Lecture Notes in Artificial Intelligence 5177

Edited by R. Goebel, J. Siekmann, and W. Wahlster

Subseries of Lecture Notes in Computer Science

Ignac Lovrek Robert J. Howlett
Lakhmi C. Jain (Eds.)

Knowledge-Based Intelligent Information and Engineering Systems

12th International Conference, KES 2008
Zagreb, Croatia, September 3-5, 2008
Proceedings, Part I

Springer

Volume Editors

Ignac Lovrek
University of Zagreb, Faculty of Electrical Engineering and Computing
Unska 3, 10000 Zagreb, Croatia
E-mail: ignac.lovrek@fer.hr

Robert J. Howlett
KES International
2nd Floor, 145-157 St. John Street, London EC1V 4PY, UK
E-mail: rjhowlett@kesinternational.org

Lakhmi C. Jain
University of South Australia, School of Electrical and Information Engineering
Mawson Lakes Campus, Adelaide, SA 5095, Australia
E-mail: lakhmi.jain@unisa.edu.au

Library of Congress Control Number: 2008933454

CR Subject Classification (1998): I.2, H.4, H.3, J.1, H.5, K.6, K.4

LNCS Sublibrary: SL 7 – Artificial Intelligence

ISSN 0302-9743
ISBN-10 3-540-85562-9 Springer Berlin Heidelberg New York
ISBN-13 978-3-540-85562-0 Springer Berlin Heidelberg New York

Springer is a part of Springer Science+Business Media

springer.com

Printed in Germany

Typesetting: Camera-ready by author, data conversion by Scientific Publishing Services, Chennai, India
Printed on acid-free paper SPIN: 12511303 06/3180 5 4 3 2 1 0

Preface

Delegates and friends, we are very pleased to extend to you a warm welcome to this, the 12th International Conference on Knowledge-Based and Intelligent Information and Engineering Systems organised by the Faculty of Electrical Engineering and Computing at the University of Zagreb, in association with KES International.

For over a decade, KES International has provided an annual wide-spectrum intelligent systems conference for the applied artificial intelligence research community. Having originated in Australia and been held there during 1997–99, the conference visited the UK in 2000, Japan in 2001, Italy in 2002, the UK in 2003, New Zealand in 2004, Australia in 2005, the UK in 2006, Italy in 2007, and now in Zagreb, Croatia in 2008. It is planned that KES 2009 will be held in Santiago, Chile before returning to the UK in 2010. The KES conference is mature and regularly attracts several hundred delegates. As it encompasses a broad range of intelligent systems topics, it provides delegates with an opportunity to mix with researchers from other groups and learn from them. The conference is linked to the International Journal of Intelligent and Knowledge-Based Systems, published by IOS Press under KES editorship. Extended and enhanced versions of the best papers presented at the KES conference may be published in the Journal.

In addition to the annual wide-range intelligent systems conference, KES has run successful symposia in several specific areas of the discipline. Agents and Multi-Agent Systems is a popular area of research. The first KES symposium on Agents and Multi-Agent Systems took place in 2007 in Wroclaw, Poland (KES-AMSTA 2007) followed in 2008 by a second in Incheon, Korea (KES-AMSTA 2008). The third in the series is planned to be held in the historic city of Uppsala in Sweden (June 3–5, 2009). Intelligent Multi-Media is a second area of focus for KES. The first KES symposium on Intelligent and Interactive Multi-Media Systems and Services (KES IIMSS 2008) will be held in Athens, Greece, in 2008, followed by a second in Venice, Italy, in 2009 (dates to be notified). A third area of interest supported by KES is Intelligent Decision Support Technologies, and the first KES symposium on this subject (KES IDT 2009) is planned for Hyogo in Japan for next year (April 23–25, 2009). Over time, each of these areas will be supported by a KES focus group of researchers interested in the topic, and if appropriate, by a journal. To this end, the International Journal of Intelligent Decision Support Technologies is published by IOS Press under the editorship of a developing KES IDT focus group.

For the future we have plans and a vision for KES. Firstly, we describe the plans.

We plan to maintain and increase the quality of KES publications. The KES quality principle is that we do not seek to expand KES activities by publishing inferior papers. However, equally, we do not believe it serves authors or the research community to reject good papers on the basis of an arbitrary acceptance / rejection ratio. Hence papers for KES conferences and symposia, and the KES Journal, will be rigorously reviewed by experts in the field, and published only if they are of a sufficiently high level, judged by international research standards.

We will further develop KES focus groups, and where appropriate, we will adopt journals and symposia, where this supports and helps us maintain our quality principle.

We introduced the concept of KES membership several years ago to provide returning KES delegates with discounted conference fees. We plan to supplement the benefits of KES membership by launching a profile page system, such that every KES member will have their own profile page on the KES web site, and be able to upload a description of themselves, their research interests and activities. The member site will act as a contact point for KES members with common interests and a potential channel to companies interested in members' research.

Printing technology is changing and this will have an effect on publishers and publications. We are conducting trials with rapid publication technology that makes it possible to print individual copies of a book on demand. The KES Rapid Research Results book series will make it convenient to publish books appealing to niche markets, for example specialised areas of research, in a way that would not have been economic before.

Many KES members and supporters have research interests outside intelligent systems. In fact, intelligent systems may just be a tool used for an application which is the main interest. A significant number of those involved in KES have interests in environmental matters. In 2009, KES will address the issue of sustainability and renewable energy through its first conference in this area, Sustainability in Energy and Buildings (SEB 2009), which will be held in Brighton, UK, April 30–May 1, 2009. SEB 2009 will address a broad spectrum of sustainability issues relating to renewable energy and the efficient use of energy in domestic and commercial buildings. Papers on the application of intelligent systems to sustainability issues will be welcome. However, it will not be compulsory that papers for SEB 2009 have significant intelligent systems content (as is a criterion for other KES conferences and symposia).

In addition to the firm plans for KES there is a longer term vision. In the time that it has been in existence, KES International has evolved from being the organiser of just a single annual conference, to a provider of an expanding portfolio of support functions for the research community. Undoubtedly, KES will continue to develop and enhance its knowledge transfer activities. The KES community consists of several thousand members, and potentially it could play a significant role in generating synergy and facilitating international research co-operation. A long term vision for KES is for it to evolve into an international academy providing the means for its members to perform international collaborative research projects. At the moment we do not have the means to turn this vision into a reality, but we will work towards this aim.

The annual KES conference continues to be a major feature of the KES organisation, and KES 2008 will continue the tradition of excellence in KES conferences.

The papers for KES 2008 were submitted either to Invited Sessions, chaired and organised by respected experts in their fields, or to General Sessions, managed by Track Chairs. Each paper was thoroughly reviewed by two members of the International Review Committee, and also inspected by a Track Chair or Invited Session Chair. A decision about whether to publish the paper was made, based on the KES quality principle described above. If the paper was judged to be of high enough quality to be accepted, the Programme Committee then decided on oral or poster presentation, based on the subject and content of the paper. All papers at KES 2008 are considered to be of equal weight and importance, no matter whether they were oral or poster presentations. This has resulted in the 316 high-quality papers included in these proceedings.

Thanks are due to the very many people who have given their time and goodwill freely to make the conference a success.

We would like to thank the KES 2008 International Programme Committee for their help and advice, and also the International Review Committee, who were essential in providing their reviews of the papers. We are very grateful for this service, without which the conference would not have been possible. We thank the high-profile keynote speakers for providing interesting expert talks to inform and inspire subsequent discussions.

An important feature of KES conferences is the Invited Session Programme. Invited Sessions give both young and established researchers an opportunity to organise and chair a set of papers on a specific topic, presented as a themed session. In this way, new topics at the leading edge of intelligent systems can be presented to interested delegates. This mechanism for feeding new ideas into the research community is very valuable. We thank the Invited Session Chairs who have contributed in this way.

The conference administrators, and the local organising committee, have all worked extremely hard to bring the conference to a high level of organisation. In this context, we would like to thank Mario Kusek, Kresimir Jurasovic, Igor Ljubi, Ana Petric, Vedran Podobnik and Jasna Slavinic (University of Zagreb, Croatia); Peter Cushion, Nicola Pinkney and Antony Wood (KES Operations, UK).

A vital contribution was made by the authors, presenters and delegates without whom the conference could not have taken place. So finally, but by no means least, we thank them for their involvement.

June 2008

Bob Howlett
Ignac Lovrek
Lakhmi Jain

Organisation

KES 2008 Conference Organisation

KES 2008 was organised by KES International, Innovation in Knowledge-Based and Intelligent Engineering Systems, and the University of Zagreb, Faculty of Electrical Engineering and Computing.

KES 2008 Conference Chairs

General Chair	Ignac Lovrek, University of Zagreb, Croatia
Executive Chair	Robert J. Howlett, University of Brighton, UK
Invited Sessions Chair	Lakhmi C. Jain, University of South Australia, Australia
Local Chair	Mario Kusek, University of Zagreb, Croatia
Award Chair	Bogdan Gabrys, University of Bournemouth, UK

KES Conference Series

KES 2008 is a part of the KES Conference Series.

Conference Series Chairs	Lakhmi C. Jain and Robert J. Howlett
KES Executive Chair	Robert J. Howlett, University of Brighton, UK
KES Founder	Lakhmi C. Jain, University of South Australia, Australia

Local Organising Committee

Kresimir Jurasovic, Igor Ljubi, Ana Petric, Vedran Podobnik, Jasna Slavinic (University of Zagreb, Croatia), Peter Cushion (KES Operations, UK)

International Programme Committee

Abe, Akinori	ATR Knowledge Science Laboratories, Japan
Adachi, Yoshinori	Chubu University, Japan
Angulo, Cecilio	Technical University of Catalonia, Spain
Apolloni, Bruno	University of Milan, Italy
Baba, Norio	Osaka Kyoiku University, Japan
Balachandran, Bala M.	University of Canberra, Australia
Dalbelo Basic, Bojana	University of Zagreb, Croatia
Beristain, Andoni	Universidad del Pais Vasco, Spain

Bianchini, Monica	Universita degli Studi di Siena, Italy
Castellano, Giovanna	University of Bari, Italy
Chen, Yen-Wei	Ritsumeikan University, Japan
Cheng, Jingde	Saitama University, Japan
Corchado, Emilio	University of Burgos, Spain
Cuzzocrea, Alfredo	University of Calabria, Italy
Damiani, Ernesto	University of Milan, Italy
Di Noia, Tommaso	Technical University of Bari, Italy
Esposito, Floriana	University of Bari, Italy
Gabrys, Bogdan	University of Bournemouth, UK
Gao, Kun	Zhejiang Wanli University, China
Hartung, Ronald L.	Franklyn University, USA
Hakansson, Anne	Uppsala University, Sweden
Holmes, Dawn	University of California Santa Barbara, USA
Howlett, Robert J.	University of Brighton, UK
Ishida, Yoshiteru	Toyohashi University of Technology, Japan
Ishii, Naohiro	Aichi Institute of Technology, Japan
Jain, Lakhmi C.	University of South Australia, Australia
Jevtic, Dragan	University of Zagreb, Croatia
Karny, Miroslav	Academy of Sciences of the Czech Republic, Czech Republic
Kunifuji, Susumu	Japan Advanced Institute of Science and Technology, Japan
Lee, Hsuan-Shih	National Taiwan Ocean University, Taiwan
Lim, Chee-Peng	University of Science, Malaysia
Liu, Honghai	University of Portsmouth, UK
Lovrek, Ignac	University of Zagreb, Croatia
Maojo, Victor	Universidad Politecnica de Madrid, Spain
Markey, Mia	University of Texas at Austin, USA
Mumford, Christine	Cardiff University, UK
Munemori, Jun	Wakayama University, Japan
Nakamatsu, Kazumi	University of Hyogo, Japan
Nakano, Ryohei	Nagoya Institute of Technology, Japan
Nakao, Zensho	University of the Ryukyus, Japan
Nauck, Detlef	BT Intelligent Systems Research Centre, UK
Negoita, Mircea Gh.	KES International
Nguyen, Ngoc Thanh	Wroclaw University of Technology, Poland
Nicoletti, Maria do Carmo	Federal University of Sao Carlos, Brazil
Nikolos, Ioannis K.	Technical University of Crete, Greece
Nishida, Toyoaki	Kyoto University, Japan
Nuernberger, Andreas	University of Magdeburg, Germany
Palade, Vasile	University of Oxford, UK
Park, Gwi-Tae	Korea University, Seoul, Korea
Pham, Tuan	James Cook University, Australia
Phillips-Wren, Gloria	Loyola College in Maryland, USA

Sharma, Dharmendra	University of Canberra, Australia
Shkodirev, Viacheslaw	St Petersburg State Polytechnic University, Russia
Sik Lanyi, Cecilia	University of Pannonia, Hungary
Sunde, Jadranka	Defence Science and Technology Organisation, Australia
Tagliaferri, Roberto	University of Salerno, Italy
Taki, Hirokazu	Wakayama University, Japan
Tsuda, Kazuhiko	University of Tsukuba, Japan
Turchetti, Claudio	Università Politecnica delle Marche, Italy
Vassanyi, Istvan	University of Pannonia, Hungary
Veganzones, Miguel A.	Universidad del Pais Vasco, Spain
Vellido, Alfredo	Technical University of Catalonia, Spain
Watada, Junzo	Waseda University, Japan
Watanabe, Toyohide	Nagoya University, Japan
Yamashita, Yoshiyuki	Tokio University of Agriculture and Technology, Japan

International Review Committee

Abe, Akinori	ATR Knowledge Science Laboratories, Japan
Abe, Jair	University of Sao Paulo, Brazil
Abu Bakar, Rohani	Waseda University, Japan
Abulaish, Muhammad	Jamia Millia Islamia, India
Adachi, Yoshinori	Chubu University, Japan
Adli, Alexander	University of the Ryukyus, Japan
Akama, Seiki	C-Republics, Japan
Al-Hashel, Ebrahim	University of Canberra, Australia
Alquezar, Ren	Technical University of Catalonia, Spain
Angelov, Plamen	Lancaster University, UK
Anguita, Davide	University of Genoa, Italy
Angulo, Cecilio	Technical University of Catalonia, Spain
Anisetti, Marco	University of Milan, Italy
Aoyama, Kouji	Fujitsu Laboratories Limited, Japan
Apolloni, Bruno	University of Milan, Italy
Appice, Annalisa	University of Bari, Italy
Aritsugi, Masayoshi	Kumamoto University, Japan
Arroyo-Figueroa, Gustavo	Instituto de Investigaciones Electricas, Mexico
Azzini, Antonia	University of Milan, Italy
Baba, Norio	Osaka Kyoiku University, Japan
Balachandran, Bala M.	University of Canberra, Australia
Balas, Marius	Aurel Vlaicu University of Arad, Romania
Balas, Valentina E.	Aurel Vlaicu University of Arad, Romania
Balic, Joze	University of Maribor, Slovenia
Bandini, Stefania	University of Milan Bicocca, Italy
Baruque, Bruno	University of Burgos, Spain

Basili, Roberto	University of Rome, Italy
Bassis, Simone	University of Milan, Italy
Bayarri, Vicente	GIM Geomatics S.L., Spain
Belanche, Lluis	Technical University of Catalonia, Spain
Bellandi, Valerio	University of Milan, Italy
Berendt, Bettina	Katholieke Universiteit Leuven, Belgium
Bianchini, Monica	Università degli Studi di Siena, Italy
Bidlo, Michal	Technical University of Brno, Czech Republic
Bielikov, Mria	Slovak University of Technology, Slovakia
Billhardt, Holger	Universidad Rey Juan Carlos I, Spain
Bingul, Zafer	Kocaeli University, Turkey
Bioucas, Jose	Instituto Superior Tecnico, Portugal
Bogdan, Stjepan	University of Zagreb, Croatia
Borghese, Alberto	University of Milan, Italy
Borzemski, Leszek	Wroclaw University of Technology, Poland
Bouchachia, Abdelhamid	University of Klagenfurt, Austria
Bouquet, Paolo	Università degli Studi di Trento, Italy
Bouridane, Ahmed	Queen's University, Belfast, UK
Bruzzone, Lorenzo	Università degli Studi di Trento, Italy
Buciu, Ioan	University of Oradea, Romania
Cabestany, Joan	Technical University of Catalonia, Spain
Calpe-Maravilla, Javier	University of Valencia, Spain
Camino-Gonzalez, Carlos Luis	Forestry and Technology Center of Catalonia, Spain
Camps-Valls, Gustavo	University of Valencia, Spain
Cao, Jiangtao	University of Portsmouth, UK
Capkovic, Frantisek	Slovak Academy of Sciences, Slovakia
Carpintero-Salvo, Irene Rosa	Universidad de Granada and Consejeria de Medio Ambiente, Spain
Castellano, Giovanna	University of Bari, Italy
Castiello, Ciro	University of Bari, Italy
Castillo, Elena	University of Cantabria, Spain
Cavar, Damir	Indiana University, USA
Ceccarelli, Michele	University of Sannio, Italy
Ceravolo, Paolo	University of Milan, Italy
Chan, Chee Seng	University of Portsmouth, UK
Chang, Chan-Chih	Industrial Technology Research Institute, Taiwan
Chang, Chuan-Yu	National Yunlin University of Science & Technology, Taiwan
Chen, Mu-Yen	National Changhua University of Education, Taiwan
Chen, Yen-Wei	Ritsumeikan University, Japan
Cheng, Jingde	Saitama University, Japan
Chetty, Girija	University of Canberra, Australia

Chi Thanh, Hoang	Hanoi University of Science, Vietnam
Ciaramella, Angelo	University of Naples Parthenope, Italy
Cicin Sain, Marina	University of Rijeka, Croatia
Claveau, Vincent	IRISA-CNRS, France
Coghill, George	University of Auckland, New Zealand
Colucci, Simona	Technical University of Bari, Italy
Corbella, Ignasi	Technical University of Catalonia, Spain
Corchado Rodriguez, Juan M.	University of Salamanca, Spain
Corchado, Emilio	University of Burgos, Spain
Costin, Mihaela	Institute for Computer Science, Romanian Academy, Romania
Cox, Robert	University of Canberra, Australia
Crippa, Paolo	Università Politecnica delle Marche, Italy
Cruz, Antonio	Federal University of Sao Carlos, Brazil
Csipkes, Gabor	Technical University of Cluj-Napoca, Romania
Curzi, Alessandro	Università Politecnica delle Marche, Italy
Cuzzocrea, Alfredo	University of Calabria, Italy
Czarnowski, Ireneusz	Gdynia Maritime University, Poland
D'Amato, Claudia	University of Bari, Italy
D'Apuzzo, Livia	University of Naples Federico II, Italy
Dalbelo Basic, Bojana	University of Zagreb, Croatia
Damiani, Ernesto	University of Milan, Italy
Davidsson, Paul	Blekinge Institute of Technology, Sweden
de Campos, Cassio Polpo	Rensselaer Polytechnic Institute, USA
De Gemmis, Marco	University of Bari, Italy
De Santis, Angela	University of Alcala, Spain
Deguchi, Toshinori	Gifu National College of Technology, Japan
Dell'Endice, Francesco	University of Zurich, Switzerland
Di Noia, Tommaso	Technical University of Bari, Italy
Di Sciascio, Eugenio	Technical University of Bari, Italy
Diani, Marco	University of Pisa, Italy
Diaz-Delgado, Ricardo	Estacion Biologica de Donana-CSIC, Spain
Dobsa, Jasminka	University of Zagreb, Croatia
Dorado, Julian	Universidad de la Coruna, Spain
Dujmic, Hrvoje	University of Split, Croatia
Dumitriu, Luminita	Dunarea de Jos University, Romania
Duro, Richard	Universidade da Coruna, Spain
Edman, Anneli	Uppsala University, Sweden
Erjavec, Tomaz	Josef Stefan Institute, Slovenia
Esposito, Anna	University of Naples Federico II and IIASS, Italy
Esposito, Floriana	University of Bari, Italy
Etchells, Terence	Liverpool John Moores University, UK
Lee, Eun-Ser	Soong Sil University, Korea
Fang, H.H.	Taipei College of Maritime Technology, Taiwan

Fasano, Giovanni	University of Venice, Italy
Feng, Jun	Hohai University, China
Fernandez-Caballero, Antonio	Universidad de Castilla-La Mancha, Spain
Fras, Mariusz	Wroclaw University of Technology, Poland
Frati, Fulvio	University of Milan, Italy
Fuchino, Tetsuo	Tokyo Institute of Technology, Japan
Fujimoto, Taro	Fujitsu Laboratories Limited, Japan
Fujinami, Tsutomu	Japan Advanced Institute of Science and Technology, Japan
Fukue, Yoshinori	Fujitsu Laboratories Limited, Japan
Gallego-Merino, Miren Josune	Universidad del Pais Vasco, Spain
Gamberger, Dragan	Rudjer Boskovic Institute, Croatia
Gao, Kun	Zhejiang Wanli University, China
Gao, Ying	Saitama University, Japan
Garcia-Sebastian, Maite	Universidad del Pais Vasco, Spain
Gastaldo, Paolo	University of Genoa, Italy
Gendarmi, Domenico	University of Bari, Italy
Georgieva, Petia	University of Aveiro, Portugal
Gianfelici, Francesco	Università Politecnica delle Marche, Italy
Gianini, Gabriele	University of Milan, Italy
Giordano, Roberto	Federal University of Sao Carlos, Brazil
Giorgini, Paolo	University of Trento, Italy
Gledec, Gordan	University of Zagreb, Croatia
Gold, Hrvoje	University of Zagreb, Croatia
Goldstein, Pavle	University of Zagreb, Croatia
Gomez-Dans, Jose Luis	University College London, UK
Goto, Yuichi	Saitama University, Japan
Grana, Manuel	Universidad del Pais Vasco, Spain
Greenwood, Garrison	Portland State University, USA
Gu, Dongbing	University of Essex, UK
Guo, Huawei	Shanghai Jiao Tong University, China
Hakansson, Anne	Uppsala University, Sweden
Halilcevic, Suad	University of Tuzla, Bosnia & Herzegovina
Hammer, Barbara	Clausthal University of Technology, Germany
Hanachi, Chihab	University Toulouse 1 and IRIT Laboratory, France
Hara, Akira	Hiroshima City University, Japan
Harada, Koji	Toyohashi University of Technology, Japan
Harris, Irina	Cardiff University, UK
Harris, Richard	University of Lancaster, UK
Hartung, Ronald L.	Franklin University, USA
Hasegawa, Shinobu	Japan Advanced Institute of Science and Technology, Japan
Hayashi, Hidehiko	Naruto University of Education, Japan
Hernandez, Carmen	Universidad del Pais Vasco, Spain

Herrero, Alvaro	University of Burgos, Spain
Hildebrand, Lars	Technical University of Dortmund, Germany
Hiroshi, Mineno	Shizuoka University, Japan
Handa, Hisashi	Okayama University, Japan
Hocenski, Zeljko	University of Osijek, Croatia
Hori, Satoshi	Monotsukuri Institute of Technologists, Japan
Howlett, Robert J.	University of Brighton, UK
Hruschka, Eduardo	University of Sao Paulo, Brazil
Huang, Xu	University of Canberra, Australia
Huljenic, Darko	Ericsson Nikola Tesla, Croatia
Ichimura, Takumi	Hiroshima City University, Japan
Inuzuka, Nobuhiro	Nagoya Institute of Technology, Japan
Ioannidis, Stratos	University of the Aegean, Greece
Ishibuchi, Hisao	Osaka Prefecture University, Japan
Ishida, Yoshiteru	Toyohashi University of Technology, Japan
Ishii, Naohiro	Aichi Institute of Technology, Japan
Ito, Hideaki	Chukyo University, Japan
Ito, Kazunari	Aoyama University, Japan
Ito, Sadanori	Tokio University of Agriculture and Technology, Japan
Itou, Junko	Wakayama University, Japan
Iwahori, Yuji	Chubu University, Japan
Jacquenet, Francois	University of Saint-Etienne, France
Jain, Lakhmi C.	University of South Australia, Australia
Jarman, Ian	Liverpool John Moores University, UK
Jatowt, Adam	Kyoto University, Japan
Jevtic, Dragan	University of Zagreb, Croatia
Jezic, Gordan	University of Zagreb, Croatia
Jiang, Jianmin	University of Bradford, UK
Jimenez-Berni, Jose Antonio	IAS-CSIC, Spain
Johnson, Ray	Defence Science and Technology Organisation, Australia
Ju, Zhaojie	University of Portsmouth, UK
Jung, Jason	Yeungnam University, Korea
Juszczyszyn, Krzysztof	Wroclaw University of Technology, Poland
Kaczmarek, Tomasz	Poznan University of Economics, Poland
Karny, Miroslav	Academy of Sciences of the Czech Republic, Czech Republic
Karwowski, Waldemar	University of Central Florida, USA
Katarzyniak, Radoslaw	Wrocaw University of Technology, Poland
Kato, Shohei	Nagoya Institute of Technology, Japan
Katsifarakis, Konstantinos	Aristotelian University of Thessaloniki, Greece
Kazienko, Przemyslaw	Wroclaw University of Technology, Poland
Kecman, Vojislav	University of Auckland, New Zealand
Keysers, Daniel	Google Switzerland, Switzerland

Kim, Dongwon	Korea University, Korea
Kimura, Masahiro	Ryukoku University, Japan
Kojiri, Tomoko	Nagoya University, Japan
Kolaczek, Grzegorz	Wroclaw University of Technology, Poland
Koukam, Abder	Université de Technologie de Belfort Montbeliard, France
Kovacic, Zdenko	University of Zagreb, Croatia
Krajnovic, Sinisa	Nippon Ericsson K.K., Japan
Krol, Dariusz	Wroclaw University of Technology, Poland
Kubota, Naoyuki	Tokyo Metropolitan University, Japan
Kucheryavskiy, Sergey	Aalborg University Esbjerg, Denmark
Kume, Terunobu	Fujitsu Laboratories Limited, Japan
Kunifuji, Susumu	Japan Advanced Institute of Science and Technology, Japan
Kuroda, Chiaki	Tokyo Institute of Technology, Japan
Kusek, Mario	University of Zagreb, Croatia
Lanjeri, Siham	University of Cordoba, Spain
Le, Kim	University of Canberra, Australia
Lee, Hsuan-Shih	National Taiwan Ocean University, Taiwan
Lee, Huey-Ming	Chinese Culture University, Taiwan
Lee, Malrey	ChonBuk National University, Korea
Lennox, Barry	University of Manchester, UK
Leray, Philippe	University of Nantes, France
Lhotska, Lenka	Czech Technical University, Czech Republic
Lisboa, Paulo	Liverpool John Moores University, UK
Liu, Honghai	University of Portsmouth, UK
Liu, Yonghuai	Aberystwyth University, UK
Loia, Vincenzo	University of Salerno, Italy
Lonetti, Francesca	University of Pisa, Italy
Loo, Chu-Kiong	Multimedia University, Malaysia
López, Beatriz	Universitat de Girona, Spain
Lops, Pasquale	University of Bari, Italy
Lovrek, Ignac	University of Zagreb, Croatia
Ma, Wanli	University of Canberra, Australia
MacDonald, Bruce	University of Auckland, New Zealand
Magnani, Lorenzo	University of Pavia, Italy
Malchiodi, Dario	University of Milan, Italy
Maldonado-Bautista, Jose O.	Universidad del Pais Vasco, Spain
Maojo, Victor	Universidad Politecnica de Madrid, Spain
Maratea, Antonio	University of Naples Parthenope, Italy
Margaris, Dionissios	University of Patras, Greece
Marinai, Simone	University of Florence, Italy
Markovic, Hrvoje	Tokyo Institute of Technology, Japan
Marrara, Stefania	University of Milan, Italy
Martin, Luis	Universidad Politecnica de Madrid, Spain

Martin-Sanchez, Fernando	Institute of Health Carlos III, Spain
Masulli, Francesco	University of Genoa, Italy
Matijasevic, Maja	University of Zagreb, Croatia
Matsuda, Noriyuki	Wakayama University, Japan
Matsudaira, Kazuya	Shizuoka University, Japan
Matsui, Nobuyuki	University of Hyogo, Japan
Matsumoto, Hideyuki	Tokyo Institute of Technology, Japan
Matsushita, Mitsunori	NTT Communication Science Labs, Japan
McCormac, Andrew	Alpha Data Ltd, UK
Mencar, Corrado	University of Bari, Italy
Meng, Qinggang	University of Loughborough, UK
Menolascina, Filippo	Technical University of Bari, Italy
Mera, Kazuya	Hiroshima City University, Japan
Minazuki, Akinori	Kushiro Public University of Economics, Japan
Mineno, Hiroshi	Shizuoka University, Japan
Minoru, Fukumi	Tokushima University, Japan
Misue, Kazuo	University of Tsukuba, Japan
Mitsukura, Yasue	Tokyo University of Agriculture and Technology, Japan
Mituhara, Hiroyuki	Tokushima University, Japan
Miura, Hirokazu	Wakayama University, Japan
Miura, Motoki	Japan Advanced Institute of Science and Technology, Japan
Miyadera, Youzou	Tokyo Gakugei University, Japan
Mizuno, Tadanori	Shizuoka University, Japan
Mohammadian, Masoud	University of Canberra, Australia
Moraga, Claudio	University of Dortmund, Germany
Mukai, Naoto	Tokyo University of Science, Japan
Mumford, Christine	Cardiff University, UK
Munemori, Jun	Wakayama University, Japan
Nachtegael, Mike	Ghent University, Belgium
Nakada, Toyohisa	Japan Advanced Institute of Science and Technology, Japan
Nakamatsu, Kazumi	University of Hyogo, Japan
Nakamura, Tsuyoshi	Nagoya Institute of Technology, Japan
Nakano, Ryohei	Nagoya Institute of Technology, Japan
Nakao, Zensho	University of the Ryukyus, Japan
Napolitano, Francesco	University of Salerno, Italy
Nara, Shinsuke	Saitama University, Japan
Nara, Yumiko	The Open University of Japan, Japan
Nascimiento, Jose	Instituto Superior de Engenharia de Lisboa, Portugal
Nebot, Angela	Technical University of Catalonia, Spain
Negoita, Mircea Gh.	KES International

Ng, Wilfred	Hong Kong University of Science and Technology, China
Nguyen, Ngoc Thanh	Wroclaw University of Technology, Poland
Nicolau, Viorel	Dunarea de Jos University of Galati, Romania
Nicoletti, Maria do Carmo	Federal University of Sao Carlos, Brazil
Nicosia, Giuseppe	University of Catania, Italy
Nijholt, Anton	University of Twente, The Netherlands
Nishida, Toyoaki	Kyoto University, Japan
Nishimoto, Kazushi	Japan Advanced Institute of Science and Technology
Nobuhara, Hajime	University of Tsukuba, Japan
Nowe, Ann	VUB, Belgium
Nowostawski, Mariusz	University of Otago, New Zealand
O'Grady, Michael	University College Dublin, Ireland
Okamoto, Takeshi	Kanagawa Institute of Technology, Japan
Oltean, Gabriel	Technical University of Cluj-Napoca, Romania
Ortega, Juan	Universidad de Sevilla, Spain
Ozawa, Seiichi	Kobe University, Japan
Palade, Vasile	University of Oxford, UK
Pan, Dan China	Mobile Group Guangdong Branch, China
Pan, Jeng-Shyang	National Kaohsiung University of Applied Sciences, Taiwan
Pandzic, Igor S.	University of Zagreb, Croatia
Papathanassiou, Stavros	National Technical University of Athens, Greece
Park, Gwi-Tae	Korea University, Korea
Parra-Llanas, Xavier	Technical University of Catalonia, Spain
Pasero, Eros	Politecnico di Torino, Italy
Paz, Abel Francisco	University of Extremadura, Spain
Pedrycz, Witold	University of Alberta, Canada
Pehcevski, Jovan	MIT Faculty of Information Technologies, Macedonia
Perez del Rey, David	Universidad Politecnica de Madrid, Spain
Perez, Rosa M.	University of Extremadura, Spain
Perez-Lopez, Carlos	Technical University of Catalonia, Spain
Pessa, Eliano	University of Pavia, Italy
Pham, Tuan	James Cook University, Australia
Phillips, Phil	Office for National Statistics, UK
Phillips-Wren, Gloria	Loyola College in Maryland, USA
Picasso, Francesco	University of Genoa, Italy
Pieczynska, Agnieszka	Instytut Informatyki Technicznej, Poland
Pirrone, Roberto	University of Palermo, Italy
Plaza, Antonio	University of Extremadura, Spain
Popa, Rustem	Dunarea de Jos University of Galati, Romania
Popescu, Daniela	University of Oradea, Romania

Ragone, Azzurra	Technical University of Bari, Italy
Raiconi, Giancarlo	University of Salerno, Italy
Raimondo, Giovanni	Politecnico di Torino, Italy
Ramon, Jan	Katholieke Universiteit Leuven, Belgium
Ranawana, Romesh	ClearView Scientific, UK
Razmerita, Liana	Copenhagen Business School, Denmark
Reghunadhan, Rajesh	Bharathiar University, India
Remagnino, Paolo	Kingston University, UK
Resta, Marina	University of Genoa, Italy
Reusch, Bernd	Technical University of Dortmund, Germany
Rizzo, Donna	University of Vermont, USA
Rohani, Bakar	Waseda University, Japan
Romero, Enrique	Technical University of Catalonia, Spain
Rosic, Marko	University of Split, Croatia
Rovas, Dimitrios	Technical University of Crete, Greece
Rozic, Nikola	University of Split, Croatia
Saito, Kazumi	University of Shizuoka, Japan
Sakai, Hiroshi	Kyushu Institute of Technology, Japan
Sakamoto, Ryuuki	Wakayama University, Japan
Sanchez, Eduardo	Logic Systems Laboratory IN-Ecublens, Switzerland
Sassi, Roberto	University of Milan, Italy
Sato-Ilic, Mika	University of Tsukuba, Japan
Scarselli, Franco	University of Siena, Italy
Schanda, Janos	University of Veszprem, Hungary
Schwenker, Friedhelm	University of Ulm, Germany
Sebillot, Pascale	IRISA/INSA de Rennes, France
Semeraro, Giovanni	University of Bari, Italy
Sergiadis, Georgios	Aristotelian University of Thessaloniki, Greece
Serra-Sagrista, Joan	Universitat Autonoma Barcelona, Spain
Sharma, Dharmendra	University of Canberra, Australia
Shiau, Yea-Jou	China University of Technology, Taiwan
Shin, Jungpil	University of Aizu, Japan
Shinagawa, Norihide	Tokyo University of Agriculture and Technology, Japan
Shizuki, Buntarou	University of Tsukuba, Japan
Shkodirev, Viacheslaw	St. Petersburg State Polytechnic University, Russia
Sidhu, Amandeep	Curtin University of Technology, Australia
Signore, Oreste	Istituto di Scienza e Tecnologie dell' Informazione A. Faedo, Italy
Sikic, Mile	University of Zagreb, Croatia
Sinkovic, Vjekoslav	University of Zagreb, Croatia
Smuc, Tomislav	Rudjer Boskovic Institute, Croatia
Snajder, Jan	University of Zagreb, Croatia

Sobecki, Janusz	Wroclaw University of Technology, Poland
Sohn, Surgwon	Hoseo University, Korea
Somol, Petr	Institute of Information Theory and Automation, Czech Republic
Staiano, Antonino	University of Naples Federico II, Italy
Stamou, Giorgos	National Technical University of Athens, Greece
Stankov, Slavomir	University of Split, Croatia
Stecher, Rodolfo	L3S Research Center, Germany
Stellato, Armando	University of Rome, Italy
Stoermer, Heiko	University of Trento, Italy
Strahil, Ristov	Rudjer Boskovic Institute, Croatia
Sugihara, Taro	Japan Advanced Institute of Science and Technology, Japan
Sugiyama, Kozo	Japan Advanced Institute of Science and Technology, Japan
Sulaiman, Ross	University of Canberra, Australia
Supek, Fran	Rudjer Boskovic Institute, Croatia
Surjan, Gyoergy	National Institute for Strategic Health Research, Hungary
Suzuki, Nobuo	KDDI Corporation, Japan
Tabakow, Iwan	Wroclaw University of Technology, Poland
Tadanori, Mizuno	Shizuoka University, Japan
Tadic, Marko	University of Zagreb, Croatia
Tagawa, Takahiro	Kyushu University, Japan
Tagliaferri, Roberto	University of Salerno, Italy
Takahashi, Osamu	Future University-Hakodate, Japan
Takahashi, Shin	University of Tsukuba, Japan
Takeda, Kazuhiro	Shizuoka University, Japan
Takenaka, Tomoya	Shizuoka University, Japan
Taki, Hirokazu	Wakayama University, Japan
Tamani, Karim	University of Savoie, France
Tanahashi, Yusuke	Nagoya Institute of Technology, Japan
Tateiwa, Yuichiro	Nagoya University, Japan
Tesar, Ludvik	Academy of Sciences of the Czech Republic, Czech Republic
Thai, Hien	University of the Ryukyus, Japan
Ting, Hua Nong	University of Malaya, Malaysia
Ting, I-Hsien	National University of Kaohsiung, Taiwan
Tohru, Matsuodani	Debag Engineering Ltd, Japan
Tonazzini, Anna	Istituto di Scienza e Tecnologie dell' Informazione A. Faedo, Italy
Torsello, Maria Alessandra	University of Bari, Italy
Tran, Dat	University of Canberra, Australia
Trentin, Edmondo	University of Siena, Italy

Trzec, Krunoslav	Ericsson Nikola Tesla, Croatia
Tsourveloudis, Nikos	Technical University of Crete, Greece
Tsumoto, Shusaku	Shimane University, Japan
Turchetti, Claudio	Università Politecnica delle Marche, Italy
Tweedale, Jeffrey	Defence Science and Technology Organisation, Australia
Uchino, Eiji	Yamaguchi University, Japan
Ugai, Takanori	Fujitsu Laboratories Limited, Japan
Ushiama, Teketoshi	Kyushu University, Japan
Vaklieva-Bancheva, Natasha	Bulgarian Academy of Sciences, Bulgaria
Vassanyi, Istvan	University of Pannonia, Hungary
Vega, Miguel	University of Granada, Spain
Veganzones, Miguel Angel	Universidad del Pais Vasco, Spain
Vellido, Alfredo	Technical University of Catalonia, Spain
Vitabile, Salvatore	University of Palermo, Italy
Vohland, Michael	Trier University, Germany
Wang, Jin-Long	Ming Chuan University, Taiwan
Wang, Yang	University of Portsmouth, UK
Watada, Junzo	Waseda University, Japan
Watanabe, Toyohide	Nagoya University, Japan
Watanabe, Yuji	Nagoya City University, Japan
Weber, Cornelius	Frankfurt Institute for Advanced Studies, Germany
Whitaker, Roger	Cardiff University, UK
Xia, Feng	Queensland University of Technology, Australia
Yamada, Kunihiro	Tokai University, Japan
Yamashita, Yoshiyuki	Tokyo University of Agriculture and Technology, Japan
Yasuda, Takami	Nagoya University, Japan
Yip, Chi Lap	University of Hong Kong, China
Yi-Sheng, Huang	Chung Cheng Institute of Technology, Taiwan
Yoshida, Kenichi	University of Tsukuba, Japan
Yoshida, Kouji	Shonan Institute of Technology, Japan
Yoshiura, Noriaki	Saitama University, Japan
Younan, Nick	Mississippi State University, USA
Yu, Donggang	James Cook University, Australia
Yu, Zhiwen	Kyoto University, Japan
Yuizono, Takaya	Japan Advanced Institute of Science and Technology, Japan
Yukawa, Takashi	Nagaoka University of Technology, Japan
Zalili, Musa	Waseda University, Japan
Zare, Alina	University of Florida, USA
Zebulum, Ricardo	NASA Jet Propulsion Laboratory, USA
Zeng, An	Guangdong University of Technology, China
Zeng, Xiangyan	University of California Davis, USA

Zhang, Bailing	Xi'an Jiaotong-Liverpool University, China
Zhu, Goupu	Sun Yat-Sen University, China
Zippo, Antonio	University of Milan, Italy
Zunino, Rodolfo	University of Genoa, Italy

General Track Chairs

Artificial Neural Networks and Connectionist Systems
Bruno Apolloni, University of Milan, Italy

Fuzzy and Neuro–Fuzzy Systems
Bernd Reusch, Technical University of Dortmund, Germany

Evolutionary Computation
Zensho Nakao, University of the Ryukyus, Japan

Machine Learning and Classical AI
Floriana Esposito, University of Bari, Italy

Agent Systems
Ngoc Thanh Nguyen, Wroclaw University of Technology, Poland

Knowledge Based and Expert Systems
Anne Håkansson, Uppsala University, Sweden

Hybrid Intelligent Systems
Vasile Palade, University of Oxford, UK

Intelligent Vision and Image Processing
Tuan Pham, James Cook University, Australia

Knowledge Management, Ontologies and Data Mining
Bojana Dalbelo Basic, University of Zagreb, Croatia

Web Intelligence, Text and Multimedia Mining and Retrieval
Andreas Nuernberger, University of Magdeburg, Germany

Intelligent Signal Processing
Miroslav Karny, Academy of Sciences of the Czech Republic, Czech Republic

Intelligent Robotics and Control
Honghai Liu, University of Portsmouth, UK

Invited Session Chairs

Advanced Groupware
Jun Munemori (Wakayama University, Japan), Hiroshi Mineno (Shizuoka University, Japan)

Advanced Knowledge-Based Systems
Alfredo Cuzzocrea (University of Calabria, Italy)

Advanced Neural Processing Systems
Monica Bianchini, Marco Maggini, Franco Scarselli (Università degli Studi di Siena, Italy)

Agent and Multi-Agent Systems: Technologies and Applications
Bala M. Balachandran, Dharmendra Sharma (University of Canberra, Australia)

Ambient Intelligence
Cecilio Angulo (Technical University of Catalonia, Spain), Honghai Liu (University of Portsmouth, UK)

Application of Knowledge Models in Healthcare
István Vassányi (University of Pannonia, Hungary), Gyoergy Surjan (National Institute for Strategic Health Research, Hungary)

Artificial Intelligence Driven Engineering Design Optimization
Ioannis K. Nikolos (Technical University of Crete, Greece)

Biomedical Informatics: Intelligent Information Management from Nanomedicine to Public Health
Victor Maojo (Universidad Politecnica de Madrid, Spain)

Chance Discovery
Akinori Abe (ATR Knowledge Science Laboratories, Japan), Yukio Ohsawa (University of Tokyo, Japan)

Communicative Intelligence
Ngoc Thanh Nguyen (Wroclaw University of Technology, Poland), Toyoaki Nishida (Kyoto University, Japan)

Computational Intelligence for Image Processing and Pattern Recognition
Yen-Wei Chen (Ritsumeikan University, Japan)

Computational Intelligence in Human Cancer Research
Alfredo Vellido (Technical University of Catalonia, Spain), Paulo J.G. Lisboa (Liverpool John Moores University, UK)

Computational Intelligence Techniques for Knowledge Discovery
Claudio Turchetti, Paolo Crippa, Francesco Gianfelici (Università Politecnica delle Marche, Italy)

Computational Intelligence Techniques for Web Personalization
Giovanna Castellano, Alessandra Torsello (University of Bari, Italy)

Computational Intelligent Techniques for Bioprocess Modelling, Monitoring and Control
Maria do Carmo Nicoletti, Teresa Cristina Zangirolami, Estevam Rafael Hruschka Jr. (Federal University of Sao Carlos, Brazil)

Contributions of Intelligent Decision Technologies (IDT)
Gloria Phillips-Wren (Loyola College in Maryland, USA), Lakhmi C. Jain (University of South Australia, Australia)

Engineered Applications of Semantic Web – SWEA
Tommaso Di Noia, Marco Degemmis, Giovanni Semeraro, Eugenio Di Sciascio (University of Bari, Italy)

Enhance Secure User Authentication Through Intelligent and Strong Techniques
Ernesto Damiani, Antonia Azzini (University of Milan, Italy)

Evolutionary Multiobjective Optimization
Christine Mumford (Cardiff University, UK)

Evolvable Hardware and Adaptive Systems – Advanced Engineering Design Methodologies and Applications
Mircea Gh. Negoita (KES International), Sorin Hintea (Technical University of Cluj-Napoca, Romania)

Evolvable Hardware Applications in the Area of Electronic Circuits Design
Mircea Gh. Negoita (KES International), Sorin Hintea (Technical University of Cluj-Napoca, Romania)

Hyperspectral Imagery for Remote Sensing: Intelligent Analysis and Applications
Miguel A. Veganzones, Manuel Grana (Universidad del Pais Vasco, Spain)

Immunity-Based Systems
Yoshiteru Ishida (Toyohashi University of Technology, Japan)

Innovation-oriented Knowledge Management Platform
Toyohide Watanabe (Nagoya University, Japan), Taketoshi Ushiama (Kyushu University, Japan)

Innovations in Intelligent Multimedia Systems
Cecilia Sik Lanyi (University of Pannonia, Hungary), Lakhmi C. Jain (University of South Australia, Australia)

Innovations in Virtual Reality
Cecilia Sik Lanyi (University of Pannonia, Hungary), Lakhmi C. Jain (University of South Australia, Australia)

Intelligent Computing for Grid
Kun Gao (Zhejiang Wanli University, China)

Intelligent Data Processing in Process Systems and Plants
Yoshiyuki Yamashita (Tokyo University of Agriculture and Technology, Japan), Tetsuo Fuchino (Tokyo Institute of Technology, Japan)

Intelligent Environment Support for Collaborative Learning
Toyohide Watanabe, Tomoko Kojiri (Nagoya University, Japan)

Intelligent Systems in Medicine and Healthcare
Chee-Peng Lim, (University of Science Malaysia, Malaysia), Lakhmi C. Jain (University of South Australia, Australia), Robert F. Harrison (University of Sheffield, UK)

Intelligent Systems in Medicine: Innovations in Computer–Aided Diagnosis and Treatment
Mia Markey (University of Texas at Austin, USA), Lakhmi C. Jain (University of South Australia, Australia)

Intelligent Utilization of Soft Computing Techniques
Norio Baba (Osaka Kyoiku University, Japan)

Knowledge-Based Interface Systems [I]
Naohiro Ishii (Aichi Institute of Technology, Japan), Yuji Iwahori (Chubu University, Japan)

Knowledge-Based Interface Systems [II]
Yoshinori Adachi (Chubu University, Japan), Nobuhiro Inuzuka (Nagoya Institute of Technology, Japan)

Knowledge Interaction for Creative Learning
Toyohide Watanabe, Tomoko Kojiri (Nagoya University, Japan)

Knowledge-Based Creativity Support Systems
Susumu Kunifuji (Japan Advanced Institute of Science and Technology, Japan), Kazuo Misue (University of Tsukuba, Japan), Motoki Miura (Japan Advanced Institute of Science and Technology, Japan), Takanori Ugai (Fujitsu Laboratories Limited, Japan)

Knowledge-Based Multi-criteria Decision Support
Hsuan-Shih Lee (National Taiwan Ocean University, Taiwan)

Knowledge-Based Systems for e-Business
Kazuhiko Tsuda (University of Tsukuba, Japan)

Neural Information Processing for Data Mining
Ryohei Nakano (Nagoya Institute of Technology, Japan), Kazumi Saito (University of Shizuoka, Japan)

Neural Networks in Image Processing
Monica Bianchini, Marco Maggini, Franco Scarselli (Università degli Studi di Siena, Italy)

New Advances in Defence and Security Systems in Intelligent Environments
Jadranka Sunde (Defence Science and Technology Organisation, Australia), Lakhmi C. Jain (University of South Australia, Australia)

Novel Foundation and Applications of Intelligent Systems
Valentina E. Balas (Aurel Vlaicu University of Arad, Romania), Chee-Peng Lim, (University of Science Malaysia, Malaysia), Lakhmi C. Jain (University of South Australia, Australia)

Reasoning-Based Intelligent Systems
Kazumi Nakamatsu (University of Hyogo, Japan)

Relevant Reasoning for Discovery and Prediction
Jingde Cheng, Yuichi Goto (Saitama University, Japan)

Skill Acquisition and Ubiquitous Human Computer Interaction
Hirokazu Taki (Wakayama University, Japan), Satoshi Hori (Monotsukuri Institute of Technologists, Japan)

Soft Computing Approach to Management Engineering
Junzo Watada (Waseda University, Japan), Huey-Ming Lee (Chinese Cultural University, Taiwan), Taki Kanda (Bunri University of Hospitality, Japan)

Smart Sustainability
Robert J. Howlett (University of Brighton, UK)

Spatio-temporal Database Concept Support for Organizing Virtual Earth
Toyohide Watanabe (Nagoya University, Japan), Jun Feng (Hohai University, Japan), Naoto Mukai (Tokyo University of Science, Japan)

Unsupervised Clustering for Exploratory Data Analysis
Roberto Tagliaferri (University of Salerno, Italy), Michele Ceccarelli (University of Sannio, Italy)

Use of AI Techniques to Build Enterprise Systems
Ronald L. Hartung (Franklin University, USA)

XML Security
Ernesto Damiani, Stefania Marrara (University of Milan, Italy)

3D Approaches for Visual Facial Expression and Emotion Dynamics Recognition in a Real Time Context
Andoni Beristain, Manuel Grana (Universidad del Pais Vasco, Spain)

Sponsoring Institutions

Ministry of Science, Education and Sports of the Republic of Croatia
University of Zagreb, Faculty of Electrical Engineering and Computing
Ericsson Nikola Tesla, Zagreb, Croatia
Croatian National Tourist Board
Zagreb Tourist Board

Table of Contents – Part I

Keynote Lectures

Assisting Human Decision Making with Intelligent Technologies 1
Gloria Phillips-Wren

Driven by Compression Progress 11
Jürgen Schmidhuber

Evolvable Hardware Architectures and Their Role in Fault Tolerant Computational Systems 12
Andy Tyrrell

Legal Issues for Military Intelligent Decision-Making Technologies 13
Anthony Finn

New Topics from Recent Interactive Evolutionary Computation Researches 14
Hideyuki Takagi

Virtual Humans: From Researcher's Fascination to Mass-Market Technology 15
Igor S. Pandzic

How Many Clusters Are There? An Essay on the Basic Notions of Clustering 16
Bernd Reusch

I General Tracks

Artificial Neural Networks and Connectionist Systems

An Intelligent Traffic Light Control Based on Extension Neural Network 17
Kuei-Hsiang Chao, Ren-Hao Lee, and Meng-Hui Wang

Design of an Auto-associative Neural Network by Using Design of Experiments Approach 25
Božidar Bratina, Nenad Muškinja, and Boris Tovornik

A New Implementation for High Speed Normalized Neural Networks in Frequency Space 33
Hazem M. El-Bakry and Mohamed Hamada

A New SOM Initialization Algorithm for Nonvectorial Data 41
Antonino Fiannaca, Riccardo Rizzo, Alfonso Urso, and Salvatore Gaglio

Accelerated Gradient Learning Algorithm for Neural Network Weights Update 49
Zeljko Hocenski, Mladen Antunovic, and Damir Filko

Neural Model-Based Segmentation of Image Motion 57
Lucia Maddalena and Alfredo Petrosino

Fuzzy and Neuro-Fuzzy Systems

Classification on Soft Labels Is Robust against Label Noise 65
Christian Thiel

Prediction of the Collapse Index by a Mamdani Fuzzy Inference System 74
Kivanc Zorlu and Candan Gokceoglu

Evolutionary Computation

A Cross-Entropy Based Population Learning Algorithm for Discrete-Continuous Scheduling with Continuous Resource Discretisation 82
Piotr Jędrzejowicz and Aleksander Skakovski

Improving the Responsiveness of NSGA-II in Dynamic Environments Using an Adaptive Mutation Operator – A Case Study 90
Alvaro Gomes, C. Henggeler Antunes, and A. Gomes Martins

Machine Learning and Classical AI

Structural Analysis of Promoter Sequences Using Grammar Inference and Support Vector Machine 98
Robertas Damaševičius

A Metric to Discriminate the Selection of Algorithms for the General ATSP Problem 106
Jorge A. Ruiz-Vanoye, Ocotlán Díaz-Parra, and Vanesa Landero N.

Classification Method Utilizing Reliably Labeled Data 114
Kouta Nakata, Shigeaki Sakurai, and Ryohei Orihara

A Manifolded AdaBoost for Face Recognition 123
C. Lu, J. Jiang, G. Feng, and C. Qing

On Lipschitz Embeddings of Graphs 131
Kaspar Riesen and Horst Bunke

An Experiment on Human Face Recognition Performance for Access Control 141
Marcus Butavicius, Chloë Mount, Veneta MacLeod, Robyn Vast, Ian Graves, and Jadranka Sunde

Agent Systems

A First Step Towards Argument Mining and Its Use in Arguing Agents and ITS 149
Safia Abbas and Hajime Sawamura

Optimizing Service Distributions Using a Genetic Algorithm 158
Kresimir Jurasovic and Mario Kusek

Agent-Based User Personalization Using Context-Aware Semantic Reasoning 166
Fran Frkovic, Vedran Podobnik, Krunoslav Trzec, and Gordan Jezic

Performance Evaluation of a Mobile Agent Network Using Network Calculus 174
Vjekoslav Sinkovic, Mario Kusek, Gordan Jezic, and Ignac Lovrek

Using Generalized Learning Automata for State Space Aggregation in MAS 182
Yann-Michaël De Hauwere, Peter Vrancx, and Ann Nowé

Recommendation of Multimedia Objects Based on Similarity of Ontologies 194
Przemysław Kazienko, Katarzyna Musiał, and Krzysztof Juszczyszyn

Toward Emotional E-Commerce: The Customer Agent 202
Nicolae Jascanu, Veronica Jascanu, and Severin Bumbaru

A Method for Integration of WordNet-Based Ontologies Using Distance Measures 210
Trong Hai Duong, Ngoc Thanh Nguyen, and Geun Sik Jo

The Multiagent System for Channel Selection and Topology Control on 802.11 Based Mesh Networks 220
Ante Prodan, Vinod Mirchandani, and John Debenham

EasyLife: A Location-Aware Service Oriented Mobile Information System 229
J.M. Shen, M.J. O'Grady, and G.M.P. O'Hare

A New Concept of Trust Modeling and Management in Complex Networks 237
Grzegorz Kołaczek

Extending Intelligent Tutoring Systems to Mobile Devices 245
Vlado Glavinić, Marko Rosić, and Marija Zelić

Integration of Knowledge in Disjunctive Structure on Semantic Level ... 253
Trong Hieu Tran and Ngoc Thanh Nguyen

Three Parameters Refinement for Distributed 3D-Image Rendering 262
Dariusz Król and Adam Bogdał

Knowledge Based and Expert Systems

CBR to Explain Medical Model Exceptions 270
Olga Vorobieva and Rainer Schmidt

Reasoning for Incomplete Authorizations 278
Yun Bai

A Place Judgment System Based on an Association Mechanism for Natural Language Interaction 286
Seiji Tsuchiya, Eriko Yoshimura, Ren Fuji, Hirokazu Watabe, and Tsukasa Kawaoka

Knowledge Maturity Based on Processing and Creation of Information 294
Sabah S. Al-Fedaghi

On Attaining Higher Level of Certainty in Evaluation Processes 304
Sylvia Encheva and Sharil Tumin

A User Interface for the User-Centred Knowledge Model, t-UCK 312
Anne Håkansson

Focus Support Interface Based on Collaborative Learning Activity 322
Yuki Hayashi, Tomoko Kojiri, and Toyohide Watanabe

Grammar-Based Argument Construction 330
Jörn Sprado and Björn Gottfried

Towards the Knowledge Capitalisation and Organisation in the Model of Conformity-Checking Process in Construction 341
Anastasiya Yurchyshyna, Catherine Faron-Zucker, Nhan Le Thanh, and Alain Zarli

A Web-Based Customer Support Knowledge Base System 349
S. Wood and R.J. Howlett

Incremental Determinization of Finite Automata in Model-Based Diagnosis of Active Systems 362
Gianfranco Lamperti, Marina Zanella, Giovanni Chiodi, and Lorenzo Chiodi

A Knowledge Based Approach for Capturing Rich Semantic Representations from Text 375
Peter Z. Yeh, Daniel R. Farina, and Alex Kass

Ontology-Based Design Pattern Recognition 384
Damir Kirasić and Danko Basch

Intelligent Vision and Image Processing

Cognitive Systems for Medical Pattern Understanding and Diagnosis 394
Lidia Ogiela

Detection and Analysis of Cell Nuclear Phases 401
Donggang Yu, Tuan D. Pham, and Xiaobo Zhou

Simple Perceptually-Inspired Methods for Blob Extraction 409
Paolo Falcoz

A Study on Self Practice Navigation System for the "Kana" Strings 417
Kenta Kouyama, Naoto Hara, Toshiaki Kuroiwa, Seiu Yamashita, Akiko Ono, Isao J. Ohsugi, Junichi Yoshino, and Hiroshi Ichimura

A Fuzzy Thresholding Circuit for Image Segmentation 425
Angel Barriga and Nashaat M. Hussein

A Method for Reading a Resistor by Image Processing Techniques 433
Yoshihiro Mitani, Yuuki Sugimura, and Yoshihiko Hamamoto

Networked Virtual Marionette Theater 440
Daisuke Ninomiya, Kohji Miyazaki, and Ryohei Nakatsu

A Generic Methodology for Classification of Complex Data Structures in Automotive Industry 448
Dymitr Ruta

Knowledge Management, Ontologies and Data Mining

Uncovering the Deep Web: Transferring Relational Database Content and Metadata to OWL Ontologies 456
Damir Jurić, Marko Banek, and Zoran Skočir

Accessing the Distributed Learner Profile in the Semantic Web 464
Mourad Ouziri

A New Travel Time Prediction Method for Intelligent Transportation Systems 473
Hyunjo Lee, Nihad Karim Chowdhury, and Jaewoo Chang

Measuring Sequence Similarity Trough Many-to-Many Frequent Correlations 484
Gianluigi Greco and Giorgio Terracina

IRPS – An Efficient Test Data Generation Strategy for Pairwise Testing 493
Mohammed I. Younis, Kamal Zuhairi Zamli, and Nor Ashidi Mat Isa

A Procedure Ontology for Advanced Diagnosis of Process Systems 501
Katalin M. Hangos, Erzsébet Németh, and Rozália Lakner

Congenital Heart Disease: An Ontology-Based Approach for the Examination of the Cardiovascular System 509
M. Esposito

ODDI: Ontology-Driven Data Integration 517
Paolo Ceravolo, Zhan Cui, Ernesto Damiani, Alex Gusmini, and Marcello Leida

Enhancing Recommendations through a Data Mining Algorithm 525
Alexis Lazanas and Nikos Karacapilidis

Using Sense Recognition to Resolve the Problem of Polysemy in Building a Taxonomic Hierarchy 533
Lei Liu, Sen Zhang, Lu Hong Diao, Shu Ying Yan, and Cun Gen Cao

A Framework for Context-Awareness in Artificial Systems 541
Fulvio Mastrogiovanni, Antonio Sgorbissa, and Renato Zaccaria

The Proposal of the Association Response Based on Commonsense in Conversation System 549
Eriko Yoshimura, Seiji Tsuchiya, Hirokazu Watabe, and Tsukasa Kawaoka

Ontological Knowledge Maintenance Methodology 557
Rim Djedidi and Marie-Aude Aufaure

A Hybrid Approach for Data Preprocessing in the QSAR Problem 565
Adina Cocu, Luminita Dumitriu, Marian Craciun, and Cristina Segal

Modeling of the Students Scenario on a Learning Course 573
Yuji Shinoda, Kenji Yoshida, and Hirotaka Nakayama

Sensor Network-Based Nonlinear System Identification 580
Giorgio Biagetti, Paolo Crippa, Francesco Gianfelici, and Claudio Turchetti

Web Intelligence, Text and Multimedia Mining and Retrieval

Automatic Generation of Computer Animation Conveying Impressions of News Articles 588
Tadahiko Kumamoto, Akiyo Nadamoto, and Katsumi Tanaka

Data Mining for Navigation Generating System with Unorganized Web Resources 598
Diana Purwitasari, Yasuhisa Okazaki, and Kenzi Watanabe

Exploring Robustness Enhancements for Logic-Based Passage Filtering 606
Ingo Glöckner and Björn Pelzer

Comparing Document Classification Schemes Using K-Means Clustering 615
Artur Šilić, Marie-Francine Moens, Lovro Žmak, and Bojana Dalbelo Bašić

Describing and Generating Web-Based Affective Human-Agent Interaction 625
Xia Mao, Zheng Li, and Haiyan Bao

Scalable Content-Based Ranking in P2P Information Retrieval 633
Maroje Puh, Toan Luu, Ivana Podnar Zarko, and Martin Rajman

A New Algorithm for Community Identification in Linked Data 641
Nacim Fateh Chikhi, Bernard Rothenburger, and Nathalie Aussenac-Gilles

Intelligent Robotics and Control

GrasSmart: An Intelligent Robotic System for Continuous Area Coverage 650
Miri Weiss-Cohen, Igal Sirotin, and Erez Rave

Stator Resistance Tuning Based on a Neural Network in an Indirect Rotor Field Oriented Control System of an Induction Motor 658
Dinko Vukadinovic, Mateo Basic, and Ljubomir Kulisic

A Simple Goal Seeking Navigation Method for a Mobile Robot Using Human Sense, Fuzzy Logic and Reinforcement Learning 666
Hamid Boubertakh, Mohamed Tadjine, and Pierre-Yves Glorennec

Fuzzy Control of a Real Time Inverted Pendulum System 674
Selcuk Kizir, Zafer Bingul, and Cuneyt Oysu

Construction and Evaluation of a Robot Dance System 682
Kuniya Shinozaki, Akitsugu Iwatani, and Ryohei Nakatsu

Development and Evaluation of a Centaur Robot 692
Satoshi Tsuda, Kuniya Shinozaki, and Ryohei Nakatsu

Collaboration Options for Small and Medium Size Enterprises 701
Sylvia Encheva and Sharil Tumin

Service Robot System Using Personal Attribute and Acquisition Preference Attribute with Networked Robots 709
Kota Nakamura, Yoshiharu Yoshida, Toru Yamaguchi, and Eri-Shimokawara-Sato

Self-Protecting Session Initiation Protocol Stack 717
Zoran Rusinovic and Nikola Bogunovic

Early Forest Fire Detection with Sensor Networks: Sliding Window Skylines Approach 725
Krešimir Pripužić, Hrvoje Belani, and Marin Vuković

Analysing Flight Data Using Clustering Methods 733
Christopher Jesse, Honghai Liu, Edward Smart, and David Brown

Visual-Based View-Invariant Human Motion Analysis: A Review 741
Xiaofei Ji, Honghai Liu, Yibo Li, and David Brown

Generalized Extreme Value for Smooth Component Analysis in Prediction Improvement 749
Ryszard Szupiluk, Piotr Wojewnik, and Tomasz Ząbkowski

Context Cookies 757
Ramón Hervás, Gabriel Chavira, Salvador W. Nava, Vladimir Villarreal, and José Bravo

Smart Environment Vectorization: An Approach to Learning of User Lighting Preferences 765
Alejandro Fernández-Montes, Juan A. Ortega, Luis González, Juan A. Álvarez, and Manuel D. Cruz

Author Index 773

Table of Contents – Part II

II Intelligence Everywhere

Artificial Intelligence Driven Engineering Design Optimization

Design of Row-Based Flexible Manufacturing System with Evolutionary Computation 1
Mirko Ficko and Joze Balic

Ant Colony System-Based Algorithm for Optimal Multi-stage Planning of Distribution Transformer Sizing 9
Eleftherios I. Amoiralis, Pavlos S. Georgilakis, Marina A. Tsili, and Antonios G. Kladas

On the Evolutionary-Fuzzy Control of WIP in Manufacturing Systems 18
Nikos C. Tsourveloudis

Numerical Study and Optimal Blade Design of a Centrifugal Pump by Evolutionary Algorithms 26
Vasilis Grapsas, Fotis Stamatelos, John Anagnostopoulos, and Dimitris Papantonis

Groundwater Numerical Modeling and Environmental Design Using Artificial Neural Networks and Differential Evolution 34
Ioannis K. Nikolos, Maria Stergiadi, Maria P. Papadopoulou, and George P. Karatzas

Biomedical Informatics: Intelligent Information Management from Nanomedicine to Public Health

Using Hierarchical Task Network Planning Techniques to Create Custom Web Search Services over Multiple Biomedical Databases 42
Miguel García-Remesal

Building an Index of Nanomedical Resources: An Automatic Approach Based on Text Mining 50
Stefano Chiesa, Miguel García-Remesal, Guillermo de la Calle, Diana de la Iglesia, Vaida Bankauskaite, and Víctor Maojo

Annotation of Colorectal Cancer Data Using the UMLS Metathesaurus 58
Marcos Martínez, José M. Vázquez, Javier Pereira, and Alejandro Pazos

A Primer in Knowledge Management for Nanoinformatics in Medicine 66
Fernando Martín-Sanchez, Victoria López-Alonso, Isabel Hermosilla-Gimeno, and Guillermo Lopez-Campos

Communicative Intelligence

Towards Natural Head Movement of Autonomous Speaker Agent 73
Marko Brkic, Karlo Smid, Tomislav Pejsa, and Igor S. Pandzic

Automatic Translation in Two Phases: Recognition and Interpretation 81
Osvaldo Cairó and Silvia Guardati

Representation of Uncertain Knowledge in Probabilistic OLAP Model 89
Maciej Kiewra

Local Topology of Social Network Based on Motif Analysis 97
Krzysztof Juszczyszyn, Przemysław Kazienko, and Katarzyna Musiał

A Semantic Language for Querying Anonymous Web Sources 106
François Pinet and Michel Schneider

CDNs with Global Adaptive Request Distribution 117
Leszek Borzemski and Krzysztof Zatwarnicki

Computational Intelligence for Image Processing and Pattern Recognition

Digital Video Watermarking Based on RGB Color Channels and Principal Component Analysis 125
Hanane Mirza, Hien Thai, and Zensho Nakao

A Watermarking Approach for MIDI File Based on Velocity and Duration Modulation 133
Alexander Adli, Hanane Mirza, and Zensho Nakao

Application of Interactive Genetic Algorithms to Boid Model Based Artificial Fish Schools 141
Yen-Wei Chen, Kanami Kobayashi, Hitoshi Kawabayashi, and Xinyin Huang

Classification of High-Resolution Satellite Images Using Supervised Locality Preserving Projections 149
Yen-Wei Chen and Xian-Hua Han

Computational Intelligence in Human Cancer Research

Downsizing Multigenic Predictors of the Response to Preoperative Chemotherapy in Breast Cancer 157
René Natowicz, Roberto Incitti, Roman Rouzier, Arben Çela, Antônio Braga, Euler Horta, Thiago Rodrigues, and Marcelo Costa

Classification of Sporadic and BRCA1 Ovarian Cancer Based on a Genome-Wide Study of Copy Number Variations 165
Anneleen Daemen, Olivier Gevaert, Karin Leunen, Vanessa Vanspauwen, Geneviève Michils, Eric Legius, Ignace Vergote, and Bart De Moor

Rule-Based Assistance to Brain Tumour Diagnosis Using LR-FIR 173
Àngela Nebot, Félix Castro, Alfredo Vellido, Margarida Julià-Sapé, and Carles Arús

Are Model-Based Clustering and Neural Clustering Consistent? A Case Study from Bioinformatics 181
Davide Bacciu, Elia Biganzoli, Paulo J.G. Lisboa, and Antonina Starita

Exploratory Characterization of Outliers in a Multi-centre ^{1}H-MRS Brain Tumour Dataset 189
Alfredo Vellido, Margarida Julià-Sapé, Enrique Romero, and Carles Arús

Feature Selection in *in vivo* ^{1}H-MRS Single Voxel Spectra 197
Félix F. González-Navarro and Lluís A. Belanche-Muñoz

Statistical Assessment of MSigDB Gene Sets in Colon Cancer 206
Angela Distaso, Luca Abatangelo, Rosalia Maglietta, Teresa Maria Creanza, Ada Piepoli, Massimo Carella, Annarita D'Addabbo, Sayan Mukherjee, and Nicola Ancona

Stratification of Severity of Illness Indices: A Case Study for Breast Cancer Prognosis 214
Terence A. Etchells, Ana S. Fernandes, Ian H. Jarman, José M. Fonseca, and Paulo J.G. Lisboa

Computational Intelligence Techniques for Web Personalization

Categorization of Web Users by Fuzzy Clustering 222
Giovanna Castellano and Maria Alessandra Torsello

Fuzzy User Profiling in e-Learning Contexts 230
Corrado Mencar, Ciro Castiello, and Anna Maria Fanelli

Personalized Web Search by Constructing Semantic Clusters of User Profiles 238
John Garofalakis, Theodoula Giannakoudi, and Agoritsa Vopi

A Comparison of Different Rating Based Collaborative Filtering Algorithms 248
Fabián P. Lousame and Eduardo Sánchez

Computational Intelligent Techniques for Bioprocess Modelling, Monitoring and Control

Application of Agent-Based System for Bioprocess Description and Process Improvement 257
Ying Gao, Katie Kipling, Jarka Glassey, Mark Willis, Gary Montague, Yuhong Zhou, and Nigel Titchener-Hooker

A Dynamic Bayesian Network to Represent a Ripening Process of a Soft Mould Cheese 265
Cédric Baudrit, Pierre-Henri Wuillemin, Mariette Sicard, and Nathalie Perrot

Nonlinear and Neural Networks Based Adaptive Control for a Wastewater Treatment Bioprocess 273
Emil Petre, Dan Selişteanu, Dorin Şendrescu, and Cosmin Ionete

Epileptic Seizure Classification Using Neural Networks with 14 Features 281
Rui P. Costa, Pedro Oliveira, Guilherme Rodrigues, Bruno Leitão, and António Dourado

Intelligent Computing for Grid

Predicting Grid Performance Based on Novel Reduct Algorithm 289
Kun Gao

Discussing Redundancy Issues in Intelligent Agent-Based Non-traditional Grids 297
Versavia Ancusa, Razvan Bogdan, and Mircea Vladutiu

A Uniform Parallel Optimization Method for Knowledge Discovery Grid 306
Kun Gao

E-Service Delivery Pattern for Knowledge Discovery Grid 313
Meiqun Liu

Design and Implementation to Intelligent Examination System for e-Business Application Operation ... 318
Xin Jin and YanLin Ma

Application Study in Decision Support with Fuzzy Cognitive Map ... 324
Yue He

Combination Study of Multi Fuzzy Cognitive Map ... 332
Chun-Mei Lin

A Hermeneutic Approach to the Notion of Information in IS ... 340
Su-Fen Wang

Intelligent Security Techniques

Access Control Labeling Scheme for Efficient Secure XML Query Processing ... 346
Dong Chan An and Seog Park

A Secure Mediator for Integrating Multiple Level Access Control Policies ... 354
Isabel F. Cruz, Rigel Gjomemo, and Mirko Orsini

Domain Account Model ... 363
Giulio Galiero, Paolo Roccetti, and Andrea Turli

Impostor Users Discovery Using a Multimodal Biometric Continuous Authentication Fuzzy System ... 371
Antonia Azzini and Stefania Marrara

Intelligent Utilization of Soft Computing Techniques

A Learning Automata Approach to Multi-agent Policy Gradient Learning ... 379
Maarten Peeters, Ville Könönen, Katja Verbeeck, and Ann Nowé

Coordinated Exploration in Conflicting Multi-stage Games ... 391
Maarten Peeters, Ville Könönen, Katja Verbeeck, Sven Van Segbroeck, and Ann Nowé

Incremental Learning Method of Simple-PCA ... 403
Tadahiro Oyama, Stephen Karungaru, Satoru Tsuge, Yasue Mitsukura, and Minoru Fukumi

Utilization of Soft Computing Techniques for Making Environmental Games More Exciting – Toward an Effective Utilization of the COMMONS GAME ... 411
Norio Baba, Kenta Nagasawa, and Hisashi Handa

Petri Net-Based Simulation and Analysis of the Software Development Process 418
Gordan Topic, Dragan Jevtic, and Marijan Kunstic

Learning Grouping and Anti-predator Behaviors for Multi-agent Systems 426
Koichiro Morihiro, Haruhiko Nishimura, Teijiro Isokawa, and Nobuyuki Matsui

Network Simulations for Relationality Design – An Approach Toward Complex Systems 434
Katsunori Shimohara

Dynamical Effect on Tick-Wise Price Predictions 442
Mieko Tanaka-Yamawaki and Keita Awaji

Time-Series Models of the EEG Wearing Overcorrected Glasses 450
Yohei Tomita, Shin-ichi Ito, Yasue Mitsukura, Minoru Fukumi, and Taketoshi Suzuki

Feature Extraction System for Age Estimation 458
Hironobu Fukai, Hironori Takimoto, Yasue Mitsukura, and Minoru Fukumi

Reasoning-Based Intelligent Systems

Improving EEG Analysis by Using Paraconsistent Artificial Neural Networks 466
Jair Minoro Abe, Helder F.S. Lopes, and Kazumi Nakamatsu

Transitive Reasoning of Before-After Relation Based on Bf-EVALPSN 474
Kazumi Nakamatsu, Jair Minoro Abe, and Seiki Akama

A Conceptual Model for Guiding the Clustering Analysis 483
Wagner F. Castilho, Gentil J. Lucena Filho, Hércules A. do Prado, Edilson Ferneda, and Margarete Axt

An Image-Based Integration System for Real-Time Dispatching of Multi-robots 491
Li-Che Chen, Tien-Ruey Hsiang, Yu Fu, and Sheng-Luen Chung

Relevant Reasoning for Discovery and Prediction

A Fast Duplication Checking Algorithm for Forward Reasoning Engines 499
Takahiro Koh, Yuichi Goto, and Jingde Cheng

Formal Definition of Relevant Logical Symbol Occurrence 508
Noriaki Yoshiura

Deontic Relevant Logic as the Logical Basis for Representing and Reasoning about Legal Knowledge in Legal Information Systems 517
Jingde Cheng

A General Forward Reasoning Algorithm for Various Logic Systems with Different Formalizations 526
Yuichi Goto, Takahiro Koh, and Jingde Cheng

Spatio-Temporal Database Concept Support for Organizing Virtual Earth

Map-Oriented Regional Information Management for Data Broadcasting Contents 536
Masahiro Ura, Takami Yasuda, Masashi Yamada, Mamoru Endo, Shinya Miyazaki, and Koji Nakamura

Continuous Clustering of Moving Objects in Spatial Networks 543
Wenting Liu, Zhijian Wang, and Jun Feng

Index Method for Tracking Network-Constrained Moving Objects 551
Jun Feng, Jiamin Lu, Yuelong Zhu, and Toyohide Watanabe

Adaptive Routing of Cruising Taxis by Mutual Exchange of Pathways 559
Kosuke Yamamoto, Kentaro Uesugi, and Toyohide Watanabe

Route Optimization Using Q-Learning for On-Demand Bus Systems 567
Naoto Mukai, Toyohide Watanabe, and Jun Feng

Sharing Event Information in Distributed Environment Based on Three-Layered Structure 575
Masakazu Ikezaki, Toyohide Watanabe, Tomoko Kojiri, and Taketoshi Ushiama

III Knowledge Everywhere

Advanced Knowledge-Based Systems

An Introduction to Authorization Conflict Problem in RDF Access Control 583
Jaehoon Kim, Kangsoo Jung, and Seog Park

Improving Text Summarization Using Noun Retrieval Techniques 593
Christos Bouras and Vassilis Tsogkas

Providing Flexible Queries over Web Databases 601
Xiangfu Meng, Z.M. Ma, and Li Yan

Fast Cryptographic Privacy Preserving Association Rules Mining on Distributed Homogenous Data Base 607
Mahmoud Hussein, Ashraf El-Sisi, and Nabil Ismail

SD-Core: Generic Semantic Middleware Components for the Semantic Web 617
Ismael Navas-Delgado and José F. Aldana-Montes

A Knowledge-Based Approach for Answering Fuzzy Queries over Relational Databases 623
Z.M. Ma and Xiangfu Meng

An Algorithm for Discovering Ontology Mappings in P2P Systems 631
Giuseppe Pirrò, Massimo Ruffolo, and Domenico Talia

Privacy Risks in Trajectory Data Publishing: Reconstructing Private Trajectories from Continuous Properties 642
Emre Kaplan, Thomas B. Pedersen, Erkay Savaş, and Yücel Saygın

A Novel Approach for Practical Semantic Web Data Management 650
Giorgio Gianforme, Roberto De Virgilio, Stefano Paolozzi, Pierluigi Del Nostro, and Danilo Avola

Distance-Based Classification in OWL Ontologies 656
Claudia d'Amato, Nicola Fanizzi, and Floriana Esposito

Kernel Methods for Graphs: A Comprehensive Approach 662
Francesco Camastra and Alfredo Petrosino

Event-Based Compression and Mining of Data Streams 670
Alfredo Cuzzocrea and Sharma Chakravarthy

Context Enabled Semantic Granularity 682
Riccardo Albertoni, Elena Camossi, Monica De Martino, Franca Giannini, and Marina Monti

Modeling Spatial Relations between Places to Support Geographic Information Retrieval 689
Alia I. Abdelmoty and Baher A. El-Geresy

Towards Relational Inconsistent Databases with Functional Dependencies 695
Sergio Greco and Cristian Molinaro

Chance Discovery

A Bank Run Model in Financial Crises 703
Katsutoshi Yada, Takashi Washio, Yasuharu Ukai, and Hisao Nagaoka

Logic of Discovery and Knowledge: Decision Algorithm 711
Sergey Babenyshev and Vladimir Rybakov

Chances, Affordances, Niche Construction 719
Lorenzo Magnani

Free Association Versus Recognition: Sensitivity for Chance Discovery in Cross-Cultural Color Design 727
Gyoung Soon Choi, Ruediger Oehlmann, and David Cottington

A Cross-Cultural Study on Attitudes toward Risk, Safety and Security 734
Yumiko Nara

Insight or Trial and Error: Ambiguous Items as Clue for Discovering New Concepts in Constrained Environments 742
Jun Nakamura and Yukio Ohsawa

Exceptions as Chance for Computational Chance Discovery 750
Akinori Abe, Norihiro Hagita, Michiko Furutani, Yoshiyuki Furutani, and Rumiko Matsuoka

Analyzing Behavior of Objective Rule Evaluation Indices Based on a Correlation Coefficient 758
Hidenao Abe and Shusaku Tsumoto

Innovation-Oriented Knowledge Management Platform

Face Image Annotation Based on Latent Semantic Space and Rules 766
Hideaki Ito, Yuji Kawai, and Hiroyasu Koshimizu

A Spam Filtering Method Learning from Web Browsing Behavior 774
Taiki Takashita, Tsuyoshi Itokawa, Teruaki Kitasuka, and Masayoshi Aritsugi

A Method of Graphics Composition Using Differential SVG Documents 782
Motohiro Matsuda, Kazunari Ito, Martin J. Dürst, and Kôiti Hasida

Externalization Support of Key Phrase Channel in Presentation Preparation 790
Koichi Hanaue and Toyohide Watanabe

X-Web: A Data Model for Managing Personal Contents Based on User Experiences 798
Taketoshi Ushiama and Toyohide Watanabe

Knowledge-Based Creativity Support Systems

Understanding Creativity-Technique Based Problem Solving Processes 806
Florian Forster and Michele Brocco

Visualisation Tool for Cooperative Relations and Its Verification with a Case Study 814
Takanori Ugai and Kouji Aoyama

Participatory Workshop as a Creativity Support System 823
Kosuke Numa, Kiyoko Toriumi, Katsuaki Tanaka, Mina Akaishi, and Koichi Hori

Readable Representations for Large-Scale Bipartite Graphs 831
Shuji Sato, Kazuo Misue, and Jiro Tanaka

Evaluation of a Prototype of the Mimamori-care System for Persons with Dementia 839
Taro Sugihara, Kenichi Nakagawa, Tsutomu Fujinami, and Ryozo Takatsuka

Aware Group Home Enhanced by RFID Technology 847
Motoki Miura, Sadanori Ito, Ryozo Takatsuka, and Susumu Kunifuji

A Tabletop Interface Using Controllable Transparency Glass for Collaborative Card-Based Creative Activity 855
Motoki Miura and Susumu Kunifuji

Random Display of Hints and Its Effect on Generating Ideas in Brain-Writing Groupware 863
Ujjwal Neupane, Kazushi Nishimoto, Motoki Miura, and Susumu Kunifuji

Visual Analysis Tool for Bipartite Networks 871
Kazuo Misue

Knowledge-Based Interface Systems [I]

A Model for Measuring Cognitive Complexity of Software 879
Sanjay Misra and Ibrahim Akman

Efficient Tracking with AdaBoost and Particle Filter under Complicated Background 887
Yuji Iwahori, Naoki Enda, Shinji Fukui, Haruki Kawanaka, Robert J. Woodham, and Yoshinori Adachi

Analog VLSI Layout Design and the Circuit Board Manufacturing of Advanced Image Processing for Artificial Vision Model 895
Masashi Kawaguchi, Takashi Jimbo, and Naohiro Ishii

Classification of Local Surface Using Neural Network and Object Rotation of Two Degrees of Freedom 903
Takashi Kojima, Yuji Iwahori, Tsuyoshi Nakamura, Shinji Fukui, Robert J. Woodham, and Hidenori Itoh

Control of Hypothesis Space Using Meta-Knowledge in Inductive Learning 911
Nobuhiro Inuzuka, Hiroyuki Ishida, and Tomofumi Nakano

Knowledge-Based Interface Systems [II]

On Capacity of Memory in Chaotic Neural Networks with Incremental Learning 919
Toshinori Deguchi, Keisuke Matsuno, and Naohiro Ishii

Improved Accuracy by Relearning and Combining Distance Functions 926
Naohiro Ishii, Takahiro Yamada, and Yongguang Bao

Influence of Character Type of Japanese Hiragana on Writer Recognition 934
Yoshinori Adachi, Masahiro Ozaki, and Yuji Iwahori

Learning Algorithm for Study Support in Web Study – Design of Prototype Model 942
Masahiro Ozaki, Yoshinori Adachi, and Naohiro Ishii

Knowledge-Based Multi-criteria Decision Support

Municipal Solid Waste Site Location Using a Fuzzy Logic Approach 950
Beatriz S.L.P. de Lima, Maria C.M. Alves, Alexandre G. Evsukoff, and Ian N. Vieira

Fuzzy-Based Call Admission Control Scheme for Mobile Networks 958
Jing-Long Wang and Shu-Yin Chiang

Routing Analysis Using Fuzzy Logic Systems in Wireless Sensor Networks 966
Shu-Yin Chiang and Jing-Long Wang

Prioritization of Incomplete Fuzzy Preference Relation 974
Hsuan-Shih Lee, Pei-Di Shen, and Wen-Li Chyr

Fuzzy Multi-criteria Decision Making Based on Fuzzy Preference Relation 980
Hsuan-Shih Lee and Chen-Huei Yeh

Knowledge-Based Systems for e-Business

The Development of the Financial Learning Tool through Business Game 986
Yasuo Yamashita, Hiroshi Takahashi, and Takao Terano

A Method for Ensuring Consistency of Software Design Information in Retrospective Computer Validation 994
Masakazu Takahashi, Satoru Takahashi, and Yoshikatsu Fujita

Egocentrism Presumption Method with N-Gram for e-Business 1002
Nobuo Suzuki and Kazuhiko Tsuda

Generating Dual-Directed Recommendation Information from Point-of-Sales Data of a Supermarket 1010
Masakazu Takahashi, Toshiyuki Nakao, Kazuhiko Tsuda, and Takao Terano

Extraction of the Project Risk Knowledge on the Basis of a Project Plan 1018
Yasunobu Kino, Kazuhiko Tsuda, and Tadashi Tsukahara

Real-Valued LCS Using UNDX for Technology Extraction 1026
Setsuya Kurahashi

Author Index 1035

Table of Contents – Part III

IV Intelligent Processing

Intelligent Data Processing in Process Systems and Plants

Development of Activity Models of Integrated Safety and Disaster Management for Industrial Complex Areas 1
Yukiyasu Shimada and Hossam A. Gabbar

LCA of the Various Vehicles in Environment and Safety Aspect 9
Kazuhiro Takeda, Shingo Sugioka, Yukiyasu Shimada, Takashi Hamaguchi, Teiji Kitajima, and Tetsuo Fuchio

Plant Model Generation for Countermeasures Planning 17
Takashi Hamaguchi, Kazuhiro Takeda, Yukiyasu Shimada, and Yoshihiro Hashimoto

Trend Analysis for Decision Support in Control Actions of Suspension Polymerization 25
Hideyuki Matsumoto, Masaya Honda, and Chiaki Kuroda

An Operational Model and a Computer Support Environment for Batch Plants Based on Adaptive Scheduling 33
Hisaaki Yamaba, Makoto Hirosaki, and Shigeyuki Tomita

Neural Information Processing for Data Mining

Improving Search Efficiency of Incremental Variable Selection by Using Second-Order Optimal Criterion 41
Kazumi Saito, Nobuaki Mutoh, Tetsuo Ikeda, Toshinao Goda, and Kazuki Mochizuki

A Biphase-Bayesian-Based Method of Emotion Detection from Talking Voice 50
Jangsik Cho, Shohei Kato, and Hidenori Itoh

EM Algorithm with PIP Initialization and Temperature-Based Selection 58
Yuta Ishikawa and Ryohei Nakano

Prediction of Information Diffusion Probabilities for Independent Cascade Model 67
Kazumi Saito, Ryohei Nakano, and Masahiro Kimura

Reducing SVR Support Vectors by Using Backward Deletion 76
Masayuki Karasuyama, Ichiro Takeuchi, and Ryohei Nakano

Soft Computing Approach to Management Engineering

Structure Analysis of Fuzzy Node Fuzzy Graph and Its Application to Sociometry Analysis 84
Hiroaki Uesu

Consumer Behaviors in Taiwan Online Shopping—Case Study of A Company 92
Lily Lin, Huey-Ming Lee, and Li-Hsueh Lin

A New Fuzzy Risk Assessment Approach 98
Huey-Ming Lee and Lily Lin

Making Scale with Units by Using Order Statistics from the Normal Distribution in AHP 106
Taki Kanda

Placement Problem in an Industrial Environment 111
Shamshul Bahar Yaakob and Junzo Watada

Motion Tracking Using Particle Filter 119
Zalili Binti Musa and Junzo Watada

Regression Model Based on Fuzzy Random Variables 127
Shinya Imai, Shuming Wang, and Junzo Watada

Biological Clustering Method for Logistic Place Decision Making 136
Rohani Binti Abu Bakar and Junzo Watada

V Intelligent Systems

Advanced Groupware

Web Application Construction by Group Work and Practice 144
Kouji Yoshida, Isao Miyaji, Hiroshi Ichimura, and Kunihiro Yamada

Effect Analysis in Implementing IRM (Information Resources Management) to Enterprise 152
Kazuya Matsudaira, Teruhisa Ichikawa, and Tadanori Mizuno

Automatic GUI Generation for Meta-data Based PUCC Sensor Gateway 159
Masatoshi Ogura, Hiroshi Mineno, Norihiro Ishikawa, Tomoyuki Osano, and Tadanori Mizuno

The Pictograph Chat Communicator II 167
Jun Munemori, Taro Fukuda, Moonyati Binti Mohd Yatid, and Junko Itou

Network Forensics on Mobile Ad-Hoc Networks 175
Akira Otaka, Tsuyoshi Takagi, and Osamu Takahashi

Efficient Reliable Data Transmission Using Network Coding in MANET Multipath Routing Environment 183
Tomonori Kagi and Osamu Takahashi

Validation at a Small Building with the Mutual Complement Network by Wireless and Wired 193
Kunihiro Yamada, Kakeru Kimura, Takashi Furumura, Masanori Kojima, Kouji Yoshida, and Tadanori Mizuno

Analysis of Relationships between Smiley and Atmosphere during Chat with Emotional Characters 201
Junko Itou and Jun Munemori

Agent and Multi-agent Systems: Technologies and Applications

Logic of Plausibility for Discovery in Multi-agent Environment Deciding Algorithms 210
Sergey Babenyshev and Vladimir Rybakov

An Agent-Oriented Dynamic Adaptive Threshold Transmission for XML Data on Networks 218
Xu Huang and Dharmendra Sharma

An Agent-Oriented Quantum Key Distribution for Wi-Fi Network Security 227
Xu Huang and Dharmendra Sharma

Developing Intelligent Agent Applications with JADE and JESS 236
Bala M. Balachandran

An Agent Based Multifactor Biometric Security System 245
Girija Chetty, Dharmendra Sharma, and Bala M. Balachandran

Engineered Applications of Semantic Web – SWEA

Classification and Retrieval through Semantic Kernels 252
Claudia d'Amato, Nicola Fanizzi, and Floriana Esposito

Semantic Bookmarking and Search in the Earth Observation Domain 260
Francesca Fallucchi, Maria Teresa Pazienza, Noemi Scarpato, Armando Stellato, Luigi Fusco, and Veronica Guidetti

VSB: The Visual Semantic Browser 269
Ismael Navas-Delgado, Amine Kerzazi, Othmane Chniber, and José F. Aldana-Montes

Lexical and Semantic Resources for NLP: From Words to Meanings 277
Anna Lisa Gentile, Pierpaolo Basile, Leo Iaquinta, and Giovanni Semeraro

A Semantic Similarity Measure for the SIMS Framework 285
Roberto Pirrone, Giuseppe Russo, Pierluca Sangiorgi, Nunzio Ingraffia, and Claudia Vicari

Fuzzy Bilateral Matchmaking in e-Marketplaces 293
Azzurra Ragone, Umberto Straccia, Fernando Bobillo, Tommaso Di Noia, and Eugenio Di Sciascio

Evolvable Hardware and Adaptive Systems – Advanced Engineering Design Methodologies and Applications

Machine Vision Application to the Detection of Micro-organism in Drinking Water 302
Hernando Fernandez-Canque, Sorin Hintea, Gabor Csipkes, Allan Pellow, and Huw Smith

Adaptive and Evolvable Hardware and Systems: The State of the Art and the Prospectus for Future Development 310
Mircea Gh. Negoita, Lukas Sekanina, and Adrian Stoica

Component Adaptation Architectures. A Formal Approach 319
Andreea Vescan

Adaptive Microarray Image Acquisition System and Microarray Image Processing Using FPGA Technology 327
Bogdan Belean, Monica Borda, and Albert Fazakas

Evolvable Hardware Applications in the Area of Electronic Circuits Design

A Digitally Reconfigurable Low Pass Filter for Multi-mode Direct Conversion Receivers 335
Gabor Csipkes, Sorin Hintea, Doris Csipkes, Cristian Rus, Lelia Festila, and Hernando Fernandez-Canque

Novel Image Rejection Filter Based on Neural Networks 343
Botond Sandor Kirei, Marina Topa, Irina Dornean, and Albert Fazakas

Towards Reconfigurable Circuits Based on Ternary Controlled Analog Multiplexers/Demultiplexers 351
Emilia Sipos, Lelia Festila, and Gabriel Oltean

Analog Multiplying/Weighting VLSI Cells for SVM Classifiers 360
Lelia Festila, Lorant Andras Szolga, Mihaela Cirlugea, and Robert Groza

Log-Domain Binary SVM Image Classifier 368
Robert Groza, Lelia Festila, Sorin Hintea, and Mihaela Cirlugea

An Adaptive GP Strategy for Evolving Digital Circuits 376
Mihai Oltean and Laura Dioşan

Computational Intelligence in Analog Circuits Design 384
Gabriel Oltean, Sorin Hintea, and Emilia Sipos

Hyperspectral Imagery for Remote Sensing: Intelligent Analysis and Applications

Dimensionality Reduction and Linear Discriminant Analysis for Hyperspectral Image Classification 392
Qian Du and Nicolas H. Younan

Endmember Extraction Methods: A Short Review 400
Miguel A. Veganzones and Manuel Graña

Super Resolution of Multispectral Images Using TV Image Models 408
Miguel Vega, Javier Mateos, Rafael Molina, and Aggelos K. Katsaggelos

Classification of Hyperspectral Images Compressed through 3D-JPEG2000 416
Ian Blanes, Alaitz Zabala, Gerard Moré, Xavier Pons, and Joan Serra-Sagristà

A New Band Selection Strategy for Target Detection in Hyperspectral Images 424
Marco Diani, Nicola Acito, Mario Greco, and Giovanni Corsini

Immunity-Based Systems

Fatal Mutations in HIV Population as an Influential Factor for an Onset of AIDS 432
Kouji Harada and Yoshiteru Ishida

Asymmetric Interactions between Cooperators and Defectors for Controlling Self-repairing 440
Yoshiteru Ishida and Masahiro Tokumitsu

A Note on Space-Time Interplay through Generosity in a Membrane Formation with Spatial Prisoner's Dilemma 448
Yoshiteru Ishida and Yuji Katsumata

Dynamic Updating of Profiles for an Immunity-Based Anomaly Detection System 456
Takeshi Okamoto and Yoshiteru Ishida

Innovations in Intelligent Multimedia Systems and Virtual Reality

OGRE-Multimedia: An API for the Design of Multimedia and Virtual Reality Applications 465
Paulo N.M. Sampaio, Roberto Ivo C. de Freitas, and Gonçalo Nuno P. Cardoso

Eighty-Color-Channel Lighting 473
Zoltán Márton and Cecília Sik Lányi

Virtual Reality Supporting Environmental Planning Processes: A Case Study of the City Library in Gothenburg 481
Kaj Sunesson, Carl Martin Allwood, Ilona Heldal, Dan Paulin, Mattias Roupé, Mikael Johansson, and Börje Westerdahl

Intelligent Environment Support for Collaborative Learning

A Prototype for Human-Like e-Learning System 491
Kenji Yoshida, Yuji Shinoda, Koji Miyazaki, Kayoko Nakagami, and Hirotaka Nakayama

A Cross-LMS Communication Environment with Web Services 499
Takashi Yukawa and Yoshimi Fukumura

A Web-Based Asynchronous Discussion System and Its Evaluation 507
Hisayoshi Kunimune, Kenzou Yokoyama, Takeshi Takizawa, Takuya Hiramatsu, and Yasushi Fuwa

CSCL Data Structurization and Inter-LMS Sharing with Use of Web Services 515
Yasuhisa Tamura, Kazuya Sumi, Takeshi Yamamuro, and Masashi Maejima

Adventures in the Boundary between Domain-Independent Ontologies and Domain Content for CSCL 523
Seiji Isotani and Riichiro Mizoguchi

Intelligent Systems in Medicine and Healthcare

Mammographic Image Contrast Enhancement Through the Use of Moving Contrast Sweep 533
Zailani Mohd Nordin, Nor Ashidi Mat Isa, Umi Kalthum Ngah, and Kamal Zuhairi Zamli

Detection of Sprague Dawley Sperm Using Matching Method 541
Mohd Fauzi Alias, Nor Ashidi Mat Isa, Siti Amrah Sulaiman, and Kamal Zuhairi Zamli

Application of the Fuzzy Min-Max Neural Networks to Medical Diagnosis 548
Anas Quteishat and Chee Peng Lim

SVD Based Feature Selection and Sample Classification of Proteomic Data 556
Annarita D'Addabbo, Massimo Papale, Salvatore Di Paolo, Simona Magaldi, Roberto Colella, Valentina d'Onofrio, Annamaria Di Palma, Elena Ranieri, Loreto Gesualdo, and Nicola Ancona

Prediction of Crohn's Disease by Profiles of Single Nucleotide Polymorphisms 564
Roberto Colella, Annarita D'Addabbo, Anna Latiano, Orazio Palmieri, Vito Annese, and Nicola Ancona

Decision Analysis of Fuzzy Partition Tree Applying AIC and Fuzzy Decision 572
Kimiaki Shinkai, Shuya Kanagawa, Takenobu Takizawa, and Hajime Yamashita

Knowledge Interaction for Creative Learning

Learning Environment for Improving Critical Thinking Skills Based on New Synthesis Theory 580
Hiroki Satake, Haruna Marumoto, Kazuhisa Seta, Motohide Umano, and Mitsuru Ikeda

English Grammar Learning System Based on Knowledge Network of Fill-in-the-Blank Exercises 588
Takuya Goto, Tomoko Kojiri, Toyohide Watanabe, Takeshi Yamada, and Tomoharu Iwata

The Planetary Simulator for Generalized Understanding of Astronomical Phenomena from Various Viewpoints 596
Masato Soga, Yasunori Nakanishi, and Kohe Tokoi

A Discussion Model for System Design Novices 604
Shoich Nakamura, Miho Watanabe, Atsuo Hazeyama, Setsuo Yokoyama, and Youzou Miyadera

Novel Foundation and Applications of Intelligent Systems

Information Fusion for Uncertainty Determination in Video and Infrared Cameras System 616
Cornel Barna

Various Speaker Recognition Techniques Using a Special Nonlinear Metric 622
Tudor Barbu and Mihaela Costin

A Collaborative Information System Architecture for Process-Based Crisis Management 630
Omar Tahir, Eric Andonoff, Chihab Hanachi, Christophe Sibertin-Blanc, Frédérick Benaben, Vincent Chapurlat, and Thomas Lambolais

A Role-Based Framework for Multi-agent Teaming 642
Jinsong Leng, Jiuyong Li, and Lakhmi C. Jain

About the Buildings' Danger Control Systems Vulnerabilities 650
Daniela E. Popescu and Marcela F. Prada

Traffic Management by Constant Time to Collision 658
Valentina E. Balas and Marius M. Balas

Skill Acquisition and Ubiquitous Human Computer Interaction

Structuring Spatial Knowledge and Fail-Safe Expression for Directive Voice Navigation 664
Taizo Miyachi, Jens J. Balvig, Ipei Kuroda, and Takeshi Suzuki

Ubiquitous Earthquake Observation System Using Wireless Sensor Devices 673
Hirokazu Miura, Yosuke Shimazaki, Noriyuki Matsuda, Fumitaka Uchio, Koji Tsukada, and Hirokazu Taki

An Effectiveness Study of Pictogram Elements for Steps in Manufacturing Procedures 680
Atsuko K. Yamazaki, Kenji Goto, Hirokazu Taki, and Satoshi Hori

Analysis of Continuous Tracks of Online Aerial Handwritten Character Recognition 687
Kazutaka Ogura, Yoshihiro Nishida, Hirokazu Miura, Noriyuki Matsuda, Hirokazu Taki, and Norihiro Abe

Interactive Learning Environment for Drawing Skill Based on Perspective 695
Yoshitake Shojiguchi, Masato Soga, Noriyuki Matsuda, and Hirokazu Taki

Smart Sustainability

Intelligent GPS-Based Vehicle Control for Improved Fuel Consumption and Reduced Emissions 701
S.H. Lee, S.D. Walters, and R.J. Howlett

Agent-Based Decision Making through Intelligent Knowledge Discovery 709
Marina V. Sokolova and Antonio Fernández-Caballero

Conceptual Modeling in a Meta-model of Sustainability Indicators 716
Alfredo Tolón-Becerra and Fernando Bienvenido

Unsupervised Clustering for Exploratory Data Analysis

Using Global Optimization to Explore Multiple Solutions of Clustering Problems 724
Ida Bifulco, Loredana Murino, Francesco Napolitano, Giancarlo Raiconi, and Roberto Tagliaferri

Robust Clustering by Aggregation and Intersection Methods 732
Ida Bifulco, Carmine Fedullo, Francesco Napolitano, Giancarlo Raiconi, and Roberto Tagliaferri

Comparison of Genomic Sequences Clustering Using Normalized Compression Distance and Evolutionary Distance 740
Massimo La Rosa, Riccardo Rizzo, Alfonso Urso, and Salvatore Gaglio

A One Class Classifier for Signal Identification: A Biological Case Study 747
Vito Di Gesù, Giosuè Lo Bosco, and Luca Pinello

A Fuzzy Extension of Some Classical Concordance Measures and an Efficient Algorithm for Their Computation 755
Michele Ceccarelli and Antonio Maratea

An Algorithm to Assess the Reliability of Hierarchical Clusters in Gene Expression Data 764
Roberto Avogadri, Matteo Brioschi, Francesca Ruffino, Fulvia Ferrazzi, Alessandro Beghini, and Giorgio Valentini

Use of AI Techniques to Build Enterprise System

Integrating Web Services and Intelligent Agents in Supply Chain for Securing Sensitive Messages 771
Esmiralda Moradian

The User Centred Knowledge Model - t-UCK 779
Anne Håkansson

Information Coordination Using Meta-agents in Information Logistics Processes 788
Mats Apelkrans and Anne Håkansson

Ontology for Enterprise Modeling 799
Ronald L. Hartung, Jay Ramanathan, and Joe Bolinger

Author Index 809

Assisting Human Decision Making with Intelligent Technologies

Gloria Phillips-Wren

The Sellinger School of Business and Management
Loyola College in Maryland, Baltimore, MD 21210 USA
gwren@loyola.edu

Abstract. The human decision making process has been characterized as relatively sequential, limited by cognitive processes, and influenced by previous experience. Under conditions associated with real-time decisions, humans can experience emotional intensity, information overload, and other stressors that often affect their choices. Decision making becomes more complex with distributed teams, collaboration across global boundaries, the speed of information refresh, and the quantity of information available through networked sources. Intelligent decision support systems attempt to address these issues and assist human decision making by developing systems that integrate capabilities from the human user and computational intelligence. This keynote address presents background and research directions needed to advance intelligent decision support systems, and proposes an architectural framework to guide design and evaluation.

Keywords: intelligent decision support systems, decision making, artificial intelligence, evaluation, intelligent agents.

1 Introduction

One of the world's leading futurists, Raymond Kurzweil, envisions a future in which machine intelligence, often called computational intelligence or artificial intelligence (AI), combines with human intelligence to accelerate human evolution (1999). He foresees a continuation of the exponential growth of information technologies and a doubling of their power (such as speed or storage) every year. By contrast, human beings change very slowly. Although humans possess unique abilities among mammals to learn, solve problems and make decisions based on reason and logic, they are relatively poor data processors, can remember only a few relationships between data at one time, reason fairly sequentially, and can be strongly influenced by emotion. Simply put, humans are poorly equipped to deal with complexity, and yet the problems and decisions facing society and business are increasingly complex [20]. Today's environment is characterized by disparate and conflicting information sources, global issues with increasing numbers of factors that must be considered, complex relationships between factors, overwhelming amounts of data, and data that dynamically change. It is no wonder that organizations, governments, and individuals look to technology to aid them in making decisions.

I. Lovrek, R.J. Howlett, and L.C. Jain (Eds.): KES 2008, Part I, LNAI 5177, pp. 1–10, 2008.

Internet-based, distributed systems have become essential aids in modern organizations. When combined with computational techniques such as intelligent agents and neural networks, such systems can become powerful aids to decision makers by mimicking human behavior in some limited yet meaningful way [15]. These newer intelligent systems can assist users with complex situations such as real-time decision making, time-pressured decisions, multiple information flows, dynamic data, information overload, inaccurate data, difficult-to-access data, collaborative or coordinated decisions, and highly uncertain decision environments. As a class, they are called intelligent decision support systems (IDSS), and they attempt to assist the decision maker overcome cognitive limitations in making a decision while possibly creating and learning useful knowledge for the future [2].

2 Models of Decision Making

A number of models have been developed to describe human decision making in "normal" situations. For example, expected utility theory formalizes the probability of different decision outcomes. A model that has proven more useful, especially as related to technology support for decision making, is to consider the process of decision making instead of the outcomes [21]. Sometimes referred to as the rationalist approach, the best decisions are seen as flowing from a largely sequential set of steps [20]. Nobel Laureate Herbert Simon is best known for this approach in his seminal work describing the decision making process as consisting of four phases: intelligence, design, choice and implementation [26]. The decision maker acquires information and develops an understanding of the problem during the intelligence phase. During the design phase the user identifies criteria, develops the decision model, and investigates alternatives. An alternative is selected during choice, and the user acts on the decision during the implementation phase. The role of information in the first two steps is crucial, and iteration may occur between the phases. Technology can support one or several steps in the process.

A similar four-step decision making process is used by researchers in military applications and is called the Observe – Orient – Decide - Act (OODA) loop or four box method [28]. The cycle was developed from Boyd's studies of air-to-air combat in the Korean War to assist pilots to achieve knowledge superiority and avoid information overload in order to win a battle.

Such rational models do not seem adequate to describe decision makers in extremely high-stress situations that require immediate action. Studies of these types of situations led to a theory of recognition-primed decision making [9]. First responders such as emergency personnel do not have time to design alternative solutions; instead they quickly mentally compare the event to their previous experiences, construct a best case solution, and simulate it in their mind to test its robustness [21]. The importance of experience is key to success, and decision makers choose a path that 'satisficies' (comes closest to a desirable outcome within the available time or resources for the decision) rather than 'maximizes' (produces the absolutely best outcome).

Decision making has also been studied in terms of what can go wrong. Often referred to as cognitive biases, some of these are the Anchoring Trap – bias from using a reference point; Status-Quo Trap – desire to maintain current conditions; Sunk-Cost Trap - perpetuating a decision path based on past investment; Confirming Evidence Trap - looking for evidence to support a decision that has already been made and discounting contradictory information; Framing Trap – bias from posing the decision in terms of risk and emotion; and Estimating and Forecasting Trap – bias from over-confidence in ability to predict [6]. Intelligent decision support systems can help overcome these traps by, for example, providing several perspectives for the user, alerting the user to threatening events, presenting information neutrally and without emotion, including past events in the analysis, and giving recommendations based on fact [20].

There is much that we do not understand about human decision making, and this area continues to be an active research field. Researchers are studying multi-criteria decision making, collaborative decision making, a pattern recognition and look ahead model, and case-based decision making, to mention only a few [21, 20]. Regardless of the decision situation or model used to understand it, the role of intelligent decision support systems is to aid the user or users to improve both the process of, and outcomes from, decision making.

3 Intelligence in Decision Support Systems

What is intelligence in the context of decision support? Turban and Aronson (1998) defined abilities that are indicative of 'intelligent behavior' as: "learn or understand from experience; make sense out of ambiguous or contradictory messages; respond quickly and successfully to a new situation; use reasoning in solving problems; deal with perplexing situations; understand and infer in ordinary, rational ways; apply knowledge to manipulate the environment; think and reason; recognize the relative importance of different elements in a situation."

As the AI community shifts from "inward-looking to outward-looking", it is apparent that research over the past several decades can deliver intelligent behaviors as posed by Turban and Aronson (1998) within decision support systems. As an example, early research focused on AI to aid design processes such as computer-assisted drafting in engineering designs of building. Current efforts use AI as "the glue that holds larger systems together using reasoning systems that represent or manage processes, information, and interaction devices that use conventional procedural programming; effectively blurring the boundaries between AI and non-AI" [11].

Intelligent agents, in particular, are mature enough to allow agent-based computing to move into the mainstream commercial sector [25]. Agents and multi-agent systems are described by researchers as: autonomous - able to make decisions; adaptive - can learn and change behavior in response to changing circumstances; proactive - ability to take an initiative; reactive - responds to changes; communicative - can communicate with other systems, agents and the user; cooperative - can act in coordination with other agents; mobile - ability to travel over networks; goal-directed - can work toward a specific goal; and persistent - can maintain state over long periods of time [3], [7], [8], [32], [23], [14]. Multi-agent systems are said to create an "artificial social system" involving agent architecture, cooperation among agents and with humans, human-like learning, and trust [32].

Agents and multi-agent systems can assist a user with the decision process. Table 1 compares agent characteristics given by the community with Simon's (1977) model of decision making. The phases can be further refined into decision steps to clarify the actions that take place in each phase [29] with one such refinement [12], [17] shown in Table 1. The extensive capabilities of agents combined with maturing research make them ideal for use in intelligent decision support systems. The user is central to the system and has the final decision on accepting or not accepting the recommended course of action and implementing it, as well as interacting with the system to provide user-specific domain knowledge or preferences.

Table 1. Decision Process and Steps vs. Intelligent Agent and Multi-Agent System Characteristics (Phillips-Wren, 2008)

DECISION PROCESS	DECISION STEPS	INTELLIGENT AGENT AND MULTI-AGENT SYSTEM CHARACTERISTICS								
		AUTONOMOUS	ADAPTIVE	PROACTIVE	REACTIVE	COMMUNICATIVE	COOPERATIVE	MOBILE	GOAL-DRIVEN	PERSISTENT
Intelligence	Problem Detection	√		√		√	√	√		√
	Data Gathering	√		√			√		√	√
	Problem Formulation		√	√	√	√			√	
Design	Model Classification		√			√				
	Model Building		√		√	√		√	√	
	Model Validation	√								
Choice	Evaluation				√	√			√	
	Sensitivity Analysis		√		√					
	Selection	√		√	√	√		√		
Implementation	Result Presentation			√	√		√		√	
	Task Planning		√		√			√		
	Task Monitoring	√							√	√

In addition to agents, other computational techniques can assist the decision maker. Artificial neural networks (ANN) use a framework inspired by human brain functioning rooted in learning. One of their useful characteristics is that ANNs can find and represent nonlinear relationships naturally since they do not pre-suppose a particular structure. This can be particularly helpful in the design phase of decision making. Fuzzy logic provides a way to represent imprecise and uncertain knowledge, helpful in both the intelligence and design phases. Evolutionary computing such as genetic algorithms benefits decision making models in several ways: by helping to choose an optimal structure for a model in the design phase; by simultaneously optimizing the multiple criteria in many decision problems; and by allowing the decision maker to view suboptimal solutions as well as the optimal one during choice [15]. Expert systems can supplement the decision maker's knowledge and provide assistance in all decision phases [1]. Spatial methods such as self organized feature maps are able to represent dimensionality assist both design and choice by providing clusters that aggregate information or processes for monitoring and analysis [4]. Computational intelligence permits humans to deal with complexity in all decision phases.

4 An Example

Many processes are essentially decision tasks. For example, a comparison of the information search process used in information retrieval and the decision making process [31] is shown in Table 2. The similarities in the processes together with characteristics of intelligent agents and multi-agent systems suggest that an intelligent decision support system implemented with intelligent agents may be able to aid the user in locating the desired information.

Table 2. Steps in the decision making process compared to the information search process [31]

Decision-Making Process	Description	Information Search Process
Intelligence	Recognize problem; Gain problem understanding; Seek and acquire information	Task Initiation Topic Selection
Design	Develop criteria; Specify relationships; Explore alternatives	Prefocus Exploration Focus Formulation
Choice	Evaluate alternatives; Develop recommendations; Make decision	Information Collection
Implementation	Weigh consequences; Implement decision	Search Closure

Agents were implemented in an Internet-based, suicide prevention website [22] developed for the U.S. National Institute of Mental Health to assist a non-expert user such a parent locate technical medical information in the U.S. National Library of Medicine [13]. Agents persist in autonomously and proactively locating remote information by helping the user identify needed resources, translating the need into medical terms, retrieving information, and communicating with the user in real-time or asynchronously [31], [16].

The difficulty for users is that the databases are large (over 15 million references from more than 5,000 biomedical journals published in the United States and 80 other countries), mix various types of medical information, are constantly updated, and are catalogued according to medical terminology using a MeSH® system whose keywords are developed, and whose articles are categorized, by medical subject area specialists [13]. Users, on the other hand, are generally non-experts without medical training who are generally unfamiliar with the specialized terminology. Access to needed information on suicide is greatly hampered and likely impossible without assistance. Intelligent agents provide appropriate assistance in identifying information and provide access to the databases for this category of non-expert users.

Intelligent agents proactively search for new information for registered users when the user is off-line. Agents are goal-driven to find information to meet the user's stated intention in accordance with the MeSH®, and agents contact the user via email. The intelligent agents are unique in information retrieval since they infer the user's technical medical interests based on a non-technical description.

5 Design and Evaluation of Intelligent Decision Support Systems

Emerging research in computational intelligence and pragmatic applications demonstrate that the community is developing an understanding of aiding human decision makers in complex situations. Why should the research community expend so much time and talent on this field, and why should organizations invest in expensive applications? To begin answering this question and to provide guidance for system design, we have begun to explore multi-criteria evaluation of intelligent decision support systems that include both the process of, and outcomes from, decision making – and that connect the criteria down to the computational method [18], [17].

Most evaluations of decision support systems are based on a single criteria such as outcome (i.e. improvement in the decision outcome such as increased profit, decreased cost, accuracy of prediction) or process (improvement in the way the decision was made such as increased efficiency, user satisfaction, time savings). The Analytic Hierarchy Process (AHP) is a multi-criteria model that synthesizes criteria into a single numeric value for comparison of different systems by structuring criteria into a hierarchy for comparing alternatives [17]. Pairwise comparisons of criteria at the lowest level of the hierarchy are entered by the user.

A conceptualized framework to guide the design and evaluation of intelligent decision support systems is shown in Figure 1. It includes four levels to connect the decision making process to the technical implementation. The top level shows the four phases of decision making, and it is connected to a decisional service-task level, e.g. the Newell's Knowledge Level of Task, Method, and Subtask. The architectural-capability level accounts for the user interface, data and knowledge, and processing capabilities. At the bottom level we have the computational symbol-program level to account for the specific artificial intelligence computational mechanism used in the intelligent decision support system. At the lowest level, the primary interests for researchers are input-output issues and computational efficiencies. The next two levels should be addressed jointly with the user to provide optimal value.

The architecture can be implemented into a multi-criteria model as shown in Figure 2 and can be extended for collaborative decision support. The branches can weighted to indicate importance to the decision maker, and the resulting numerical comparison of alternatives provides guidance on the overall best decision system. One of the benefits of the implementation is that specific contributions of different computational techniques can be precisely known down to individual process detail.

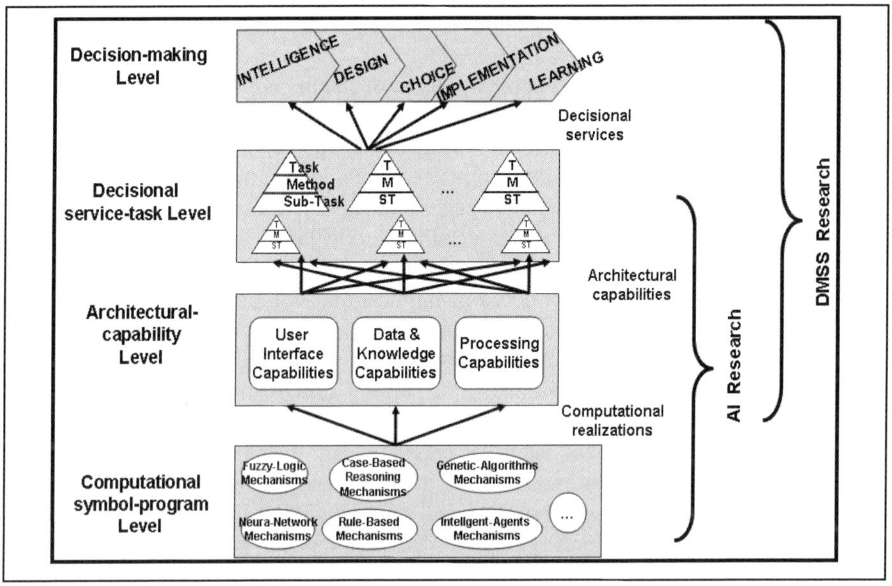

Fig. 1. Conceptual framework for the design and evaluation of intelligent decision support systems [18]

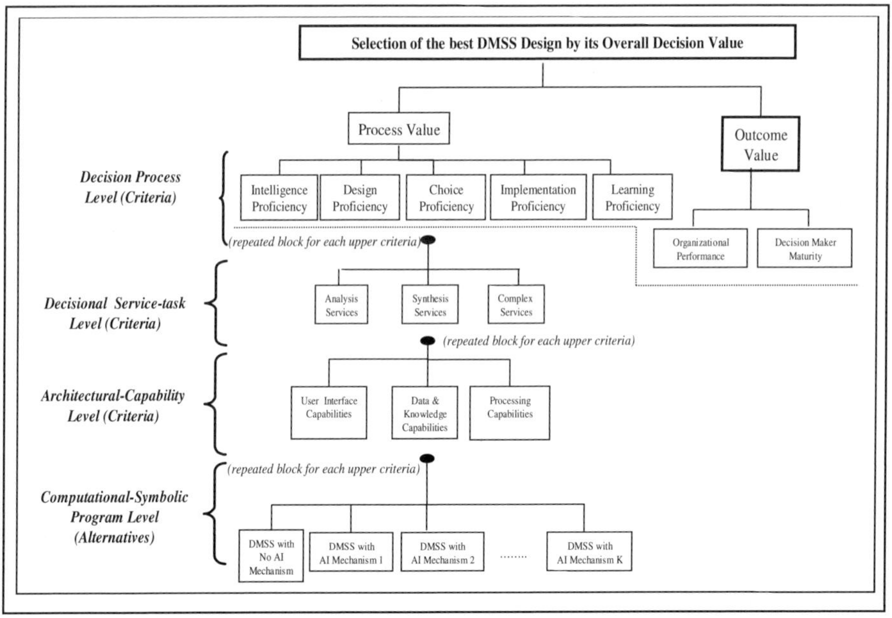

Fig. 2. Multi-criteria model implementation to evaluate intelligent decision support systems [18]

6 Future Trends

Computational technologies and applications in decision support systems offer many interesting opportunities for research. Such systems are not currently widespread. Although some technologies such as intelligent agents are advanced enough to be implemented in practical applications, other technologies are not ready for widespread application. General frameworks and generic infrastructures such as BDI (Belief-Desire-Intention) for intelligent agents need to be developed for technologies such as neural networks. Such architectures will enable specialized applications.

Partnerships need to develop between humans and computers as complementors of each other, with humans strong in areas such as communication and learning, and computers strong in areas such as memory and computational reasoning [20]. Computers are fast, accurate, unemotional and parallel processors. By contrast, human decision makers can be slow, inaccurate, emotional, and single-focused, especially in stressful situations. Intelligent agent researchers call this approach human-centric [27]. Advances in agent coordination, collaboration, and learning will be major steps forward. Agents will need social ability and communication in order to interact with humans as well as other agents. Adaptive decision support systems that personalize for different users, recognize context, and perceive user intention are on the horizon.

One of the biggest challenges for real applications is trust in autonomous systems in general, and in intelligent decision support systems in particular. Future research will need to address questions such as: What decisions are humans willing to allow machines to make autonomously? What decisions and actions should we allow them to take without supervision and under what conditions? What checkpoints need to be implemented in such systems? Do we really trust them to act in our best interests?

In closing, I would like to invite all of you to the First International Symposium on Intelligent Decision Technologies, sponsored by KES International, and to be held in Himeji, Japan, in April 2009. Research papers on theory, computational methods applied to decision making, or applications are encouraged for publication in *Intelligent Decision Technologies: An International Journal (IDT).* Kurzweil's (1999) vision of the merging of human and machine intelligence may be within reach.

Acknowledgements

Portions of this work were funded by iTelehealth, Inc., and Consortium Research Management, Inc., under a Small Business Innovation Research contract # N44MH22044 from the National Institutes of Mental Health for an Intermediary-Based Suicide Prevention Website Development Project. Other portions were funded by research grants from The Sellinger School of Business and Management.

References

1. Ahmad, A.-R., Basir, O., Hassanein, K., Azam, S.: An intelligent expert systems' approach to layout decision analysis and design under uncertainty. In: Phillips-Wren, G., Ichalkaranje, N., Jain, L. (eds.) Intelligent Decision Making: An AI-Based Approach, pp. 321–364. Springer, Berlin (2008)

2. Burnstein, F.: Foreword. In: Phillips-Wren, G., Ichalkaranje, N., Jain, L. (eds.) Intelligent Decision Making: An AI-Based Approach, pp. IX–XI. Springer, Berlin (2008)
3. Bradshaw, J.: Software Agents. The MIT Press, Cambridge (1997)
4. Ceglowski, A., Churilov, L.: Using self organizing feature maps to unravel process complexity in a hospital emergency department: A decision support perspective. In: Phillips-Wren, G., Ichalkaranje, N., Jain, L. (eds.) Intelligent Decision Making: An AI-Based Approach, pp. 365–385. Springer, Berlin (2008)
5. Design-Ireland, Accessed on 1 February (2007), `http://www.design-ireland.net/index.php?http%3A//www.design-ireland.net/internet/browsing-13.php`
6. Hammond, J., Keeney, R., Raiffa, H.: The hidden traps in decision making. Harvard Business Review 79(5), 1–10 (1998)
7. Huhns, M., Singh, M.: Readings in Agents. Morgan Kaufmann Publishers Inc., San Francisco (1998)
8. Jennings, N., Woolridge, M.: Agent Technology: Foundations, Applications and Markets. Springer, Berlin (1998)
9. Klein, G.: A recognition-primed decision (RPD) model of rapid decision making. In: Klein, G., Orasanu, J., Calderwood, R. (eds.) Decision Making in Action: Model and Methods. Ablex Publishing, New York (1993)
10. Kurzweil, R.: The Age of Spiritual Machines. Viking, Penguin Group, New York (1999)
11. Maher, M.L.: Blurring the boundaries. Artificial Intelligence for Engineering Design, Analysis and Manufacturing 21, 7–10 (2007)
12. Mora, M., Forgionne, G., Cervantes, F., Garrido, L., Gupta, J., Gelman, O.: Toward a comprehensive framework for the design and evaluation of Intelligent Decision-making Support Systems (i-DMSS). Journal of Decision Systems 14(3), 321–344 (2005)
13. NLM - National Library of Medicine Gateway (2007), `http://www.nlm.nih.gov/`
14. Padgham, L., Winikoff, M.: Developing Intelligent Agent Systems. John Wiley & Sons Ltd., West Sussex (2004)
15. Pedrycz, W., Ichalkaranje, N., Phillips-Wren, G., Jain, L.: Introduction to computational intelligence for decision making. In: Phillips-Wren, G., Ichalkaranje, N., Jain, L. (eds.) Intelligent Decision Making: An AI-Based Approach, pp. 79–96. Springer, Berlin (2008)
16. Phillips-Wren, G.: Agent-Enabled Decision Support for Information Retrieval in Technical Fields. In: Gabrys, B., Howlett, R.J., Jain, L.C. (eds.) Knowledge-Based Intelligent Information and Engineering Systems 10th International Conference Proceedings, pp. 508–514. Springer, Berlin (2006)
17. Phillips-Wren, G.: Intelligent agents in decision support systems. Encyclopedia of Decision Making and Decision Support Technologies. IGI Global, Hershey, PA, pp. 505–513 (2008)
18. Phillips-Wren, G., Mora, M., Forgionne, G.: Evaluation of intelligent decision support systems. Encyclopedia of Decision Making and Decision Support Technologies, IGI Global, Hershey, PA, pp. 320–328 (2008)
19. Phillips-Wren, G., Mora, M., Forgionne, G., Gupta, J.: An integrative evaluation framework for intelligent decision support systems. European Journal of Operational Research (forthcoming, 2008)
20. Pohl, J.: Cognitive Elements of Human Decision Making. In: Phillips-Wren, G., Ichalkaranje, N., Jain, L. (eds.) Intelligent Decision Making: An AI-Based Approach, pp. 41–76. Springer, Berlin (2008)
21. Pomerol, J.-C., Adam, F.: In: Phillips-Wren, G., Ichalkaranje, N., Jain, L. (eds.) Intelligent Decision Making: An AI-Based Approach, pp. 3–40. Springer, Berlin (2008)

22. PSN, Preventing Suicide Network. Accessed on 1 February 1, http://www.preventingsuicide.com
23. Russell, S., Norvig, P.: Artificial Intelligence: A Modern Approach, 2e. Prentice-Hall Inc., Upper Saddle River (2003)
24. Saaty, T.L.: A scaling method for priorities in hierarchical structures. Journal of Mathematical Psychology 15, 234–281 (1977)
25. Sedacca, B.: Best-kept secret agent revealed. ComputerWeekly (12 October) (2006), http://www.computerweekly.com/Articles/2006/10/12/219087/best-kept-secret-agent-revealed.htm
26. Simon, H.: Administrative Behavior, 4th edn. The Free Press, New York (1997) (Original publication date 1945)
27. Tweedale, J., Ichalkaranje, N., Sioutis, C., Jarvis, B., Consoli, A., Phillips-Wren, G.: Innovations in multi-agent systems. Journal of Network & Computer Applications 30(3), 1089–1115 (2007)
28. Tweedale, J., Sioutis, C., Phillips-Wren, G., Ichalkaranje, N., Urlings, P., Jain, L.: Future directions: Building a decision making framework using agent teams. In: Phillips-Wren, G., Ichalkaranje, N., Jain, L. (eds.) Intelligent Decision Making: An AI-Based Approach, pp. 387–408. Springer, Berlin (2008)
29. Turban, E., Aronson, J.: Decision Support Systems and Intelligent Systems. A. Simon and Schuster Company, Upper Saddle River (1998)
30. Wang, Y.D.: A Decision Theoretic Approach to the Evaluation of Information Retrieval Systems, unpublished Ph.D. dissertation, University of Maryland Baltimore County, Baltimore, MD (2006)
31. Wang, Y.D., Phillips-Wren, G., Forgionne, G.: E-delivery of personalized healthcare information to intermediaries for suicide prevention. International Journal of Electronic Healthcare 1, 396–412 (2006)
32. Wooldridge, M.: An Introduction to MultiAgent Systems. John Wiley & Sons, Ltd., West Sussex (2002)

Driven by Compression Progress

Jürgen Schmidhuber

IDSIA - Instituto Dalle Molle di Studi sull'Intelligenza Artificiale, Lugano, Switzerland & Cognitive Robotics Lab, Technische Universität München, Germany
juergen@idsia.ch

Abstract. I argue that data becomes temporarily interesting by itself to some self-improving, but computationally limited, subjective observer once he learns to predict or compress the data in a better way. Curiosity is the desire to create or discover more data that allows for compression progress. This drive motivates exploring infants, pure mathematicians, composers, artists, dancers, comedians, yourself, and recent artificial systems.

References

1. Schmidhuber, J.: Simple Algorithmic Principles of Discovery, Subjective Beauty, Selective Attention, Curiosity & Creativity. In: Corruble, V., Takeda, M., Suzuki, E. (eds.) Proc. 10th Intl. Conf. on Discovery Science (DS 2007). LNCS (LNAI), vol. 4755, pp. 26–38. Springer, Heidelberg (2007); Also In: Hutter, M., Servedio, R. A., Takimoto, E. (eds.) Proc. 18th Intl. Conf. on Algorithmic Learning Theory (ALT 2007). LNCS (LNAI), vol.4754, p. 32. Springer, Heidelberg (2007)
2. Schmidhuber, J.: Developmental Robotics, Optimal Artificial Curiosity, Creativity, Music, and the Fine Arts. Connection Science 18(2), 173–187 (2006)
3. Schmidhuber, J.: Hierarchies of generalized Kolmogorov complexities and nonenumerable universal measures computable in the limit. International Journal of Foundations of Computer Science 13(4), 587–612 (2002)
4. Schmidhuber, J.: Exploring the Predictable. In: Ghosh, S.T. (ed.) Advances in Evolutionary Computing, pp. 579–612. Springer, Heidelberg (2002)
5. Schmidhuber, J.: Low-Complexity Art. Leonardo, Journal of the International Society for the Arts, Sciences, and Technology 30(2), 97–103 (1997)
6. Schmidhuber, J.: A possibility for implementing curiosity and boredom in model-building neural controllers. In: Meyer, J.A., Wilson, S.W. (eds.) Proc. of the International Conference on Simulation of Adaptive Behavior: From Animals to Animats, pp. 222–227. MIT Press/Bradford Books (1991)
7. Schmidhuber, J.: Curious model-building control systems. In: Proc. International Joint Conference on Neural Networks, Singapore, vol. 2, pp. 1458–1463. IEEE, Los Alamitos (1991)

I. Lovrek, R.J. Howlett, and L.C. Jain (Eds.): KES 2008, Part I, LNAI 5177, p. 11, 2008.

Evolvable Hardware Architectures and Their Role in Fault Tolerant Computational Systems

Andy Tyrrell

University of York, UK
amt@ohm.york.ac.uk

Abstract. Biological inspiration in the design of computing machines finds its source in essentially three biological models: phylogenesis, the history of the evolution of the species, ontogenesis, the development of an individual as directed by his genetic code, and epigenesis, the development of an individual through learning processes influenced both by their genetic code and by the environment. These three models share a common basis: a one-dimensional description of the organism, the genome. If one would like to implement some or all of these ideas in hardware can we use COTS or do we need specifically designed-for-purpose devices? This talk will consider some historical work on bio-inspired architectures before moving on to consider a new device designed and built specifically for bio-inspired work. It will consider some of the novel features present in this device, such as self-configuration and dynamic routing, which assist the implementation of ontogenetic capabilities such as development, self-repair and self-replication.

I. Lovrek, R.J. Howlett, and L.C. Jain (Eds.): KES 2008, Part I, LNAI 5177, p. 12, 2008.

Legal Issues for Military Intelligent Decision-Making Technologies

Anthony Finn

Defence Science and Technology Organisation, Department of Defence,
Edinburgh, Australia
anthony.finn@dsto.defence.gov.au

Abstract. Intelligent decision-making technologies (IDT) will soon have the capacity to control weapons in a manner that includes decisions regarding target identification and engagement. If permitted, this would provide these systems with the ability to decide which targets to prosecute with lethal force, without any operator intervention. This would mark a sea-change in the role of technology in warfare as the human can be removed from the decision-making loop. There are complex technological and legal issues regarding the development, deployment and exploitation of such systems. This paper outlines our obligations in respect of target discrimination under the Law of Armed Conflict (LOAC) and then uses these principles to discuss the allocation of roles and responsibilities between a human supervisor and the associated IDT.

I. Lovrek, R.J. Howlett, and L.C. Jain (Eds.): KES 2008, Part I, LNAI 5177, p. 13, 2008.

New Topics from Recent Interactive Evolutionary Computation Researches

Hideyuki Takagi

Kyushu University, Japan
http://www.design.kyushu-u.ac.jp/~takagi/

Abstract. First, we review wide varieties of IEC applications. They include artistic applications such as generating computer graphics, music, and editorial design, acoustic and image signal processing, hearing aid fitting, data mining, architectural design, virtual reality, and others.

Secondly, we introduce new type of IEC applications. Major IEC applications are optimizing target systems and creating graphics, images, shapes, sounds, vibrations, and others. We introduce two new types of IEC applications. The first one is measuring human characteristics. IEC is an optimization method based on human subjective evaluation. Likely reverse engineering, we may measure the evaluation characteristics or mental conditions of an IEC user by analyzing the outputs from the target system optimized by the user. The second one is extension of IEC evaluation. Usually IEC optimizes a target system based on IEC user's subjective evaluation, i.e. psychological evaluation. We may extend the evaluation from psychological one to physiological one. We show the framework of the extended IEC.

Thirdly, we overview the researches that try to reduce IEC user fatigue and show our latest research in this area. Several approaches have been proposed to reduce IEC user's fatigue; some of them are improving input/output interface, accelerating EC search, allowing human intervention into EC search, estimating human evaluations, and others. Here, we introduce our latest research and show our view.

I. Lovrek, R.J. Howlett, and L.C. Jain (Eds.): KES 2008, Part I, LNAI 5177, p. 14, 2008.

Virtual Humans: From Researcher's Fascination to Mass-Market Technology

Igor S. Pandzic

University of Zagreb, Faculty of Electrical Engineering and Computing, Croatia
igor.pandzic@fer.hr

Abstract. The talk will give a snapshot of virtual humans research and their applications. The fundamental driving force of scientific progress is the researchers' fascination and deep passion for their research topic. In few fields is this so obvious as in virtual humans: what can be more fascinating than trying to reproduce the image of ourselves - walking, talking, communicating, intelligent human-like beings. It is therefore no wonder that efforts in this direction have been going on since decades. So what do we have to show after all this work? In terms of graphics and animation, some of today's virtual characters are quite amazing. However, high levels of expressivity and naturalness require huge amounts of manual work, so fully autonomous real-time characters lag far behind such creations. When it comes to perception, most of today's virtual humans are blind (i.e. computer vision or other sensors are used only in a small number of systems), hard of hearing (i.e. speech recognition is unreliable) and the other senses are barely explored, if at all. While natural language processing, dialog systems and theory of mind have all made progress, it is fair to say that cognitive and conversational capabilities of intelligent virtual agents are nowhere near those of real people. All things considered, while huge progress has been made, there is a tremendous amount of work still in front of us on the road to truly interactive and believable virtual humans. At the same time, we have reached the stage of development where numerous applications for virtual people are springing up. In parallel with the research work, we are witnessing an increased entrepreneurial activity in this field in the recent years. Beyond movies and games, virtual humans are appearing as tutors, advisors, receptionists, personal avatars or companions in a whole array of application fields including health care, finance, retail, communications, entertainment and others.

I. Lovrek, R.J. Howlett, and L.C. Jain (Eds.): KES 2008, Part I, LNAI 5177, p. 15, 2008.

How Many Clusters Are There? – An Essay on the Basic Notions of Clustering

Bernd Reusch

Technische Universität Dortmund
Computer Science I
Otto-Hahn-Str. 16, 44227 Dortmund, Germany
bernd.reusch@udo.edu

Abstract. This lecture starts with two remarks:

1. Many people, papers, books claim „the number of clusters has to be given in advance"
2. Even good texts do not provide a formal definition of "cluster" or "clustering".

Starting with remark 1 we show, that any formal definition of "clustering" uniquely defines the number of clusters. For remark 2 we prove that relations called "similarities" define nicely "clustering". We also show that every clustering is defined by some similarity, although it may be given only implicitly. Finding clusters given by some similarity is the same problem as the "maximal clique problem". Hence it is NP-complete. It is also very clear that the very old techniques of "reduction of partial automate" is the same. In both areas algorithms are known, for maximal cliques even good heuristics. Coming back to remark 1. The problem that people really address, is the case where the desired similarity is not given explicitly, but by some of its properties. Among these, there may be the number of clusters. Very many of the known algorithms have the following structure:

1. Generate some similarity Pi
2. Is it "good"?
   Yes: Stop here and use Pi for clustering
   No: generate Pi+1 and continue.

For generating new similarities many techniques are used: distances, probabilities, etc. and also fuzzy theory. Finally we show that using "representatives" for clustering is not the same as using similarities.

I. Lovrek, R.J. Howlett, and L.C. Jain (Eds.): KES 2008, Part I, LNAI 5177, p. 16, 2008.

An Intelligent Traffic Light Control Based on Extension Neural Network

Kuei-Hsiang Chao, Ren-Hao Lee, and Meng-Hui Wang

Department of Electrical Engineering, National Chin-Yi University of Technology, Taichung, Taiwan, R.O.C.
{Kuei-Hsiang Chao,chaokh}@ncut.edu.tw,
{Ren-Hao Lee,ricklzh}@yahoo.com.tw,
{Meng-Hui Wang,wangmh}@ncut.edu.tw

Abstract. This paper presents an intelligent traffic light control method based on extension neural network (ENN) theory for crossroads. First, the number of passing vehicles and passing time of one vehicle within green light time period are measured in the main-line and sub-line of a selected crossroad. Then, the measured data are adopted to construct an estimation method based on ENN for recognizing the traffic flow of a standard crossroad. Some experimental results are made to verify the effectiveness of the proposed intelligent traffic flow control method. The diagnostic results indicate that the proposed estimated method can discriminate the traffic flow of a standard crossroad rapidly and accurately.

Keywords: Extension neural network theory, traffic light system, traffic flow control.

1 Introduction

The conventional traffic light control methods include fix-time control, time-of-day control, vehicle actuated control, semi-actuated control, green wave control, area static control and area dynamic control [1]. However, there is no system meeting the adaptive characteristic. Although some significant artificial intelligence (AI) methods such as fuzzy logic [2,3], neural network [4], evolutionary algorithms [5,6] and reinforcement learning [7,8] have been proposed to tune the cycle length and splits adaptively, the success in timing optimization and convergence rate are still limited. The cycle length and splits could be determined by using the fuzzy control method, and thus that could shorten the queue, and reduce total traffic delay. However, most researchers work at controlling an isolated intersection with the fuzzy control method. Few apply this method to the coordinated control of area traffic because it is a complex large-scale system. There are many interaction factors, and it is difficult to describe the whole system using some qualitative knowledge. This is the limitation of fuzzy control methods. The applying effect of artificial neural network (ANN) depends on its generalization capability. So the samples should be ergodic and the learning process should converge to the global optimal point. In fact, it is hard to meet these conditions for a real application. The evolutionary algorithms such as genetic algorithm, ant algorithm and particle swarm optimization are all biomimetic methods

I. Lovrek, R.J. Howlett, and L.C. Jain (Eds.): KES 2008, Part I, LNAI 5177, pp. 17–24, 2008.

for global optimization. Therefore, evolutionary algorithms are not likely to be trapped in local optima because of their characteristics of random search and implicit parallel computing. Also, when meeting a large-scale problem, these methods will spend much time to converge to the optima. It is disadvantageous for on-line optimization of area traffic coordinated control. In addition, the convergence rate is sensitive to parameters selected, which depend on practical problems to be solved. Thus, applying the evolutionary algorithm to area traffic coordinated control is limited. The advantage of reinforcement learning is that it is not necessary to set up the mathematic model for the external environment. However, there is also the disadvantage of converging slowly.

To satisfy the requirements of timing optimization and convergence rate for urban traffic light control, an intelligent control method based on extension neural network theory is proposed in this paper. The proposed traffic signal control method can adjust various traffic signal control parameters adaptively in response to varying traffic demand. The proposed ENN method has the advantages of less learning time, higher accuracy and less memory consumption.

2 Extension Neural Network

Extension neural network [9] uses a combination of neural networks and extension theory. The extension theory [10] provides a novel distance measurement for classification processes, while the neural network can embed the salient features of parallel computation power and learning capability. The schematic structure of the ENN is depicted in Fig. 1. It includes both the input layer and the output layer. The nodes in the input layer receive an input feature pattern and use a set of weighted parameters to generate an image of the input pattern. In this network, there are two connection values (weights) between input nodes and output nodes, one connection represents the lower bound for this classical domain of the features, and the other connection represents the upper bound. The connections between the j-th input node and the k-th output node are W_{kj}^L and W_{kj}^U. Only one output node in the output layer remains active to indicate a classification of the input pattern.

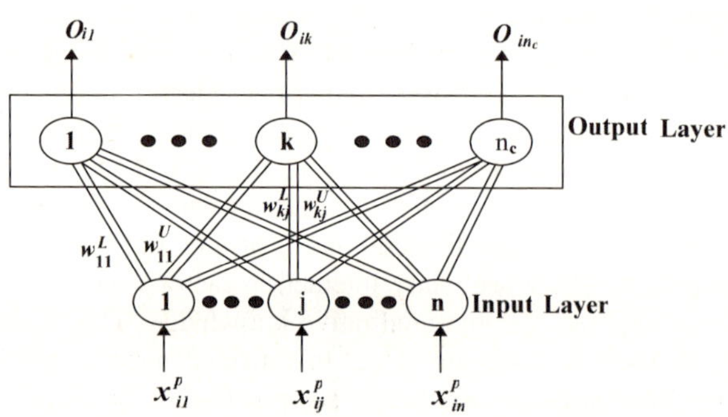

Fig. 1. The structure of extension neural network

2.1 Learning Algorithm of the ENN

The learning of the ENN can be seen as supervised learning, and its purpose is to tune the weights of the ENN to achieve good clustering performance or to minimize the clustering error. Before the learning, several variables have to be defined. Let training pattern set be $X \equiv \{X_1, X_2, ..., X_{NP}\}$, where N_P is the total number of training patterns. The i-th pattern is $X_i^p \equiv \{x_{i1}^p, x_{i2}^p, ..., x_{in}^p\}$, where n is the total number of the feature of patterns, and the cluster of the i-th pattern is p. To evaluate the clustering performance, the total error number is set as N_m, and the total error rate E_τ is defined below:

$$E_\tau = \frac{N_m}{N_p} \tag{1}$$

The detailed supervised learning algorithm can be described as follows:

Step 1: Set the connection weights between input nodes and output nodes. The range of classical domains can be either directly obtained from the previous requirement, or determined from training data as follows:

$$w_{kj}^L = \min_{i \in N}\{x_{kj}^k\} \tag{2}$$

$$w_{kj}^U = \max_{i \in N}\{x_{kj}^k\} \tag{3}$$

Step 2: Calculate the initial cluster center of every cluster.

$$Z_k = \{z_{k1}, z_{k2}, ..., z_{kn}\} \tag{4}$$

$$z_{kj} = \left(w_{kj}^L + w_{kj}^U\right)/2, \quad \text{for } k = 1, 2...n_c \text{ ; } j = 1, 2, ...n \tag{5}$$

Step 3: Read the i-th training pattern and its cluster number p.

$$X_i^p = \{x_{i1}^p, x_{i2}^p, ..., x_{in}^p\}, \quad p \in n_c \tag{6}$$

Step 4: Use the proposed extension distance (ED) to calculate the distance between the training pattern X_i^p and the k-th cluster, as follows:

$$ED_{ik} = \sum_{j=1}^{n}\left[\frac{\left|x_{ij}^p - z_{kj}\right|}{\left|\left(w_{kj}^U - w_{kj}^L\right)/2\right|} + 1\right], \quad k = 1, 2, ..., n_c \tag{7}$$

The proposed distance is a modification of extension distance [9], and it can be graphically presented as in Fig. 2. It can describe the distance between the x and a range $\langle w^L, w^U \rangle$.

Step 5: Find the k^*, such that $ED_{ik^*} = \min\{ED_{ik}\}$, If $k^* = p$ then go to Step 7, otherwise Step 6.

Step 6: Update the weights of the p-th and the k^*-th clusters as follows:
(a) Update the centers of the p-th and the k^*-th clusters.

$$z_{pj}^{new} = z_{pj}^{old} + \eta\left(x_{ij}^{p} - z_{pj}^{old}\right) \tag{8}$$

$$z_{k^*j}^{new} = z_{k^*j}^{old} - \eta\left(x_{ij}^{p} - z_{k^*j}^{old}\right) \tag{9}$$

(b) Update the weights of the p-th and the k^*-th clusters.

$$\begin{cases} w_{pj}^{L(new)} = w_{pj}^{L(old)} + \eta(x_{ij}^{p} - z_{pj}^{old}) \\ w_{pj}^{U(new)} = w_{pj}^{U(old)} + \eta(x_{ij}^{p} - z_{pj}^{old}) \end{cases} \tag{10}$$

$$\begin{cases} w_{k^*j}^{L(new)} = w_{k^*j}^{L(old)} - \eta(x_{ij}^{p} - z_{k^*j}^{old}) \\ w_{k^*j}^{U(new)} = w_{k^*j}^{U(old)} - \eta(x_{ij}^{p} - z_{k^*j}^{old}) \end{cases} \tag{11}$$

where η is a learning rate. The result of tuning two clusters' weights shown in Fig. 3, which clearly indicates the change of ED_A and ED_B. The cluster of pattern x_{ij} is changed from cluster A to B because $ED_A > ED_B$. From this step, we can clearly see that the learning process is only to adjust the weights of the p-th and the k^*-th clusters. Therefore, the proposed method has a rapid speed advantage over other supervised learning algorithms and can quickly adapt to new and important information.

Step 7: Repeat Step 3 to Step 6, and if all patterns have been classified then a learning epoch is finished.

Step 8: Stop if the clustering process has converged or the total error rate E_τ has arrived at a preset value; otherwise, return to Step 3.

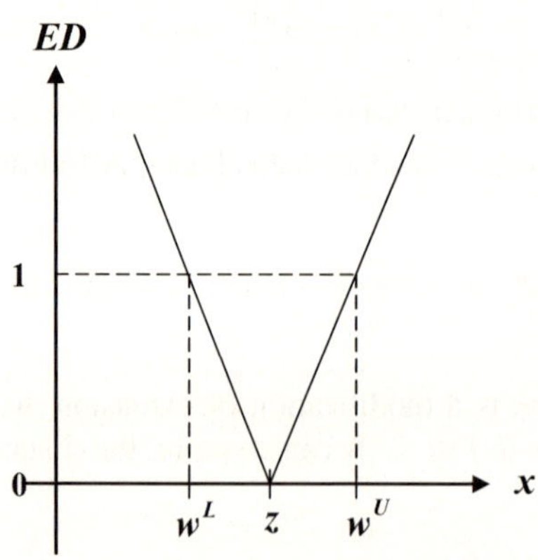

Fig. 2. The proposed extension distance

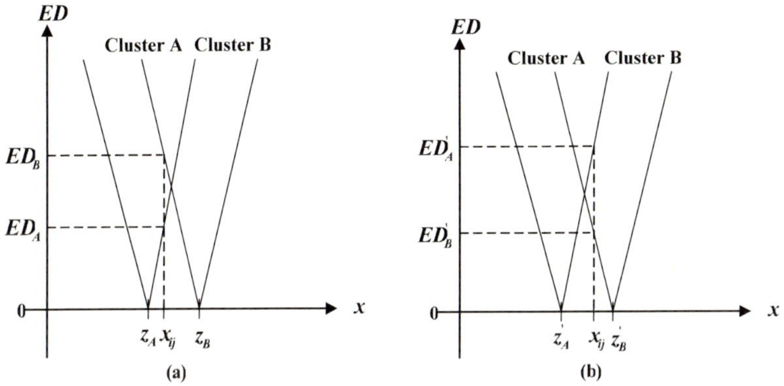

Fig. 3. The results of tuning cluster weights: (a) original condition; (b) after Tuning

It should be noted that the proposed ENN can take input from human expertise before the learning, and it can also produce meaningful output after the learning, because the classified boundaries of the features are clearly determined.

2.2 Operation Process of ENN

There can be recognition or sorting the cluster clearly when the ENN completes a learning procedure and its operation procedure is summarized as follows:

Step 1: Read the weighting matrix of ENN.
Step 2: Calculate the initial cluster centers of every cluster by using equation (4) and equation (5).
Step 3: Read the test pattern.

$$X_t = \{x_{t1}, x_{t2}, \ldots, x_{tn}\} \tag{12}$$

Step 4: Use the proposed extension distance (ED) to calculate the distance between the tested pattern and every existing cluster by equation (7).
Step 5: Find the k^*, such that $ED_{ik^*} = \min\{ED_{ik}\}$, and set the $o_{ik^*} = 1$ to indicate the cluster of the tested pattern.
Step 6: Stop, if all the test patterns have been classified, otherwise go to Step 3.

3 The Proposed Traffic Light Control Method

We can divide traffic flow into nine categories according to the number of passing vehicles and the passing time of vehicles during a green light time period. The represented symbols of these categories are described below:

- TF_1: High traffic flow in main-line and high traffic flow in sub-line.
- TF_2: High traffic flow in main-line and medium traffic flow in sub-line.
- TF_3: High traffic flow in main-line and low traffic flow in sub-line.

- TF_4: Medium traffic flow in main-line and high traffic flow in sub-line.
- TF_5: Medium traffic flow in main-line and medium traffic flow in sub-line.
- TF_6: Medium traffic flow in main-line and low traffic flow in sub-line.
- TF_7: Low traffic flow in main-line and high traffic flow in sub-line.
- TF_8: Low traffic flow in main-line and medium traffic flow in sub-line.
- TF_9: Low traffic flow in main-line and low traffic flow in sub-line.

The actual measured 900 data of different flows at certain crossroads are used to train the ENN proposed in the previous section. To obtain higher and precise convergence rate, the learning rate η and total error rate E_τ are set to be 0.2 and 0.1%, respectively. After the training procedure, one can find that the total error rate is 0% and only learning times 16 is needed.

The proposed ENN method can calculate the distance with respect to each cluster, and accordingly the traffic flow cluster and green light time in next period can be determined. To increase the sensitivity and adaptive capability, the green light time G^*_{time} of next period in each line is determined as follows:

$$G^*_{time} = G_{time,r} + (G_{time,n} - G_{time,r}) \times \frac{ED_r}{(ED_r + ED_n)} \tag{13}$$

Where $G_{time,r}$ and ED_r are the nominal green light time and extended distance of the judged traffic flow cluster. Whereas $G_{time,n}$ and ED_n are the nominal green light time and extended distance next to the judged traffic flow cluster. The nominal green light time of high, medium, and low traffic flow are indicated as GH_n, GM_n and GL_n, respectively.

4 Experimental Results

To prove the effectiveness of the ENN traffic light control method, the traffic flow records at certain crossroads are first selected to test. Table 1 lists the 18 tested data selected arbitrarily from the traffic flow records. The passing time among the vehicles passing through the main-line (c_1) and sub-line (c_3) within the green light time period of one traffic light cycle can be calculated by using the infra-red timer. The number of passing vehicles in the main-line (c_2) and sub-line (c_4) within the green light time period of one traffic light cycle can be counted by using the infra-red counter. Table 2 shows the identified results of the proposed method. Compared to the test data listed in Table 1, it shows the proposed method can correctly recognize the traffic flow cluster. For instance, in tested number 3, the ED_{TF2} (2.02) is the minimum value for the traffic flow cluster TF_2. It signals the crossroad is now toward high traffic flow in the main-line and medium traffic flow in the sub-line. Besides, the ED of other traffic flow cluster are all above 2.02, which means the possibility of the other traffic flow cluster is much lower than the traffic flow cluster TF_2. Letting $GH_n = 25\text{sec}$, $GM_n = 15\text{sec}$, and $GL_n = 8\text{sec}$, the green light time period G^*_{time} of main-line and sub-line found from (13) are 24sec and 18sec, respectively.

Table 1. The tested traffic flow data selected from the records at certain crossroad

Test no.	c_1	c_2	c_3	c_4
1(TF_1)	5.74132	0.775133	5.99438	1.66561
2(TF_1)	4.20955	0.648859	6.3288	2.28554
3(TF_2)	4.068	2.79041	2.80516	0.44863
4(TF_2)	5.76158	2.20405	1.95535	0.488244
5(TF_3)	3.25161	1.77464	1.15812	0.007883
6(TF_3)	5.55277	1.87215	1.12699	0.008835
7(TF_4)	1.67236	0.301306	4.58024	1.55529
8(TF_4)	3.49254	0.274962	4.14906	1.28237
9(TF_5)	1.91049	0.371855	1.69464	0.208063
10(TF_5)	2.76221	0.41706	1.50794	0.152139
11(TF_6)	3.4022	0.467625	1.65596	0.076918
12(TF_6)	1.86082	0.29613	0.022523	0.009581
13(TF_7)	0.427486	0.024916	5.85175	2.09693
14(TF_7)	1.475	0.039962	4.19514	2.90234
15(TF_8)	0.5038	0.091482	1.8569	0.111365
16(TF_8)	1.122	0.061527	2.73535	0.383965
17(TF_9)	1.24789	0.033883	0.519273	0.006247
18(TF_9)	0.435039	0.076354	1.2066	0.020695

Table 2. The identified results of the proposed traffic flow control method

Test no.	The distance between i-th tested pattern and k-th cluster									Judged cluster	Green Light time period	
	ED_{TF1}	ED_{TF2}	ED_{TF3}	ED_{TF4}	ED_{TF5}	ED_{TF6}	ED_{TF7}	ED_{TF8}	ED_{TF9}		Main -line (sec.)	Sub-line (sec.)
1	1.91	8.20	35.63	4.96	12.95	39.48	18.86	30.47	62.69	TF_1	22	23
2	2.66	11.93	47.10	3.62	14.30	49.04	14.50	29.31	71.49	TF_1	21	23
3	5.03	2.02	11.10	16.23	12.66	23.04	53.96	61.24	71.52	TF_2	24	18
4	4.11	1.41	9.93	14.73	12.24	21.05	46.22	52.26	61.62	TF_2	24	18
5	6.23	4.39	2.20	11.93	9.55	9.03	36.51	41.35	40.00	TF_3	23	10
6	4.90	3.12	1.19	13.78	11.81	11.32	41.01	46.38	45.10	TF_3	23	10
7	4.68	9.31	34.22	1.36	8.22	32.51	5.71	13.68	43.20	TF_4	17	24
8	4.09	6.94	27.90	1.79	5.85	26.62	7.91	13.77	38.86	TF_4	18	23
9	7.55	4.98	7.29	4.35	2.73	4.91	9.93	9.22	12.76	TF_5	19	19
10	7.16	4.85	5.54	4.41	2.67	3.24	11.89	11.70	13.43	TF_5	19	12
11	6.68	4.69	3.99	5.00	3.12	2.49	13.53	13.84	13.96	TF_6	19	11
12	8.93	7.00	4.59	5.18	4.69	2.81	9.64	9.96	8.38	TF_6	19	11
13	5.32	13.29	46.09	3.04	13.84	46.12	2.09	12.90	50.84	TF_7	11	24
14	5.03	15.02	57.30	3.26	15.31	57.54	2.00	15.61	65.81	TF_7	11	24
15	8.61	6.23	6.91	5.70	4.67	5.54	4.80	2.89	3.96	TF_8	11	12
16	7.43	4.23	12.32	4.75	3.07	11.11	3.06	0.98	10.45	TF_8	10	17
17	9.18	7.21	4.63	6.48	5.90	4.11	5.24	4.63	2.25	TF_9	10	10
18	9.18	7.09	5.13	6.29	5.73	4.55	5.06	3.66	2.19	TF_9	10	11

5 Conclusions

In this paper, an intelligent traffic flow estimation method based on the extension neural network theory for a standard crossroad was proposed. The experimental results show the proposed traffic flow diagnosis method can easily recognize the main traffic flow cluster and determine the green light time period of main-line and sub-line in next cycle. The good features of the proposed traffic flow diagnosis method include less learning time, higher accuracy and less memory consumption. When the traffic flow of the selected crossroad changes, only a fractional amount of the data should be adjusted, thus the update interval may be much reduced. Therefore, the proposed traffic light control method will be easy to implement in a real-time traffic flow detecting device or a portable instrument. It is also has good economic benefits to apply the proposed traffic light control method to the coordinated control of area traffic.

References

1. Papageorgiou, M., Diakaki, C., Dinopoulou, V., Kotsialos, A., Wang, Y.: Review of Road Traffic Control Strategies. Proceeding of the IEEE 91(12), 2043–2067 (2003)
2. Pappis, C., Mamdani, E.: A Fuzzy Logic Controller for a Traffic Junction. IEEE Transactions on Systems, Man, and Cybernetics 7, 707–717 (1977)
3. Li, Y., Fan, X.: Design of Signal Controllers for Urban Intersections Based on Fuzzy Logic and Weightings. In: 6th IEEE Conference on Intelligent Transportation Systems, vol. 1, pp. 867–871. IEEE Press, New York (2003)
4. Srinivasan, D., Choy, M.C., Cheu, R.L.: Neural Networks for Real-Time Traffic Signal Control. IEEE Transactions on Intelligent Transportation Systems 7(3), 261–272 (2006)
5. Dong, C., Liu, Z., Liu, X.: Chaos-Particle Swarm Optimization Algorithm and Its Application to Urban Traffic Control. International Journal of Computer Science and Network Security 6(1B), 97–101 (2006)
6. Chang, S.C., Tsai, M.W., Huang, G.W.: A GA Based Intelligent Traffic Signal Scheduling Model. In: IEEE Symposium on Computational Intelligence in Scheduling, pp. 93–97. IEEE Press, New York (2007)
7. Littman, M., Szepesvari, C.: A Generalized Reinforcement Learning Model: Convergence and applications. In: 13th International Conference Machine Learning, pp. 310–318. IEEE Press, New York (1996)
8. Li, Z., He, F., Yao, Q., Wang, F.Y.: Signal Controller Design for Agent-Based Traffic Control System. In: IEEE International Conference on Networking, Sensing and Control, pp. 199–204. IEEE Press, New York (2007)
9. Wang, M.H., Hung, C.P.: Extension Neural Network. In: Proceedings of the International Joint Conference on Neural Networks, vol. 1, pp. 399–403 (2003)
10. Cai, W.: Extension Set and Incompatible Problems. Science Exploration 3(1), 83–97 (1983)

Design of an Auto-associative Neural Network by Using Design of Experiments Approach

Božidar Bratina, Nenad Muškinja, and Boris Tovornik

University of Maribor, Faculty of Electrical Engineering and Computer Science, Smetanova ulica 17, 2000 Maribor, Slovenia
{bozidar.bratina,nenad.muskinja,boris.tovornik}@uni-mb.si

Abstract. Data driven computational intelligence methods have become popular in Fault detection and isolation (FDI) due to relatively quick design and not so difficult implementation on real systems. In this paper a research work on a Taguchi DoE approach for training the auto-associative neural network to extract non-linear principal components of a system, is presented. Design of such network was first proposed by Kramer however for achieving robustness to unspecified parameters such as noise level and disturbances, a design of experiments methodology can be used to optimally define network structure and parameters.

Keywords: Fault detection and isolation, nonlinear principle components, neural networks, Design of Experiments.

1 Introduction

Faults in technical processes are unavoidable. To be able to reduce plant down-time, minimize maintenance costs and quickly react to malfunctions systems that are able to identify malfunctions in the process are needed. Decades of research and development served a lot of methods for fault detection, isolation and diagnosis (FDID), however usually with predefined assumptions or limitations. No single method is universally applicable therefore development of various hybrid approaches became popular, and which include data-driven, analytical, and/or knowledge-based methods. Information about most used ones can be found in review papers [1], [2].

In case of large scale systems (plants) a model of the system is difficult to obtain so data driven and statistical methods are preferred. For example principle component analysis (PCA) which is practical but unfortunately due to its linear nature, not very appropriate for highly nonlinear systems [3]. To improve the fitting between the model and the system many derivates of PCA were developed, which uses computational intelligence methods. In 1991, Kramer presented nonlinear PCA method by using auto-associative neural network (AANN) along with recommendations for determining optimal number of neurons for neural network. To achieve adequate network model of the system a compromise between many network parameters is important (number of neurons, training algorithm, etc.). Properly designed AANN is capable of capturing nonlinearities in the system therefore a good data-driven model

I. Lovrek, R.J. Howlett, and L.C. Jain (Eds.): KES 2008, Part I, LNAI 5177, pp. 25–32, 2008.

can be obtained to be used in FDI, where less false alarms are produced that affect the reliability of the system.

Instead of Kramer's recommendations an experimental approach based on Taguchi Design of Experiments was used to achieve optimal parameters for the neural network model, where the structure and main parameters were taken into consideration. Direct comparison with Kramer's results is difficult because types of training algorithm, types of activation function, amount and quality of data samples, etc., were also considered in the presented approach.

2 Data Driven Fault Detection

Computational intelligence methods and multivariate statistical methods are more and more used due to their easy implementation on real systems. Secondly they are appropriate to obtain the model of nonlinear system, which improves the reliability of the FDI scheme by reducing the number of false alarms. As the practical use of statistical methods in FDI is increasing, neural networks are their main competition due to easy model derivation. Artificial neural networks exists for more than 60 years however their practical value became important in the 1980's as they can be very successful for solving many issues in modeling, identification, control, etc. So far around thirty types of neural network structures exist, but four of them are dominant: ART networks, multilayered perceptrons (multilayered feed forward neural network), recurrent neural networks and self-organizing maps. The structure and parameters of the network are usually set up according to the project task and presented method is perhaps an interesting way to choose them.

2.1 Nonlinear Principle Components Analysis

Development of neural networks has delivered a special case of a network - an auto-associated structure which found its use in many different areas (dimensionality reduction, signal processing, compression, etc). It's especially suitable for nonlinear systems where a nonlinear technique for multivariate data analysis should be used. Kramer first presented a feed-forward neural network to perform identity mappings, where network inputs are reproduced at the output layer [4]. Later it was used in many applications [5]-[8]. Such non-linear PCA enables nonlinear mappings in hidden layers of neural network (Fig.1). Next to the input layer is the encoding layer, followed by the bottleneck layer. The network layers are mirrored, where first and third hidden layers are encoding and decoding layers, respectively. In the middle there is the bottle-neck layer with reduced number of neurons and nonlinear model of the system. So transfer function f_l maps from x, the input column vector of length l, to the encoding layer, represented by $h^{(x)}$, a column vector of length m, with elements,

$$h_k^{(x)} = f_1\left(\left(W^{(x)}x + b^{(x)}\right)_k\right) \tag{1}$$

where, $b^{(x)}$ is a column vector of length m containing the bias parameters, $W^{(x)}$ is an $m \times l$ weight matrix.

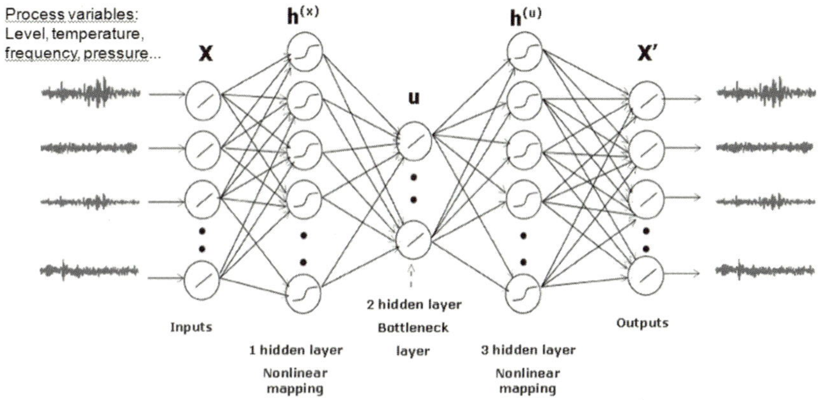

Fig. 1. Auto-associative neural network structure (AANN)

A transfer function f_2 maps from the encoding layer to the bottleneck layer containing a reduced number of neurons, which represents nonlinear principle component(s) u,

$$u = f_2\left(W^{(x)} h^{(x)} + \bar{b}^{(x)}\right) \tag{2}$$

The transfer function f_1 is generally nonlinear, while f_2 can also be the identity function. The transfer function f_3 maps from u to the final hidden layer $h^{(u)}$,

$$h_k^{(u)} = f_3\left(\left(W^{(u)} u + b^{(u)}\right)_k\right) \tag{3}$$

followed by f_4 mapping from $h^{(u)}$ to x', the output column vector of length l, with

$$x'_i = f_4\left(\left(W^{(u)} h^{(u)} + \bar{b}^{(u)}\right)_i\right) \tag{4}$$

The cost function $J = \left\langle \|x - x'\|^2 \right\rangle$ is minimized to solve the weights and offset parameters of the AANN, meaning finding the optimal values of $W^{(x)}$, $b^{(x)}$, $w^{(x)}$, $\bar{b}^{(x)}$, $w^{(u)}$, $b^{(u)}$, $W^{(u)}$ and $\bar{b}^{(u)}$. Desired minimum square error between the neural network output and the original data is thus minimized. According to necessary compromise when choosing the network parameters, Kramer recommends optimal selection of mapping nodes by using final prediction error (FPE) and information theoretic criterion (AIC) function. The number of neurons is important for the complexity of the nonlinear functions that can be generated by the network therefore in case of a small number accuracy might be low due to limited representational capacity of the network. On the other hand, if there are too many, the network can become over-fitted. To choose between network parameters that affect on networks operation and model quality, we used the Taguchi DoE method and respective experimental results to determine which combination of the structure and parameter settings are optimal for desired study case. To achieve better results the influences between structure of input and output data, sizes of training and testing sets, number of hidden neurons in each layer and initial weight settings was inspected. [9]

2.2 Design of Experiments

The DoE method is known for 80 years but it was rarely used in practice (economy, industry) until improved by Genichi Taguchi. With the philosophy of designing experiments he started a small revolution as the method could be used for solving many issues with final result of increasing the quality of the project task. This can be achieved by trial and error approach, DoE or Taguchi DoE, where the latter one involves conducting planned experiments based on orthogonal arrays (OA), Signal/Noise ratio calculations and correlation analysis between variables. Advantage of OA is in less conducted experiments to achieve optimal set up of the system parameters, where upon Signal/Noise ratio calculations optimization can be performed. By revealed correlations between variables, also robustness of the system can be addressed.

Selection of control and noise factors (variables) with respective variation levels is very important as this implies the selection of the OA type. Control factors of a neural network can be the number of hidden neurons, number of layers, type of neuron's activation function, type of training algorithm, etc. Included noise factors show how disturbances and noise affects the system. With each combination of set up parameter values, the accuracy of the trained network is analyzed, where more than one experiment (three) is conducted for each combination to avoid measurement issues hence an average value is calculated. The goal is to find such structure and parameter values of the network, to optimally fit the process behavior under optimum computational demands. A delicate decision in practice is which factors can be treated as control and which as noise, as it is a subjective decision based on operator's experience and type of the process. When best combination is selected the network is validated by conducting a final experiment to check if desired goals are met. To find correlated influences between factors further analysis by using an ANOVA method can be conducted. More about Taguchi DoE can be found in [10].

The best network structure and parameters can be used to obtain a good system model with possibility to detect small sensor and actuator faults and with minimized number of false alarms. The implementation of real-time FDI scheme consisting of neural network model was performed on a laboratory hydraulic model.

3 Application to a Laboratory Model

The process flow-sheet of the three-tank laboratory plant is depicted in Fig. 2. The upright tanks T_1 and T_2 are mounted above the tank T_3, hence, the inlet to the tanks also depends on the level (hydrostatic pressure) in the tanks T_1 and T_2, respectively (the pumps P_1 and P_2 are not an ideal generators to the system). Also, the outlet pipes are mounted at the bottom of the tank T_3 hence the amount of water in tank T_3 affects the outlet and the inlet flow of the upper tanks. The following faults can be introduced to the system: displacement of the level sensors in the tank T_1 and T_2, and pipeline of the pumps P_1 and P_2 can be partially clogged (partially closing the inlet valves).

By implementing FDI method to a real process it must be considered that result highly depends on the quality of data acquisition and data extraction from the noise correlated signals. In order to set up as much as typical industrial environment, an OPC standard together with TCP/IP protocol was used. The laboratory model was

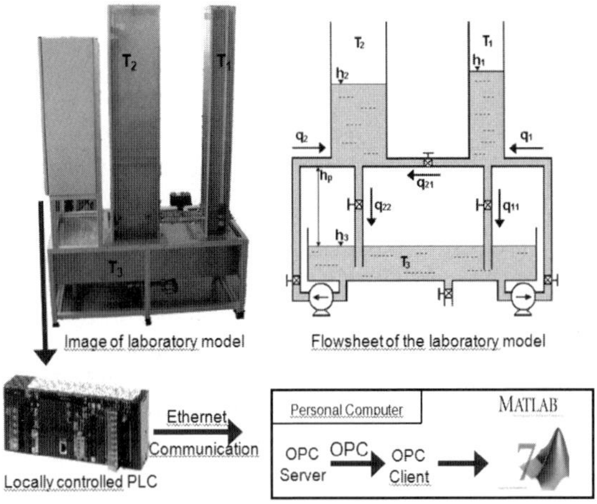

Fig. 2. The laboratory hydraulic plant

controlled locally by a PLC, while the process variables (inputs and outputs of the model) were processed in Matlab/ Simulink.

3.1 AANN Design by Using Taguchi DoE

An important step before actual deign procedure is data preparation (centered and scaled values). Direct measurements from the process were used (levels and frequencies of the control pumps) as an input and training matrix with 5000 samples.

The DoE method was defined with appropriate number of factors and respective variation levels which is an operator's subjective decision and usually case oriented task. In our case we focused on parameters that affect the operation of the network (Table 1) upon which an orthogonal array of L_{27} (3^{13}) was selected to define planned experiments.

Table 1. Selected factors and respective levels

Factor	Level 1	Level 2	Level 3
A	5	10	15
B	1	2	3
C	Linear	Sigmoid	Tanh
D	LM	GD	GDX
E	10%	30%	50%

Factor A: number of hidden neurons in encoding/decoding layer.
Factor B: number of neurons in bottle-neck layer.
Factor C: activation function.
Factor D: back-propagation learning method:
LM (Levenberg-Marquardt),
GD (gradient descent)
GDX (momentum gradient descent with adaptation)
Factor E: size of learning data against size of complete data.

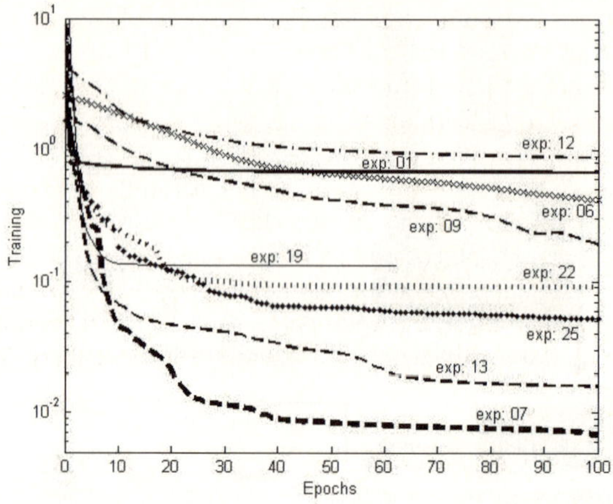

Fig. 3. Trained AANN with respective accuracy (mse)

To complete the task of experiments each experiment should be conducted more than ones (in our case 3x27=81 experiments; classic DoE=3x5^3 experiments) and in random order to achieve mean values of respective results. According to the OA combinations, a neural network structures with different parameter settings were generated and tested in Matlab/Simulink. Training was supervised with appropriate initial weights and rich data samples. The rest of the parameters such as speed of training, number of iterations and minimum error, were defined according to standard Matlab default values. All together 81 different neural networks were trained, where a mean-square-error function was used to serve as an output result.

After conducted experiments, Signal/Noise ratios were calculated to show how factor variations affect to the network accuracy. Fig. 4 shows calculated S/N ratios for all

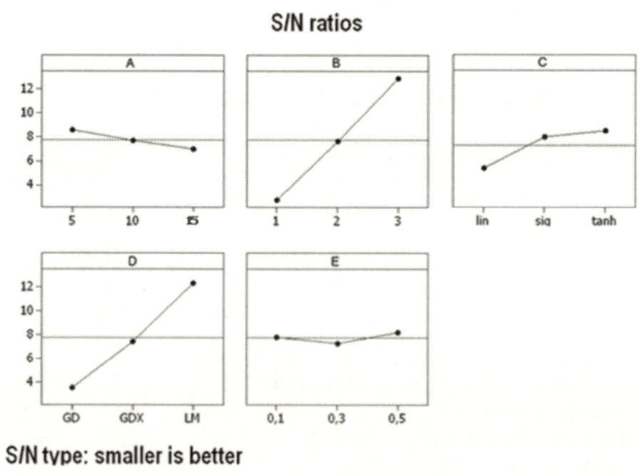

Fig. 4. Signal-to-noise ratios

five factors. The increasing number of hidden neurons didn't show much influence to the speed of training however number of neurons in the bottle-neck layer, the selected type of learning function and activation function were important for adequately trained network. The size of learning samples against the complete data samples had minor effect to the network training procedure. As it can be seen in Fig. 4, better trained neural network can be achieved by designing the neural network with parameter values that give higher S/N ratios. By conducting a final experiment with setting the "winning" combination of parameter values and best structure, desired results were achieved. In case of the laboratory hydraulic model, the winning combination shows exp. 7 with 5 neurons in the coding/decoding layer, 3 neurons in the bottle-neck layer, the use of "tangensh" activation function and Levenberg-Marquardt learning algorithm (Fig. 3).

According to chosen parameters, neural network structure 4-5-3-5-4 was used for FDI realization in Matlab/ Simulink. With such neural network structure fault AANN models were trained, where also nonlinear principle components could be extracted out of the bottle-neck layer to detect deviations in the process operation.

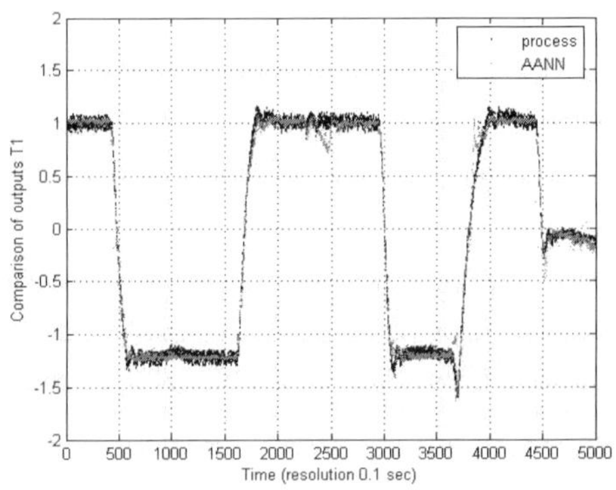

Fig. 5. Comparison of the process and best AANN for variable "level in tank 2"

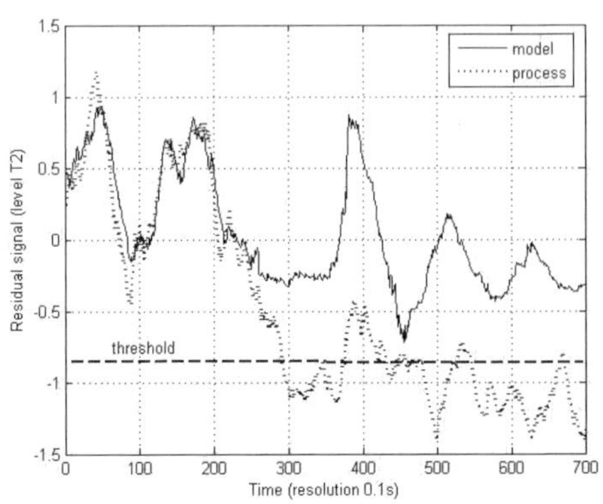

Fig. 6. Sensor drift detection of level sensor in Tank 2

Predefined faults could be detected and isolated upon process and AANN output data comparison. With appropriate threshold settings a 4% shift detection of the level sensors could be identified. Also a test for sensor drift was conducted, although the controller hides small deviations in the process (Fig. 6). The noise on the measurement signals was around 2-3%, however for better results it was slightly filtered out to improve detection results.

4 Conclusion

By applying the Taguchi DoE method instead of using Kramer's recommendations for selection of optimal neural network parameters, a reasonable AANN structure for process model could be achieved. Good model is usually essential for successful FDI results since a large number of false alarms can be reduced. DoE design procedure proved to be relatively easy however in case of poor OA definition, factor selection or neglecting hidden influences in the process, obtained results are hardly satisfying. In the paper 81 experiments (neural networks) were conducted (trained) to obtain proper structure of an auto-associative neural network. Obtained AANN model of the process enabled fault identification under close-loop conditions even when relatively small faults were introduced. When searching for optimal neural network parameter values the presented approach can be useful, however the use of the AANN model depends on the case specifics that a neural network should solve.

References

1. Venkatasubramanian, V., Raghunathan, R., Yin, K., Kavuri, N.S.: A review of process fault detection and diagnosis. Computers and Chemical Engineering 27, 293–326, Part I, Part II, Part III (2003)
2. Uraikul, V., Chan, C.W., Tontiwachwuthikul, P.: Artificial intelligence for monitoring and supervisory control of process systems. Engin. App. of Artificial Intelligence 20, 115–131 (2007)
3. Jolliffe, I.T.: Principal Component Analysis, 2nd edn. Springer, New York (2004)
4. Kramer, M.A.: Nonlinear principal component analysis using auto-associative neural networks. AIChE Journal 37, 233–243 (1991)
5. Hsieh, W.W.: Nonlinear principal component analysis by neural networks. Tellus 53A, 599–615 (2001)
6. Malthouse, E.C.: Limitations of nonlinear PCA as performed with generic neural networks. IEEE transactions on neural networks 9, 165–173 (1998)
7. Hines, J.W., Uhrig, R.E., Wrest, D.J.: Use of Autoassociative Neural Networks for Signal Validation. Journal of Intelligent and Robotic Systems 21, 143–154 (1998)
8. Jia, F., Martin, E.B., Morris, A.J.: Non-linear Principal Components Analysis for Process fault detection. Computers and Chemical Engineering 20, 851–854 (1998)
9. Kim, Y., Yum, B.: Robust design of multilayer feedforward neural networks: an experimental approach. Engin. App. of Artificial Intelligence 17, 249–263 (2004)
10. Taguchi, G., Chowdhury, S., Wu, Y.: Taguchi's Quality Engineering Handbook. John Wiley&Sons, New Jersey (2004)

A New Implementation for High Speed Normalized Neural Networks in Frequency Space

Hazem M. El-Bakry[1] and Mohamed Hamada[2]

[1] Faculty of Computer Science & Information Systems,
Mansoura University, EGYPT
helbakry20@yahoo.com
[2] University of Aizu
Aizu Wakamatsu, Japan
Hamada@u-aizu.ac.jp

Abstract. Neural networks have shown good results for detection of a certain pattern in a given image. In our previous work, a fast algorithm for object/face detection was presented. Such algorithm was designed based on cross correlation in the frequency domain between the input image and the weights of neural networks. In this paper, a simple design for solving the problem of local subimage normalization in the frequency domain is presented. This is done by normalizing the weights in the spatial domain off line. Furthermore, it is proved that local subimage normalization by normalizing the weights is faster than subimage normalization in the spatial domain. Moreover, the overall speed up ratio of the detection process is increased as the normalization of weights is done off line.

Keywords: Fast Pattern Detection, Neural Networks, Cross Correlation, Image Normalization.

1 Introduction

Pattern detection is a fundamental step before pattern recognition. Its reliability and performance have a major influence in a whole pattern recognition system. Nowadays, neural networks have shown very good results for detecting a certain pattern in a given image [5,8]. But the problem with neural networks is that the computational complexity is very high because the networks have to process many small local windows in the images [4]. In our pervious papers, we presented fast neural networks based on applying cross correlation in the frequency domain between the input image and the input weights of neural networks. It was proved that the speed of these networks is much faster than conventional neural networks [1-3]. It was also proved that fast neural networks introduced by previous authors [7,9,10] are not correct. The reasons for this were given in [2].

The problem of subimage (local) normalization in the Fourier space was presented in [6]. Here, a simple method for solving this problem is presented. By using the proposed algorithm, the number of computation steps required for weight normalization becomes less than that needed for image normalization. Furthermore, the effect of

I. Lovrek, R.J. Howlett, and L.C. Jain (Eds.): KES 2008, Part I, LNAI 5177, pp. 33–40, 2008.

weight normalization on the speed up ratio is theoretically and practically discussed. Mathematical calculations prove that the new idea of weight normalization, instead of image normalization, provides good results and increases the speed up ratio. This is because weight normalization requires fewer computation steps than subimage normalization. Moreover, for neural networks, normalization of weights can be easily done off line before starting the search process. In section II, fast neural networks for pattern detection are described. Subimage normalization in the frequency domain through normalization of weights is presented in section III. The effect of weight normalization on the speed up ratio is presented in section IV.

2 Theory of Fast Neural Networks Based on Cross Correlation in the Frequency Domain for Pattern Detection

Finding a certain pattern in the input image is a search problem. Each subimage in the input image is tested for the presence or absence of the required pattern. At each pixel position in the input image each subimage is multiplied by a window of weights, which has the same size as the subimage. The outputs of neurons in the hidden layer are multiplied by the weights of the output layer. A high output implies that the tested subimage contains the required pattern and vice versa. Thus, we may conclude that this searching problem is cross correlation between the image under test and the weights of the hidden neurons.

The convolution theorem in mathematical analysis says that a convolution of *f* with *h* is identical to the result of the following steps: let *F* and *H* be the results of the Fourier transformation of f and h in the frequency domain. Multiply *F* and *H* in the frequency domain point by point and then transform this product into spatial domain via the inverse Fourier transform [1]. As a result, these cross correlations can be represented by a product in the frequency domain. Thus, by using cross correlation in the frequency domain a speed up in an order of magnitude can be achieved during the detection process [1,2,3,5,12].

In the detection phase, a subimage *X* of size mxn (sliding window) is extracted from the tested image, which has a size *P*x*T*, and fed to the neural network. Let W_i be the vector of weights between the input subimage and the hidden layer. This vector has a size of mxn and can be represented as mxn matrix. The output of hidden neurons *h(i)* can be calculated as follows:

$$h_i = g\left(\sum_{j=1}^{m}\sum_{k=1}^{n} W_i(j,k)X(j,k) + b_i\right) \tag{1}$$

where *g* is the activation function and *b(i)* is the bias of each hidden neuron *(i)*. Eq.1 represents the output of each hidden neuron for a particular subimage *I*. It can be computed for the whole image *Z* as follows:

$$h_i(u,v) = g\left(\sum_{j=-m/2}^{m/2}\sum_{k=-n/2}^{n/2} W_i(j,k)\, Z(u+j, v+k) + b_i\right) \tag{2}$$

Eq. (2) represents a cross correlation operation. Given any two functions f and g, their cross correlation can be obtained by:

$$f(x,y)\otimes g(x,y)=\left(\sum_{m=-\infty}^{\infty}\sum_{n=-\infty}^{\infty}f(m,n)g(x+m,y+n)\right) \tag{3}$$

Therefore, Eq. (2) can be written as follows:

$$h_i = g\left(W_i \otimes Z + b_i\right) \tag{4}$$

where h_i is the output of the hidden neuron *(i)* and $h_i(u,v)$ is the activity of the hidden unit *(i)* when the sliding window is located at position (u,v) in the input image Z and $(u,v)\in[P-m+1,T-n+1]$.

Now, the above cross correlation can be expressed in terms of the Fourier Transform:

$$W_i \otimes Z = F^{-1}\left(F(Z)\bullet F^*\left(W_i\right)\right) \tag{5}$$

(*) means the conjugate of the *FFT* for the weight matrix. Hence, by evaluating this cross correlation, a speed up ratio can be obtained comparable to conventional neural networks. Also, the final output of the neural network can be evaluated as follows:

$$O(u,v)=g\left(\sum_{i=1}^{q} W_o(i)\,h_i(u,v)+b_o\right) \tag{6}$$

where q is the number of neurons in the hidden layer. $O(u,v)$ is the output of the neural network when the sliding window located at the position (u,v) in the input image Z. W_o is the weight matrix between hidden and output layer.

The complexity of cross correlation in the frequency domain can be analyzed as follows:

1. For a tested image of NxN pixels, the *2D-FFT* requires a number equal to $N^2log_2N^2$ of complex computation steps. Also, the same number of complex computation steps is required for computing the *2D-FFT* of the weight matrix for each neuron in the hidden layer.

2. At each neuron in the hidden layer, the inverse *2D-FFT* is computed. So, q backward and $(1+q)$ forward transforms have to be computed. Therefore, for an image under test, the total number of the *2D-FFT* to compute is $(2q+1)N^2log_2 N^2$.

3. The input image and the weights should be multiplied in the frequency domain. Therefore, a number of complex computation steps equal to qN^2 should be added.

4. The number of computation steps required by the faster neural networks is complex and must be converted into a real version. It is known that the two dimensions Fast Fourier Transform requires $(N^2/2)log_2N^2$ complex multiplications and $N^2log_2N^2$

complex additions [11]. Every complex multiplication is realized by six real floating point operations and every complex addition is implemented by two real floating point operations. So, the total number of computation steps required to obtain the *2D-FFT* of an *NxN* image is:

$$\rho=6((N^2/2)log_2N^2) + 2(N^2 log_2N^2) \quad (7)$$

which may be simplified to:

$$\rho=N^2 log_2N^2 \quad (8)$$

Performing complex dot product in the frequency domain also requires $6qN^2$ real operations.

5. In order to perform cross correlation in the frequency domain, the weight matrix must have the same size as the input image. So, a number of zeros = (N^2-n^2) must be added to the weight matrix. This requires a total real number of computation steps = $q(N^2-n^2)$ for all neurons. Moreover, after computing the *2D-FFT* for the weight matrix, the conjugate of this matrix must be obtained. So, a real number of computation steps $=qN^2$ should be added in order to obtain the conjugate of the weight matrix for all neurons. Also, a number of real computation steps equal to *N* is required to create butterflies complex numbers $(e^{-jk(2\Pi n/N)})$, where $0<K<L$. These (N/2) complex numbers are multiplied by the elements of the input image or by previous complex numbers during the computation of the *2D-FFT*. To create a complex number requires two real floating point operations. So, the total number of computation steps required for the faster neural networks becomes:

$$\sigma=(2q+1)(5N^2 log_2N^2) +6qN^2+q(N^2-n^2)+qN^2 +N \quad (9)$$

which can be reformulated as:

$$\sigma=(2q+1)(5N^2 log_2N^2) +q(8N^2-n^2) +N \quad (10)$$

6. Using a sliding window of size nxn for the same image of *NxN* pixels, $q(2n^2-1)(N-n+1)^2$ computation steps are required when using traditional neural networks for face/object detection process. The theoretical speed up factor *B* can be evaluated as follows:

$$B=\frac{q(2n^2-1)(N-n+1)^2}{(2q+1)(5N^2 log_2 N^2)+q(8N^2-n^2)+N} \quad (11)$$

The theoretical speed up ratio (Eq. 11) with different sizes of the input image and different in size weight matrices is listed in Table 1. Practical speed up ratio for manipulating images of different sizes and different in size weight matrices is listed in Table 2 using 700 MHz processor and *MATLAB ver 5.3*.

In practical implementation, the multiplication process consumes more time than the addition one. The effect of the number of multiplications required for conventional neural networks in the speed up ratio (Eq. 11) is more than the number of of multiplication steps required by the faster neural networks. In order to clear this, the following equation (B_m) describes the relation between the number of multiplication steps required by conventional and faster neural networks:

$$B_m = \frac{qn^2(N-n+1)^2}{(2q+1)(3N^2 log_2 N^2)+6qN^2} \tag{12}$$

The results listed in Table 3 prove that the effect of the number of multiplication steps in case of conventional neural networks is more than faster neural networks and this the reason why practical speed up ratio is larger than theoretical speed up ratio.

For general fast cross correlation the speed up ratio becomes in the following form:

$$Bg = \frac{q(2n^2 - 1)N^2}{(2q+1)(5(N+\tau)^2 log_2(N+\tau)^2) + q(8(N+\tau)^2 - n^2) + (N+\tau)} \tag{13}$$

where τ is a small number depends on the size of the weight matrix. General cross correlation means that the process starts from the first element in the input matrix. The theoretical speed up ratio for general fast cross correlation Eq. (13) is shown in Table 4. Compared with *MATLAB* cross correlation function *(xcorr2)*, experimental results show that the our proposed algorithm is faster than this function as shown in Table 5.

3 Subimage Normalization in the Frequency Space

In [5], the authors stated that image normalization to avoid weak or strong illumination could not be done in the frequency space. This is because the image normalization is local and not easily computed in the Fourier space of the whole image. Here, a simple method for image normalization is presented. Normalizing the image can be obtained by centering and normalizing the weights as follows:

Let be the zero-mean centered subimage located at (r,c) in the input image ψ:

$$\overline{X}_{rc} = X_{rc} - \overline{x}_{rc} \tag{14}$$

where, $\overline{x}_{rc}$ is the mean value of the sub image located at position (r,c). We are interested in computing the dot multiplication between the subimage $\overline{X}_{rc}$ and the weights W_i the of hidden layer as follows:

$$\overline{X}_{rc} \bullet W_i = X_{rc} \bullet W_i - \overline{x}_{rc} \bullet W_i \tag{15}$$

where,

$$\overline{x}_{rc} = \frac{\sum_{k,\ j=1}^{n} X_{rc}(k,\ j)}{n^2} \tag{16}$$

The dot multiplication denoted by (•) is not a matrix multiplication but is but is done element-wise (multiply each element in the first matrix by its corresponding element at the same position in the second matrix and sum up the results to obtain a one final value).

Combining Eq. (15) and Eq. (16), we get the following expression:

$$\overline{X}_{rc} \bullet W_i = X_{rc} \bullet W_i - \frac{\sum_{k,j=1}^{n} X_{rc}(k,j)}{n^2} \bullet W_i \tag{17}$$

For any two matrices with the same size, multiplying the first matrix dot by the mean of the second and summing the results the same as multiplying the second matrix dot by the mean of the first one and summing the results of multiplication. Therefore, Eq. (17) can be written as:

$$\overline{X}_{rc} \bullet W_i = X_{rc} \bullet W_i - X_{rc} \bullet \frac{\sum_{k,j=1}^{n} W_i(k,j)}{n^2} \tag{18}$$

The zero mean weights are given by:

$$\overline{W_i} = W_i - \frac{\sum_{k,j=1}^{n} W_i(k,j)}{n^2} \tag{19}$$

Also, Eq. (18) can be written as:

$$\overline{X}_{rc} \bullet W_i = X_{rc} \bullet \left(W_i - \frac{\sum_{k,j=1}^{n} W_i(k,j)}{n^2} \right) \tag{20}$$

So, we may conclude that:

$$\overline{X}_{rc} \bullet W_i = X_{rc} \bullet \overline{W}_i \tag{21}$$

which means that multiplying a normalized image with a non-normalized weight matrix dot multiplication is equal to the dot multiplication of the non – normalized image with the non-normalized weight matrix.

4 Conclusion

Normalized neural networks for fast pattern detection in a given image have been presented. It has been proved mathematically and practically that the speed of the detection process becomes faster than conventional neural networks. This has been accomplished by applying cross correlation in the frequency domain between the input image and the normalized input weights of the neural networks. Furthermore, a new general formulas for fast cross correlation as well as the speed up ratio have been given. Moreover, the problem of local subimage normalization in the frequency space has been solved. Simulation results have confirmed the theoretical computations by using *MATLAB*. The proposed approach can be applied to detect the presence/absence of any other object in an image.

References

[1] Klette, R., Zamperon,: Handbook of image processing operators. John Wiley & Sons ltd., Chichester (1996)

[2] El-Bakry, H.M.: Comments on Using MLP and FFT for Fast Object/Face Detection. In: Proc. of IEEE IJCNN 2003, Portland, Oregon, July 20-24, pp. 1284–1288 (2003)

[3] Lin, J.Y., Cheng, C.T., Chau, K.W.: Using support vector machines for long-term discharge prediction. Hydrological Sciences Journal 51(4), 599–612 (2006)

[4] Srisuk, S., Kurutach, W.: A New Robust Face Detection in Color Images. In: Proc. of IEEE Computer Society International Conference on Automatic Face and Gesture Recognition (AFGR 2002), Washington D.C., USA, May 20-21, pp. 306–311 (2002)

[5] El-Bakry, H.M.: Automatic Human Face Recognition Using Modular Neural Networks. Machine Graphics & Vision Journal (MG&V) 10(1), 47–73 (2001)

[6] Feraud, R., Bernier, O., Viallet, J.E., Collobert, M.: A Fast and Accurate Face Detector for Indexation of Face Images. In: Proceedings of the Fourth IEEE International Conference on Automatic Face and Gesture Recognition, Grenoble, France, 28-30 March (2000)

[7] Ben-Yacoub, S., Fasel, B., Luettin, J.: Fast Face Detection using MLP and FFT. In: Proc. of the Second International Conference on Audio and Video-based Biometric Person Authentication (AVBPA 1999) (1999)

[8] Baluja, S., Rowley, H.A., Kanade, T.: Neural Network - Based Face Detection. IEEE Trans. on Pattern Analysis and Machine Intelligence 20(1), 23–38 (1998)

[9] Beat F.: Fast Multi-Scale Face Detection, IDIAP-Com 98-04 (1998)

[10] Ben-Yacoub, S.: Fast Object Detection using MLP and FFT, IDIAP-RR 11, IDIAP (1997)

[11] Lewis, J.P.: Fast Normalized Cross Correlation, `http://www.idiom.com/~zilla/Papers/nvisionInterface/nip.html`

[12] El-Bakry, H.M.: New Faster Normalized Neural Networks for Sub-Matrix Detection using Cross Correlation in the Frequency Domain and Matrix Decomposition. Applied Soft Computing Journal 8(2), 1131–1149 (2008)

Appendix: Tables 1 to 5

Table 1. The theoretical speed up ratio (SUR) for images with different sizes

Image size	SUR (n=20)	SUR (n=25)	SUR (n=30)
100x100	3.67	5.04	6.34
200x200	4.01	5.92	8.05
300x300	4.00	6.03	8.37
400x400	3.95	6.01	8.42
500x500	3.89	5.95	8.39
600x600	3.83	5.88	8.33
700x700	3.78	5.82	8.26
800x800	3.73	5.76	8.19
900x900	3.69	5.70	8.12
1000x1000	3.65	5.65	8.05

Table 2. Practical SUR for images with different sizes Using MATLAB ver 5.3

Image size	SUR (n=20)	SUR (n=25)	SUR (n=30)
100x100	7.88	10.75	14.69
200x200	6.21	9.19	13.17
300x300	5.54	8.43	12.21
400x400	4.78	7.45	11.41
500x500	4.68	7.13	10.79
600x600	4.46	6.97	10.28
700x700	4.34	6.83	9.81
800x800	4.27	6.68	9.60
900x900	4.31	6.79	9.72
1000x1000	4.19	6.59	9.46

Table 3. A comparison between the number of multiplication steps required for conventional and faster neural networks to manipulate images with different sizes (n=20, q=30)

Image size	Conventional Neural Nets	Fast Neural Nets	SUR (η_m)
100x100	7.8732e+007	2.6117e+007	3.01
200x200	3.9313e+008	1.1911e+008	3.30
300x300	9.4753e+008	2.8726e+008	3.29
400x400	1.7419e+009	5.3498e+008	3.26
500x500	2.7763e+009	8.6537e+008	3.21
600x600	4.0507e+009	1.2808e+009	3.16
700x700	5.5651e+009	1.7832e+009	3.12
800x800	7.3195e+009	2.3742e+009	3.08
900x900	9.3139e+009	3.0552e+009	3.05
1000x1000	1.1548e+010	3.8275e+009	3.02

Table 4. The theoretical SUR for the general fast cross correlation algorithm

Image size	SUR (n=20)	SUR (n=25)	SUR (n=30)
100x100	5.59	8.73	11.95
200x200	4.89	7.64	10.75
300x300	4.56	7.12	10.16
400x400	4.35	6.80	9.68
500x500	4.20	6.56	9.37
600x600	4.08	6.38	9.13
700x700	4.00	6.24	8.94
800x800	3.92	6.12	8.77
900x900	3.85	6.02	8.63
1000x1000	3.79	5.93	8.51

Table 5. Simulation results of the SUR for the general fast cross correlation compared with the MATLAB cross correlation function (xcorr2)

Image size	SUR (n=20)	SUR (n=25)	SUR (n=30)
100x100	10.14	13.05	16.49
200x200	9.17	11.92	14.33
300x300	8.25	10.83	13.41
400x400	7.91	9.62	12.65
500x500	6.77	9.24	11.77
600x600	6.46	8.89	11.19
700x700	5.99	8.47	10.96
800x800	5.48	8.74	10.32
900x900	5.31	8.43	10.66
1000x1000	5.91	8.66	10.51

A New SOM Initialization Algorithm for Nonvectorial Data

Antonino Fiannaca[1,2], Riccardo Rizzo[2],
Alfonso Urso[2], and Salvatore Gaglio[1,2]

[1] Dipartimento di Ingegneria Informatica, Universitá di Palermo, Italy
[2] ICAR-CNR, Consiglio Nazionale delle Ricerche, Palermo, Italy

Abstract. Self Organizing Maps (SOMs) are widely used mapping and clustering algorithms family. It is also well known that the performances of the maps in terms of quality of result and learning speed are strongly dependent from the neuron weights initialization. This drawback is common to all the SOM algorithms, and critical for a new SOM algorithm, the Median SOM (M-SOM), developed in order to map datasets characterized by a dissimilarity matrix. In this paper an initialization technique of M-SOM is proposed and compared to the initialization techniques proposed in the original paper. The results show that the proposed initialization technique assures faster learning and better performance in terms of quantization error.

Keywords: Median SOM, initialization, pairwise data.

1 Introduction

The self-organizing maps (SOMs) are unsupervised neural networks used to support the exploration of a set of multidimensional patterns in clustering and classification applications. It is also well known that SOM results are heavily influenced by the initialization of the neurons configuration in the input space, because a proper initialization can reduce the time required for the training phase and the quantization error or network performances.

The initialization techniques can be grouped in two main categories: *random initialization techniques* that are not linked to the input pattern set, and *dataset techniques based on the analysis of the training data* trying to exploit some regularity or manifold of the input data. In random initialization the neural weights are chosen usually inside an hypercube that contains the input dataset[1]; the most common dataset techniques are sample and linear initialization: in the sample initialization[2], weight values are a random selection of input patterns, while in linear initialization[1,3] weight vectors are defined along the subspace spanned by the principal eigenvectors of the input dataset.

For many sets of real objects it is difficult to obtain a vector space representation, but it is possible to calculate a pairwise dissimilarity matrix. For example it is easy to compare DNA or Protein sequences using alignment techniques, or compute an edit distance for two generic strings, but it is difficult to set up a

I. Lovrek, R.J. Howlett, and L.C. Jain (Eds.): KES 2008, Part I, LNAI 5177, pp. 41–48, 2008.

method to identify a set of meaningful features and to build a vector space in order to use the Euclidean distance. From a dissimilarity matrix it is possible to obtain a SOM map using a Median SOM (*M-SOM*), a modification of the SOM algorithm proposed in [4]. In [4,5,6] these maps were used to visualize and cluster nonvectorial data. Notice that an efficient initialization algorithm, as well as a fast learning algorithm, plays a crucial role in computation of large datasets. For example in the genomic field [8] gene expression datasets and microarray collections are modeled using a pairwise data matrix. This representation becomes relevant for biological data, because specific distance measures have been created, also knows as evolutionary distances. In this paper, a new technique based on the fundamental idea of linear initialization will be introduced. Compared with the previous proposed technique[4], our initialization method does not require a conversion of data into vectorial representations and does not introduce extra auxiliary model vectors associated with the nodes.

The paper has the following structure: the next section reports an overview of the Median SOM algorithm and describes the batch learning process; the section 3 shows in detail the proposed initialization algorithm; the section 4 reports the evaluation criteria and the experimental results. Finally some conclusions are reported in section 5.

2 SOM Learning Algorithms for Nonvectorial Data

SOM neural networks can be trained using two classes of learning algorithms: the on-line and the batch learning; advantages and drawbacks of these algorithms has been discussed in [10]. The Median SOM is made by a lattice of $N = m \times n$ neurons (called models in Median SOM). Each neuron is not a weighted vector, but it is an element corresponding to an input pattern. In this work a batch learning algorithm trained by epochs is used, and its pseudo-code is reported in table 1. The batch learning algorithm for a Median SOM begins with a random assignment of an input pattern to each unit of the model grid (*sample initialization*).

In (step 3.(b).i) the best matching unit (*bmu*) selection is performed by the *affectation phase*, where each input is associated with a model, using the distance defined in the input pairwise matrix; this way, each map unit i collects a list of pattern to whom the reference model of unit i is the nearest reference pattern. In (step 3.(c)) the update process is performed by the *representation phase*, where the prototype of each model is updated [4]; this way each map unit i takes for the new reference pattern the median over the union of the lists that belongs to the topological neighbourhood of unit i [5]. The topological neighbourhood is computed using a gaussian function kernel around the best matching unit, and its neighbourhood radius $\sigma(t)$ is a decreasing function of time.

Obviously the proposed algorithm will take place in (step 1) of Table 1, and advantages of its use will be measured, epoch by epoch, at (step 3.(d)) of Table 1, when the evolution of the neural network is evaluated with the Quantization

Table 1. Median SOM batch learning algorithm

1. Initialize the N neurons;
2. Set step counter $t = 1$, epoch counter $p = 1$, maximum number of epochs $MaxP$;
3. While the stop condition ($p \geq MaxP$) is not verified,
   (a) Define *RndDataset* as a list where input patterns are randomly ordered;
   (b) While *RndDataset* is not empty,
      i. Get a pattern $x(t)$ from *RndDataset*;
      ii. Find the best matching unit $bmu(x(t))$;
      iii. Remove $x(t)$ from *RndDataset*;
      iv. $t = t + 1$;
   (c) For all N neurons update reference model;
   (d) Calculate Quantization Error at epoch p;
   (e) $p = p + 1$, $t = 1$;
4. End of learning after $MaxP$ epochs.

Error (QE). The local QE of a M-SOM is defined as the pairwise distance between a data vector input and its best matching unit.

3 The Initialization Algorithm for Median SOM

As the authors pointed out in [4,6] the SOM and M-SOM algorithms are heavy dependent from initialization. In the above papers the authors observed experimentally that the convergence of the SOM algorithm is significantly faster and safer, if the initial models are at least roughly ordered in two dimensions. For this reason, they propose a vectorial initialization method that uses a projection of the pairwise data into a vectorial space, and then a standard SOM algorithm execution to find models.

The main idea of the proposed initialization algorithm is the projection, over the Kohonen map, of some distinctive patterns of the dataset, preserving theirs mutual relationships. In order to perform this projection, a totally connected undirect graph $G(V, E)$ (the "*initGraph*") is built using some selected patterns of the dataset. In this graph V is the set of vertices of the graphs that are selected among the most distant elements of the input data and E is the set of edges, where the length of each edge $E(i, j)$ is closer to dissimilarity between patterns i, j, in the pairwise matrix. The constraint "closer to" instead of "equal to" is used because the geometry of the graph must be respected. The introduced algorithm resolves some mutual conflicts through "elastic edges" that would balance all pairwise relationships using a mean square error.

The *initGraph* plays a fundamental role in our initialization technique for two reasons: mainly it performs a bridge between original space of dataset and two-dimensional space of Kohonen map, secondly it can preserve most of mutual relations among selected patterns.

Table 2. Pseudo code of inizialization algorithm. Step 1.

1. Set k = number of biggest values to take into account in pairwise matrix;
2. Let p = number of rows and columns that contain the k biggest values ($p \leq 2 \times k$);
3. Mark the greater k values in pairwise matrix;
4. Mark the p elements related to k marked distance;
5. Extract the pairwise sub-matrix S ($p \times p$) with only marked elements;

The initialization is made in the following three steps:

1. *Selection* of the k most distant patterns (elements) of the input dataset; these patterns are selected using the biggest values (distances) in the pairwise dissimilarity matrix;
2. *Arrangement* of selected patterns into a 2D space using a 2D projection of the *initGraph*;
3. *Fitting* the *initGraph* projection into the Median SOM.

The first step is the retrieval of the most distant elements for the dataset, whose identification is done by exploration of input pairwise matrix. The aim is to obtain a subset of patterns that carry out information about the patterns set. First of all, a pairwise sub-matrix is pulled out from the pairwise matrix input. Selection of patterns, used to generate the pairwise sub-matrix, depends on values in pairwise dissimilarity matrix. These values can be regarded as distances between patterns. Greater distances will be taken into account to build an initialization map where the most distant elements will be placed in opposite areas of the network, according to the *initGraph*. Pseudo code for *step*1 is reported in Table 2.

The second step is the devoted to build a model able to translate relation among selected patterns into a two dimensional structure; this model must observe the geographic distance among areas where selected patterns should be located. Pseudo code for *step*2 is reported in Table 3.

The third step is the adaptation of initialization graph on the Kohonen map. Obviously the graph will be properly stretched and scaled before being overlapped on the map. After that, selected patterns will be located over some neurons, which set their models equal to the corresponding representative patterns. Remaining neurons will be initialized depending on the previous ones; in fact each neuron into the nearest neighbourhood of a neuron selected at previous step, will be initialized with its model. Remaining neurons will be initialized with random models, pulled out from input patterns. In this manner, during learning process, samples that are similar to selected patterns will fall into the neighbourhood of them. The implemented graph is eventually located in the Kohonen map, properly scaled, in order to assing models to the neurons. Pseudo code for *step*3 is reported in Table 4.

Table 3. Pseudo code of initialization algorithm. Step 2.

1. Set *tol* = tolerance adopted for elastic edges;
2. Generate a totally connected undirect graph $G(V, E)$ where vertices V are the first three marked elements of S;
3. Set lengths of edges $E(i, j)$ equal to value (distance) between features i, j in pairwise dissimilarity matrix (with respect to geometry of graph);
4. Calculate the *Mean Square Error* among all edges $E(i, j)$ and theirs theoretical distances (the pairwise dissimilarity between patterns i, j in pairwise matrix S) according to $MSE(||E(i, j) - S(i, j)||)$;
5. While the tolerance adopted for elastic edges is not satisfied ($MSE \geq tol$),
   (a) Add a random little value ϵ_1 to each edge $E(i, j)$, in order to get each edge closer to its theoretical measure (according to pairwise matrix);
   (b) Decrease parameter *tol* with a random little value ϵ_2 (*tol* = *tol* - ϵ_2);
6. For $c = 4$ to p (for each remaining $p - 3$ rows or columns of the matrix S),
   (a) Add a new vertex v, corresponding to element c of S, and $c(c-1)/2$ new edges;
   (b) Calculate measures of new edges using S matrix (with respect to geometry of graph);
   (c) Calculate the *Mean Square Error* among all edges $E(i, j)$ and theirs theoretical distances, as in step 4;
   (d) While the tolerance adopted for elastic edges is not satisfied ($MSE \geq tol$),
      i. Add a random little value ϵ_1 to each edge $E(i, j)$, in order to get each edge closer to its theoretical measure (according to pairwise matrix);
      ii. Decrease parameter *tol* with a random little value ϵ_2 (*tol* = *tol* - ϵ_2);
7. Resulting graph is the "initialization graph".

Table 4. Pseudo code of initialization algorithm. Step 3.

1. Let m = height and n = width of M-SOM grid, thus number of neurons is $m \times n$;
2. Build a bounding box R for the *initGraph*;
3. Scale the box R in order to obtain a rectangle $m \times n$;
4. For $s = 1$ to p (for each vertex of the graph),
   (a) Find the neuron overlapped with vertex s and assign to its prototype the model corresponding to vertex s;
   (b) Assign the same model of the selected neuron to each neuron into its neighbourhood with radius = 1.
5. Each remaining neuron will be initialized with a model selected randomly.

4 Experimental Results

In this section the comparison among the proposed, the sample and the vectorial initialization method is evaluated. In order to reach this goal, three M-SOMs are

Table 5. Average execution time measured in sec. of M-SOM learning process for the three initialization algorithms

Average Execution Time of training processes			
	Proposed	**Sample**	**Vectorial**
M-SOM Initialization	0.029	–	152.610
M-SOM Learning process	192.827	193.342	193.315

implemented with the same learning algorithm, but with different initialization algorithms. The result of training process will be evaluated using the evolution of quantization error.

4.1 Quality Criteria

The Quantization Error is used to compare the effectiveness of the different initialization methods. This criterion is commonly used in evaluation of neural networks resolution. Moreover this popular measure is easy to compute and well defined for M-SOM. The evolution of QE during training process was monitored over 50 experiments for each initial configuration and means of achieved values will be analyzed. The significance level of analyzed means are evaluated by the t-Test, that checks if the means of compared groups are statistically different from each others.

4.2 Evaluation of the Proposed Algorithm

In order to compare the M-SOM results after the training phase it is necessary to select a suitable number of epochs; after several trials it was seen that 20 epochs are enough to have a fixed configuration of models in the M-SOM. All the maps have a 20×20 square lattice. The training phase for each epoch is done with a neighbourhood radius function that decreases exponentially from $\sigma_{max} = 5$ to $\sigma_{min} = 1$.

In the vectorial initialization algorithm, a projection of input data into a 50-dimensional space is performed using the Sammon Projection [7]; then a SOM with a square lattice 20×20 neurons, a learning rate function that decreases exponentially from $\alpha_{MAX} = 0.75$ to $\alpha_{MIN} = 0.15$, and a neighbourhood radius function, that decreases exponentially from $\sigma_{MAX} = 4$ to $\sigma_{MIN} = 1$, is used in order to find models. These values offer the best result for the used dataset.

In the proposed algorithm, the $k = 6$ biggest values in pairwise matrix have been taken into account. These parameter values are those that give the best results for several datasets widely used in literature.

4.3 Validation of the Proposed Initialization Process

The validation of the M-SOM has been carried out using a real world dataset, [8], here called *Garrity-RNA*, made of 1436 small subunit ribosomal RNA sequences,

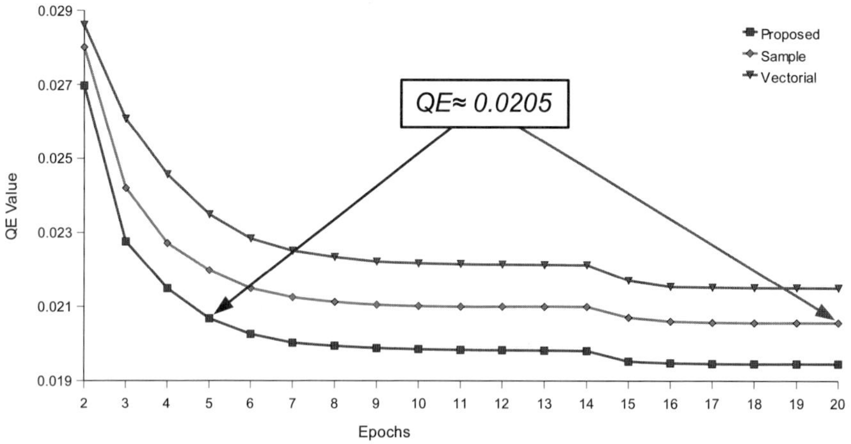

Fig. 1. Quantization error versus number of epochs for Garrity-RNA dataset. QE values at first steps are not shown. Starting values for porposed, sample and vectorial algortihms are respectively 0.0535, 0.0571 and 0.0336. The Sample initialization algorithm reaches the stop condition at the epoch $p = 20$ with a $QE \approx 0.0205$. Almost the same QE value was reached by proposed initialization algorithm at epoch $p = 5$.

that were initially classified (based on the GenBank [9] annotation) as belonging to the *Gammaproteobacteria*.

Figure 1 shows the average evolution of quantization error during the training process of *Garrity-RNA* dataset for the three initialization algorithm. The chart clearly shows the effectiveness of the proposed algorithm in terms of resolution of the map, in fact the map trained with the proposed algorithm reaches the lowest QE value among the maps. Moreover the sample initialization algorithm reaches the stop condition at the epoch $p = 20$ with a $QE \approx 0.0205$, whereas the proposed one reaches almost the same QE value at epoch $p = 5$. Notice that the evolution of M-SOM learning process, initialized with the vectorial method, starts with the lower value (epoch 1 not reported in figure 1), with respect to the other maps, but it ends with a greater QE value: this means that the M-SOM learning process with the vectorial initialization is fallen out in a local minimum for the network. Since the curves in the figure seem very close and have almost the same shape, and it should be possible that they come from the same distribution, the t-Test has been calculated. The test rejects the null hypothesis with a level of significance = 0.17%.

Table 5 reports average execution times for SOMs initialized with the three techniques. All quoted times are derived from tests on a machine having a 3.00GHz Pentium IV processor, 1014Mb of RAM, Windows Vista Business 32bit operating system. Our technique spends 0.038 sec to execute the Median SOM process, whereas vectorial initialization spends 131.878 sec. We assume the time of the sample initialization technique to be equal 0 sec. The time percentage of the proposed algorithm is about 0.015% of the learning process. This is a very small value with respect to the advantages shown in previous figures.

5 Conclusions

In this paper a new initialization algorithm for SOMs with non-vectorial data input is proposed. Unlike previous initialization methods, the proposed one does not require a conversion of data into vectorial representations, but it introduces a totally connected undirect graph (here called *initGraph*) that connects some key patterns; the edges of *initGraph* take into account the pairwise dissimilarity among all key patterns. Finally, a projection of the *initGraph* over the neuron grid reports the distance informations into the map.

This method has been compared with both sample and vectorial initialization techniques. Results of experimental tests, carried out on a real biological dataset, demonstrate the good performances obtained by our technique in terms of resolution of the map and execution time.

References

1. Kohonen, T.: Self-Organizing Maps, 3rd edn. Springer, Berlin (2001)
2. Varsta, M.: Self organizing maps in sequence processing. Dissertation, Department of Electrical and Communications Engineering, Helsinki University of Technology (2002)
3. Vesanto, J., Alhoniemi, E.: Clustering of the Self-Organizing Map. J. IEEE-NN 11(3), 586–600 (2000)
4. Kohonen, T., Somervuo, P.: How to make large self-organizing maps for nonvectorial data. Neural Network, 945–952 (2002)
5. Kohonen, T., Somervuo, P.: Self-organizing maps of symbol strings, pp. 413–418. IEEE Press, Piscataway (1998)
6. Oja, M., Somervuo, P., Kaski, S., Kohonen, T.: Clustering of human endogenous retrovirus sequences with median self-organizing map. In: Workshop on Self-Organizing Maps (2003)
7. Sammon, J.W.: A nonlinear mapping for data structure analysis. J. IEEE Transactions on Computers 18, 401–409 (1969)
8. Garrity, G.M., Lilburn, T.G.: Self-organizing and self-correcting classifications of biological data. J. Bioinformatics 21(10), 2309–2314 (2005)
9. National Center for Biotechnology Information, Entrez Nucleotide query, `http://www.ncbi.nlm.nih.gov/entrez/query.fcgi?db=Nucleotide`
10. Fort, J.C., Letrrmy, P., Cottrell, M.: Advantages and drawbacks of the Batch Kohonen algorithm. In: Verleysen, M. (ed.) ESANN 2002, pp. 223–230 (2002)

Accelerated Gradient Learning Algorithm for Neural Network Weights Update

Zeljko Hocenski, Mladen Antunovic, and Damir Filko

University J.J. Strossmayer, Faculty of Electrical Engineering, Kneza Trpimira 2b, 31000 Osijek, Croatia
{Zeljko.Hocenski,Mladen.Antunovic,Damir.Filko}@etfos.hr

Abstract. This work proposes decomposition of gradient learning algorithm for neural network weights update. Decomposition enables parallel execution convenient for implementation on computer grid. Improvements are reflected in accelerated learning rate which may be essential for time critical decision processes. Proposed solution is tested and verified on MLP neural network case study, varying a wide range of parameters, such as number of inputs/outputs, length of input/output data, number of neurons and layers. Experimental results show time savings in multiple thread execution.

Keywords: time critical processes, algorithm decomposition, neural network, weights update, gradient learning method, parallel processing.

1 Introduction

Each neural network is determined by its' structure and learning algorithm. Learning algorithm adjusts network weights, θ, in a way that after each iteration its' behavior improves. In identification and control of nonlinear dynamic processes desired behavior is usually known. Whenever the desired behavior is known, learning algorithms are based on output error, e. Output error is the difference between the neural network output, y, and desired output, y_d and is a function of θ. Quality of system performance is based on a criteria function $\Im(\theta)$ which may be any positive function dependant on network weights [1], [2], [3].

2 Gradient Methods

Most convenient criteria function is the L_2 norm of output error, (1).

$$\Im(\theta)=\frac{1}{2}\sum_{i=1}^{N} e_i^2(\theta)=\frac{1}{2}\,\boldsymbol{e}^{\mathrm{T}}(\theta)\cdot\boldsymbol{e}(\theta). \tag{1}$$

Optimization is done on N-pair input-output data set acquired on real process. The goal is to minimize criteria function. If θ^* minimize $\Im(\theta)$, than conditions (2) must be fulfilled.

I. Lovrek, R.J. Howlett, and L.C. Jain (Eds.): KES 2008, Part I, LNAI 5177, pp. 49–56, 2008.

$$\nabla\Im\Big|_{\theta=\theta^*} = 0\,; \quad \Delta\theta^T \cdot \boldsymbol{H} \cdot \Delta\theta > 0, \tag{2}$$

where $H(\theta)$ is Hessian matrix of criteria function, defined as matrix of second partial derivatives of $\Im(\theta)$.

$$\boldsymbol{H}(\theta) = \nabla^2\Im(\theta). \tag{3}$$

Numerical methods for finding θ^* that satisfies equation (2) converge to a local minimum, but not necessary to a global minimum. This means that calculated values may not be optimal. But for most practical application it is not even necessary to find θ^* that give global minimum. Satisfactory is to find any set of θ for which criteria function becomes small enough, less than ε, where $\varepsilon \in \mathrm{R}$ is small positive constant. There are two different types of learning algorithms – batch and on-line. In batch learning, weights are changed using all data block, while on-line training use vector by vector of acquired (measured) data in adjusting weights. Whenever all data are available before learning process, batch algorithm is advised. In this article batch algorithm is used. Most batch algorithms are based on classical methods of non-linear optimization in finding minimum. Goal function that is to be minimized is criteria function $\Im(\theta)$. Learning algorithms minimize $\Im(\theta)$ by adjusting neural networks' weights. Most common batch algorithms are based on iterative procedure, (4)

$$\theta(k+1) = \theta(k) + \alpha(k)\cdot \boldsymbol{s}_d(k), \tag{4}$$

where $\boldsymbol{s}_d(k)$ is direction of searching for minimum and $\alpha(k)$ is learning coefficient in k^{th} iteration.

Gradient methods use either (5) or (6) to calculate $\boldsymbol{s}_d(k)$.

$$\boldsymbol{s}_d(k) = -\nabla\Im(\theta(k)), \tag{5}$$

$$\boldsymbol{s}_d(k) = -\nabla\Im(\theta(k)) + \beta(k)\cdot \boldsymbol{s}_d(k\text{-}1). \tag{6}$$

Gradient methods that use (5) are also known as steepest descent method. In dependence on α there are many variations of algorithm defined by equations (4) and (5). If too small learning is too slow. If too big, the algorithm may oscillate in minimum neighborhood or even diverge. Thus if constant, α should be selected carefully. Main advantage of (5) when compared to (6) is that it is quite simple and requires less computer power and memory space. Another advantage is that it is inherently parallel, i.e. for each neuron a separate formula is used to adjust weights. There are many variations of (5) described in literature [6], [7], [8] and [9]. Algorithm (6) is extended from (5) and had additional part that is factor of $\boldsymbol{s}_d$ in previous iteration, where β is a factor selected to provide perpendicular property of $\boldsymbol{s}_d$ and $\boldsymbol{H}(\theta)$. This property gives faster convergence compared to (5). Calculation of $\nabla\Im$ and $\boldsymbol{H}(\theta)$ is numerically demanding, specially in networks with great number of neurons. This work gives improvement of gradient methods. Suggested improvement results in alternative method that converge in less iteration and is inherently parallel which enables implementation on computer grid. Method is described in next two sections, and finally experimental results on computing efficiency in dependence on multiple parameters are given in last section.

3 Square Approximation Gradient Method

Square approximation learning algorithm is one of gradient methods, but it uses, not only first partial derivative of $\Im(\theta)$, but also the second. In another words, square approximation method approximates $\Im(\theta)$ with square function in the neighborhood of $\theta(k)$. The approximation is

$$\Im(\theta) \cong \Im(\theta(k)) + \nabla^T \Im(\theta(k))(\theta - \theta(k)) + \frac{1}{2}(\theta - \theta(k))^T \nabla^2 \Im(\theta(k))(\theta - \theta(k)). \quad (7)$$

Condition for minimum is that first derivative of $\Im(\theta)$ equals zero,

$$\frac{\partial \Im}{\partial \theta} = \nabla \Im(\theta(k)) + \nabla^2 \Im(\theta(k))(\theta - \theta(k)) = 0. \quad (8)$$

Substituting θ with $\theta(k+1)$ emerges recursive relation

$$\theta(k+1) = \theta(k) - \left[\nabla^2 \Im(\theta(k))\right]^{-1} \nabla \Im(\theta(k)). \quad (9)$$

Compared to (4) the direction of searching is

$$s_d(k) = -\left[\nabla^2 \Im(\theta(k))\right]^{-1} \nabla \Im(\theta(k)) = -\boldsymbol{H}^{-1} \nabla \Im(\theta(k)). \quad (10)$$

With respect to (4) $s_d(k)$ will have direction towards minimum if $\boldsymbol{H}$ is positive-definite. $\boldsymbol{H}$ is positive-definite for strictly convex functions. If $\Im(\theta)$ is not strictly convex function, the algorithm (9) may diverge. It is important to select initial values $\theta(1)$ to be close enough to minimum to ensure positive-definity of $\boldsymbol{H}$.

For criteria function defined by (1), gradient vector and Hessian matrix are:

$$\nabla \Im(\theta) = \boldsymbol{J}^T(\theta) \cdot \boldsymbol{e}(\theta), \quad (11)$$

$$\boldsymbol{H}(\theta) = \nabla^2 \Im(\theta) = \boldsymbol{J}^T(\theta)\, \boldsymbol{J}(\theta) + \sum_i e_i(\theta) \cdot \nabla^2 e_i(\theta). \quad (12)$$

where $\boldsymbol{J}(\theta)$ is Jacobean matrix,

$$\boldsymbol{J}(\theta) = \frac{\partial e(\theta)}{\partial \theta}, \quad (13)$$

In relation (12) in each iteration it is required to calculate elements of matrix $\nabla^2 e(\theta)$ which is numerically demanding. Thus calculation of $\boldsymbol{H}$ can be modified. Here approximations of $\boldsymbol{H}$ are labeled as $\tilde{H}$ and $\breve{H}$. Modification should satisfy condition of positive definity, square convergence, be computationally efficient and at the same time be good approximation of $\boldsymbol{H}$ [10], [11], [12], [13].

Vector $\boldsymbol{e}(\theta)$ in the neighborhood of $\theta(k)$ can be substituted with its' linear approximation,

$$e(\theta) \cong \tilde{e}(\theta) = e(\theta(k)) + \nabla e(\theta(k)) \cdot (\theta - \theta(k)). \quad (14)$$

Now, instead of (1), we minimize approximation of $\Im(\theta)$

$$\tilde{\Im}(\theta)=\frac{1}{2}\tilde{e}^T \cdot \tilde{e} \cdot \tag{15}$$

From condition (2) we get relation that minimize function (15)

$$\boldsymbol{J}^T(\theta(k))\cdot\boldsymbol{J}(\theta(k))\cdot(\theta-\theta(k))+ \boldsymbol{J}^T(\theta(k))\cdot\boldsymbol{e}(\theta(k)). \tag{16}$$

Combining (11) with (16) and adding learning coefficient $\alpha(k)$ we get recursive relation for calculation of network parameters

$$\theta(k+1)=\theta(k)-\alpha(k)\left[J^T(\theta(k))\cdot J(\theta(k))\right]^{-1} J^T(\theta(k))e(\theta(k))\cdot \tag{17}$$

Comparing (9) with (17) one can see that Hessian matrix is substituted with approximation (18) which is equal to the first part of relation (10).

$$\tilde{H}(\theta(k))= \boldsymbol{J}^T(\theta(k))\cdot\boldsymbol{J}(\theta(k)). \tag{18}$$

Matrix $\tilde{H}$ is positive semi-definite which enables algorithm convergence, but if measuring data are not informative enough or neural network has too many neurons, matrix $\tilde{H}$ can be bad conditioned, i.e. almost singular, which may be cause of numerical instability. To overcome this problem it is necessary to modify $\tilde{H}$ in order to get positive definite matrix for all input space of criteria function. It is shown in [12] that when a constant μ is added to the elements on main diagonal of $\tilde{H}$, resulting matrix $\breve{H}$ (19) is good conditioned.

$$\breve{H}(\theta(k))= \boldsymbol{J}^T(\theta(k))\cdot\boldsymbol{J}(\theta(k)) + \mu\mathbf{I}. \tag{19}$$

On-line training algorithm finally becomes

$$\theta(k+1)=\theta(k)-\alpha(k)\cdot \breve{H}^{-1} \cdot J^T(\theta(k))\cdot e(\theta(k))\cdot \tag{20}$$

4 Parallel Decomposition

Square approximation learning algorithm described in previous section has faster convergence compared to gradient methods, but is computationally complex. Full occupancy of Hessian matrix and its' approximations (18) and (19), algorithm (20) requires centralized calculation of new network parameters for each iteration k. Due to full occupancy of $\boldsymbol{H}$ it is not possible to implement parallel calculation of neural network parameters. To overcome this disadvantage it is necessary to find matrix that can enable parallel calculation of $\theta(k+1)$ and at the same time retain square convergence property. In this section a proposal is made how to decompose algorithm for calculation of network parameters in nn independent relations, where nn is total number of neurons.

If vector of network parameters θ, that has $n(\theta)$ elements, is split into nn vectors θ_i,, each with $n(\theta_i)$ elements, Jacobean matrix, (13) consists of sub matrices (21) and (22).

$$\boldsymbol{J}(\theta) = [\boldsymbol{J}_1(\theta)\ \ \boldsymbol{J}_2(\theta)\ \ \ldots\ \ \boldsymbol{J}_{nn}(\theta)]. \tag{21}$$

$$J_i(\theta)=\frac{\partial e(\theta)}{\partial \theta_i}=\begin{bmatrix} \frac{\partial e_1(\theta)}{\partial \theta_{i,1}} & \frac{\partial e_1(\theta)}{\partial \theta_{i,2}} & \cdots & \frac{\partial e_1(\theta)}{\partial \theta_{i,n(\theta)}} \\ \frac{\partial e_2(\theta)}{\partial \theta_{i,1}} & \frac{\partial e_2(\theta)}{\partial \theta_{i,2}} & \cdots & \frac{\partial e_2(\theta)}{\partial \theta_{i,n(\theta)}} \\ \vdots & \vdots & \ddots & \vdots \\ \frac{\partial e_{N_e}(\theta)}{\partial \theta_{i,1}} & \frac{\partial e_{N_e}(\theta)}{\partial \theta_{i,2}} & \cdots & \frac{\partial e_{N_e}(\theta)}{\partial \theta_{i,n(\theta)}} \end{bmatrix}. \tag{22}$$

where N_e is number of input data and $n(\theta)$ number of network weights.

Combining (19) and (21) it follows

$$\breve{H}(\theta(k))=\begin{bmatrix} \breve{H}_{1,1} & \breve{H}_{1,2} & \cdots & \breve{H}_{1,nn} \\ \breve{H}_{2,1} & \breve{H}_{2,2} & \cdots & \breve{H}_{2,nn} \\ \vdots & \vdots & \ddots & \vdots \\ \breve{H}_{nn,1} & \breve{H}_{nn,2} & \cdots & \breve{H}_{nn,nn} \end{bmatrix}, \tag{23}$$

where $\breve{H}_{i,j}$ has $n(\theta_i)$ rows and $n(\theta_j)$ columns.

$$\breve{H}_{i,j} = \boldsymbol{J}_i^T \cdot \boldsymbol{J}_j + \mu \mathbf{I}. \tag{24}$$

if all sub matrices $\breve{H}_{i,j}$, $i \neq j$ are ignored, as a result emerges quasi-diagonal matrix $\overline{H}$,

$$\overline{H}(\theta)=\begin{bmatrix} J_1^T J_1+\mu I & 0 & \cdots & 0 \\ 0 & J_2^T J_2+\mu I & \cdots & 0 \\ \vdots & \vdots & \ddots & \vdots \\ 0 & 0 & \cdots & J_{nn}^T J_{nn}+\mu I \end{bmatrix}. \tag{25}$$

Combining (20) with (25) on-line training algorithm finally becomes

$$\theta_i(k+1)=\theta_i(k)-\left[J_i^T(\theta(k))J_i(\theta(k))+\mu I\right]^{-1}\cdot J_i^T(\theta(k))e(\theta(k)), \qquad i=1,2,\ldots,nn. \tag{26}$$

5 Experimental Verification

Parallel execution of square approximation learning algorithm was implemented and tested on standard MLP neural network.

Parallel execution is accomplished in splitting one Jacobean for whole neural network into t Jacobeans, one for each neuron, (21) and (22). This enables parallel execution of recursive formula for calculating neural weights θ. Instead of one formula for central calculation of all network weights, there are t formulas for θ, one for each neuron, (26). Thus, algorithm execution can be parallel in that part. Additional time saving is accomplished due to smaller dimensions of matrix that it is to be inversed $J_{t,\bullet,\bullet,k}^T J_{t,\bullet,\bullet,k}+\mu I$, as matrix inversion is numerically demanding, and increases exponentially by the dimension.

Practical implementation of the algorithm was done in C programming language. Standard ANSI C libraries and legacy code are mostly used, with an exception of additional windows libraries used for simpler time measurement and thread organization. Program was developed in MS Visual Studio IDE, the ANSI C code was used for the purpose of easier conversion for future work in MPI interface and the Grid/Cluster environment.

Experiments consist of measuring algorithm execution time in dependence of network parameters such as number of neurons and layers and size of input/output data vector.

Experiments were performed in sequence on a Dell Inspiron 6400 laptop with the following configuration: Intel Core Duo T2300@1.66GHz (two cores), 1GB DDR2@667Mhz, 80GB HDD. Algorithm is executed using one, two, three and four treads respectively and the results are given in figures 1 and 2.

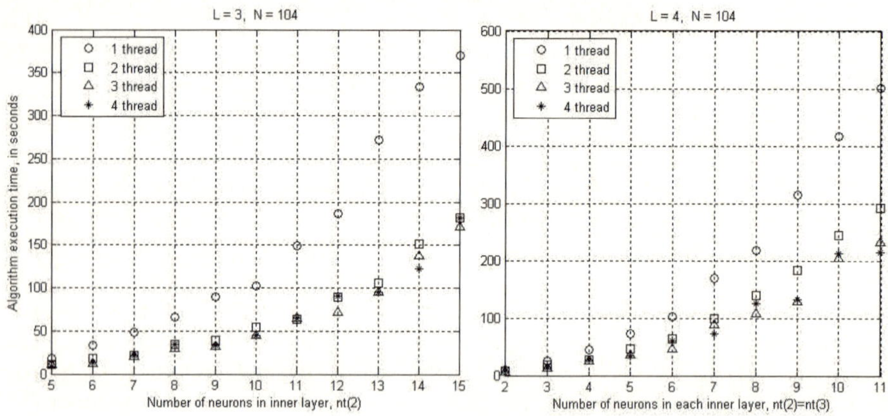

Fig. 1. Algorithm execution time over number of neurons and layers for MLP NN

It is important to note that for every single execution of the application, random initial values $\theta(1)$ were generated, in that way execution times for identical neural networks do not have the same time values in the end of execution. All experiments were executed 10 times, and the score in the end is arithmetic mean of the time values needed for 10 iterations of the algorithm for neural network weights calculation to pass, per each parameter set.

As it is seen in the diagrams in figures 1 and 2, the time values get reduced as the number of worker threads used for calculation are increased. The drop in time is about 50% comparing 1 and 2 threads, which is logical since in that case calculations are distributed on two available cores. Additional increase in thread number gives roughly the same results since all the threads occupy the same amount of available resources and the score can be only worse because of the switching between executing threads. This is especially true in the simple networks where it takes very little time to execute the program on a single thread, parallelizing it only aggravate execution time because of the need for time consuming thread initializations and data distribution.

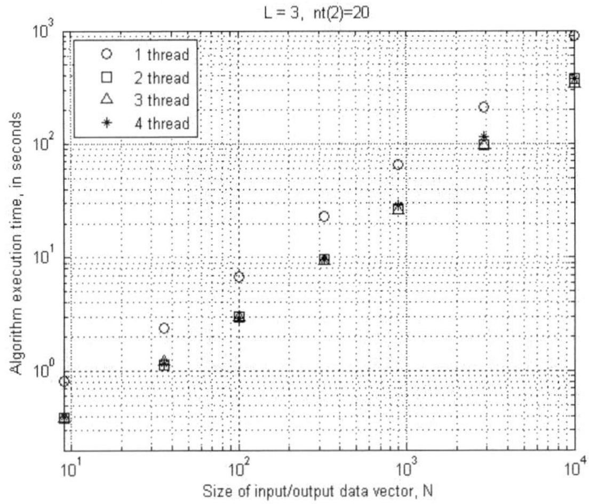

Fig. 2. Algorithm execution time over size of input/output data vector for MLP NN with 20 neurons in single inner layer

Analyzing CPU usage while the application is running shows that the processor is only used at about 50%(of both cores) in case of the single thread execution. Where in case of multiple thread execution CPU usage increases to 100%(of both cores). Reason for that behavior is obvious, when we have single thread, its work is constantly shifted from one core to the other and we have idle state when one core has no thread to execute, whereas multiple threads execution, every core is dedicated a single thread all the time and there is no idle time. But since the used processor in all measurement is dual core, only two threads are run simultaneously and if we use 3 or 4 threads in our calculation, execution of threads are rotated on the available resources (two cores).

6 Conclusion

Square approximation learning algorithm has fast convergence, but is computationally complex. In this work a proposal is made how to decompose algorithm for calculation of network parameters in *nn* independent relations, where *nn* is total number of neurons in order to adapt the algorithm for parallel execution. Parallel execution of square approximation learning algorithm was implemented and tested for standard MLP neural network. Experimental results show time savings in multiple thread execution for a wide range of MLP neural network parameters, such as size of input/output data matrix, number of neurons and layers. The drop in time is about 50% comparing 1 and 2 threads. Since the used processor in all measurement is dual core, no time savings exist for 3 and 4 threads. Future work will be use of this algorithm in the Grid/Cluster environment where, using MPI interface, execution can be parallelized and distributed on numerous nodes which in turn is expected to improve execution time in real-time systems with strict time requirements.

References

1. Brent, R.P.: Fast Training Algorithms, for Multilayer Neural Nets. IEEE Transactions on Neural Networks 2(3), 346–353 (1991)
2. Barnard, E.: Optimization for training neural nets. IEEE Transactions on Neural Networks 3(2), 232–241 (1992)
3. Cichocki, A., Umbehauen, R.: Neural Networks for Optimization and Signal Processing. Wiley, New York (1993)
4. Himmelblau, D.M.: Applied nonlinear programming mathematical theory. McGraw-Hill, New York (1972)
5. Turk, S., Budin, L.: Computer Analysis, Školska knjiga, Zagreb (in Croatian) (1989)
6. Choi, J.J., Oh, S., Marks II, R.J.: Training Layered Perceptrons Using Low Accuracy Computations. In: Proc. Int'l Joint Conf. Neural Networks, pp. 554–559. IEEE, Piscataway (1991)
7. Chen, D.S., Jain, R.C.: A robust backpropagation learning algorithm for function approximation. IEEE Transactions on Neural Networks 5(3), 467–479 (1994)
8. Piche, S.W.: Steepest descent algorithms for neural network controllers and filters. IEEE Transactions on Neural Networks 5(2), 198–212 (1994)
9. Ergenzinger, S., Thomsen, E.: An accelerated learning algorithm for multilayer perceptrons: optimization layer by layer. IEEE Trans. Neural Networks 6(1), 31–42 (1995)
10. Dennis Jr., J.E., More, J.J.: Quasi-Newton Methods, Motivation and Theory. SIAM Review 19(1), 46–89 (1977)
11. Marquardt, D.W.: An Algorithm for Least-Squares Estimation of Nonlinear Parameters. Journal of the Society for Industrial and Applied Mathematics 11(2), 431–441 (1963)
12. Levenberg, K.: A method for the solution of certain non-linear problems in least squares. Quart. Appl. Math. 2, 164–168 (1944)
13. Petrović, I., Baotić, M., Perić, N.: An Efficient Newton-type learning Algorithm for MLP Neural Networks. In: Proceedings of the International ICSC/IFAC Symposium on Neural Computation - NC 1998, pp. 551–557 (1998)

Neural Model-Based Segmentation of Image Motion

Lucia Maddalena[1] and Alfredo Petrosino[2]

[1] ICAR - National Research Council
Via P. Castellino 111, 80131 Naples, Italy
lucia.maddalena@na.icar.cnr.it
[2] DSA - University of Naples Parthenope
Centro Direzionale, Isola C/4, 80143 Naples, Italy
alfredo.petrosino@uniparthenope.it

Abstract. Besides enabling the segmentation of video streams into moving and background components, detecting moving objects provides a focus of attention for recognition, classification, and activity analysis, making these later steps more efficient. We propose a novel model for image sequences based on self organization through artificial neural networks, that is used both for background modeling, allowing to handle scenes containing moving backgrounds or gradual illumination variations, and for stopped foreground modeling, helping in distinguishing between moving and stopped foreground regions and leading to an initial segmentation of scene objects. Experimental results are presented for real video sequences.

Keywords: background modeling, foreground modeling, image sequence modeling, neural network, self organization.

1 Introduction

Although primary aim of moving objects detection in image sequences is the segmentation of video streams into foreground and background components, it also provides a focus of attention for recognition, classification, and activity analysis, making these later steps more efficient, since only moving pixels need be considered. For many of these tasks, it is necessary to build representations of the appearance of objects in the scene [2]. This paper presents a model for image sequences that is used for two issues: to model the scene background in a way that supports sensitive detection of moving objects, and to model the foreground objects in a way that supports their tracking.

Moving object detection is known to be a significant and difficult problem [7]. The most common and efficient method to tackle it for scenes taken from stationary cameras is background subtraction, that is based on the comparison of each sequence frame with a reference background model, including information on the scene without moving objects (see [6]). This method is independent on moving object velocity and it is not subject to the foreground aperture problem, but it is extremely sensitive to dynamic scene changes due to lighting and extraneous events.

I. Lovrek, R.J. Howlett, and L.C. Jain (Eds.): KES 2008, Part I, LNAI 5177, pp. 57–64, 2008.

Apart from background modeling, also building representations for foreground objects is essential for tracking them and maintaining their identities throughout the image sequence [2]. Modeling the color distribution of homogeneous regions has been used successfully to track nonrigid bodies (e.g. [8]), and a variety of parametric and nonparametric statistical techniques have been adopted (see [2]).

Our objective is to construct a system for motion detection based on the background and the foreground model automatically generated by a self-organizing method without prior knowledge of the pattern classes. The approach, firstly proposed for background modeling [4,5], consists in using biologically inspired problem-solving methods to solve motion detection tasks, typically based on visual attention mechanisms [1]. The aim is to obtain the objects that keep the users attention in accordance with a set of predefined features. By learning the trajectories and features of moving objects using a self-organizing approach, the background and foreground models are built up. Such models, together with a layering mechanism, allow to construct a system able to detect motion and distinguish foreground objects into moving or stopped objects, even when they appear superimposed.

The paper is organized as follows. In §2 we describe a self-organizing approach to image sequence modeling. In §3 and §4 we describe how such model is adopted for background modeling and foreground modeling, respectively. §5 presents preliminary results obtained with the implementation of the proposed approach, while §6 includes concluding remarks.

2 Image Sequence Modeling

In this section we describe a model for image sequences taken from stationary cameras that automatically adapts to scene changes in a self-organizing manner and without a priori knowledge. The idea is to build the model by learning in a self-organizing manner image sequence variations, seen as trajectories of pixels in time. A neural network mapping method is proposed to use a whole trajectory incrementally in time fed as an input to the network.

The adopted artificial neural network is organized as a 2D flat grid of neurons and, similarly to Self-Organizing Maps [3], allows to produce representations of training samples with lower dimensionality, at the same time preserving topological neighborhood relations of the input patterns. Each neuron computes a function of the weighted linear combination of incoming inputs, where weights resemble the neural network learning. Each neuron can be therefore represented by a weight vector, obtained collecting the weights related to incoming links. An incoming pattern is mapped to the neuron whose set of weight vectors is "most similar" to the pattern, and weight vectors in a neighborhood of such node are updated. Therefore the network behaves as a competitive neural network that implements a winner-take-all function with an associated mechanism that modifies the local synaptic plasticity of the neurons, allowing learning to be restricted spatially to the local neighborhood of the most active neurons.

For each image pixel we consider a neuronal map consisting of $n \times n$ weight vectors, that are all initialized with the pixel value. The complete set of weight vectors for all pixels of an image I with N rows and M columns is represented as a neuronal map A with $n \times N$ rows and $n \times M$ columns, where adjacent blocks of $n \times n$ weight vectors correspond to adjacent elements in image I. An example of such neuronal map structure for a simple image I with $N = 2$ rows and $M = 3$ columns obtained choosing $n = 3$ is given in Fig. 1. As depicted, the upper left pixel a of I (Fig. 1 - (a)) has weight vectors $(a_1, \ldots, a_9)$ stored into the 3×3 elements of the upper left part of neuronal map A (Fig. 1 - (b)), and analogous relations exist for each pixel of I and corresponding weight vectors storage. It appears evident that the neuronal map A can be seen as an initial image sequence model, enlarged 3×3 times.

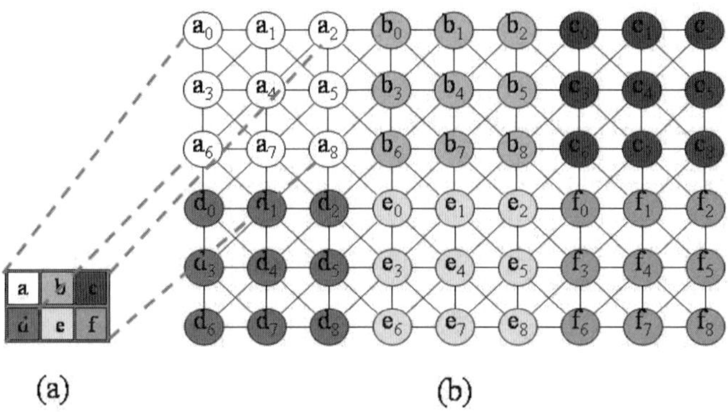

Fig. 1. A simple image (a) and the modeling neuronal map (b)

Subsequent learning of the neuronal map allows to adapt the image sequence model to scene modifications. The learning process consists in updating the model by changing the neural weights according to a visual attention mechanism of reinforcement. Specifically, temporally subsequent samples are fed to the network. Each incoming pixel p_t of the t-th sequence frame I_t is compared to the current pixel weight vectors $(c_1, c_2, \ldots, c_{n^2})$ to determine the weight vector c_m that best matches it:

$$d(c_m, p_t) = \min_{i=1,\ldots,n^2} d(c_i, p_t) \tag{1}$$

where the metric $d(\cdot)$ is suitably chosen according to the specific color space being considered. The best matching weight vector c_m is used as the pixel encoding approximation, and, together with its $n \times n$ neighborhood, is reinforced according to weighted running average:

$$A_t(i,j) = (1 - \alpha_{i,j})A_{t-1}(i,j) + \alpha_{i,j}p_t, \quad \begin{array}{l} i = \overline{x} - \lfloor \frac{n}{2} \rfloor, \ldots, \overline{x} + \lfloor \frac{n}{2} \rfloor \\ j = \overline{y} - \lfloor \frac{n}{2} \rfloor, \ldots, \overline{y} + \lfloor \frac{n}{2} \rfloor \end{array} \tag{2}$$

where $c_m = A_t(\overline{x}, \overline{y})$. Here $\alpha_{i,j} = \alpha\ w_{i,j}$, where α represents the learning factor, that depends on the scene variability, while $w_{i,j}$ are Gaussian weights in the $n \times n$ neighborhood of c_m, that well correspond to the lateral inhibition activity of neurons. Therefore, such updating allows to take into account spatial relationships between incoming pixel and its surrounding.

The neuronal network obtained as described gives at each time instant t a fairly compact representation of the image sequence $I_0, \ldots, I_t$. Specifically, it can be seen as a nonlinear projection of the probability density function of the high-dimensional input data onto the two-dimensional network, and therefore weight vectors give a discrete approximation of training samples distribution [3]. Such distribution can be seen as a mixture of distributions, one for modeling just the background and one for modeling the stopped foreground of an image sequence. This allows us to achieve moving object detection with discrimination between stopped and moving foreground objects, as described in the following sections.

3 Modeling the Background

The main problem of the background subtraction approach to moving object detection is its extreme sensitivity to dynamic scene changes due to lighting and extraneous events. Although these are usually detected, they leave behind "holes" where the newly exposed background imagery differs from the known background model. While the background model eventually adapts to these "holes", they generate false alarms for a short period of time. Therefore, our aim is to devise an approach to moving object detection based on a background model that automatically adapts to changes in a self-organizing manner and without a priori knowledge. To this end we use the self-organizing image sequence model described in §2, that is indeed a generalization of the one already adopted in [4,5] for background modeling, as briefly described in the following.

In order to represent each weight vector of the neuronal map, we choose the HSV color space, that allows to specify colors in a way that is close to human experience of colors, relying on the hue, saturation and value properties of each color. Each weight vector $c_i, i = 1, \ldots, n^2$, is therefore a 3D vector initialized to the HSV components of the corresponding pixel of the first sequence frame I_0, assuming this frame as initial approximation of the background.

Learning of the neuronal map in this case consists in updating the background model using *selective* weighted running average, in order to adapt the model to slight scene modifications without introducing the contribution of pixels that do not belong to the background scene. Specifically, for each incoming pixel p_t of the t-th sequence frame I_t, weighted running average of eqn. (2) is applied only if the best match c_m to p_t is close enough to the background model, i.e. only if

$$d(c_m, p_t) \leq \epsilon, \tag{3}$$

where ϵ allows to distinguish between foreground and background pixels. Otherwise, if no acceptable best matching weight vector exists, then p_t is detected as belonging to a moving object.

The above described background subtraction and update procedure for each pixel can be sketched as in the following algorithm:

Algorithm SOBS (Self-Organizing Background Subtraction)
Input: pixel value p_t in sequence frame I_t, $t = 0, \ldots, LastFrame$
Output: background/foreground binary mask value $B(p_t)$

```
Initialize weight vectors C for pixel p_0 and store it into A
for t=1, LastFrame
  Find best match c_m in C to current sample p_t satisfying eqn. (3)
  if (c_m found) then
    B(p_t) = 0   //background
    Update A in the neighborhood of c_m as in eqn. (2)
  else
    B(p_t) = 1   //foreground
```

A thorough discussion about the metric adopted in eqn. (1) and about the selection of parameters α in eqn. (2) and ϵ in eqn. (3) can be found in [5]. There the proposed approach has been shown to outperform several existing moving object detection methods, in particular in handling cases as moving backgrounds and gradual illumination changes.

4 Modeling the Foreground

The binary mask B obtained by SOBS algorithm for each sequence frame I_t allows to distinguish incoming pixels p_t into background pixels (i.e. pixels modeled by the background model, for which $B(p_t) = 0$) and foreground pixels (i.e. pixels not modeled by the background model, for which $B(p_t) = 1$). However, due to the selective nature of SOBS background update process, objects that enter the scene and stop are always detected as moving foreground objects, even if they are not really *moving*. An example is given in Fig. 2-(b), showing that both the moving and the stopped cars are detected as moving foreground in the mask B computed by the SOBS algorithm.

In order to distinguish between moving and stopped foreground objects, we introduce a layering mechanism, where each layer models a single stopped foreground object. If a pixel is detected as a foreground pixel for several consecutive frames, then it is considered as belonging to a stopped object, and it is inserted into a stopped foreground layer.

Stopped foreground pixels are modeled by a neuronal map analogous to the one described in §2. When a stopped foreground pixel is first detected, the corresponding weight vectors are initialized to the stopped foreground pixel itself. For subsequent sequence frames, weight vectors for stopped foreground pixels are updated using selective running average as in eqn. (2), where now A indicates the stopped foreground model and pixel $(\overline{x}, \overline{y})$ of A contains the best match c_m for the incoming stopped pixel (x, y) satisfying eqn. (3).

The foreground modeling algorithm for an incoming pixel value p_t in sequence frame I_t not belonging to the background model allows to obtain a binary mask Mov indicating moving foreground pixels and a set of stopped foreground layers $S = \{S_i, i = 1, \ldots, NumOfStoppedLayers\}$ as follows:

Foreground modeling algorithm

Find best match c_m in S to current sample p_t satisfying eqn. (3)
if (c_m found in layer S_i) **then**
  $Mov(p_t) = 0$ //not moving
  update S_i in the neighborhood of $c_m = A(\overline{x}, \overline{y})$ as in eqn. (2)
else
  $Mov(p_t) = 1$//moving

Foreground modeling together with the layering strategy allow to keep an up-to-date representation of all stopped pixels and an initial classification of scene objects, which can be used to support subsequent tracking and classification phases. Moreover stopped foreground modeling allows to distinguish moving and stopped foreground objects when it happens they are located in the same region of the image. Indeed, in this case the availability of an up-to-date model of all stopped objects allows to discern whether pixels that are not modeled as background belong to one of the stopped objects or to moving objects. An example is given in Fig. 2, where the stopped foreground model reported in Fig. 2-(d) allows to distinguish the stopped and the moving cars, as shown in Fig. 2-(f).

5 Experimental Results

Experimental results for moving object detection using the proposed approach have been produced for several image sequences. Here we report results obtained for sequence *Dataset*1 of PETS2001 sequences available on the web[1]. This is an outdoor sequence consisting of 2688 frames of 768 × 576 spatial resolution. The scene consists of a street crossing with moving cars and people, where one of the cars stops in a parking lot. One representative frame together with obtained results is reported in Fig. 2. Here we report the original sequence frame no. 824 (Fig. 2-(a)), where the first car has already stopped, while the second car is passing in front of it. The corresponding moving object detection mask computed by the SOBS algorithm choosing $n = 3$ (Fig. 2-(b)) shows that all moving objects are well detected. However the parked car is detected as a moving object too, due to the selective update of SOBS algorithm, and therefore the detection mask does not allow to distinguish between the stopped car and the moving one.

Fig. 2-(c) shows the background model A computed by the SOBS algorithm described in §3, that appears to be a quite accurate representation of the real background. We would remark that the background model A is represented by a neuronal map whose size is nine times that of the original image I; in the

[1] http://www.cvg.cs.rdg.ac.uk/PETS2001/pets2001-dataset.html

(a) (b) (c) (d) (e) (f)

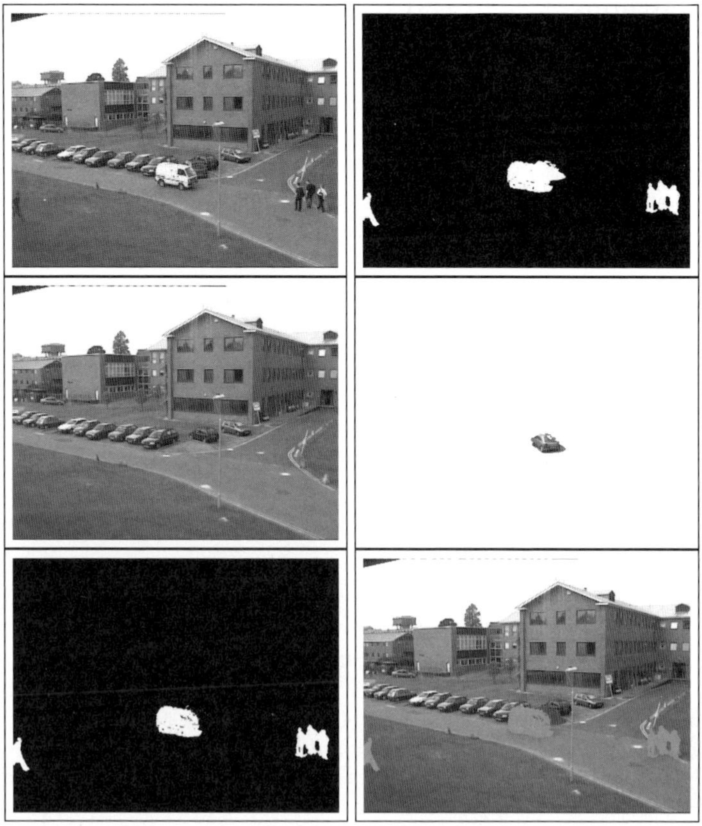

Fig. 2. Results of background and foreground modeling for *Dataset*1 sequence: (a) original frame; (b) moving object detection mask computed by SOBS algorithm; (c) background model; (d) stopped foreground model; (e) moving foreground mask; (f) original frame with moving (green) and stopped (red) foreground objects

reported figures they appear to have the same size only for space constraints and for an easier comparison.

In Fig. 2-(d) we show the first stopped foreground layer obtained as explained in §4, that represents the parked car. Such stopped foreground model allows to distinguish the stopped and the moving cars, as shown by the mask containing only the moving foreground reported in Fig. 2-(e) and by the final result reported in Fig. 2-(f), where we show the original frame with moving foreground objects (in green) and stopped foreground objects (in red).

It can be observed that the neuronal maps modeling background (Fig. 2-(c)) and stopped foreground (Fig. 2-(d)), together with the moving foreground mask (Fig. 2-(e)), give a quite accurate and up-to-date representation of current sequence frame (Fig. 2-(a)), that can be used for an initial segmentation of the scene.

6 Conclusions

We described a novel model for image sequences that automatically adapts to scene changes in a self-organizing manner and without a priori knowledge. The model is then targeted for modeling background and foreground by learning motion patterns and so allowing the separation of background, moving foreground and stopped foreground in scenes taken from stationary cameras, strongly required in video surveillance systems. The adopted background modeling is able to handle scenes containing moving backgrounds or gradual illumination variations, achieving robust detection for different types of videos taken with stationary cameras. Moreover, we showed that the proposed modeling of stopped foreground pixels helps in distinguishing between moving and stopped foreground regions, leading also to an initial segmentation of scene objects. Experimental results on real data demonstrate the effectiveness of the proposed system and its ability to capture both moving and stopped foreground objects.

References

1. Cantoni, V., Marinaro, M., Petrosino, A. (eds.): Visual Attention Mechanisms. Kluwer Academic/Plenum Publishers, New York (2002)
2. Elgammal, A., Duraiswami, R., Harwood, D., Davis, L.S.: Background and Foreground Modeling Using Nonparametric Kernel Density Estimation for Visual Surveillance. Proceedings of the IEEE 90(7), 1151–1163 (2002)
3. Kohonen, T.: Self-Organizing Maps, 3rd edn. Springer, Heidelberg (2001)
4. Maddalena, L., Petrosino, A.: A Self-Organizing Approach to Detection of Moving Patterns for Real-Time Applications. In: Mele, F., Ramella, G., Santillo, S., Ventriglia, F. (eds.) BVAI 2007. LNCS, vol. 4729, pp. 181–190. Springer, Heidelberg (2007)
5. Maddalena, L., Petrosino, A.: A Self-Organizing Approach to Background Subtraction for Visual Surveillance Applications. IEEE Transactions on Image Processing 17(7), 1168–1177 (2008)
6. Radke, R.J., Andra, S., Al-Kofahi, O., Roysam, B.: Image change detection algorithms: a systematic survey. IEEE Trans. Image Process 14(3), 294–307 (2005)
7. Toyama, K., Krumm, J., Brumitt, B., Meyers, B.: Wallflower: Principles and Practice of Background Maintenance. In: Proc. of the Seventh IEEE Conference on Computer Vision, vol. 1, pp. 255–261 (1999)
8. Wren, C., Azarbayejani, A., Darrell, T., Pentland, A.: Pfinder: Real-Time Tracking of the Human Body. IEEE Trans. on PAMI 19(7), 780–785 (1997)

Classification on Soft Labels Is Robust against Label Noise

Christian Thiel

Institute of Neural Information Processing, University of Ulm, 89069 Ulm, Germany
christian.thiel@uni-ulm.de

Abstract. In a scenario of supervised classification of data, labeled training data is essential. Unfortunately, the process by which those labels are obtained is not error-free, for example due to human nature. The aim of this work is to find out what impact noise on the labels has, and we do so by artificially adding it. An algorithm for the noising procedure is described. Not only individual classifiers are studied, but also ensembles of classifiers whose answers are combined, increasing the overall performance. Also, we will answer the question if classifiers trained on soft labels are more resilient to label noise than those trained on hard labels.

1 Introduction

In supervised classification, we naturally can not work without labels that are associated with our training data. Obtaining labels, hard or soft, is prone to errors, human or otherwise. That means that a classification algorithm has falsely labeled data in his training set, which, in extreme cases, might render it useless. In this paper, we will evaluate the impact that label noise has on the accuracy of single classifiers, and also multiple classifier schemes.

The training data available for a specific classification task must not necessarily be labeled hard. That is, a training sample might belong, to different degrees, to multiple classes simultaneously. For example, multiple experts might not agree on the diagnosis for the sample, or hear different emotions in a spoken sentence [1,2]. In fact, problems in the field of medical or life sciences, like predicting the secondary structure of proteins [3], often produce and require soft labels. We will compare the noise-resilience of hard-trained versus soft-trained classifiers.

In the following section, we review previous work on the topic of label noise, before presenting a model to artificially add distinct levels of noise. After describing the experimental setup, we present our results. A summary wraps up this article.

2 Previous Works

There is not much literature on how label noise should be modeled and dealt with. One exception is Anluin and Lairds paper [4] that details how to embed

I. Lovrek, R.J. Howlett, and L.C. Jain (Eds.): KES 2008, Part I, LNAI 5177, pp. 65–73, 2008.

training data with flipped labels into the framework of Probably Approximately Correct (PAC) learning [5], even allowing malicious noise. They give a lower bound on the number of training samples necessary. The analysis is only applicable to a "flip" kind of noise and does not hold for more than two classes. Then, some works exist that do present algorithms that can deal with noise, but mostly with the restrictions mentioned above. One example is [6], where the approach is to learn the parameters of the model that generates the noise. Closely related, in [7] the label flip probabilities are incorporated into the training target criterion. The thrust of that paper is different, however, as the setting is semi-supervised learning, where no noise is actively added, but only a faction of the training data has labels associated. Being an extention of the methods proposed by McLachlan in [8], the algorithm changes the labels attributed to the unlabeled training data with each iteration, minimising the modified *Classification Maximum Likelihood* criterion.

We have ourselves previously investigated the impact of noise on classification accuracy, with a focus on fuzzy labels [9]. Our current work uses a more intricate noise model and does not limit the scope of classifiers to Fuzzy KNN.

3 Modelling Label Noise

We are taking the training data from our sensors as it is, but the class information associated with each object may be erroneous. This we call label noise. To experimentally determine the impact of label noise on classification accuracy, we need to artificially add noise according to a certain model. In a two-class case, a given portion of the training data would get randomly selected and the associated label flipped to the opposite class. This methods extends to the multi-class case, with the label being flipped to one of the other classes in a random manner (as employed in [9]). But since we are dealing with fuzzy labels, where not only one class is given in the label, those noise models are not applicable. Thus, we employ the procedure described in Algorithm 1:

Algorithm 1. Adding noise to labels

Input: Normalised labels $label_i$ ($i = 1 \ldots \#$ samples), desired *noise* level
for all $label_i$ **do**
  % Generate random label
    $rlabel$ = for each class, draw from uniform distribution in $(0, 1)$
    $rlabel = normalise\_label(rlabel)$
  % Mix original and random label
    $label_i = label_i * (1 - noise) + rlabel * noise$
end for
Output: Normalised labels $label_i$ with added noise

This approach models the noise as it could appear in real-world scenarios, and can be understood intuitively. Note that it is important to first normalise the random label and then combine it with the original one using a weighted sum. Normalising only in the end would seriously flatten the fuzzy labels and decrease their variance[1]. It is obviously essential that the original label and the random label be normalised in the same way, or their combination would yield unpredictable results. In our experiments, we simply made the labels to sum up to one.

For certain applications, it might be useful to use a different noise model than the uniform one we employed. For example, one could treat each class label differently, either by assigning a special variance for the randomisation, or using a different noise level for each. A totally different approach to generate the new noised labels would be to see each label as a point on the hyperplane of possible (regarding the normalisation) labels. To add noise, one would advance on this hyperplane into a random direction, with the length of this vector being determined by the desired noise level.

4 Experimental Setup

We want to evaluate the impact that noise on the training labels has on the accuracy of single classifiers and multiple classifier architectures. To this end, different levels of artificial noise (see Section 3) are added to the labels. For each level, four basic classifiers, based on different features, are trained, and their decisions combined. Essentially, we are interested in the behaviour of classification accuracy as we increase the level of label noise. The entire experiment is run twice, once with the fuzzy labels, once with hard labels that have been derived from the fuzzy labels. This allows us to see whether it is beneficial to use soft labels, or if hard labels are to be preferred. In all our experiments, conversion of soft labels to hard labels, often called *hardening*, is done via the *winner takes all* rule, also known as *maximum membership* rule. It works by assigning the class with the highest membership value all the weight, and zero to the rest. For example, hardening [0.3 0.4 0.1 0.2] would result in [0 1 0 0].

The fruits data set employed comes from a robotic environment, consisting of 840 pictures from 7 different classes (apples, oranges, plums, lemons,... [10]). Each image was divided evenly into multiple parts (indicated by PxP in the following), the feature values calculated independently for each part and then concatenated. Using the results in [11], we selected the four individual features on which classifiers had the highest accuracy, albeit with the restriction that the features should be fundamentally different. The features selected are (described in detail in [11], dimensions given in brackets): *Colour Histograms* (3x3, 216 dim) in the RGB space. Orientation histograms on the edges in a greyscale version of the picture. Here we used both the *Sobel* operator (4x4, 128 dim) and the *Canny*

[1] We did an experiment on 700 fuzzy labels. Without noise, 428 of them had a class membership with a value above 0.5. After adding 30% noise ($noise = 0.3$), this dropped to 250, flattening the labels a bit. Had we used the normalising only at the end, this value would have significantly dropped to 38. Looking at the mean variance of the labels, the original ones had 5.4e-2, the noised ones 2.7e-2, and for the end-normalised ones it dropped down to 1.2e-2.

(3x3, 72 dim) algorithm to detect edges. As weakest feature, colour histograms in the black-white opponent colour space were calculated (*APQBW*, 2x2, 32 dim). All results were obtained using 5-fold cross validation.

As the data set initially only had hard labels (a banana is quite clearly a banana), we had to convert them to soft labels first. This was done using the fuzzy K-Means clustering [12] algorithm. First, we clustered the data, then assigned a label to each cluster centre according to the hard labels of the samples associated with it to varying degrees. Then, each sample was assigned a new soft label as a sum of products of its cluster memberships with the centres' labels[2].

The basic classifiers we used were Radial Basis Function networks (for the Sobel and Colour Histogram features) and Fuzzy-Input Fuzzy-Output Support Vector Machines (for Canny and APQBW) which we will call F^2-SVMs. Those two classifier types can handle soft training labels, give soft answers and generally show a good classification performance.

The number of kernels in the Gaussian RBF network [13] was set to 47 using a simple heuristic formula, their position determined by running a fuzzy c-means algorithm [12] on the training data. The individual variance of the kernels was set using an experimental observation of Breimann [14], which allows to have only one parameter to optimise for the whole net[3].

The Fuzzy-Input Fuzzy-Output Support Vector Machines [2] employed are, as their name suggests, SVMs able to take training data with fuzzy labels, and to give a fuzzy reply to test samples. The choice of the parameter C common to all machines, which controls the sensitivity to class memberships, had to be determined beforehand using cross validation. The overall architecture is One-Against-One [15], the kernels were polynomials of degree three.

The combination of the classifier's decisions is accomplished using several established Multiple Classifier System architectures in parallel. See [16] for an introduction, and [17] for a comparison of the performance of many methods. We tested the following schemes: Minimum, Median, Average, Dempster-Shafer orthogonal sum rule [18,19,20], Decision Templates ([17], using measure S_1), simple probabilistic product, and an optimal least squares solution calculated using the pseudoinverse. A theoretical comparison of the last three ones can be found in a previous paper of ours [21].

The basic performance measure in our experiments is the classifier accuracy. That is, we harden the soft labels and soft outputs, to find out the agreement between them. There are plenty of other comparison metrics available, a good survey can be found in [22]. The authors of that paper also conceived and tested

---

[2] Clustering is done for all classes together, using the APQBW feature. After looking at the resulting labels we chose a fuzzifier of 1.3. Hardened fuzzy labels agree with the hard labels in about 80% of the cases.

[3] Using a kernel of the form $f_i(x) = \frac{1}{(\alpha_k d_{i,k})^2} K\left(\frac{x-c_i}{\alpha_k d_{i,k}}\right)$, good results can be obtained if the following ratio including the free parameter α_k is kept constant: $\frac{\alpha_k \overline{d_k}^2}{\sigma(d_k)}$. Here $\overline{d_k}$ is the mean of the kth nearest neighbour distances and $\sigma(d_k)$ their standard deviation.

a very promising new measure, but had to conclude that "the best course of action to obtain all the accuracy information is to support the interpretation of the descriptive measures with a detailed inspection of the full fuzzy error matrix". So, as long as there are no other widely accepted methods for comparison, we decided to use classifier accuracy for our graphs.

5 Results

The most important findings of our experiments can be seen in Figure 1: Even the individual classifiers hold up to noise incredibly well. The classifier fusion step is always able to improve over the individual results. Most importantly, classifiers working with the soft labels have higher accuracy.

A more detailed analysis of the performance of the soft-trained versus the hard-trained classifiers is shown in Figure 2. The individual soft classifiers have, in most cases, a higher accuracy (shown as gain in percent points) than their hard counterpart. Taking the mean value, soft wins. But this advantage gets less and less noticeable once more and more noise is added. The better performance of soft trained classifiers comes, in our opinion, from the ability of the classifiers to take advantage of interdependencies that are encoded in the fuzzy labels, for which we found experimental evidence in separate F^2-SVM-experiments [2] on a dataset of emotional speech [23].

As can be seen rather clearly in Figure 1, the extra fusion step is really worth doing (reasons for this can be found in [16]). The combined answer is always more accurate than even the best single classifier. But of course not all fusion schemes are equally powerful, and when investigating this issue, it turns out one has to make the distinction between the hard-trained and soft-trained setups. In the case with only hard labels, Median, Product, Decision Templates, and Pseudoinverse fusion rules are in the top group, a clear winner for all noise

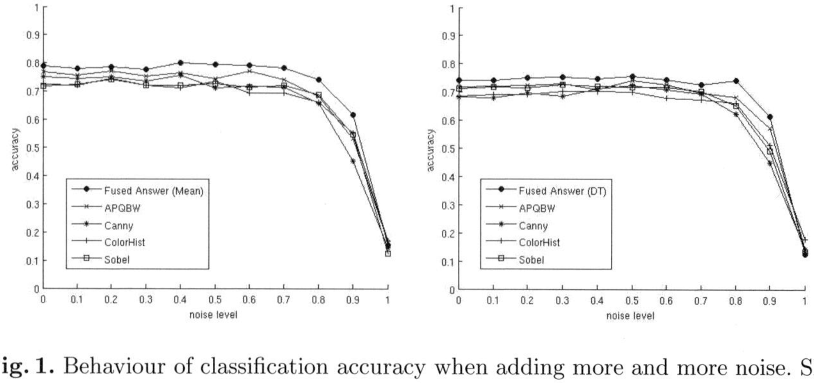

Fig. 1. Behaviour of classification accuracy when adding more and more noise. Shown are plots for the four basic classifiers and their fused answer (method given in brackets). The classifiers for the left plot were trained on soft labels, whereas only hard labels were provided to the ones in the right plot.

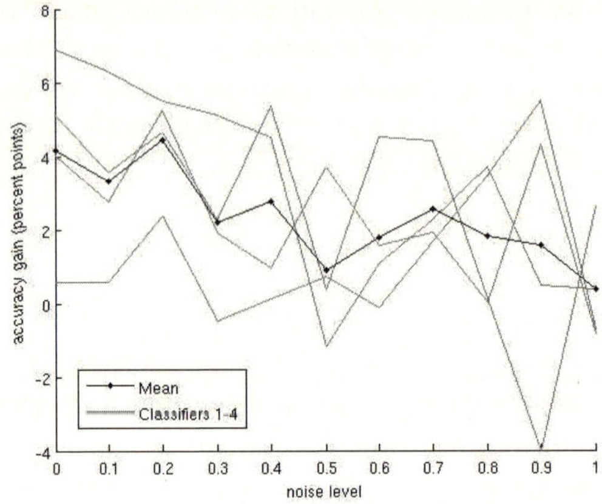

Fig. 2. Accuracy gain in percent points of the four soft-trained classifiers over their hard-trained counterparts. The dotted black line gives the mean value.

levels[4] can not be declared. For all crossvalidation runs and noise levels, the worst result of the top group is only 6.0 percent points away from the best fusion result for this run (MaxDistanceToBest = 6.0). As for the soft-trained case, only Median and Decision Templates form the top group, with Median being the most stable. The Product rule had to be discarded, as in some single cross validation runs it has outlier drops in accuracy worse than 10 percent points. The MaxDistanceToBest here is very low, only 3.6 percent points.

One observations strikes as particularly surprising: the high resilience of even single classifiers, trained on hard or soft labels, to added noise. Revisiting Figure 1, we see that despite adding 80% noise on the training labels, the best single classifier still has the very high accuracy of 68%. To put this into perspective, we shall look at the effects of noise on the labels from another angle. If we did not treat them as fuzzy labels, but are only interested in the class with the highest probability, the behaviour shown in Figure 3 comes up. The hardened noisy labels agree pretty much with the hardened original, not noised labels. For example, at the above-mentioned noise level of 80%, there still is agreement of 56.4%, meaning more than half of the samples associated (via hardening) with one class are originally from that class. This seems to still be enough for the classifiers to train reasonably well. So, the fuzzy labels can take quite some amount of such noise before it poses problems in our classification setting.

A short note on fuzzy versus hard in this context: the findings that the classifiers are quite resilient to high label noise levels is only valid for the soft labels. As shown, the noise added does deteriorate a fuzzy label gracefully, and it will take high noise levels until the winning class with the highest probability changes. As the hard labels in this experiment have been derived by hardening the available soft

[4] For the fusion architecture selection, we disregarded the noise levels from 80-100% noise, for those cases are not relevant in practical applications and exhibit very volatile behaviour.

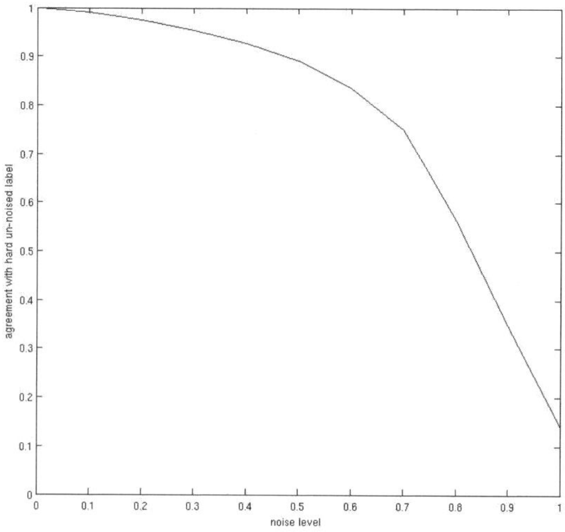

Fig. 3. The x axis gives the level of noise added to the fuzzy labels, as described in Section 3. The y axis shows what portion of the hardened noised labels is still the same as the hardened un-noised labels.

labels, they share this property. Any noise added directly on a hard label, which is only possible using the "flip" rule, would instantly change its winning class. Unfortunately, a direct comparison of such noise and our model is not possible.

As a more general note, experiments undertaken for this paper suggest that when classifiers are trained with hard labels, their responses should also be taken to be hard, and fused with corresponding schemes, to achieve the highest accuracy.

6 Summary

We investigated the effects of noise that was added to the training labels. Noising was accomplished by calculating the new label as a weighted sum of the original and a completely random label. It turned out that even individual classifiers hold up very well to high noise levels (see Figure 1). Combining several classifiers improves the overall accuracy further, but the right architecture has to be selected. Comparing classifiers trained on soft versus hard labeled data, it turned out that the soft approach is more resistant to noise.

References

1. Steidl, S., Levit, M., Batliner, A., Nöth, E., Niemann, H.: "Of all things the measure is man" - Automatic Classification of Emotions and Inter-Labeler Consistency. In: International Conference on Acoustics, Speech, and Signal Processing, ICASSP 2005, pp. 317–320 (2005)

2. Thiel, C., Scherer, S., Schwenker, F.: Fuzzy-Input Fuzzy-Output One-Against-All Support Vector Machines. In: Apolloni, B., Howlett, R.J., Jain, L. (eds.) KES 2007, Part III. LNCS (LNAI), vol. 4694. Springer, Heidelberg (2007)
3. Bondugula, R., Duzlevski, O., Xu, D.: Profiles and Fuzzy K-Nearest Neighbor Algorithm for Protein Secondary Structure Prediction. In: Chen, Y.P.P., Wong, L. (eds.) Proceedings of the 3rd Asia-Pacific Bioinformatics Conference, pp. 85–94. World Scientific, Singapore (2005)
4. Angluin, D., Laird, P.: Learning from Noisy Examples. Machine Learning 2, 343–370 (1988)
5. Valiant, L.G.: A Theory of the Learnable. Commun. ACM 27, 1134–1142 (1984)
6. Lawrence, N.D., Schölkopf, B.: Estimating a kernel Fisher discriminant in the presence of label noise. In: Proceedings of the 18th International Conference on Machine Learning, pp. 306–313. Morgan Kaufmann, San Francisco (2001)
7. Amini, M.R., Gallinari, P.: Semi-supervised learning with an explicit label-error model for misclassified data. In: IJCAI 2003 (2003)
8. McLachlan, G.J.: Discriminant Analysis and Statistical Pattern Recognition. John Wiley & Sons, Chichester (1992)
9. El Gayar, N., Schwenker, F., Palm, G.: A study of the robustness of KNN classifiers trained using soft labels. In: Schwenker, F., Marinai, S. (eds.) ANNPR 2006. LNCS (LNAI), vol. 4087, pp. 67–80. Springer, Heidelberg (2006)
10. Fay, R., Kaufmann, U., Schwenker, F., Palm, G.: Learning Object Recognition in a NeuroBotic System. In: Groß, H.M., Debes, K., Böhme, H.J. (eds.) 3rd Workshop on SelfOrganization of AdaptiVE Behavior SOAVE 2004. Fortschritt-Berichte VDI, Reihe 10, vol. 743, pp. 198–209. VDI (2004)
11. Fay, R.: Feature Selection and Information Fusion in Hierarchical Neural Networks for Iterative 3D-Object Recognition. PhD thesis, University of Ulm, Germany (2007)
12. MacQueen, J.B.: Some methods for classification and analysis of multivariate observations. In: 5th Berkeley Symposium on Mathematical Statistics and Probability, pp. 281–298. University of California Press (1967)
13. Powell, M.J.D.: Radial basis functions for multivariate interpolation: A review. In: Mason, J.C., Cox, M.G. (eds.) Algorithms for Approximation, pp. 143–168. Clarendon Press, Oxford (1987)
14. Breiman, L., Meisel, W., Purcell, E.: Variable Kernel Estimates of Multivariate Densities. Technometrics 19, 135–144 (1977)
15. Kahsay, L., Schwenker, F., Palm, G.: Comparison of multiclass SVM decomposition schemes for visual object recognition. In: Kropatsch, W.G., Sablatnig, R., Hanbury, A. (eds.) DAGM 2005. LNCS, vol. 3663, pp. 334–341. Springer, Heidelberg (2005)
16. Kuncheva, L.I.: Combining Pattern Classifiers: Methods and Algorithms. Wiley, Chichester (2004)
17. Kuncheva, L.I., Bezdek, J.C., Duin, R.P.W.: Decision templates for multiple classifier fusion: An experimental comparison. Pattern Recognition 34, 299–314 (2001)
18. Shafer, G.: Dempster-Shafer Theory (2002), `http://www.glennshafer.com/assets/downloads/articles/article48.pdf`
19. Dempster, A.P.: A generalization of Bayesian inference. Journal of the Royal Statistical Society 30, 205–247 (1968)
20. Shafer, G.: A Mathematical Theory of Evidence. University Press, Princeton (1976)

21. Schwenker, F., Dietrich, C., Thiel, C., Palm, G.: Learning decision fusion mappings for pattern recognition. ICGST International Journal on Artificial Intelligence and Machine Learning (AIML) 6, 17–21 (2006)
22. Binaghi, E., Brivio, P.A., Ghezzi, P., Rampini, A.: A fuzzy set-based accuracy assessment of soft classification. Pattern Recognition Letters 20, 935–948 (1999)
23. Strauss, P.M., Hoffmann, H., Minker, W., Neumann, H., Palm, G., Scherer, S., Schwenker, F., Traue, H., Walter, W., Weidenbacher, U.: Wizard-of-oz data collection for perception and interaction in multi-user environments. In: International Conference on Language Resources and Evaluation (LREC) (2006)

Prediction of the Collapse Index by a Mamdani Fuzzy Inference System

Kivanc Zorlu[1,*] and Candan Gokceoglu[2]

[1] Mersin University, Engineering Faculty, Department of Geological Engineering, Ciftlikkoy, Mersin, Turkey
[2] Hacettepe University, Engineering Faculty, Department of Geological Engineering, Beytepe, Ankara, Turkey
{Kivanc Zorlu,kivancgeo}@mersin.edu.tr,
{Candan Gokceoglu,cgokce}@hacettepe.edu.tr

Abstract. Determination of collapse potential of collapsible grounds is an important problem for civil engineers. However, this requires extensive field and laboratory works. For this reason, prediction tools for this purpose are highly attractive for engineers. Considering this difficulty, development of a Mamdani fuzzy inference system for prediction of collapse index is the main purpose of the study. The fuzzy inference system developed in the study includes two inputs, one output and 25 lingusitic if-then rules. The performace of the fuzzy inference system is checked by various indices and these indices reveal that the fuzzy inference system has a significant prediction performance.

Keywords: fuzzy inference system, caliche, collapsible ground, collapse index.

1 Introduction

Economy and safety are two main componens of an engineering design. All civil engineering structures are constructed on natural ground. However, some grounds exhibit a considerable volume decrease under stresses. Collapsible grounds are characterized by considerable volume decreases under constant stress if they are saturated. Such type materials exhibit high strength when they are dry. However, if they are saturated, they can lose their strength because of their low unit weight and highly porous structure. Collapse occurs if the intergranular stress is higher than intergranular bonding strength provided by bridging. This collapse mechanism is triggered by increasing in stress or dennundation or both. Increasing in overburden thickness sourced from sedimentation, dynamic loadings created by earthquakes and surcharges of engineering structures are some causes of increases in stress [1]. To avoid a sudden failure of a structure, the collapse potential of a ground should be considered before construction on a collapsible ground. However, determination of collapse potential of a ground is highly difficult. For this reason, the purpose of the

[*] Please note that the LNCS Editorial assumes that all authors have used the western naming convention, with given names preceding surnames. This determines the structure of the names in the running heads and the author index.

I. Lovrek, R.J. Howlett, and L.C. Jain (Eds.): KES 2008, Part I, LNAI 5177, pp. 74–81, 2008.

present study is to propose a Mamdani fuzzy inference system to predict the collapse index of a collapsible ground.

Before the preparation of the fuzzy inference system, an extensive field work to collect representative samples is carried out. In the second stage of the study, a series of laboratoru tests are performed on the samples collected from the field. Finally, a Mamdani fuzzy inference system is developed to predict the collapse index.

2 Collapse Mechanism and Data Structure

The sampling for laboratory tests is conducted on the clayey level of the caliche profile in the Adana Organized Industrial Site. A total of 20 undisturbed samples from the silty, sandy levels are extracted using 35x198 mm mould for oedometer tests to describe the collapse potential.

For the determination of collapse index (C_p) of solis, two types of laboratory tests are employed. One of them is double oedometer collapse test and the other one is single oedometer collapse test [2]. In the double oedometer collapse test, under different stress levels, the deformation differences of partially and fully saturated samples are determined. However, the purpose of the present study is to determine the volume decrease depending on saturation under constant stress. In addition, due to highly heterogenous nature of the geomaterial studied, preparation of two samples having similar void ratio for the double oedometer collapse tests is almost impossible. For these reasons, in this study, the single oedometer collapse test is preferred. In this procedure, the tests are conducted on one undisturbed sample having natural water content. During the single oedometer testing program, the procedure suggested by ASTM [3] is followed. At the end of the tests, the collapse index of the samples is determined by the following equation:

$$C_p=\Delta e_c/(1+e_1) \quad (1)$$

where; C_p is the collapse index; $\Delta e_{c,}$ is the change of void ratio depending on saturation and e_1 is the original void ratio.

This equation may be used to determine the collapse potential, Ic, of soil at particular vertical stress or the collapse index, $C_{p,}$ at an applied vertical stress of 200 kPa. C_p for smaller applied vertical stress may be estimated assuming that the soil does not swell after inundation at smaller applied vertical stress [3]. In other words, the parameter of C_p reflects the special conditions, because the samples are non-swell nature depending on increase in water content, and the constant load employed during the tests is 200 kPa. However, the term of collapse potential is a general concept. For this reason, instead of collapse potential, use of the term of collapse index reflecting a special condition is preferred in the study. In this study, after the dennundation under 200 kPa loading level, the tests are terminated as soon as the collapse occurred. Theoretically, the change in void ratio after the dennundation can be determined by combining the void ratio at the beginning (original void ratio) and the void ratio after the test. The list of the collapse index values obtained from the single oedometer tests and the inputs (fine content and void ratio) are given in Table 1.

Table 1. The parameters used in the fuzzy inference system [1]

No	i	e_0	C_p
L1	27	0.354	1.77
L2	66	0.417	2.58
L3	53	0.427	2.53
L4	65	0.529	2.73
L5	12	0.345	1.62
L6	47	0.411	2.22
L7	64	0.476	2.68
L8	18	0.348	1.82
L9	55	0.489	3.13
L10	82	0.720	3.79
L11	77	0.599	3.38
L12	52	0.400	2.82
L13	39	0.453	2.32
L14	28	0.312	1.67
L15	54	0.604	2.68
L16	45	0.439	2.32
L17	42	0.381	2.27
L18	48	0.478	2.47
L19	52	0.537	2.78
L20	46	0.410	2.37

i : fine content (%); e_o: void ratio; C_p: collapse index.

3 Fuzzy Inference System

Sometimes, due to the difficulties encountered during sample collection and preparation, some projects can be prepared employing limited number of tests. To overcome these limitations, some empirical relationships for indirect determination of the collapse index are investigated by Zorlu and Kasapoglu [1]. However, the empirical equations developed by Zorlu and Kasapoglu [1] are based on regression techniques. To determine the collapse index by regression equations requires numerical data. However, the linguistic based fuzzy inference systems can be used by not only numerical data but also linguistic terms such as "low", "high" etc. For this reason, a Mamdani fuzzy inference system to predict the collapse index is developed in the present study. An interesting and perhaps the most attractive characteristic of fuzzy models compared with other conventional methods commonly used in geosciences, such as statistics, is that they are able to describe complex and nonlinear multivariable problems in a transparent way [4]. In literature, commonly two different types of fuzzy inference system are used. These are the Mamdani and the Takagi-Sugeno-Kang algorithms. Among them, particularly the Mamdani algorithm is mostly preferred in engineering geology studies [5]. This algorithm was first developed by Mamdani and Assilian [6] to use in a steam machine control. They put forward their expert opinion on the use of this machine using "if-then" rules. Considering the conventional mathematical techniques, integration of this expert opinion to an indirect model evidently seems to be impossible [5]. Alvarez Grima [5] mentioned that the Mamdani algorithm constitutes one of the most efficient techniques to solve the complex engineering geological problems. The main reason for this evaluation is that the materials studied in engineering geology are commonly natural, and hence they involve a high level of uncertainty. In this study, as the previous studies related to the engineering geology and geology [7-12], the Mamdani fuzzy inference system is considered to introduce a prediction model for the collapse index. The model includes two inputs (the fine content, i and the void ratio, e) and one output (the collapse index, C_p).

For inference in a rule-based fuzzy model, the fuzzy propositions need to be represented by an implication function called a fuzzy if-then rule or a fuzzy conditional statement [5]. A fuzzy set is a collection of paired members consisting of members and degrees of "support" or "confidence" for those members. A linguistic variable whose values are words, phrases or sentences are labels of fuzzy sets [13]. In literature, many methods such as intuition, rank ordering, angular fuzzy sets, genetic algorithms, inductive reasoning, soft partitioning, etc. exist for the membership value assignment [e.g. 14-16]. Although, a fuzzy model is built generally by using expert knowledge in the form of linguistic rules, recently, there is an increasing interest in obtaining fuzzy models from measured data [4]. Designing membership functions is the most difficult, laborious and critical stage of building a fuzzy model, particularly when the available data is limited [5]. In such a case, the best alternative is simply to partition the numerical domain of the fuzzy input/output variables into a specified number equally spaced membership functions [17]. In this study, the fuzzy sets of the membership functions are obtained from the data partition. The type of membership functions used to represent the fuzzy terms of the inputs and the outputs are piecewise triangular membership functions. Before obtaining the membership functions, to provide standardization among the inputs and the outputs, all data is normalized in a close interval of [0, 1] using the following equation:

$$X_{norm} = (X-X_{min})/(X_{max}-X_{min}) \quad (2)$$

Table 2. Linguistic "if-then" rules used in fuzzy inference system

If	Fine content is VL and void ratio is VL	*then*	Collapse potential is VL
If	Fine content is VL and void ratio is L	*then*	Collapse potential is VL
If	Fine content is VL and void ratio is M	*then*	Collapse potential is L
If	Fine content is VL and void ratio is H	*then*	Collapse potential is M
If	Fine content is VL and void ratio is VH	*then*	Collapse potential is H
If	Fine content is L and void ratio is VL	*then*	Collapse potential is VL
If	Fine content is L and void ratio is L	*then*	Collapse potential is L
If	Fine content is L and void ratio is M	*then*	Collapse potential is L
If	Fine content is L and void ratio is H	*then*	Collapse potential is M
If	Fine content is L and void ratio is VH	*then*	Collapse potential is H
If	Fine content is M and void ratio is VL	*then*	Collapse potential is L
If	Fine content is M and void ratio is L	*then*	Collapse potential is L
If	Fine content is M and void ratio is M	*then*	Collapse potential is M
If	Fine content is M and void ratio is H	*then*	Collapse potential is M
If	Fine content is M and void ratio is VH	*then*	Collapse potential is H
If	Fine content is H and void ratio is VL	*then*	Collapse potential is L
If	Fine content is H and void ratio is L	*then*	Collapse potential is M
If	Fine content is H and void ratio is M	*then*	Collapse potential is H
If	Fine content is H and void ratio is H	*then*	Collapse potential is H
If	Fine content is H and void ratio is VH	*then*	Collapse potential is VH
If	Fine content is VH and void ratio is VL	*then*	Collapse potential is M
If	Fine content is VH and void ratio is L	*then*	Collapse potential is M
If	Fine content is VH and void ratio is M	*then*	Collapse potential is H
If	Fine content is VH and void ratio is H	*then*	Collapse potential is VH
If	Fine content is VH and void ratio is VH	*then*	Collapse potential is VH

Where; X_{norm} is the normalized value of the measured variable; X is the measured variable; X_{min} is the minimum value of the measured variable in the data; X_{max} is the maximum value of the measured variable in the data.

The graphs of the membership functions are given in Figure 1. In the fuzzy inference system, a total of 25 linguistic rules are used (Table 2). The control surface of the model is given in Figure 2. In the model, "min" and "max" are employed as "and" and "or", respectively. The final output of the Mamdani fuzzy model is also a fuzzy set. However, numerical values are commonly desired in practice. For this reason, a defuzzification procedure is required. Defuzzification is briefly defined as the transformation of a fuzzy

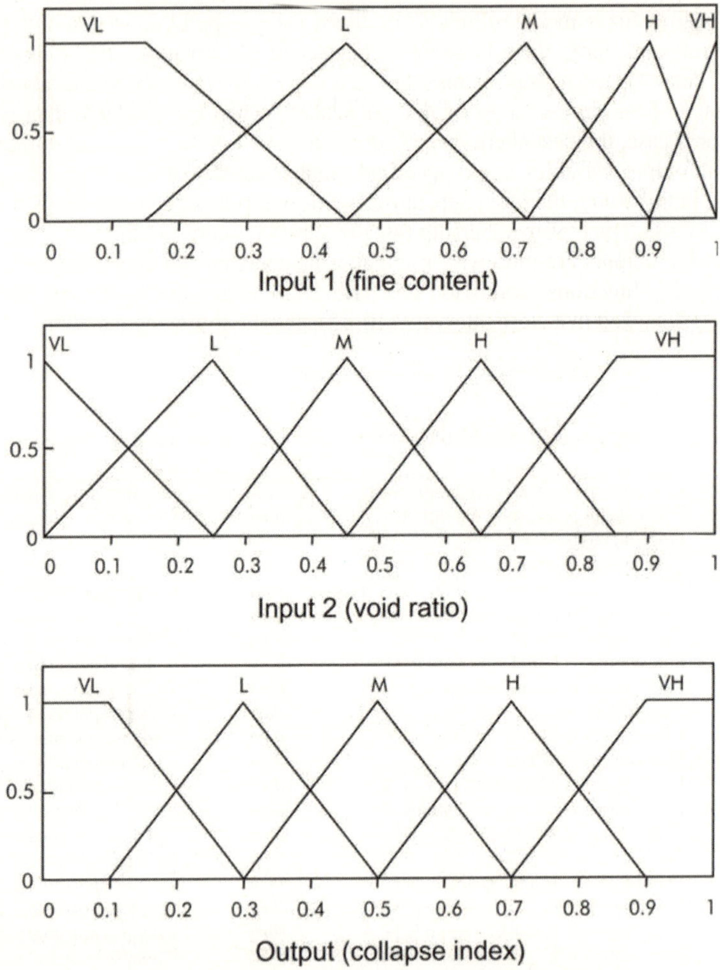

Fig. 1. Input and output fuzzy membership functions

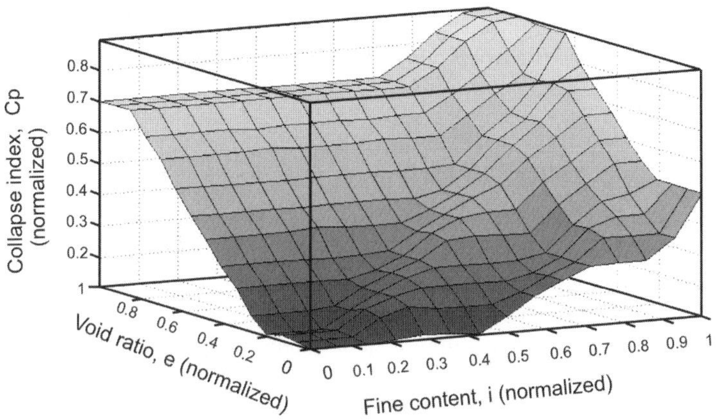

Fig. 2. Control surface of the fuzzy inference system

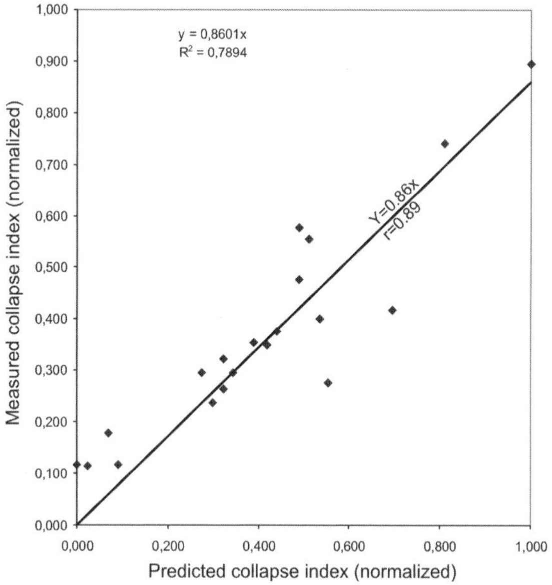

Fig. 3. Graph of cross-correlation between the predicted and the measured collapse indices

set into a numerical value. In literature, many defuzzification methods are proposed [18]. However, the center of gravity (COG) method is mostly preferred among them because the calculation stage is simple and it produces plausible results. By running the FIS constructed for the description of the collapse index, the collapse index of each specimen is predicted. The prediction performance of the constructed model is assessed by some prediction indices such as variance account for (VAF) and root mean square error (RMSE). Theoretically, if VAF and RMSE are equal to 100% and 0 respectively, the model produces excellent results. The VAF and RMSE values for

20 specimens are calculated as 57.2% and 0.11 respectively. The coefficient of cross-correlation between the predicted and the measured collapse potential is obtained as 0.89 (Figure 3).

4 Results and Discussion

The following results and discussions can be drawn from the present study:

Even though its crucial importance for the safety of structures, the direct determination of collapse potential of collapsible grounds is highly difficult. For this reason, indirect determination of such feature is an attractive subject for civil engineers especially. In literature, some limited studies to indirect determination of the collapse index exist. However, these approaches are based on regression techniques. Considering this lack, in the present study, a Mamdani fuzzy inference system is developed. The performance analyses show that the developed inference system has a sufficient prediction capacity for civil engineering projects. Moreover, the fuzzy inference system can be used by employing "if-then" linguistic rules.

References

1. Zorlu, K., Kasapoglu, K.E.: Determination of Geomechanical Properties and Collapse Potential of a Caliche by In-Situ and Laboratory Tests. Environmental Geology (2008), doi:10.1007/s00254-008-1239-7
2. Lutenegger, A.J., Saber, R.T.: Determination of Collapse Potential of Soils. Geotechnical Testing Journal 11(3), 173–178 (1988)
3. ASTM (American Society of Testing and Materials): Annual Book of ASTM Standards. 04.08, 1391–1393 (1992)
4. Setnes, M., Babuska, R., Verbruggen, H.B.: Rule-Based Modeling: Precision and Transparency. IEEE Transactions on Systems, Man and Cybernetics. Part C, Applications and Reviews 28, 165–169 (1998)
5. Alvarez Grima, M.: Neuro-Fuzzy Modeling in Engineering Geology. A.A. Balkema, Rotterdam, p. 244 (2000)
6. Mamdani, E.H., Assilian, S.: An Experiment in Linguistic Synthesis with a Fuzzy Logic Controller. Int. Journal of Man-Machine Studies 7(1), 1–13 (1975)
7. Gokceoglu, C.: A Fuzzy Triangular Chart to Predict the Uniaxial Compressive Strength of the Ankara Agglomerates from their Petrographic Composition. Engineering Geology 66(1–2), 39–51 (2002)
8. Sonmez, H., Gokceoglu, C., Ulusay, R.: An Application of Fuzzy Sets to the Geological Strength Index (GSI) System Used in Rock Engineering. Engineering Applications of Artificial Intelligence 16, 251–269 (2003)
9. Sonmez, H., Gokceoglu, C., Ulusay, R.: A Mamdani Fuzzy Inference System for the Geological Strength Index and its Use in Slope Stability Assessments. International Journal of Rock Mechanics and Mining Sciences 41, 513–514 (2004)
10. Nefeslioglu, H.A., Gokceoglu, C., Sonmez, H.: A Mamdani Model to Predict the Weighted Joint Density. In: Palade, V., Howlett, R.J., Jain, L. (eds.) KES 2003. LNCS, vol. 2773, pp. 1052–1057. Springer, Heidelberg (2003)

11. Nefeslioglu, H.A., Gokceoglu, C., Sonmez, H.: Indirect Determination of Weighted Joint Density (wJd) by Empirical and Fuzzy Models: Supren (Eskisehir, Turkey) Marbles. Engineering Geology 85(3/4), 251–269 (2006)
12. Gokceoglu, C., Zorlu, K.: A Fuzzy Model to Predict the Uniaxial Compressive Strength and the Modulus of Elasticity of a Problematic Rock. Engineering Applications of Artificial Intelligence 17, 61–72 (2004)
13. Zadeh, L.A.: Outline of a New Approach to the Analysis of Complex Systems and Decision Processes. IEEE Transactions on Systems, Man, and Cybernetics SMC-3, 28–44 (1973)
14. Zadeh, L.A.: A Rationale for Fuzzy Control. Journal of Dynamic Systems, Measurement and Control Transaction ASME 94, 3–4 (1972)
15. Hadipriono, F., Sun, K.: Angular Fuzzy Set Models for Linguistic Values. Civil Engineering Systems 7(3), 148–156 (1990)
16. Karr, C.L., Gentry, E.J.: Fuzzy Control of pH Using Genetic Algorithms. IEEE Transaction on Fuzzy Systems 1(1), 46–53 (1993)
17. Babuska, R.: Fuzzy Modeling and Identification. PhD thesis, Delft University of Technology, Delft, The Netherlands, p. 294 (1996)
18. Hellendoorn, H., Thomas, C.: Defuzzification if fuzzy controllers. Journal of Intelligent Fuzzy Systems 1, 109–123 (1993)

A Cross-Entropy Based Population Learning Algorithm for Discrete-Continuous Scheduling with Continuous Resource Discretisation

Piotr Jędrzejowicz[1] and Aleksander Skakovski[2]

[1,2] Gdynia Maritime University,
[1] Chair of Information Systems,
[2] Department of Navigation,
[1,2] ul. Morska 83F, 81-225 Gdynia, Poland
{P.Jedrzejowicz,askakow}@am.gdynia.pl

Abstract. The problem of scheduling nonpreemtable tasks on parallel identical machines under constraint on discrete resource and requiring, additionally, renewable continuous resource to minimize the schedule length is considered in the paper. A continuous resource is divisible continuously and is allocated to tasks from given intervals in amounts unknown in advance. Task processing rate depends on the allocated amount of the continuous resource. The considered problem can be solved in two steps. The first step involves generating all possible task schedules and second - finding an optimal schedule among all schedules with optimal continuous resource allocation. To eliminate time consuming optimal continuous resource allocation, a problem Θ_Z with continuous resource discretisation is introduced. Because Θ_Z is NP-hard a population-learning algorithm (PLA2) is proposed to tackle the problem. PLA2 belongs to the class of the population-based methods. Experiment results proved that PLA2 excels known algorithms for solving the considered problem.

Keywords: discrete-continuous scheduling, evolutionary computation, population learning algorithm.

1 Introduction

A problem of scheduling jobs on multiple processors under constraint on discrete resource and requiring, additionally, renewable continuous resource to minimize the schedule length is considered in the paper. In the problem two types of resources are considered: discrete and continuous. A discrete resource is divisible discretely, for example a set of processors or a set of mechanical or pumping machines. A continuous resource is divisible continuously and is allocated to the jobs from given intervals in amounts unknown in advance. In practice a continuous resource may be limited in amount - for example power (electric, pneumatic, hydraulic) supplying a set of machines, limited gas flow intensity supplying forge furnaces in a steel plant, or limited fuel flow intensity in refueling terminals.

The problem of scheduling jobs on multiple processors under constraint on discrete resource and requiring, additionally, renewable continuous resource was intensively

I. Lovrek, R.J. Howlett, and L.C. Jain (Eds.): KES 2008, Part I, LNAI 5177, pp. 82–89, 2008.

explored in [8], [9], [10], [11], and we define the problem in the same way. Namely, we consider n independent, nonpreemptable jobs, each of them simultaneously requiring for its processing at time t a machine from a set of m parallel, identical machines (the discrete resource) and an amount (unknown in advance) $u_i(t) \in [0, 1]$, $i = 1, 2, \ldots, n$, of a continuous renewable resource. The job model is given in the form:

$$\dot{x}_i(t) = \frac{dx_i(t)}{d(t)} = f_i[u_i(t)],\ x_i(0) = 0,\ x_i(C_i) = \tilde{x}_i\ , \tag{1}$$

where $x_i(t)$ is the state of job i at time t, f_i is an increasing continuous function, $f_i(0) = 0$, C_i is (unknown in advance) completion time of job i, and $\tilde{x}_i$ is its processing demand (final state). We assume, without loss of generality, that $\sum_{i=1}^{n} u_i(t) = 1$ for every t. The problem is to find a sequence of jobs on machines and, simultaneously, a continuous resource allocation that minimizes the given scheduling criterion. The problem is NP-hard [10]. The defined problem can be decomposed into two interrelated sub problems: (i) to find a feasible sequence of jobs on machines, and (ii) to allocate the continuous resource among jobs already sequenced. The notion of a feasible sequence is of crucial importance. According to [9] a feasible schedule can be divided into $p \leq n$ intervals defined by completion times of consecutive jobs. Let Z_k denote the combination of jobs processed in parallel in the k-th interval. Thus, in general, a feasible sequence FS of combinations Z_k, $k = 1, 2, \ldots, p$, can be associated with each feasible schedule. Feasibility of such a sequence requires that the number of elements in each combination does not exceed m and that each job appears exactly in one or in consecutive combinations in FS (nonpreemptability). It has been shown in [8] that for concave job models and the schedule length minimization problem, it is sufficient to consider feasible sequences of combinations Z_k, $k = 1, 2, \ldots, n - m + 1$, composed of exactly m jobs each. For a given feasible sequence FS of jobs on machines, we can find an optimal continuous resource allocation, i.e. an allocation that leads to a schedule minimizing the given criterion from among all feasible schedules generated by FS. At this point, a convex mathematical programming problem has to be solved, in the general case (see [8]). An optimal schedule for a given feasible sequence (i.e. a schedule resulting from an optimal continuous resource allocation for this sequence) is called a semi-optimal schedule. In consequence, a globally optimal schedule can be found by solving the continuous resource allocation problem optimally for all feasible sequences. Unfortunately, in general, the number of feasible sequences grows exponentially with the number of jobs. Therefore it is justified to apply some approximation algorithm or metaheuristic.

Because finding an optimal allocation of a continuous resource to a feasible schedule requires using specialized and time-consuming solver, an idea of continuous resource discretisation was proposed in [11]. We use the same approach in the paper. Namely, we assume that the number of possible continuous resource allocations to a task J_i is D_i, i.e. is fixed, and the amount of the continuous resource for each $l_i = 1, 2, \ldots, D_i$ is known in advance (in the original problem there was infinite number of the continuous resource allocations to a task and the amount of the continuous resource to be allocated was not known in advance). Because a different amount of the continuous resource is allocated to task J_i for each l_i, l_i is called a processing mode of task J_i. Such discretisation of the continuous resource allows treating it as a discrete resource.

The problem of scheduling jobs on multiple processors under additional continuous resource with continuous resource discretisation still remains NP-hard [11]. A population-learning algorithm (PLA) first proposed in [6] is used to tackle the problem, since it was effective in solving other scheduling problems considered in [5], [3], [4]. Promising results obtained by the proposed in [7] version of PLA - PLA1 proved the approach for solving Θ_Z to be effective and caused the design of PLA2. PLA2 uses three procedures: cross-entropy (CE) described in Section 3.1, Tabu Search (TS) procedure and island-based evolution algorithm (IBEA) thoroughly described in [7].

2 Problem Formulation

We define a problem Θ_Z in the same way as in [11]. Namely, let $\boldsymbol{J} = \{J_1, J_2, \ldots, J_n\}$ be a set of nonpreemtable tasks, with precedence relations and ready times $r_i = 0$, $i = 1, 2, \ldots, n$, and $\boldsymbol{P} = \{P_1, P_2, \ldots, P_m\}$ be a set of parallel and identical machines, and there is one additional renewable discrete resource in amount $U = 1$ available. A task J_i can be processed in one of the modes $l_i = 1, 2, \ldots, D_i$ (D_i – the number of processing modes of task J_i), for which J_i requires a processor from $\boldsymbol{P}$ and amount of the additional resource known in advance. The processing mode of J_i cannot change during the processing. For each task two vectors are defined: a processing times vector $\tau_i = \{\tau_i^1, \tau_i^2, \ldots, \tau_i^{D_i}\}$, where $\tau_i^{l_i}$ is the processing time of task J_i in mode $l_i = 1, 2, \ldots, D_i$ and a vector of additional resource quantities allocated in each processing mode $u_i = \{u_i^1, u_i^2, \ldots, u_i^{D_i}\}$. The problem is to find processing modes for tasks from $\boldsymbol{J}$ and their sequence on processors from $\boldsymbol{P}$ such that schedule length $Q = \max\{C_i\}$, $i = 1, \ldots, n$ is minimized.

3 Population Learning Algorithm

Population learning algorithm proposed in [6] has been inspired by analogies to a social phenomenon rather than to evolutionary processes. The population learning algorithm takes advantage of features that are common to social education systems:

- A generation of individuals enters the system.
- Individuals learn through organized tuition, interaction, self-study and self-improvement.
- Learning process is inherently parallel with different schools, curricula, teachers, etc.
- Learning process is divided into stages.
- More advanced and more demanding stages are entered by a diminishing number of individuals from the initial population (generation).
- At higher stages more advanced education techniques are used.
- The final stage can be reached by only a fraction of the initial population.

General idea of the present implementation of PLA2 is shown in the following pseudo code:

```
Procedure PLA2
Begin
  Create an initial population P0 of the size x0 - 1 using
  procedure cross-entropy (CE).
  Create an individual TSI in which all tasks Ji are to be
  executed in mode li = 1 (a mode characterized by minimal
  quantity of additional resource u_i^1 and maximal task
  processing time τ_i^1, 1 ≤ i ≤ n).
  Improve the individual TSI with the tabu search (TS)
  procedure.
  Create population P1 = P0 + TSI.
  Improve all individuals in P1 with the IBEA.
  Output the best solution to the problem.
End.
```

In the procedure PLA2 $x_0 = K \cdot PS$, where K – the number of islands and PS – the population size on an island defined in procedure IBEA.

All individuals (solutions) used in the PLA2 procedure can be characterized in the following manner:

- an individual (a solution) is represented by an n-element vector $S = \{c_i | \ 1 \le i \le n\}$,
- all processing modes of all tasks are numbered consecutively. Thus processing mode l_b of task J_b has the number $c_b = \sum_{i=1}^{b-1} D_i + l_b$,
- all S representing feasible solutions are potential individuals;
- each individual can be transformed into a schedule by applying LSG, which is a specially designed list-scheduling algorithm for discrete-continuous scheduling;
- each schedule produced by the LSG can be directly evaluated in terms of its fitness.

3.1 A Cross-Entropy Algorithm

Because TS and IBEA algorithms are thoroughly described in [7], we only present CE procedure. In PLA2 the proposed CE procedure is perceived as the procedure preparing some solution basis for further improvement by procedure IBEA. In CE procedure a cross-entropy method first proposed in [12] is used since it was effective in solving various difficult combinatorial optimization problems [1]. Because in CE procedure a solution is viewed as a vector of n jobs, we would like to know the probability of locating job J_i on a particular place j in the vector. For this reason we introduce two success probability vectors $\hat{p}_j$ and $\hat{p}'_{ji}$ related to each job J_i and its place j in solution S. Vector $\hat{p}_j = \{p_{ji} \,|\, 1 \le i \le n\}$, $1 \le j \le n$ contains p_{ji} values, which is the probability that on place j there will be located job i. Vector $\hat{p}'_{ji} = \{p_{jil} \,|\, 1 \le l \le D_i\}$, $1 \le j \le n$, $1 \le i \le n$ contains p_{jil} values, which is the probability that on place j task i

will be executed in mode l. A procedure CE using cross-entropy method for combinatorial optimization described in [1] and modified for solving Θ_Z problem is shown in the following pseudo code:

Procedure CE

Begin

Set ic:= 1 (ic - iteration counter), ic^{stop} - maximal number of iterations, a:= 1.

Set $\hat{p}_j = \{p_{ji} = 1/n \mid 1 \le i \le n\}, 1 \le j \le n.$

Set $\hat{p}'_{ji} = \{p_{jil} = 1/D_i \mid 1 \le l \le D_i\}, 1 \le j \le n, 1 \le i \le n.$

While $ic \bullet ic^{stop}$ do

Generate a sample S_1, S_2, ... , S_s, ... , S_n of solutions with success probability vectors $\hat{p}_j$ and $\hat{p}'_{ji}$.

Order S_1, S_2, ... , S_s, ... , S_n by nondecreasing values of their fitness function.

Set $\gamma = \lceil \rho \cdot N \rceil, \rho \in (0,1)$.

Set

$$\hat{p}_j = \left\{ p_{ji} = \frac{\sum_{s=1}^{\gamma} I(S_s(j) = i)}{\gamma} \middle| 1 \le i \le n \right\}, \tag{2}$$

$1 \le j \le n$, $I(S_s(j) = i) = 1$, $I(S_s(j) \bullet i) = 0$, where $S_s(j)$ - number of the task located on j-th place in s-th solution S.

Set

$$\hat{p}'_{ji} = \left\{ p_{jil} = \frac{\sum_{s=1}^{\gamma} I(S_s(ji) = l)}{\gamma} \middle| 1 \le l \le D_i \right\}, \tag{3}$$

$1 \le j \le n$, $1 \le i \le n$, $I(S_s(ji) = l) = 1$, $I(S_s(ji) \bullet l) = 0$, where $S_s(ji)$ - an execution mode of task i located on j-th place in s-th solution S.

Save the first $h = \lceil K \cdot PS / ic^{stop} \rceil$ best solutions from the ordered sample into P_0 under address a. Set a:= $a + h$.

Set ic:= ic + 1.

EndWhile.

EndProcedure.

In the presented pseudo code parameters K – the number of islands and PS – the population size defined in procedure IBEA.

4 Computational Experiments

The proposed Cross-Entropy Based population learning algorithm for solving discrete-continuous scheduling problems with continuous resource discretisation (PLA2) was implemented and tested. Results were compared to the best known obtained by a genetic GAVRdyskr, tabu search, simulated annealing algorithms [11], IBEA [2], and PLA1 [7] (GAVRdyskr denoted in our paper as G_{dskr} and PLA1 were used for results comparison as the ones of the same nature). For testing purposes three combinations of n x m were considered (n – the number of tasks and m – the number of machines): 10x2, 10x3, and 20x2. For each combination n x m 100 instances of a problem Θ_Z were generated and three discretisation levels were considered: 10, 20, and 50, which makes 900 instances of the problem. Each instance was tested 24 times. Relative error (RE) of the solutions found by PLA2 compared to best-known solutions was used to evaluate the quality of PLA2. Value of the RE calculated according to the formulae RE = $(Q_{PLA2} - Q_{best\text{-}known})/Q_{best\text{-}known}$ for each instance was used to find mean and maximum relative errors. RE_{mean} and RE_{max} of the solutions found by PLA2, PLA1, and G_{dskr} are presented in Table 1. The qualities of the solutions found by PLA2 in general are better than the qualities of the solutions found by PLA1 and G_{dskr}. For example, for case 20x2, D = 20 RE_{mean} = -0.23%, which means that the schedule length of all schedules yielded by PLA2 was 0.23% shorter on average than the best-known. For the same case, RE_{max} = 7.23% means that the longest schedule among all schedules yielded by PLA2 was 7.23% longer than the best-known. In our tests PLA2 improved 80.33% of 300 the best known solutions for combinations 10x2, 10x3, and 20x2, and reduced average of RE_{mean} compared to PLA1 and G_{dskr}. Table 2 shows

Table 1. Comparison of the results obtained by PLA2, PLA1 and G_{dskr} for problem Θ_z

nxm		D=10			D=20			D=50		
		PLA2	PLA1	G_{dskr}	PLA2	PLA1	G_{dskr}	PLA2	PLA1	G_{dskr}
10x2	RE_{mean}	2.75%	1.82%	7.59%	1.53%	2.52%	9.18%	2.25%	2.93%	10.41%
	RE_{max}	9.29%	9.39%	15.34%	5.94%	9.00%	15.84%	8.32%	12.21%	17.54%
10x3	RE_{mean}	2.99%	3.76%	14.58%	2.20%	3.32%	15.79%	2.19%	3.96%	17.11%
	RE_{max}	10.66%	11.58%	25.18%	8.70%	12.40%	23.15%	33.54%	14.31%	27.52%
20x2	RE_{mean}	2.70%	2.49%	13.25%	-0.23%	3.11%	15.42%	2.44%	3.42%	17.65%
	RE_{max}	9.21%	9.48%	19.40%	7.23%	9.65%	24.02%	9.03%	9.65%	26.87%

Table 2. The average of PLA2's RE_{mean} reduction in comparison to average RE_{mean} of PLA1 and G_{dskr}

nxm	PLA1	G_{dskr}
10x2	0.25%	6.89%
10x3	1.22%	13.36%
20x2	1.37%	13.80%

how many percents on average the solutions found by PLA2 were better than the solutions found by PLA1 and G_{dskr}.

Mean time required by PLA2 and PLA1 to find a solution for 10x2 on Pentium (R) 4 CPU 3.00GHz compiled with aid of Borland Delphi Personal v.7.0 was 5s. G_{dskr} was implemented in C++. Mean time required by G_{dskr} to find a solution for 10x2 on supercomputer Silicon Graphics Power Challenge XL designed in 64-bit SMP (Symmetrical Multi Processing) architecture on twelve RISC MIPS R8000 processors using 1GB RAM and 20GB disc memory was 33s. Such results make PLA2 quite an effective algorithm for solving problem Θ_Z.

5 Conclusion

In his paper we propose a hybrid population-based approach to solving the problem of scheduling nonpreemtable tasks on parallel identical machines under constraint on discrete resource and requiring, additionally, renewable continuous resource to minimize the schedule length. Since the discussed problem is computationally difficult it has been possible to obtain only approximate solutions within a reasonable time. In such case the validation of the approach seems only possible through computational experiment and comparison of the results with those obtained by applying other approaches. Such experiment which results are discussed in the paper proved that combining cross-entropy, tabu search and island-based evolutionary algorithm within the population based scheme is an effective approach. Further research should focus on further reducing computation time needed to obtain satisfactory results.

References

1. De Boer, P.-T., Kroese, D.P., Mannor, S., Rubinstein, R.Y.: A Tutorial on the Cross-Entropy Method. Annals of Operations Research 134(1), 19–67 (2005)
2. Czarnowski, I., Gutjahr, W.J., Jędrzejowicz, P., Ratajczak, E., Skakovski, A., Wierzbowska, I.: Scheduling Multiprocessor Tasks in Presence of Correlated Failures. Central European Journal of Operations Research 11(2), 163–182 (2003); Luptaćik, M., Wildburger U.L (ed.), pp.163–182. Physika-Verlag, A Springer-Verlag Company, Heidelberg (2003)
3. Jędrzejowicz, J., Jędrzejowicz, P.: Population–Based Approach to Multiprocessor Task Scheduling in Multistage Hybrid Flowshops. In: Palade, V., Howlett, R.J., Jain, L. (eds.) KES 2003. LNCS, vol. 2773, pp. 279–286. Springer, Heidelberg (2003)
4. Jędrzejowicz, J., Jędrzejowicz, P.: PLA–Based Permutation Scheduling. Foundations of Computing and Decision Sciences 28(3), 159–177 (2003)
5. Jędrzejowicz, J., Jędrzejowicz, P.: New Upper Bounds for the Permutation Flowshop Scheduling Problem. In: Moonis, A., Esposito, F. (eds.) IEA/AIE 2005. LNCS (LNAI), vol. 3533, pp. 232–235. Springer, Heidelberg (2005)
6. Jędrzejowicz, P.: Social Learning Algorithm as a Tool for Solving Some Difficult Scheduling Problems. Foundation of Computing and Decision Sciences 24, 51–66 (1999)
7. Jędrzejowicz, P., Skakovski, A.: A Population Learning Algorithm for Discrete-Continuous Scheduling with Continuous Resource Discretisation. In: Chen, Y., Abraham, A. (eds.) ISDA 2006 Special session: Nature Imitation Methods Theory and practice (NIM 2006), 16-18 October, vol. II, pp. 1153–1158. IEEE Computer Society, Jinan (2006)

8. Józefowska, J., Węglarz, J.: On a methodology for discrete-continuous scheduling. European J. Oper. Res. 107(2), 338–353 (1998)
9. Józefowska, J., Mika, M., Różycki, R., Waligóra, G., Węglarz, J.: Solving discrete-continuous scheduling problems by Tabu Search. In: 4th Metaheuristics International Conference MIC 2001, Porto, Portugal, July 16-20, pp. 667–671 (2001)
10. Józefowska, J., Różycki, R., Waligóra, G., Węglarz, J.: Local search metaheuristics for some discrete-continuous scheduling problems. European J. Oper. Res. 107(2), 354–370 (1998)
11. Różycki, R.: Zastosowanie algorytmu genetycznego do rozwiązywania dyskretno-ciągłych problemów szeregowania. PhD dissertation, Institute of Computing Science, Poznań University of Technology, Piotrowo 3A, 60-965, Poznań, Poland (2000)
12. Rubinstein, R.Y.: Optimization of computer simulation models with rare events. European Journal of Operations Research 99, 89–112 (1997)

Improving the Responsiveness of NSGA-II in Dynamic Environments Using an Adaptive Mutation Operator – A Case Study

Alvaro Gomes[1,2], C. Henggeler Antunes[1,2], and A. Gomes Martins[1,2]

[1] Department of Electrical Engineering and Computers, University of Coimbra, Portugal
[2] INESC Coimbra, R Antero de Quental, 199, Coimbra, Portugal
{agomes,ch,amartins}@deec.uc.pt

Abstract. This paper presents a comparative analysis of the results obtained with two different implementations of the NSGA-II genetic algorithm in the framework of load management activities in electric power systems. The multiobjective real-world problem deals with the identification and the selection of suitable control strategies to be applied to groups of electric loads aimed at reducing maximum power demand, maximize profits and minimize user discomfort. It is shown that the algorithm performance is improved when the NSGA-II mutation operator is adaptively changed to incorporate information about the results of the search process and transfer this "knowledge" to the population.

Keywords: Evolutionary Multiobjective Optimization, Real-world applications.

1 Introduction

The complexity of real world optimization problems stems often from its combinatorial nature namely whenever multiple, incommensurate and conflicting objectives are at stake. The shape and the size of the search spaces and an irregular distribution of solutions throughout the search space can also be sources of complexity. Besides, real world multiobjective optimization problems often present a dynamic behavior, in the sense that the coefficients and/or the structure of the mathematical model may change over time, thus changing the location of the non-dominated (Pareto) front. In face of these issues, optimization tools must be able to adapt and react to the changes occurring in the environment. Such non-stationary problems ask for an extra effort to the optimization tools since these must be able to track the non-dominated front moving from one region to another when the changes in the environment occur. Several techniques to deal with non-stationarity have been reported in the literature [1].

The identification and the selection of a suitable value or range of values for every parameter become essential steps when using meta-heuristics and particularly GAs. A way to continuously adapt the parameters to the environment is through adaptive control, that is, instead of being constant over each simulation the values of different parameters may vary with time [2]. Adaptive control can be a privileged manner of incorporating knowledge about the evolutionary process into the search, thus potentially contributing for the GA to work more effectively. Mutation and crossover

I. Lovrek, R.J. Howlett, and L.C. Jain (Eds.): KES 2008, Part I, LNAI 5177, pp. 90–97, 2008.

operators are used as a way for discovering new solutions, by contributing to focus the search into new regions of the search space.

In this paper, the influence of using adaptive control of the mutation operator in the NSGA II algorithm for a non-stationary multiobjective problem is analyzed. The real-world problem herein reported deals with the identification and the selection of suitable control strategies to be applied to groups of electric loads, in the framework of load management in electric power systems. In section 2, the direct load control problem is presented, while in section 3 the design of the mutation operator is described. The results obtained applying this approach are presented in section 4. In section 5, some conclusions are drawn about the merits of the proposed adaptive behaviour of the mutation operator.

2 A Direct Load Control Problem

With the restructuring and unbundling of the electricity sector, a common scenario for buying and selling electricity is the one in which electricity retailers buy the electricity in the wholesale market and sell the electricity to the end-users in the retail markets. Generally, the electricity prices at the wholesale market change more frequently and more intensely than at the retail market. In such scenarios, an interesting situation, from a marketer point of view, is having the capability of changing the shape and level of the customers demand through the implementation of direct load control actions; for instance, in such a way that reduces the maximum peak demand and at the same time increases the profits by taking advantage of the differences between the retail market and the wholesale market electricity prices. The aim is to identify adequate on/off periods (load control actions that change the regular working cycle of end-use loads) to be implemented on a daily basis that allow, at the same time, to reduce the maximum power demand and to increase profits without imposing a severe discomfort to the end-users. These competing objectives make the design and selection of the direct load control actions a hard combinatorial multiobjective problem that can be tackled by EAs [3][4]. However, since the prices at the wholesale market can change frequently and intensely during the day and most prices at the retail market are fixed over that period of time, the marketer's profits with the reselling of electricity can also present a high frequency of variation and sometimes display unpredictable changes. Changes occurring during the identification of direct load control actions, for instance changes in profits forecasts or in weather forecasts, which in turn influence demand, must be taken into consideration.

Since most consumers buy electricity at fixed costs (over a period of time),if retailers have the ability to appropriately change their customer's load then they can take advantage of the difference of prices between the wholesale market and the retail market. Energy profits depend on the amount of energy sold and the maximum peak demand. The application of control strategies to some end-use loads changes demand patterns and levels and thus changes the profits per unit of energy sold. The objective functions considered in the proposed model are:

- Minimization of peak power demand (PD). Peak reduction enables the distribution utility/retailer to postpone the investment in new capacity, reduces the electricity costs (by reducing transmission/distribution charges) and provides the capability of continuously exploiting the differences between purchasing and selling prices.

- Maximization of profits (PR). Profits in the selling of electricity are generally influenced by the amount of electricity sold and the time of day/season/year. In the presence of demand and wholesale price forecasts, the distribution utility/retailer can design adequate load shedding actions in order to maximize profits once retail prices are fixed.

- Minimization of discomfort caused to customers (MI). The electricity service provided by loads under control is changed, possibly postponed, when load management actions are implemented. These changes can eventually cause some discomfort to customers that must be minimized, so that those actions become also attractive from the customers' point of view (with eventual reduction in their electricity bill) and/or at least not decrease their willingness to accept them. In this particular situation the loads under control are air conditioners and discomfort is evaluated through an objective function related with the time the room temperature is over a pre-specified threshold level. The objective function to be minimized is the maximum continuous time interval in which this situation has occurred.

The mathematical formulation of a similar model can be found in [3].

3 Adaptive Mutation Operator

NSGA-II, proposed by Deb et al. [5], combines two populations, offspring and parent, each one of size N, creating a population of size 2N. Then a non-domination sorting is performed in the resulting population. Each individual solution is assigned a fitness (or rank) equal to its non-domination level. Once the non-dominance sorting procedure is over, a new population is created using solutions of different non-dominated fronts. The first individuals to be selected for making part of the new population come from the first (best) non-dominated front and continues with the second non-dominated front, followed by the third non-dominated front, and so on. If it is necessary to decide which solutions must be selected from the same front then a niche strategy is used, choosing the solutions that reside in the least crowded region in that front.

In this work, in order to better adapt NSGA-II to the problem at hand, the mutation operator has been changed in such a way that it allows transferring information about the search process to the population. This adaptive mutation operator was firstly implemented in a specific genetic algorithm to deal with the direct load control problem [6]. The operator, with adaptive behavior, has now been adapted and implemented in the NSGA-II algorithm to be used in the design and selection of direct load control actions in a dynamic environment. The information is collected by analyzing the results obtained by every individual in the population in the phenotype space. Next, this information is used to compute mutation rates for every gene in every individual in the population. Thus, the mutation operator becomes a function of the performance of each individual in every objective as described below. The dynamic behavior of this operator, changing from one individual to another and from one generation to another, allows the search to be effectively guided towards regions of the search space in which more interesting compromise solutions can be found. In order to distinguish between the "original" NSGA-II and the "modified" algorithm in which the mutation operator has an adaptive behavior, we call this latter algorithm "NSGA-II with knowledge".

In the problem under study the binary alphabet is used:

"0" - is the normal operation state of the devices (normal on period)
"1" - is the "no operation state" (off period)

So, there are two different possible mutations that may occur in each gene: "normal operation state" ("0") mutates to "no operation state" ("1") (*pm_0_1*) and an "off period" ("1") mutates to a "on period" ("0") (*pm_1_0*). In order to better adapt the mutation operator to the environment, the two mutations that may occur in each gene may present different probability values. Changes occurring in the environment are made visible to the algorithm by the mutation operator, since usually when the environment changes the performance of each individual also varies leading to changes into the mutation operator values. The influence of the environment in the mutation operator is translated in the following way [6].

The objective associated with the minimization of peak power demand presents a time varying behaviour (within each simulation and from one run to another). The contributions of this objective ($pm^1\_0\_1; pm^1\_1\_0$) to the mutation operator are also time dependent, in a way closely related to the variation in time of the objective, and limited by thresholds (specified by an analyst assisting the decision maker): *pm_max* and *pm_min*. The way each contribution changes is a function of the normalized difference between the current value of the objective function in time interval i (D[i]) and a fraction, α, of its maximum value (MD).

$$pm^1\_0\_1[i] = \frac{\max(0; D[i] - \alpha * MD)}{\max_i (D[i] - \alpha * MD)} * pm\_\max$$

$$pm^1\_1\_0[i] = \left[1 - \frac{\max(0; D[i] - \alpha * MD)}{\max_i (D[i] - \alpha * MD)}\right] * pm\_\max$$

$$0 \le pm^1\_0\_1 \le pm\_\max$$

$$0 \le pm^1\_1\_0 \le pm\_\max$$

The level α is determined empirically and depends on the amount of controllable load (more load under control allows higher reductions in peak power demand; 90% has been used in the experiments whose results are shown in the following section).

D[i] is the power demand at interval i and *MD* is the original maximum peak demand. $pm^1\_0\_1$ ($pm^1\_1\_0$) is the contribution of this objective to the corresponding mutation probability. When the difference is negative (the actual demand is below the threshold level) the value 0 is considered for $pm^1\_0\_1$, meaning that, from this dimension's point of view, there is no interest in a mutation 0 to 1. For $pm^1\_1\_0$, the value *pm_max* is considered. That is, when power demand is low there is no interest in having load curtailments.

The aspects related with clients' comfort participate in the construction of the mutation operator with fixed values during the simulation. The contribution is based on the *rationale* that in order to minimize the clients' discomfort the number of load curtailments should be reduced. The values of the individual contribution for this objective ($pm^2\_0\_1$ and $pm^2\_1\_0$) are:

$$pm^2\_0\_1[i] = pm\_\min,$$

$$pm^2\_1\_0[i] = pm\_\max$$

Profits are in the range [*Min_profit*, *Max_profit*] according to the forecasts and, in each time interval *i*, the contribution for *pm* is given by

$$pm^3\_0\_1[i] = \frac{Max\_profits - \Pr ofits[i]}{Max\_profit - Min\_profit} * pm\_\max$$

$$pm^3\_1\_0[i] = \left[1 - \frac{Max\_profits - \Pr ofits[i]}{Max\_profit - Min\_profit}\right] * pm\_\max$$

where *Profits[i]* is the value for profits in interval i, based on daily forecasts.

The aggregation of all contributions is:

$$pm\_0\_1[i] = \sum_{t=1}^{3} \omega_t . pm^t\_0\_1[i] \text{ and } pm\_1\_0[i] = \sum_{t=1}^{3} \omega_t . pm^t\_1\_0[i]$$

This aggregation process is just a way to implement the operator based on the performance of every individual in each objective. The mutation operator still remains a probabilistic and blind operator.

4 The Case Study

An electric distribution utility facing some capacity pressure at a sub-station wants to analyze if it is possible to postpone the investment for increasing capacity by implementing direct load control actions without decreasing its profits. In order to proceed with this analysis the utility performed a survey on end-use appliances among its customers. It was possible to identify 1150 air conditioners (ACs) available for control. These loads have been grouped in 14 groups.

The peak power demand at the sub-station (SS) is 18377.5 kW (at 15:00h) and the maximum demand of the controllable load is about 2917 kW. At this time the controllable load correspond to 15.9% of the total peak. Figure 1 shows the demand at SS level as well as the demand of the loads under control.

When dealing with thermostatic loads one must be aware that changing their regular working cycle may impose some discomfort to the consumers, then decreasing their willingness to participate in such kind of activities. Also, the maximum power demand can become higher than it was with no control actions, instead of lower as intended, if the reestablishment of power to loads after curtailments is too coincident. Thus, the load curtailment actions must be properly designed in order to achieve the objectives pursuit by the retailer: minimize maximum power demand (PD); maximize profits (PR) and minimize eventual discomfort caused to the end-users (MI). Discomfort is perceived as the maximum interval of time (MI) in which the inside temperature is above a threshold level previously established. MI is measured in intervals of 5 minutes. The profits forecast is shown in figure 2.

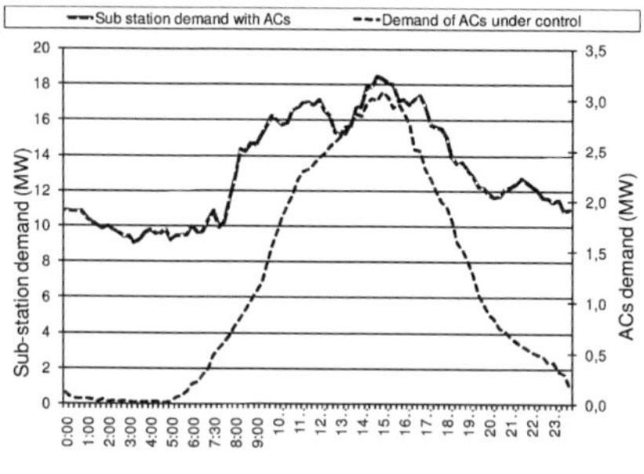

Fig. 1. SS and controllable demand

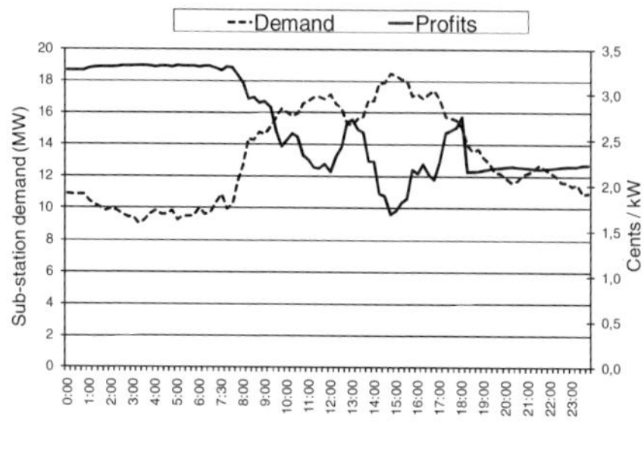

Fig. 2. Profits forecast

5 Results

In this section a comparative analysis between the results obtained with NSGA-II and NSGA-II "with knowledge" is made. The main characteristics of the implementations are:

Population size: 32 individuals;

Stop condition: maximum number of generations or when the DM is satisfied with the results obtained;

Crossover probability: 0.1;

Mutation probability:

NSGA-II: 0.001;

NSGA-II with adaptive mutation probability: [0.00008, 0.004].

All the parameters but the mutation operator are equal for both algorithms. The results are based on 7 runs for each algorithm.

The solutions found with NSGA-II and with NSGA-II "with knowledge" are displayed in Figures 3-4. 2D graphs enable to evaluate qualitatively each non-dominated front. The results obtained for objective function MI are integer values that are multiple of 5 (5, 10, 15…). Solutions obtained for MI= 10 minutes are shown in Figure 5.

The solutions obtained with NSGA-II "with knowledge" dominate all the solutions obtained with NSGA-II (this is also true for the case MI=5minutes).

In Figure 4 the best values of objective function PD in each generation are shown. Also NSGA-II "with knowledge" converges more rapidly than the standard NSGA-II.

The results at load diagram level can be seen in figure 5. The maximum global demand decreases about 780 kW, representing about 4% of the total demand (maximum peak demand decreases from 18377.5 kW to 17597.8 kW). The maximum demand of controllable load is 2917 kW, being the amount of reduction obtained about 27% of controllable load.

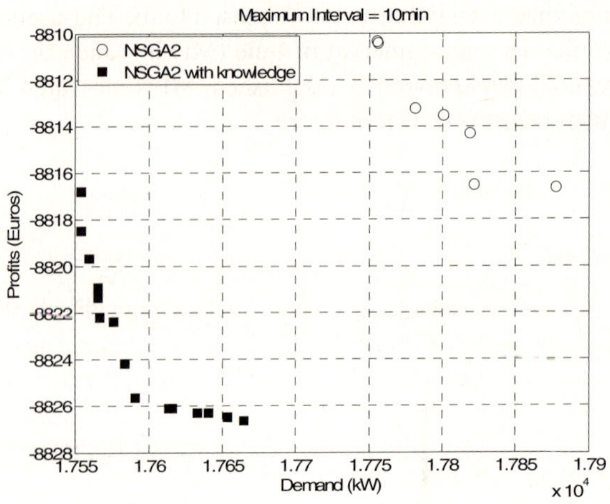

Fig. 3. Non-dominated front for MI=10 minutes

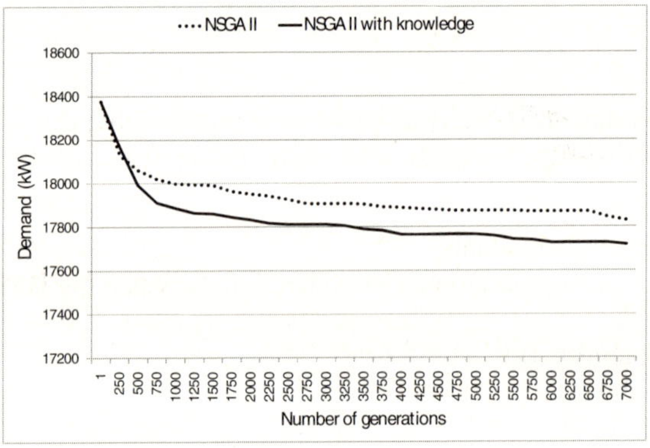

Fig. 4. Best value in objective function PD in each generation, when MI= 10 minutes

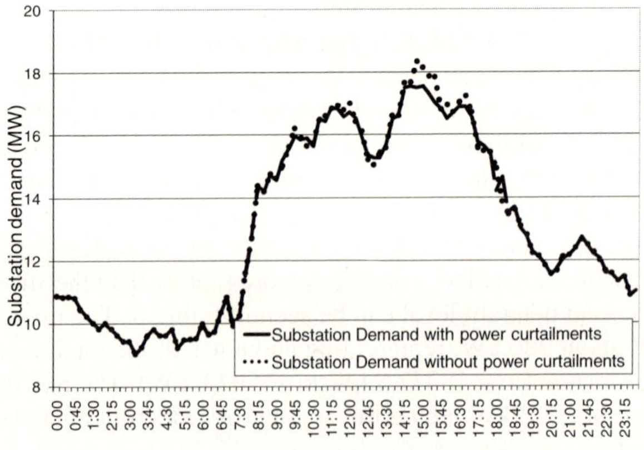

Fig. 5. SS demand without and with load curtailments

NSGA-II "with knowledge" leads to better results both for the power demand objective (17597.8 kW) and the profits objective (8820.3 €) in comparison with the standard NSGA-II (17853.9 kW and 8809.6 €, respectively). Also the number of load curtailments is much lower in NSGA-II with knowledge (469) than with NSGA-II without knowledge (1261).

6 Conclusions

In this paper a case study has been presented dealing with the application of load control strategies to groups of end-use loads aimed at reducing maximum demand without reducing profits. For the identification of a suitable load control strategy the NSGA-II algorithm has been implemented. The mutation operator has been changed in order to take into consideration the results of the individuals in each objective function, thus passing into the algorithm information about the search process. Is has been shown that, in this particular problem, NSGA-II with adaptive mutation operator (that is, "with knowledge" about the search) performs better that NSGA-II with a traditional (constant value) mutation operator (thus not using any knowledge about the search process), finding solutions that allow lower peak demand, and higher profits with fewer load curtailments.

References

1. Branke, J.: Evolutionary Approaches to Dynamic Optimization Problems – Updated Survey. In: GECCO Workshop on Evolutionary Algorithms for Dynamic Optimization Problems, pp. S.27–30 (2001)
2. Michalewicz, Z., Fogel, D.: How to Solve it: Modern Heuristics. Springer, Berlin (2000)
3. Gomes, A., Antunes, C.H., Martins, A.G.: A multiple objective evolutionary approach for the design and selection of load control strategies. IEEE Transactions on Power Systems 19(2), 1173–1180 (2004)
4. Yao, L., Chang, W., Yen, R.: An iterative deepening genetic algorithm for scheduling of direct load control. IEEE Trans. Power Systems 20(3), 1414–1421 (2005)
5. Deb, k., Pratap, A., Agarwal, S., Meyarivan, T.: A fast Elitist Non-Dominated Sorting Genetic Algorithm for Multi-Objective Optimization: NSGA-II. IEEE Trans. Evolutionary Computation 6(2), 181–197 (2002)
6. Gomes, A., Antunes, C.H., Martins, A.G.: Design of an Adaptive Mutation Operator in an Electrical Load Management Case Study. Computers and Operations Research 35(9), 2925–2936 (2007)

Structural Analysis of Promoter Sequences Using Grammar Inference and Support Vector Machine

Robertas Damaševičius

Software Engineering Department, Kaunas University of Technology,
Studentų 50-415, LT-51368, Kaunas, Lithuania
robertas.damasevicius@ktu.lt

Abstract. Promoters are short regulatory DNA sequences located upstream of a gene. Structural analysis of promoter sequences is important for successful gene prediction. Promoters can be recognized by certain patterns that are conserved within a species, but there are many exceptions which makes the structural analysis of promoters a complex problem. Grammar rules can be used for describing the structure of promoter sequences; however, derivation of such rules is not trivial. In this paper, stochastic L-grammar rules are derived automatically from known drosophila and vertebrate promoter and non-promoter sequences using genetic programming. The fitness of grammar rules is evaluated using a machine learning technique, called Support Vector Machine (SVM). SVM is trained on the known promoter sequences to obtain a discriminating function which serves as a means of evaluating a candidate grammar (a set of rules) by determining the percentage of generated sequences that are classified correctly. The combination of SVM and grammar rule inference can mitigate the lack of structural insight in machine learning approaches such as SVM.

1 Introduction

Promoters are short regulatory DNA sequences that precede the beginnings of genes. They are common both in prokaryotic and eukaryotic genomes. Analysis of promoter structure is important for our understanding of gene regulation mechanisms and genome evolution process, elucidation of the mechanisms for transcriptional activation of genes, annotation of transcriptional regulatory elements, and development of efficient promoter prediction programs [1]. The crucial obstacle in analyzing promoters is that they usually do not share extensive sequence similarity even when they are functionally correlated, which prevents their detection using sequence similarity-based search methods such as BLAST or FASTA. Though there are conserved patterns in the promoter sequences of a species, numerous exceptions make the promoter recognition a complex problem [2].

Modern promoter recognition tools use machine learning techniques such as Naive Bayes, Decision Trees, Hidden Markov Models, Neural Networks or Support Vector Machine (SVM) [3]. Best machine-learning based promoter recognition methods allow achieving up to 98% accuracy [4], however they do not provide any insight on the internal structure of the promoter sequences. Analysis of some promoters suggests

I. Lovrek, R.J. Howlett, and L.C. Jain (Eds.): KES 2008, Part I, LNAI 5177, pp. 98–105, 2008.

that promoter sequences may have a modular structure. The structural models of promoter sequences can be analyzed using computational approaches such as *adaptive quality-based clustering* [5] and the iterative expectation-maximization algorithm-based method (MEME) [6].

Formal grammars can provide a means of describing complex repeatable structures such as DNA. The structure of promoter sequences can be described using a formal system such as *L-grammar* (or *L-system*) [7], and the problem of promoter recognition can be replaced by grammar induction [8]. Koza [9] demonstrates the possibility of discovering the rewrite rule for L-systems and the state transition rules for cellular automata using genetic programming techniques. For the study of protein formation, Marcus [10] considers so-called *semi-Lindenmayer* systems and discusses an isomorphism between the genetic and natural language. Jimenéz-Montano [11] proposes an algorithm to construct a short context-free grammar that generates a given sequence. Infante-Lopez and de Rijke [12] describe the inference of regular language grammar rules based on *n-grams* and minimization of the *Kullback-Leibler* divergence. O'Neill *et al.* [13] generate regular expressions for promoter recognition problem using *Grammatical Swarm* technique. Each individual swarm particle represents choices of program construction rules, where these rules are specified using a *Backus-Naur Form* (BNF) grammar. Denise *et al.* [14] generate genomic sequences and basic RNA secondary structures according to a given probability distribution and syntactical (grammatical) parameters. Stochastic context-free grammars constructed from sample sets of sequences are also considered for modeling RNA sequences [15, 16].

The aim of this paper is describe a method for structural analysis of promoter sequences which combines SVM classification with automatic derivation of stochastic L-grammar rules using genetic programming. The structure of the paper is as follows. Section 2 provides an introduction into L-grammar. Section 3 describes the principles of SVM classification. Section 4 considers grammar induction problem. Section 5 describes derivation of stochastic L-grammar rules for drosophila and vertebrate promoter datasets. Finally, Section 6 evaluates the results and presents conclusions.

2 Introduction into L-Systems

Treating genome as a language can allow to generalize the structural information contained in biological sequences and to investigate it using formal language theory methods. From the biological point of view, the components of any biological organism evolve simultaneously, so we cannot expect that biological processes could be modeled using a sequential approach. It is more likely that the cells that can reproduce simultaneously would be modeled by a mechanism that is based on the same behavioral principles. Formal language theory can be used for modeling biological phenomena in general and genetic mechanisms, in particular [17]. Some aspects of formal languages are similar to biological processes, e.g.:

- *Pure grammars* do not differentiate between *terminal* and *non-terminal* symbols, so that all words generated from the grammar rules are put into the generated language. This notion is well-motivated biologically, because all symbols in a DNA sequence have similar role.
- *Erasing rule* $A \to \varepsilon$ models a delete mutation in a DNA sequence.

- *Chain rule* $A \to B$ reflects a single nucleotide mutation in DNA.
- *Repetition rule* $A \to AA$ models highly repetitive DNA sequences such as tandem repeats.
- *Growing rule* $A \to BC$ models the growth of a DNA sequence.
- *Stochastic rule* $p : A \to B$ models random mutations of DNA sequences.

L-systems or L-grammars are a special class of grammars that can model the growth of living organisms, e.g. plant development, but also are able to model the morphology of a variety of organisms [18]. They produce sentences that can be interpreted graphically to produce images of fractals or organisms. Recently L-grammars also have been applied for modeling DNA sequences [19-23].

In L-systems the production rules are applied in parallel and may simultaneously replace all letters in a given word. Also there is no distinction between terminal and non-terminal symbols. All symbols that appear in the grammar are valid in the final string, and any symbol in the alphabet can head a rule [24]. The recursive nature of the L-system rules leads to self-similarity and fractal-like forms which is also a property of DNA sequences [25].

For modeling DNA sequences we use a *context-free stochastic L-grammar*, which is defined as a tuple:

$$G = \{V, \omega, R, P\} \tag{1}$$

here:

$V = \{A, C, G, T\}$ is a set of symbols containing elements (nucleotide types, in our case) that can be replaced (the alphabet).

$\omega = V^K$ is a K-length string of symbols that define the initial state of the system.

$R \subseteq V^1 \times V^L$ is a finite set of production rules defining the way an individual nucleotide can be replaced with combinations of other nucleotides. A rule consists of two strings - the *predecessor* and the *successor*.

P is a set of probabilities $p_j \in [0,1]$ that a production rule $r_j \in R$ will be applied.

3 Classification Using Support Vector Machine

Support Vector Machine (SVM) [26] is a structural risk minimization-based method for creating binary classification functions from a set of labeled training data. SVM requires that each data instance is represented as a vector of real numbers in *feature space*. Hence, if there are categorical attributes, we first have to convert them into numeric data. First, SVM implicitly maps the training data into a (usually higher-dimensional) feature space. A *hyperplane* (decision surface) is then constructed in this feature space that bisects the two categories and maximizes the margin of separation between itself and those points lying nearest to it (the *support vectors*). This decision surface can then be used as a basis for classifying vectors of unknown classification.

Consider an input space X with input vectors $x_i \in X$, a target space $Y = \{1, -1\}$ with $y_i \in Y$ and a training set $T = \{(x_1, y_1), \ldots, (x_N, y_N)\}$. In SVM classification, separation of the two classes $Y = \{1,-1\}$ is done by means of the *maximum margin* hyperplane, i.e. the hyperplane that maximizes the distance to the closest data points and guarantees the best generalization on new examples. In order to classify a new point x_j, the classification function $g(x_j)$ is used:

$$g(x_j) = \operatorname{sgn}\left(\sum_{x_i \in SV} \alpha_i y_i K(x_i, x_j) + b \right) \tag{2}$$

here: SV are the support vectors, $K(x_i, x_j)$ is the kernel function, α_i are weights, and b is the offset parameter.

If $g(x_j) = +1$, x_j belongs to the *Positive* (P) class, if $g(x_j) = -1$, x_j belongs to the *Negative* (N) class, and if $g(x_j) = 0$, x_j cannot be classified.

For evaluating the efficiency of the classification function $g(x_j)$ we use a simple *Accuracy* (ACC) metric, which is a measure of how well a binary classification test correctly identifies or excludes a condition.

$$ACC = \frac{n_{TP} + n_{TN}}{n_P + n_N} \cdot 100\% \tag{3}$$

here: n_{TP} and n_{TN} is the number of *True Positives* (TP) and *True Negatives* (TN) in the classification results, respectively.

4 Problem of Grammar Inference

Derivation of L-system grammar rules, also known as *grammatical induction* or *grammar inference,* refers to the process of inducing a formal grammar (usually in the form of production rules) from a set of observations using the machine learning techniques. The result of grammar inference is a model that reflects the characteristics of the observed objects. Here, for inference of grammar rules we use a "*trial-and-error*" method [27]. The method suggests successively guessing grammar rules (productions) and testing them against positive and negative observations. The best ruleset is then mutated randomly following the ideas of genetic programming until no improvement in accuracy is found within a certain number of iterations.

The rules of the L-system grammar are applied iteratively and simultaneously starting from the initial string. Let us denote $L(G)$ all those strings over V that can be generated by starting with the start string ω and then applying the production rules in R with probabilities P. We describe the grammar inference problem as a classical optimization problem as follows. Determine G, where V is constant, from a given set of input sequences $X = V^N$ such as to achieve the best classification $c(X')$:

$$c(X') = \max \sum_{X'} g(x_j) \tag{4}$$

here: $x_j, x_j \in X' \subset L(G)$ are strings produced by production rules R, and $g(x_j)$ is the classification function trained on a set of input sequences $x_i \in X$.

5 Case Study: Derivation of L-Grammar Rules for Promoters

We use 2 datasets: 1) The 2002 collection of data of drosophila core promoter regions [28]. The test file contains 6500 examples (1842 promoters, 1799 introns (non-coding sections of DNA), and 2859 protein-coding sequences), each 300 bp length. 2) The collection of human and additional eukaryotic (vertebrate) promoter regions extracted from the Eukaryotic Promoter Database rel. 50; the negative set contains coding and non-coding sequences from the 1998 GENIE data set [29]. The test file contains 5800 examples (565 promoters, 4345 introns, and 890 protein-coding sequences), each 300 bp length.

The L-grammar rule generation process is performed as follows:

1) *Derivation of a learned classifier.* The promoter dataset is used for training a SVM classifier (we use SVM^{light} [30]) with a linear kernel function. Nucleotide sequences are mapped onto feature space using *orthogonal encoding*, where nucleotides are represented by 4-dimensional orthogonal binary vectors: $\{A \to 0001, C \to 0010, G \to 0100, T \to 1000\}$. The result of training is a model of a learned classifier that can separate promoter and non-promoter sequences.

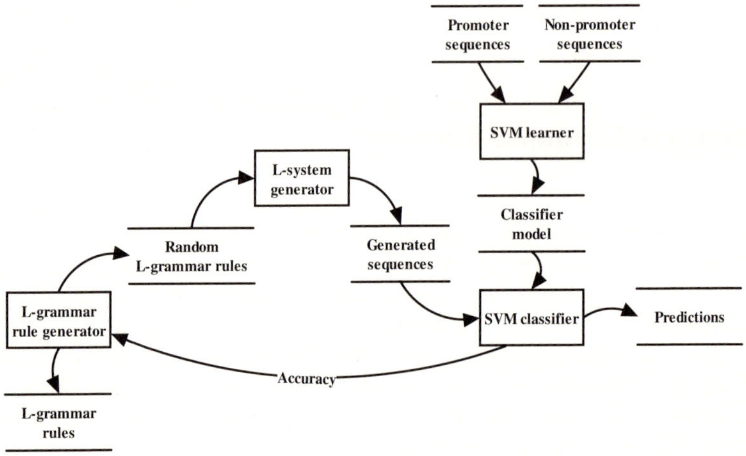

Fig. 1. Structure of the L-grammar induction system

2) *L-grammar rule induction.* The classifier model is used for evaluating the fitness of L-grammar rules. Rules are generated by L-grammar rule generator using a directed random search method: the best ruleset so far is mutated randomly and used to generate 1000 of 300 bp length L-system strings. The mutation parameters are the start string, successor strings and production rule probabilities. The number of rules as well as the predecessor strings is fixed. There are four rules for each type of nucleotide: A-rule, C-rule, G-rule and T-rule. The generated strings are fed to the trained

SVM classifier and the accuracy of the classification is obtained. If the accuracy of the mutated ruleset is better than the accuracy of the previous best ruleset, the mutated ruleset is saved as the best ruleset; otherwise the previous best ruleset is retained. The process is continued until the required accuracy is achieved. The structure of the L-grammar induction system is summarized in Fig. 1.

The classification was done in two stages: 1) SVM was trained using promoters vs. non-promoters, and the trained classifier was used 2) to classify random sequences labeled as promoters, and artificially generated (from induced L-grammar rules) sequences labeled as promoters. The classification results are presented in Table 1.

We can see that artificially generated sequences can be classified vs. non-promoter sequences almost as good as real (natural) promoters. That suggests that induced L-grammar rules indeed capture some essential dataset properties of promoter sequences. The results are worse for the vertebrate dataset, because the vertebrate promoters have more complex patterns with more irregularities and exceptions.

Table 1. Classification results

Classified sequences	No. of sequences	Classification accuracy	
		Drosophila dataset	Vertebrate dataset
Promoters vs. non-promoters	6500/5800	99.74%	94.67%
Random sequences	1000	3.3%	1.2%
Artificially generated sequences	1000	93.40%	92.10%

The induced L-grammar rules are presented in Fig. 2. Note that in drosophila grammar rules C and in vertebrate grammar rules T symbols are missing from the successor side of production rules.

```
Variables: A, C, G, T
Start:     AAACTAAT
Rules:     0.85: (A -> TATA),
           0.94: (C -> TA),
           0.91: (G -> TAG),
           0.10: (T -> TGA)
```

```
Variables: A, C, G, T
Start:     A
Rules:     0.50: (A -> CGGAA),
           0.86: (C -> CCCCG),
           0.19: (G -> ACGG),
           0.50: (T -> AA)
```

Fig. 2. Stochastic L-grammar rules for generating drosophila promoter-like sequences (left) and vertebrate promoter-like sequences (right)

6 Evaluation and Conclusions

The lack of structural insight is one of the major drawbacks of machine learning approaches such as SVMs. Since the task of promoter recognition is already solved sufficiently by the underlying SVM (99.7% accuracy on drosophila test sequences), the derived grammars are used for structural analysis of promoter sequences.

The classification results of the artificial promoter sequences generated using the derived L-grammar rules are almost as accurate as that of the natural promoters, thus showing a great deal of similarity between the discriminating features of both types of sequences. The analysis of the derived L-grammar production rules also allows identifying common promoter sequence elements (so called "boxes").

The drosophila promoter production rules feature the TATA successor string, which matches the TATA-box (TATAA or TATAAA) typical for the drosophila promoters. Other subsequences typical for the promoter sequences are produced by the successive application of the production rules, e.g., the *Pribnow* box (TATAAT) is produced by two successive applications of the A-rule.

The vertebrate promoters are typically more complex than the drosophila promoters, because there are many TATA-less promoters, which are characterized by other more complex subsequences. These subsequences also can be produced from the induced grammar rules, e.g., the CACG subsequence characteristic to the *E-box* (CACGTG) is produced by the successive application of the A-rule and G-rule. The AACC subsequence characteristic to the *Y-box* (GGGTAACCGA) is produced by successive application of the G-rule, A-rule, and C-rule. Human promoter sequences are also characterized by the occurrence of the DPE (*downstream promoter element*), a distinct 7-nucleotide subsequence (A/G)G(A/T)CGTG that is similar to the G-rule.

The issue open for further research is the completeness of the derived grammar. The construction of promoter sequences is limited by the size of the DNA grammar and properties of L-systems itself: the ruleset consists of only four rules, one for each nucleotide, which can appear on the left-hand side, the corresponding probabilities, and a start word. Though such small grammars can produce reliable results (93.4% for the drosophila dataset), it is difficult to evaluate the extent of the structural characteristics of promoter sequences that can be captured by such grammars.

Future work may include the extension of grammar ruleset for context sensitivity, which could reveal more interesting structural details of promoter sequences.

References

1. Bajic, V.B., Choudhary, V., Hock, C.K.: Content analysis of the core promoter region of human genes. Silico Biol. 4, 109–125 (2004)
2. Werner, T.: The state of the art of mammalian promoter recognition. Briefings in Bioinformatics 4(1), 22–30 (2003)
3. Monteiro, M.I., de Souto, M.C.P., Gonçalves, L.M.G., Agnez-Lima, L.F.: Machine Learning Techniques for Predicting Bacillus subtilis Promoters. In: Setubal, J.C., Verjovski-Almeida, S. (eds.) BSB 2005. LNCS (LNBI), vol. 3594, pp. 77–84. Springer, Heidelberg (2005)
4. Ranawana, R., Palade, V.: A neural network based multiclassifier system for gene identification in DNA sequences. J. of Neural Computing Applications 14, 122–131 (2005)
5. Florquin, K., Saeys, Y., Degroeve, S., Rouzé, P., Van de Peer, Y.: Large-scale structural analysis of the core promoter in mammalian and plant genomes. Nucleic Acids Res. 33(13), 4255–4264 (2005)
6. Ohler, U., Liao, G.C., Niemann, H., Rubin, G.M.: Computational analysis of core promoters in the Drosophila genome. Genome Biol. 3 (2002) RESEARCH0087
7. Lindenmayer, A.: Mathematical models for cellular interactions in development. Journal of Theoretical Biology 18, 280–315 (1968)
8. Unold, O.: Grammar-Based Classifier System for Recognition of Promoter Regions. In: Beliczynski, B., Dzielinski, A., Iwanowski, M., Ribeiro, B. (eds.) ICANNGA 2007. LNCS, vol. 4431, pp. 798–805. Springer, Heidelberg (2007)

9. Koza, J.R.: Discovery of Rewrite Rules in Lindenmayer Systems and State Transition Rules in Cellular Automata via Genetic Programming. In: Symp. on Pattern Formation (SPF 1993), Claremont, CA (1993)
10. Marcus, S.: Linguistic structures and generative devices in molecular genetics. Cahiers. Ling. Theor. Appl. 1, 77–104 (1974)
11. Jiménez-Montaño, M.A.: On the Syntactic Structure of Protein Sequences and the Concept of Grammar Complexity. Bull. Math. Biol. 46, 641–659 (1984)
12. Infante-Lopez, G., de Rijke, M.: Alternative approaches for generating bodies of grammar rules. In: Proc. of 42nd Annual Meeting of the Association for Computational Linguistics (ACL 2004), Barcelona, Spain, 21-26 July, pp. 454–461 (2004)
13. O'Neill, M., Brabazon, A., Adley, C.: The Automatic Generation of Programs for Classification Problems with Grammatical Swarm. In: Proc. of the Congress on Evolutionary Computation CEC 2004, Portland, OR, USA, June 2004, pp. 104–110 (2004)
14. Denise, A., Ponty, Y., Termier, M.: Random Generation of structured genomic sequences. In: Proc. of 7th Annual Int. Conf. on Research in Computational Molecular Biology (RECOMB 2003), Berlin, Germany, 10-13 April (2003)
15. Grate, L., Herbster, M., Hughey, R., Haussler, D.: RNA modelling using Gibbs sampling and stochastic context-free grammars. In: Proc. of the Second Int. Conf. on Intelligent Systems for Molecular Biology, vol. 2, pp. 138–146. AAAI/MIT Press (1994)
16. Sakakibara, Y., Brown, M., Hughey, R., Mian, I.S., Sjoelander, K., Underwood, R., Haussler, D.: Stochastic context-free grammars for tRNA modelling. Nucleic Acids Res. 25, 5112–5120 (1994)
17. Fernau, H.: Parallel Grammars: A Phenomenology. Grammars 6(1), 25–87 (2003)
18. Prusinkiewicz, P., Lindenmayer, A.: The Algorithmic Beauty of Plants. Springer, New York (1990)
19. Searls, D.B.: The computational linguistics of biological sequences. In: Hunter, L. (ed.) Artificial Intelligence and Molecular Biology, pp. 47–120. AAAI/MIT Press (1993)
20. Yokomori, T., Kobayashi, S.: Learning local languages and their application to DNA sequence analysis. IEEE Trans. on Pattern Analysis and Machine Intelligence 10(20), 1067–1079 (1998)
21. Mihalache, V., Salomaa, A.: Lindenmayer and DNA: Watson-Crick D0L Systems. Current Trends in Theoretical Computer Science, 740–751 (2001)
22. McGowan, J.F.: Nanometer Scale Lindenmayer Systems. In: Proc. of SPIE, vol. 4807 (2002)
23. Gheorghe, M., Mitrana, V.: A formal language-based approach in biology. Comparative and Functional Genomics 5, 91–94 (2004)
24. Prusinkiewicz, P., Hanan, J.: Lindenmayer Systems, Fractals, and Plants. Lecture Notes in Biomathematics. Springer, Heidelberg (1989)
25. Abramson, G., Cerdeira, H.A., Bruschi, C.: Fractal properties of DNA walks. Biosystems 49(1), 63–70 (1999)
26. Vapnik, V.: Statistical Learning Theory. Wiley-Interscience, New York (1998)
27. Duda, R.O., Hart, P.E., Stork, D.G.: Pattern Classification. John Wiley & Sons, Chichester (2001)
28. Berkeley Drosophila Genome Project. Drosophila promoter dataset, `http://www.fruitfly.org/seq_tools/datasets/Drosophila/promoter/`
29. Berkeley Drosophila Genome Project. Human promoter dataset, `http://www.fruitfly.org/seq_tools/datasets/Human/promoter/`
30. SVMlight, `http://svmlight.joachims.org/`

A Metric to Discriminate the Selection of Algorithms for the General ATSP Problem

Jorge A. Ruiz-Vanoye[1], Ocotlán Díaz-Parra[2], and Vanesa Landero N.[1]

[1] Centro Nacional de Investigación y Desarrollo Tecnológico
Interior internado Palmira s/n., Col. Palmira, Cuernavaca, Mexico
{ruizvanoye04c,landerov}@cenidet.edu.mx
[2] CIICAp. Universidad Autonóma del Estado de Morelos
Av. Universidad, Col. Chamilpa, 62209, Cuernavaca, Mexico
odiazp@uaem.mx

Abstract. In this paper we propose: (1) the use of discriminant analysis as a means for predictive learning (data-mining techniques) aiming at selecting metaheuristic algorithms and (2) the use of a metric for improving the selection of the algorithms that best solve a given instance of the Asymmetric Traveling Salesman Problem (ATSP). The only metric that had existed so far to determine the best algorithm for solving an ATSP instance is based on the number of cities; nevertheless, it is not sufficiently adequate for discriminating the best algorithm for solving an ATSP instance, thus the necessity for devising a new metric through the use of data-mining techniques.

Keywords: Inductive learning, genetic algorithm, discriminant analysis, data-mining techniques, machine learning.

1 Introduction

Discriminant analysis [1, 2] is a multivariate statistical technique, whose purpose is to analyze if there exist significant differences between groups of objects with respect to a set of variables measured. Linear discriminant analysis was introduced by Fisher in 1936 [1] as a statistical procedure for classification. Classification is the most common task in generic inductive learning; additionally, it is a function of predictive learning that classifies data of diverse classes [3]. In this paper we propose the use of classification as a means for predictive learning in metaheuristic algorithm selection, and a metric to improve the selection of algorithms that best solve a given instance of the Asymmetric Traveling Salesman Problem (ATSP). Metrics are mathematical formulations used for reflecting a certain situation; i.e., a relation between quantitative or qualitative variables that allows observing the situation and the tendencies of changes generated in the objects [1]. They can be used to measure the influence on algorithm performance.

The problem that will be addressed is ATSP, which can be stated as follows: given a set of nodes and distances for each pair of nodes, find a route of minimal overall length that visits each of the nodes exactly once [4]. The distance of node i to node j and the distance of node j to node i can be different (equations 1, 2, 3, 4).

I. Lovrek, R.J. Howlett, and L.C. Jain (Eds.): KES 2008, Part I, LNAI 5177, pp. 106–113, 2008.

$$\min z(x) = \sum_{j=1}^{m} \sum_{i=1}^{m} d_{ij} x_{ij} \tag{1}$$

$$\sum_{j=1}^{m} x_{ij} = 1; \qquad i = 1,\dots,m; \qquad d_{ij} \neq d_{ji} \tag{2}$$

$$\sum_{i=1}^{m} x_{ij} = 1; \qquad j = 1,\dots,m\,; \qquad 0 \le x_{ij} \le 1 \tag{3}$$

$$x_{ij} = \begin{cases} 1, & \text{if tour traverses from } i \text{ to } j \\ 0, & \text{otherwise} \end{cases} \qquad \forall i, j \tag{4}$$

The parameters of ATSP instances can be coded in a encoding scheme: L= "nc, d_1, d_2, ..., d_n", where nc represents the number of cities and d_i is the distance between pairs of cities. The purpose of this work is to propose the use of discriminant analysis as a means for predictive learning (data-mining techniques) to select the best algorithm than solves an ATSP instance.

2 Hypothesis of the Investigation

In the area of data mining, Fink [5] developed a technique of algorithm selection for decision problems, which is based on the estimation of the gain of an algorithm, obtained from the statistical analysis of its performance. The investigation group that worked in the METAL project [6], proposed a method to select an algorithm for a set

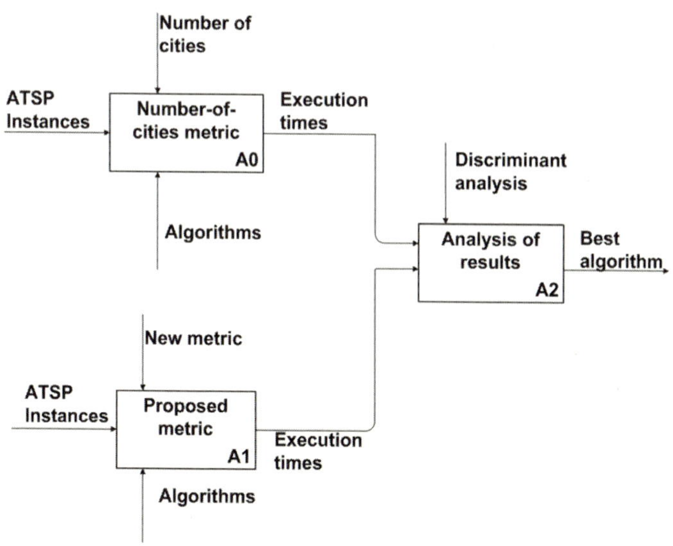

Fig. 1. Hypothesis of the investigation

of related cases. They find a set of old cases whose characteristics are the most similar to those of the new set. The algorithms performances for the old cases are known and used to predict the best algorithm for the new set of cases. Pérez-Cruz [7, 8, 9, 10, 11] proposed a methodology, based on automatic learning to systematically develop mathematical models for algorithm performance. The proposal consists of characterizing the performance of a set of metaheuristic algorithms applied to the solution of NP-hard problems, through the determination of regions of superiority of the algorithms.

The purpose of this investigation (Fig. 1) consists of developing a method to improve the prediction on the algorithm that best solves the instances of a set of benchmark ATSP instances. At present, there exists a metric based on the number of cities; however, it is not sufficiently adequate for discriminating the best algorithm. Thus the following question arises: is it possible to discriminate adequately using additional metrics?

3 Genetic Distance Metric for Algorithm Selection for the General ATSP

The purpose of this work is to develop a formal metric that helps improve algorithm selection for the general ATSP. To this end we used a concept from descriptive statistics called relative frequency. The index of similarity or genetic distance quantifies the similarity or difference in intercity distances (genetic is attributed to the quantification of the molecular markers; for example, in a jungle a lion and a leopard have some characteristics that are similar but there are also differences, which are reflected in their genes). The genetic term applied to ATSP will tell us how much variability in the intercity distances exists; i.e., the number of measurements in an interval of a frequency distribution.

The index of similarity S expresses the similarity among intercity distances, which is obtained from the frequencies of the distances, where similarity of distances will exist as long as the frequencies of the distances are larger than 1. In general, the cases with many similar frequencies are considered less complex to solve by a certain type of algorithm.

$$S = \begin{cases} \sum Frequency(d_i) > 1, & \text{if } Frequency(d_i) > 1 \quad \text{for some } i \\ 0, & \text{otherwise} \end{cases} \tag{5}$$

Where S = index of similarity, d_i = intercity distance, and $Frequency(d_i)$ = number of distances equal to d_i.

4 Experimentation

The experimentation was carried on an Acer Travelmate 2330LC computer with an Intel Celeron processor at 1.5GHz, 512 MB, 80 GB hard disk. The metaheuristic algorithms tested were: a generic Genetic algorithm (GA), standard Tabu Search (STS), Random Search (RS), and Scatter Search (SS) [12, 13, 14, 15, 16].

The parameters for GA were: selection operator=roulette, crossover operator=OX, crossover rate=1, mutation rate=0.05, generations=1000, population size=100, replacement strategy=elitism, n-elitism=1, tournament group size=2. For STS the parameters were: stop criterion (iterations=1000, evaluations=10000), tabu list tenure=10, neighborhood size=10, selection operator=standard, neighborhood operator =simple inversion. The parameter for RS was: rounds=1000. For SS the parameters were: stop criterion (steps=50, evaluations=50000), PSize=40, HQ ref. set size=5, div. ref. set Size=3, diversification=none, similarity operator=Hamming, improvement method=simple inversion, improvement cycles =10, neighborhood op=simple inversion, neighborhood size=10, ref. set update method=standard (more diversity), subset generator=pairs, solution combinator= OX. The instances were obtained from the TSPLIB benchmark [17].

Table 1. Some results obtained for algorithms GA and STS

Instance	*GA*		*STS*	
	Time	*Ratio*	*Time*	*Ratio*
br17	0.13019	1.000	1.15166	1.000
ft53	0.78112	0.674976	1.44207	0.478915
ft70	2.2933	0.791053	1.33192	0.705725
ftv33	1.99287	0.867746	1.35194	0.612381
ftv35	2.18314	0.764798	1.42204	0.617869
ftv38	2.17312	0.785018	1.37197	0.572177
ftv44	2.23321	0.755504	1.36196	0.528333
ftv47	2.26325	0.736929	2.28328	0.511226
ftv55	2.12305	0.590959	1.44207	0.432491
ftv64	2.44351	0.485352	1.45209	0.393874
ftv70	2.33336	0.460014	1.4621	0.39196
ftv170	2.88415	0.250295	1.75252	0.17178
kro124p	2.54366	0.555479	1.50216	0.471131
p43	2.24323	0.966965	1.35194	0.990658
rbg323	3.78544	0.330508	1.57226	0.254169
rbg358	4.29618	0.24771	1.54222	0.193254
rbg403	4.39632	0.460317	1.59229	0.366597
rbg443	5.01721	0.436878	1.74251	0.369214
ry48p	2.17312	0.905222	1.4621	0.685066

Table 1 shows some results obtained for GA and STS; while Table2 shows some results for RS and SS.

The discriminant analysis implemented in the SPSS software was used [18]. This kind of analysis is used as a machine-learning method to find the relation between the characteristics of a problem and the performance of an algorithm [3].

The performance results obtained were the theoretical ratio and run time. For each instance the list of algorithms that best solve it was determined; to this end the following criterion for algorithm evaluation was defined: the algorithm with the smallest value of the run time divided by the theoretical ratio is considered the best for the instance.

Table 2. Some results obtained for algorithms RS and SS

Instance	*RS*		*SS*	
	Time	*Ratio*	*Time*	*Ratio*
br17	0.9714	0.433333	1.21174	1.000
ft53	1.06153	0.321746	3.44954	0.509031
ft70	1.07154	0.588415	5.10734	0.730575
ftv33	1.12161	0.373729	2.18314	0.681505
ftv35	1.02147	0.374714	2.5036	0.757326
ftv38	1.09157	0.355897	2.61376	0.726496
ftv44	1.0415	0.331961	2.98429	0.558711
ftv47	1.19171	0.309948	3.30475	0.508737
ftv55	0.99143	0.256705	3.81549	0.554865
ftv64	1.0415	0.257419	4.7268	0.420151
ftv70	1.02147	0.235821	5.06729	0.418545
ftv170	1.0415	0.117645	17.2949	0.181596
kro124p	1.02147	0.221605	7.74113	0.547736
p43	1.01145	0.474302	3.12449	0.997515
rbg323	1.12161	0.227561	53.9476	0.271055
rbg358	1.22176	0.179199	01:08.6	0.209173
rbg403	1.28184	0.339065	01:17.1	0.400357
rbg443	1.392	0.348763	01:31.8	0.400825

Table 3 shows the list of the best algorithms for the sample instances, as well as the calculation of the proposed metric S for each of the TSPLIB instances [17]. The values of table 3 were used as input for the discriminant analysis (where n is the number of cities and S is the index of similarity), and Table 4 shows the results from the discriminant analysis.

In order to obtain the classification criterion, two indicators were used which were used as independent variables, and the name of the best algorithm as dependent variable.

Table 3. Results obtained using the metric on the solution of the ATSP instances

Instance	*n*	*S*	*Best Algorithm*
br17	17	0.53	GA
ft53	53	0.32	GA
ft70	70	0.24	RS
ftv33	33	0.22	STS
ftv35	35	0.36	STS
ftv38	38	0.26	GA
ftv44	44	0.3	STS
ftv47	47	0.48	GA
ftv55	55	0.56	STS
ftv64	64	0.65	STS
ftv70	70	0.71	STS
ftv170	170	1.7	RS
kro124p	124	1.0	STS
p43	43	1.03	STS
rbg323	323	49.28	RS
rbg358	358	81.15	RS
rbg403	403	121.04	RS
rbg443	443	894	RS
ry48p	48	0.12	STS
ftv170	170	1.7	RS
p43	43	1.03	STS

Table 4. Results from the discriminant analysis

Group Origin	*Group Destiny*			*Total*
	GA	*STS*	*RS*	
GA	7	0	0	7
	100 %	0%	0%	100%
STS	6	0	0	6
	100%	0%	0%	100%
RS	1	0	5	6
	16.7 %	0%	83.3%	100%

Table 5. Example of instances with their characteristics and the best algorithm predicted

Instances	*Characteristics* *n*	*S*	*Best Algorithm*	*Squared Mahalanobis Distance to Centroid*	*Function 1*	*Function 2*
ftv160	160	1.61	GA	2.468	0.549	-0.510
ftv110	110	1.1	GA	0.702	-0.145	-0.275
ftv120	120	1.21	GA	0.969	-0.006	-0.322
dc112	112	1.12	GA	0.752	-0.117	-0.284
dc126	126	1.26	STS	1.236	0.77	-0.350
dc134	134	1.34	STS	1.510	0.188	-0.388

A criterion of classification called group discriminant function will be used later for each new observation in the corresponding group. The percentage of the new observations correctly classified is an indicator of the effectiveness of the discriminant functions. If the functions are effective for the training sample, they will classify new instances correctly. To validate the effectiveness of the discriminant classifier, we used other ATSP instances. Table 5 shows the results for a fraction of the instances. For each instance, it displays the result of the proposed metric, the prediction of the best algorithm that solves the given instance, the discriminant Mahalanobis distance to the centroid, and discriminant functions 1 and 2. The results were obtained for validating the effectiveness of the discriminant classifier. The prediction of the classifier was of 58.9 %.

5 Conclusions

The main contribution of this work is the proposal of a similarity metric or genetic distance metric that quantifies the similarity or difference among intercity distances (through the distribution of distance frequencies). A method devised to improve the prediction on the algorithm that best solves a benchmark set of ATSP instances, which can adequately discriminate the best solution algorithm. From the experimental results, it follows that it is possible to discriminate adequately adding another metric to the ATSP problem. A continuation of this work would aim at devising additional metrics to increase the prediction percentage on the algorithm that best solves a given instance. The results were obtained for validating the effectiveness of the discriminant classifier. The prediction of the classifier was 58.9 %.

References

1. Kirkpatrick, S., Gelatt, C., Vecci, M.: Optimization by Simulated Annealing. Science 220(4598) (1983)
2. Rutenbar, R.: Simulated Annealing Algorithms: An Overview. IEEE Circuits and Devices Magazine 5(5), 19–26 (1989)
3. Kantardzic, M.: Data Mining: Concepts, Models, Methods, and Algorithms. John Wiley & Sons, Chichester (2003)
4. Cirasella, D.S., Cirasella, J., Johnson, D.S., McGeoch, L.A., Zhang, W.: The asymmetric traveling salesman problem: Algorithms, instance generators, and tests. In: Buchsbaum, A.L., Snoeyink, J. (eds.) ALENEX 2001. LNCS, vol. 2153, pp. 32–59. Springer, Heidelberg (2001)
5. Fink, E.: How to Solve it Automatically, Selection among Problem-solving Methods. In: Proceedings of the Fourth International Conference on AI Planning Systems AIPS 1998, pp. 128–136 (1998)
6. Soares, C., Brazdil, P.: Zoomed Ranking, Selection of Classification Algorithms Based on Relevant Performance Information. In: Zighed, D.A., Komorowski, J., Żytkow, J. (eds.) PKDD 2000. LNCS (LNAI), vol. 1910, pp. 126–135. Springer, Heidelberg (2000)
7. Pérez, J., Pazos, R.A., Frausto, J., Rodriguez, G., Romero, D., Cruz, L.: A Statistical Approach for Algorithm Selection. In: Proceedings of III International Workshop on Experimental an Efficient Algorithms, Brazil. LNCS. Springer, Heidelberg (2004)
8. Pérez, J., Pazos, R.A., Frausto, J., Rodriguez, G., Cruz, L.: Self-Tuning Mechanism for Genetic Algorithms Parameters an Application to Data-Object Allocation in the Web. In: Laganá, A., Gavrilova, M.L., Kumar, V., Mun, Y., Tan, C.J.K., Gervasi, O. (eds.) ICCSA 2004. LNCS, vol. 3046. Springer, Heidelberg (2004)
9. Pérez, J., Pazos, R.A., Fraire, H., Cruz, L., Pecero, J.: Adaptive Allocation of Data-Objects in the Web using Neural Networks. LNCS, vol. 2829. Springer, Heidelberg (2003)
10. Pérez, J., Pazos, R.A., Frausto, J., Rodriguez, G., Cruz, L.: Comparison and Selection of Exact and Heuristic Algorithm. In: Laganá, A., Gavrilova, M.L., Kumar, V., Mun, Y., Tan, C.J.K., Gervasi, O. (eds.) ICCSA 2004. LNCS, vol. 3045. Springer, Heidelberg (2004)
11. Pérez, J., Pazos, R.A., Fraire, H., Cruz, L., Santiago, E., García, N.E.: A Machine Learning Ap-proach for Modeling Algorithm Performance Predictors. In: Torra, V., Narukawa, Y. (eds.) MDAI 2004. LNCS (LNAI), vol. 3131. Springer, Heidelberg (2004)
12. Wagner, S., Affenzeller, M.: The HeuristicLab Optimization Environment. Technical Report. Institute of Formal Models and Verification. Johannes Kepler University Linz. Austria (2004)
13. Affenzeller, M.: New Generic Hybrids Based Upon Genetic Algorithms. In: Garijo, F.J., Riquelme, J.-C., Toro, M. (eds.) IBERAMIA 2002. LNCS (LNAI), vol. 2527, pp. 329–339. Springer, Heidelberg (2002)
14. Wagner, S., Affenzeller, M.: HeuristicLab: A Generic and Extensible Optimization Environment. In: Adaptive and Natural Computing Algorithms. Springer Computer Science, pp. 538–541. Springer, Heidelberg (2005)
15. Wagner, S., Affenzeller, M.: HeuristicLab Grid - A Flexible and Extensible Environment for Parallel Heuristic Optimization. In: Proceedings of the 15th International Conference on Systems Science. Oficyna Wydawnicza Politechniki Wroclawskiej, vol. 1, pp. 289–296 (2004)
16. Affenzeller, M.: New Variants of Genetic Algorithms Applied to Problems of Combinatorial Optimization. In: Proceedings of the EMCSR 2002, vol. 1, pp. 75–80 (2002)

17. Reinelt, G.: TSPLIB - A Traveling Salesman Problem Library. ORSA Journal on Computing 3, 376–384 (1991)
18. SPSS, Inc. Headquarters, Chicago, Illinois (2008), http://www.spss.com/es/
19. Witten, I.H., Frank, E.: Data Mining Practical Machine Learning Tools an Techniques. Morgan Kaufmann Publishers, Elsevier (2005)
20. Mitchell, T.M.: Machine Learning. McGraw-Hill, New York (1997)

Classification Method Utilizing Reliably Labeled Data

Kouta Nakata[1], Shigeaki Sakurai[2], and Ryohei Orihara[3]

System Engineering Laboratory, Corporate Research & Developement Center, Toshiba Corporation

Abstract. Making an accurate classifier needs accurate labeling, and accurate labeling needs accurate domain knowledge, experience and criteria, that is, experts to label. In reality, having such experts label all data that we need is often impossible because it requires of the high cost, and sometimes we have to make use of 'cheaper' data labeled by non-experts. In such case, experts' and non-experts' data are not discriminated in learning, even if mislabeled data in non-experts' data may make the resultant classifier poor. In this paper, we propose a classification method utilizing reliably labeled data. We utilize the previous knowledge of how reliable persons have given the labels, and set the degrees of label confidence on non-experts' data based on neighboring reliable experts data. The degrees of confidence are reflected in learning as data with higher confidence make a greater contribution to the classifier. With these assumptions, the results of experiments with publicly available data suggest that our method can make a more precise classifier than the conventional method that adopts all data equally.

1 Introduction

In the field of supervised learning, accurate labeling is essential for accurate learning. Since such labeling requires the domain knowledge, enough experience and the fairness and consistency of judgements, it is desirable to have experts of the domain give labels in order to obtain accurate training data.

But in reality, having such experts for all data is often impossible because of its economic and temporal costs or unlikelihood of being able to call upon the service of capable experts. In view of these practical limitations, training data of experts' labeling are often in short supply for actual problems, and, in that case, we have to make use of non-experts' labeling in order to obtain a sufficient amount of training data. While lower costs and ubiquity of non-experts are favorable for obtaining a large training set, the accuracy of labeling becomes a problem: since non-experts' labeling is not so reliable as experts', it is possible that a large amount of training data contains proportional mislabels. We frequently face a dilemma of fine/small and coarse/large training data.

In traditional learning, such background is not taken into account. each data is dealt with equally as a member of a single training set, no matter whether experts or non-experts are responsible for its labeling. But let us reconsider that

I. Lovrek, R.J. Howlett, and L.C. Jain (Eds.): KES 2008, Part I, LNAI 5177, pp. 114–122, 2008.

data of non-experts' labeling are dominant in terms of quantity. In that case, fine data of experts' labeling would be submerged and likely to have little influence on classifiers. On the other hand, as the ratio of non-experts' labeling becomes large, their accompanying mislabeled data would proportionally make up some part of the training set, and they may make the resultant classifier poor. It is highly likely that data of expensive labeling are not fully utilized.

In this paper, we propose a method utilizing reliably labeled data by discriminating data of experts' and non-experts' labeling. Our purpose is to make classifiers more precise than those generated by the traditional method that adopts the data equally. We take advantage of the previous knowledge of how reliable persons have given the labels, and set the degrees of label confidence on non-experts' data based on neighboring reliable experts' data. The degrees of confidence are reflected in learning, as data with higher confidence make a greater contribution to training of the classifier. Chapter 2 describes our method in detail. Chapter 3 shows the results of experiments with publicly available data and discusses how our method works. We summarize this paper in Chapter 4.

2 Methodology

We propose a method utilizing reliably labeled data. Our first assumption is that experts' labeling is definitive but non-experts' labeling accompanies mislabeling. For simplicity, we call the former "expert data" and the latter "non-expert data".

Our method emphasizes two main points. Firstly, 'credit', the degree of label confidence, is set on each non-expert data. In this paper, credits are calculated in an intuitive way, while various alternatives are potentially possible. We introduce the estimation of credits in the next section. Secondly, the credits of non-expert data are reflected in learning as data with higher credits make a greater contribution to training. As an actual learner, we adopt AdaBoost ([7]) that is favorable for reflecting credits. The latter part of this section describes how credits are reflected in learning.

2.1 Credits on Non-expert Data

In this paper, our experiments calculate credits on non-expert data based on neighboring expert data. This approach has something in common with that of Semi-Supervised Learning (e.g. [10]), but differs especially in our assumption that we have all data already labeled and set credits on unreliably labeled data.

A 2-dimensional illustration is shown in Fig.1, where two different labels are distributed on both sides of a broken line. It is assumed that some non-expert data (squares) are mislabeled and present in the wrong side, while expert data (circles) are correctly labeled and kept accurately on their own sides.

Credit c_j of a non-expert instance $(x_j^{\text{ne}}, y_j^{\text{ne}})$ is estimated by (1).

$$c_j = a \sum_{i=1, y_i^{\text{ex}}=y_j^{\text{ne}}}^{k} \frac{1}{l_i}, \tag{1}$$

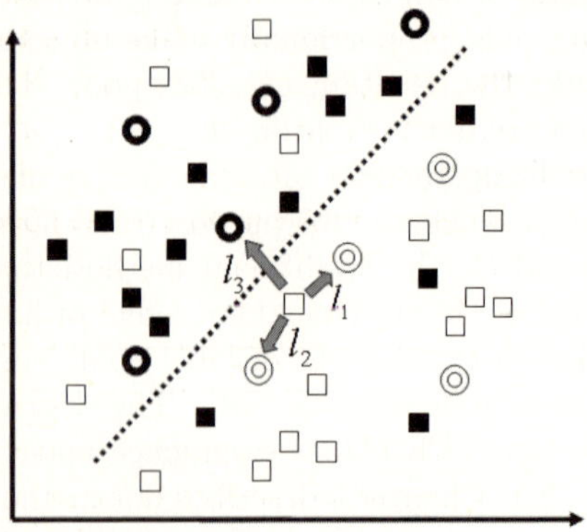

Fig. 1. Illustration of non-expert data and neighboring expert data

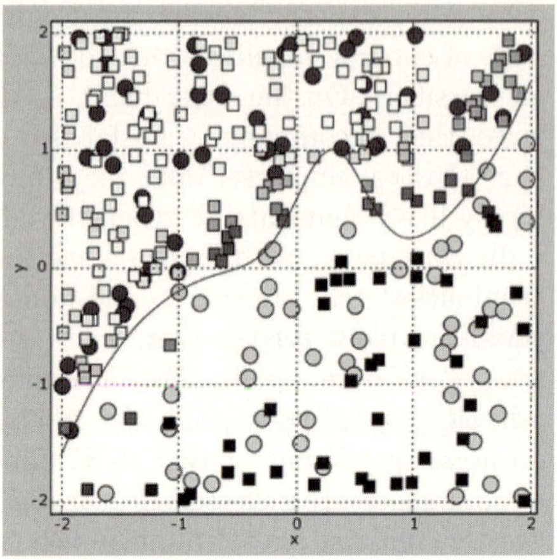

Fig. 2. Distribution of Credits (artificial, $\beta = 20\%$)

where x represents the feature vector, y represents the label, 'ex' represents 'experts', and $a = 1/\sum_{i=1}^{k} \frac{1}{l_i}$ is a normalization factor.

Each non-expert data refers to the k-nearest neighbors of expert data ($k = 3$ in Fig.1). It can take credits that are inversely proportional to the distance as the nearer instance gives the more credit, but can take only from the expert data of the same label. In Fig.1, the target non-expert instance takes credits from the experts 1 and 2, but fails to take it from the expert 3 whose label is not the same.

Our experiments define the credits of expert data as 1, and all of expert data contribute equally to training classifiers. An instance of non-expert data obtains the highest credit of $c_j = 1$ if all of the neighboring expert data agree, and it contributes to training as much as expert data. To the contrary, if none of the neighboring expert data agrees, it obtains no credit of $c_j = 0$ and becomes equivalent to being excluded from training set. The numerical magnitude of $c_j = [0, 1]$ indicates how much it affects learning.

For l_i in (1), our experiments use Euclidean distance for continuous features and Hamming distance for categorical features. Values in continuous features are normalized in a standard way where the average is set to 0 and the variance is set to 1.

2.2 Credit-Sensitive Learning

We reflect the credits of non-expert data in learning by extending the algorithm of AdaBoost ([7]). One reason for adopting AdaBoost is its high performance for various problems. Furthermore its compatibility with cost-sensitive learning is also favorable for our experiments ([6], [9]). The method of cost-sensitive learning can be applied to our 'credit-sensitive' learning as we accordingly incorporate credits of non-expert data into weights in the AdaBoost algorithm.

Algorithm 1 shows the learning algorithm. Our only extension is step 2(a), where credits are incorporated into the data weights. As shown in the previous work on cost-sensitive learning, the way of altering weights varies and the best way seems still open to question ([8]). Thus, we limit ourselves to making the simplest modification by multiplying weights by credits as $D_t^{'}(i) = c_i D_t(i)$. New weight $D_t^{'}(i)$ is incorporated in generating weak learners in the next step. Data with smaller c_i is obviously utilized less for learning however large the value of the original weight $D_t(i)$ becomes. As the weak learner, decision tree C4.5, a proven classifier for AdaBoost, is adopted.

Algorithm1. Credit-Sensitive AdaBoost

1. Start with weights $D_1(i) = 1/N, i = 1, 2, ..., N$.
2. Repeat for $t = 1, 2, ...,$:
   (a) Incorporate credits c_i into weights $D_t(i)$ and obtain new weights $D_t^{'}(i)$.
   (b) Train weak learner $h_t(x)$ using weights $D_t^{'}(i)$ on the training data.
   (c) Set $\beta_t = \log \frac{\epsilon_t}{1-\epsilon_t}$, where ϵ_t is error function, $\epsilon_t = \sum_{y_i \neq h_t(x_i)} D_t(i)$.
   (d) Set $D_{t+1}(i) = D_t(i) \exp(\beta_t)$, $i = 1, 2, ..., N$, when $y_i \neq h_t(x_i)$, and renormalize weights as $\sum_i D_t(i) = 1$.
3. Output the final classifier: $sign(\sum_{t=1}^{T} \beta_t h_t(x))$

3 Experiments

In this section, performance of our method is evaluated by the experiments with artificial and publicly available data. We simulate the sets of expert and non-expert data by dividing the original data. We briefly illustrate the distribution of credits with 2-dimensional artificial data, and then show the results of accuracy experiments. We discuss the improvement of our method at the end of the section.

3.1 Datasets

Our experiments use one artificial dataset and seven UCI datasets ([1]). Properties of the datasets are summarized in Table 1. The dataset of *artificial* is made as follows. We prepare the 2-dimensional field $(-2 < x < 2, -2 < y < 2)$ and the border $f(x) = 0.2x^3 + \exp(-6.0(x - 0.3)^2)$. Random 1000 points are generated on the field and each of them is labeled '0' if it falls on the area above the border $(y >= f(x))$, and labeled '1' otherwise.

Obviously, these 'ready-made' datasets have little or no information about the persons or situations of labeling. Thus, we have to simulate the sets of expert and non-expert data by dividing the original data into subsets. First, we keep α% of the original data as training data and use the rest as test data as in the traditional validation process. The training set is again divided into simulated expert and non-expert data with the ratio of β% and $(100 - \beta)$%, respectively. Mislabeling is simulated by randomly flipping the labels of γ% of non-expert data. Our experiments uses $\alpha = 90\%$ (10-fold cross-validation) and $\gamma = 20\%$ hereafter.

Table 1. Properties of the Datasets

	Dataset	Cases	Feature		Class	Err. of Original		
			Numerical	Categorical		C4.5	AdaBoost	Bagging
1	artifcial	1000	2	-	2	5.2	2.4	3.8
2	votes	435	-	16	2	4.8	5.5	4.3
3	breast-cancer	683	10	-	2	5.9	3.4	3.8
4	ionosphere	351	34	-	2	12.7	6.7	8.7
5	tic-tac-toe	958	-	9	2	11.6	1.2	3.6
6	mushroom	5644	-	22	2	0.6	0.2	0.3
7	hypothyroid	2000	6	18	2	2.4	1.9	1.9
8	kr-vs-kp(chess)	3196	-	36	2	4.6	1.8	3.9

The experiments have assumed two criteria of "binary labels" and "high classification accuracy" in choosing datasets. The former is to simplify the way of simulating mislabeled data. We have only to flip the labels without considering which label the original should be altered to. The latter is essential to our assumption that expert data are given correct labels. High accuracy confirms that consistent labeling has been done to the dataset, and also to the expert data that is its pure subset. We examine the second criterion by the performances of C4.5 and its boosted and bagged classifiers. As shown in Table.1, classification errors with AdaBoost and Bagging are estimated $< 10\%$ for all datasets by 10-fold cross-validation.

3.2 Distribution of Credits

We briefly illustrate the distribution of credits before showing the results of accuracy experiments, An example of an *artificial* dataset is shown in Fig.2, where black and gray circles represent expert data of label 0 and label 1, respectively, and squares represent non-expert data of label 0. Non-expert data should be present only in the label 0 region of $y \geq f(x)$, but squares also appear in $y < f(x)$ due to the manual noises. Note that the tones of squares represent the values of credits, not labels.

The distribution simply explains the nature of credits: 1) In region far above $y \geq f(x)$, $c_i = 1$ (white) since the neighboring expert data all agree with the target non-expert instance. 2) In region far below $y < f(x)$, $c_i = 0$ (black) since the expert data all disagree with the non-expert instance. 3) Near the border of $y = f(x)$, $c_i = [0, 1]$ (gray) since some expert data agree and others disagree. Such distribution suggests that trusted non-expert data can enforce the learning while doubtful data are downplayed or eliminated, to make the resultant classifier more precise.

3.3 Accuracy

In this section, we show the results of running our method. For *artificial* data, the test error rates of our boosting model (AdaBoost with Credits, CAB) and the traditional AdaBoost (TAB) are shown in Fig.3. The horizontal axis represents β, the ratio of expert data, and the vertical axis represents the error rate. The results of the conventional Bagging method ([4], TB) and our Bagging model (Bagging with credits, CB) are also presented for comparison, and they are discussed at the end of this section.

The results show the distinct superiority of CAB in its performance. The error of CAB is far lower than that of TAB in all ranges of β, and it decrease more rapidly as β increases. These properties hold true for another dataset of *votes* as shown in Fig.4. Although *votes* are represented with 16 categorical features, CAB works as well as with numerical *artificial* dataset.

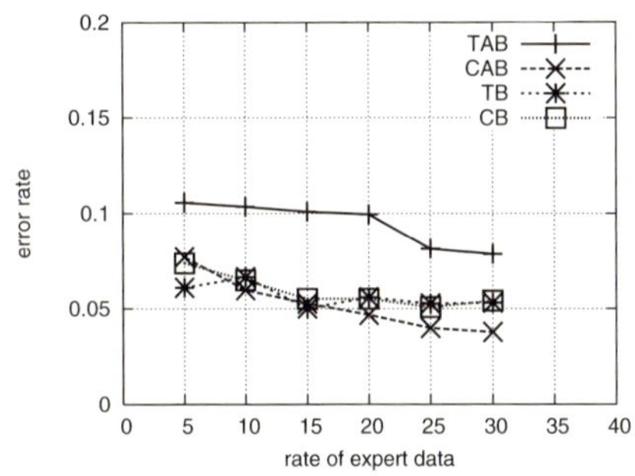

Fig. 3. Error Rates for *artificial*

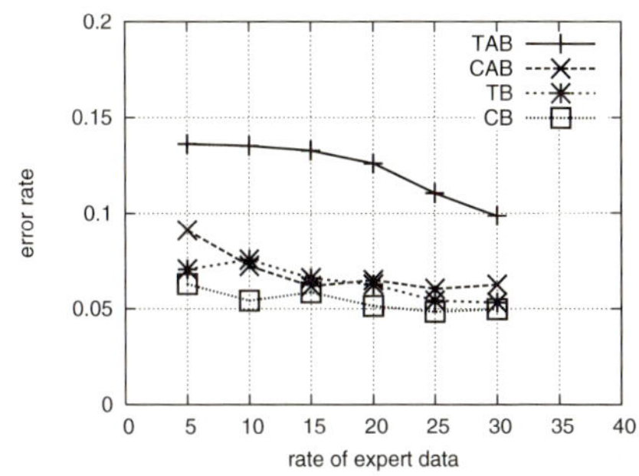

Fig. 4. Error Rates for *votes*

The performances of CAB and TAB are summarized in Fig.5, where black and white bars represent error rates of CAB and TAB at $\beta = 20\%$. The horizontal axis represents each dataset, which is marked with the symbol * if the difference between CAB and TAB is statistically significant. Remarkable improvement with CAB is confirmed for 6 of 8 datasets. As seen in Table.1, these 6 sets vary in their sizes and features, suggesting that CAB can cover a broad range of datasets. To the contrary, the degradation with CAB is also confirmed for *ionosphere* and *tic-tac-toe*. The degradation is not trivial and discussed in the next section.

The performances of CB and TB are shown in Fig.6. Unlike the results of CAB and TAB, CB fails to distinguish itself, only showing performance similar to that of the conventional method of TB. These results suggest that bagging is not sensitive to the credits and not favorable for our method.

3.4 Discussion

CAB distinctively outperforms TAB for 6 datasets. Since these 6 datasets varies in their sizes or features, the results suggest that our method are applicable to datasets with various properties, and also are supportive of our intention to employ proposed method on real-world data that are labeled actually by experts and non-experts.

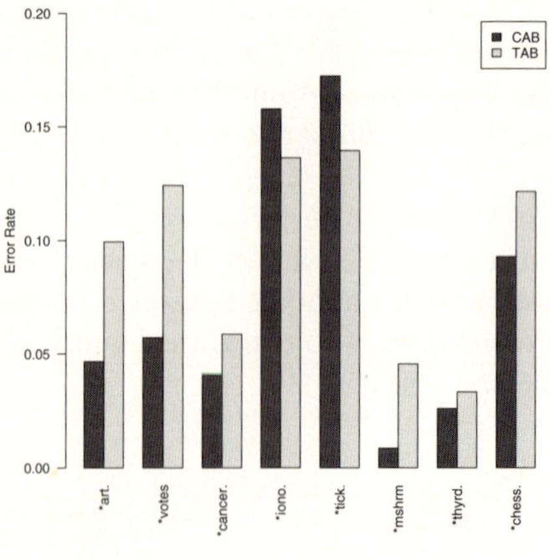

Fig. 5. Comparison of the CAB and TAB performances

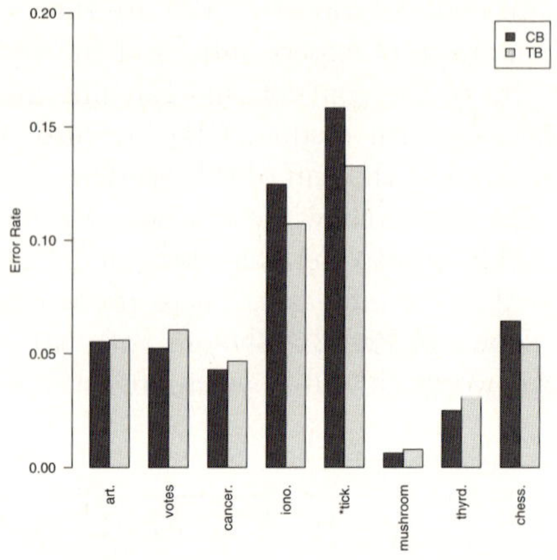

Fig. 6. Comparison of the CB and TB performances

In contrast with the AdaBoost method, there is no significant difference between the performances of CB and TB, suggesting that credits are hardly reflected in the learning in the Bagging algorithm. This may be due to the nature of Bagging that repeatedly samples the instances with replacement. For example, consider that some mislabeled instances are not sampled into a training set by the Bagging algorithm, and such sampling may be equivalent to giving $c_i = 0$ to the instances. If the number of weak learners is sufficiently large, bootstrap sampling of Bagging may counteract the effect of credits.

It is remarkable that CAB is significantly inferior to TAB for *ionosphere* and *tic-tac-toe*. We suppose that these degrations are due to the specialty of their features: the properties of the features prevent the accurate estimation of distance, the nearest neighbors based on the distance are also supposed to be inaccurate, and the resultant credits and classifiers are not reliable. We futher examine this presumption by manipulating features of both datasets.

The *ionosphere* dataset is studied well in the field of feature selection or reduction. The previous works suggest that *ionosphere* contains many irrelevant features, and k-NN classifiers that are sensitive to irrelevant features often suffer severe degradation due to them ([3], [2]). These irrelevant features may harm the estimation of credits where distance is used in a similar way to the k-NN method. In order for accurate estimation of distance, we choose 3 features (feature 3, 5 and 14) that seem to be effective on k-NN classification based on the work of [5]. The feature subset uses only 3 of 34 features, but a small set containing feature 3 and 5 is known to work better for some classification methods.

The *tic-tac-toe* dataset describes the end condition of the board game. The player 'x' is assumed to play first. The labels are binary: 'positive' is given when the player 'x' wins and 'negative' otherwise. As illustrated in Fig.7, The feature x_i represents the player that takes the cell i: $x_i = x$ if it is taken by the player 'x', $x_i = o$ if it is taken by the player 'o', and $x_i = b$ if it is a blank. In the

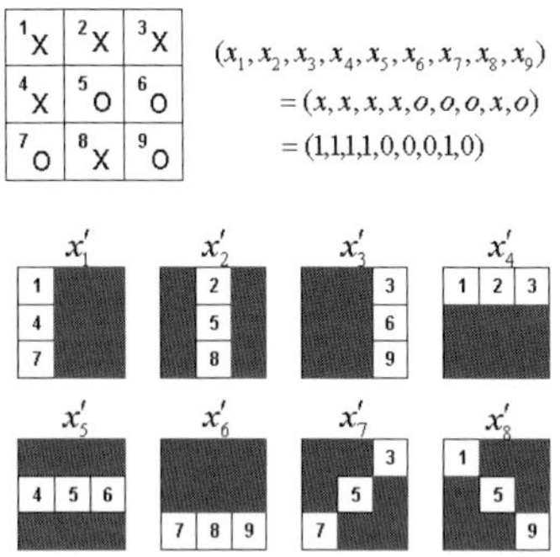

Fig. 7. Illustration of 'cell' and 'three-in-a-row' features

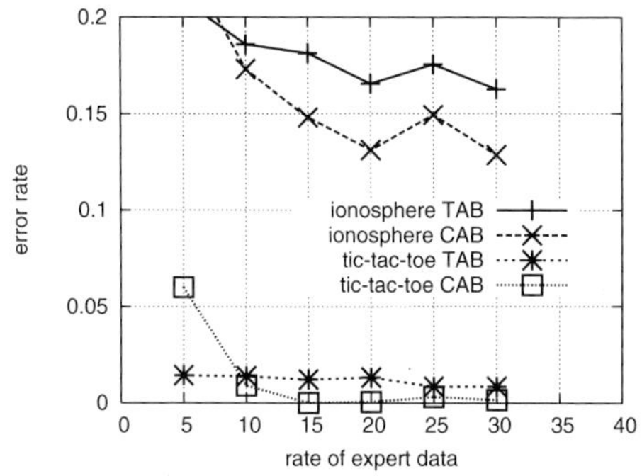

Fig. 8. Comparison of the CAB and TAB performances (ionosphere and tic-tac-toe)

calculation of Euclidean distance, $x_i = x$ is converted into $x_i = 1$ and $x_i = o, b$ into $x_i = 0$.

These 'cell' features are not favorable for estimation of the distance between games. For example, let us consider two instances of $X_1 = (x, x, x, x, o, o, o, x, o)$ and $X_2 = (o, x, x, x, o, o, b, x, o)$. Two games look much different since the player 'x' wins in $X1$ and 'o' wins in $X2$, but they are the nearest neighbours in terms of Euclidean distance. In order to avoid such discrepancy, we introduce 8 'three-in-a-row' features $(x'_1, ..., x'_8)$ that are specialized for *tic-tac-toe*. The illustration is shown in Fig.7. Each feature is square of sum of the value in white cells, as $x'_1 = (x_1 + x_4 + x_7)^2$, for instance. X_1 and X_2 are converted into $X'_1 = (4, 4, 1, 9, 1, 1, 1, 1)$ and $X'_2 = (1, 4, 1, 4, 1, 1, 1, 0)$, respectively. The large distance between X'_1 and X'_2 seems to reflect the difference of two games well.

Our method is re-evaluated with the improvements of feature subset (*ionosphere*) and three-in-a-row features (*tic-tac-toe*). The results of CAB and TAB are shown in Fig.8. In both datasets, CAB is stably superior to TAB when $\beta \geq 10\%$, showing no degradation as the original features might cause. The results show that accurate estimation of distance is essential to our method, as it leads to setting accurate credits on non-expert data and finally to producing superior classifiers.

4 Conclusion

We have proposed a classification method utilizing reliably labeled data, with the assumption that small training data of experts' labeling and large training data of non-experts' labeling are both obtained. We set credits, the degrees of label confidences, on non-expert data based on the neighboring expert data. The credits of non-expert data are reflected in learning as data with higher credits make a greater contribution to training.

The experiments with artificial and UCI datasets have shown that the boosting model of our method performs better than the conventional method that adopts all training data equally. They have also suggested that accurate estimation of distance leads to better classifiers. These results indicate the possibility that our method is applicable to datasets with various properties. In future work, we intend to apply proposed method to the real-world data that are actually labeled by experts and non-experts.

References

1. Newman, D.J., Asuncion, A.: UCI machine learning repository (2007)
2. Akkus, A., Guvenir, H.A.: K nearest neighbor classification on feature projections. In: Proc. 13th International Conf. on Machine Learning, pp. 12–19 (1996)
3. Bay, S.D.: Combining nearest neighbor classifiers through multiple feature subsets. In: Proc. 15th International Conf. on Machine Learning, pp. 37–45 (1998)
4. Breiman, L.: Bagging predictors. Machine Learning 24(2), 123–140 (1996)
5. Dy, J.G., Brodley, C.E.: Feature selection for unsupervised learning. J. Mach. Learn. Res. 5, 845–889 (2004)
6. Fan, W., Stolfo, S.J., Zhang, J., Chan, P.K.: AdaCost: misclassification cost-sensitive boosting. In: Proc. 16th International Conf. on Machine Learning, pp. 97–105 (1999)
7. Freund, Y., Schapire, R.E.: Experiments with a new boosting algorithm. In: Proc. 13th International Conf. on Machine Learning, pp. 148–156 (1996)
8. Masnadi-Shirazi, H., Vasconcelos, N.: Asymmetric boosting. In: Proc. 24th International Conf. on Machine Learning, pp. 609–619 (2007)
9. Ting, K.M.: A comparative study of cost-sensitive boosting algorithms. In: Proc. 17th International Conf. on Machine Learning, pp. 983–990 (2000)
10. Wang, F., Zhang, C., Shen, H.C., Wang, J.: Semi-supervised classification using linear neighborhood propagation. CVPR 1, 160–167 (2006)

A Manifolded AdaBoost for Face Recognition

C. Lu[1,*], J. Jiang[2], G. Feng[1], and C. Qing[2]

[1] School of Mathematics and Computational Science,
Sun Yat-Sen University, China, 510275
[2] School of Informatics, University of Bradford, BD7 1DP, UK
luchuny@mail2.sysu.edu.cn, j.jiang1@bradford.ac.uk,
mcsfgc@mail.sysu.edu.cn, c.qing@bradford.ac.uk

Abstract. Manifold learning is an effective dimension reduction method to extract nonlinear structures from high dimensional data. Recently, manifold learning started to attract attention within the research communities of image analysis, computer vision, and document data analysis. In this paper, we propose a Manifolded AdaBoost algorithm towards automatic 2D face recognition by using AdaBoost to fold the manifold space dimension and exploit the strength of both techniques. Experimental results support that the proposed algorithm improve over existing benchmarks in terms of stability and recognition precision rates.

Keywords: manifold learning, face recognition, nonlinear dimension reduction, Manifolded AdaBoost.

1 Introduction

As one of the mostly targeted applications in machine learning, face recognition has recently received significant attention, especially during the past two decades. A great amount of face recognition algorithms have been proposed, reported, and detailed surveys of the developments in this area have been published in the literature [1]. The present work can be summarized into three major categories, including Holistic matching methods, Feature-based (structural) matching methods and Hybrid methods. The most well-known methods are based on Eigenfaces [2] and Fisherfaces [3]. Both of them can be classified as holistic matching techniques. These approaches generally operate directly on the whole facial region. In this way, the two-dimensional (2D) recognition system can avoid difficulties of 3D face data acquisition, expensive computation, and sensitiveness to strong lighting sources or to reflective surfaces. Nowadays the sensitivity to illumination expressions and pose variations are the main problems researchers have to address in their new attempts, and recent publication in the literature started to show this trend in addressing some of these problems [23].

Most of the existing techniques use illumination normalization as a pre-processing method for illumination invariant face recognition and assume that the face images lie on a linearly embedded manifold to preserve the global Euclidean structure of the

* Corresponding author.

I. Lovrek, R.J. Howlett, and L.C. Jain (Eds.): KES 2008, Part I, LNAI 5177, pp. 123–130, 2008.

image space. However, a lot of researches have shown that facial images possibly lie on a nonlinear sub-manifold [4], [5], [6].

Since 2000, manifold learning attracted more and more attention in signal processing since it is an effective dimension reduction methodology for extracting nonlinear structures from high dimensional data. Now it has become increasingly more popular technique in the area of machine learning. Manifold learning is generally divided into global and local embedding algorithms. ISOMAP [4], as a global algorithm, presumes that isometric properties should be preserved in both the observation space and the intrinsic embedding space. On the other hand, Locally Linear Embedding (LLE) [5] and Laplacian Eigenamps [6] focus on the preservation of local neighboring structure. Recently, manifold learning starts to be applied in image analysis, computer vision, and documents data analysis [10], [11], [12].

As one of the most important machine learning algorithms, AdaBoost, short for Adaptive Boosting, formulated by Yoav Freund and Robert Schapire [13], is an algorithm for constructing a "strong" classifier as linear combination of simple weak classifiers. It can be used in conjunction with many other learning algorithms to improve their performance and solve many complicated learning problems. At present, AdaBoost has been widely used for feature selection [16] and classifier learning, examples of which include object detection [15] and face recognition [14]. AdaBoost learning can minimizes the upper bound on both training and generalization errors [24]. In our paper, we folded the manifold space dimension by using AdaBoost to select discriminating projection directions.

The rest of this paper is structured into three further sections, where we describe the background and the Manifolded AdaBoost algorithms in section 2, and we present the experimental results and the analysis in Section 3. Finally, we give the conclusions in Section 4.

2 Manifolded AdaBoost

People working with high-dimensional data regularly confront the problem of dimension reduction without losing valuable information. When the data of interest lies on an embedded nonlinear manifold, which is part of the higher dimensional space, we need nonlinear dimensionality reduction for its effective and efficient processing. Since manifold learning algorithms such as ISOMAP, LLE and Laplacian eigenmaps operate in a batch mode, i.e., when new data arrives, the entire algorithms need to be rerun with the original data augmented by the new arrivals, they are not suitable for input data that sequentially arrives at their inputs. Linear form manifold learning has been proposed to solve this problem for application problems.

Denote the sample set as $X = \left[\vec{x_1}, \vec{x_2}, \ldots \vec{x_N}\right], x_i \in R^D$, W be an affinity matrix, that is, the N-dimensional matrix with the (i,j)-th element W_{ij} being the affinity between x_i and x_j. ω is the projection matrix. For linear form manifold learning, the nonlinear structure preserving criterion as follows:

$$\min_{X^T \omega B \omega^T X = b} \sum_{i \neq j} \left(\omega^T x_i - \omega^T x_j\right)^2 W_{ij} \tag{1}$$

Essentially, the minimization problem is equivalent to finding:

$$\underset{X^T \omega B \omega^T X = b}{\arg\min} \ \omega^T XLX^T \omega \tag{2}$$

where b is a constant and B is the constraint matrix defined to avoid a trivial solution of the objective function. L is the Laplacian matrix defined as $L = S - W$, where S is the N-dimensional diagonal matrix with the i-th diagonal element being: $S_{ii} = \sum_j W_{ij}$.

The objective function of manifold learning is an attempt to ensure that it looks for a projection matrix ω such that the nonlinear data structure can be preserved. (2) can be solved by a generalized eigenvalues decomposition problem. The columns of transform matrix span a face space defined by manifold learning.

Linear form manifold learning has many advantages for face recognition, which include: (i) It operates much faster than the implicit expression manifold learning algorithms; (ii) When we get the explicit expression of the manifold learning algorithm, we find that linear form manifold is also a incremental form manifold; (iii) Linear form manifold is easier to be combined with other classification methods or to be modified for face recognition. Many other research areas are also benefited from the linear form manifold, but almost all of them change the affinity matrix W by considering the discriminative information as a constraint condition of manifold learning, including Neighborhood Preserving Embedding (NPE) [9] which we used in this paper. Manifolded AdaBoost is based on linear form manifold learning.

Following the above analysis and discussion, we introduce AdaBoost into the linear form manifold learning to help selecting transformation matrix ω and projecting the high-dimensional raw data into low-dimensional ones, and yet maintains the effective characterization of original data samples towards face recognition.

It is known that AdaBoost is originally developed to support binary classification tasks. There are several methods of extended AdaBoost to multiclass cases. The most straightforward generalization, called AdaBoost.M1 [13]. However, this method fails if the weak learner can't achieve at least 50% accuracy when run these hard distributions. For the latter case, several more sophisticated methods have been developed. Representative examples include AdaBoost.MH [17] and AdaBoost.M2 [13], which is a special case of AdaBoost.MR [17].

To improve the performance of an existing face recognition algorithm, we exploit the strength of AdaBoost.M2 that it not only learns from hard-to-classify examples, but also from the incorrect labels, where the mislabeled distribution is introduced to enhance the communication between the learner and the booster. By using AdaBoost as a projection direction selection tool, the most discriminating projection directions can be extracted by selecting only those projection directions that can best discriminate among classes. AdaBoost works under the principle that each of its iterations tracks the minimum error hypothesis to pick up the most discriminative projection direction vector.

By combining AdaBoost with linear form manifold, we are able to select the most discriminating projection and information for classification and pattern recognition. By boosting the most discriminating information from the linear form manifold, the proposed algorithm is able to exploit the strength of both AdaBoost and manifold learning to improve the performance of face recognition.

Let $\{\vec{x_1}, \vec{x_2}, ...\vec{x_N}\}$ be a data set of D-dimensional vectors, which is to be classified into C classes. We use linear form manifold learning dimension reduction algorithm to map $\{\vec{x_1}, \vec{x_2}, ...\vec{x_N}\}$ into a new sample set $Z = \{\vec{z_1}, \vec{z_2}, ...\vec{z_N}\}$ with d-dimensions, where $d << D$. Essentially, linear form manifold learning is to determine a projection matrix $[\omega]_{D\times m}$ to complete the dimension reduction and transformation:

$$X = [\omega]_{D\times m} Z^T \tag{3}$$

where $\omega = (\omega_1, \omega_2, ..., \omega_m)$ is learnt by linear form manifold learning ($d \le m < D$).

However, existing manifold learning assumes $d=m$ and fails to select the best possible projection which is beneficial to pattern recognition and classification. To make an appropriate balance and trade-off between pattern classification and data structure preservation, we use the AdaBoost [25] to eliminate the m-d non-effective projections and select the most effective d-dimensional projection by setting $d<m$.

Following the t-th iteration, specifically, the projection selected can be denoted as: $\{\omega_{j_1}, \omega_{j_2}, ..., \omega_{j_t}; j_i \in \{1, ..., m\}\}$. As a result, the remaining candidates to be selected can be represented as: $\omega_j : j \in \{[1,2,...m] - [j_1, ...j_t]\}$, and the selection process at the $(t+1)$-th iteration is minimize the following formula:

$$\varepsilon_j^{t+1} = \min_j \varepsilon_{t+1} = \frac{1}{2}\sum_{i=1}^{N} D_t(i)\left(1 - h_j(x_i, y_i) + \sum_{y\neq y_i} q_{i,y}^t h_t(x_i, y)\right) \tag{4}$$

where $D_t(i)$ is the sample distribution, $q_{i,y}^t$ is the label weighting function derived at the t-th iteration, and $h(x, y)$ is a weak learner used in AdaBoost [25].

To ensure that the new projection is discriminatingly informative, the following parameter is calculated:

$$\beta_{t+1} = \frac{\varepsilon_{t+1}}{1 - \varepsilon_{t+1}} \tag{5}$$

the $(t+1)$ project is thus selected by:

$$\omega^{t+1} = (\ln\left(\frac{1}{\beta_1}\right)\omega_{j_1}, \ln\left(\frac{1}{\beta_2}\right)\omega_{j_2}, ..., \ln\left(\frac{1}{\beta_{t+1}}\right)\omega_{j_{t+1}}) \tag{6}$$

Therefore, the proposed Manifolded AdaBoost map, which is helped by AdaBoost to boost its discriminating information and classification power, can be summarized as:

$$g_{new} : X \subset R^D \rightarrow Z_{new} \subset R^d \tag{7}$$

where $z_{new,i} = \omega_{new}^T x_i$, $\omega_{new} = \omega^d = (\ln\left(\frac{1}{\beta_1}\right)\omega_{j_1}, \ln\left(\frac{1}{\beta_2}\right)\omega_{j_2}, ..., \ln\left(\frac{1}{\beta_d}\right)\omega_{j_d})$.

A key feature of this algorithm is that it is a task-oriented flexible boosting method, which can be easily extended to solving other application problems. As an example, boosting can be selected to reduce the lighting effect or expression effect in face recognitions.

3 Empirical Studies

In order to evaluate the performance of our Manifolded AdaBoost, we have designed a number of experimental phases and structured this section into three further subsections, where Section 3.1 describes our test data sets, Sections 3.2 evaluate the proposed algorithms in terms of classification accuracy, and finally, we give analysis of the results in Section 3.3.

3.1 Datasets and Experimental Design

Our experiments are performed on two face databases: 1) The Yale database [19]; 2) The YaleB database [20] to compare different algorithms. In the Yale database, 165 frontal face images are included, covering 15 individuals taken under 11 different conditions. Each individual has different facial expressions, illumination conditions and small occlusion. There are 10 individuals under 64 different lighting conditions for 9 poses in the database of the YaleB Database. Since we are only concerned with the illumination problem, frontal face images under varying lighting conditions are primarily used for our experiments.

In order to be consistent with the experiments given in [18], pre-processing is applied to locate the faces. Original images were normalized and cropped. The size of each cropped image is 32×32 pixels.

In addition, we average the results over T random splits to obtain the recognition rates. A random subset with k images per individual was taken with labels to form the training set, and the rest of the database was considered to be the testing set. In the Yale database we set $T=10$, $k=6$, and $T=5$, $k=40$ for the YaleB database.

3.2 Results of Manifolded AdaBoost

In order to evaluate the proposed Manifolded AdaBoost algorithm, we compare the recognition accuracy with three existing algorithms, including Locality Preserving Projections (LPP) [7], [8], Neighborhood Preserving Embedding (NPE) [9] and Principle Component Analysis (PCA) [2] in the experiments. In our paper, we are not focus on pattern classifier and the specificities of our algorithm are not dependent on classifier, so we apply the nearest-neighbor classifier for its simplicity.

The experiments is designed and implemented with three steps, which include: (i) low-dimensional linear form manifold is calculated and extracted from the training face samples by using LPP, NPE and PCA; (ii) the linear form manifold is modified into a folded discriminating manifold by our proposed method; and (iii) testing faces are recognized by using a specified pattern classifier.

Fig. 1. shows the graphs of recognition rates versus the number of projection directions. LPP, NPE and our method are tested on the Yale database. PCA and our method are tested on the YaleB database. Table 1. shows the recognition rates of LPP, NPE and our method versus different projection directions on the Yale database.

It is easy to see that our method is more stable on the Yale database and illumination invariant on the YaleB database.

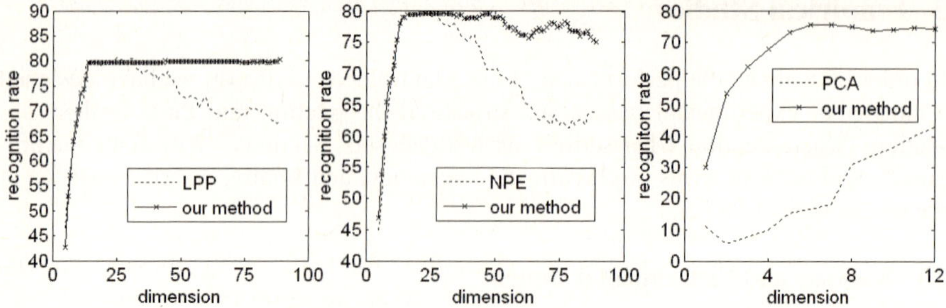

Fig. 1. Recognition rate versus dimension of reduced space

Table 1. Recognition rate versus different dimensions of reduced space on Yale database

dimensions	LPP vs our method	NPE vs our method
80	70.2% vs 79.6%	62.2% vs 76.8%
50	73% vs 79.8%	70.6% vs 79%
20	79.2% vs 79.6%	79.8% vs 79.4%
14	79.8% vs 79.6%	79% vs 79%

Table 1., the left and middle graphs clearly show that our method improve over LPP and NPE, and when the dimension rise to *C*-1 (*C* is the number of classes) or large than *C*-1, the recognition rate nearly reaches the maximum by our method and the recognition rate is very stable. But other methods are unstable. And moreover, it is obvious that when the dimension rises to *C*-1, the recognition rate reaches the maximum. We can determine the number of classes by our method.

In addition, the right graph in Fig. 1. illustrates that, while PCA is sensitive to illumination effects since YaleB database images we used are under varying lighting conditions, our method is almost illumination invariant. So the average recognition rate of our method is greatly higher than the average recognition rate of PCA.

3.3 Analysis of the Results

The main observations from the performance comparisons can be made as follows:

(i) Recognition performance will suffer from insufficient information if dimension is underestimated. On the other hand, an over-estimate of dimension will introduce noisy components which also reduce its performances [21]. The number of dimensions in the face space affects the speed and accuracy in processing large face databases. Studies in Dimensionality of Face Space [21] reveals that appropriate dimension of face space is in the range of 100-200 eigen-features, on average. With manifold learning strategy, however, we can reduce the dimension of face space to nearly *C*-1 without compromising the quality of its face recognition performances. This means that we can project images to a very small subspace (usually smaller than the intrinsic dimension of the face image but bigger than *C*-1) and retain the most discriminative information.

(ii) From our experimental result on YaleB database, it can be seen that the proposed algorithm is still effective for face recognition with variable lighting effects. In comparison with PCA, LPP and NPE, it clearly shows significant advantages in terms of the stability over the recognition performances.
(iii) The analysis and discuss of our method indicate that the proposed Manifolded AdaBoost retains the structure of input data well.
(iv) The proposed Manifolded AdaBoost method is linear and defined on all the training and testing samples, which makes it suitable for practical classification problems. Similar to KPCA [22] generalized from PCA, the proposed algorithm can also be generalized into a kernel Manifolded AdaBoost algorithm.

4 Conclusions

In this paper, we proposed a Manifolded AdaBoost, which exploits the strength of both manifold learning and AdaBoost for face recognition. While manifold learning provides a good framework for representing data in lower dimensions, combination of AdaBoost enables its projection towards more discriminating information and power that is useful for face recognition. Extensive experiments show that the proposed algorithm improves over relevant algorithms including PCA, LPP and NPE. This proves the concept that discriminative information selection process can improve the performance of the linear form manifold learning for face recognition.

Further research can be identified to include: (i) by focusing on the image classification space, new manifold learning classification algorithms can be developed to improve its performances in face recognition; (ii) Global and local features can be integrated and considered together to quantify their contributions to classifications, and hence better balance can be achieved to develop new algorithms for further improvement; and (iii) the reduced feature set could be further enabled by manifold learning to improve its processing speed and thus new algorithms can be developed that are suitable for real-time classification and recognition.

Acknowledgement. This work is partially supported by the National Natural Science Foundation in China under grant No. 60675007 and the Education Ministry of China under key grant No. 104145.

References

1. Zhao, W., Chellappa, R., Rosenfeld, A., Phillips, P.J.: Face recognition: a literature survey. ACM Comput. Surv. 35(4), 399–458 (2003)
2. Turk, M., Pentland, A.: Eigenfaces for recognition. Journal of Cognitive Neuroscience 3(1), 71–86 (1991)
3. Belhumeur, P.N., Hepanha, J.P., Kriegman, D.J.: Eigenfaces vs. fisherfaces: recognition using class specific linear projection. IEEE. Trans. PAMI 19(7), 711–720 (1997)
4. Tenenbaum, J.B., Silva, V., Langford, J.C.: A Global Geometric Framework for Nonlinear Dimensionality Reduction. Science 290, 2319–2323 (2000)
5. Roweis, S.T., Saul, L.K.: Nonlinear Dimensionality Reduction by Locally Linear Embedding. Science 290, 2323–2326 (2000)

6. Belkin, M., Niyogi, P.: Laplacian Eigenmaps and Spectral Techniques for Embedding and Clustering. Advances in NIPS 15, Vancouver, British Columbia, Canada (2001)
7. He, X., Niyogi, P.: Locality Preserving Projections. Advances in NIPS Vancouver, Canada (2003)
8. He, X., Yan, S., Hu, Y., Niyogi, P., Zhang, H.J.: Face Recognition using Laplacianfaces. IEEE. Trans. PAMI 27(3), 328–340 (2005)
9. He, X., Cai, D., Yan, S., Zhang, H.J.: Neighborhood preserving embedding. In: Proceedings of the Tenth IEEE International Conference on Computer Vision, pp. 1208–1213 (2005)
10. Chang, H., Yeung, D.Y., Xiong, Y.: Super-resolution through neighbor embedding. In: Proc. of CVPR 2004, pp. 275–282 (2004)
11. Elgammal, A., Lee, C.S.: Inferring 3D Body Pose from Silhouettes using Activity manifold learning. In: IEEE Computer Society Conference on CVPR, pp. 681–688 (2004)
12. Lebanon, G.: Metric learning for text documents. IEEE Trans. PAMI 28, 497–508 (2006)
13. Freund, Y., Schapire, R.E.: A decision-theoretic generalization of online learning and an application to boosting. J. Comp. & Sys. Sci. 55(1), 119–139 (1997)
14. Viola, P., Jones, M.J.: Robust real-time face detection. Int. J. Comput. Vis. 57, 137–154 (2004)
15. Viola, P., Jones, M.J.: Rapid object detection using a boosted cascade of simple features. In: Proc. IEEE Conf. on CVPR, pp. 511–518 (2001)
16. Silapachote, P., Karuppiah, D.R., Hanson, A.R.: Feature Selection Using Adaboost For Face Expression Recognition. In: Proceedings of the Fourth IASTED International Conference on Visualization, Imaging, and Image Processing, pp. 84–89 (2004)
17. Schapire, R.E., Singer, Y.: Improved boosting algorithms using confidence-rated predictions. In: Proceedings of the Eleventh Annual Conference on Computational Learning Theory, pp. 80–91 (1998)
18. Cai, D., He, X., Han, J.: Using Graph Model for Face Analysis. Technical Report, UIUCDCS-R-2005-2636, UIUC (2005)
19. Yale University, `http://cvc.yale.edu/projects/yalefaces/yalefaces.html`
20. Yale University, `http://cvc.yale.edu/projects/yalefacesB/yalefacesB.html`
21. Meytlis, M., Sirovich, L.: On the Dimensionality of Face Space. IEEE Trans. PAMI 29(7), 1262–1267 (2007)
22. Scholkopf, B., Smola, A., Muller, K.-R.: Kernel principal component analysis. In: Scholkopf, B., Burges, C., Smola, A. (eds.) Advances in Kernel Methods— Support Vector Learning, pp. 327–352. MIT Press, Cambridge (1999)
23. Liu, D., Lam, K.M., Shen, L.S.: Illumination invariant face recognition. Pattern Recognition 38(10), 1705–1716 (2005)
24. Freund, Y., Schapire, R.: A short introduction to boosting. Journal of Japanese Society for Artificial Intelligence 14(5), 771–780 (1999)
25. Qing, C., Jiang, J.: Recognition of jpeg compressed face images based on adaboost. In: The 2nd Int. Conf. on Semantics And digital Media Technologies (SAMT), pp. 272–275 (2007)

On Lipschitz Embeddings of Graphs

Kaspar Riesen and Horst Bunke

Institute of Computer Science and Applied Mathematics, University of Bern,
Neubrückstrasse 10, CH-3012 Bern, Switzerland
{riesen,bunke}@iam.unibe.ch

Abstract. In pattern recognition and related fields, graph based representations offer a versatile alternative to the widely used feature vectors. Therefore, an emerging trend of representing objects by graphs can be observed. This trend is intensified by the development of novel approaches in graph based machine learning, such as graph kernels or graph embedding techniques. These procedures overcome a major drawback of graphs, which consists in a serious lack of algorithms for classification and clustering. The present paper is inspired by the idea of representing graphs by means of dissimilarities and extends previous work to the more general setting of Lipschitz embeddings. In an experimental evaluation we empirically confirm that classifiers relying on the original graph distances can be outperformed by a classification system using the Lipschitz embedded graphs.

1 Introduction

Representing objects or patterns by feature vectors $\mathbf{x} \in \mathbb{R}^n$ offers a number of useful properties. In particular, due to the mathematical wealth of operations available in a vector space, a huge amount of algorithms for classification, clustering, and analysis of objects given in terms of feature vectors have been developed in recent years [1].

Yet, the use of feature vectors implicates two limitations. First, as vectors always represent a predefined set of features, all vectors in a particular application have to preserve the same length regardless of the size or complexity of the corresponding objects. Furthermore, there is no direct possibility to describe binary relationships among different parts of an object. It is well known that both constraints can be overcome by graph based representations [2]. As a matter of fact, graphs allow us to adapt their size to the size and complexity of the underlying object. Futhermore, graphs offer a convenient possibility to describe structural relationships among different parts of an object.

A major drawback of graphs, however, is that they offer little mathematical structure, i.e. most of the basic mathematical operations are not available or not defined in a standardized way for graphs. Examples are the sum, product, average, etc. of a set of graphs. Nevertheless, since the concept of kernel machines has been extended from vectors to symbolic data structures, and in particular to graphs [3], this drawback can be overcome. The key idea of graph kernels is

I. Lovrek, R.J. Howlett, and L.C. Jain (Eds.): KES 2008, Part I, LNAI 5177, pp. 131–140, 2008.

that rather than defining handcrafted mathematical operations in the original graph domain, the graphs are implicitly mapped into a vector space where all those operations are readily available. Obviously, by means of graph kernels one can benefit from both the high representational power and flexibility of graphs and the wide range of pattern recognition algorithms for feature vectors.

The present paper is closely related to graph kernels. However, rather than defining a kernel function on graphs, mapping pairs of graphs implicitly to dot products in a vector space, the graphs are individually embedded in a vector space. To this end we apply Lipschitz embeddings to graphs [4]. The basic idea of Lipschitz embeddings is that an object o is transformed into an n-dimensional vector $\mathbf{x} = (x_1, \ldots, x_n)$ such that each of the components x_i corresponds to the distance of o to a predefined reference set. The intuition behind such an embedding is that important information of an object is captured by its distances to the reference sets. Moreover, and this is our major motivation, graphs are transformed explicitly into a vector of real numbers. Hence, sophisticated algorithms for classification, originally developed for feature vectors, become applicable to graphs.

The present paper is a successor of recent work [5], where singleton reference sets, i.e. reference sets with a sole member, are used for graph embedding. Note that the methodology presented in [5] is crucially inspired by the work about dissimilarity representation [6]. Obviously, this is a special case of a Lipschitz embedding. However, in the present paper we extend the approach in [5] towards general Lipschitz embeddings. By using a set of subsets rather than singletons, we increase the likelihood that the original distance between two graphs is captured adequately by the distance in the embedding space [7]. In the experimental evaluation we emprically confirm that this extension leads to further improvements compared to the previous approach [5] of the classification accuracy on five different graph sets of quite diverse nature.

2 Graphs and Graph Edit Distance

Generally, a graph g is given by a finite set of nodes V, a finite set of edges E, and their corresponding labeling functions. Let L_V and L_E be finite or infinite sets of labels for nodes and edges, respectively.

Definition 1 (Graph). *A graph g is defined by the four-tuple $g = (V, E, \mu, \nu)$, where V is the finite set of nodes, $E \subseteq V \times V$ is the set of edges, $\mu : V \rightarrow L_V$ is the node labeling function, and $\nu : E \rightarrow L_E$ is the edge labeling function.*

The definition given above allows us to handle arbitrary graphs with unconstrained labeling functions. For example, the label alphabet can be given by the set of integers, the vector space $\mathbb{R}^n$, or a set of symbolic labels. Moreover, unlabeled graphs can be handled by assigning the same label l to all nodes and edges. Edges are defined by pairs of nodes (u, v), where $u \in V$ denotes the source node and $v \in V$ the target node of a directed edge. Undirected graphs can be modeled by inserting a reverse edge $(v, u) \in E$ for each edge $(u, v) \in E$ with $\nu(u, v) = \nu(v, u)$.

In order to apply Lipschitz embeddings to graphs (further details will be presented in Section 3), a distance model for graphs has to be introduced. In recent years a number of graph matching techniques have been introduced in the literature [2]. One of the most flexible methods for error-tolerant graph matching is based on the edit distance of graphs [8]. The key idea of graph edit distance is to define the dissimilarity, or distance, of graphs by the minimum amount of distortion that is needed to transform one graph into another. A standard set of distortion operations is given by *insertions*, *deletions*, and *substitutions* of nodes and edges.

Given two graphs, the source graph g_1 and the target graph g_2, the idea of graph edit distance is to delete some nodes and edges from g_1, relabel (substitute) some of the remaining nodes and edges, and insert some nodes and edges in g_2, such that g_1 is finally transformed into g_2. A sequence of edit operations $e_1, \ldots, e_k$ that transform g_1 into g_2 is called an *edit path* between g_1 and g_2. Obviously, for every pair of graphs (g_1, g_2), there exist a number of different edit paths transforming g_1 into g_2. Let $\Upsilon(g_1, g_2)$ denote the set of all such edit paths. To find the most suitable edit path out of $\Upsilon(g_1, g_2)$, one introduces a cost for each edit operation, measuring the strength of the corresponding operation. The idea of such cost functions is to define whether or not an edit operation represents a strong modification of the graph. Hence, between two similar graphs, there should exist an inexpensive edit path, representing low cost operations, while for different graphs an edit path with high costs is needed. Consequently, the *edit distance* of two graphs is defined by the minimum cost edit path between two graphs.

Definition 2 (Graph Edit Distance). *Assume that a graph domain $\mathcal{G}$ is given. Let $g_1 = (V_1, E_1, \mu_1, \nu_1) \in \mathcal{G}$ be the source graph and $g_2 = (V_2, E_2, \mu_2, \nu_2) \in \mathcal{G}$ be the target graph. The graph edit distance between g_1 and g_2 is a mapping $d : \mathcal{G} \times \mathcal{G} \to \mathbb{R}$ defined by*

$$d(g_1, g_2) = \min_{(e_1, \ldots, e_k) \in \Upsilon(g_1, g_2)} \sum_{i=1}^{k} c(e_i),$$

where $\Upsilon(g_1, g_2)$ denotes the set of edit paths transforming g_1 into g_2, and c denotes the edit cost function measuring the strength $c(e_i)$ of edit operation e_i.

The edit distance of graphs can be computed, for example, by a tree search algorithm [8] or by faster, suboptimal methods which have been proposed recently [9]. Note that graph edit distance, in particular when computed with suboptimal methods, does not guarantee to be metric in general. However, any non-metric distance matrix can be postprocessed such that it becomes metric [6].

3 Lipschitz Embeddings of Graphs

3.1 Definition and Basic Properties

Originally, Lipschitz embeddings [4] were proposed to embed metric spaces into other ones with low distortion. In [7], for instance, Lipschitz embeddings were used for similarity searching in metric spaces. In the context of the present paper, however, the paradigm of Lipschitz embedding is adopted in order to transform a graph domain $\mathcal{G}$ into a vector space $\mathbb{R}^n$.

Let us assume that a set of graphs $\mathcal{G} = \{g_1, \dots, g_N\}$ is given. A Lipschitz embedding defines a vector space $\mathbb{R}^n$ such that each dimension corresponds to a reference set of graphs drawn from $\mathcal{G}$. Formally, we first define a set $\mathcal{S} = \{\mathcal{P}_1, \dots, \mathcal{P}_n\}$ that consists of n subsets of $\mathcal{G}$. The n subsets $\mathcal{P}_i \subset \mathcal{G}$ define the *reference sets* of the Lipschitz embedding. The extended graph edit distance function between graphs and reference sets is defined as $d(g, \mathcal{P}_i) = \min_{p \in \mathcal{P}_i}\{d(g,p)\}$.

Definition 3 (Lipschitz Embedding of Graphs). *The Lipschitz embedding with respect to $\mathcal{S} = \{\mathcal{P}_1, \dots, \mathcal{P}_n\}$ is a function $\varphi_\mathcal{S} : \mathcal{G} \to \mathbb{R}^n$ defined as*

$$\varphi_\mathcal{S}(g) = (d(g, \mathcal{P}_1), \dots, d(g, \mathcal{P}_n)).$$

Obviously, the range of function $\varphi_\mathcal{S}$ is a vector space where each dimension corresponds to a subset $\mathcal{P}_i \subset \mathcal{G}$ and the coordinate values of the embedded graph g are the distances from g to the nearest element in $\mathcal{P}_i$.

Given d is a metric, we have $|d(g_1, p) - d(g_2, p)| \leq d(g_1, g_2)$ (triangle inequality). This argument can be easily extended to the case of subsets $\mathcal{P}_i$, i.e. $|d(g_1, \mathcal{P}_i) - d(g_2, \mathcal{P}_i)| \leq d(g_1, g_2)$. Let $p_1 \in \mathcal{P}_i$ and $p_2 \in \mathcal{P}_i$ be the nearest elements in $\mathcal{P}_i$ to g_1 and g_2, respectively, i.e. $d(g_1, \mathcal{P}_i) = d(g_1, p_1)$ and $d(g_2, \mathcal{P}_i) = d(g_2, p_2)$. Obviously, $d(g_1, p_1) \leq d(g_1, p_2)$ and $d(g_2, p_2) \leq d(g_2, p_1)$ and therefore $|d(g_1, \mathcal{P}_i) - d(g_2, \mathcal{P}_i)| = |d(g_1, p_1) - d(g_2, p_2)|$. Note that $d(g_1, p_1) - d(g_2, p_2)$ can be both positive or negative. Consequently,

$$|d(g_1, p_1) - d(g_2, p_2)| \leq \max\{|d(g_1, p_1) - d(g_2, p_1)|, |d(g_1, p_2) - d(g_2, p_2)|\}$$

From the triangle inequality we conclude

$$\max\{|d(g_1, p_1) - d(g_2, p_1)|, |d(g_1, p_2) - d(g_2, p_2)|\} \leq d(g_1, g_2)$$

Hence, $|d(g_1, \mathcal{P}_i) - d(g_2, \mathcal{P}_i)|$ is a lower bound on $d(g_1, g_2)$.

In [7,10] a slight modification of the Lipschitz embedding is used. The authors propose that each coordinate value is divided by a factor depending on n, where n is the number of reference sets. That is, the Lipschitz embedding is defined as

$$\varphi_\mathcal{S}(g) = (d(g, \mathcal{P}_1)/q, \dots, d(g, \mathcal{P}_n)/q),$$

where $q = \sqrt{n}$. Because of this modified definition and particularly due to the relation $(|d(g_1, \mathcal{P}_i) - d(g_2, \mathcal{P}_i)| \leq d(g_1, g_2))$ established above, we obtain

$$\begin{aligned} ||\varphi_\mathcal{S}(g_1) - \varphi_\mathcal{S}(g_2)|| &= \left(\sum_{i=1}^{n} \left(\frac{d(g_1, \mathcal{P}_i) - d(g_2, \mathcal{P}_i)}{\sqrt{n}} \right)^2 \right)^{\frac{1}{2}} \\ &\leq \left(n \cdot \frac{d(g_1, g_2)^2}{n} \right)^{\frac{1}{2}} = d(g_1, g_2). \end{aligned}$$

That is, an upper bound of the Euclidean distance of a pair of graphs in the embedding space is given by the graph edit distance between the corresponding graphs. Consequently, by means of the normalization, the influence of the number of reference sets is reduced. Regardless of the number of reference sets, it is guaranteed that distances in the resulting embedding space are bounded by $d(g_1, g_2)$.

Since the computation of graph edit distance is exponential in the number of nodes for general graphs, the complexity of this graph embedding is exponential as well. However, one can use efficient approximation algorithms for graph edit distance (e.g. [9] with cubic time complexity). Consequently, given n predefined reference sets of size m each, the embedding of one particular graph is established by means of $n \cdot m$ distance computations within polynomial time.

3.2 Defining the Subsets

One crucial question in the proposed embedding method is an adequate definition of the sets $\mathcal{S} = \{\mathcal{P}_1, \ldots, \mathcal{P}_n\}$ of reference sets $\mathcal{P}_i$. In [10] a definition of $\mathcal{S}$ based on random subsets of variable size is intoduced. In this work, $\mathcal{S}$ consists of $O(\log^2 N)$ randomly selected subsets, where each group of $O(\log N)$ subsets is of size 2^i with $i = 1, \ldots, O(\log N)$. Note that N is the size of the original set to be embedded. This approach establishes not only an upper bound as shown above, but also a lower bound on the distance of two embedded objects with respect to their original distance. The authors prove that given their definition of the reference sets, the relative amount of deviation of the distance values in the embedding space with respect to the original distance values is $O(\log N)$, with high probability.

In the present paper, however, we do not use the procedure proposed in [10] as it is not our primary goal to preserve the original graph edit distances up to a certain distortion level. Rather, we are interested in a vectorial description of the underlying graphs which enables us to outperform traditional classifiers operating directly on the graph distances. We use two different methods to define our reference sets, viz. a random selection (RandSel) and a more advanced technique based on k-medoids clustering for graphs (k-MedSel).

In the random method we randomly select n subsets $\mathcal{S} = \{\mathcal{P}_1, \ldots, \mathcal{P}_n\}$ each of size m. After drawing a graph g from $\mathcal{G}$, g is put back such that the same graph can have multiple occurences in $\mathcal{P}_i$, and can occur in different sets $\mathcal{P}_i$ and $\mathcal{P}_j$ of $\mathcal{S}$ as well.

For the second method we try to choose n subsets from $\mathcal{G}$ such that they are approximately evenly distributed with respect to the dissimilarity information given by d. To this end, k-medoids clustering for graphs [11] is applied on $\mathcal{G}$. This results in n disjoint subsets $\mathcal{P}_i \subset \mathcal{G}$ of different size. Next, we iteratively remove the set marginal graphs of $\mathcal{P}_i$ until m graphs remain in $\mathcal{P}_i$. The set marginal graph $g_{mrg} \in \mathcal{P}_i$ is located at the border of a given graph set. Formally, the set marginal graph is the graph whose sum of distances to all other graphs in $\mathcal{P}_i$ is maximal. Note that no graphs are removed whenever the size of a set $\mathcal{P}_i$ is smaller than, or equal to, m. Note that in contrast with the random selection this method leads to disjoint subsets.

4 Experimental Evaluation

The purpose of the experimental evaluation is twofold. First, a comparison between classifiers based on Lipschitz embeddings and those relying on the original graph distances is carried out. The aim of these experiments is to show that the former are able to outperform the latter. Secondly, the influence of the size m of the individual subsets is analyzed. The crucial question is if it is possible to further improve the classification accuracy by using subsets of size $m \geq 1$ as reference for the embedding rather than singletons as proposed in [5]. As base classifier a support vector machine (SVM) is used.

The pattern classification tasks considered in this paper involves a total of five different graph data sets. Because of lack of space, we can give only a short description of the data. For a more thorough description we refer to [5] where the data sets are discussed in greater detail. Note that each of our graph sets is divided into three disjoint subsets, viz. a training, a validation, and a test set.

The first database used in the experiments consists of graphs representing distorted letter drawings out of 15 classes (Letter). Next we apply the proposed method to graphs representing images out of 100 categories from the COIL-100 database [12] (COIL). The third data set is given by graphs representing fingerprint images of the NIST-4 database [13] out of the four classes *arch*, *left*, *right*, and *whorl* (Fingerprint). The fourth graph set is constructed from the AIDS Antiviral Screen Database of Active Compounds [14]. Graphs from this data set represent molecules out of two classes (*active*, *inactive*). The last data set consists of graphs representing webpages [15] that originate from 20 different categories (*Business*, *Health*, *Politics*, ...) (Webgraph).

Note that the graph datasets used in our experiments are of quite different nature, comming from a variety of applications. Furthermore, the graph sets differ in their characteristics, such as the number of available graphs ($|G|$), the number of different classes ($|\Omega|$), and the average and maximum number of nodes and edges per graph ($\varnothing|V|$, $\varnothing|E|$, $\max|V|$, $\max|E|$). In Table 1 a summary of all graph datasets and their corresponding characteristics is given.

Table 1. Graph dataset characteristics

Database	$\|G\|$	$\|\Omega\|$	$\varnothing\|V\|$	$\varnothing\|E\|$	$\max\|V\|$	$\max\|E\|$
Letter	2250	15	4.7	4.5	9	9
COIL	2700	100	21.4	53.9	79	228
Fingerprint	2800	4	5.4	4.4	26	24
AIDS	2000	2	9.5	10.0	85	328
Webgraph	2340	20	186.1	104.6	834	596

4.1 Experimental Set Up

We use three reference systems to compare our novel approach with. The first reference system is a trivial similarity kernel in conjunction with an SVM. This approach basically turns the existing graph edit distance into similarities by mapping low distance values to high similarity values and vice versa. To this end we use a simple monotonically decreasing transformation. Given the edit distance $d(g, g')$ of two graphs g and g', the similarity kernel is defined as $\kappa(g, g') = -d(g, g')^2$. The meta parameter C for the SVM, i.e. the weighting parameter which controls whether the maximization of the margin or the minimization of the error is more important, is the only additional parameter that has to be tuned on the validation set. We refer to this method as GED-SVM.

For the second and third reference system we use a special case of Lipschitz embedding where we define each of the reference sets $\mathcal{P}_1, \ldots, \mathcal{P}_n$ as a singleton ($m = 1$). This special case is equal to the approach presented in [5]. In order to define the reference sets, we use both RandSel and k-MedSel with $m = 1$[1]. In contrast to the first reference system, this approach leads to explicitly defined feature vectors. Consequently, a standard kernel function can be applied to the embedded graphs $\varphi_S(g)$. We make use of an RBF kernel function $k(\varphi_S(g), \varphi_S(g')) = exp\left(-\gamma||\varphi_S(g) - \varphi_S(g')||^2\right)$, where $\gamma > 0$. That is, for both selection methods three meta parameters have to be validated on the validation set, viz. the number of subsets n, and the parameter pair (C, γ) for the RBF-kernel SVM. We refer to these methods as RandSel(1) and k-MedSel(1), respectively.

Our novel approach establishes a generalization of the second and third reference system described above. That is, we allow the reference sets $\mathcal{P}_i$ to have size larger than, or equal to, one ($m \geq 1$). Hence, besides the parameter tripel (n, C, γ) a fourth parameter, namely the size of the subsets m, has to be optimized.

4.2 Results and Discussion

In Table 2 the classification accuracies of the different systems on the independent test sets are given. Also the optimized parameter value m is indicated for both construction methods (RandSel and k-MedSel). Regarding the random construction of the reference sets one observes that on four out out of five data

Table 2. Classification accuracy in the original graph domain and in the embedded vector space

	Reference Methods			Lipschitz Embeddings					
Database	GED-SVM	RandSel(1)	k-MedSel(1)	RandSel	(m)		k-MedSel	(m)	
Letter	92.3	91.3	92.4	91.3	(1)		92.7	(3)	②
COIL	91.4	90.6	91.5	90.6	(1)	❸	92.0	(9)	②
Fingerprint	79.4	80.3	77.7	81.3	(1)	①③	81.9	(5)	①③
AIDS	93.6	97.4	97.7	97.1	(7)	①❸	98.2	(7)	①②
Webgraph	84.6	81.7	81.9	81.7	(1)	❶	82.3	(3)	❶

Z-test for testing statistical significance ($\alpha = 0.05$):

①/②/③ Stat. sig. improvement over the first/second/third reference system.
❶/❷/❸ Stat. sig. deterioration compared to the first/second/third reference system.

[1] Note that $\mathcal{S}$ is defined on the training set.

sets the tuned size of the subsets m is one. Hence, no additional benefit is gained by allowing $m \geq 1$ with RandSel. However, regarding the Lipschitz embedding based on k-MedSel with $m \geq 1$, we observe that the optimized size m of the reference sets is greater than one on all data sets. In Fig. 1 this part of the validation procedure is illustrated on the AIDS data set. The figure shows the classification accuracy on the validation set as a function of n and m. In this particular example, the additional benefit of the generalization is clearly observable, i.e. the best accuracy is achieved with a subset size $m > 1$.

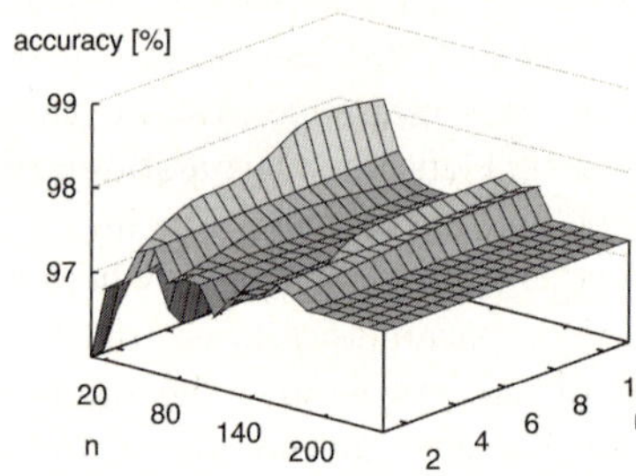

Fig. 1. AIDS data set: Classification accuracy plotted as a function of the dimensionality n and size of the subsets m

Regarding the classification accuracies given in Table 2, three main findings can be reported. First the proposed Lipschitz embedding based on k-MedSel outperforms the embedding based on RandSel on all data sets. That is, the construction of the reference sets by means of an advanced technique (k-medoids clustering in the present case) rather than just a random definition shows superior performance. Secondly, the generalized Lipschitz embedding based on k-MedSel outperforms all reference systems on all data sets except for one case (GED-SVM on Webgraphs). Six out of the 14 improvements are statistically significant. That is, an excellent performance and a high degree of robustness and flexibility of the proposed approach is empirically verified. Thirdly, allowing general subsets greater than one, rather than singletons, is a clear benefit. Comparing the results achieved by k-MedSel with $m = 1$ to those with $m \geq 1$, it turns out that the lattter outperforms the former on all data sets (once with statistical significance).

5 Conclusions

In the present paper a novel approach to graph embedding using Lipschitz mappings is proposed. The basic idea of the embedding method is to describe a graph by means of n distances to predefined reference sets of graphs. Hence, a graph g is mapped explictly to the n-dimensional real space $\mathbb{R}^n$ by representing the edit distances of g to all of the n reference sets as a vector.

For our experimental evaluation, five data sets with quite different characteristics are used. The datasets vary with respect to graph size, edge density, type of labels for the nodes and edges, and meaning of the underlying objects. From the experimental evaluation one can draw the following conclusions. A definition of the reference sets based on an elaborated method rather than random selection is preferable. Classifiers using the Lipschitz embedded graphs outperform classification systems using the original graph edit distances. Finally, the generalization of the Lipschitz embedding to the case where the subsets are not necessarily singletons ($m \geq 1$) is clearly beneficial.

In future work we will study other embedding techniques related to Lipschitz embeddings. Also the application of other distance functions to sets of graphs seems to be an attractive topic for further investigation. Finally, the application of other clustering methods which allow overlapping clusters and therefore non-disjoint reference sets could be a rewarding avenue to be explored.

Acknowledgements

This work has been supported by the Swiss National Science Foundation (Project 200021-113198/1).

References

1. Duda, R., Hart, P., Stork, D.: Pattern Classification, 2nd edn. Wiley-Interscience, Chichester (2000)
2. Conte, D., Foggia, P., Sansone, C., Vento, M.: Thirty years of graph matching in pattern recognition. Int. Journal of Pattern Recognition and Artificial Intelligence 18(3), 265–298 (2004)
3. Gärtner, T.: A survey of kernels for structured data. SIGKDD Explorations 5(1), 49–58 (2003)
4. Bourgain, J.: On Lipschitz embedding of finite metric spaces in Hilbert spaces. Israel Journal of Mathematics 52(1-2), 46–52 (1985)
5. Riesen, K., Neuhaus, M., Bunke, H.: Graph embedding in vector spaces by means of prototype selection. In: Escolano, F., Vento, M. (eds.) GbRPR. LNCS, vol. 4538, pp. 383–393. Springer, Heidelberg (2007)
6. Duin, R., Pekalska, E.: The Dissimilarity Representations for Pattern Recognition: Foundations and Applications. World Scientific, Singapore (2005)
7. Hjaltason, G., Samet, H.: Properties of embedding methods for similarity searching in metric spaces. IEEE Trans. on Pattern Analysis ans Machine Intelligence 25(5), 530–549 (2003)
8. Bunke, H., Allermann, G.: Inexact graph matching for structural pattern recognition. Pattern Recognition Letters 1, 245–253 (1983)
9. Riesen, K., Neuhaus, M., Bunke, H.: Bipartite graph matching for computing the edit distance of graphs. In: Escolano, F., Vento, M. (eds.) GbRPR. LNCS, vol. 4538, pp. 1–12. Springer, Heidelberg (2007)
10. Linial, N., London, E., Rabinovich, Y.: The geometry of graphs and some of its algorithmic applications. Combinatorica 15, 215–245 (1995)

11. Kaufman, L., Rousseeuw, P.: Finding Groups in Data: An Introduction to Cluster Analysis. John Wiley & Sons, Chichester (1990)
12. Nene, S., Nayar, S., Murase, H.: Columbia Object Image Library: Coil-100. Technical report, Department of Computer Science, Columbia University, New York (1996)
13. Watson, C., Wilson, C.: NIST Special Database 4, Fingerprint Database. National Institute of Standards and Technology (1992)
14. DTP, D.T.P.: AIDS antiviral screen (2004), `http://dtp.nci.nih.gov/docs/aids/aids_data.html`
15. Schenker, A., Bunke, H., Last, M., Kandel, A.: Graph-Theoretic Techniques for Web Content Mining. World Scientific, Singapore (2005)

An Experiment on Human Face Recognition Performance for Access Control

Marcus Butavicius, Chloë Mount, Veneta MacLeod, Robyn Vast, Ian Graves, and Jadranka Sunde

Defence Science and Technology Organisation (DSTO), PO Box 1500, Edinburgh, SA, 5111, Australia
{Marcus.Butavicius,Chloe.Mount,Veneta.MacLeod,Robyn.Vast, Ian.Graves,Jadranka.Sunde}@DSTO.Defence.gov.au

Abstract. An experiment was conducted on human face recognition performance in an access control scenario. Ten judges compared fifty individuals to security ID style photos where 20% of the photos were of different people, assessed to look similar to the individual presenting the photo. Performance was better than that observed in the only other comparable live-to-photo experiment [1] with a false match rate of 9% [$CI_{95\%}$: 2%, 16%] in this study compared to 66% [$CI_{95\%}$: 50%, 82%] and a false reject rate of 5% [$CI_{95\%}$: 0%, 11%] compared to 14% [$CI_{95\%}$: 0.3%, 28%]. These differences were attributed to divergences in experimental methodology, especially with regards to the distractor tasks used. It is concluded that the figures provided in the current study are more appropriate estimates of performance in access control scenarios. Substantial individual variation in face matching abilities, response time and confidence ratings was observed.

Keywords: access control, security, face recognition, human factors.

1 Introduction

Since the events of September 11, there has been a rapid increase in the implementation of automated biometric solutions, such as face recognition, to authenticate the identity of individuals in access control settings [2]. However, there has been relatively little effort to undertake performance evaluation of face recognition in operational scenarios (for exceptions see [3], [4], [5], [6]). In addition, an even more fundamental issue has not been investigated - how well do the current systems involving human operators work in access control settings? The data gained from such research could be used to gauge the security of current systems as well as a baseline for considering the introduction of automated biometric systems. In fact, even in settings with automated systems, human face recognition will still play an important role in authentication where the system creates an alarm that must be verified and acted upon by a human operator.

A literature review [7] found only one published paper where human face recognition performance, comparing photos to participants presented in a live scenario, had been investigated in a controlled experiment. In [1], a study was conducted investigating how

I. Lovrek, R.J. Howlett, and L.C. Jain (Eds.): KES 2008, Part I, LNAI 5177, pp. 141–148, 2008.

well six cashiers matched the photos on credit cards to forty six shoppers in real life under two conditions. The first condition included cases where the photo was of the shopper but with different clothes, hairstyle and accessories consistent with normal day-to-day variation. The second condition included cases where the photos depicted a different person who was judged to look very similar to the shopper. In this study the cashiers incorrectly accepted the false credit cards 66% of the time while they incorrectly rejected the correct credit cards 14% of the time.

However, while the performance figures in [1] are relevant to the credit card fraud scenario, their experimental paradigm is different in many ways to an access control setting. Most importantly, the cashiers' duties included verification of signatures in addition to making assessments on the basis of face information. This task is not present in an access control scenario and there is empirical evidence that performance on face matching performance is influenced by the demands of an additional task, known in the psychological literature as a distractor task [8]. In addition, because the signatures always matched in [1], this additional task always provided evidence that the shopper was holding a correct credit card, even when the photo did not match them. Given the different scenarios and the potential influence on performance they could have, it was decided that an experiment should be conducted to measure human face recognition performance in an access control scenario.

1.1 Participants

60 participants (42 males and 18 females) took part in the experiment. Of these, 50 DSTO staff members (33 males and 17 females) acted as 'card holders'. The remaining ten participants were Defence Force personnel who acted as 'judges'. To ensure that the judges were not familiar with the card holders, the participants in each group were selected from different work sites. The age range of participants was 21 to 42 years (M = 32.14).

1.2 Materials

An 'ID card' was made for each card holder featuring a true photograph of the individual taken for the purpose of this experiment. These photographs were headshots taken against a white background at a distance of one metre. Card holders had neutral facial expressions and were asked to remove security passes and glasses only if they interfered with the photograph. The fraudulent photographs were chosen by two research investigators from a database of headshot photographs available for public use, for 19 of the participants [9]. Each investigator was asked to rank them in order of similarity with the original photo and the ten fraudulent photos with the highest average similarity scores, were selected for use in the experiment.

Each ID card also featured correct details of the card holder including country of origin, name, date of birth, gender, an expiry date and a code. Each card was placed in an envelope with the card holder's code and a list of room numbers (identifying the order that the card holder was to visit the judges) on the front. In order to record the judge's response choices, confidence ratings and response times, a computer program was written for the experiment. Each judge was provided with a laptop computer on which this program was run. A video camera was set up in each of the ten rooms to capture both the judge and the card holders in order to measure response times using Noldus software [10].

1.3 Procedure

Judges were shown to their individual rooms where they would remain for the experiment. Each room contained a desk, a chair and a laptop set up in such a way that judges would sit facing the entrance to the room and the computer screen was not visible to card holders. Judges were given practice trials on the computer program to familiarise themselves with the software for the experiment. Following this practice, the judges were instructed not to leave the room or open the door, so that card holders were kept from their view before the start of the experiment proper.

Each card holder presented themselves once to each judge, so that each judge processed a total of 50 faces in the experiment. The experiment was divided into 10 rounds in which a judge viewed five card holders. To counteract order effects, the following procedure was followed. Firstly, for each round, card holders were randomly allocated to one of the ten rooms such that there were five card holders at each room at any one time, and each card holder visited each room once and once only. In addition, the order of the five card holders entering each room in each round was randomised. A research investigator monitoring each room verified the presence and order of the participants for each round. When all rooms were ready, the first card holder for each room was instructed to enter the room and hand the judge his/her envelope.

Judges were required to enter the code into the computer. This distractor task was carried out in addition to the main task of interest (i.e., face matching). Judges clicked on 'accept' on the interface if they believed that the photo was of the person presenting to them (or 'reject' otherwise). They were then required to rate their confidence in their decision on a seven point Likert scale by clicking a box corresponding to a number from one ('not at all confident') to seven ('extremely confident'), place the card back in the envelope, and return it to the bearer. They were not allowed to ask the bearer any questions about their appearance, or request other forms of identification. The card holders were instructed not to look at their ID photos and were not informed whether or not they had a fraudulent ID card.

Card holders were not able to see the judge's computer screen or the decision the judge had made. The card holder left the room and lined up outside the next room as per the order on their envelope. The next card holder in the round was only permitted to enter once the previous card holder had left the room. When the five card holders who had originally lined up outside each room had presented themselves to that judge, the experiment was paused so that investigators could check that all card holders were at the correct second room, and could re-order them if necessary. This procedure continued until all card holders had presented themselves once to each of the judges.

2 Results

There was considerable variation in the performance of the ten judges in terms of how many errors they made (ranging from 0 to 12 errors, $M = 2.7$, $SD = 3.5$) and the average time taken to make a decision (ranging from 5.85 – 27.48 seconds, $M = 15.8$ seconds,

SD = 7). However, the average confidence rating for all participants was very high (M = 6.1, SD = 0.74).

Overall, 95% of decisions made in the experiment were correct. Of all the ID cards that were presented to judges, 62% were accepted at least once, with 60% of the fraudulent cards being accepted at least once. In addition, 40% of the judges falsely rejected at least one card. Of the total number of incorrect decisions made about card holders, 26% were about females, who made up 34% of the sample. Of all the decisions that were made about female card holders, 4% proved to be incorrect whereas out of all decisions made about male card holders, 6% were incorrect.

The accuracy of each judge was determined by examining their true match, false match, true reject and false reject error rates. In cases where the photo presented depicts the card holder, a true match (TM) occurs when the judge responds correctly (i.e., 'accept') and a false reject (FR) occurs when the judge responds incorrectly (i.e., 'reject'). In cases where the photo presented does not depict the card holder, a true reject (TR) occurs when the judge responds correctly (i.e., 'reject') and a false match (FM) occurs when the judge responds incorrectly (i.e., 'accept'). As can be seen in Table 1, overall matching performance was very good with two of the judges (Judges E and G) responding correctly in all 50 cases. The overall false match rate (FMR) was 9% ($CI_{95\%}$: 2%, 16%) and the false reject rate was only 5% ($CI_{95\%}$: 0%, 11%).

An error rate plot based on this human data is presented in Figure 1. In this graph, each judge's error rates (i.e., the probabilities of true and false matches) are plotted as separate points. In cases where there was more than one person with the same error rates, the size of the point has been increased in proportion to the number of judges with the same rates.

Examination of the error data indicates that the participants differ in the way they match faces in more than just a decision threshold, i.e., they are not simply using the same perceptual / cognitive technique at different thresholds, but that there are more fundamental differences across individuals in the manner they are judging similarity.

Table 1. The true match, false match, true reject and false reject rates for each judge

Judge	True Match %(N/Total)	False Match %(N/Total)	True Reject %(N/Total)	False Reject %(N/Total)
A	100 (40/40)	10 (1/10)	90 (9/10)	0 (0/40)
B	100 (40/40)	10 (1/10)	90 (9/10)	0 (0/40)
C	98 (39/40)	20 (2/10)	80 (8/10)	2 (1/40)
D	93 (37/40)	0 (0/10)	100 (10/10)	7 (3/40)
E	100 (40/40)	0 (0/10)	100 (10/10)	0 (0/40)
F	100 (40/40)	30 (3/10)	70 (7/10)	0 (0/40)
G	100 (40/40)	0 (0/10)	100 (10/10)	0 (0/40)
H	93 (37/40)	0 (0/10)	100 (10/10)	8 (3/40)
I	100 (40/40)	10 (1/10)	90 (9/10)	0 (0/40)
J	73 (29/40)	10 (1/10)	90 (9/10)	28 (11/40)
AVERAGE	96 (38.2/40)	9 (0.9/10)	91 (9.1/10)	5 (2.3/40)

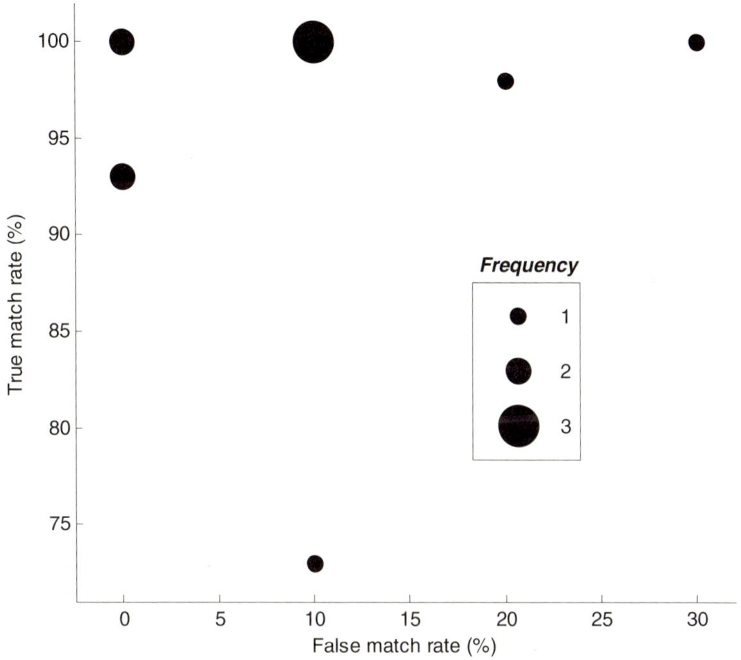

Fig. 1. True match and false match rates. The size of the symbol is proportional to the number of judges with identical error rates.

At the very least, there are individual differences in the effectiveness of face matching strategies, i.e., that discrimination, and not just decision bias, varies across individuals (for a more detailed discussion of bias and discrimination in human face matching see [11]).

Examination of the raw data indicates that considerable individual differences can be observed in the errors that judge participants made. There was little evidence that card holder effects were consistent across different judges. The maximum number of judges to accept the same fraudulent card was only 3 and, of the 18 false rejects of match participants, three participants were falsely rejected twice each while the remaining 12 false rejects were for 12 different individuals.

Response times tended to be longer when the judge made an incorrect decision than when they made a correct decision. A repeated measures Analysis of Variance (ANOVA) was conducted on the data, but revealed no significant differences between the response times for different response types ($F(3,3) = 0.728$, $p = .55$, Greenhouse-Geisser corrected). However, the observed effect size ($\eta_p^2 = .421$) was small to medium in size according to [12] and there were only ten judges, with large individual variation observed in reaction times. Therefore a larger sample of judges would be required to determine whether there is a statistically significant difference between the response times for the decision classes.

While in general confidence ratings were high, there was variation in these ratings based on the type of decisions that were made. For correct decisions, a confidence rating of 7 (‘extremely confident’) was by far the most common response recorded. However, for incorrect decisions, much more variability in confidence ratings was noted, with the three equally most common confidence ratings being 2, 4 and 6. This effect was most apparent for false reject decisions. However, the difference in mean confidence ratings for the different response types was not statistically significant ($F(3,3) = 2.046$, $p = .388$, Greenhouse-Geisser corrected), although the effect size was relatively large ($\eta_p^2 = .672$), and there was considerable individual variation between judges’ ratings. As mentioned previously, a larger sample of judges would be necessary to examine the reliability of such differences.

There was evidence that longer response times were associated with lower confidence ratings and this was largely independent of whether the decision was correct. For all the trials, the correlation between confidence and response time was -0.176 using Spearman’s ρ ($N = 500$, $CI_{95\%}$: -.285, -.062) which is considered a medium-sized effect [12].

3 Discussion

The aim of the current study was to investigate human face recognition performance in an access control setting. Our analysis revealed that the differences between the human judges could not be characterised by differences in some kind of match threshold whereby humans may differ only in the stringency of their decision-making. In other words, our findings suggest that the judges used fundamentally different face matching strategies. In addition, the human-photo combinations on which judges made mistakes varied markedly between individuals.

Match accuracy was better than expected from previously published results, with two judges responding correctly to all card holders. However, there was a great deal of individual variation in performance, with one judge making as many as 12 out of 50 possible errors (76% correct). Although a definitive relationship between response time and the accuracy of decisions was not found, it was apparent that overall response times tended to be longer when judges incorrectly rejected card holders with true photographs. In addition, it appears that for some judges a relationship existed between accuracy and response time, with incorrect decisions (false rejects and false matches) often corresponding to longer response times. However, further investigation in studies with larger samples will be required to ensure that this observation is generally valid. In general, the finding of individual variation is consistent with previous evidence of individual differences in face verification performance using photo-to-photo matching ([11], [8]).

In comparison to the only previous study to investigate live-to-photo matching [1], we found that face matching performance was superior in our scenario. In the current study, judges made substantially more correct decisions (95%) compared to [1] (67%). Additionally, four out of six cashiers in the previous study made at least one false reject, whereas in the current study only four out of ten judges made this type of error. The biggest difference between the studies was in the frequency with which different errors were made. On the one hand, the false reject rates were moderately comparable in the two studies with 5% [$CI_{95\%}$: 0%, 11%] in the current study and 14% [$CI_{95\%}$: 0.3%, 28%] in [1]. That is, the ability to judge cases where the photo indicates the same person is similar in the two studies. On the other hand, the ability to judge cases where a person is different from their photo varied markedly between studies. In our study the false match rate was only 9% [$CI_{95\%}$: 2%, 16%] compared to 66% [$CI_{95\%}$: 50%, 82%] in [1].

A gender effect was found in [1] whereby cashiers had more difficulty detecting fraudulent females than they did in detecting males. The current study found the opposite effect, with more errors being made about male than female card holders. It should, however, be noted that the shoppers in [1] were equally split for gender, whereas in the current study, only 34% of card holders were female. Furthermore, all the cashiers were females, while in this study only one out of the ten judges was female. It is possible that both these facts may have affected the results of both studies.

We propose that the performance differences between our study and [1] are due to the influence of the experimental set-up and methodology – while their scenario resembled a retail environment, our scenario resembled an access control setting. In general, there is a large amount of variability between estimates of human performance in face recognition tasks across different experimental paradigms [7]. More specifically, there is also evidence that the different tasks performed concurrently with face matching influence performance and it is probable that this underlies a large amount of variability between the two studies [8]. Not only did the distractor tasks differ in the skills required (i.e., entering a number into a computer versus comparing two signatures) but they differed in the influence on the decision to accept or reject an individual. In our study, the distractor task was unrelated to the task of identifying an individual. In contrast, the distractor task in [1] always provided evidence that the individual was holding a correct credit card because the signatures always matched. This information was consistent with face matching when correct credit cards were presented but inconsistent when incorrect cards were presented. The contradictory information present in only the fraudulent trials in [1] may explain why the false match rates were higher than what we have found in this study and also why the false reject rate remained consistent.

We suggest that further research is needed to directly compare human and automated performance in an access control scenario. While previous research has sought to compare human and automated face recognition performance, such studies have involved memory or identification tasks rather than the face matching or authentication tasks ([13], [14]). While we believe that the results from this study are indicative of human performance, what is needed is a controlled study where both human and automated systems are presented with the same stimuli or data in an access control scenario to better compare performance between the two approaches.

References

1. Kemp, R., Towell, N., Pike, G.: When seeing should not be believing: Photographs, credit cards and fraud. App. Cog. Psych. 11, 211–222 (1997)
2. Blackburn, T., Butavicius, M., Graves, I., Hemming, D., Ivancevic, V., Johnson, R., Kaine, A., McLindin, B., Meaney, K., Smith, B., Sunde, J.: Biometrics Technology Review 2002. DSTO-GD, 0359 (2002)
3. Butavicius, M.: Evaluating and predicting the performance of an identification face recognition system in an operational setting. Aust. Soc. for Op. Res. Bull. 25(2), 2–13 (2006)

4. Kaine, A.: The Impact of Facial Recognition Systems on Business Practices within an Operational Setting. In: Proc. 25th conference on Inf. Techn. Interfaces (ITI 2003), pp. 315–320 (2003)
5. McLindin, B., Butavicius, M., Meaney, K.: Gallery Image Effects on Facial Recognition Systems. In: Proc. EC-VIP, 4th EURASIP conference on Video/Image Processing and Multimedia Communications, vol. 2, pp. 445–460 (2003)
6. Sunde, J., Butavicius, M., Graves, I., Hemming, D., Ivancevic, V., Johnson, R., Kaine, A., McLindin, B.A., Meaney, K.A.: Methodology for evaluating the Operational Effectiveness of Facial Recognition Systems. In: Proc. EC-VIP, 4th EURASIP conference on Video/Image Processing and Multimedia Communications, vol. 2, pp. 441–448 (2003)
7. Vast, R., Butavicius, M.: A Literature Review of Face Recognition for Access Control: Human Versus Machine Solutions. DSTO-TR, 1747 (2005)
8. Lee, M.D., Vast, R.L., Butavicius, M.A.: Face matching under time pressure and task demands. In: Sun, R., Miyake, N. (eds.) Proc. of the 28th Annual Conference of the Cog. Sci. Soc., pp. 1675–1680. Cog. Sci. Soc., Vancouver (2006)
9. Phillips, P.J., Moon, H., Rizvi, S.A., Rauss, P.J.: The FERET evaluation methodology for face recognition algorithms. IEEE Trans. Pattern Analysis and Machine Intelligence 22, 1090–1104 (2000)
10. Noldus: The Observer Quick Start Guide, Version 5.0. Noldus Information Technology (2003)
11. Fletcher, K., Butavicius, M.A., Lee, M.D.: The effects of external feature similarity and time pressure on unfamiliar face matching. British Journal of Psychology (in press)
12. Cohen, J.: Statistical power analysis for the behavioral sciences, 2nd edn. Lawrence Erlbaum, Hillsdale (1988)
13. Burton, M.A., Miller, P., Bruce, V., Hancock, P.J.B., Henderson, Z.: Human and automatic face recognition: a comparison across image formats. Vision Research 41, 3185–3195 (2001)
14. Luckman, A.J., Allinson, N.M., Ellis, A.W., Flude, B.M.: Familiar face recognition: A comparative study of a connectionist model and human performance. Neurocomputing 7, 3–27 (1995)

A First Step towards Argument Mining and Its Use in Arguing Agents and ITS

Safia Abbas[1] and Hajime Sawamura[2]

[1] Graduate School of Science and Technology, Niigata University
8050, 2-cho, Ikarashi, Niigata, 950-2181 Japan
safia@cs.ie.niigata-u.ac.jp
[2] Institute of Natural Science and Technology,
Academic Assembly, Niigata University
8050, 2-cho, Ikarashi, Niigata, 950-2181 Japan
sawamura@ie.niigata-u.ac.jp

Abstract. Argumentation is an interdisciplinary research area that incorporates many fields such as artificial intelligence, multi-agent systems, and collaborative learning. Although different argumentation tools have been developed, a structured data representation format has been missing. Recent researches have focused on applying mining techniques to find meaningful knowledge from these unstructured textual data. This paper reports work in progress on building Relational Argument DataBase(RADB) for argument mining and its use in arguing agents and ITS. The RADB depends on the Argumentation Interchange Format Ontology (AIF) using "Walton Theory" for argument analysis. Our aim is to present a preliminary attempt to support argument construction for agents and/or humans from structured argument database together with different mining techniques. We also discuss the usage of relational argument database in agent-based intelligent tutoring system(ITS) framework.

1 Introduction

Argumentation theory is considered as an interdisciplinary research area. Its techniques and results have found a wide range of applications in both theoretical and practical branches of artificial intelligence, education, and computer science [7, 8, 12]. Among other things, we are mainly concerned with argument mapping, analysis and formal computational argumentation frameworks, where many efforts have been most devoted as far as we see in the literature[10].

Argument mapping (e. g., Compendium, Araucaria, Rationale, etc.) aims at improving our ability to articulate, comprehend and communicate reasoning, by supporting the production of reasoning diagrams and argumentation especially for complex arguments and debates. It is greatly anticipated that it helps students developing their critical thinking as well as promoting critical thinking in daily life. On the other hand, the main concern in formal computational argumentation frameworks is to formalize methods in which the final statuses of arguments are to be decided semantically and/or dialectically [8].

I. Lovrek, R.J. Howlett, and L.C. Jain (Eds.): KES 2008, Part I, LNAI 5177, pp. 149–157, 2008.

One deficient point in both areas of argumentation-conscious work is the lack of retrieving, reusing, and processing (such as modifying, adjusting, etc.) pre-existing arguments for an argument to be newly constructed. Put it differently, such an idea of constructing arguments from a large argument database is missing. Constructing arguments are considered as one of the most important creative activities since we argue all the time in our daily life, scientific communities, parliaments, courts, online discussion boards and so on. That is, we would say "living is arguing in the age of globalization just as living is eating in the days of plenty". Argument mining and argument discovery technology are particularly needed to allow us to create new ideas, opinions and thoughts by finding out meaningful arguments and reusing them from such large-scale argument repositories. Then, a common format for argument representation, like Argument Markup Language (AML) [4, 13], will be required and plays an important role in argument mining as well as argument discovery technology.

In this paper, we present a preliminary attempt to support argument construction for agents and/or humans from argument database together with different mining techniques. In doing so, we employ a classical but firm approach to organizing argument database and their usage, that is, a relational database approach for arguments. Then, arguments are supposed to be analyzed and constructed with Compendium, an argument mapping system. The paper is organized as follows. Section 2 illustrates the building of relational argument databases and represents some argument mining techniques that used to mine these argument data. Section 3 reviews the relevant work performed in this field. Finally, concluding remarks and future work are presented in Section 4.

2 Relational Argument DataBase

A relational database can be defined as a set of information reformulated and categorized into a set of files (tables) that can be accessed, gathered (queried), and manipulated in different manners. According to the AIF ontology, arguments can be represented semantically in the form of nodes connected with directed edges in a directed graph known as argument network[1]. If the cyclic problem (the same information node (I-node) refines more than one scheme node (S-node)) is avoided, the arguments can semantically be represented as directed tree. This presentation can be structured in the form of well-established relational database, and annotated as relational argument database. This is properly needed to facilitate the mining process in the proposed framework.

2.1 Building Relational Argument DataBases

This subsection describes the building blocks of the relational argument database. We consider the AIF ontology [1, 6] with some restrictions (such that no edge emanates from I-node to I-node), and Walton schemes [5] for arguments analysis. Any argument scheme based on Walton theory can be presented as shown in *Fig*.1. which represents a general skeleton for the different schemes description in

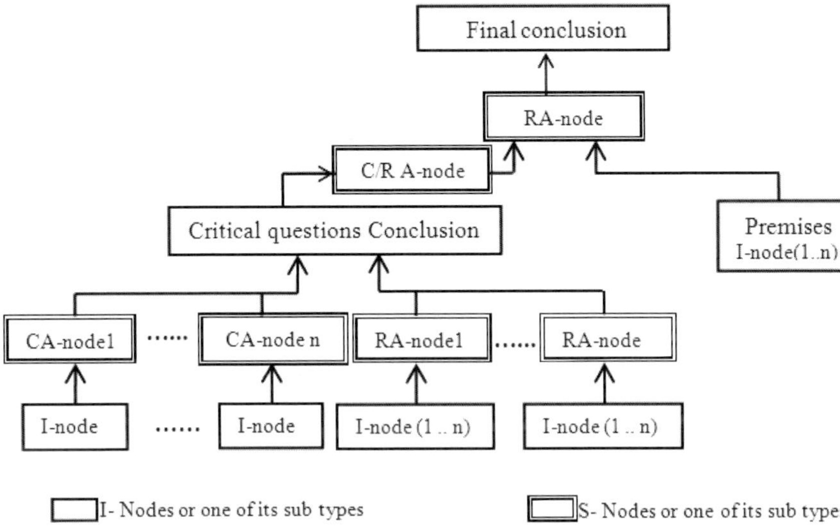

Fig. 1. Argument network representation for different Walton schemes

Walton theory[5]. The premises block gathers the different premises types (majors, minors). The critical question's conclusion block assembles the result of the different critical questions together with the results of the different presumption questions that are to be exposed in a specific scheme.

Considering the canonical representation for the schemes, we pose some sentiments about the relational database baselines. In our vision, we gather the different scheme information into three basic files (tables): Scheme_TBL, Scheme_Struct TBL and Data_TBL. First, the scheme kind is formulated in Scheme_TBL ($Table1$), in which rows act as records of data for different schemes, and columns as features (attributes) of records. The Scheme_TBL creates an identification number(ID) for each scheme name ($Scheme\_Name$), where this ID plays a role of primary key for the table and foreign key in the others. In addition, any ID attribute will stand for the same function in all files. Scheme_Struct_TBL exemplifies the details of each scheme, where rows represent records for the different information associated with the schemes, and columns reveal ID, $Scheme\_Id$ that stands for the foreign key of $Scheme\_TBL$, indicating the scheme concerned. The $Content$ field contains the

Table 1. The main schemes (Scheme_TBL)

ID	Scheme_Name
1	Expert Opinion
2	Popular Opinion
3	Verbal Classification
4	Inference
..	...

ID	Scheme_Id	Type	Content
1	1	P	Source E is an expert in the subject domain X containing proposition B.
2	1	P	E asserts that proposition B in domain X is true.
3	1	C	B may plausibly be taken to be true.
4	1	CC	Critical argumentation conclusion
5	1	CQ	*Expertise Question*: How credible is expert E as an expert source?
6	1	CQ	*Field Question*: Is E an expert in the field that the B is in?
7	1	CQ	*Opinion Question*: Does E's assertions imply B?
8	1	CQ	*Trustworthiness Question*: Is E reliable as source?
9	1	CQ	*Consistency Question*: Does B consistent with the assertions of other experts?
10	1	CQ	*Backup Evidence Question*: are there any evidences sustain B?
11	4	RA	*RA-Node*
12	5	CA	*CA-Node*
13	6	PA	*PA-Node*

Fig. 2. The structure of schemes (Scheme_Struct_TBL)

details of the associated information (premises, conclusion, critical questions, etc.). The *Type* field has four values, P for premises, C for conclusion, CQ for critical question and CC for critical argumentation conclusion. For instance, the expert opinion scheme[5] can be represented in *Fig.* 2.

The Data_TBL table contains all users' transactions. This table gathers all the analysis done by the user for specific argument contexts. The table consists of the *ID* attribute, defined as before, the *Stru_Id* attribute, which serves as foreign key for the *ID* in the Scheme_Struct_TBL to refer to a specific part of the scheme details, the *Content* attribute contains a portion of the analyzed text that fulfills the referred fixed scheme part, the *Type* attribute, which holds three values only, 1 for the supported node, 0 for rebuttal node, -1 for undetermined value that denotes neither supported nor rebuttal node. One of these values is to be given to the final conclusion of the analysis. Since we consider any argument network as a kind of directed rooted trees, the *Child_Of* attribute points to the parent of each node, whereas the root node has no parents (0 refers to no parent). The *level* attribute refers to the level of each node in the tree, such that the value 0 indicates the root node of the argument. Finally, the *argumentation_no* attribute contains the number of the analyzed argument context. For example, the following context from Araucaria repository database [3, 4, 13] is reanalyzed based on the expert opinion scheme as in *fig.* 3. *"Eight monthold Kyle Mutch's tragic death was not an accident and he suffered injuries consistent with a punch or a kick, a court heard yesterday. The baby, whose stepfather denies murder, was examined by pathologist Dr James Grieve shortly after his death. Dr. Grieve told the High Court at For far the youngest was covered in bruises and had suffered a crushed intestine as well as severe internal bleeding. When asked by Advocate Depute Mark Stewart, prosecuting, if the bruises could have been caused by an accident, he said "No. Not in a child that is not walking, not toddling and has not been in a motor car." Dr. Grieve said the injuries had happened "pretty quickly" and would be "difficult for an infant to cope with". The lecturer in forensic medicines at Aberdeen University told the jury that the bruises could have been caused by a single blow from a blunt instrument, like a closed hand. Death, not accident, court told, "Evening Telegraph", Monday, September 13, 2004, p.11"*

ID	Stru_ID	Content	Type	Child_of	level	argumentation_no
1	3	Kyle Mutch's tragic death was not an accident and he suffered injuries	-1	0	0	argument_602
2	7	RA-Node	1	1	1	argument_602
3	2	Dr.James told the high court at Forfar the youngest death could not be cau	1	2	2	argument_602
4	1	Dr.James Grieveis a Pathologists who examne the baby shortly after his (	1	2	2	argument_602
5	4	Death, not an accident, court told	1	2	2	argument_602
6	7	RA-Node	1	5	3	argument_602
7	6	Field Question: Dr.James Grieve is an exper. in pathology	1	6	4	argument_602
8	7	Opinion Question: the baby is a child that is not walking,	1	6	4	argument_602
9	10	Backup Evidence Question: the lecturer in forensic medicines	1	6	4	argument_602
10	12	CA-Node	0	5	3	argument_602
11	9	conflict from inconsistent testimony: the baby's stepfather denies murder	0	10	4	argument 602

Fig. 3. The analysis of the above context based on expert opinion scheme

2.2 Some Techniques for Argument Mining

Since we consider the RADB as a forest of rooted unordered directed tree, where each tree expresses an analysis of a specific argument context, the different tree mining techniques could be applied to mine this forest. Yun Chi et al. [14] surveyed the current algorithms used for mining frequent subtrees from databases. They focused on two main components of these algorithms, the candidate generation step and the support counting step. Also they presented thorough performance studies on a representative family of algorithms, and revealed that there is no single best tree mining algorithm. Some algorithms offer a better time efficiency, while others require less memory. So every time we manipulate the proposed RADB we will consider the time and memory consuming.

We draw a preliminary vision for retrieving and mining the RADB, using a framework with ITS component incorporated. The framework as depicted in *Fig*.4 consists of three main components: the *parser* module, the mining *classifier* agent, and the *ITS* program. The *parser* module receives a statement S from the intended users such as students or agents and divides it into tokens, then reduces the number of tokens, and sends the final crucial set of words { w_1 w_2... w_n } to the *classifier* agent. The tokens are reduced if they belong to a look up table containing the set of all unnecessary words like the articles { a, an, the }, pronouns { I, he, she, it} and others, otherwise it is added to the set of tokens which is to be sent to the *classifier* agent. The importance of the *parser* module lies in reducing the set of tokens which in turn will reduce the number of iterations done by the *classifier* agent, and improve the complexity of the used mining algorithms. The *classifier* agent classifies the retrieved contexts depending on the students specification. The agent can classify the retrieved arguments by priority, polarity, scheme name, premises (with/against), and by conclusion. The priority aims to show the retrieved contexts organized by the maximum support number based on the specific classification mining techniques, polarity classifies the retrieved arguments in to two classes, support class and against class, using the text mining techniques, scheme name retrieves the desired contexts depending on a specific scheme name determined by the student, premises (with/against) retrieves arguments by searching only in the different premises, and conclusion retrieves and classifies the arguments by searching only in the different conclusions. The *classifier* agent receives the

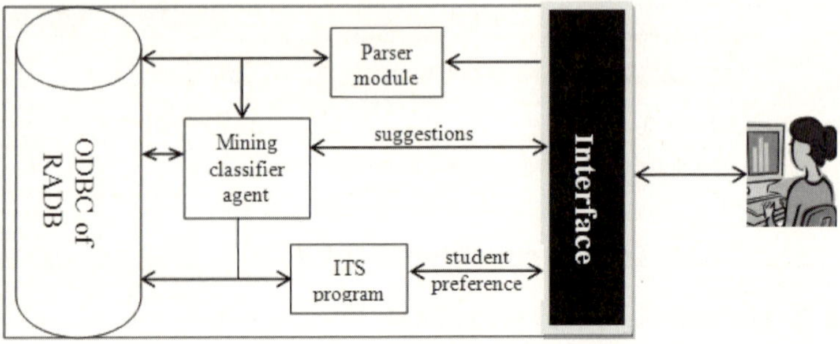

Fig. 4. Framework outline

set of crucial words { w_1 w_2...w_n } from the *parser* module and the search type from the student, then retrieves and classifies the documents that are relevant to the student statement from the database. After the *classifier* agent exposed the pertinent contexts to the student, the student picks up one context among them. The student preference then delegates to the *ITS* program. The program exposes the corresponding context, and gives the student the ability to analyze the selected argument based on a specific chosen scheme. Finally the program negotiates with the student about the way of analysis through adaptive learning techniques to cultivate the student analysis ability.

2.3 Mining Arguments by Querying RADB: An Example

Suppose a student wants information about Iraq war, so he/she writes "the destructive war in Iraq". First the *parser* will divide the statement into tokens such that w_1=the, w_2=destructive, w_3=war, w_4=in, w_5=Iraq. Then, the parser access to the database to reduce the number of tokens by checking the lookup table. So, the output will be the items set {destructive, war, Iraq}. Now the *classifier* should find a set of documents D= $\{d \parallel \{w_1, w_2, w_3\} \subseteq d\}$. Assume that conclusion is the search criterion. The *classifier* will use the mining AprioriTid algorithm [11] to make all possible combinations of the different items set. Then, the result is calssified depending on the support number for each combination. Firstly, the algorithm will calculate the support number for each single token, and select the tokens that has the support number greater than *minsup*, we will take the *minsup* = 1, such that any token appears at least once will be considered. The support number for each token can be counted by the number of transactions resulted from the following SQL statements.

Select argument_no from Data_TBL where
Stru_id = 3 *and Content like '%destructive%'*;
Select argument_no from Data_TBL where
Stru_id = 3 *and Content like '%war%'*....;

The output will be the set of ordered pairs L_1 = {(destructive, 5), (war, 10), (Iraq, 20)}, where this pairs is of the form (the token, the support number), and

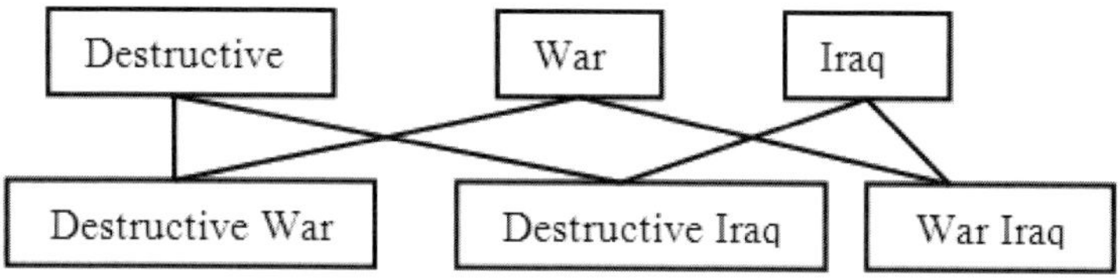

Fig. 5. The super set C_2 of the singleton token set L_1

the set of arguments A_1={ argument_801, argument_509,...} that contains those tokens. Secondly, the algorithm consequently builds the super set C_k = apriori_gen(L_{k-1}) for all possible combinations of the tokens. *Fig*.5 shows the first iteration for C_2=apriori_gen(L_1). Then the support number for each combination is checked through the set A_1. Suppose that the support number for the item set "War Iraq" is 0, so this item set is neglected. The output of this iteration will be L_2= {(Destructive war, 3), (Destructive Iraq, 5)}, and the set A_2={argument_509,...}. Finally, the last iteration of our example will out put the set L_3={(Destructive war Iraq, 1)} and the set of arguments A_3={ argument_509}. If A_3 has more than one argument_no the arguments will be ordered depending on the possessed counter arguments such that the argument with more cons is the weakest.

Furthermore, the ITS will negotiate with the student partially (step by step hints) or totally (compare the student whole analysis with the original one retrieved from the repository) using some mining techniques in order to cultivate his/her intellectual and analysis skills.

3 Related Work

A number of argument mark-up languages and mining weblogs have been proposed to facilitate data preprocessing. The argument-markup language (AML, XML based language) behind the Araucaria system [3, 4, 12, 13], and the classification problem of mining the legal reasoning or the informal reasoning considering the law [2, 9], are examples of these trials. However, current approaches either retrieve data without mining, as Araucaria search tool, or mine the textual data, which is intractable to be processed, as opinion mining in legal blogs that considers specific field like law[2]. This paper, introduces a novel approach to retrieve the information using mining techniques based on RADB. This structure facilitates fast interaction, and enjoys general applicability since it does not require a specialized knowledge. The idea is to mine the pre-existing arguments in order to (i) direct the search towards hypotheses that are more relevant to the users needs, even with more than one word in the search statement, (ii) add flexibility to the retrieving process to suit the users aims.

I. Rahwan presents the ArgDf system [1, 6], through which users can create, manipulate, and query arguments using different argumentation schemes. Comparing ArgDf system to our approach, both of them sustain creating new arguments based on existing argument schemes. Details of the selected argumentation scheme are retrieved from the repository, and the generic form of the argument is displayed to the user to guide the creation of the premises and conclusion. For example, querying the ArgDF repository to extract the name of the schemes can be done through the following RQL statement:

$$SelectScheme,\ PresumptiveInferenceScheme - hasSchemeName$$
$$FromScheme\!:\!kb : PresumptiveInferenceScheme\ kb : hasSchemeName....$$

whereas, in our approach, querying the RADB to extract the name and the details of the schemes is done through the following SQL statments:

$$SELECT\ SCH\_Name,\ ID\ FROM\ [Scheme\_TBL]$$
$$SELECT\ Content\ FROM\ [Scheme\_Struct\_TBL]\ WHERE$$
$$Id\_of\_sel\_scheme = \ [Scheme\_Struct\_TBL].SCH\_ID.$$

Moreover, the ArgDf guides the user during the creation process based on the scheme structure only. However, in our approach, the user is not only guided by the scheme structure but also by crucial hints devolved through mining techniques. So, the creation process is restricted by comparing the contrasting reconstruction of the user's analysis and the pre-existing one. Such restriction helps in refining the user's underlying classification.

In ArgDf, searching arguments is revealed by specifying text in the premises or the conclusion, as well as the type of relationship between them. Whereas, in our approach, searching the existing arguments is not only done by specifying text in the premises or the conclusion but also by providing different strategies based on different mining techniques (as explained in subsection 2.2). Which guarantees the retrieval of the most convenient hypotheses relevant to the subject of search.

4 Concluding Remarks and Future Work

In this paper we have described a novel approach of building a highly structured argument repertoire (RADB) with managing tools. We use different mining techniques to support argument analysis, retrieval, construction, and re-usage. Also, the paper introduced an educational framework that utilizes the RABD. The presented framework aims to:(i)facilitate the search process by providing a different search criteria that reveals a convenient result relevant to the search issue, (ii)guide the analysis process to refine the user's underlying classification, (iii)deepen the users' understanding of negotiation, decision making, and critical thinking.

One deficient point in the computational argumentation research is that they lack a means of retrieving, reusing, and processing (such as modifying, adjusting, etc.) arguments for an argument to be newly constructed. Put it differently, such an idea of argument construction from a large argument database is missing. Moreover, the on-line textual data (i.e., unstructured or semi-structured) is intractable to be processed. So in this paper, we spot the light on another dimension for argument analysis and representation to support argument construction for agents and/or humans based on structured/relational argument database (RADB). The proposed attempt enjoys certain advantages when compared to others, especially with respect to the search of pre-existing arguments. A relevant and convenient result is obtained when the search statement is in this form: "the destructive war in Iraq".

In the future, user's mistakes will be handled by the ITS component. Arguments are supposed to be analyzed and constructed with Compendium, an argument mapping system, taking in consideration the presence of cycles and/or more than one scheme. In addition, refining the search ability by finding out arguments in which specific facts or rules are used.

References

1. Modgil, S., Rahawan, I., Reed, C., Chesnevar, C., McGinnis, J., et al.: Towards an argument interchange format. In: The Knowledge Engineering Review, vol. 00(0), pp. 1–25. Cambridge University Press, Cambridge (2007)
2. Conrad, J., Schilder, F.: Opinion mining in legal blogs. In: ICAIL 2007, Palo Alto, California USA, June 4-8, 2007, pp. 231–236 (2007)
3. Walton, D., Rowe, G., et al.: Araucaria as a tool for diagramming arguments in teaching and studying philosophy. Teaching Philosophy 29, 111–124 (2006)
4. Reed, C., Rowe, G., Katzav, J.: Araucaria: Making up argument. In: European Conference on Computing and Philosophy (2003)
5. Godden, M., Walton, D.: Argument from expert opinion as lgal evidence: Critical questions and admissibility criteria of expert testimony in the american legal system. In: Ratio Juris, vol. 19, pp. 261–286 (2006)
6. Zablith, F., Rahawan, I., Reed, C.: The foundation for a world wide argument web. In: Artificial Intelligence Conference (AAAI), April 04 (2007); published in the Artificial Intelligence Journal
7. Baker, M., Andriessen, J., Suthers, D.: Arguing to learn confronting cognitions in computer-supported collaborative learning environments, vol. 1. Kluwer Academic Publishers, Dordrecht (2003)
8. Reed, C., Katzav, J., Rowe, G.: Argument research corpus. In: Huget, M.-P. (ed.) Communication in Multiagent Systems. LNCS (LNAI), vol. 2650, pp. 269–283. Springer, Heidelberg (2003)
9. Boiy, E., Moens, M., Mochales, R., Reed, C.: Automatic detection of arguments in legal texts. In: ICAIL 2007, Palo Alto, California, USA, June 4-8, 2007, pp. 225–230 (2007)
10. Prakken, H., Vreeswijk, G.: Logical systems for defeasible argumentation. In: Gabbay, D., Guenther, F. (eds.) Handbook of Philosophical Logic, pp. 219–318. Kluwer, Dordrecht (2002)
11. Srikant, R., Agrawal, R.: Fast algorithms for mining rules. In: Proceeding of the 20th VLDB Conference Santiago, Chile (1994)
12. Rahawan, I., Sakeer, P.V.: Representing and querying arguments on semantic web. In: Dunne, P.E., Bench-Capon, T.J.M. (eds.) Computational Models of Argument. IOS Press, Amsterdam (2006)
13. Reed, C., Rowe, G.: Araucaria: Software for argument analysis, diagramming and representation. International Journal on Artificial Intelligence Tools 13, 983 (2004)
14. Muntz, R.R., Chi, Y., et al.: Frequent subtree mining-an overview. In: Fundamenta Informaticae, pp. 1001–1038 (2001)

Optimizing Service Distributions Using a Genetic Algorithm

Kresimir Jurasovic and Mario Kusek

University of Zagreb
Faculty of Electrical Engineering and Computing
Unska 3, HR-10000, Zagreb, Croatia
{kresimir.jurasovic,mario.kusek}@fer.hr

Abstract. This paper deals with optimizing the distribution of services performed by agents in a Mobile Agent Network (MAN). The MAN model consists of agents, capable of performing remote software management services, and of various network elements and nodes used by agents during service execution. In order to verify various service distributions a Mobile Agent Network Simulator was developed. This simulator allows the user to define the services which need to be performed by the agents and describe the physical network used in the process. In this paper, an approach using a genetic algorithm which optimizes the distribution of services with the objective to reduce the network load and total execution time is presented.

Keywords: multi-agent system, mobile agent network, genetic algorithm, simulation.

1 Introduction

In recent years, telecommunication systems have been transforming from traditional telecommunication systems, consisting only of network providers with basic telecommunication services, into systems based on Next Generation Network (NGN) principles [1]. The Next Generation Network (NGN) consists of different types of networks, nodes and terminals, all aimed at providing an appropriate environment for advanced services emerging from the convergence of Internet information services and traditional telecoms services (e.g. telephony). This convergence results in the creation of new innovative services which can be offered to the users [2] and which have to be sold to the user [3]. Discovery mechanisms which enables the user to find an appropriate service is also important [4]. Since the NGN integrates different networks, terminals and technologies used to create new services, it is becoming increasingly difficult to deploy new services in the network. Service deployment frameworks now have to take into account the characteristics of nodes were the service is installed (OS, installed software components, deployment environment, terminal characteristics), as well as the deployment procedures specific to the service [5].

To cope with these problems, we have designed a system called the Multi–Agent Remote Maintenance Shell (MA–RMS) [6]. The MA–RMS is an agent–based

I. Lovrek, R.J. Howlett, and L.C. Jain (Eds.): KES 2008, Part I, LNAI 5177, pp. 158–165, 2008.

framework used for remote software service execution and maintenance. It is based on a formal model of a multi–agent system called the Mobile Agent Network (MAN) [7] and is organized as a team oriented multi–agent system i.e. it consists of a team of agents. One of these agents is the planning agent responsible for distributing services to other agents in the team.

The drawback of the MA–RMS is that it currently uses predefined strategies which define how services are distributed among agents that execute them. In complex systems, like the MA–RMS it is difficult to verify system's properties formally or on a real system. Thus, in order to analyze behavior of a multi–agent system with various parameters, such as different agent coordination strategies, creating a simulation is the only viable approach. Such an approach can simulate the functionality of the system and can therefore be used to perform faster analysis of the system. The MAN Simulator we created is capable of simulating agent–based systems based on the MAN model.

To increase the effectiveness of service execution, a genetic algorithm was developed. Our solution approach uses this algorithm to find a distribution of services aimed at reducing the total execution time. The fitness function of the genetic algorithm uses simulation results from the MAN simulator to evaluate the efficiency of individual chromosomes.

The paper is organized as follows: Section 2 describes the Mobile Agent Network Simulator and the solution approach for service distribution based on a genetic algorithm. Simulation results where we compared the efficiency of the algorithm with previous service distributions is presented in Section 3. Section 4 concludes the paper.

2 Optimizing Distribution Strategies

2.1 The Mobile Agent Network Simulator

When deploying software components at a large number of remote locations using software agents, one of the major obstacles faced is determining the number of agents to use and distributing the required services among the agents. Executing software component deployment with an inefficient distribution of services can result in substantially larger deployment times.

In order to try and solve the problem we designed a MAN Simulator [8]. The MAN Simulator is capable of simulating different agent coordination strategies. It also conforms to the MAN model which allows us to analyze the performance of different service distributions for software deployment. Using the the MAN Simulator, we can perform system analysis much faster than with the MA–RMS. Related work on the MAN Simulator and the MA–RMS can be found in the [8,6,7]

Both the MAN Simulator and the MA-RMS use several predefined service distributions. Those distributions are:

- R1: a single agent executes all services on all nodes;
- R2: an agent executes a single service on one node only;
- R3: an agent executes all services on one node only;

- R4: an agent executes a specific service on all nodes;
- R5: an agent executes a specific service only once on all nodes;
- R6: services are assigned to the agents in order to exploit maximal parallelism in service execution. Mutually independent services are assigned to different agents, in order to execute them simultaneously on nodes with parallel execution supported;
- R7: a hybrid solution combining R4 and R3. An agent is responsible for a specific service on all nodes; all other agents execute all other services, each on a different node;
- R8: a hybrid solution combining R5 and R3 (specialization of R7 in the way R5 is specialization of R4).

2.2 Current Issues

The problem with the current approach is that the MAN Simulator and the MA-RMS currently use predefined service distributions which are optimal for only some network topologies. For example the R3 distribution is only optimal when used in network topologies where all nodes are connected with the links of the same bandwidth. In reality this is often not the case. Executing remote software maintenance services on a large number of nodes can be very time consuming if not performed with the an efficient distribution of services among agents. As a result it is critical to try to find the best possible distribution before starting the execution. The optimal solution can be found by running a simulation for every possible distribution of services. However, as the number of nodes and services increases, the time needed to find the optimal solution grows exponentially. Thus, we propose a heuristic approach, that uses the genetic algorithm, to find an efficient distribution of services, based on the network topology and the services that are to be executed. To goal is to find an suboptimal solution that would give better results than previously used distribution strategies in a reasonable time.

Designing a genetic algorithm involves three basic steps: defining a chromosome, designing a fitness function and configuring the parameters of the genetic algorithm. To apply the genetic algorithm principles to our problem domain we needed a framework that would be capable of executing our genetic algorithm. The API we chose to execute our genetic algorithm was JGAP [9].

2.3 Designing the Genetic Algorithm

The first step in designing our genetic algorithm was to define the structure of the chromosomes. We define a chromosome to be a par $\{N, AS_k\}$, where N represents the number of agents which will be used to execute services at remote locations, while AS_k represents the set of services to be executed. Each AS_k is defined as $AS_k = \{g_1, g_2, \ldots, g_i, \ldots, g_p\}$, where g_i is a par $g_i = \{es_{k,nes}, agent_k\}$. Here $es_{k,nes}$ represents the elementary service that is to be executed and $agent_k$ represents the agent which will be executing that service. N and every g_p from AS_k represents one gene in the chromosome. The first gene, N, restricts the number of agents which will be used to execute the services. This number depends on the number of nodes and services which must be executed at each node. Therefore, N varies from 1 to the sum of all services executed on every node. The remaining genes denote the agents that will perform individual services. In order words, the second gene defines the agent that will execute service 1, the third that will execute service 2 and so on up until the n_{th} gene which represents the agent which will perform service n-1. $agent_k$ from $\{es_{k,nes}, agent_k\}$ is represented in JGAP as an Integer Gene. The value of $agent_k$ is limited by the first gene.

The fitness of a chromosome is calculated in the following way: in every cycle of the genetic algorithm, a fitness function is called and its value is calculated using the chromosome as a parameter. Based on the genes contained in the chromosome, the fitness function configures the Mobile Agent Network Simulator. The first gene is used to determine the number of agents which are to be created in the simulation, while the remaining genes are used to assign services to agents. After the fitness function configures the simulation, it starts execution of the MAN Simulator. The result of the simulation, i.e. the final fitness value, represents the total execution time needed to execute all the services at the remote nodes. Naturally, the lower the execution time, the more efficient the distribution of services is. Chromosomes which correspond to lower fitness values have a higher chance of continuing their existence in the next generation. The genetic algorithm uses the following parameters:

- The size of the population was 50;
- The number of evolutions the algorithm performed before returning the optimal distribution of services was 50;
- The only operator used by the genetic algorithm was the crossover operator. The crossover probability was 50%;
- The best chromosome from every evolution went on into the next generation unchanged (i.e. an elitistic approach was used).

When defining which parameters will be used for the genetic algorithm it was essential to find the parameters that could generate good results in different network topologies and a set of services that are to be executed. For that reason we have performed a series of testing in which we changed the parameters of the genetic algorithm and analysed how good was the distribution of services from the genetic algorithm compared to other distributions. This testing was performed on a variety of different network topologies and a set of services. The parameters selected produce best results in a reasonable time.

When designing the genetic algorithm, we encountered a problem with the crossover operator. The cause of the problem was the fact that the number of agents which are available to execute services is limited by the first gene. Figure 1 shows how a corrupted chromosome can be generated. Namely, during crossover, the last gene from chromosome 1 and chromosome 2 exchanged their positions. This resulted in the creation of a new service distribution where $task_4$ was to be executed by $agent_4$. However, this new chromosome can use only two agents according to its first gene, and yet services are assigned to three different agents.

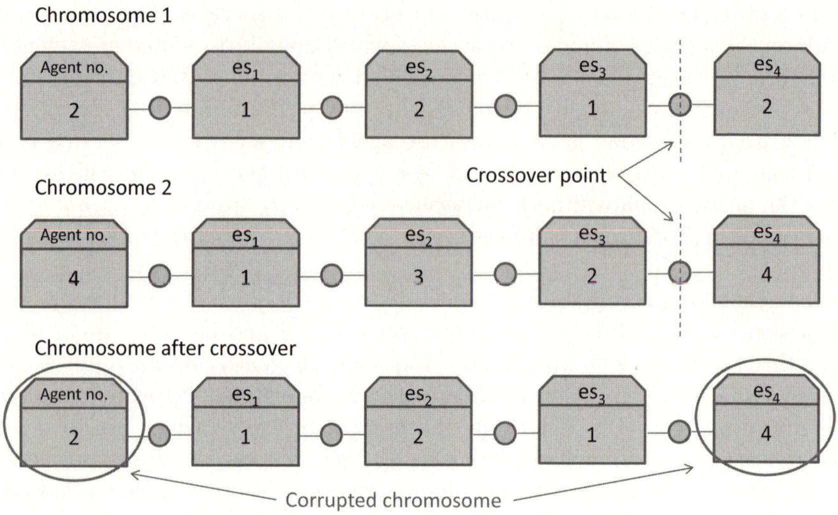

Fig. 1. A corrupted chromosome

This problem was solved in the following way: the invalid chromosome was left unchanged in the population, but the fitness function was modified in such a way as to create a new chromosome if an invalid chromosome was detected. In this chromosome, the agent numbers assigned to tasks which were greater than the agent number in the first gene were replaced by the value of $agent_k$ modulo N. The newly created chromosome was then used to configure the MAN Simulator.

3 Result Verification

In order to verify the efficiency of the proposed genetic algorithm, we performed simulations on three different network topologies. Figure 2 shows the overall network topology. In the first series of simulations, the bandwidth of all links was set first to 512 Kbit/s, than to 1 Mbit/s and finally to 10Mbit/s in . In the second series of simulations, the link between switch SW6 and SW2 was kept constant at 512 Kbit/s, while the remaining link bandwidths were set first to 1 Mbit/s and then to 10 Mbit/s. In the last simulation, link bandwidths between SW3 and SW5 and SW2 and SW4 were kept constant at 512 Kbit/s, while the remaining link bandwidths were set to 1 Mbit/s and then to 10 Mbit/s. The results of the simulations are shown in Fig. 3.

In the first series of simulations (Fig. 3a), the best distribution strategy in all simulations was the R3 distribution strategy which distributes all services on one node to one agent. The service distribution generated had (in average for all the simulations in the series) 27% larger total execution time when compared to the R3. The worst strategy in all the simulations was the R1 distribution with had 734% larger results in average. For this topology, the optimal service

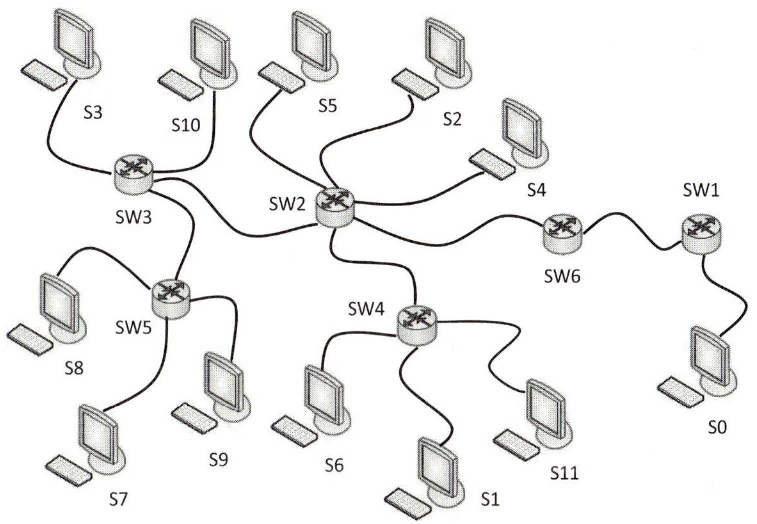

Fig. 2. Network topology used in the simulations

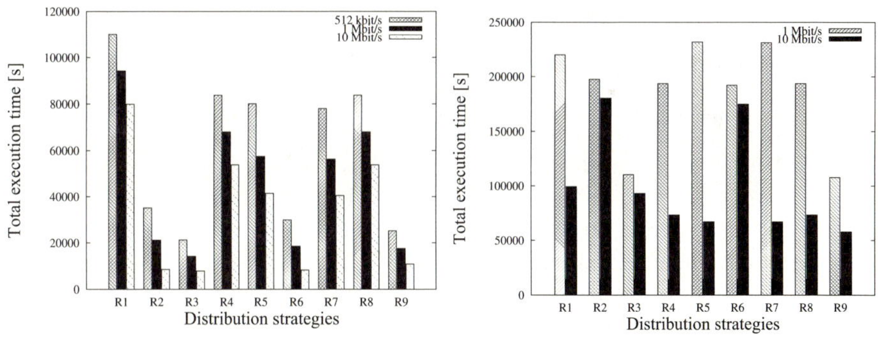

a) Results from simulation series 1 b) Results from simulation series 2

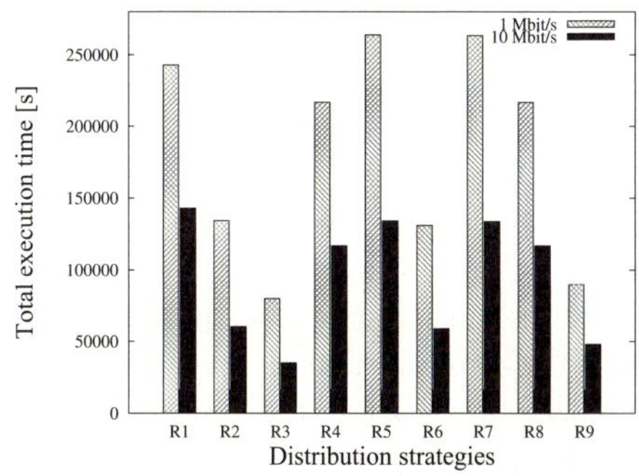

c) Results from simulation series 3

Fig. 3. Results from the simulations

distribution strategy is the R3. The reason for this is that the all link bandwidth in all cases are large enough so that all the agents can migrate to the remote nodes and start executing the services almost in parallel. For the topology from the second series of simulations (Fig. 3b), the best results were obtained by the genetic algorithm in both simulations in this series. The fastest distribution strategy after the genetic algorithm was the R3 which had 32% larger results in average. The worst strategy was the R2 strategy with average 248% larger results. The reason the genetic algorithm was the best in this series is because of the bottleneck between switches SW2 and SW6. The first agents that migrated on the remote location finished their execution before some of the agents could migrate. This allows the genetic algorithm to generate a distribution where the first agents would execute more services that the ones that migrated the last. In the third simulation (Fig. 3c), the best distribution of strategies was the R3. The genetic algorithm had 25% larger results in average.

From the simulation results, we can conclude that the distribution of services generated by the genetic algorithm generates better results on network topologies where there are bottlenecks between two network subnets. This also means that it generates better results on network topologies where there are variations between link bandwidths since in these cases the mechanisms of the genetic algorithm can produce service distribution that could maximize service execution parallelism.

4 Conclusion and Future Work

In this paper, we presented a solution approach for using a genetic algorithm for the optimization of service distribution strategies which are used by the MA–RMS while performing services at remote locations. In its every iteration the genetic algorithm calculates the effectiveness of the current chromosomes based on the results of simulations it performed by our Mobile Agent Network Simulator which is capable of simulating different agent coordination strategies. The best chromosomes evolve into the next generation. In the performance analysis, we compared the results with those of the existing predefined distribution already used by the MA–RMS. The analysis showed that in a very short time, the algorithm can generate service distributions with very effective total execution times.

In all our series of simulations, the service distribution generated by the genetic algorithm was among the best distributions. The second series of simulation have shown that the genetic algorithm generates better service distributions on network topologies where there are a variety of network bandwidths on links connecting remote locations. This is especially true when there are bottlenecks in the network. The cause of these bottlenecks is that the time needed for agent migration is larger than the time needed to execute the service. This situation allows the genetic algorithm to create a distribution where the first agents that migrate execute more service that the one that migrate the last. The consequence is the increase of parallelism in service execution. The future work will include trying to add more genetic operators and improve existing ones which should increase the effectiveness of this approach.

Acknowledgments

This work was carried out within research project 036-0362027-1639 "Content Delivery and Mobility of Users and Services in New Generation Networks", supported by the Ministry of Science, Education and Sports of the Republic of Croatia.

References

1. Yoon, J.L.: Telco 2.0: a new role and business model. IEEE Communications Magazine 45(1), 10–12 (2007)
2. Sherif, M.H., Ho, S.: Evolution of operation support systems in public data networks. In: Proceedings of the 5th IEEE Symposium on Computers & Communications, pp. 72–77 (2000)
3. Podobnik, V., Petric, A., Jezic, G.: The CrocodileAgent: Research for efficient agent-based cross-enterprise processes. In: Meersman, R., Tari, Z., Herrero, P. (eds.) OTM 2006 Workshops. LNCS, vol. 4277, pp. 752–762. Springer, Heidelberg (2006)
4. Podobnik, V., Trzec, K., Jezic, G.: An auction-based semantic service discovery model for e-commerce applications. In: Meersman, R., Tari, Z., Herrero, P. (eds.) OTM 2006 Workshops. LNCS, vol. 4277, pp. 97–106. Springer, Heidelberg (2006)
5. Houssos, N., Alonistioti, A., Merakos, L., Mohyeldin, E., Dillinger, M., Fahrmair, M., Schoenmakers, M.: Advanced adaptability and profile management framework for the support of flexible mobile service provision. Special Issue on (R)Evolution towards 4G Mobile Communication Systems 10(4) (August 2003)
6. Jezic, G., Kusek, M., Desic, S., Caric, A., Huljenic, D.: Multi–agent remote maintenance shell for remote software operations. In: Palade, V., Howlett, R.J., Jain, L. (eds.) KES 2003. LNCS, vol. 2774, pp. 675–682. Springer, Heidelberg (2003)
7. Kusek, M., Jezic, G., Jurasovic, K., Sinkovic, V.: Network simulation in a fragmented mobile agent network. In: Apolloni, B., Howlett, R.J., Jain, L. (eds.) KES 2007. LNCS (LNAI), vol. 4694, pp. 214–221. Springer, Heidelberg (2007)
8. Mario, K., Kresimir, J., Ana, P.: Simulation of mobile agent network. In: 9th International Conference on Telecommunications, 2007. ConTel 2007, 13-15 June 2007, pp. 49–56 (2007)
9. Rotstan, N., Meffert, K.: JGAP (January 23, 2008), `http://jgap.sourceforge.net/`

Agent-Based User Personalization Using Context-Aware Semantic Reasoning

Fran Frkovic[1], Vedran Podobnik[1], Krunoslav Trzec[2], and Gordan Jezic[1]

[1] University of Zagreb, Faculty of Electrical Engineering and Computing, Croatia
[2] Ericsson Nikola Tesla, R&D Center, Croatia
{fran.frkovic,vedran.podobnik,gordan.jezic}@fer.hr,
krunoslav.trzec@ericsson.com

Abstract. The future of mobile telecommunications is aimed at creating a user-centric wireless world which takes into account user's preferences, as well as communication context (e.g., network and terminal heterogeneity or user location). In order to enable context-aware personalization of communication services offered by next-generation mobile networks, we propose an agent-based approach in combination with semantic reasoning techniques from the Semantic Web. In particular, we use ontology-based user profiles to create an agent-based context-aware service which supports personalization according to user's preferences. An ontology is created which contains knowledge regarding terminal capabilities and user preferences, as well as a software agent which manages user personalization according to the context extracted from the user profile.

Keywords: Software Agents, User Personalization, Context-aware Semantic Reasoning, Ontology-based User Profiles.

1 Introduction

Third-generation (3G) mobile networks and beyond are characterized by a large number of content-rich services delivered by an infrastructure based on the Internet Protocol (IP). From a user's perspective, efficient content selection and presentation on mobile devices are tasks difficult to achieve, mainly due to the insensitivity of service offers to communication context or device capabilities [1]. Consequently, adaptive support for delivering services to mobile users is needed. This could enable more efficient service behavior based on communication context, as well as user preferences. Combining the Semantic Web and agent technologies provides a promising solution for the automation of user personalization tasks which offers both semantic-aware and context-aware information processing related to content-rich mobile services [2].

The level at which modern computers manage information can be described as the *data* level in the DIKW (*Data, Information, Knowledge and Wisdom*) information hierarchy [3] shown in Fig. 1. *Data* is the most basic level which comes in the form of raw observations or measurements, while *information* adds semantics to *data*. *Knowledge* tells us *how* to use it, and *wisdom* tells us *when* to use it.

I. Lovrek, R.J. Howlett, and L.C. Jain (Eds.): KES 2008, Part I, LNAI 5177, pp. 166–173, 2008.

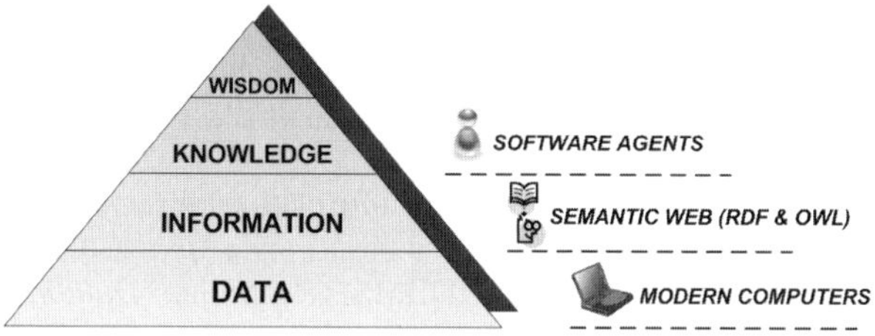

Fig. 1. The DIKW hierarchy. Modern computers process at the *data* level. Semantic Web technologies bring computers to the *information* level, while software agents introduce the *knowledge* layer [4].

The development of the idea of semantic reasoning has resulted in a large number of data models and languages. Among them are RDF[1] (*Resource Description Framework*), RDFS[2] (*RDF Schema*) and OWL[3] (*Web Ontology Language*).

This paper is organized as follows. In Section 2, we present the technologies of the Semantic Web which enable semantic reasoning. Section 3 describes related work regarding user personalization issues in mobile networks. In Section 4, we present ontology-based user profiles which contain terminal capabilities and user preferences, as well as case studies which demonstrate agent-based user personalization. Section 5 proposes ideas for future research work and concludes the paper.

2 The Semantic Web

Currently, Web accessible resources are mainly described using HTML, and presented to human users via Web browsers. HTML, however, does not enable computers to fully interpret the information. Internet pioneer Tim Berners-Lee speaks of a "dream" of the future in which computers are truly capable of analyzing data on the Web [5] and presenting it in a human-friendly way. The Semantic Web is a vision in which knowledge is organized into conceptual spaces according to meaning, and keyword-based searches are replaced by semantic query answering [6].

Formally, an ontology is a statement of logical theory. An ontology in the context of information science is a *data model* which represents certain concepts within a domain of interest, as well as relationships between these concepts. Ontologies are used in the areas of AI, the Semantic Web, etc.

RDF is a family of W3C specifications generally used for modeling information. The RDF model is based on statements or triples, which include a subject, a verb and an object (SVO). A collection of RDF statements is represented by a labeled, directed pseudo-graph. While RDF allows users to describe resources using their own vocabulary and does not make assumptions on any particular domain, RDFS is used to define the semantics of a domain. The main RDFS constructs are *Class* and *subClass* relations, as well as the ability to define domains and a range of properties.

[1] http://www.w3.org/RDF/
[2] http://www.w3.org/TR/rdf-schema/
[3] http://www.w3.org/TR/owl-features/

OWL can be considered an evolution of RDF/RDFS in its ability to represent machine-processable semantic content. OWL adds a number of features to RDF/RDFS, such as the local scope of properties, disjointness of classes, cardinality restrictions and special characteristics of properties.

Semantic queries are the main means of information retrieval used in current research in this area. Inspiration for a query-based style of reasoning stems directly from the widespread propagation of RDBMS (*Relational Database Management Systems*). Semantic query languages have a number of features in which they differ from SQL queries due to Semantic Web knowledge, which can be either *asserted* (explicitly stated) or *inferred* (implicit), being *network structured,* rather than *relational.* Also, the Semantic Web assumes an OWM (*Open World Model*) in which the failure to derive a fact does not imply the opposite [7], in contrast to *closed world* reasoning where all relations that can not be found are considered false [8].

3 Related Work

The W3C is working on the CC/PP[4] (*Composite Capabilities / Preferences Profile*), an RDF-based specification which describes device capabilities and user preferences used to guide the adaptation of content presented to that device. It is structured to allow a client to describe its capabilities by reference to a standard profile, accessible to an origin server or other sender of resource data, and a smaller set of features that are in addition to or different than the standard profile. A set of CC/PP attribute names, permissible values and associated meanings constitute a CC/PP vocabulary.

OMA's (*Open Mobile Alliance*) UAProf[5] (*User Agent Profile*) specification, based on the CC/PP, is concerned with capturing classes of mobile device capabilities which include the hardware and software characteristics of the device. Such information is used for content formatting, but not for content selection purposes. The UAProf specification does not define how the user preferences part of the profile is structured.

An intelligent software agent capable of performing semantic reasoning is supposed to manage user personalization tasks (i.e., both content selection and formatting) according to user profiles described by an UAProf schema-based OWL ontology. The concept of software agents appeared in the mid-1990's [9] and resulted in the application of an agent-based computing paradigm in various research domains [10, 11, 12, 13]. However, multi-agent systems have recently become very relevant with the advent of the Semantic Web.

4 User Personalization

An explicit user request or an event triggered in a *user's device* (i.e., terminal) describing e.g., its battery status or location initiates a reaction from the agent managing the course of service provisioning. RDQL[6] (*RDF Data Query Language*) and SeRQL[7] (*Sesame RDF Query Language*) semantic queries are constructed and performed on

---

[4] http://www.w3.org/Mobile/CCPP/
[5] http://www.openmobilealliance.org/
[6] http://www.w3.org/Submission/2004/SUBM-RDQL-20040109/
[7] http://www.openrdf.org/doc/sesame/users/ch06.html

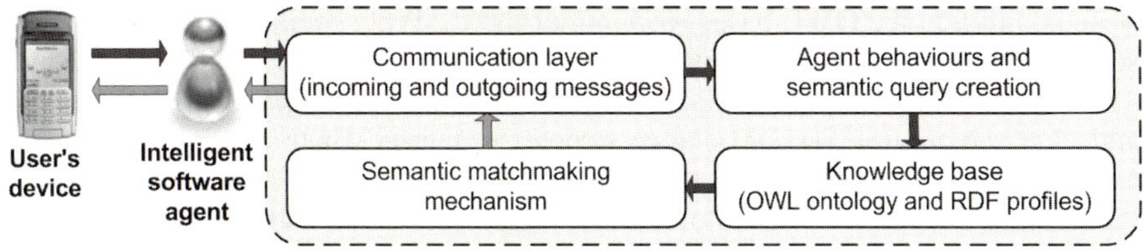

Fig. 2. User personalization overview. An event from the *user's device* is processed by the *intelligent software agent* and required *semantic queries* are performed upon *knowledge base*. *Semantic matchmaking mechanism* determines appropriate response.

two types of information: general device capabilities and individual user preferences. An overview of the agent's desired functionality is shown in Fig. 2.

Using the retrieved information, the appropriate content can be selected, formatted, and presented to the user. In order to achieve proof-of-concept agent-based contextual user personalization, CC/PP compliant user profiles are created according to the OWL ontology based on the UAProf schema.

4.1 Ontology Modeling

According to the CC/PP and UAProf schema, the main elements of a user profile are components and their corresponding attributes. With reference to that fact, a class named *Component*, intended to be an upper-class for every component created, was placed in the UAProf schema-based profile ontology and extended by components representing the software and hardware capabilities of a user's device (classes *BrowserUA*, *HardwarePlatform*, *NetworkCharacteristics*, *PushCharacteristics*, *SoftwarePlatform*, *WapCharacteristics*), as well as individual user preferences (class *UserPreferences*), which are not included in the UAProf schema. A class *Profile* was created with the constraint that it must have a relation with at least one component. A fragment of the created ontology's taxonomy is shown in Fig. 3. Also, a great number of properties representing attributes from the CC/PP and UAProf schema were created and linked to the appropriate components.

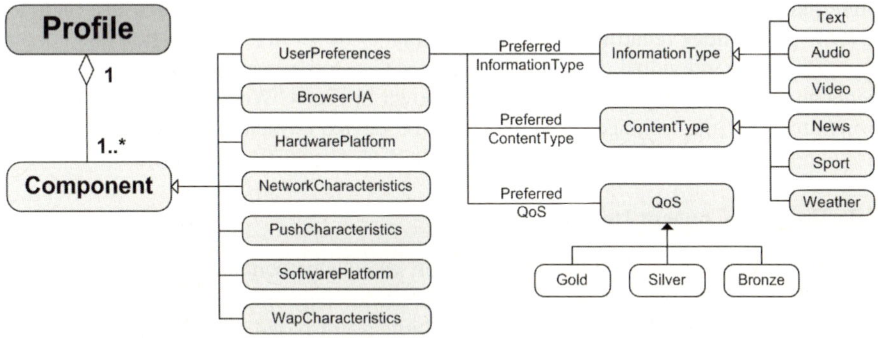

Fig. 3. A fragment of a class taxonomy of our OWL ontology including three attributes which have the *UserPreferences* class as the domain, and the *InformationType*, *ContentType* and *QoS* classes as the range. Additionally, three instances of the *QoS* class are shown.

In addition to the OWL-based mapping of the UAProf specification, classes representing content, content type, information type, and quality of service were added. The *Content* class contains available content while the *InformationType*, *ContentType* and *QoS* (*Quality of Service*) classes contain instances which are used to describe available content. For example, a specific instance of the *Content* class can represent the weather forecast in the form of a low resolution streaming video.

4.2 User Profiles

A great number of user profiles which are in compliance with UAProf specifications can be found in the W3Development's UAProf repository[8]. However, these profiles had to be processed in order to comply with the OWL ontology described in subsection 4.1. The UAProf schema was expanded to support user preference description, which enables us to not only format, but also select the appropriate information for each user. Profile components, attributes, and values refer to an ontology, while the profile itself is an instance of a schema.

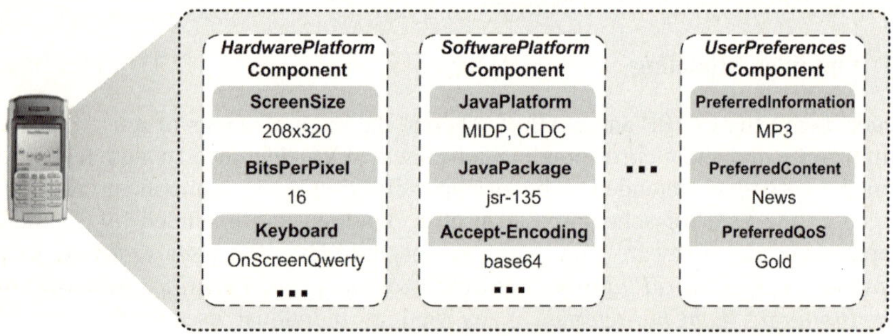

Fig. 4. A fragment of a user profile showing three components: *HardwarePlatform*, *Software-Platform* and *UserPreferences*, each with three related attributes

The profiles we created to test our approach covered a variety of mobile devices, ranging from older text and audio-only phones, to newer models capable of reproducing high quality multimedia content. The user preference parts of the profiles also included diverse possibilities regarding the requested quality of service, information and content type. Part of an individual profile, showing how the profile is built from various components and a set of designated attributes, is shown in Fig. 4. We can see how the profile brings together the technical capabilities of the device (which affect content presentation) and the user preferences (which direct the content selection process).

4.3 Proof-of-Concept Implementation

In order to demonstrate the inter-play between knowledge contained in the OWL ontology and individual profiles composed of the device capabilities and user preferences (i.e., context information), we conducted multiple agent-based user personalization case studies. Information was retrieved by means of RDQL and SeRQL queries. A Sesame

[8] http://w3development.de/rdf/uaprof_repository/

[14, 15] repository with OWL support [16] was utilized to store the required knowledge. The program component implemented in Java provides an interface for repository management, querying and the processing of results.

Retrieving and Ranking Content for a Particular User. Considering that user profiles store device capabilities and user preferences, a service provider should be able to deliver the most suitable content in the appropriate form for each user. Semantic queries are used to retrieve the available content which is then ranked according to user preferences and, finally, delivered to the user, as shown in Fig. 5.

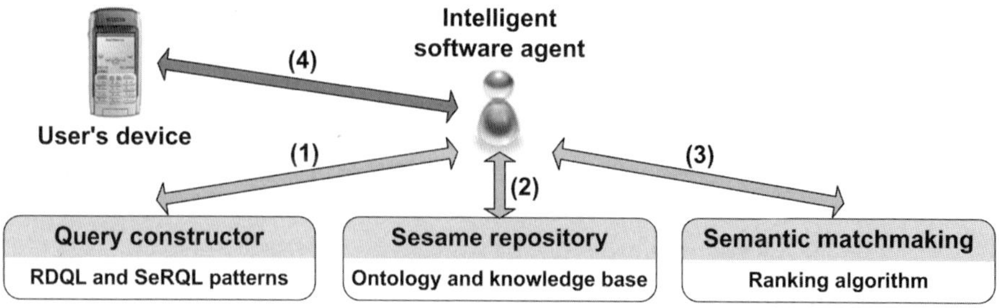

Fig. 5. An *Intelligent agent* communicates with the *user's device*. Queries are created (1) and sent to the *Sesame repository* (2). The *Matchmaking algorithm* ranks the acquired information (3) according to user preferences. Finally, the preferred content is sent to the user (4).

Consider the following example. Suppose it is necessary to retrieve information regarding all users who are interested in the weather forecast and their device's screen size. A simple RDQL query can be formed as follows:

```
SELECT ?x,?z WHERE (?x, <p:PreferredContent>, ?y),
(?x, <p:ScreenSize>, ?z) AND ?y="WeatherForecast"
```

An important role in this process is played by the semantic matchmaking mechanism which enables a more expressive ranking of content by taking into account the meaning, relation and semantic similarity of resources introduced by utilizing Semantic Web languages for information representation.

Content Formatting for Users. Context information and ontology-based user profiles are first used to find users interested in a certain category of content, and then to format the appropriate content making it presentable to each user. It is also verified whether the discovered users have the minimum required technical capabilities. For example, semantic queries can be generated to find all those users who are interested in any kind of images (low or high resolution, different number of colors) and are able to display them. Available images are processed to fit each user's terminal depending on their technical specifications, as shown in Fig. 6.

Finding Users Interested in Specific Content. This scenario supposes that we need to find users who might be interested in a given specific content. Users are chosen if they prefer the same content type, information type or quality of service that the particular content has. Content properties are discovered through semantic query

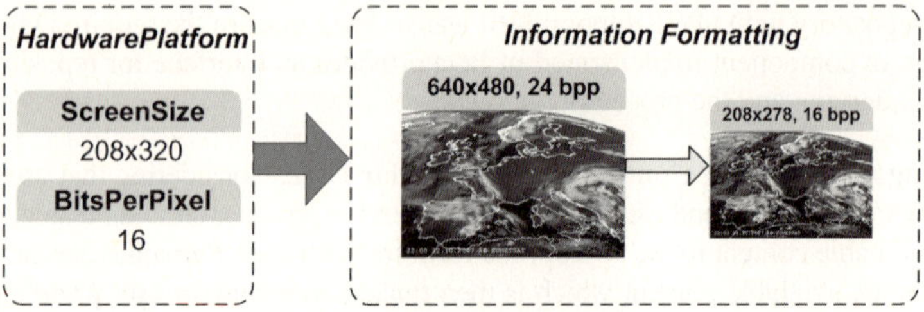

Fig. 6. Information formatting demonstrated on an image file representing the weather forecast. Device capabilities stored in the *HardwarePlatform* component of a profile guide user personalization. Media information is processed to match application-specific requirements.

creation and ontology querying similar to that in the previous example, while user preferences are found in the RDF-based knowledge database. The case study is expanded by using the semantic matchmaking algorithm in order to find a wider base of users who might be interested in the information with slightly reduced probability.

5 Conclusion and Future Work

In this paper, we describe how context-aware semantic reasoning can affect the course of telecommunication service provisioning and enable agent-based contextual user personalization. The UAProf schema was mapped to an OWL ontology, while features for user preferences support were added. User profiles, describing a number of different mobile devices and user preferences, were created. Finally, case studies demonstrating the advantages of combining the Semantic Web and agent technologies were implemented and described.

Future research will include further development of the terminal capabilities ontology by expanding it to provide more context information. Another important step will be to find better and more precise algorithms for semantic matchmaking and ranking of eligible resources which should result in major improvements with respect to telecommunication service provisioning.

Acknowledgements. This work was carried out within research projects 036-0362027-1639 "Content Delivery and Mobility of Users and Services in New Generation Networks", supported by the Ministry of Science, Education and Sports of the Republic of Croatia, and "Agent-based Service & Telecom Operations Management", supported by Ericsson Nikola Tesla, Croatia.

References

1. Podobnik, V., Trzec, K., Jezic, G.: Auction-Based Semantic Service Discovery Model for E-Commerce Applications. In: Meersman, R., Tari, Z., Herrero, P. (eds.) OTM 2006 Workshops. LNCS, vol. 4277, pp. 97–106. Springer, Heidelberg (2006)
2. Podobnik, V., Trzec, K., Jezic, G.: Context-Aware Service Provisioning in Next-Generation Networks: An Agent Approach. International Journal of Information Technology and Web Engineering 2(4), 41–62 (2007)

3. Ackoff, R.L.: From Data to Wisdom. Journal of Applied System Analysis 16, 3–9 (1989)
4. Dimkovski, M., Deeb, K.: Knowledge Technology through Functional Layered Intelligence. Future Generation Computer Systems 23(3), 295–303 (2007)
5. Berners-Lee, T., Fischetti, M.: Weaving the Web, Harper San Francisco, New York, USA (1999)
6. Antoniou, G., van Harmelen, F.: Semantic Web Primer. MIT Press, Cambridge (2004)
7. Walton, C.: Agency and the Semantic Web. Oxford University Press, New York (2007)
8. Grimm, S., Motik, B.: Closed World Reasoning in the Semantic Web through Epistemic Operators. In: Proc. of the Workshop OWL: Experiences and Directions (OWLED), Galway, Ireland (2005)
9. Nwana, H.S.: Software Agents: An Overview. Knowledge and Engineering Review 11(3), 205–244 (1996)
10. Trzec, K., Lovrek, I., Mikac, B.: Agent Behaviour in Double Auction Electronic Market for Communication Resources. In: Gabrys, B., Howlett, R.J., Jain, L.C. (eds.) KES 2006. LNCS (LNAI), vol. 4251, pp. 318–325. Springer, Heidelberg (2006)
11. Petric, A., Podobnik, V., Jezic, G.: The CrocodileAgent: Designing a Robust Trading Agent for Volatile E-Market Conditions. In: Nguyen, N.T., Grzech, A., Howlett, R.J., Jain, L.C. (eds.) KES-AMSTA 2007. LNCS (LNAI), vol. 4496, pp. 597–606. Springer, Heidelberg (2007)
12. Trzec, K., Lovrek, I.: Field-based Coordination of Mobile Intelligent Agents: An Evolutionary Game Theoretic Analysis. In: Apolloni, B., Howlett, R.J., Jain, L.C. (eds.) KES 2007, Part I. LNCS (LNAI), vol. 4692, pp. 198–205. Springer, Heidelberg (2007)
13. Jurasovic, K., Kusek., M.: Verification of Mobile Agent Network Simulator. In: Nguyen, N.T., Grzech, A., Howlett, R.J., Jain, L.C. (eds.) KES-AMSTA 2007. LNCS (LNAI), vol. 4496, pp. 520–529. Springer, Heidelberg (2007)
14. Broekstra, J., Kampman, A., van Harmelen, F.: Sesame: A Generic Architecture for Storing and Querying RDF and RDF Schema. In: Horrocks, I., Hendler, J. (eds.) ISWC 2002. LNCS, vol. 2342, pp. 54–68. Springer, Heidelberg (2002)
15. Broekstra, J., Kampman, A.: RDF(S) Manipulation, Storage and Querying Using Sesame. In: Demo Proc. of the 3rd Int. Semantic Web Conference (ISWC), Hiroshima, Japan (2004)
16. Kiryakov, A., Ognyanov, D., Manov, D.: OWLIM - A Pragmatic Semantic Repository for OWL. In: Dean, M., Guo, Y., Jun, W., Kaschek, R., Krishnaswamy, S., Pan, Z., Sheng, Q.Z. (eds.) WISE 2005 Workshops. LNCS, vol. 3807, pp. 182–192. Springer, Heidelberg (2005)

Performance Evaluation of a Mobile Agent Network Using Network Calculus

Vjekoslav Sinkovic, Mario Kusek, Gordan Jezic, and Ignac Lovrek

University of Zagreb
Faculty of Electrical Engineering and Computing
Unska 3, HR-10000, Zagreb, Croatia
{vjekoslav.sinkovic,mario.kusek,gordan.jezic,ignac.lovrek}@fer.hr

Abstract. This paper deals with performance evaluation of a mobile agent network using network calculus. A mobile agent network includes a multi-agent system and a set of processing nodes connected by a communication network where agents reside and operate as a team. We describe such a network as a queuing system where an agent represents an information unit to be served, while nodes represent servers capable of hosting, executing and communicating with agents. In a case study, a simulation-based method for performance evaluation is proposed. Various parameters for a given scenario are analysed and the experimental results are presented.

1 Introduction

New generation networks integrate different types of networks and enable full mobility of users with seamless service provisioning. In such an environment, everything is mobile, including both users and their equipment. Services should be provided on the move, be it in the same network or while changing networks during a session. Furthermore, personalisation, communication context, and user communities are becoming increasingly important to users, as well as service providers. Complex relationships between users and service providers should remain or become manageable, which is the major role of intelligent software agents representing them in the network. In an "all-mobile" network, the mobility of agents can improve network functionality and service performance. A mobile agent representing a user can migrate autonomously from node to node in order to perform certain tasks on behalf of the user.

In this paper, we analyse Mobile Agent Network performance by applying network calculus [1]. A Mobile Agent Network (MAN) consists of mobile and stationary agents organized as an agent team responsible for performing operations in a network. Network calculus is deterministic calculus that constrains input flow. A MAN includes a multi-agent system and a set of processing nodes connected by a communication network where agents reside and operate as a team. We describe such a network as a queuing system where an agent represents an information unit to be served. The nodes represent servers capable for hosting, executing and communicating with agents. We simulated a Mobile

I. Lovrek, R.J. Howlett, and L.C. Jain (Eds.): KES 2008, Part I, LNAI 5177, pp. 174–181, 2008.

Agent Network with different input parameters and used network calculus to evaluate its capabilities of executing user requests. In a case study, simulations with different parameters for a given scenario are analysed and experimental results are presented.

The paper is organised as follows: The Mobile Agent Network is described in Section 2, while Section 3 elaborates upon performance analysis and network calculus. Section 4 presents a case study, experiments and simulation results, and Section 5 concludes the paper.

2 The Mobile Agent Network

The Mobile Agent Network (MAN) is used for modelling agent organization and coordination in an agent team. The idea is that the user sends a request to the system. The request is then intelligently decomposed and executed by software agents. The MAN is represented by a triple $\{A, S, N\}$, where A represents a multiagent system consisting of cooperating and communicating mobile agents that can migrate autonomously from node to node; S is a set of ns nodes in which the agents perform operations; and N is a network that connects nodes and assures agent mobility [2].

Each processing node S_i has a unique $address_i$ from the set of addresses, $address = \{address_1, address_2, \ldots, address_i, \ldots, address_{ns}\}$. An agent is defined by a triple, $agent_k = \{name_k, address_k, service_k\}$, where $name_k$ defines the agent's unique identification, $address_k$ represents the current agent location and $service_k$ denotes the functionality the agent provides in the form of $service_k = \{es_{k,1}, es_{k,2}, \ldots, es_{k,i}, \ldots, es_{k,nes}\}$ representing a set of assigned elementary services. When $agent_k$ is hosted by node S_i then $address_k = address_i$.

A network N is represented by an undirected graph, $N = (S, E)$ which denotes network connections and assures agent mobility. The set of processing nodes is denoted as $S = \{S_1, S_2, \ldots, S_i, \ldots, S_{ns}\}$. E represents the set of links $E = \{e_1, e_2, \ldots, e_m, \ldots, e_{ne}\}$. For example $e_m = \{S_i, S_j\}$ represents a link between nodes S_i and S_j. Figure 1a shows the network considered in this paper, while Figure 1b shows its matrix representation.

A user request is decomposed into a set of elementary services that are presented by a directed acyclic graph $G = (T, L)$, where T denotes the list of elementary services, while L represents the set of directed edges that define precedence relations between elementary services [3]. Each elementary service is represented as a type of elementary service with input and output parameters. An elementary service from list T is defined as $es_i = \{i, s_{esi}, address_i\}$, where element i represents the elementary service number, s_{esi} its type, and $address_i$ denotes the address of the node where the elementary service should be executed. If the address is not defined then the elementary service can be executed on an arbitrary node that can execute elementary service s_{esi}. The agent, according to its knowledge and intelligence, decides where the elementary service will be executed. Each s_i is defined by $s_i = \{I_i, O_i\}$ where I_i represents a set of input data $I_i = \{i_1, i_2, \cdots, i_{ni}\}$ and O_i a set of output data $O_i = \{o_1, o_2, \cdots, o_{no}\}$. A set of directed edges L is defined

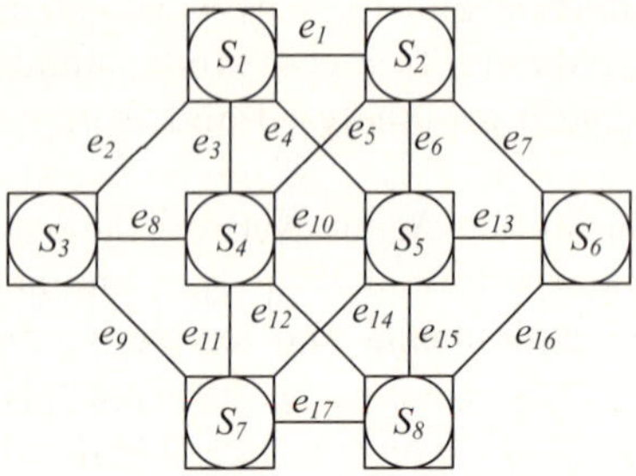

a) Graphical Representation

$$E = \left\{ \begin{matrix} \{S_1, S_2\} \\ \{S_1, S_3\} \\ \{S_1, S_4\} \\ \{S_1, S_5\} \\ \{S_2, S_4\} \\ \{S_2, S_5\} \\ \{S_2, S_6\} \\ \vdots \\ \{S_7, S_8\} \end{matrix} \right\}$$

b) Matrix Representation

Fig. 1. The test network graph

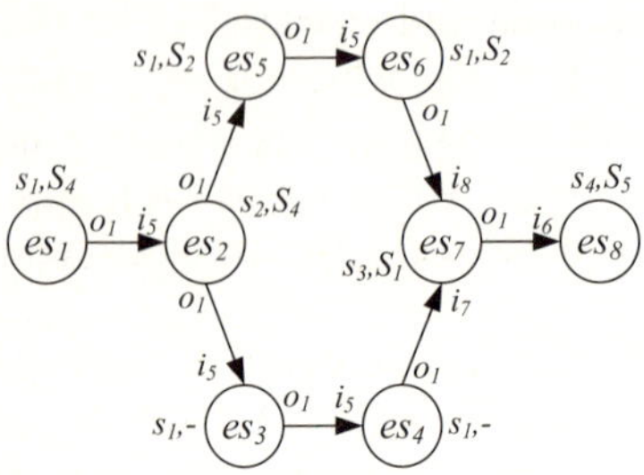

a) Graphical Representation

$$T = \left\{ \begin{matrix} \{1, 1, 4\} \\ \{2, 2, 4\} \\ \{3, 1, -\} \\ \{4, 2, -\} \\ \{5, 1, 2\} \\ \{6, 2, 2\} \\ \{7, 3, 1\} \\ \{8, 4, 5\} \end{matrix} \right\} \qquad L = \left\{ \begin{matrix} \{1, 1, 2, 5\} \\ \{2, 1, 3, 5\} \\ \{2, 1, 5, 5\} \\ \{3, 1, 4, 5\} \\ \{4, 1, 7, 7\} \\ \{5, 1, 6, 5\} \\ \{6, 1, 7, 8\} \\ \{7, 1, 8, 6\} \end{matrix} \right\}$$

b) Matrix Representation

Fig. 2. The operations graph

as $L = \{l_1, l_2, \cdots, l_i, \cdots, l_{nl}\}$. Each l_i is defined as $l_i = \{t_{io}, o_i, t_{ii}, i_i\}$ where t_{io} is the operation number of the output parameter o_i and t_{ii} is the task number of input parameter i_i. If the operation receives input parameters during creation, the edge is not presented in the operations graph. An example of an elementary service graph is shown in Figure 2, displaying both graphical (Figure 2a) and matrix representation (Figure 2b). Consider the edge connecting output o_1 of elementary service es_1 with input i_5 of elementary service es_2. This edge is represented by the first row of matrix L. Elementary service es_2 is of type s_2 and it should be executed at node S_4. This is represented by the second row of matrix T. In the third row, there is dash in place of the execution node denoting that elementary service es_3 does not have a predefined node where it should be executed.

To execute all the operations included in the graph, there is a team of agents. The process of distributing elementary services includes determining the number of agents in the team, as well as deciding which agent will execute which elementary service [2]. Consequently, intelligent agents can choose different distributions depending on their knowledge of the network topology, service execution ability, current network and node load, etc. Consider the following example of a simple distribution: $agent_1$ will execute the first two elementary services: es_1 and es_2 and, thus, $service_1 = \{es_1, es_2\}$. $agent_2$ will execute the next two $service_2 = \{es_3, es_4\}$. For the next two agents, their corresponding services are: $service_3 = \{es_5, es_6\}$ and $service_4 = \{es_7, es_8\}$. It is evident that $agent_1$ will first execute elementary service es_1 and then the es_2 on the same node (S_4). However, $agent_4$ will first execute elementary service es_7 on node S_1 and then it will migrate to node S_5 where it will execute es_8. $agent_2$ can execute its services at any node since it is not specified.

In the prototype the agents are created on the agent platform JADE [4]. The JADE agent platform conforms FIPA standards which defines that the agent and their location is tracked and any agent can find out the current location of any other agent in the platform.

3 Performance Analysis and Network Calculus

In order to introduce performance evaluation capabilities, the mobile agent network is described as a queuing system where the agents represent information units to be served, competing to run a service on the same node (Figure 3).

Node S_i with processing capacity B_i receives input agent flow G_i and network agent flow X_{ji}. These flows are summarised as L_i and sent to queue Q_i organised according to the domain principle. When it is $agent_k$'s turn in queue Q_i, the processor executes es_{kl}. After completion, $agent_k$ either executes es_{kn+1} on the same node (flow U_i) or migrates to another node and executes its next elementary service there. The migrating agent is placed in the migration queue, TQ_i. When it is $agent_k$'s turn to migrate, it is serialised and transferred to node S_j. If $agent_k$ has completed its last elementary service, it is placed in output flow R_i.

When an agent is created (agent birth), its elementary service list is set and the agent migrates to the node where the first elementary service should be executed. After elementary service execution, the agent tries to send the result to the team agent(s) that execute(s) successor elementary service. The actual communication

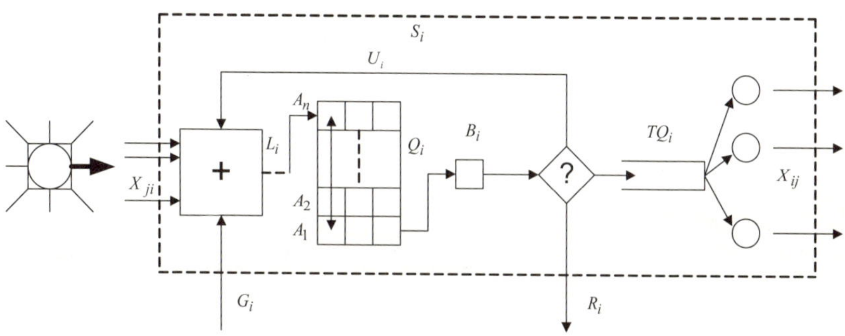

Fig. 3. Node structure

can be internal (inside of one agent), local (between agents on the same node) or global (between agents on different node). After communication, the agent takes the next elementary service from its list. If the subsequent elementary service is to be executed on a different node, the agent migrates to the node in question. This is repeated for all elementary services on its list ($service_k$). The final elementary service is the agent's death by which the agent is disposed of. In case of unpredicted situations, the agent tries to solve the problem by itself. If it fails, it contacts other agents and they cooperatively solve the problem. In a situation of node disconnection, user requests will not be served, i.e. they will be lost.

The research in the area of performance evaluation and coordination of agent systems includes different theoretical frameworks such as Markov processes [5], Petri nets [6] and classical queuing theory considering stochastic processes and average quantities in an equilibrium state [7,8]. In this paper, agent performance evaluation tools are extended to network calculus, a theory of deterministic queuing systems [9]. Some earlier work also applied network calculus to communicating agents [10].

Comparing to other system approaches, network calculus is based on Min-Plus algebra, with the underlying idea to a) regulate the input flow of bits, packets or generally information units, or in our case agents carrying elementary services, and b) use deterministic scheduling in order to achieve service guarantees. The network calculus constraints input flow with an arrival curve that can be viewed

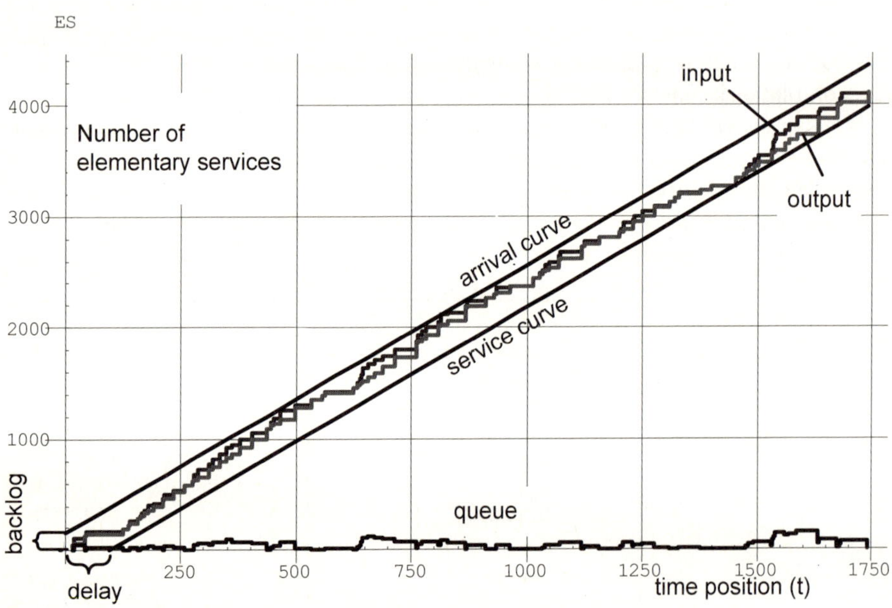

Fig. 4. Example graph

as an abstraction of a regulating mechanism, in our case defined by the maximum number of agents in a burst submitted to execution, maximum number of elementary services per agent and burst interarrival time. Output is bounded by a service curve as an abstraction of the scheduling, defining in our case the number of completed elementary services. An example showing analysis of the test network from the Figure 1 is shown in the Figure 4.

In our case study, we used basic simulation parameters and made variations therein. The basic simulation parameters follow. The considered network was composed of 8 nodes. Each node had the same processing speed of 3 elementary services per Δt, where Δt is a discrete time unit. 100 agent bursts were simulated with a mean burst interarrival time of $20\Delta t$. Inside of each burst, agent arrival intensity was 50. Each agent had a random number (up to 10) of elementary services to execute. The node where each elementary service was to be executed was also chosen randomly. In order to migrate from one node to another, the agents were assumed to require $1\Delta t$ time for each hop.

Figure 4 shows the results for simulations with the basic parameters for one node in the network. Input and output flows were constrained with arrival and service lines. The queue at the node in question is also shown, seeming to always return to 0 after some time. That means that the system is well designed and that there is no rejecting execution.

Three basic performance measures are the following: backlog, delay and agent rate. The backlog $backlog(t)$ is defined as a number of elementary services held inside the node at some position t on discrete time scale. It is expressed by a vertical deviation between input and output. The delay $delay(t)$ at position t is the delay experienced by some elementary service arriving at t if all elementary services received before it are served before it. The delay is the horizontal deviation between input and output function. Agent rate r is represented by the total number of elementary services (ES total) submitted to execution up to t. Linear arrival and service curves are defined as follows:

$$r = \frac{EStotal}{t} = \frac{3876}{1608} = 2.41 \quad (1)$$

$$arrival = backlogmax + r \times t = backlog(1608) + r \times t = 153 + 2.41xt \quad (2)$$

$$service = r \times (t - delaymax) = r \times (t - delay(1538)) = 2.41(t - 94) \quad (3)$$

4 Case Study

In each of the following experiments, we changed only one of the basic parameters. In the first experiment (Figure 5a), we changed the number of processing nodes. We can see that the backlog, delay and bitrate decrease as the number of nodes increases. When the number of nodes is 5 or higher, all the values decrease a bit, which means that the system is within normal parameters (i.e., can execute everything that comes into the system). In the second experiment (Figure 5b), the processing speed of all nodes was examined. As seen in the figure, the backlog and delay are high for processing speed below 3 elementary services per Δt.

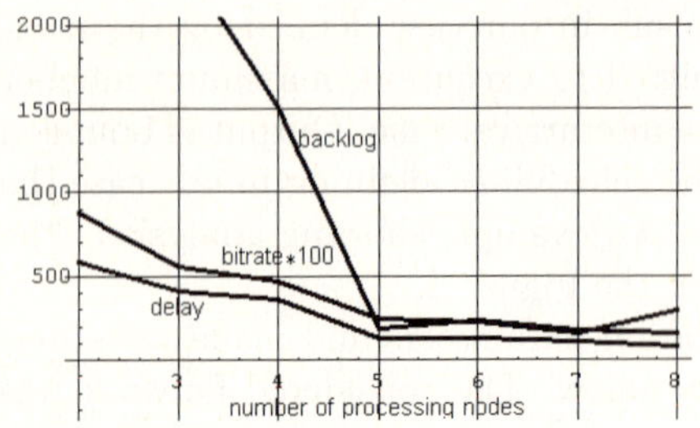

a) Number of Processing Nodes

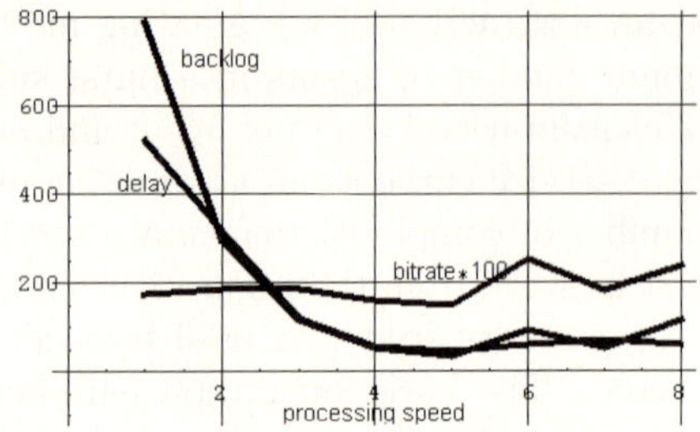

b) Processing Speed

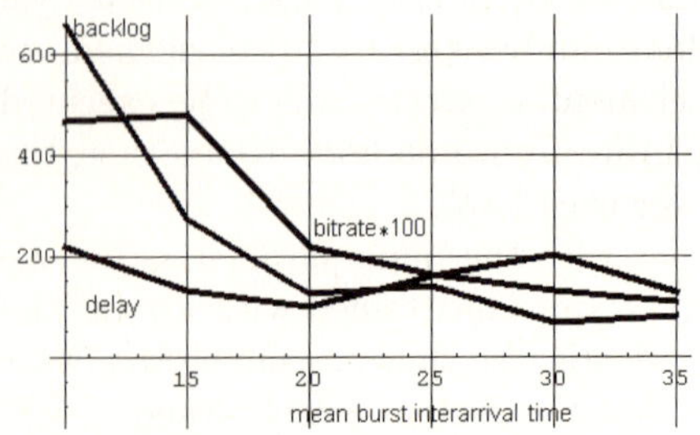

c) Mean Burst Interarrival Time

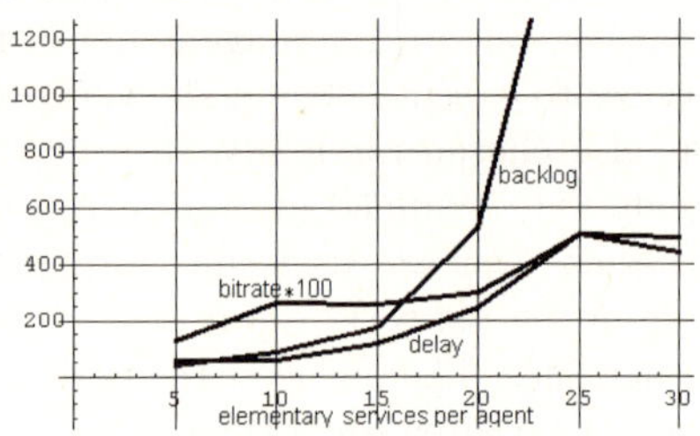

d) Maximum Number of Elementary Services per Agent

Fig. 5. Simulation results

Thus, the minimal processing speed of all nodes should be greater than 3. In the third experiment (Figure 5c), we changed the mean burst interarrival time (TA). The results show that for cases when agent bursts arrive faster (smaller TA), the backlog, delay and bitrate are high. After increasing TA to over $20\Delta t$, the system works within normal parameters. In the last experiment (Figure 5d), the maximal number of elementary services per agent was evaluated. Increasing this number to over 8 generates a jump in backlog and delay. That means that the system is within normal performance for cases where the number of elementary services per agent is below 8.

5 Conclusion

A mobile agent network is described as a queuing system and evaluated by applying network calculus. An agent represents an information unit to be served, and nodes represent servers capable of hosting, executing and communicating with agents. A simulation-based method for performance analysis is proposed. By using network calculus, we evaluated the capabilities of our Mobile Agent Network to execute complex user requests. In our experiments, various parameters for the given scenario are analysed and the results are elaborated upon.

References

1. Boudec, J.Y.L., Thiran, P.: Network calculus: a theory of deterministic queuing systems for the internet. Springer, New York (2001)
2. Kusek, M., Lovrek, I., Sinkovic, V.: Agent team coordination in the mobile agent network. In: Khosla, R., Howlett, R.J., Jain, L.C. (eds.) KES 2005. LNCS (LNAI), vol. 3681, pp. 240–245. Springer, Heidelberg (2005)
3. Lovrek, I., Sinkovic, V., Jezic, G.: Communicating agents in mobile agent network. In: Sixth International Conference on Knowledge-Based Intelligent Engineering Systems & Allied Technologies (KES), Crema, Italy, pp. 126–130 (2002)
4. JADE: Java Agent DEvelopment Framework (2008), http://jade.tilab.com/
5. Boutilier, G.: Sequential optimality and coordination in multiagent systems. In: Proceedings of the Sixteenth International Joint Conference on Artificial Intelligence, pp. 478–485. Stockholm, Sweden (1999)
6. Koriem, S.: Development, analysis and evaluation of performance models for mobile multi-agent networks. The Computer Journal 49(6), 685–709 (2006)
7. Sinkovic, V., Lovrek, I.: Generic model of a mobile agent network suitable for performance evaluation. In: Fourth International Conference on Knowledge–Based Intelligent Engineering Systems & Allied Technologies (KES), Brighton, UK, pp. 675–678 (2000)
8. Sinkovic, V., Lovrek, I.: A model of massively parallel call and service processing in telecommunications. Journal of Systems Architecture 43(6-7), 479–490 (1997)
9. Pandit, K., Schmitt, J., Steinmetz, R.: Network calculus meets queueing theory -a simulation based approach to bounded queues. In: Proceedings IWQOS 2004 Twelfth IEEE International Workshop on Quality of Service, Montreal, Canada, pp. 114–120 (2004)
10. Schiøler, H., Jessen, J.J., Dalsgaard, J., Larsen, K.G.: Network calculus for real time analysis of embedded systems with cyclic task dependencies. In: Computers and Their Applications, pp. 326–332 (2005)

Using Generalized Learning Automata for State Space Aggregation in MAS

Yann-Michaël De Hauwere, Peter Vrancx*, and Ann Nowé

Computational Modeling Lab - Vrije Universiteit Brussel
{ydehauwe,pvrancx,anowe}@vub.ac.be

Abstract. A key problem in multi-agent reinforcement learning remains dealing with the large state spaces typically associated with realistic distributed agent systems. As the state space grows, agent policies become more and more complex and learning slows. One possible solution for an agent to continue learning in these large-scale systems is to learn a policy which generalizes over states, rather than trying to map each individual state to an action.

In this paper we present a multi-agent learning approach capable of aggregating states, using simple reinforcement learners called learning automata (LA). Independent Learning automata have already been shown to perform well in multi-agent environments. Previously we proposed LA based multi-agent algorithms capable of finding a Nash Equilibrium between agent policies. In these algorithms, however, one LA per agent is associated with each system state, as such the approach is limited to discrete state spaces. Furthermore, when the number of states increases, the number of automata also increases and the learning speed of the system slows down. To deal with this problem, we propose to use Generalized Learning Automata (GLA), which are capable of identifying regions within the state space with the same optimal action, and as such aggregating states. We analyze the behaviour of GLA in a multi-agent setting and demonstrate results on a set of sample problems.

1 Introduction

Reinforcement learning (RL) has already been shown to be a powerful tool for solving single agent Markov Decision Processes (MDPs). Basic RL techniques are not suited for problems with very large state spaces, however, as they mostly rely on a tabular representation for policies and numerating all possible state-action pairs is not feasible (the so called *curse of dimensionality*). Because of these issues, several extensions have been proposed to reduce the complexity of learning. The use of temporally extended actions has recently been introduced in the RL community as a possible solution [1,2]. Other methods for representing

* Yann-Michaël De Hauwere and Peter Vrancx are both funded by a Ph.D grant of the Institute for the Promotion of Innovation through Science and Technology in Flanders (IWT Vlaanderen).

I. Lovrek, R.J. Howlett, and L.C. Jain (Eds.): KES 2008, Part I, LNAI 5177, pp. 182–193, 2008.

the agent's policy such as neural networks, decision trees and other regression techniques are already widely used.

The non-stationary environment agents experience, and the uncertainty about the other agents' goal make the problem of large state spaces even more pertinent in Multi Agent Systems (MAS). Relatively little work has been done, however, on extending these RL techniques for large state spaces to MAS.

When the state space does not have such huge dimensions, a commonly used formalization for learning is the Multi-agent Markov Decision Process (MMDP). The problem with this theoretic framework is that the state-action space grows exponentially with the number of agents. This results in longer learning times, and learning can even become impossible in a reasonable time frame. It is clear that some form of generalization over the state space is necessary, to solve this problem. With this generalization, a certain amount of accuracy must be traded in for acceptable learning times and memory considerations.

In 1996, Boutilier had already pointed out the interest and difficulties for using structured problem representations and generalization techniques in multiagent environments [3] after having shown the usefulness of using Bayesian Networks (BN) to compactly represent the state transition function [4]. More recently, Guestrin et al. [5] introduced an algorithm for planning in cooperative MAS, using factored value functions and a simple message passing scheme among the agents to coordinate their actions.

However, all these approaches often assume that agent actively collaborate with each other or have full knowledge of the underlying transition and reward functions. Most research done so far towards learning to aggregate states in the RL problem has focused on learning the structure of the underlying problem [6,7,8]. Using this learned model, conventional techniques such as Dynamic Programming (DP) can be used to solve the problem.

In this paper we present a system where agents act completely individually, without knowledge of other agent's rewards or actions and without trying to model the problem at hand.

2 Background

2.1 Multi-agent MDPs

An extension of single agent Markov decision problems (MDPs) to the cooperative multi-agent case can be defined by Multi-agent MDPs (MMDPs) [9]. Formally, we can define an MMDP as a five-tuple, $M = \langle \mathbb{A}, \{A\}_{\forall i \in \mathbb{A}}, \mathbb{S}, T, \mathbb{R} \rangle$, where:

- $\mathbb{A}$ is the set of agents participating in the game,
- $\{A\}_{\forall i \in \mathbb{A}}$ is the set of actions available to agent i,
- $\mathbb{S}$ is the set of states (same set as an MDP),
- $T(s, \overrightarrow{a}, s)$ is the transition function stating the probability that a joint-action a will lead the agents from state s to state s',
- and $R : S \times A_1 \times \ldots A_{|\mathbb{A}|} \rightarrow \mathbb{R}$ is the reward function denoting the reward the agents get for performing a joint action in the current state.

Because the agents share the same transition and reward function, one can think of the collection of agents being a single super agent with joint actions at its disposal and whose goal is to learn the optimal policy for the joint MDP. Since the agents' individual action choices may be jointly suboptimal, the added problem in MMDP's is for the agents to learn to coordinate their actions so that joint optimality is achieved.

2.2 Factored Representations

In Multi-agent Factored MDPs [10] system states are described using a set of random variables $X = \{X_1, \ldots, X_n\}$ where each state variable X_i can assume values in a finite domain $Dom(X_i)$. Each possible system states corresponds to a value assignment $x_i \in Dom(X_i)$ for every state variable X_i.

The state transition function in such systems is described by a *Dynamic Bayesian Network (DBN)*. This is a two-layer directed acyclic graph G_a where the nodes are $\{X_1, \ldots, X_n, X'_1, \ldots, X'_1\}$. In this graph, the parents of X'_i are denoted by $Parents_a(X'_i)$. With every node $X'_i \in G_a$, a Conditional Probability Distribution $CPD^a_{X_i}(X'_i|Parents_a(X'_i))$ is associated quantifying the DBN. This method benefits from the dependencies that exist (or don't exist) between the variables of the network.

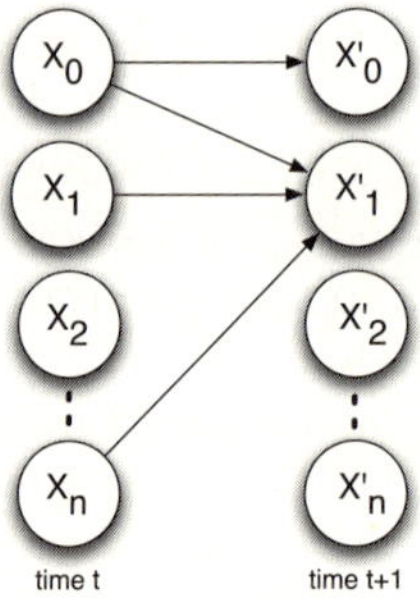

Fig. 1. DBN for action a_i

Figure 1 shows the dependencies between the variables at time t and time $t+1$ when action a_i is executed. According to this network, the Parents of X'_1 are X_0, X_1 and X_n.

2.3 Learning Automata

Learning Automata are simple reinforcement learners which attempt to learn an optimal action, based on past actions and environmental feedback. Formally, the automaton is described by a tuple $\{A, \beta, p, T\}$ where $A = \{a_1, \ldots, a_r\}$ is the set of possible actions the automaton can perform, p is the probability distribution over these actions, β is a random variable between 0 and 1 representing the evironmental response, and T is a learning scheme used to update p.

A single automaton is connected in a feedback loop with its environment. Actions chosen by the automaton are given as input to the environment and the environmental response to this action serves as input to the automaton. Several automaton update schemes with different properties have been studied. In this paper we use the so called Linear Reward Inaction (L_{R-I}) scheme:

$$p_m(t+1) = p_m(t) + \lambda\beta(t)(1 - p_m(t)) \quad (1)$$
$$\text{if } a_m \text{ is the action taken at time } t$$
$$p_j(t+1) = p_j(t) - \lambda\beta(t)p_j(t) \quad (2)$$
$$\text{if } a_j \neq a_m$$

Where $\lambda \in [0, 1]$ is a constant called the reward parameter or learning rate.

2.4 Generalized Learning Automata

A Generalized Learning Automaton (GLA) is an associative reinforcement learning unit. The purpose of a GLA is to learn a mapping from given inputs or contexts to actions. At each time step the GLA receives an input which describes the current system state. Based on this input and its own internal state the unit then selects an action. This action serves as input to the environment, which in turn produces a response for the GLA. Based on this response the GLA then updates its internal state.

Formally, a GLA can be represented by a tuple (X, A, β, u, g, T), where X is the set of inputs to the GLA and $A = \{a_1, \ldots, a_r\}$ is the set of outputs or actions the GLA can produce. $\beta \in [0, 1]$ again denotes the feedback the automaton receives for an action. The real vector u represents the internal state of the unit. It is used in conjunction with the probability g to determine the action probabilities, given an input $x \in X$:

$$P\{a(t) = a|u, x\} = g(x, a, u) \quad (3)$$

where g has to satisfy following conditions:

$$g(x, a, u) \geq 0 \; \forall x, a, u$$
$$\sum_a g(x, a, u) = 1 \quad \forall x, u$$

T is a learning algorithm which updates $\mathbf{u}$, based on the current value of $\mathbf{u}$, the given input, the selected action and response β. In this paper we use a modified version of the REINFORCE [11] update scheme. In vector notation this update scheme can be described as follows:

$$\begin{aligned}\mathbf{u}(t+1) = \mathbf{u}(t) + \lambda\beta(t)\tfrac{\delta ln\, g}{\delta \mathbf{u}}(\mathbf{x}(t), a(t), \mathbf{h}(\mathbf{u}(t))) \\ +\lambda\mathbf{K}(\mathbf{h}(\mathbf{u}(t)) - \mathbf{u}(t))\end{aligned} \quad (4)$$

where $\mathbf{h}(\mathbf{u}) = [h_1(u_1), h_2(u_2), \ldots h_r(u_r)]$, with each h_i defined as:

$$h_i(\eta) = \begin{cases} L_i & \eta \geq L_i \\ 0 & |\eta| \leq L_i \\ -L_i & \eta \leq -Li \end{cases} \tag{5}$$

In this update scheme λ is the learning rate and $L_i, K_i > 0$ are constants. The update scheme can be explained as follows. The first term added to the parameters is a gradient following term, which allows the system to locally optimize the action probabilities. The next term uses the $h_i(u)$ functions to keep parameters u_i bounded within predetermined boundaries $[-L_i, L_i]$. This term is added since the original REINFORCE algorithm can give rise to unbounded behavior. In [12] it is shown, that the adapted algorithm described above, converges to local maxima of $f(\mathbf{u}) = E[\beta|\mathbf{u}]$, showing that the automata find a local maximum over the mappings that can be represented by the internal state in combination with the function g.

3 GLA for State Space Aggregation

As was explained in the previous section, GLAs can be used to learn a mapping of inputs to actions. Originally these systems were proposed for classification problems, in which the context vectors represent features of objects to be classified and the GLA output represents class labels. We propose to use the same techniques in factored MMDPs. In such a system each agent internally uses a set of GLA to learn the different regions in the state space where different actions are optimal.

We use the following set-up for the GLA. With every action $a_i \in A$ the automaton can perform, it associates a vector $\mathbf{u_i}$. This results in an internal state vector $\mathbf{u} = [\mathbf{u_1}^\tau \dots \mathbf{u_r}^\tau]$ (where τ denotes the transpose). With this state vector we use the Boltzmann distribution as probability generating function:

$$g(x, a_i, \mathbf{u}) = \frac{e^{\frac{x^\tau u_i(a_i)}{T}}}{\sum_j e^{\frac{x^\tau u_j(a_i)}{T}}} \tag{6}$$

Of course, since this function is fixed in advance and the environment in general is not known, we have no guarantee that the GLA can represent the optimal mapping. For instance, when using the function given in Equation 6 with a 2-action GLA, the internal state vector represents a hyperplane. This plane separates context vectors which give a higher probability to action 1 from those which action 2. If the sets of context vectors where different actions are optimal, are not linearly separable the GLA cannot learn an optimal mapping.

To allow a learner to better represent the desired mapping from context vectors to actions, we can utilize systems composed of multiple GLA units. For instance the output of multiple 2-action GLAs can be combined to allow learners to build a piecewise linear approximation of regions in the space of context

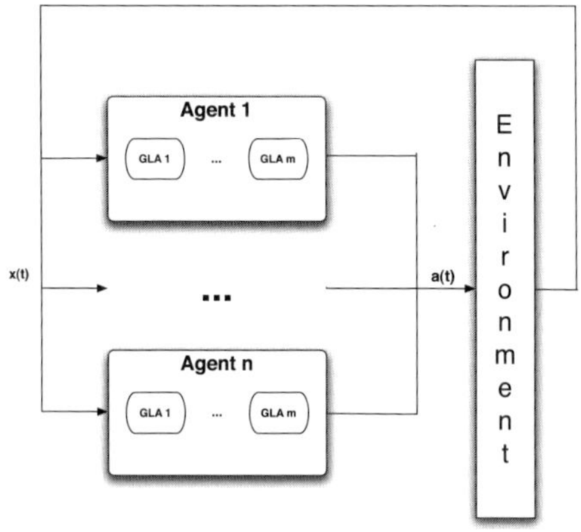

Fig. 2. Learning set-up. Each agent receives factored state representation as input. GLA decide action to be performed.

vectors. In general, we can use systems which are composed of feedforward structured networks of GLA. In these networks, automata on one level use actions of the automata on the previous level as inputs. If the feedforward condition is satisfied, meaning that the input of a LA does not depend on its own output, convergence to local optima can still be established [13].

Figure 2 shows the general agent learning set-up. Each time step t a vector $\mathbf{x(t)}$ giving a factored representation of the current system state is generated. This vector is given to each individual agent as input. The agents internally use a set of GLA to select an action corresponding to the current state. The joint action $\mathbf{a(t)}$of all agents serves as input to the environment, which responds with a feedback $\beta(t)$ that agents use to update the GLA. One of the main advantages of this approach is that convergence guarantees exist for general feedforward GLA structures. In the common interest problems under study in this paper, a group of agents each internally using one or more GLA can be viewed as a single large network of GLA, thus ensuring convergence to a local optimum.

4 Experimental Results

In this section we demonstrate our approach in a number of relatively simple experiments. Our basic experimental set-up is shown in Figure 3. Two agents A and B move on a line between $[-1, 1]$. Each time step both agents select action *left (L)* or *right (R)*, move and then receive a reward based on their original joint location and the joint action they chose. Each agent then updates using only the reward signal and the joint location, without any knowledge of the action selected by the other agent. The GLA use a continuous state space, whereas for the comparison with the traditional LA (section 4.3), we used a discretisation of our real line into 201 distinct locations, to ensure the

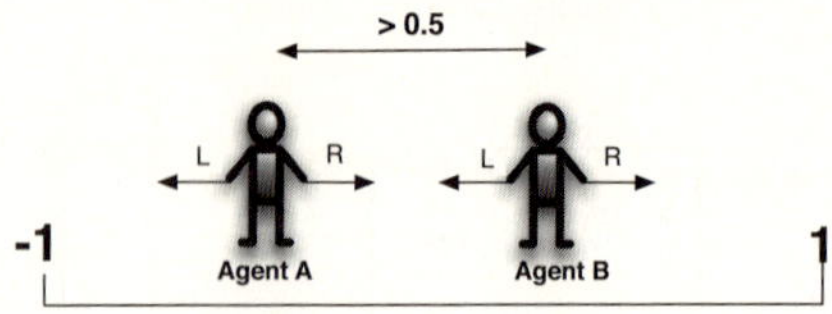

Fig. 3. Experimental set-up. Two agents move around on a line between positions −1 and 1. Each time step both agents take a step left or right.

Markov property on our environment. This yields in a state space of 201×201 — $40,401$ possible joint locations.

4.1 Agents Using a Single GLA

In this experiment the state space is divided in three regions, as shown in Figure 4(a). In region 1 Agent A is left of Agent B. In the second, Agent A is to the right of Agent B. The third region encapsulates all the states where the absolute value of the distance between the two agents is less than 0.5. Each agent has two possible actions, i.e. *Left* or *Right*. The reward scheme is as follows:

1. Region 1: A reward of +1 is given, when both agents choose action *Left*, 0 otherwise.
2. Region 2: A reward of +1 is given, when both agents choose action *Right*, 0 otherwise.
3. Region 3: A reward of +1 is given, when both agents move apart from each other, 0 otherwise.

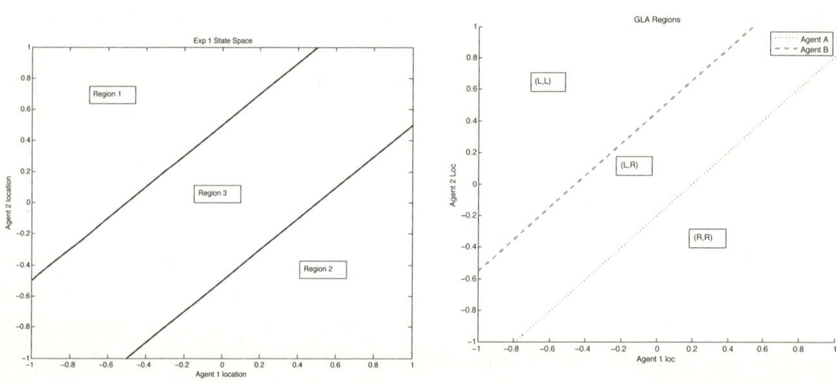

(a) (b)

Fig. 4. Experiment 1. (a) State space regions for experiment 1. (b) Typical result learnt by GLA. Lines separate regions where agents prefer different actions. Joint actions with highest probability are given in each region. Parameter settings where $\lambda = 0.01, K_i = L_i = 1$, $T = 0.5$.

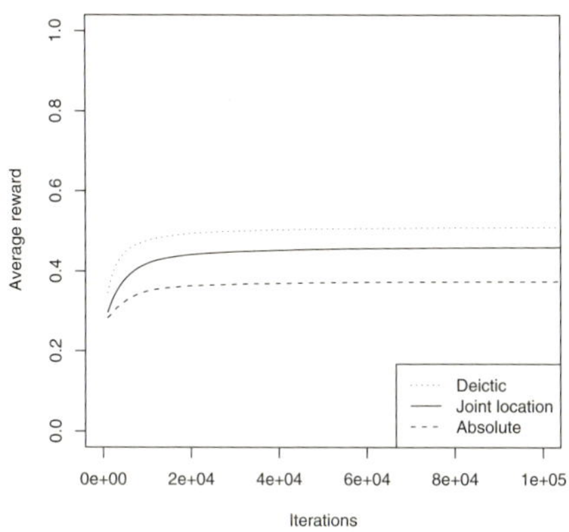

Fig. 5. Comparison of the influence of the state information given to the GLA

For this experiment each agent uses a single GLA with 2 actions corresponding to the agent actions L and R. Each time step we give both agents an input vector $\mathbf{x} = [x1\ x2\ 1]$, where $x1$ is the position of agent A and $x2$ is the position of agent B. The GLA use a vector $\mathbf{u_i} = [u_{i1}\ u_{i2}\ u_{i3}]$ for each action i. The learning process of a GLA can then be seen as moving the line $(\mathbf{u_1} - \mathbf{u_2})^T\mathbf{x}$ which separates regions in the state space where the GLA prefers action 1(L) from those where it prefers action 2(R). Typical results obtained with this system can be seen in Figure 4(b). This result was obtained running the experiment for 100000 iterations. Each iteration consists of a single action choice and update for both agents. After each move and subsequent learning update, the agents were reset to new random positions and the game was restarted. This was done to avoid the undersampling problem which occurs easily when dealing witch such large state spaces.

Since GLA take context vectors as input, it is possible to present the state information in different forms to the agent. Figure 5 shows a comparison of the average reward obtained, with three distinct ways of information. We compared the use of the joint location described above, to an absolute distance metric $(AbsoluteValue(Pos(AgentA) - Pos(AgentB)))$ and a deictic distance metric $(Pos(AgentA) - Pos(AgentB))$. This experiment was run without tuning of the exploration of the Boltzmann action selection method, so these values are not necessarily measures for optimal performance of the GLA, but rather serve as a criterion to compare the influence of the information given in the context vectors.

The absolute distance metric clearly performs the worst due to the inability of making a distinction between different positions of the other agent. When presenting the agent with a deictic information to the position of the other agent, it outperforms agents using a joint location based state information. We performed this experiment to show that, even though the same information is used, presenting it in different forms to the agent clearly benefits the learning results.

4.2 Agents Using Multiple GLA

In the second experiment we examine a situation where the different regions in the state space are not linearly separable. In such a case the agents cannot exactly represent the optimal mapping, but rather have to approximate it using hyperplanes. We use the same set-up as in the previous experiment, but now we consider two regions, as given in Figure 6(a). In region I, given by the inside of the parabola action (L, L) is optimal with a reward of 0.9. When the joint location of the agents falls outside the parabola, however, action (R, R) is optimal with reward 0.5. In both cases all other joint actions have a pay-off of 0.1.

Both agents use a system consisting of 2 GLA, connected by an AND operation. Both GLA have 2 actions: 0 and 1. If the automata both choose 1 the agents performs its first action L else it performs action R. Figure 6(a) shows 2 typical results for the boundaries that the agents learn to approximate the parabola. Figure 6(b) shows for both agents the evolution of probability of the optimal action L in region I. The probabilities in this plot where obtained by generating 100 points in the region with uniform probability and calculating the average probability over these points.

While it can be seen from the results in Figure 6 that the agents are able to approximate the desired regions, this experiment also demonstrated limits to out approach. As was mentioned in the previous section the GLA are only guaranteed to converge to a local optimum. This means that the agent can get stuck in suboptimal solutions. Such a situation was observed when the reward for the optimal action in region II is increased. In this case it is possible for the agents to get stuck in a situation where they both always prefer the optimal action for region II and neither agent has a good approximation of the region inside the parabola. Since the rewards of both agents are based on their joint

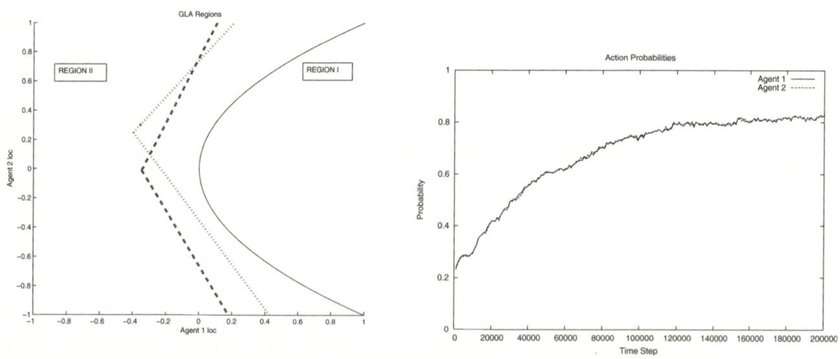

(a) (b)

Fig. 6. Experimental results for the second experiment. (a) Typical results for approximations for parabola learnt by agents. (b) Probabilities of optimal action in region I for both agents (average over 100 runs). Parameter settings where $\lambda = 0.005$,$K_i = L_i = 1$, $T = 0.5$.

action, no agent can get out of this situation on its own. The agents can only improve their pay-off by switching actions inside region I together. In such a situation with multiple local optima, the final result obtained by the agents is depended on their initialization.

4.3 GLA vs. LA

In this experiment we compare results obtained by GLA and traditional learning automata using the problem described in experiment 1. To be able to perform a fair comparison of the GLA vs LA in a multi-agent setting, we gave both techniques the same information, i.e. the joint location of the agents. For the LA this means that each agent has an equal amount of LA to the amount of possible joint locations, which is, as mentioned earlier, 40,401. For the GLA, we used this joint location as the context vectors, which means each agent only needs *one* generalized learning automaton for this problem.The experiments where performed 20 times and the results averaged. Each experiment ran for 10^7 iterations.

Figures 7(a) and 7(b) give the probability of the agents playing the optimal action in regions 1 and 2 for both techniques. As in our second experiment, the probabilities were obtained using 100 generated points in every region with uniform probability and calculating the average probability over these points. We see that the LA manage to achieve a higher convergence ratio than the GLA but at a very hight cost. The LA need nearly 6 million iterations to converge to a policy where they play the optimal action in almost 100% of the cases, whereas the GLA need only 15,000 iterations to achieve a level of 85%. This is

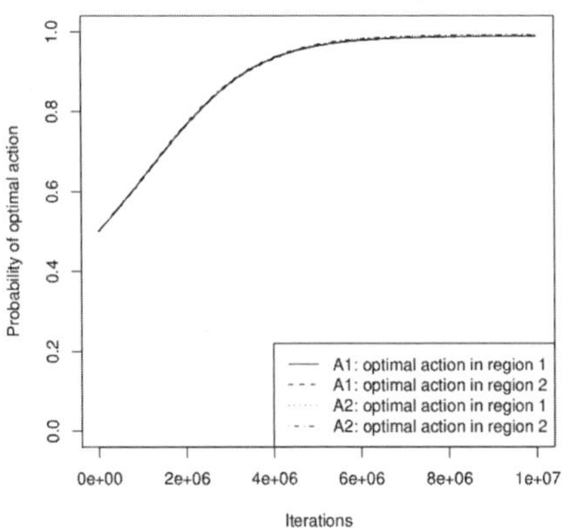

(a)

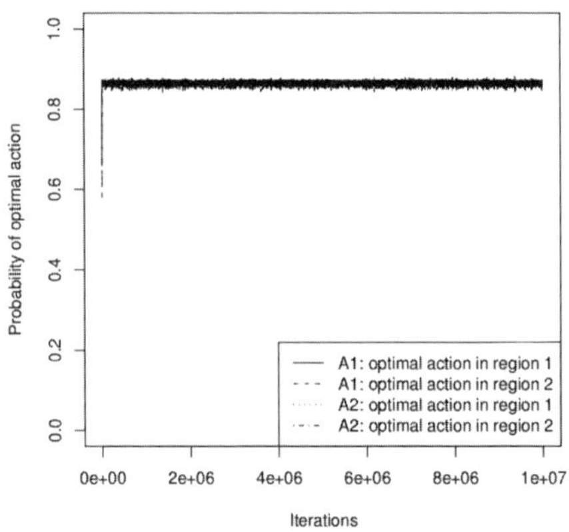

(b)

Fig. 7. Convergence to the optimal action of the LA-agents (a) and GLA-agents (b). The LA, converge to almost 100% of the optimal action, but need roughly 6.10^6 iterations to achieve this result, whereas the GLA converge to approximately 85% of the optimal action, and, even though it is unclear from the graph, the GLA reach this convergence level after only 15,000 iterations

400 times faster than the LA. If we were to increase the difficulty of the problem, for instance letting the agents move on a line between $[-10, 10]$ with increments of 0.01, we would have $2001^2 \approx 4.10^6$ possible states and an equal amount of LA per agent. When using GLA-agents, we would however still need only one GLA per agent and learning should not be affected by this increase of the state space.

5 Discussion and Future Work

When dealing with large state spaces the learning problem becomes difficult and state aggregation can offer a solution. When generalizing over states, the issue of converging to the optimal solution, must be traded of with the time needed to find this solution. In this paper we compared the use of GLA to LA. Altough the LA were able to converge with a higher probability to the optimal action, they needed 400 times the time needed for the GLA to converge and this factor only increases with the size of the state space, whereas the GLA are not subject to this problem. We also demonstrated the importance of the way the state information is represented, and that the inclusion of domain knowledge in this information can vastly improve the performance of the system.

While the system proposed in this paper is able to achieve good results in some problems, it still is possible for agents to converge to suboptimal solutions. To this end it is important to explore alternative learning schemes for the automata. Existing LA update schemes inspired by simulated annealing are known to be able to escape local optima and offer global convergence in some cases.

Another issue arises when we want to learn non-linear boundaries in the state space. The structure of the GLA network, which must be determined in advance, introduces a language bias and limits the mappings an agent can learn. In most cases information about the best structure to be used will not be available. A possible solution to this problem would be adapt existing tree-like GLA algorithms, which add automata at run-time and can build general piecewise linear approximations.

References

1. Sutton, R.S., Precup, D., Singh, S.P.: Between MDPs and semi-MDPs: A framework for temporal abstraction in reinforcement learning. Artificial Intelligence 112(1-2), 181–211 (1999)
2. Stolle, M., Precup, D.: Learning options in reinforcement learning. In: Koenig, S., Holte, R.C. (eds.) SARA 2002. LNCS (LNAI), vol. 2371, pp. 212–223. Springer, Heidelberg (2002)
3. Boutilier, C.: Planning, learning and coordination in multiagent decision processes. In: Theoretical Aspects of Rationality and Knowledge, pp. 195–201 (1996)
4. Boutilier, C., Dearden, R., Goldszmidt, M.: Exploiting structure in policy construction. In: Mellish, C. (ed.) Proceedings of the 14th International Joint Conference on Artificial Intelligence, pp. 1104–1111. Morgan Kaufmann, San Francisco (1995)
5. Guestrin, C., Koller, D., Parr, R.: Multiagent planning with factored mdps. In: 14th Neural Information Processing Systems (NIPS-14) (2001)

6. Degris, T., Sigaud, O., Wuillemin, P.H.: Learning the structure of factored markov decision processes in reinforcement learning problems. In: Proceedings of the 23rd International Conference on Machine learning, New York, NY, USA, pp. 257–264 (2006)
7. Strehl, A.L., Diuk, C., Littman, M.L.: Efficient structure learning in factored-state mdps. In: AAAI, pp. 645–650. AAAI Press, Menlo Park (2007)
8. Abbeel, P., Koller, D., Ng, A.Y.: Learning factor graphs in polynomial time and sample complexity. Journal of Machine Learning Research 7, 1743–1788 (2006)
9. Boutilier, C.: Planning, Learning and Coordination in Multiagent Decision Processes. In: Proceedings of the Sixth Conference on Theoretical Aspects of Rationality and Knowledge table of contents, pp. 195–210 (1996)
10. Guestrin, C., Hauskrecht, M., Kveton, B.: Solving factored MDPs with continuous and discrete variables. In: Proceedings of the 20th conference on Uncertainty in artificial intelligence, pp. 235–242 (2004)
11. Williams, R.: Simple Statistical Gradient-Following Algorithms for Connectionist Reinforcement Learning. Reinforcement Learning 8, 229–256 (1992)
12. Thathachar, M., Sastry, P.: Networks of Learning Automata: Techniques for Online Stochastic Optimization. Kluwer Academic Pub., Dordrecht (2004)
13. Phansalkar, V., Thathachar, M.: Local and global optimization algorithms for generalized learning automata. Neural Computation 7(5), 950–973 (1995)

Recommendation of Multimedia Objects Based on Similarity of Ontologies

Przemysław Kazienko, Katarzyna Musiał, and Krzysztof Juszczyszyn

Wrocław University of Technology, Wyb.Wyspiańskiego 27, 50-370 Wrocław, Poland
{kazienko,katarzyna.musial,krzysztof}@pwr.wroc.pl

Abstract. A new framework for recommendation of multimedia objects based on individual ontologies is presented in the paper. The recommendation process takes into account similarities calculated both between objects' and users' ontologies that respect the social and semantic features existing in the system. The system was developed for the use in the *Flickr* multimedia sharing system.

1 Introduction

Flickr or *YouTube* are typical examples of multimedia sharing systems (MSSs). that enable their users to upload, manage, and browse multimedia contents such as photos, videos, etc. called in this paper multimedia objects (MOs). Each of the multimedia object can be tagged by its author. In other words, users can describe their MOs with one or more short phrases that are most meaningful for the authors. However, the tags are proposed solely by authors even though they can browse the tags used by others. For that reason, tags do not have to be understandable for all other users. Moreover, users can comment the items added by others, include them to their favorites, etc. They also have the opportunity to set up direct relationships with other system users, establish groups of collective interests and directly enumerate their friends.

The vast amount of data about both multimedia objects and user activities can also be exploited to create complex ontologies that would provide the comprehensive view onto the multimedia objects existing within the system, the relationships between them as well as the users operations connected with these multimedia objects. Next, the knowledge built into these ontologies can be utilized by the recommender system to suggest to the active user the items, which are the most suitable for them.

Recommender systems help people to make decision, what items to buy, which movie to watch or even who they can invite to their social network [9]. They are especially useful in the environments with the vast amount of information since they cope with selection of a small subset of items that appears to fit to the users' needs [1, 13, 15]. The recommender systems are usually divided into three main categories: collaborative filtering, content-based filtering, and hybrid recommendation [1, 13]. In the collaborative filtering, the system recommends products or people that have been positively evaluated by others, whose ratings and tastes are similar to the preferences of the current user who will receive recommendations [1]. In the content-based filtering the items recommended to the user are similar to the items picked and rated high by this user in the past [14]. The hybrid method combines two previously enumerated approaches [7].

In Web 2.0 applications we can use domain ontologies in order to generate recommendations [17]. The problem of finding similarities between ontology elements and

I. Lovrek, R.J. Howlett, and L.C. Jain (Eds.): KES 2008, Part I, LNAI 5177, pp. 194–201, 2008.

also between entire ontologies as complex structures plays an important role in managing the content of Web portals [6] such as *Flickr*. One of the first systems that applied the idea of ontology for recommendation were Quickstep and Foxtrot proposed in [12] in which the collections of research papers were classified using ontological classes. Tan and Lambrix proposed the method useful to recommend the best alignment strategy for ontologies [16]. In another example, the view-based search method developed within the information retrieval community was combined with the ontology-based annotations [5]. Ontologies were also utilized to address the cold-start and interest acquisition problems [11].

A typical recommender framework processes the data gathered by the system and generates some suggestions to users whose have no influence on the recommendation process. The character of MSS requires a new method of recommendation, in which users would be able to change relationships between MOs generated by the system. However, these relationships result from either semantic or social links. The former include common tags and similar descriptions, whereas the latter are consequences of relations between system users and can be derived from lists of favourites, groups, contact lists and comments to the same MOs. The exaggerated independence of the recommender system can be weakened by the introduction of the automatically created individual ontologies that could be manually changed by the users.

2 Recommendation of Multimedia Objects Based on Similarity of Ontologies

2.1 Individual Ontologies

Overall, the domain knowledge consists of the two types of ontologies, representing knowledge about users and multimedia objects. Both define the domain concepts and the basic relations met in proposed system. Their basic structure is predefined, however the individual set of concepts for each user and multimedia object may differ depending on the users' actions and their history. Nevertheless, these ontologies used all extracted concepts whereas in the real system every individual ontology can contains different types of concepts. Moreover, most of the concepts are optional, so for example the user does not have to possess favourite MOs. The same situation appears regarding the multimedia object.

The individual user ontology represents knowledge about users' activity:

- authored, favourite and commented MOs. Note that, these three activities can overlap e.g. user can both add the given MO to the favourites and in comment it,
- the tags used by the user to annotate MOs,
- the descriptions made by users in order to provide more information about MOs,
- users included in contact list and the fact of being included in others contact list,
- the fact of being a member or an administrator of user group.

On the other hand, the concepts within the individual multimedia object ontology reflect the knowledge about MOs uploaded to the system:

- the users who authored, favourite or commented given object,
- description attached to MO by the author,

- the tags which describe MO,
- the groups that this MO belongs to.

The ontologies are created in the moment when the user or MO appears in the system for the first time. When the MO is added by the user then the individual MO's ontology is created based on such information as tags, description and authorship of the photo. On the other hand, when a new user registers to the system then the empty ontology for this person is created. The process of ontologies creation is not the trivial one, however, it is not the most important component of this research so it will not be described in details. Obviously, the ontologies must be revised continuously. The changes will come both from the users who want to update their ontologies and from the system itself. Such considerations provokes the formation of two layers of ontology: the system and the user ontology. The former will be managed by the system itself and the latter one will be changed by the user of MSS. Moreover, each person can maintain only their own ontology and the ontologies of photos they authored. The final ontology that will be presented by the system is the product of these two enumerated ontologies. In order to facilitate the process of updating the individual user ontology the appropriate mechanisms, which support the users in their activities, ought to be developed. One of them is to guarantee the user the access to the dictionaries such as WordNet, which enable to introduce the unified tags as the keywords for their photos. Furthermore, some ontology visualization tools need to be available in order to support users in their activities regarding their ontologies.

2.2 Ontology Similarity Measure

In order to compute ontology-based similarities between users and MOs, an *ontology similarity measure* is introduced and applied to individual user's and MO's ontologies. It should be noticed that in majority of researches, addressing problems of ontology similarity or merging and alignment of ontologies, only similarity between the elements of ontological structures is considered [2]. There are only few works which deal with comparing ontologies as a whole knowledge structures [3, 10].

We use a *Taxonomic Precision* (*TP*), a similarity measure based on the notion of *semantic cotopy* (see def. 2) recently presented and analysed in [4]. The reason to chose this measure was to take advantage of its ability to compare ontologies as whole structures. The values of *TP* are from the range [0,1]. As stated in [4] this definition of taxonomic precision may be influenced by the lexical term layer in the case of significant differences in domain models. However, in our approach, most of the concepts used in individual ontologies come from global sets (user, group and object names, tags), so this issue is not expected to appear. In our approach terms like tags, user names etc. are directly identified with concepts. Moreover, we do not distinguish between relations in our ontologies, when applying similarity measures we treat them as taxonomies with root concepts user and MO.

Definition 1. The ontology O is a structure $O := (C, root, \leq_C)$ where C is a set of concept identifiers and *root* is a designated root concept for the partial order $\leq_C$ on C.

Definition 2. *Semantic Cotopy* $sc(c,O)$ of a concept c from ontology O is a set containing c and all super- and subconcepts of c in O, excluding root concept $root(O)$.

Note that the above modification of the standard definition (exclusion of the root concept) comes from the specific features of our system. In our case, when comparing ontologies, the root concepts will always be different.

Definition 3. *Taxonomic Precision* $tp(c,O_1,O_2)$ of concept c and two ontologies O_1 and O_2 such that $c\in O_1$ and $c\in O_2$ is defined as:

$$tp(c,O_1,O_2)=\frac{\left|sc(c,O_1)\cap sc(c,O_2)\right|}{\left|sc(c,O_1)\right|} \tag{1}$$

Definition 4. *Global Taxonomic Precision* $TP(O_1,O_2)$ of the two ontologies O_1 and O_2 is defined as:

$$TP(O_1,O_2)=\frac{1}{|C_1|}\sum_{c\in C_1}\begin{cases} tp(c,O_1,O_2) & if \quad c\in C_2 \\ 0 & if \quad c\notin C_2 \end{cases}, \tag{2}$$

where C_1, C_2 – the sets of concepts of O_1 and O_2 respectively.

2.3 Ontology Similarity Assessment

In order to decide whether the given user or MO is similar to another one, their individual ontologies need to be processed. Since individual ontologies of users do not represent information about the features of the processed MOs or users in the contact list (the same concerns MOs' ontologies) we postulate their extension by adding relevant subconcepts. This action is performed only for the purpose of computing similarities. The Ontology Similarity Algorithm (OSA) for the two users' or MOs' ontologies (from here on denoted as O_1 and O_2) looks as follows:

The ontology similarity algorithm – OSA

Input:

- Ontologies O_1 and O_2 to be compared. Note: we assume that O_1 and O_2 are of the same type, i.e. user's or MO's individual ontologies, as defined in sec 4.2.

Output:

- The value of similarity $TP(O_1^*,O_2^*)$ between O_1 and O_2 from the range [0,1].

1. begin
2. $O_1^* = O_1$, $O_2^* = O_2$ /* create extensions O_1^* and O_2^* of O_1 and O_2, respectively*/
3. for (each *user* concept $C_i \leq_C$ *root* in O_1^*) do begin
4. find ontology O_i such that $root(O_i) = C_i$
5. attach all subconcepts of C_i from O_i as subconcepts of C_i in O_1^*
6. end
7. for (each MO concept $C_j \leq_C$ *root* in O_1^*) do begin
8. find ontology O_j such that $root(O_j) = C_i$
9. attach all subconcepts of C_j from O_j as subconcepts of C_j in O_1^*
10. end
11 repeat steps 3-10 for O_2^*
12. calculate $TP(O_1^*, O_2^*)$ according to Def. 4.
13. return $TP(O_1^*, O_2^*)$
14. end

In order to compute the similarities between the O_1 and O_2, they will be extended by attaching concepts from individual ontologies met in O_1 and O_2. The motivation is to take into account their characteristic features that could be omitted otherwise. For example, two different MOs in ontologies of two users are not signs of their similarity, but if they are tagged in the same way, by the same users and have similar descriptions – it should have positive influence on similarity between these users.

2.4 Recommendation Process

Based on the gathered information from the individual users and MOs ontologies we have built the recommender framework that enables users to view, comment, add to the list of favourites the MOs that they will be keenly interested in. Moreover, if one finds the recommended MO interesting then this person can find the author of the photo and set up a new relationship with this user. By combining the different data sources, the method facilities a bootstrap user to find interesting content in the MSS .

The overall view of the recommender framework for the MSS is presented in Fig. 1. Before the recommendation process for the given person is launched the individual ontologies for all people as well as for all MOs are created. These ontologies can be changed by both the system itself as well as each user can maintain their own ontology as well as ontologies of MOs added by them to the system (see sec. 2.1)

The first step of the recommendation process is to capture the user context, i.e. that both the user who browse the MOs and the MO selected by this person to browse are identified. In order to facilitate the further explanation of the process let's assume that the user u_x is watching MO a_i. The individual ontologies serve as the input data for the whole process and are utilized in the ontology-based similarities calculation phase. These ontologies are periodically recalculated in order to assure their validity. Based on individual MO ontologies the k-nearest MOs that are close to MO a_i are selected and list L_{MO} of these objects is established. List L_{MO} contains the weights that reflects the level of similarity between MO a_i and k-nearest MOs. This element can be seen as the content-based filtering, whereas the method of list L_U creation is also called social-based filtering. List L_U is obtained by comparing the given user u_x's ontology with all other users' ontologies. In this list the weights that reflects the level of similarity between user u_x and all other users are stored.

After that, the user context filtering is performed. During the recommendation process some of the MOs must be rejected from the list of candidates (L_{MO}) in order to avoid the situation in which the user have already seen the particular MO. The MOs that ought to be omitted and in consequence rejected from the list L_{MO} are: the objects owned by user u_x as well as the objects that have already been commented or added to favourites by user u_x. Moreover, the weight of MOs that have been already viewed by user u_x should be made smaller. The level of reduction basically depends on two elements, i.e. how often was the particular MO viewed in the past and the second one when was this object browsed for the last time by the given user.

The next step is the integration of lists L_{MO} and L_U. The main idea is to create the final recommendation list L_F and top N elements from this list are presented to u_x. List L_F is obtained by verifying L_{MO} according to MOs' authors. It means, that for each

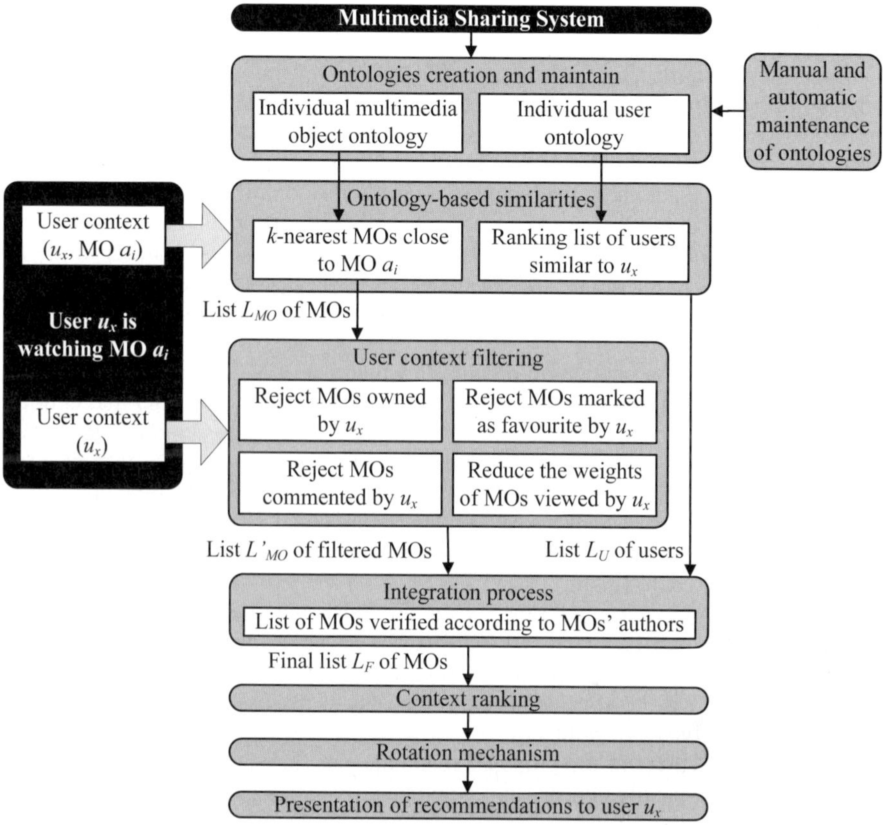

Fig. 1. Ontology-based Recommendation Process in Multimedia Sharing Systems

MO from L_{MO} its author's weight from list L_U is taken and these weights are summed up. For both of these weights the importance coefficients is assigned (see formula 3).

$$w^{final}(a_i, a_j, u_x) = \alpha \cdot w^{MO}(a_i, a_j) + \beta \cdot w^{A}(u_x, u_j), \tag{3}$$

where $w^{final}(a_i, a_j, u_x)$ – the final weight for the MO a_j from the list L_{MO} of items most similar to MO a_i viewed by u_x; $w^{MO}(a_i, a_j)$ – the weight for the MO a_j from the list L_{MO} of items most similar to MO a_i from the range [0,1]; $w^{A}(u_x, u_j)$ – the weight for the author u_j of MO a_j from the list L_U of users most similar to u_i from the range [0,1]; α, β – importance coefficients with values from the range [0,1].

Constants α and β are used to simulate and adjust the influence of the weights from lists L_1 and filtered L_2. For example, if α is low and β is high then the author's weight is more significant than MO's weight. Since values of both components are from the range [0,1], the value of final weight belongs to the range [0,2]. After the integration process the list L_F is sorted and finally top N selected MOs from L_{MO} are suggested to person u_x. The rotary mechanism is used, to prevent the same MOs to be recommended to user u_x all the time [8].

2.5 Discussion

During the development of the recommender process several issues, which need to be addressed, have appeared. One of the most important concerns is the high complexity of the performed calculations. In order to overcome this shortcoming the entire processes (see sec. 2.4) can be divided into two separate categories, i.e. the tasks that can be executed offline and these that ought to be performed online.

The efficiency problems are also related to the list of authors that are similar to the given one (L_U). The whole long list is stored and processed by the system separately for each user. These lists should be shortened, i.e. only m-nearest authors and their weights can be retained while one common, small system weight ought to be assigned for the rest users.

Another issue is the integration process in which list of MOs is verified according to MOs' authors. In the proposed method the two weights, one for MO that is in the list L_{MO} and another one for the author of this object from L_U, are summed up. Nevertheless, some other merging functions can be applied as well. For example, one weight can be multiply by the second one. However, in this case, the outcome will be much more diversified as well as the weights cannot equal zero. This problem can be addressed by establishing the minimal non-zero value of the weight or by adding small number ε to each weight.

The descriptions which are represented in MOs' and users' ontologies are expressed in the natural language. To increase the accuracy of text comparison some advanced NLP tools can be utilized. In this case, the concepts *Description* from different ontologies will be considered the same only if the tool returns text similarity value above the given threshold.

An important feature of the proposed recommender framework is lack of the cold-start problem. It typically appears in recommender systems based on collaborative or content-based filtering. In the proposed recommender framework, a hybrid approach is used. If a new user registers and starts navigating within the system then only the similarities between the just viewed MO and other MOs from the system are calculated. On the other hand, if a new MO is uploaded then the system automatically creates the individual ontology for this object.

3 Conclusions and Future Work

The proposed concept of recommender system utilizes the new method that compares ontologies as a whole structures to assess similarity between multidimensional profiles of users and multimedia objects. The ontologies provide the comprehensive view about the information gathered in the multimedia sharing system. As a result, we can execute the recommendation process, which takes into account many distinct features of system users and multimedia objects that are created and annotated by them.

Future work will include research on *Flickr* – the photo sharing system, in order to prove the effectiveness of ontology-based recommendation and show the synergy effect that results from the joint use of recommender system with ontology-based user and multimedia object assessment. Note that there are many ways of further developments of the proposed scheme. They lay in more sophisticated mechanisms of ontology extension before similarity computation (see sec. 2.2) as well as providing users with some advanced visual interfaces and conversational modules, which will make use of the underlying ontological structures. The challenge is also to develop the effective method of ontologies creation and maintenance.

Acknowledgments. The work was supported by The Polish Ministry of Science and Higher Education, grant no. N516 037 31/3708.

References

1. Adomavicius, G., Tuzhilin, A.: Toward the Next Generation of Recommender Systems: A Survey of the State-of-the-Art and Possible Extensions. IEEE Transactions on Knowledge and Data Engineering 17(6), 734–749 (2005)
2. Andrea, M., Egenhofer, M.: Determining Semantic Similarity Among Entity Classes from Different Ontologies. IEEE Trans. on Knowledge and Data Engineering 15, 442–456 (2003)
3. Croitoru, M., Hu, B., Dashmapatra, S., Lewis, P., Dupplaw, D., Xiao, L.: A Conceptual Graph Based Approach to Ontology Similarity Measure. In: Priss, U., Polovina, S., Hill, R. (eds.) ICCS 2007. LNCS (LNAI), vol. 4604, pp. 154–164. Springer, Heidelberg (2007)
4. Dellschaft, K., Staab, S.: On How to Perform a Gold Standard Based Evaluation of Ontology Learning. In: Cruz, I., Decker, S., Allemang, D., Preist, C., Schwabe, D., Mika, P., Uschold, M., Aroyo, L.M. (eds.) ISWC 2006. LNCS, vol. 4273, pp. 228–241. Springer, Heidelberg (2006)
5. Hyvonen, E., Saarela, S., Vilijanen, K.: Ontogator: Combining View- and Ontology-Based Search with Semantic Browsing. In: XML Finland 2003, Kuopio (2003)
6. Juszczyszyn, K.: Virtual Communities and the Alignment of Web Ontologies. In: Dasgupta, S. (ed.), Hershey, USA, pp. 497–499. Idea Group Reference (2006)
7. Kazienko, P., Kołodziejski, P.: Personalized Integration of Recommendation Methods for E-commerce. Int. Journal of Computer Science & Applications 3(3), 12–26 (2006)
8. Kazienko, P., Adamski, M.: AdROSA - Adaptive Personalization of Web Advertising. Information Sciences 177(11), 2269–2295 (2007)
9. Kazienko, P., Musiał, K.: Recommendation Framework for Online Social Networks. In: AWIC 2006. Studies in Computational Intelligence, pp. 111–120. Springer, Heidelberg (2006)
10. Maedche, A., Staab, S.: Measuring Similarity between Ontologies. In: Gómez-Pérez, A., Benjamins, V.R. (eds.) EKAW 2002. LNCS (LNAI), vol. 2473, pp. 251–263. Springer, Heidelberg (2002)
11. Middleton, S.E., Alani, H., Shadbolt, N.R., De Roure, D.C.: Exploiting Synergy Between Ontologies and Recommender Systems. In: Proc. of the 11th International World Wide Web Conference WWW, Int. Workshop on the Semantic Web, pp. 41–50 (2002)
12. Middleton, S., Shadbolt, N., De Roure, D.: Ontological User Profiling in Recommender Systems. ACM Transactions on Information Systems 22(1), 54–88 (2004)
13. Montaner, M., López, B., de la Rosa, J.L.: A Taxonomy of Recommender Agents on the Internet. Artificial Intelligence Review 19(4), 285–330 (2003)
14. Pazzani, M., Billsus, D.: Learning and revising user profiles: The identification of interesting web sites. Machine Learning 27, 313–331 (1997)
15. Perguini, S., Goncalves, M.A., Fox, E.A.: Recommender systems research: A Connection-Centric Survey. Journal of Intelligent Information Systems 23(2), 107–143 (2004)
16. Tan, H., Lambrix, P.: A Method for Recommending Ontology Alignment Strategies. In: Aberer, K., Choi, K.-S., Noy, N., Allemang, D., Lee, K.-I., Nixon, L., Golbeck, J., Mika, P., Maynard, D., Mizoguchi, R., Schreiber, G., Cudré-Mauroux, P. (eds.) ASWC 2007 and ISWC 2007. LNCS, vol. 4825, pp. 494–507. Springer, Heidelberg (2007)
17. Ziegler, C., Lindner, W.: SemanticWeb Recommender Systems. In: Lindner, W., Mesiti, M., Türker, C., Tzitzikas, Y., Vakali, A.I. (eds.) EDBT 2004. LNCS, vol. 3268, pp. 78–89. Springer, Heidelberg (2004)

Toward Emotional E-Commerce: The Customer Agent

Nicolae Jascanu, Veronica Jascanu, and Severin Bumbaru

Department of Computer Science, University "Dunarea de Jos" of Galati, Romania
{nicolae.jascanu,veronica.jascanu,severin.bumbaru}@ugal.ro

Abstract. Experiencing and expressing emotions are integral parts of our life. In the retail area of commerce, emotion plays a fundamental role. When we choose or buy a commodity, our choice has a profound emotional dimension. We are social beings, we are influenced by others opinions, we have our opinions and we like to bargain for everything. This paper proposes a novel model, based on dimensional models of emotion, able to capture and interpret the customer emotional knowledge.

Keywords: affective computing, e-commerce, emotions.

1 Emotions and Customer Satisfaction

Evaluation of service quality can be described as a cognitive process where customers consider the goodness or badness of the commodities, by evaluating the perceived service performance only, or by comparing the service performance with some predetermined standard [1], [2], [3], [4]. On the other hand, satisfaction contains an affective dimension, without which customers responses cannot be fully accounted for. Cognitive and affective responses can be seen as distinct, and having a separate influence on satisfaction formation [5], [6], [7].

Oliver [8], Oliver and Westbrook [9] defined affect as a mediator between cognitive evaluations, such as perceived product performance, and satisfaction. In his theory, Oliver argues that it is possible, regarding to a commodity, to experience both negative and positive affects at the same time. When we speak about life quality, we consider both types of affect. Oliver finds these ideas adaptable to the field of consumer goods and service consumption. When a service is seen as consisting of several different attributes which can be evaluated by the consumer before, during and after consumption, each of these evaluations of service attributes, may also be seen as a potential source of negative or positive affect. If a product fails to accomplish the customer's needs or expectations, the response will include negative emotions . If the commodity is perceived as desirable, the customer will respond with positive emotions. The satisfaction comes from the combination of positive and negative emotions of all service attributes. Customers may have a large palette of different opinions, not simply all positive or all negative emotions. Stauss [10] has suggested that different satisfaction types may exist according to the pattern of emotions, cognitions and intentions that the customer expresses. This may explain observed weak links between satisfaction and customer loyalty. An empirical study in the financial service industry

I. Lovrek, R.J. Howlett, and L.C. Jain (Eds.): KES 2008, Part I, LNAI 5177, pp. 202–209, 2008.

indicates support for the hypothesis that a particular overall satisfaction score may be connected with different levels of an emotion [11].

In the process of acquiring a commodity, the customer interacts with the supplier by separate service episodes, the sum of these episodes forming a so-called relationship. A relationship consists of several episodes in which the post-purchase reactions serve as input into the next pre-purchase phase. In the same time, there is an evaluation mechanism, some kind of introspection, in which the customer makes an overall appraisal of the relationship in terms of both its cognitive and emotional components. In fact, some levels of emotion are present during all interactions between the customer and the service provider. Emotions can be experienced from the start of consuming a service to the termination of the service experience, and even a long time after the actual consumption has ended. They may also change for the worse or for the better during the service encounter, depending on the actions taken by the service employees.

There are many aspects and facets of emotion in customer-supplier interactions, but as a conclusion for this paragraph, ignoring the customer's emotions in the process of acquiring products or services it's a huge mistake for a supplier. If there is a computerized model, it should be able to take into account the emotions elicited during interactions between parties [12], [13].

1.1 The Circumplex Model of Emotions

Emotional judgments and affective self-ratings often are found to array in circular arrangement, referred to as a "circumplex" structure arrangement [14], [15]. A circumplex is a two-dimensional, circular structure in which single attributes correlate highly with those attributes nearby on the circumference of the circle, correlate near zero with those attributes one-quarter way around the circle, and correlate inversely with those attributes directly opposite on the circle. Using a circumplex structure for representing emotion is equivalent to making several assertions about the nature of the emotion domain. At the most fundamental level, a circumplex model means that some emotions are similar to each other yet measurably different from other emotions. Structural theories of emotion presume that emotions are not all unrelated and discrete but have certain underlying similarities and differences. Second, by using the circumplex model, a claim is made that two affect dimensions can capture the majority of emotional experience. Third, a circumplex suggests that emotions can be described in a circular fashion in two-dimensional space and that emotions do not simply aggregate together in several groupings or fall in order along two axes. The circumplex model holds that some affects will always fall between any two axes that are drawn through the two-dimension circumplex space. In addition, a circumplex implies that a very high or very low value on one dimension is accompanied by a moderate value on the other dimension. A circumplex model of emotion suggests a clear structure for the effects emotion will have on behavior and thus has large heuristic value [16]. The circumplex provides a measurement model that can be useful for understanding and organizing the many emotion measures used in research today. Emotion circumplex models are very specific in indicating mathematically testable relationships between various affects. Thus, the emotion circumplex provides a theoretical structure that can potentially advance our understanding of emotional life.

2 The Customer Agent

The model of customer agent relies on the circumplex theory, for emotional knowledge acquisition and representation. When someone explains why he or she decided to buy or wish to buy that product or service, the explanation includes both the rational and emotional components of the reasoning process. For example, when you try to rent a house in a Caribbean resort, the final decision is a pleasing one, even if the price is a little too high. The pleasure comes from the fact that the location is near the beach and the renting period is the desired one. Therefore, it is natural to make a compromise, accept the offer, and feel good about it imaging how good it will be on the next summer. Of course, if the price is too high, the offer is denied no matter how good the other issues are. When you buy something, you aim for several issues so that the final decision is a trade-off between those issues. The model assumes that each of those issues have multiple possible values, each value representing an important mark for customer. For example (see Figure 1), the value of 10 minutes on *time* issue is such a mark and it means an excellent distance to get to the beach. The 20 minutes mark is not such a good distance, and we do not "feel" so good about walking twice a day this distance to get on beach. It is hard for a human being to think more granularly and say how it feels at exactly 13 minutes away from beach. Therefore, it is naturally to collect from customer only representative marks.

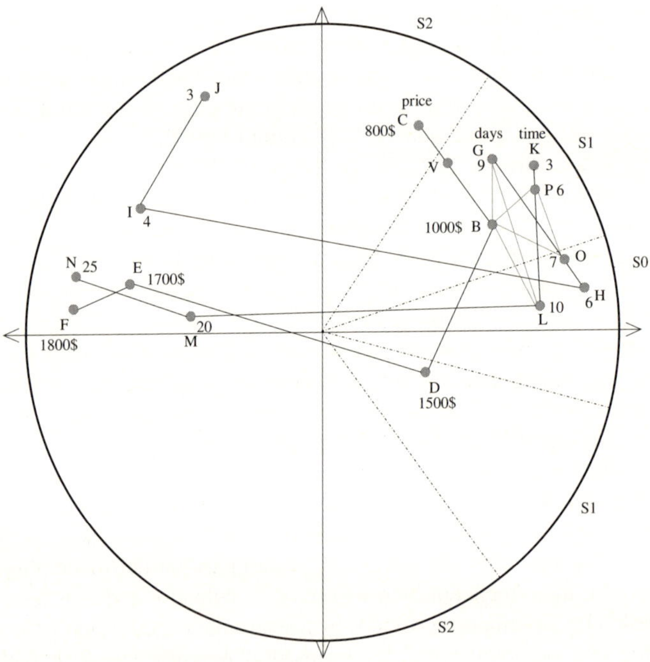

Fig. 1. Configured circumplex: price, days, time to destination

The scenario for collecting knowledge from customer is very simple: the customer defines the commodity issues and the mark points on each issue. Let us consider a negotiation for an excursion to a Caribbean resort. The selected issues are those from Figure 1: *price* of excursion, *days* to stay and *time* to get to the beach. We will explain the algorithm for selecting between following two configurations: B, G, L = 1000 \$, 9 days, 10 min; B, O, P = 1000 \$, 7 days, 6 min. Even for us is hard to decide which configuration is better. We could anticipate that the [B, O, P] configuration is selected over [B, G, L] configuration. Intuitively, [B, O, P] is preferred because it is not so important to stay 9 days (7 days are well enough) and because the time to beach really matters, as we can see from the Figure 1. In order to take a decision, the engine performs two inferences: a quantitative one and a qualitative one. The quantitative analyze offers an order between mark points for every issue. Therefore, we will know which mark point is the best and which is the worst. The qualitative analyze offers, for every segment between two mark points a measure named *cost*. Therefore, we will know how much we will loose if we negotiate a mark point over other.

2.1 Quantitative Inference

We define the following hypotheses:

H1: *Between two mark points with the same arousal value, will be selected that with a greater pleasure value.*

H2: *Between two mark points with the same pleasure value, will be selected that with a greater arousal value.*

H3: *Between two mark points situated at the same distance from circumplex center, will be selected that with a greater arousal value.*

For the other three quadrants, we define similar hypotheses. To implement these hypotheses, we could use a rule engine or we can imagine a three dimensional surface, ordering the mark points accordingly with z height. The general surface equation is shown below:

$$Z = \left[\left(\frac{2\beta}{\pi} \right)^b (Z_M - Z_m) + Z_m \right] \frac{r^a}{R^a} . \tag{1}$$

where: β is quadrant dependant; a and b could be integers like 2 or 3; Z_m the height at maximum pleasure (positive or negative); Z_M the height at maximum arousal

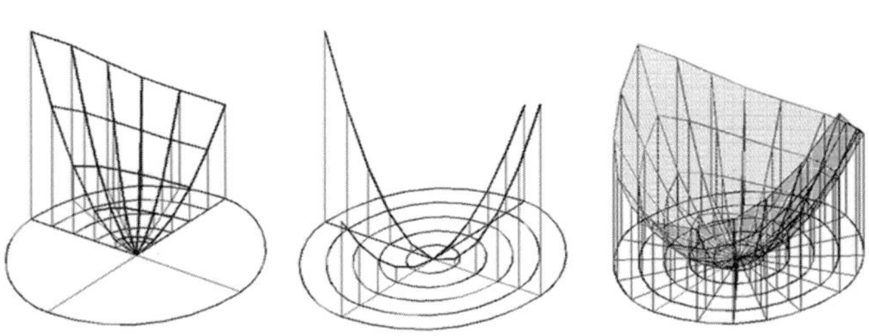

Fig. 2. Graphical representations for equation 1

(positive or negative). The chosen values for each parameter aren't so important because the surface role is to offer an order for mark points. The general equation 1, have the following graphical representations:

Appling equation 1, we will have the following orders for each issue: *price*: C, V, B, D, E, F; *days*: G, O, H, I, J; *time*: K, P, L, M, N

2.2 Qualitative Inference

Having an order between mark points, it is not sufficient because, obviously, we do not have constant costs for every segment. Even if we consider the Euclidian distance between two mark points, a long distance cost much than a short distance. The model tries to capture different qualitative aspects of these segments and estimates a cost for each one.

We will consider the following parameters for the positive qualitative analyze: negotiation profile, segment profile, segment angle (theta), segment distance, one-to-four quadrant traversal and positive-to-negative traversal. Each of these parameters is implemented as a fuzzy variable. The output of the fuzzy logic rules is the cost parameter. *Negotiation profile* is a fuzzy variable obtained from the distribution of positive mark points. There are five possible negotiation profiles: neutral, activated, extreme, normal and abnormal. From a negotiation point of view, a neutral profile means life quality. In other words, we negotiate over values that make a standard for our lifestyle and therefore, we will hardly give up on those values. An extreme profile means a negotiation over values that arouse us, but those values does not necessarily means life quality. Every type of negotiation profile has a direct influence over the cost value of each segment. *Segment profile* is calculated from the influence of each segment's mark points over sectors. It is important to know what the relation between segment profile and negotiation profile is. For example, if we have a neutral negotiation profile and the segment profile is extreme, than that mark points really means something and should be carefully considered during negotiation. The fuzzy engine is configured for all possible combination between profiles. *Segment angle* is another important parameter for cost calculation. Depending on the negotiation profile and segment profile, the angle gives an intermediate parameter, named impact. For example, if the negotiation profile is neutral and the segment profile is also neutral, than we talk about values affecting lifestyle so pleasure is more important than arousal. In other words, for this configuration a small angle (theta) gives a bigger impact than a big angle. The parameters of *distance* and *traversal* regulate the intermediary impact parameter. The result is the final cost of the positive analyze. For the negative analyze we consider a similar fuzzy model. Because the negotiation enters into the negative area, it is enough to make a rough analyze. It is obvious that no one wishes to buy products or services with negative values. Of course, sometimes you have to make compromises and accept some negative issues, but nobody will accept a product with all issues negative.

2.3 The Selection Inference

Now, we have an order and a cost between mark points. The model of selection engine is a pure geometrical one. We will consider a circle of unitary radius. The circle

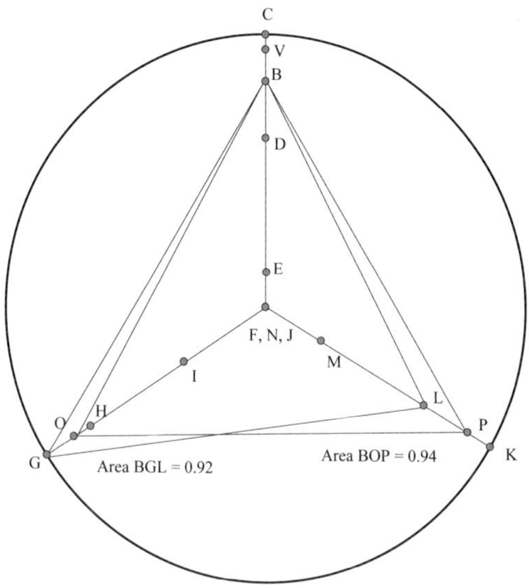

Fig. 3. Geometrical model for configuration selection

is divided into a number of sectors equally with the number of issues. In our case, we will have three radii, each one corresponding to an issue. On every radius, we will place the mark points, in the computed order, from extremity to center. Into the center, we will have from all issues the worst mark point. In Figure 3 is represented the geometrical model. In order to chose between two configurations, in our case between [B, G, L] and [B, O, P], we simply calculate the area of polygon.

The configuration with the biggest area is selected: [B, O, P]. The presented model of customer agent is versatile and is able to incorporate knowledge from its own experience. In addition, it is able to negotiate with multiple supplier agents and to choose between them.

3 Prototype

For now, we have built a prototype engine based on fuzzy theory. The engine is able to set an order between mark points, to calculate a cost for each segment and finally to order a list of combinations. A rough picture of the fuzzy engine for the positive qualitative analyze is given below:

Intuitively, the engine results are good. We are building a questionnaire to collect emotional data and to compare the engine results with the ones provided by the human counterpart. Because, the customer agent is part of a multi-agent e-commerce platform, we need to test the negotiation protocol to demonstrate that the order and cost are adequate measurements to obtain a Pareto optimal solution.

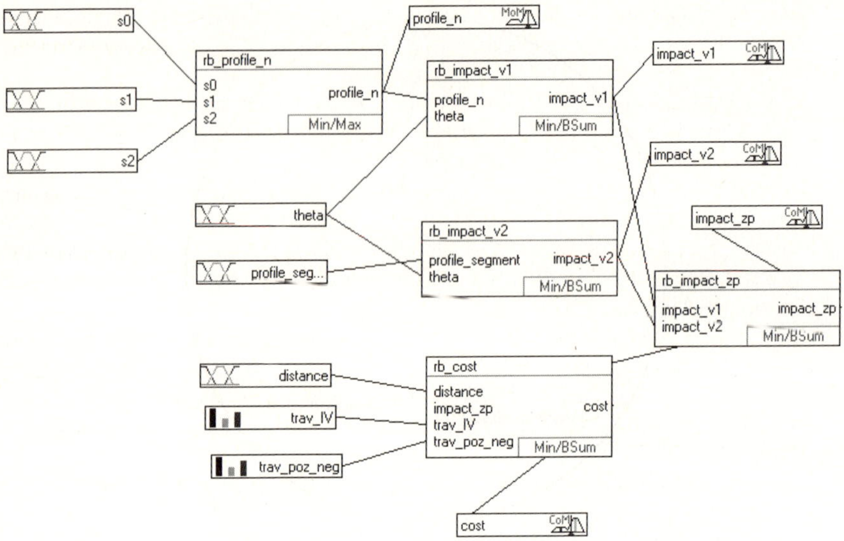

Fig. 4. Fuzzy system for positive qualitative analyze

4 Conclusions

The customer agent is part of a multi-agent e-commerce platform that involves three main actors: the customer, the supplier and the community. The novelty of the model is represented by the way, in which emotions are integrated in every aspect, from customer knowledge acquisition and representation to bilateral negotiation and supplier marketing research tools. By incorporating emotions, the model is able to capture almost naturally the rational and the emotional aspects, to learn about personal preferences, to gather and use in negotiation process the community opinions, which are essentially emotional, to offer snapshots of emotional state to negotiation partners without exposing internals, and finally to offer a more human-like experience over negotiation.

Acknowledgements. This work is partially supported by the National University Research Council CNCSIS, Romania, Project TD-182/2007.

References

[1] Militaru, D.: Consumer behavior in Electronic Commerce Environments and Fashion Effect. In: ICEC, Minneapolis, USA (2007)

[2] Oliver, R.L.: Satisfaction: A behavioral perspective on the consumer. McGraw-Hill, New York (1997)

[3] Yi, Y.: A critical review of consumer satisfaction. In: Zeithaml, V.A. (ed.) Review of marketing, Chicago, pp. 68–123 (1990)

[4] Karat, J.: Beyond task completion: evaluation of affective components of use. In: Handbook of human-computer interaction, vol. 59, pp. 1152–1164. Springer, Heidelberg (2002)
[5] Desmet, P.M.A., Tax, S.J.E.T., Overbreeke, C.J.: Designing products with added emotional value: development and application of an approach for research through design. The Design Journal 4, 32–47 (2000)
[6] Desmet, P.M.A., Hekkert, P.: The basis of product emotions. In: Pleasure with products: beyond usability, London (2002)
[7] Schwartz, N.: Feelings as information: Informational and motivational functions of affective states. In: Handbook of motivation and cognition: Foundations of social behavior, pp. 527–561. Guilford Press, New York (1990)
[8] Oliver, R.L.: Cognitive, affective, and attribute bases of the satisfaction response. Journal of Consumer Research 20 (1993)
[9] Oliver, R.L., Westbrook, R.A.: Profiles of consumer emotions and satisfaction in ownership and usage. Journal of Consumer Satisfaction, Dissatisfaction and Complaining Behavior 6, 12–27 (1993)
[10] Stauss, B.: The qualitative satisfaction model. In: Advancing Service Quality: A Global Perspective, New York, pp. 35–44 (1996)
[11] Morrin, M., Chebat, J.C.: Person-place congruency: the interactive effects of shopper style and atmospherics on consumer expenditures. Journal of Service Research 8, 181–191 (2005)
[12] Pullman, M., Gross, M.: Ability to experience design elements to elicit emotions and loyalty behaviors. Decision Sciences 35, 551–578 (2004)
[13] Vilnai, I., Rafaeli, A.: Aesthetics and professionalism of virtual servicescapes. Journal of Service Research 8, 245–259 (2006)
[14] Russell, J.A.: A circumplex model of affect. Journal of Personality and Social Psychology 39, 1161–1178 (1980)
[15] Russell, J.A.: Core affect and the psychological construction of emotion. Psychological Review 110, 145–172 (2003)
[16] Russell, J.A., Fehr, B.: Relativity in the perception of emotion in facial expression. Journal of Experimental Psychology 116, 223–237 (1987)

A Method for Integration of WordNet-Based Ontologies Using Distance Measures

Trong Hai Duong[1], Ngoc Thanh Nguyen[2], and Geun Sik Jo[1]

[1] School of Computer and Information Engineering,
Inha University, Korea
haiduongtrong@gmail.com, gsjo@inha.ac.kr
[2] Institute of Information Science and Engineering,
Wroclaw University of Technology, Poland
thanh@pwr.wrov.pl

Abstract. While there is a large body of previous work focused on WordNet-based for finding the semantic similarity of concepts and words, the application of these word oriented methods to ontology integration tasks has not been yet explored. In this paper, we propose a methodology of WordNet-based distance measures, and we apply the meaning of concepts of upper ontologies to an ontology integration process by providing semantic network called *OnConceptSNet*. It is a semantic network of concepts of ontologies in which relations between concepts derived from upper ontology WordNet. We also describe a methodology for conflict in ontology integration process.

Keywords: knowledge integration, information system, ontology integration.

1 Introduction

Ontology integration is an important task which needs to be performed when several information systems share or exchange their knowledge. Because ontology in these systems is a separated element of their knowledge bases, the knowledge integration process very often begins with ontology integration.

Basically, the ontology is defined by the following elements:

- ***C*** - a set of concepts (classes),
- ***I*** - a set of instances of concepts,
- ***R*** - a set of binary relations defined on ***C***,
- ***Z*** - a set of axioms, which can be interpreted as integrity constraints or relationships between instances and concepts.

In general, the problem of ontology integration can be formulated as follows: *For given ontologies $O_1, \ldots, O_n$ one should determine one ontology which could replace them* [3, 10]. Integration of ontologies is such a complex task, since ontologies have various characteristics and forms by nature, i.e., languages, domains, structures of ontologies may differ from each other. Therefore, authors of [4] suggested an Ontology Architecture, which provide a solid basis for studies about

I. Lovrek, R.J. Howlett, and L.C. Jain (Eds.): KES 2008, Part I, LNAI 5177, pp. 210–219, 2008.

ontology integration task. Pinto and Martins [10] identified the activities that should be performed in the ontology integration process. One of the first tools, PROMPT [8] helps in the merge process are now available. It uses labels to extent the structure of ontologies. Their focus is on ontology merging, i.e., how to create one ontology from two source ontologies. Most of the ideas for ontology integration tasks deal with upper ontologies as domain specific of the ontology [2]. The upper ontologies not only provide definitions for general-purpose terms [7, 12], but also extend them as semantic domain layering of the ontology architecture [4]. However, all these approaches often lack the specific application for ontology integration task and significant testing.

In our study, we apply the meaning of concepts of the upper ontologies to ontology integration process by providing semantic network called *OnConceptSNet*. It is a semantic network of concepts of ontologies in which relations between concepts derived from upper ontology WordNet. We propose a methodology for WordNet-based distance measures between the concepts. We also describe a methodology for conflict in ontology integration process.

2 Definitions

2.1 Basic Notions

As stated in above section, by an ontology we understand a quadruple: $(\boldsymbol{C}, \boldsymbol{I}, \boldsymbol{R}, \boldsymbol{Z})$. We assume a real world $(\boldsymbol{A}, \ \boldsymbol{V})$ where $\boldsymbol{A}$ is a finite set of attributes and $\boldsymbol{V}$ is the domain of $\boldsymbol{A}$, that is $\boldsymbol{V}$ is a set of attribute values, and $\boldsymbol{V} = \bigcup_{a \in \boldsymbol{A}} \boldsymbol{V}_a$ ($\boldsymbol{V}_a$ is the domain of attribute a). In this paper, we accept the following assumptions:

1. A concept is defined as a triple:

$$concept = (c, A^c, V^c) \tag{1}$$

where c is a unique name of the concept, $A^c \subseteq \boldsymbol{A}$ is a set of attributes describing the concept and $V^c \subseteq \boldsymbol{V}$ is the attributes' domain: $V^c = \bigcup_{a \in A^c} V_a$. Pair (A^c, V^c) is called the structure of concept c.

2. An instance of a concept c is described by the attributes from set A^c with values from set V_c. Thus, an instance of a concept c is defined as a pair:

$$instance = (id, \ v) \tag{2}$$

where id is a unique identifier of the instance in world (A, V) and v is the value of the instance, which is a tuple of type A^c. All instances of the same concept in an ontology are different with each other.

By $Ins(O, c)$ we denote the set of instances belonging to concept c in ontology O. We have

$$\boldsymbol{I} = \bigcup_{c \in C} Ins(O, c) \tag{3}$$

2.2 Similarity

We present a formal definition of similarity method which derived from [2] as follows: let x, y, z are entities, value of $sim(x, y)$ represents the semantic similarity between x and y.

1. $sim(x, y) \in [0, 1]$
2. if $sim(x, y) = 1$ then $y = x$ or x semantic equivalent y.
3. $sim(x, y) = 0$: two objects are disjoint, i.e., no common characteristics.
4. $sim(x, y) = sim(y, x)$: similarity is symmetric
5. $sim(x, z) \leq (sim(x, y) + sim(y, z))$: The triangular inequation is valid for the similarity measure

However, when we apply similarity characteristics to find similarity between two structures of concepts, we reject characteristics 2, and 4 to satisfy with overlap characteristic of two concepts which mentioned below section.

2.3 Problem

In this section, we present the ontology integration process by providing semantic network of concepts of ontologies, called *OnConceptSNet*. The *OnConceptSNet* builds or extends their representations by acquiring knowledge from WordNet-Base and *static rules*. The knowledge may change the old network by adding and deleting nodes and arcs or by modifying numerical values (similar) or type arcs (relation), called weights, associated with the arcs.

The *OnConceptSNet* is defined as a graph:

$$\boldsymbol{G} = (\boldsymbol{C}^*, \boldsymbol{R}^*) \tag{4}$$

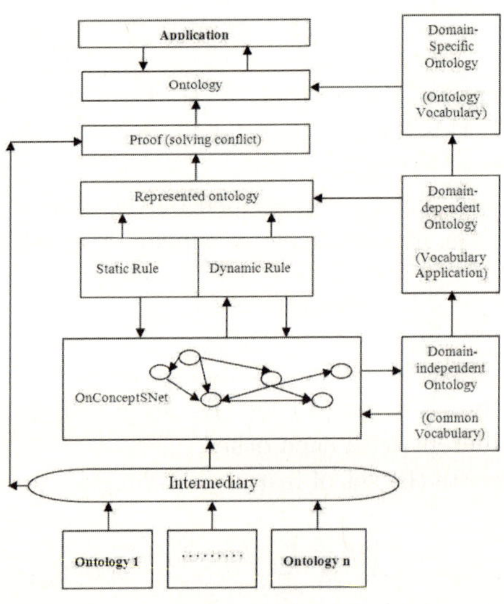

Fig. 1. Ontology integration process

$\boldsymbol{C}^*$ *is a set of nodes representing concepts that come from* $O_1, \ldots, O_n$.
$\boldsymbol{R}^*$ *is a set of arcs representing relations between concepts: semantic equivalent* ($\Leftrightarrow$), *more general* ($\sqsupseteq$), *less general* ($\sqsubseteq$), *overlap* ($\frown$). *Each arc is associated by a numerical similar value between two concepts.*

Here we denote S_i, l_i, and L_i corresponding to structure, name, and label of concept c_i, where the label of concept is either its name or its comment, or its label. We define the relations on *OnConceptSNet* as follow:

1. $c_1 \frown c_2$, that must be satisfied with one of conditions:
   - $sim(S_1, S_2) = 1$, and $sim(S_2, S_1) = 1$.
   - S_1, and S_2 are empty.
2. $c_1 \Leftrightarrow c_2$, that must be satisfied with one of conditions:
   - $sim(c_1, c_2) = 1$
   - $sim(L_1, L_2) > 0$, and $c_1 \frown c_2$.
   - $sim(L_1, L_2) > 0$, and *super-concept of* $c_1 \Leftrightarrow$ *super-concept of* c_2
   - $sim(L_1, L_2) > 0$, and *sub-concept of* $c_1 \Leftrightarrow$ *sub-concept of* c_2
3. $c_1 \sqsupseteq c_2$, that must be satisfied with one of conditions:
   - $sim(L_1, L_2) > 0$, $sim(S_2, S_1) = 1$, and $sim(S_1, S_2) < 1$
   - l_1 is a hyponym of l_2, $sim(S_2, S_1) = 1$, and $sim(S_1, S_2) < 1$
   - l_2 is a hypernym of l_1, $sim(S_2, S_1) = 1$, and $sim(S_1, S_2) < 1$
4. $c_1 \sqsubseteq c_2$, that of $c_2 \sqsupseteq c_1$.

We apply the upper ontologies as semantic domain layering. The idea of domain-independent ontologies provides basic concepts and relations to build the semantic network of concepts of ontologies which we call *OnConceptSNet*.

Represented ontology is an ontology which represents candidate ontologies. It is derived from *OnConceptSNet*.

Rules include *static rules*, and *dynamic rules*. The main *static rules* that we mentioned above to create *OnConceptSNet*. *Dynamic rules* are used to reduce or extend *OnConceptSNet* into *Represented ontology*, the rules as follows:

1. Rules for Concept:
   - if $c_1 \Leftrightarrow c_2$ then *delete* c_1
   - if $c_1 \sqsupseteq c_2 \wedge \exists\, c_1 \Leftrightarrow c_3$, where c_2 *is a sub-class* c_3 then *delete* c_1
   - if $(c_1 \sqsupseteq c_2) \wedge \neg\exists\, c_1 \Leftrightarrow c_3$, where c_2 *is a sub-class* c_3 then c_2 *sub-class* c_1
   - if $c_1 \sqsubseteq c_2 \wedge \exists\, c_1 \Leftrightarrow c_3$, where c_2 *is a sup-class* c_3 then *delete* c_1
   - if $c_1 \sqsubseteq c_2 \wedge \neg\exists\, c_1 \Leftrightarrow c_3$, where c_2 *is a sup-class* c_3 then c_2 *sup-class* c_1
2. Rules for property, the symbols differ in above-mentioned, $p_1 \Leftrightarrow p_2$ (p_1 *similarity* p_2), $\sqsubseteq$ (*hypernym/holonym*), $\sqsupseteq$ (*hyponym/meronym*), $\perp$ (*antonym*):
   - if $p_1 \Leftrightarrow p_2$ then *delete* p_2 (eg., *job* $\Leftrightarrow$ *occupation*)
   - if $p_1 \sqsubseteq p_2$ then *delete* p_1 (eg., *age* $\sqsubseteq$ *birthday*)
   - if $p_1 \sqsupseteq p_2$ then *delete* p_2 (eg., *sex* $\sqsupseteq$ *female*)
   - if $p_1 \perp p_2$ then *delete* p_1 (eg., *single* $\perp$ *married*)

Proof (solving conflict) contacts candidate ontologies, and the *represented ontology* to proof conflict in the *represented ontology*. After solving conflict, the *represented ontology* becomes an *ontology* that replaces the candidate ontologies.

Intermediary plays an intermediary role to connect the candidate ontologies and the *OnConceptSNet*. It translates the candidate ontologies into synchronous.

3 Ontology Conflict and Integration

3.1 Conflicts on Instance Level

At this level we assume that 2 ontologies differ from each other only in values of instances. That means they may have the same concepts and relations.

Definition 1. Let O_1 and O_2 be $(\boldsymbol{A}, \boldsymbol{V})$-based ontologies. Let concept (c, A^c, V^c) belong to both ontologies and let the same instance i belong to concept c in each ontology, that is $(i, v_1) \in Ins(O_1, c)$ and $(i, v_2) \in Ins(O_2, c)$. We say that a conflict takes place if $v_1 \neq v_2$.

For solving conflicts of ontologies on instance level, consensus methods seem to be very useful. Different criteria, structures of data and algorithms have been worked out [5, 6]. For this kind of conflict, the consensus problem can be defined:

Given a set of values $X = \{v_1, \ldots, v_n\}$ where v_i is a tuple of type A^c, that is:

$$v_i : A^c \rightarrow V^c \tag{5}$$

for $i = 1, \ldots, n$; $A^c \subseteq \boldsymbol{A}$ and $V = \bigcup_{a \in A^c} V_a$ one should find tuple v of type A, such that one or more selected postulates for consensus are satisfied [6].

One of very popular postulate requires minimizing the following sum.

$$\sum_{i=1}^{n} d(v, v_i) = \min_{v' \in T(A^c)} \sum_{i=1}^{n} (v', v_i) \tag{6}$$

where $T(A^c)$ is the set of all tuples of type A^c.

3.2 Conflicts on Concept Level

At this level we assume that two ontologies differ from each other in the structure of the same concept. That means they contain the same concept but its structure is different in each ontology.

Definition 2. Let O_1 and O_2 be $(\boldsymbol{A}, \boldsymbol{V})$-based ontologies. Let concept (c_1, A^{c_1}, V^{c_1}) belong to O_1 and concept (c_2, A^{c_2}, V^{c_2}) belong to O_2. We say that a conflict takes place in concept level if $c_1 = c_2$ but $A^{c_1} \neq A^{c_2}$ or $V^{c_1} \neq V^{c_2}$.

Definition 2 specifies such situations in which two ontologies define the same concept in different ways. For example, concept person in one system may be defined by attributes: *Name, Age, Address, Sex, Job*, while in another system it is defined by attributes: *Id, Name, Address, Date_of_birth, Taxpayer identification number, Occupation.*

The problem is the following: For given, a set of pairs $X = \{(A^i, V^i) : (A^i, V^i)$ is the structure of concept c in ontology O_i for $i = 1, \ldots, n\}$, it is needed to determine a pair (A^*, V^*) which at best represents the given pairs.

Words "at best" mean one or more postulates for satisfying by pair (A^*, V^*).

4 Distance Measures

4.1 WorkNet-Base Similarity between Two Words

The Palmer and Wu [9] similarity metric measures the depth of the two concepts in the WordNet taxonomy, and the depth of the least common subsumer. Resnik [11] defines the similarity between two words as the information content of the lowest superordinate in the hierarchy, defining the information content of a concept c (where a concept is the WordNet class containing the word). The Lesk similarity [1] of two concepts is defined as a function of the overlap between the corresponding glosses and those that surround it in the given context.

Our purpose is to apply similarity measure between words for ontology integration tasks. This similarity degree depends on complex candidate ontologies. So another of our approach for WorkNet-base similarity of two words is proposed as follows: the words occur together in a synset, they have the synonym relation with each other in context of a gross. For example, the words *learner* occurs in two noun synsets {*learner, scholar, ass-imilator*} and {*apprentice, learner, prentice*}; *student* occurs in two noun synsets {*student, pupil, educate* } and {*scholar, s-cholarly person, bookman, student*}. Thus, *scholar* is a common word of a *student*'s synset and a *learner*'s synset, so *student* and *learner* have a relation, if we continue finding synonym of words in student's synsets and in learner's synsets, the number of similar words occurs together with *student* and *learner* may be much larger. That means the similarity degree between *student* and *learner* is quite larger. Moreover, each word occur in many synsets that cross part of speech. For example, the word *base* occurs in 7 adjective synsets, 3 verb synsets, and 19 noun synsets, that means the similarity acrosses part of speech. For these reasons, we proposed a formulate for measuring the semantic similarity of words as follows:

$$sim(w_1, w_2) = \max_{level=1,\ldots,n} \left(\frac{\triangle + \sum_{w_i \in Syn_1 \cap E} (\sum_{w_j \in Syn_2 \cap E} Inc(w_i, w_j))}{min(size(Syn_1), size(Syn_2)) + size(E)} \right) \quad (7)$$

where

$$Inc = \begin{cases} 0 & \text{if } w_i \neq w_j \\ 1 & \text{if } w_i = w_j \end{cases}$$

If $Inc(w_1, w_2) = 1$ then $E = E \cup \{w_1\}$.
$\triangle$ is total return value of *Inc* at *level*=1..*k*-1, *k* is current *level*.

The *level* will be increased from 1 to *n*, each increasing time of *level* then,

$$Syn_1 = \bigcup_{w \in Syn_1 \cap E} Synonym(w) \text{ and } Syn_2 = \bigcup_{w \in Syn_2 \cap E} Synonym(w).$$

We experiment with the method to find out similarity between 100 pairs of words with different similarity degree and crossing part of speech, we chose *level* is equal to 3, and limit of size of array *Syn* is 1000, most of similarity between words is found out. Please note that, the more we increase level, the more similarity between words is increased. For example, similarity between *learner* and *student* at *level* 1, 2, 3 corresponding to 0.24, 0.65, 0.84.

4.2 Similarity between Two Properties

p_i is representation identification of property i,
$R_i = \{r_1, r_2, \ldots, r_n\}, r_j$ is a name/value of instance j of priperty p_i
$A_i = \{a_1, a_2, \ldots, a_k\}, a_j$ is a either single word or compound word come from r_i
$G_i = \{g_1, g_2, \ldots, g_m\}, G_i$ is set of more general words of $a_j \in A_i, j = 1, 2, \ldots, k$. The words of set G_i which come from WordNet through HYPERNYM.

$$H = \bigcup_{i=1}^{n} (G'_i) \tag{8}$$

where $G'_i \subseteq G_i$ and if $g_j \in G'_i, g_j$ exists in at least $\frac{1}{2}n$ sets $G_i, i = 1, 2, \ldots, n$

$$sim(H_1, H_2) = \frac{\sum_{a \in H_1} (max(sim_{b \in H_2}(a, b)))}{min(size(H_1), size(H_2))} \tag{9}$$

Similarity between two properties

$$sim(p_1, p_2) = max(sim(L_1, L_2), sim(H_1, H_2)) \tag{10}$$

where $sim(L_1, L_2)$ is similarity between two labels of properties p_1 and p_2.

4.3 Similarity between Two Concepts

c_i is representation identification of concept i,
$C_i = \{l_1, l_2, \ldots, l_n\}, l_i$ is a name/label of instance i of concept c_i
$A_i = \{a_1, a_2, \ldots, a_k\}, a_j$ is a either single word or compound word come from l_i
$G_i = \{g_1, g_2, \ldots, g_k\}, G_i$ are set of more general words of $a_j \in A_i$. Those words of set which come from WordNet through HYPERNYM.

$$H = \bigcup_{i=1}^{n} (G'_i) \tag{11}$$

where $G'_i \subseteq G_i$ and if $g_j \in G'_i, g_j$ exists in at least $\frac{1}{2}n$ sets $G_i, i = 1, 2, \ldots, n$

$$M = \bigcup_{i=1}^{n} (A'_i) \tag{12}$$

where $A'_i \subseteq A_i$ and if $a_k \in A'_i$ then exist at least $g_j \in G_i$ and g_j is general word of a_k.

We define similarity between 2 structures of two concepts as follows:
S is representation of Structure of concept. $S = \{p_1, p_2, \ldots\}$, where $p_i, i = 1, 2, \ldots, n$ is properties of concept

$$sim(S_1, S_2) = \frac{\sum_{p \in S_1} (max(sim_{p' \in S_2}(p, p')))}{size(S_2)} \tag{13}$$

Similarity between 2 concepts c_1 and c_2

$$sim(c_1, c_2) = max(\frac{sim(L_1, L_2) + sim(S_1, S_2)}{2}, \frac{sim(H_1, H_2) + sim(S_1, S_2)}{2}, \frac{sim(M_1, M_2) + sim(S_1, S_2)}{2}) \quad (14)$$

where $sim(L_1, L_2)$ is similarity between two labels of concepts c_1 and c_2.

5 An Algorithm for Ontology Integration

- Input: n candidate ontologies $O_1, \ldots, O_n$.
- Output: Ontology O^* that replaces $O_1, \ldots, O_n$.

Begin

1. *Intermediary* translates $O_1, \ldots, O_n$ with language $L_1, \ldots, L_n$ into L_0;
2. Create *OnConceptSNet* for $O_1, \ldots, O_n$ pass WordNet-Based;
   - Find similarity between properties of $O_1, \ldots, O_n$;
   - Find similarity between concepts of $O_1, \ldots, O_n$;
   - Create relation between concepts of $O_1, \ldots, O_n$ on *OnConceptSNet* pass *static rules* and similarity between concepts;
3. Create *dynamic rules* that have states come from *OnConceptSNet*;
4. Execute *dynamic rules* for reducing *OnConceptSNet* to *represented ontology* that best represents n candidate ontologies;
5. Execute algorithm for solving conflict in *represented ontology*;
6. Compute O^*, to build the domain-dependent, and domain-specific ontology;
7. Return O^*;

End.

6 Experiments

We used four data sets, each consisting of at least two ontologies which we refer to [2] http://www.aifb.uni-karlsruhe.de/WBS/meh/foam/ontologies.htm for evaluation purposes. From their differences, we expect a representative evaluation. Because of limited space, we only present some compares with author's result [2] (see the table 1, and table 2).

Our purpose is to integrate ontologies of the information systems with knowledge bases which have to be integrated when they want to share or exchange their knowledge. These information systems' knowledge bases include ontologies and their instances. Therefore we assume that there are enough instances for our similarity finding method. However, we have to note that our method is sufficient not only in environments with many instances but also in environments with lack of instanes. An example, although the ontologies of authors [2] are really less instances, but our results are still sufficient. In this paper, we only show some experiment results from which is stressed the role of WordNet-Based as independent-domain ontologies is stressed. Our complete system for ontology integration will be shown in our future work.

Table 1. Some results comparing between our similarity and in [2]

Property 1	*Property 2*	*Author's sim[2]*	*Our sim*
russia#music	meh://8807#music	1.0	1.0
russia#cost_money_eating	meh://8807#cost_money	0,9473	0.98
russia#include_city	meh://8807#include_town	0,9167	1.0
animalsA.owl#hasMaleParent	animalsB.owl#hasFather	1.0	1.0
animalsA.owl#hasFemaleParent	animalsB.owl#hasMother	1.0	1.0
...	...	...	...

Table 2. Some results comparing between our relation and similarity in [2]

Concept 1	*Concept 2*	*Author's sim[2]*	*Our relation*
animalsA.owl#Woman	animalsB.owl#Person	?	sub-class
animalsA.owl#HumanBeing	animalsB.owl#Man	?	sup-class
animalsA.owl#HumanBeing	animalsB.owl#Person	1.0	equivalent
animalsA.owl#TwoLeggedPerson	animalsB.owl#BipedalPerson	1.0	equivalent
animalsA.owl#TwoLeggedThing	animalsB.owl#BipedalThing	1.0	equivalent
...	...	...	...

7 Conclusions

In this paper, we built the semantic network, called OnConceptSNet which is derived from the upper ontologyWordNet to integrate multiple ontologies as reconcile semantic conflicts between the ontologies. We designed the semantic similarities between ontology elements using WordNet-Based. We also described a methodology to solve the conflict in ontology integration process. In future work, we will approach dynamic inference rules for ontology integration tasks that derive and aggregate relation between attributes, instances, concepts, and to insert, remove a derived object when the condition of the deductive rule is satisfied. We will also applied consensus theory for solving conflict in relation level and restriction level. Finally, we will build a auto-ontology integration system.

References

1. Banerjee, S., Pedersen, T.: An adapted Lesk algorithm for word sense disambiguation using Word- Net. In: Proceedings of the Third International Conference on Intelligent Text Processing and Computational Linguistics, Mexico City, Mexico, pp. 136–145 (2002)
2. Ehrig, M., Sure, Y.: Ontology mapping - an integrated approach. In: First European Semantic Web Symposium, pp. 76–91 (2004)
3. Gangemi, A., Pisanelli, D.M., Steve, G.: Ontology Integration: Experiences with Medical Terminologies. In: Guarino, N. (ed.) Formal Ontology in Information Systems, pp. 163–178. IOS Press, Amsterdam (1998)

4. Jeongsoo, L., Heekwon, C., Kwangsoo, K., Cheol-Han, K.: An Ontology Architecture for Integration of Ontologies. In: Processing The Semantic Web - ASWC, pp. 205–211 (2006)
5. Nguyen, N.T.: Processing Inconsistency of Knowledge on Semantic Level. Journal of Universal Computer Science (2), 285–302 (2005)
6. Nguyen, N.T.: Advanced Methods for Inconsistent Knowledge Management. Springer, London (2008)
7. Niles, I., Pease, A.: Towards a standard upper ontology. In: Welty, C., Smith, B. (eds.) FOIS 2001: Proceedings of the international conference on Formal Ontology in Information Systems, New York, NY, USA, pp. 2–9 (2001)
8. Noy, N.F., Musen, M.A.: PROMPT: Algorithm and Tool for Automated Ontology Merging and Alignment. In: AAAI 2000 Proceedings, pp. 450–455 (2000)
9. Palmer, M., Wu: Verb semantics and lexical selection. In: Proceedings of the 32nd Annual Meeting of the Association for Computational Linguistics, Las Cruces, New Mexico, pp. 133–138 (1994)
10. Pinto, H.S., Martins, J.P.: A Methodology for Ontology Integration. In: Proceedings of the First International Conference on Knowledge Capture, pp. 131–138. ACM Press, New York (2001)
11. Resnik: Using information content to evaluate semantic similarity. In: Proceedings of the 14th International Joint Conference on Artificial Intelligence, Montreal, Canada, pp. 448–453 (1995)
12. Semy, S.K., Pulvermacher, M.K., Obrst, L.J.: Toward the use of an upper ontology for U.S. government and U.S. military domains: An evaluation. Technical Report MTR 04B0000063, The MITRE Corporation (2004)

The Multiagent System for Channel Selection and Topology Control on 802.11 Based Mesh Networks

Ante Prodan, Vinod Mirchandani, and John Debenham

University Of Technology Sydney, NSW 2007, Australia
aprodan@it.uts.edu.au
http://qdine.it.uts.edu.au/

Abstract. To address the problem of topology control on multi-radio wireless mesh networks a distributed, lightweight, co-operative multiagent system that guarantees scalability has been developed and validated by simulation. Our goal is twofold, to select channels so to reduce interference, and improve connectivity by shortening paths between portal and client nodes. As this system is to be deployed over large networks the scalability and stability of the solution are of the main concern. The proposed algorithms have been implemented and evaluated with the Java based framework as well as NetLogo a multiagent simulation tool. The positive attributes of the algorithms are demonstrated through the comprehensive simulation result analysis.

1 Introduction

The work discussed here is based on previous work in the area of mesh networking and in particular in distributed algorithms at Columbia University, Microsoft Research, University of Maryland and Georgia Institute of Technology [1],[2]. The algorithms described in this paper leverage on the work on topology control that we have conducted in [3],[4].

Related recent work on 802.11 Mesh Networks, such as [5], is predicated on a network whose prime purpose is to route traffic to and from nodes connected to the wired network - in which case it is assumed to be no traffic between end user nodes. Although similar, our methods have, where possible, been designed for the more general clas of "wireless mesh networks".

The main motivation for the use of multi-radio routers with self-organising channel selection is to cause a decrease of the interference between the channels of two adjacent routers. A principal objective of a multiagent system is to increase the overall MR-WMN capacity by making the throughput of the links less susceptible to the channel interference and to reduce the path length between portal and client nodes.

This paper is organised as follows: In section 2 we introduce important terms and concepts; section 3 describes the non Transmit Power Control (TPC) based Topology Control; the preliminary algorithm is discussed in Section 4; enhanced topology control algorithm is explained in Section 5; the comprehensive simulation result analysis is provided in Section 6 and conclusion in Section 7.

I. Lovrek, R.J. Howlett, and L.C. Jain (Eds.): KES 2008, Part I, LNAI 5177, pp. 220–228, 2008.

2 Basic Terms and Concepts

- A *node* is an intelligent router; a set of interfaces of which most are radio interfaces but some can be wired, where each radio interface is associated with a particular channel. A node contains a lightweight agent that (intelligently we hope) assigns the radio channel and directs other nodes activities.
- A *link* is a pair of interfaces within the same communication domain where each interface is assigned the same channel. Two interfaces communicate through a shared link.
- A *portal node* is a node that contains a wired interface (e.g. Ethernet) and is used to provide connectivity between the client nodes and the Internet.
- A *client node* is a node that contains only wireless interfaces. To be able to route traffic to and from the Internet this node has to be directly or indirectly connected to at least one portal node.
- The *initialisation process* is the process though which a set of spatially distributed client and portal nodes cooperatively create links among each other.
- The term *topology control* indicates the way in which the links between mesh nodes are created. Broadly the concept of topology control of MR-WMN can be explained as a control over connectivity between the mesh nodes. Our non-TPC based topology control relies on increasing the number of shortest paths to the portal nodes which raises the system capacity.
- The term *path reduction* means the reduction in path length (measured in hops) between a client node and a closest portal node.
- *MR-WMN* is a type of a mesh network in which each node has more than one wireless interface. In contrast to this, the use of single radio nodes for a mesh network results in a progressive decreases in the throughput of the link between each hop due to the co-channel interference [6].
- The term *interference cost* (IC) of a link is used to describe the measure of bandwidth reduction that the rest of the MR-WMN is subjected to when two interfaces communicate on a specific channel.
- The term path reduction (PR) is used to describe a process that minimises a number of hops between client nodes and portal nodes (path).
- The term *network density* is used to indicate the number of nodes on a network that occupies the area that is in all experiments the same. Network densities used in our simulation are: 35, 70 and 100 nodes per network area. In addition, the scalability of all evaluated algorithms is tested on an extended area with the density of 1000 nodes.
- The *lightweight multiagent system* used in this work can be classified as: *inaccessible* - agents can not obtain complete information and accuracy cannot be guaranteed; *non deterministic* - agents actions can have unforeseen effect e.g. creation of a link between a pair of nodes can cause an increase in interference that renders some other link on a network unusable. The environment is *dynamic* - other processes can change the environment beyond agent's control. To select appropriate actions agents have to perform *information gathering*.

3 Non TPC Based Topological Control

In section 2, we had introduced the concept of topological control; as part of our further study, we have identified the need to implement this concept on MR-WMNs. This is possible by means of algorithms that increase the number of shortest paths between the client nodes and the portal nodes. Through the stochastic simulations, we conclusively show in this section that the distribution of shortest paths from the client to the portal nodes is very uneven, which motivates the development of the above stated algorithms. Such a topological control would result in a noticeable increase of the overall WMN capacity. We base this anticipated outcome on the simplistic preliminary premise that the client nodes generate the traffic load evenly.

3.1 Simulation Model

We have used a Java based framework to carry out the simulations for the results shown and discussed in this section. The key attributes of the simulation were:

- Number of radio interfaces per router was randomly selected from 3 to 5.
- Default signal strength was 100 mW (20 dBm) Signal strength for each interface was randomly generated with +/- 25% variation.
- Network site area was 750 m X 500 m.

The simulations were carried out for realistic node densities and topologies. In all, we had carried out a total of 900 simulations that includes 100 simulations per each of the node density and topology combination. The large number of simulation runs has helped us to generate 98% confidence intervals for the obtained results.

3.2 Path Length Problem

This part of study investigates the path length as a function of different topologies and densities. In this regard figure 1 gives a cumulative distribution of the number of hops for mean number of links across all network topologies and 100 node network density. It can be seen that the hop count is approximately the same across all the three topologies. The previous result of invariance with topologies is reconfirmed. In addition, we have obtained the frequency of links normalised with respect to the node densities for different hop counts across different topologies. From the results presented in this subsection, we intend to create and evaluate the algorithms for path length reduction that will preserve the interference cost and result in an increase of network system capacity.

4 Preliminary Algorithm for Path Length Reduction

The preliminary algorithm for path length reduction consists of two separate mechanisms: (i) link substitution and (ii) link addition. It is important to note

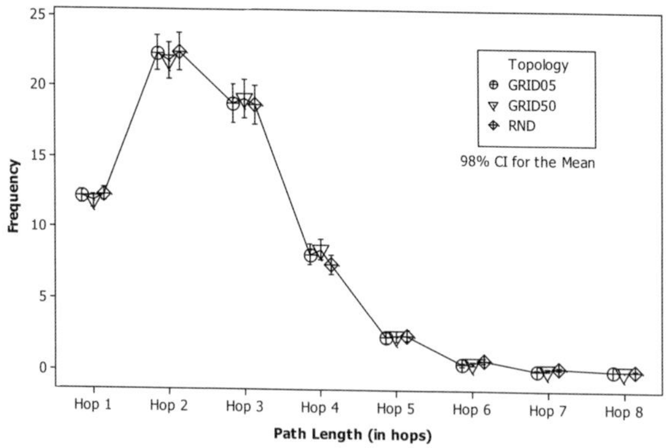

Fig. 1. Frequency distribution of the path length (in hops) for different topologies at 100 node network density - different topologies types are shown by distinct symbols

that both of these mechanisms can be triggered only when the path length to portal nodes is known. Furthermore, these mechanisms rely on node blocking, self-blocking and interference cost measurement techniques, which are in detail discussed in our earlier work [4]. The first mechanism enables a node to initiate a link substitution whenever it discovers a free radio interface in the transmission range that can provide a shorter path link. This is a preferred mechanism for path reduction as it can be deployed in a way that does not result in an increase of the overall interference.

The second mechanism used for preliminary path length reduction algorithm is link addition. Similar to link substitution mechanism, a node looks for available radio interfaces only this time new i.e. additional links are created. However, this approach almost always results in an increase of the overall interference and thus care has to be taken that the benefit of path length reduction is not nullified by an increase of interference. For this reason we use it only in conjunction with interference cost measurements.

During the simulations two important observations were made: (i) Path length reduction due to link substitution does not cause any increase in interference. (ii) Link addition did result in an increase of interference, which still was not significant enough to diminish the benefits of path length reduction. This was possible when the added links were selected from a pool of potential links by using the criteria of lowest IC.

5 Enhanced Topology Control

In this section, we explain our non-TPC based enhanced algorithm for topology control in MR-WMN. The algorithm has been implemented and tested in a multi-agent simulation tool called NetLogo. NetLogo is a cross-platform, multi-agent programmable modelling environment used for simulating natural and social phenomena. The primary purpose of NetLogo has been to provide a higher-level platform that allows modellers to build and learn from multi-agent based models. In continuation of this section results of simulations are provided to show that the enhanced algorithm produces IC reduction in MR-WMN system when compared to a non-enhanced algorithm. The enhanced algorithm stage was reached after progressively creating and examining the topology control approaches explained in previous sections.

5.1 Motivation for Using NetLogo

To evaluate the topology control approaches discussed in section 5 we had created and used a Java framework. This was also used to carry out a feasibility study of the self-organisation algorithm that we proposed in [4]. However, it soon became apparent that building and modifying algorithms within the existing framework was getting tedious due to the complexity of the tasks related to the integration of additional distributed algorithms (e.g. topology control algorithms) into an essentially sequential framework. The benefits such as interactivity and extensibility of a multi-agent simulation environment offered by NetLogo are very significant because model corresponds more realistically to the natural representation of a self-organising system.

5.2 Initialisation Algorithm

Our enhanced initialisation process is composed of two independent algorithms- (i) Portal first and (ii) Path reduction through link substitution. Due to lack of space we have only given a wide scope of our algorithm rather than describing the details of its operation. Portal first (PF) - As a matter of fundamental principle in a MR-WMN system the portal nodes are not linked with other portal nodes however each portal node is linked to its neighbouring client node until all its radio interfaces are exhausted.

5.3 New Path Reduction Algorithm

In our new path reduction (PR) algorithm only link substitution method is used which makes it different from our previous PR algorithm in which link substation is used along with link addition. Another distinction is that previously we selected only those substituted links that would not increase the IC whereas in the current algorithm we select links irrespective of their effect on the IC.

The outline of our algorithm is succinctly given here: An initiator node selects one of its available radio interfaces on the basis of strongest transmission power, the selected interface creates a list of available interfaces in its communication range (locality principle), from these interfaces those that have a path length to the portal node longer than the shortest path are filtered out. The remaining ones are short listed and an interface from these that offers the best SNIR is selected; the new link between the two interfaces is created and the previous shortest path link is switched off and its interfaces operational attributes are reset to their default values. This process occurs simultaneously across the multiagent system. A more formal description of the algorithm is given below:

For node $n_x \in N$ select one of its interfaces i_x that is free and has the strongest signal of all its free interfaces;
if $i_x = \varnothing$
    for n_x set *blockFree* ← *bc;*
    end;
else
    set *blockBusy* ← *bc*;
set I_{free} ← *communicationRange*(i_x, I) ;
if $I_{free} = \varnothing$
    endAction.
set i_y ← *bestSNR*(I_{free})
if $i_y = \varnothing$
    endAction.
$l_{i_x i_y}$ ← *createLink*(i_x, i_y)
for n_x *linkCounter linkSubCounter* + *1;*
remove ($l_{i_u i_v}$);
reset (i_u, i_v);
set p ← *newShortestPath*(n_x)
end.

N	a set of all nodes
I	a set of all interfaces
L	a set of all links
I_{free}	a subset of interfaces
i_x	an available interface.
n_x	a node which contains i_x.
i_y	an available interface.
n_y	a node which contains i_y.
i_u	an interface with shortest path on n_x
n_v	a node that contains i_v.
i_u	an interface on n_z that is linked to i_u
p	a shortest path for a n_x
$l_{i_u u_v}$	a link between i_u *and* i_v
bc	a blocking constant

endAction
    for n_x
        set *blockBusy* ← 0;
        set *blockFree* ← *bc;*
end.

- cummuncationRange(interface, set of interfaces): This function selects all interfaces from the set of interfaces that are free and within the range of interface.
- shortestPath(p, interface, set of interfaces): This function selects a set of interfaces that has a shortest path from the interface to the set of interfaces. If the path between the interface with a shortest path and the interface is not shorter than path p the function returns ∅. Otherwise it returns a set of interfaces with a shortest path.
- bestSNIR(set of interfaces): This function selects an interface with a best SNIR from the set of interfaces. If there is more than one such interface this function randomly selects one.
- createLink(interface A, interface B): This function creates a link between interface A and interface B and returns a newly created link.
- remove(link): This function removes a link from the set L.
- reset(interface A, interface B): This function resets attributes of interface A and interface B to default attributes.
- newShortestPath(node): This function returns a shortest path value for the node.
- A node is either 'locked' or 'unlocked'. A locked node is either locked because it has committed to lock itself for a period of time on request from another node, or it is 'self-locked' because it has recently instigated one of the self organisation procedures. A locked node is only locked for a 'very short' period

during the operation of each of those procedures. The length of blocking time is indicated by the value of blockBusy and blockFree attributes. This is simply to ensure that no more than one alteration is made during any one period which is necessary to ensure the stability of the procedures.

6 Results and Discussion

We present below some of the key results that we have obtained to illustrate the performance of our enhanced path reduction algorithm. In this regard Fig. 2 shows a graph of ICR vs. network density when just the algorithm for ICR has been invoked and both Path Reduction and ICR algorithms has been

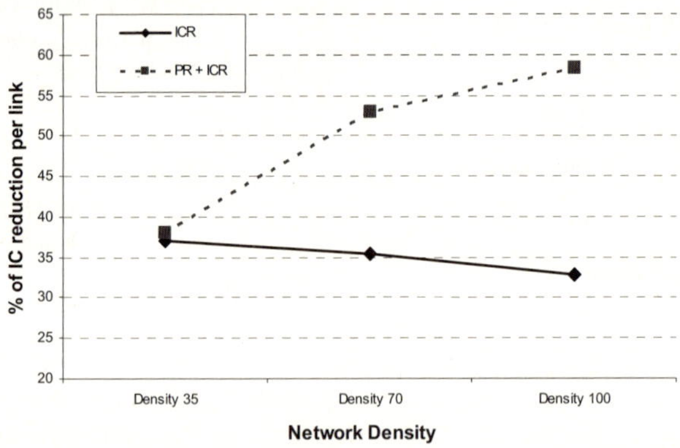

Fig. 2. Result (comparative study) showing the increase in IC reduction in comparison to ICR only algorithm

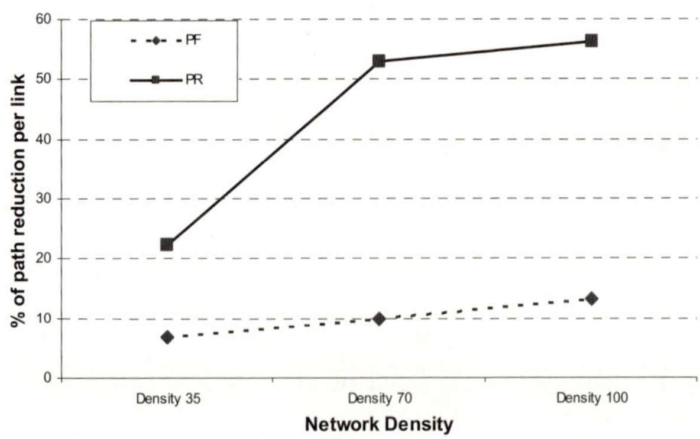

Fig. 3. Portal first (PF) versus new algorithm for path reduction (PR) through link substitution

invoked. This comparative study clearly shows that without the invocation of the PR algorithm the effect of ICR process results in much higher IC. A reason for this is twofold. Firstly, because of an increased number of interfaces that the initiator node can use and secondly, because of the additional mechanism that selects the interface based on the best SNIR value. Furthermore, as the network density increases the performance of the PR followed by ICR significantly increases whereas just the performance of the ICR algorithm on its own slightly decreases.

Fig. 3 compares the path length reduction that is achieved by using just the PF algorithm and the new path length reduction algorithm. It can be seen that the new path length reduction is much more effective at reducing the path length relative to that achieved by PF algorithm. Although PF is providing significant decrease in path length PR provides 2-5 times better improvements.

7 Conclusions

Topology control in MR-WMN is carried out by initialisation schemes to create connectivity between the mesh nodes at MR-WMN set-up or reconfiguration. This paper has conclusively shown supported by numerous simulation results that non-TPC based topological control algorithms could be developed. Such algorithms should increase the number of shortest paths to the portal nodes which will lead to an increase of system capacity. In this regard, the simulation result obtained for the proposed preliminary path reduction algorithms, based on link substitution and link addition, showed that a significant path reduction can be achieved.

Even more important are results of performance simulations of the enhanced path reduction algorithm. Invocation of this algorithm in addition to the expected level of path reduction leads to a significant IC reduction. We have also shown that there is no need for incorporation of PF algorithm since IC reduction results are better without it and it does not provide any increase in path reduction in comparison to enhanced path reduction algorithm.

Notes and Comments. This work is supported by a grant from BellLabs - Alcatel-Lucent and the Australian Research Council.

References

1. Ko, B.J., Misra, V., Padhye, J., Rubenstein, D.: Distributed Channel Assignment in Multi-Radio 802.11 Mesh Networks. In: WCNC IEEE, Hong Kong, China, pp. 3981–3986 (2007)
2. Mishra, A., Shrivastava, V., Banerjee, S.: Partially Overlapped Channels Not Considered Harmful. SIGMetrics/Performance. ACM, Saint Malo (2006)
3. Prodan, A., Mirchandani, V., Debenham, J., Green, L.: Performance Evaluation of a Self-Organising Scheme for Multi-Radio Wireless Mesh Networks. In: IEEE AusWireless. IEEE, Sydney (2007)

4. Mirchandani, V., Prodan, A., Debenham, J.: A Method and Study of Topology Control based Self-Organisation in Mesh Networks. In: AccessNets (BWIA Workshop). IEEE, Ottawa (2007)
5. Raniwala, A., Chiueh, T.-C.: Architecture and Algorithms for an IEEE 802.11-Based Multi-Channel Wireles Mesh Network IEEE Infocom, Miami, USA (2005)
6. Gupta, P., Kumar, P.R.: The capacity of wireless networks. IEEE Transactions on Information Theory 46, 388–404 (2000)

EasyLife: A Location-Aware Service Oriented Mobile Information System

J.M. Shen[1], M.J. O'Grady[2], and G.M.P. O'Hare[2]

[1] School of Computer Science & Informatics, University College Dublin (UCD), Belfield, Dublin 4, Ireland
[2] Adaptive Information Cluster (AIC), University College Dublin (UCD), Belfield, Dublin 4, Ireland
sjm2218@hotmail.com, {Michael.J.OGrady,Gregory.OHare}@ucd.ie

Abstract. Many examples of Location-aware services have been developed in recent years as the enabling technologies mature. However, these services frequently exist in isolation and address specific niche markets. The diversity of the mobile computing community suggests that enabling dynamic combinations of location-area services would be an appropriate deployment strategy. In this way, customers could subscribe to those services that address their particular needs. This paper introduces EasyLife, an agent-based architecture designed to facilitate the development of suites of location-aware services that customers can configure according to their preferences.

Keywords: Ambient Intelligence, Mobile Computing, Agent-oriented Information Systems, Location-aware computing.

1 Introduction

Location-aware computing is one of the foremost developments in mobile computing [1] as it is perceived as offering a new paradigm for service delivery that has significant economic potential. Hence, it has attracted the attention of many industrial sectors as well as various disciplines in academia. The enabling technologies are well understood – positioning systems of various hues, broadband wireless communications and sophisticated personal devices. Though there are many exemplar examples of location-aware services, one common characteristic concerns their focus on addressing a single application domain or end-user group.

Heterogeneity is one of the defining characteristics of the mobile computing community. Thus arbitrary customers may require certain combinations of services at various times, depending on the prevailing context. Meeting those customers' needs would obligate service providers to provide their services in a mix-and-match fashion. Such a scenario gives rise to technical issues as a need for new business models. This paper proposes an architecture – EasyLife, which is aimed at providing a framework that would enable service providers to deliver customised combinations of location-aware services to their subscribers.

This paper is structured as follows. In the Section II, a brief snapshot of location-aware services is provided. The architecture of EasyLife is presented in Section III

I. Lovrek, R.J. Howlett, and L.C. Jain (Eds.): KES 2008, Part I, LNAI 5177, pp. 229–236, 2008.

after which the initial implementation is described. Some future work is discussed in Section V after which the paper is concluded.

2 Related Research

The success of the Global Positioning System (GPS) and ubiquitous availability of mobile data services has made location-aware services feasible. A significant number of projects have been described in the literature, and a number of these are now discussed.

AccesSights project [2] is a multimodal location-aware mobile tourist information system. The aim is to help disabled people to get the same tourist information as sighted people and help them explore tourist destinations. By using GPS, the system knows where the user is, their orientation, and their movement, thus providing the relevant sight information to the user through sound and some accessible text. It is very benefit for those special people. Interestingly, tourism is a popular application domain for location-aware services and various other systems have been documented, for example, HIPS [3], CRUMPET [4] and Gulliver's Genie [5].

AudioGPS [6] is more than converting the GPS data retrieved as simple coordinates and depicting them onto a map. It involves building a prototype for mapping this data to non-speech spatial audio, so that the user could get the location information with less attention and eliminating the need for language recognition. The idea is attractive, and an initial prototype has been constructed and evaluated.

MStream [7] seeks to develop a mobile music streaming application that provides a location-aware audio service to the end user. Apart from the client-server model, the MStream project also uses P2P technology for large scale and interactive audio among users. It could be used for location-based conferencing or as a tour guide.

CoMPASS [8] is concerned with the delivery of context-sensitive spatial information to mobile devices. In particular, personalization is an important aspect, and the system enables the recommendation of context-aware spatial information to users as they interact with electronic maps.

Agent Channeling ContExt Sensitive Services (ACCESS) [9] is an agent-based architecture which is used for development and deployment of context sensitive services. The main focus of the project is to build an infrastructure so that the developers could focus on the implementation of the context-aware services.

From the above, we can see that there are plenty of good ideas about location-based services, and lots of effort have been put into this area. In Easy Life, we would like to use the cutting edge technology and develop a multi-tier location-aware service-oriented system. Such a system would offer a complete solution for delivering location-based services, and would leverage the following technologies:

- Location aware services that are more that just navigation aids;
- Web 2.0 services incorporated into mobile technologies;
- Heterogeneous agent technologies for modeling and delivering the different services.

Initial prototype services offered by EasyLife will include a real-time weather service, a shopping service based on shopper location and a restaurant service that can recommend nearest restaurant to the user. Though initially focusing on the issue of location-awareness, it can be extended to included context-awareness.

3 Architecture

From an architectural perspective, EasyLife consists of three key components:

- EasyLife client;
- AgentProxy_app;
- AgentServices.

The benefits for dividing the system into these three different parts are that it is loosely coupled and each component is well abstracted and easily extended. Each of these components is now described.

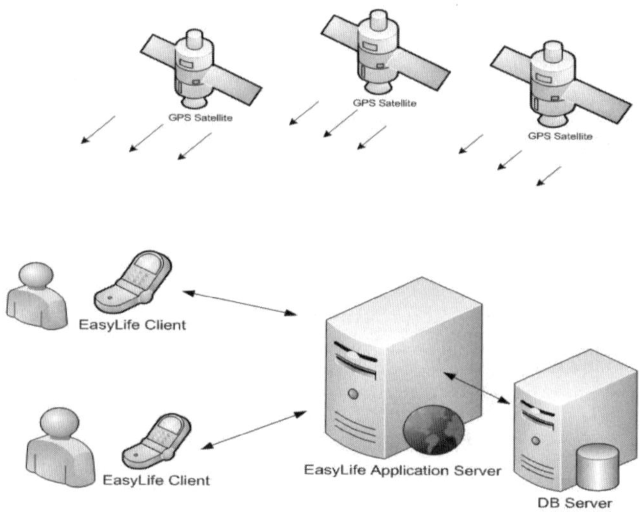

Fig. 1. Overview of EasyLife

3.1 EasyLife Client

An illustration of the architecture of the EasyLife client may be seen in Figure 2. It is composed of five key elements:

- **Agent:** controls the behavior and workflow of the EasyLife client. It also listens and handles all events generated.
- **Context:** is responsible for getting the context from the external environment. In particular, it is used to get the location information from the external Bluetooth GPS receiver. It can be extended to handle other sensors if need be.
- **Controller:** provides an access point for certain functions – one example being the UIController. It provides an API for calling different UIs. For the extensibility purpose, other controllers could be developed.
- **UIs:** provides the Graphic User Interface (GUI) for the EasyLife client.
- **Data Model:** represents an abstraction of the data from different domains, such as the GPS data model, the weather data model and so on.

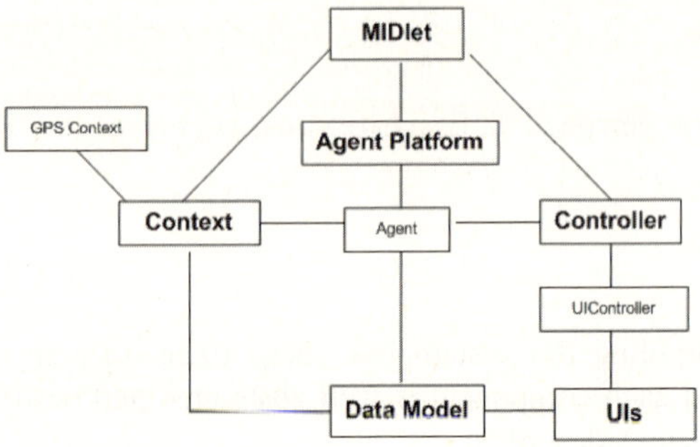

Fig. 2. EasyLife – Client Architecture

3.2 AgentProxy_app

The AgentProxy_app is best viewed as a virtual router. It has a Gateway Agent which parses service requests, and forwards them to the correspondent agent service. It also relays messages to the mobile client.

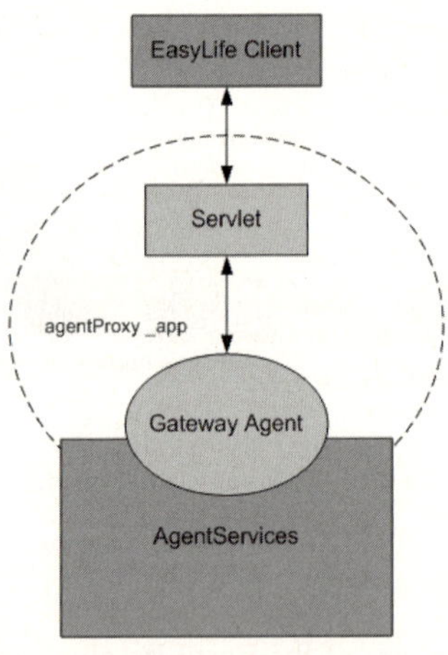

Fig. 3. AgentProxy_app architecture

From an implementation perspective, agentProxy_app is implemented as a servlet (Figure 3). It communicates with the EasyLife client via WiFi and via standard TCP/IP with the Gateway Agents which acts as the interface to Agent Services.

3.3 AgentServices

AgentServices consists of many different agents, each representing an enterprise and providing different services. There are controller agents which could be seen as service brokers interacting with other systems. The benefits of dividing the system into these three different parts are that it loosely coupled. Each component is well abstracted and used for different purposes.

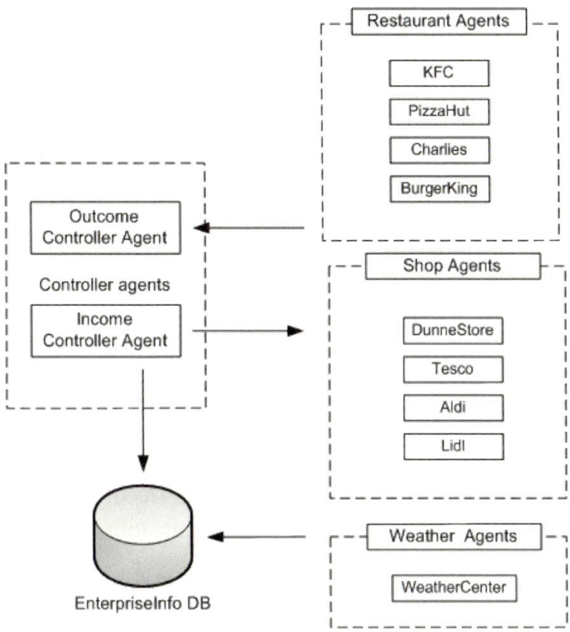

Fig. 4. AgentServices architecture

AgentServices consists of two parts, controller agents and service agents (Figure 4).

Controller Agents

- **Income Controller Agents** are the entry points for the agent service. They are responsible for parsing the request and dispatching it to the correspondent agent according to the location and the type of the service. They also inform the Outcome Controller Agent of the address of the Gateway Agent, so that the Outcome Controller Agent knows where to send the service data.
- **Outcome Controller Agent** obtains the address of the Gateway Agent from the Income Controller Agent. All the service agents send their data to the outcome agent and this dispatches the service data to the correspondent Gateway Agent.

Service Agents
In generally all the service agents would conform to similar role with the following functions.

- Event Listening;
- Service Data Constructing;
- Sending Messages.

In the case of the Restaurant service, four agents were constructed - each representing different restaurants, namely the KFC Agent, the PizzaHut Agent, the BurgerKing Agent and the Charlies Agent. The restaurant service demonstrates the location-based recommending service. Four agents for the Shop service were created, representing different shops. The Shop service demonstrates a location-based advertisement service. One agent for the weather service was created. This leverages the Yahoo weather service by calling two REST APIs - one for getting the code for the location, the other for getting weather forecast information from the Yahoo website.

4 Implementation

EasyLife harnesses a number of technologies but intelligent agents form the constituent components. Two different agent development environments were used – Jade for the resource intensive server components and AFME for the lightweight mobile devices.

JADE
Java Agent Development Environment (JADE) [10] is a robust and efficient environment for agent development. JADE agents support a thorough range of agent characteristics. From a communications perspective, JADE complies with the FIPA specifications.

Agent Factory Micro Edition
Agent Factory Micro Edition (AFME) [11] was specifically designed for deploying intentional agents on mobile devices. AFME agents follow a sense-deliberate-act cycle It has a perceptor feature which are used to monitor the state of the environment, and actuators for affected change within the environment.

4.1 EasyLife Prototype

An initial implementation of Easylife has been developed (figure 5). The server is hosted on a standard workstation and communications occurs with the client via Wifi.

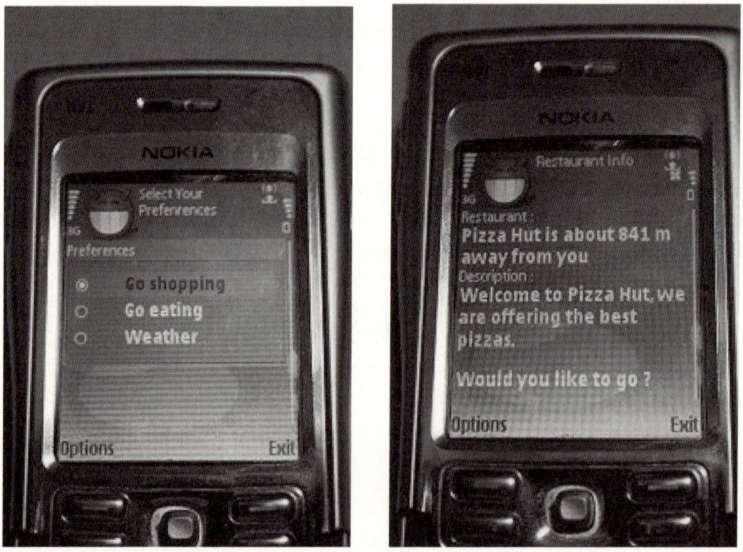

(a) (b)

Fig. 5. Subscriber selects an EasyLife service (a) and recommends a nearby restaurant (b)

Databases are implemented in mySQL and Hibernate for object persistence. The client is hosted on a Nokia N91. Position is obtained using a Bluetooth GPS receiver.

5 Future Work

EasyLife is very much a work in progress. The core architecture is in place and it is intended to use it as a base for further research. Initially, it is planned to explore information fusion through the use of mashups in mobile contexts, and the integration of Web services into mobile applications and services. A second issue concerns the use of heterogeneous agents. In theory, such agents should be capable of interoperating provided that they adhere to a recognised standard. In practice, there may be difficulties when the inherent resource limitations of mobile devices are considered, and the implications of these need to identified, and appropriate solutions identified.

6 Conclusion

As location-aware services become available, the need for robust extensible architectures will become paramount. Such architectures will harness a number of disparate technologies to enable the delivery of services to mobile users. In addition, distributed computing technologies are essential for the effective realization of mobile services. In this paper, we have presented EasyLife as an example of an architecture that encapsulates a number of characteristics essential for the construction and deployment of location-aware services.

Acknowledgments. This material is based upon works supported by the Science Foundation Ireland (SFI) under Grant No. 03/IN.3/1361.

References

1. Vasilakos, A., Pedrycz, W.: Ambient Intelligence, Wireless Networking, Ubiquitous Computing. Artec House (2006)
2. Klante, P., Krösche, J., Boll, S.: AccessSights - a multimodal location-aware mobile tourist information system. In: Proceedings of the 9th International Conference on Computers Helping People with Special Needs (ICCHP) (2004)
3. O'Grady, M.J., O'Rafferty, R.P., O'Hare, G.M.P.: A tourist-centric mechanism for interacting with the environment. In: Proceedings of the First International Workshop on Managing Interactions in Smart Environments, Dublin, Ireland, pp. 56–67. Springer, Heidelberg (1999)
4. Poslad, S., Laamanen, H., Malaka, R., Nick, A., Zipf, A.: Crumpet: Creation of user-friendly mobile services personalized for tourism. In: Proceeding of the Second IEE International Conference on 3G Mobile Communication Technologies, London, UK (2001)
5. O'Grady, M.J., O'Hare, G.M.P., Sas, C.: Mobile Agents for Mobile Tourists: A User Evaluation of Gulliver's Genie. Interacting with Computers 17(4), 343–366 (2005)
6. Holland, S., Morse, D., Gedenryd, H.: AudioGPS: Spatial Audio Navigation with a Minimal Attention Interface. Personal and Ubiquitous Computing 6(4), 253–259 (2002)

7. Liu, L.S., Zimmermann, R., Carter, W.: MStream: Position-aware Mobile Music Streaming. In: Proc. Third Annual International Conference on Mobile Systems, Applications and Services, Seattle, Washington, USA (2005)
8. Weakliam, J., Lynch, D., Doyle, J., Bertolotto, M., Wilson, D.: Delivering Personalized Context-Aware Spatial Information to Mobile Devices. In: Li, K.-J., Vangenot, C. (eds.) W2GIS 2005. LNCS, vol. 3833, pp. 194–205. Springer, Heidelberg (2005)
9. Strahan, R., O'Hare, G.M.P., Phelan, D., Muldoon, C., Collier, R.: ACCESS: An Agent based Architecture for the Rapid Prototyping of Location Aware Services. In: Proceedings of the 5th International Conference on Computational Science (ICCS 2005), Emory University Atlanta, USA (2005)
10. Bellifemine, F., Caire, G., Greenwood, D.: Developing Multi-Agent Systems with JADE. John Wiley & Sons, New Jersey (2007)
11. Muldoon, C., O'Hare, G.M.P., Collier, R., O'Grady, M.J.: Agent Factory Micro Edition: A Framework for Ambient Applications. In: Proceedings of Intelligent Agents in Computing Systems (IACS 2) Workshop held in Conjunction with International Conference on Computational Science (ICCS) 2006, Reading, UK. LNCS, vol. 3, pp. 727–734. Springer, Heidelberg (2006)

A New Concept of Trust Modeling and Management in Complex Networks

Grzegorz Kołaczek

Institute of Information Science and Engineering
Wroclaw University of Technology, Wroclaw, Poland
grzegorz@pwr.wroc.pl

Abstract. The paper presents the novel concept of trust evaluation method for autonoumous multi-agent systems. The method is based on assumption that multi-agent system constitutes social network and so it is an instantiation of the complex network. Several metrics used in description of this type of networks can be used to model trust relation between agents. In the presented model the agent's *A* trust in an agent's *B* statements is obtained as the Subjective Logic consensus of modified trust levels of all agents that trust the agent *B*. The trust level modification is performed in relation to the current position of the agent in the network.

1 Introduction

The Internet, the electrical power grid, the transportation network but also multi-agent systems can be viewed and analysied as the examples of compex networks.Two important properties displayed by many of these networks are the smallworld and scale-free properties [2,7]. Small-world networks are characterized by the clustering coefficient and the average network distance. The clustering coefficient is the probability that any two nodes are connected to each other, given that they are both connected to a common node. The average network distance measures the average minimal number of links connecting any two nodes in the network. Many regular networks have high clustering coefficients and large network distances. Random networks, on the other hand, have small network distances and low clustering coefficients [3]. Small-world networks fall somewhere in between these two extremes as they have large clustering coefficients and small average network distances [7,12]. The scale-free property is defined by an algebraic behaviour in the probability distribution $P(k)$ of the number k of links at a node.

When we consider multi-agent system all these parameters describing small-world or scale-free networks can be used to analyze the position and relations among agents within their society.

One of the most important factors in human interaction and communication is trust. Trust is also very important feature for all autonomous multi-agent systems. Performing their activities agents collaborate with other agents this means that they obtain and process data provided by them. So, the final decision or action performed by the agent strongly depends on the quality of the previously obtained data. The evaluation of the

I. Lovrek, R.J. Howlett, and L.C. Jain (Eds.): KES 2008, Part I, LNAI 5177, pp. 237–244, 2008.

risk related to the agent's decisions in relation to uncertainty about data quality provided by other agents is one of the most important problems and so interesting research area in multi-agent systems.

The trust and reputation of subjects has been typically assessed as a function of the quality of their response to requests coming from other members in the community. This approach is used in some organizational learning systems as, for example, *Answer Garden* [1] or knowledge communities [4,6,8,14]. Discussion about the different ratings that can be obtained by analyzing response quality is to be found in [5]. These systems rely on ratings provided as a feedback from the subject receiving the response to a previous demand. Subsequently these ratings are combined and finally each subject calculates its own trust value. Trust and reputation measure gives an idea of the confidence one can have on the quality of a subject's responses. The disadvantage of this type of mechanism is that it needs the explicit and frequent involvement of users that issue ratings. This implies that a good reputation calculation and maintenance depends on the involvement of users and continued contribution of ratings. Less intrusive and less demanding in terms of users involvement methods are more interesting. The problem is how reputation can be measured in the absence of any user feedback for subject's responses. In [8,9,10,14] some general discussion of trust and reputation in multi-agent is presented.

In the paper a novel concept of a trust modeling and management that addresses above mentioned problems by using the some universal measures for complex networks is presented.

Organisation of the paper is as follows. In section two some basic concepts of trust modeling in information systems are presented. Section three describes Subjective Logic as the one of the important elements of the novel trust modeling system. Section four presents the algorithm of trust level evaluation using som characteristic complex networks parameters. The last section closes by discussing results and pointing to further work.

2 Modeling Trust in Information Systems

There are two main reasons why the trust is so important aspect in contemporary information systems security. The first one is that there is no information system we could believe that is 100% secure. The consequence of this is that we cannot be completely sure about the subject identity and its true intensions. The second is that most contemporary information systems are a part of open networks. These networks allow subjects to communicate without any prior arrangements like for example organisation membership and so this also makes information authenticity difficult to verify.

The level of an autonomous agent uncertainty about other agents intensions, behaviour, etc. may be expressed and evaluated by trust level. However there are a few problems on the way to define the notion of trust formally. The first one is that there are some differences in defining of the most fundamental nature of trust. Second difficulty arises when a set of attributes which values decide about a final trust level must be defined. The third and probably the most important one is that trust will always have some subjective aspects that always create problems in formal usage. This idea was expressed by A.Jøsang in the following way: "we claim that there can be no other measure for security and authenticity than subjective trust" [4].

Apart from many controversial points about trust formal modelling there are also some elements that are commonly accepted. These aspects are as follows:

- trust is required when an external subject is to be permitted to modify the private data,
- trust is not transitive relation , it means that if A trusts B and B trusts C it does not automatically imply that A trusts C,
- high level of trust never means certainty about the subject's future behaviour,
- higher level of trust is the lower risk of a subject's bad behaviour.

The formal representation of trust should create possibility to use this notion in a context of information systems in a similar way as it is being used in reality. The most general idea about representing trust for security system purposes states that trust level must be estimated according to:

- agent's own experience,
- recommendation from other agents,
- agent's context (history, time, location, etc.).

The paper presents the concept of trust modeling which enables the autonomous agent to combine recommendation data obtained from other agents and data related to the agent's context. As the area of possible contexts of agent's interactions is very diverse we will focus on agent's position in agent's society described by its location in the social network.

3 Subjective Logic

Subjective Logic was elaborated by A. Jøsang and it is a kind of a framework for artificial reasoning [4]. Subjective Logic defines various logical operators for combining opinions. In Jøsnag approach opinion is an uncertain probability measure and so Subjective Logic can also be called as a calculus for uncertain probabilities.

Subjective Logic contains the equivalent of the traditional logical operators such as *conjunction* (AND), *disjunction* (OR) and *negation* (NOT). A novelty in Subjective Logic is an introduction of some non-traditional operators as *recommendation* and *consensus.*

Because our knowledge about reality is always incomplete or imprecise it is impossible to know with certainty whether such statement is true or false. Jøsang proposes to express this propriety of our knowledge by using a term *opinion* Opinions about facts translates into degrees of belief or disbelief as well as uncertainty which fills the void in the absence of both belief and disbelief. In formal notation it can be expressed as follows:

$$b+d+u=1, \qquad \{b,d,u\}\in[0,1] \qquad (1)$$

where b, d and u designate belief, disbelief and uncertainty respectively.

The ordered triplet $\omega=\{b,d,u\}$ is then called an opinion. An opinion may be represented as a point in the triangle (Fig.1.).

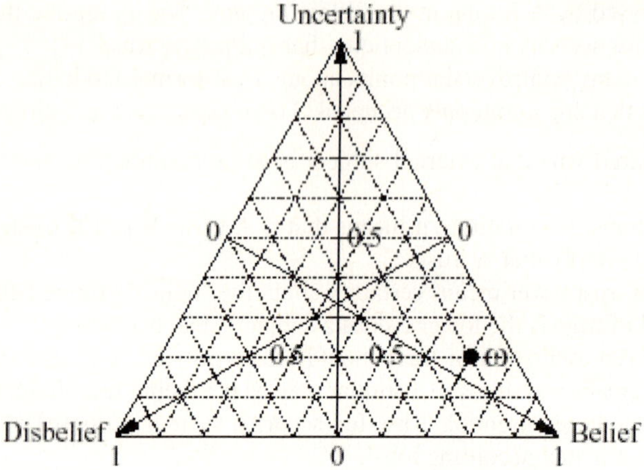

Fig. 1. Opinion triangle

In our approach we will especially use the two from numerous Subjective Logic operators: *consensus* and *recommendation*.

Assume two agents, *A* and *B*, where *A* has an opinion about *B*. Opinion expressed by *A* about other agent *B* is interpreted as opinion about proposition "*B's opinion is reliable*". We'll denote opinion expressed by agent *B* about given predicate *p* and agent's *A* opinion about *B* as ω_p^B and ω_B^A respectively. Assuming that ω_p^B and ω_B^A are known, the opinion of agent *A* about *p* is given by *discounting operator* (also known as *reputation* operator, denoted by ⊗):

$$\omega_p^{AB} = \omega_B^A \otimes \omega_p^B = \left\langle b_B^A b_p^B, b_B^A d_p^B, d_B^A + u_B^A + b_B^A u_p^B \right\rangle \tag{2}$$

It may be proved, that recommendation operator is associative but not commutative. This implies that the order of opinions in recommendation chains is significant. The joint opinion of two agents *A* and *B* about given predicate is computed by *consensus* operator, denoted by ⊕:

$$\omega_p^{AB} = \omega_p^A \oplus \omega_p^B = \left\langle \left(b_p^A u_p^B + b_p^B u_p^A\right)/k, \left(d_p^A u_p^B + d_p^B u_p^A\right)/k, u_p^A u_p^B / k \right\rangle \tag{3}$$

where: $k = u_p^A + u_p^B - u_p^A u_p^B$

4 Trust Modeling in Complex Networks

The term "complex network" refers to a network that has certain non-trivial topological features that do not occur in simple networks. Most social and technological networks can be considered complex by virtue of non-trivial topological structure (e.g., social network, computer network). Among these non-trivial features are: a heavy-tail in the degree distribution; a high clustering coefficient; assortativity or disassortativity among vertices; community structure at many scales; and evidence of a hierarchical structure [8].

In our opinion, some of these features can be very useful in a case of trust modelling and management. We selected two parameters that are frequently used to characterise complex network structure: clustering coefficient and centrality.

The clustering coefficient of a node in a network quantifies how close the node and its neighbours are from being a clique. This measure has been introduced by Duncan J. Watts and Steven Strogatz to determine whether a graph is a small-world network. At the other hand, the clique is defined as an exclusive group of people who share common interests, views, purposes, or patterns of behaviour [8,12]. A clique is a subset of individuals from a larger group, who are more closely identified with one another than the remaining members of the group, and who exchange something among themselves, such as friendship, affection, or information [13].

In a context of trust analysis existence of cliques in a trust network or a fact that a particular node is a member of the clique can be interpreted in two completely different ways. Positive interpretation is that a member of the clique has deeper knowledge about other clique members so its recommendation is more trustworthy. However in negative scenario we may assume that the clique members can collaborate to cheat other agents.

This ambiguity is the reason why in the proposed trust evaluation method the clique membership is an element of *uncertainty* component of the agent's opinion.

The numerical value of clustering coefficient can be calculated using for example following formula:

$$C_i = \frac{\left|\{e_{jk}\}\right|}{k_i(k_i-1)} : v_i, v_k \in N_i, e_{jk} \in E \tag{4}$$

where: v_i , - nodes, N_i – neighbourhood of the node v_i , e_{jk} –link between v_i and v_k, k_i – degree of the node v_i

In further discussion the parameter *CLIQUE* will be used in evaluation of agent's opinion *uncertainty* component. This parameter should be understood as the normalised value of clustering coefficient calculated using selected method (e.g. formula (4)) for a given node.

The second parameter - centrality determine the relative importance of a vertex within the graph (e.g. how important an agent is within a social network). There are four main measures of centrality that are used in network analysis: degree centrality, betweenness, closeness, and eigenvector centrality.

We propose to use one of the well known centrality measures (e.g. Google's PageRank which is a variant of the Eigenvector centrality measure [7]) to describe the level of confidence in agent's opinion. We assume that value of centrality measure is a kind of recommendation which originate from the network structure and so the opinions of the "more central" nodes are much worthy to be believed in.

In a next section the parameter *CENTRALITY* is used to refer to the normalised value of a certain centrality measure for a particular network node.

4.1 Trust Evaluation Algorithm

Let us consider a simple multi-agent network with trust relations as it is shown in the Fig.2.

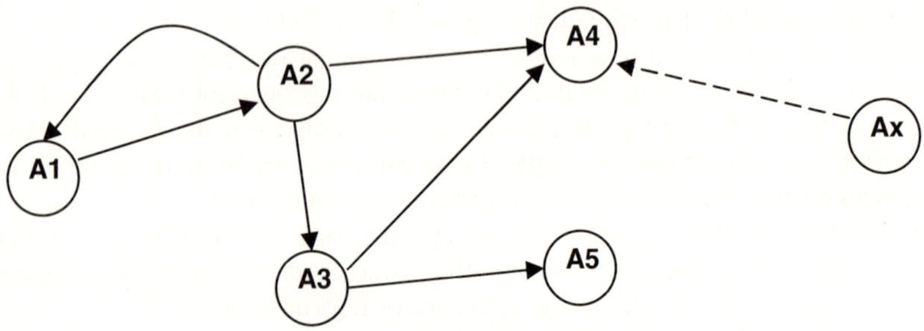

Fig. 2. Simple trust network in multi-agent system

where:

A1,A2,... - autonomous agents with established trust relations
Ax – a new agent which tries to derive trust in A4

At the early stage of each multi-agent system there is a moment where we have several agents and no trust relations. There is no historical records of interaction between agents so the agents can't use their own experience or recommendation to establish trust relations and to calculate trust levels. The only possible source of information that could be used is the agent's context. We assume that in such cases each agent will propose arbitrary some trust level with high values of uncertainty component and this level will be updated according to the further interactions among agents.

The second case, described in the Fig.2 is when a new agent joins the network. The new agent can use the existing relations between agents and their recommendations to elaborate their own opinions. In such situations agent uses some data coming from the structure of existing trust network.

We propose the following algorithm which will enable the new agent to calculate its own opinion about any other agent in the network:

Given: An agent *A* who calculate its trust level to the agent *B*, trust network – a set of agents' opinions about other agents

Result: ϖ_B^A - the agent's *A* opinion about the agent *B*

BEGIN

1. Get all opinions about the agent *B*. Let **C**={C_1, C_2,..., C_k} be a set of all agents that have opinion about *B*.
2. Measure the network context for each agent from the set **C** and present it in a form of Subjective Logic opinion $\varpi_{C_i}^{net} = (b_{netC_i}, d_{netC_i}, u_{netC_i})$. Where $\varpi_{C_i}^{net}$ is a

recommendation and it reflects the agent's C_i position in the network and is calculated as follows:

$$b_{net,C_i} = CENTRALITY \tag{5}$$

$$d_{net,C_i} = 1 - b_{C_iB} - u_{C_iB} \tag{6}$$

$$u_{net,C_i} = \min(1 - CENTRALITY, CLIQUE) \tag{7}$$

CENTRALITY and *CLIQUE* are normalised values calculated using selected complex network metrics as it was discussed at the beginning of the Section 4.

3. Calculate the modified opinion for each agent from the set **C** using context from Step 2. and Subjective Logic recommendation operation.

$$\varpi'^{C_i}_{B} = \varpi^{net}_{C_i} \otimes \varpi^{C_i}_{B} \tag{8}$$

Where opinion $\varpi^{C_i}_{B}$ is the original opinion of the agent C_i about the agent B.

4. Calculate the agent A opinion about the agent B as the Subjective Logic conensus of all modified in the step 3 opinions of agents from the set **C**.

$$\varpi^{A}_{B} = \varpi'^{C_1}_{B} \oplus \varpi'^{C_3}_{B} \oplus \ldots \oplus \varpi'^{C_k}_{B} \tag{9}$$

END

5 Conclusions

The paper presents a general concept about trust modeling and management using network based analysis of relations between autonomous agents. The method is based on assumption that multi-agent system constitutes social network and so the network specific parameters can be used to evaluate the trust levels. Proposition describes how metrics like centrality and clustering coefficient used in description of this type of networks can be used to model trust relation between agents. Within the presented model the agent's A opinion about agent's B statements is obtained as the Subjective Logic consensus of modified trust levels of all agents that trust the agent B. The trust level modification is performed in relation to the current position of the agent in the network.

The presented idea is a starting point to further practical simulations. Experiments that are intended as the next step of current research concern the following problems:

- which network measure fits the best our expectation about trust modeling,
- how much presented method is vulnerable to intentional manipulation of collaborating agents,
- how fast opinions (trust levels) can be spread within a network,
- how connectivity level influences propagation of opinions among nodes,

– how much network structure determine possibilities of opinion change,
– how distribution of values among opinion's components: believe, disbelieve and certainty, influence possibilities of opinion change.

References

1. Ackerman, M.S., McDonald, D.W.: Answer Garden 2: Merging organizational memory with collaborative help. In: Computer Supported Cooperative Work, pp. 97–105 (1996)
2. Barabasi, A.-L., Albert, R.: Emergence of scaling in random networks. Science 286, 509–512 (1999)
3. Bollobas, B.: Random Graphs. Academic Press, New York (1985)
4. Josang, A.: An Algebra for Assessing Trust in Certification Chains. In: Kochmar, J. (ed.) Proceedings of the Network and Distributed Systems Security Symposium, pp. 124–140 (1999)
5. Garcia, R.: Extensió col.laborativa del servei de localització d'expertesa. Master's thesis. Technical University of Catalonia (2001)
6. Maheswaran, M., Tang, H.C., Ghunaim, A.: Towards a Gravity-Based Trust Model for Social Networking Systems. In: Distributed Computing Systems Workshops, pp. 240–248 (2007)
7. Page, L., Brin, S., Motwani, R., Winograd, T.: The PageRank Citation Ranking: Bringing order to the Web. Technical Report, Computer Science Department, Stanford University (1998)
8. Schillo, M., Funk, P., Rovatsos, M.: Who can you Trust: Dealing with Deception. In: Castelfranchi, C., Tan, Y., Falcone, R., Firozabadi, B.S. (eds.) Proceedings of the Workshop Deception, Fraud and Trust in Agent Societies of the Autonomous Agents Conference, pp. 293–320 (1999)
9. Teacy, W.T., et al.: Travos: Trust and reputation in the context of inaccurate information sources. Autonomous Agents and Multi-Agent Systems 12(2) (2006)
10. Wang, Y., Vassileva, J.: A Review on Trust and Reputation for Web Service Selection. In: First International Workshop on Trust and Reputation Management in Massively Distributed Computing Systems (TRAM 2007), pp. 322–340 (2007)
11. Watts, D.J., Strogatz, S.H.: Collective dynamics of 'small-world' networks. Nature 393, 432–440 (1998)
12. Watts, D.J.: Small Worlds: The Dynamics of Networks between Order and Randomness. Princeton University Press, Princeton (1999)
13. Webster's Revised Unabridged Dictionary (1913)
14. Zacharias, G., Maes, P.: Trust Management Through Reputation Mechanisms. Applied Artificial Intelligence 14, 881–907 (2000)

Extending Intelligent Tutoring Systems to Mobile Devices

Vlado Glavinić[1], Marko Rosić[2], and Marija Zelić[2]

[1] Faculty of Electrical Engineering and Computing, University of Zagreb, Unska 3, 10000 Zagreb, Croatia
[2] Faculty of Natural Sciences, Mathematics and Kinesiology, University of Split, Teslina 12, 21000 Split, Croatia
vlado.glavinic@fer.hr,
{marko.rosic,marija.zelic}@pmfst.hr

Abstract. Developments in the wireless infrastructure have paved the way to a new e-learning paradigm named mobile learning (m-learning). M-learning systems aim to improve the quality of learning by providing mobile learners with an easy, contextualized and ubiquitous access to knowledge. Our discussion focuses specifically on mobile intelligent tutoring systems. Based on our previous work in the field of intelligent tutoring systems as well as agent technology we have outlined a multi-agent architecture for our intelligent tutoring system xTEx-Sys to be extended to mobile devices. Given the present absence of relevant literature and referent material we think that this paper provides software developers with some valuable guidelines.

Keywords: agents, intelligent tutoring systems, m-learning, agent-based systems, mobile intelligent tutoring systems.

1 Introduction

Developments in the wireless infrastructure have paved the way to a new e-learning paradigm named mobile learning. Mobile learning (m-learning) is nowadays attracting a lot of attention as the next kind of computer supported learning, the respective implementation issues however still being the object of more detailed further elaboration [1]. Most present-day learning systems run on desktop computers and are not designed for use on mobile devices such as mobile phones, smart phones, PDAs, etc. On the other hand m-learning systems aim to improve the quality of learning by providing mobile learners with an easy, contextualized and ubiquitous access to knowledge. These systems are intended to blend into a mobile learner's daily routine, by being at the reach of a hand, hence enabling the learner to brush up on her/his knowledge, review some referent material, take a short quiz and the like; all this while waiting for a friend at an airport, waiting in a queue at a bank, traveling by train, etc. The global intention is to make learning "a way of being".

The development of the m-learning paradigm, much like any new learning paradigm, consequently comes with a whole new set of problems and issues to be addressed. We identify two main problems that should be considered more carefully.

I. Lovrek, R.J. Howlett, and L.C. Jain (Eds.): KES 2008, Part I, LNAI 5177, pp. 245–252, 2008.

First, mobile learners usually work alone and thus have little time on hand for educational activities. With this in mind the systems should be able to "play the role of a human expert" and provide the learners with intelligent help and support. This requires personalization i.e. adaptation of the systems to the learners, their individual characteristics, abilities, learning backgrounds, preferences and so on. Obviously we are talking about intelligent tutoring systems extended to the mobile computing environment. The second problem is of a more technical nature, which is related to porting existing systems to a new environment. This possibly introduces some complex implementation issues. Namely, the wireless environment (GSM, UMTS, WLAN) is much different from the standard "wireline" one; bandwidth, delay, error rate, interference and the like, may change dramatically as the learner changes her/his location. Another thing to consider is the characteristics of different mobile devices used for accessing the system, spanning from low-performance mobile phones to high-performance laptops. Features like processing power, amount of memory and display qualities have to be taken into account.

An interesting approach to overcome the above mentioned problems, and certainly the one most talked about in the research community, is the use of agent-based technology. Agents' characteristics such as autonomy, learning capabilities, proactiveness and social skills should enable adaptation of the system to the learner as well as to the mobile computing environment. Intelligent agents are considered to be a promising approach to building complex and high-quality systems due to the fact that the agent paradigm enables modeling of systems in an intuitive and natural way, resembling to human perception of the problem domain. Even so the agent technology has not yet been widely accepted. There are a few issues holding back agent technology from a global breakthrough. Developing agent systems requires skills in many different fields such as distributed systems engineering, communication infrastructure, etc. More significantly there is no general consensus on the methodology for analysis and design of such systems: even though a lot of methodologies have been developed, none of them have reached a high level of maturity. There are also no adequate templates or guidelines for porting existing applications (as well as building completely new ones) to the world of agents. Apart from some elementary examples, software developers are "left in the dark" when trying to outline the structure of their system. This is exactly the situation we found ourselves in when we decided to extend our intelligent tutoring system xTEx-Sys (Extended Tutor Expert System) to mobile devices [2]. Our intention is to enrich the system with delivery of personalized educational contents to mobile devices. This of course would only be the main function of the system; others would include short quizzes, course notifications, adequate chat applications, and so on.

Based on our previous work in the field of intelligent tutoring systems as well as agent technology we have outlined the architecture for our new system. Given the present absence of relevant literature and referent material we think that the following discussion provides developers with some new ideas. The paper is structured as follows. Section 2 gives an overview of the xTEx-Sys system. Section 3 describes mapping of the system's functions to intelligent agents, along with a detailed discussion on the agents, their internal structure, performance and other relevant issues. Section 4 concludes the paper and outlines the guidelines for future work.

2 Overview of the xTEx-Sys System

xTEx-Sys is actually not an intelligent tutoring system itself; it is an authoring shell supporting the development of intelligent tutoring systems (ITSs). In order to gain some insight into both functionality and architecture of the system, in the following we give a short overview of xTEx-Sys. By outlining some basic ideas of the system, we can further discuss porting (a part of its) functionality to mobile devices and contemplate on the agents enabling such a migration. The detailed description of the system, its history, pedagogical framework, specifications and methodologies, evaluation, comparison to other similar systems and the like, can be found in [3].

The formalism for knowledge representation in the xTEx-Sys system is based on semantic networks with frames. Knowledge is represented by concepts (i.e. domain knowledge objects) and semantic relations between them. The concepts can have structural attributes such as textual descriptions, pictures, presentations, documents, sound and animation, URL addresses and the like. Figure 1 shows an illustration of the concept "Heat" in the course "Physics 1".

The left side of Figure 1 shows a hierarchical structure of the courseware. A course can be decomposed into the following elements: units which generally include one or more lessons, lessons which include one or more topics, and finally topics which include one or more instructional items. An instructional item is a collection of related concepts from various knowledge bases, but it can also take the form of a dynamic quiz. The system exhibits intelligent behavior by proper sequencing of course elements, adjusting the difficulty level of teaching contents to the student and providing interactive problem solving support.

The xTEx-Sys system has its shortcomings, primarily concerning the limited expressiveness of the knowledge representation formalism as well as insufficient adaptation of the system to the student. The future version will include natural language processing (NLP) techniques to provide students with more fluent and intelligible reading materials.

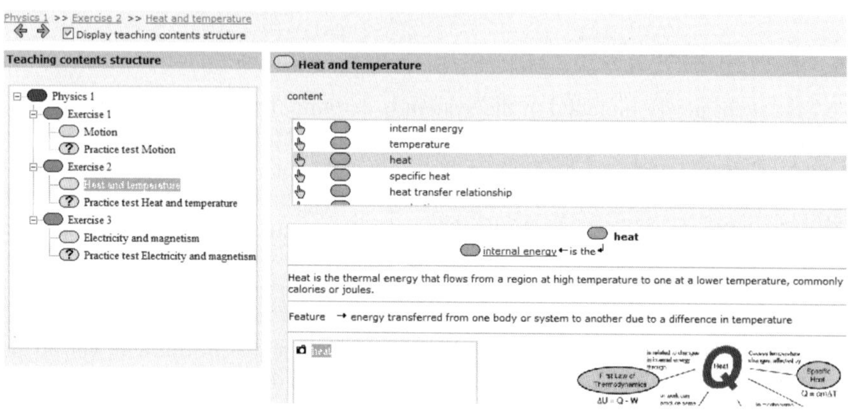

Fig. 1. Presentation of a concept in xTEx-Sys

3 Extending xTEx-Sys to Mobile Devices

The following discussion on agents enabling migration of the original xTEx-Sys system to mobile devices is actually quite general, and most of the presented ideas can be applied to different intelligent tutoring systems. The analysis aims to give some insight into a potential multi-agent architecture that would support the extension of ITSs to mobile devices. The respective system architecture is shown in Figure 2. It is based on the following agents, which are discussed in more detail in the following sections:

- the personal agent assists the student while she/he is using the system,
- the pedagogical agent guides the learning and teaching process,
- the device agent enables adaptation of content to mobile devices,
- the network agent enables adaptation of the system to different networks,
- the database agent communicates with the system's database, and
- the chat agent enables communication among students.

Identifying the agents and their responsibilities based on the functional requirements, modeling of agent acquaintances, agent refinement as well as deployment is system dependent and should be discussed in detail as the specific system requires [4].

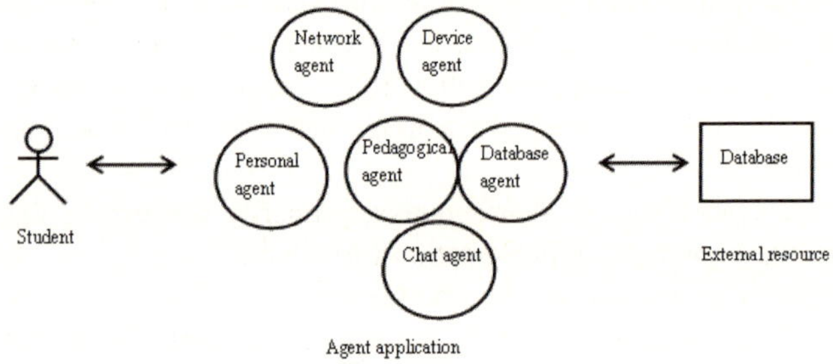

Fig. 2. New system simplified architecture

3.1 Personal Agent

The first agent newly included in the system is certainly the *personal agent*. The personal agent resides on a student's mobile device, interacting with her/him usually via a graphical user interface. This agent is responsible for assisting the student while she/he is using the system, providing intelligent help and support, updating her/his personal profile, reacting to the system's notifications, etc. It basically behaves like a secretary, accomplishing routine support tasks thus allowing the student to concentrate on her/his real job. The above specified functionality implies a complex internal architecture of this agent.

First of all, the personal agent should contain the student's profile with her/his personal information as well as preferences (e.g. beneficial display properties: fonts, colors, text size; different study habits and learning styles). The way the personal agent builds a student's profile is an area of vast and extensive research [5]. The personal agent containing a student's profile can provide her/him with a more friendly and enjoyable learning environment. Secondly, the personal agent should manage the student's calendar containing her/his appointments, meetings, events, seminars, exams, etc. The above described functions imply that the personal agent maintains its internal knowledge in some form of semantic markup. The agent should be able to manage the knowledge accumulated from different sources and draw inferences from it. Therefore it should be equipped with a reasoning engine.

The xTEx-Sys mobile version (although at a very early stage) is being developed within the JADE agent platform [6]. The agents' mental attitudes are modeled by the Jadex BDI reasoning engine [7]. Information is structured using the OWL language [8]. There are of course different agent platforms, reasoning engines and semantic markup languages, and developers should make their own choice depending on application requirements, personal preferences and the like.

Applying some of the above mentioned features to xTEx-Sys includes for example: coloring different elements of the course tree (e.g. elements the student has already reviewed), applying preferred themes and styles, alerting the student when a course notification arrives, etc. The agent should be able to communicate with other students' personal agents, collect information, recommendations and so on. Privacy and other security issues should be taken into account.

3.2 Pedagogical Agent

Closely tied to the personal agent is the *pedagogical agent* responsible for guiding the learning and teaching process. The pedagogical agent represents the heart of the system enabling students to learn from personalized teaching material. The agent is equipped with the student's profile containing information on her/his learning background, previous test results, learning goals, etc. Depending on this information the agent can suggest to the student a revision of previous lessons or some additional material, an exercise, a related test and so on. The educational contents should be customized based on the student's profile, for example a lesson might include more or less explanations, examples and such. The agent should also provide help to the student either explicitly (e.g. a "help" button) or implicitly (monitoring the student's performance and appearing when needed). There already exists a lot of research in this area, especially dealing with animated agents, peer agents, etc [9].

Our focus however is on the mobile computing environment. This automatically assumes some "downsizing" of the pedagogical agent. Viewing the xTEx-Sys mobile version as an extension of the system enabling "study on-the-go", it is clear that the student will not be engaged in an exhaustive learning process but rather in a quick review of learning matter, a short quiz, a simple educational game and the like. With this in mind, the agent should be able to parse educational contents and show the student only the relevant parts. For example, the xTEx-Sys student can initially be presented with a course tree as shown in Figure 1. After selecting a course element the associated lesson's summary might be presented along with a link to a short exercise.

Content adaptation should be done in close cooperation with the device agent that is described in the following section. It is important to understand that there is no clear boundary between these agents; they have to engage in constant communication, negotiation and information sharing in order to accomplish their tasks. For instance, the pedagogical agent might request data from the database agent, tailor the collected data to fit the student, pass it along to the device and network agents for further content adaptation, and eventually engage the personal agent to apply the appropriate display styles.

3.3 Device Agent

Mobile devices are quite restrictive in terms of screen size, memory and processing power. They typically display less than 20 lines of text, run on different operating systems of restricted functionality, support different markup languages, etc. When displayed directly on mobile devices the content designed for desktop computers is usually aesthetically unappealing and difficult to navigate. So far the content adaptation implied creating different versions of the same content for different target devices. Obviously this approach is device dependent and inflexible. The *device agent* should be used for dynamic content adaptation which includes layout changes, reconfigurations of content format and so on. This should be coupled with client-side navigation techniques enabled by the personal agent. Additionally the device profile containing physical characteristics of the device (color depth, screen size, memory) can be used, e.g. the CC/PP profile [10].

There are some simple ways of adapting content to mobile devices based on device profile and student preferences [11]:

- the "fit to screen" approach produces a scaled-down version of the content by removing or shrinking images, using smaller fonts and casting out irrelevant parts;
- the "hierarchical display" of the content is suitable for navigating a large document, providing the student with a global view of the document and enabling selection of a particular section;
- different visualization techniques for maximizing the display space (e.g. the fisheye view, the circular interface, etc.);
- methods for multimedia content adaptation (e.g. layered encoding, rate shaping, etc.).

The device agent of the xTEx-Sys mobile version can use a combination of the above mentioned techniques for content adaptation. The agent can, for example, remove the header and footer of a page, replace images of different course elements in a course tree with smaller ones, use the fisheye view to display focal content of a course tree in a larger font size and peripheral content in a smaller font size, etc. The applied techniques are obviously application dependent.

3.4 Network Agent

The device agent works in close cooperation with the *network agent*. The network agent provides information on the Quality-of-Service (QoS) values of the wireless network currently in use. WLAN, UMTS, GSM and other networks have different characteristics regarding bandwidth, bit rate, delay, roundtrip time, and so on. Other agents in the system can request these values from the network agent or subscribe to notifications about changes in the values.

Adaptation of the system to changes in the mobile environment is obviously an important issue. The deployed agents should be able to make necessary corrections in a transparent fashion. For instance, if the connection is slow, the personal agent might fetch only email headers to the mobile device and not entire emails, the device agent might use different compression methods on a document, etc. The network agent should also have knowledge about advanced error recovery methods, techniques for handling disconnected learning and so on.

In addition to the main agents there are a few other supporting agents that are discussed in the following section.

3.5 Other Agents

The *database agent* is responsible for communication with the system's database. Interaction with an external resource such as a database, file or legacy software requires some careful consideration since resources might change their status independently of the agent [12]. The xTEx-Sys system uses Web services as a database front end, therefore the *Web services agent* is deployed instead of the database agent.

The *chat agent* enables communication among students, registering different personal agents as clients. Students can engage in conversation with other students, create or join different chat rooms, form buddy lists, transfer files and the like. A useful feature of the chat agent would be forming of different interest groups based on the students' personal profiles. Figure 3 shows a simple chat module of our xTEx-Sys mobile version.

Additional agents required in the system deal with issues such as monitoring of agents, dynamic creation of agents, killing agents, discovering service providing agents and so on. These issues however are usually addressed in the agent platform itself.

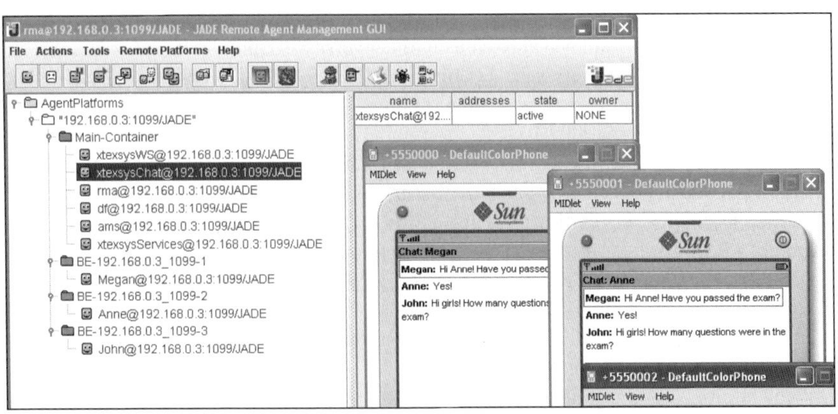

Fig. 3. Simple chat module of the xTEx-Sys mobile version

4 Conclusion and Future Work

This paper presents a mock-up of a multi-agent architecture for m-learning systems. Event though the approach is illustrated in a case-study manner, basing on experience gained during the development of the mobile version of an existing ITS (i.e. xTEx-Sys), we feel that the discussion itself is quite general and grants its application to a broader range of related systems. Since each agent needs to be carefully discussed and worked out in detail, as the specific system requires, the presented analysis should obviously be considered as a starting point only.

The main research challenge is the modeling of agents, their functionality, internal structures and mechanisms. Future work foresees writing a thorough specification of the system based on the presented analysis in order to obtain sufficient insight which would enable a relatively straightforward implementation of the system. This specification should include items like agent interaction tables, message templates, service registrations, agent behaviors, ontologies, content languages and the like. Additional issues to be addressed comprehend security, persistency, performance and scalability of the system.

Acknowledgements. This paper describes the results of research being carried out within the project 036-0361994-1995 *Universal Middleware Platform for e-Learning Systems*, as well as within the program *Intelligent Support for Ubiquity of e-Learning Systems*, both funded by the Ministry of Science, Education and Sports of the Republic of Croatia.

References

1. Glavinić, V., Rosić, M., Zelić, M.: Agents in m-Learning Systems Based on Intelligent Tutoring. In: Stephanidis, C. (ed.) UAHCI 2007. LNCS, vol. 4556, pp. 578–587. Springer, Heidelberg (2007)
2. Stankov, S.: Isomorphic Model of the System as the Basis of Teaching Control Principles in the Intelligent Tutoring System. PhD Dissertation. Faculty of Electrical Engineering, Mechanical Engineering and Naval Architecture University of Split, Split (in Croatian) (1997)
3. Stankov, S., Rosić, M., Žitko, B., Grubišić, A.: TEx-Sys model for building intelligent tutoring systems. Computers & Education (in press, 2007), doi:10.1016/j.compedu.2007.10.002
4. Glavinić, V., Rosić, M., Zelić, M.: Analysis of an Agent-based m-Learning System. In: Proc. 14th IEEE Mediterranean Electrotechnical Conf. Ajaccio. CD-ROM Proc, pp. 1–6 (2008)
5. Kuflik, T., Shoval, P.: User profile generation for intelligent information agents. In: Proc. 2nd Int'l Workshop on Agent-Oriented Information Systems, Stockholm, pp. 63–72 (2000)
6. Bellifemine, F., Poggi, A., Rimassa, G.: JADE: a FIPA2000 compliant agent development environment. In: Proc. 5th Int'l Conf. on Autonomous Agents, Montreal, pp. 216–217 (2001)
7. Braubach, L., Lamersdorf, W., Pokahr, A.: Jadex: Implementing a BDI-Infrastructure for JADE Agents. EXP – in search of innovation 3, 76–85 (2003)
8. Web Ontology Language OWL, `http://www.w3.org/2004/OWL`
9. Sklar, E., Richards, D.: The Use of Agents in Human Learning Systems. In: Proc. 5th Int'l Joint Conf. on Autonomous Agents and Multiagent Systems, Hakodate, pp. 767–774 (2006)
10. Composite Capabilities/Preferences Profile Home Page, `http://www.w3.org/Mobile/CCPP`
11. Zhang, D.: Web Content Adaptation for Mobile Handheld Devices. Communications of the ACM 50, 75–79 (2007)
12. Nikraz, M., Caire, G., Bahri, P.A.: A methodology for the analysis and design of multiagent systems using JADE. Int'l J. of Comp. Systems Science and Eng. 21, 99–116 (2006)

Integration of Knowledge in Disjunctive Structure on Semantic Level

Trong Hieu Tran and Ngoc Thanh Nguyen

Institute of Information Science and Engineering,
Wroclaw University of Technology, Poland
{Trong.Hieu.Tran,Ngoc-Thanh.Nguyen}@pwr.wroc.pl

Abstract. Knowledge integration is a very important task in Knowledge Management field. The aim of this paper is to present an algorithm for knowledge integration based on disjunctive structures. A distance function between disjunctions is analyzed and some postulates for knowledge integration in this logic structure are introduced. Key properties of these postulates and aspects of algorithm are also investigated.

Keywords: knowledge integration, consensus methods.

1 Introduction

Integration processes are of great importance in knowledge-based systems. The main task of knowledge integration is to create new pieces of knowledge from a given set of initial pieces of knowledge by some such processes as merging, extracting and eliminating inconsistencies. Knowledge integration becomes essential when one wants some knowledge systems to cooperate with each other, or to unify some independent ones. That mission, however, is difficult because not only the autonomy of each system but also non-deterministic mechanics for knowledge processing. The most likely and unwanted consequence is the inconsistency of knowledge.

Therefore, handling inconsistency is an important process in knowledge integration even if there is no inconsistence, for example in such situation that the knowledge pieces refer to different subjects. To this end, there have been several known approaches for knowledge integration such as using relation [16, 18] and logic [9, 11]. Consensus methods have also been introduced and proved to be powerful tool for dealing with inconsistency [16] and integration tasks [10].

The scheme of a knowledge integration method based on Consensus Theory consists of at least two elements: i) the definition of the structure of knowledge and ii) a set of postulates (criteria) which define the aim of the integration process. In general, the purpose of a criterion for a consensus-based knowledge integration method is based on the requirement that the result of integration should best represent the states of knowledge which are to be integrated. This methodology was applied in some works, for example knowledge integration using conjunctive and disjunctive structures [17, 19], using fuzzy structure [15]. In this paper, the authors concentrate on disjunctive structure with the original contribution being based on the defining of new distance function between the disjunctions and proposition of an effective semantic-oriented algorithm for knowledge integration process.

I. Lovrek, R.J. Howlett, and L.C. Jain (Eds.): KES 2008, Part I, LNAI 5177, pp. 253–261, 2008.

Accordingly, the paper is organized as follows. After the introduction Section 2 reviews the basic related notions. Section 3 contains the definition of the distance function between disjunctions. The outline of conflict and consensus technique is analyzed in Section 4. Section 5 presents postulates for knowledge integration. Section 6 will analyze these postulates and bring in integration algorithm. At last, some conclusions are included in Section 7.

2 Basic Notions

In this work we assume that an agent uses a finite set L of symbols to represent the truth of facts and events in the real world where this agent acts. Each symbol has a value of either *True* or *False*. A *literal* is defined as an expression a or $\neg a$ where a is a symbol in L. A literal with symbol "¬" is called a *negative literal*, otherwise it is called a *positive literal.*

A conjunctive formula of literals (*conjunction* in short) is a logic formula which is represented by the following expression:

$$t_1 \wedge t_2 \wedge \ldots \wedge t_k$$

where t_i is a literal (positive or negative) for $i = 1, 2, \ldots, k$.

Similarly, a disjunctive formula of literals (*disjunction* in short) is a logic formula which represented by the following expression:

$$t_1 \vee t_2 \vee \ldots \vee t_k$$

where t_i is a literal (positive or negative) for $i = 1, 2, \ldots, k$.

In a conjunction or disjunction we assume that there is no repetition or inconsistency among literals. It means that each literal occurs at most once in a formula. Moreover, we also assume that all formulae satisfy the Closed World Assumption, it means that the literals which do not occur in a formula are implied that they are negative literals. It means we can eliminate all negative literals from a formula but preserving the meaning. So, in this work each formula can be represented by only a set of positive literals and when literals are mentioned in a formula, it is implied that they are positive literals.

At the semantic level, a conjunction has only one way to be understood that it achieves true value if and only if all its literals are satisfied. Now we can represent a conjunction by a set of symbols which are occurred in it. For a disjunction, it is more complicated, a disjunction will achieves true value if one or more literals are satisfied. It means that a disjunction is true if and only if there exists a conjunction which is constructed from the set of literals of that disjunction is true. Basing on this idea, a disjunction will be understood as a set of conjunctions of which literals are in set of literals occurred in that disjunction. Further, a disjunction can be represented by a set of all subset of literals occurred in this disjunction.

Let $S_c(x)$ be the set of conjunctions which can be constructed from the literals occurred in a disjunction x and $S_L(x)$ be the set of all subset of these literals.

For example, if $x = \alpha \vee \beta \vee \chi$ then

$S_c(x) = \{\alpha, \beta, \chi, \alpha\wedge\beta, \alpha\wedge\chi, \beta\wedge\chi, \alpha\wedge\beta\wedge\chi\}$
$S_L(x) = \{\{\alpha\}, \{\beta\}, \{\chi\}, \{\alpha, \beta\}, \{\alpha, \chi\}, \{\beta, \chi\}, \{\alpha, \beta, \chi\}\}$
Let *Clause*(L) be the set of all disjunctions can be built from the literals in L.

In this paper we will use algebraic sets with repetitions given by Lipski and Marek [13]. Some main ideas of this algebra are as follows:

- Set $A = (3{*}x,\ 1{*}y,\ 2{*}z)$ be a set with repetitions with cardinality equal to 6 in which element x appears 3 times, y appears once and z appears twice
- If $A = (3{*}x,\ 1{*}y,\ 2{*}z)$ and $B = (2{*}x,\ 2{*}y)$, then

  $A \overset{*}{\bigcup} B = (3{*}x, 2{*}y, 2{*}z)$, $A \overset{+}{\bigcup} B = (5{*}x, 3{*}y, 2{*}z)$ and $A \overset{*}{\bigcap} B = (2{*}x, 1{*}y)$.

3 Distance Functions between Disjunctions

Distance functions are generally understood as tools which allow to measure the difference between objects in the same space. They have a lot of important applications in Artificial Intelligence such as to determine how close a new input to stored instances and to predict the output class correspond to that input in machine learning[7], self-organizing maps or competitive learning in neural network[12], statistics[2], and pattern recognition[6]. In the sub-field of Knowledge Management distance functions also have some significant applications. For instance, they are used to define the macrostructure of a universe in consensus problems [17], determine the median of elements and the distance between semilattices [14], and measure the inconsistency in knowledge bases [8], or the distance between equivalence relations [4].

In this paper, with the aim at determining knowledge integration of disjunction set, we define a distance function between disjunctions in general way as follows:

Definition 1
By a distance function between disjunction from set Clause (L) we understand the following function:

$$d: Clause(L) \times Clause(L) \rightarrow R^+$$

where R^+ denotes the set of non–negative real numbers.

In this paper we define a distance function for disjunctions which uses the sets of material literals of disjunctions. This function not only mentions about the occurrences of literals but it also considers to the contribution of a literal in each disjunction.

Definition 2
For $x \in Clause(L)$, let Material(x) be a repetition set defined as follows:

$$Material(x) = \overset{+}{\bigcup_{V \in S_L(X)}} V$$

Example 1
For $\alpha, \beta, \gamma \in L$. Let $x = \alpha \vee \beta \vee \gamma$ we have

$$S_L(x) = \{\{\alpha\},\{\beta\},\{\gamma\},\{\alpha,\beta\},\{\alpha,\gamma\},\{\beta,\gamma\},\{\alpha,\beta,\gamma\}\}.$$

Thus $Material(x) = \{4{*}\alpha, 4{*}\beta, 4{*}\gamma\}$

Proposition 1
For each disjunction, we can determine only one repetition set of literals and for each repetition set of literals, we only can determine at most one disjunction.

We define the distance function between disjunctions as follows:

Definition 3
Let $x, y \in Clause(L)$

$$d(x, y) = 1 - \frac{|Material(x) \overset{*}{\bigcap} Material(y)|}{|Material(x) \overset{*}{\bigcup} Material(y)|}$$

Example 2. For $\alpha, \beta, \gamma, \delta \in L$.

Let $x = \alpha \vee \beta$, $y = \alpha \vee \gamma \vee \delta$ we have $Material(x) = \{2_*\alpha, 2*\beta\}$ and $Material(y) = \{4*\alpha, 4*\gamma, 4*\delta\}$. So, $d(x, y) = 1 - 2/14 = 6/7$.

Assume that we have two disjunctions $x = \alpha_1 \vee \alpha_2 \vee \ldots \vee \alpha_n \vee \beta_1 \vee \beta_2 \vee \ldots \vee \beta_m$ and $y = \alpha_1 \vee \alpha_2 \vee \ldots \vee \alpha_n \vee \gamma_1 \vee \gamma_2 \vee \ldots \vee \gamma_k$. Without loss of generality let us assume that $m \le k$. Then, we have

$Material(x) = (2^{m+n-1} * \alpha_1, \ldots, 2^{m+n-1} * \alpha_n, 2^{m+n-1} * \beta_1, \ldots, 2^{m+n-1} * \beta_m)$

$Material(y) = (2^{n+k-1} * \alpha_1, \ldots, 2^{n+k-1} * \alpha_n, 2^{n+k-1} * \gamma_1, \ldots, 2^{n+k-1} * \gamma_k)$

Thus we have

$$d(x,y) = 1 - \frac{n * 2^{m+n-1}}{(k+n) * 2^{k+n-1} + m * 2^{m+n-1}} = 1 - \frac{n}{(k+n)2^{k-m} + m} \tag{1}$$

Based on the *Proposition 1* and *Definition 3* we can easily prove that the distance function d is a metric.

4 Outline of Conflict and Consensus Technique

One of the most common problems in integrating processes is to cope with conflicts caused by different sources of knowledge. In [17] authors defined the simplest conflict taking place when two bodies bear different opinions on the same subject. In general, a conflict includes the following main components [8]:

- *Conflict body* specifies the direct participants of the conflict.
- *Conflict subject* specifies the issues to which the conflict refers.
- *Conflict content* specifies the opinions of the participants on the conflict subject.

Within a conflict we can determine several conflict profiles. A conflict profile is a multi-set (i.e. with repetitions) of opinions which are generated by the agents regarding an issue. Information system theory [18] has been proved to be very suitable for managing this kind of conflicts.

Some common methods for resolving the conflict problems are presented in [5, 17]. In this paper we define conflict in the similar way in which we assume that a conflict profile is a multi-set of disjunctions and we will use consensus methods to resolve the conflict.

In general, Consensus Theory deals with problems of data analysis in order to extract valuable information. Consensus methods enable us to determine a version of data which i) should best represent a set of given data and ii) should be a good compromise acceptable for parties that are in conflict because of their authorship of the original data.

The consensus problem for resolving inconsistency of a set of opinions should be formulated as follows: *Given n opinions* $O_1, O_2, \ldots, O_n$, *one should determine one which could best represent these ones.*

The chosen opinion is called a *consensus* of the given opinions.

Consensus technique is a powerful tool used in the alternatives ranking problem [1] or the committee election problem [3].

In this paper, the consensus technique is used to determine the knowledge integration of agents in which the knowledge of each agent is represented as a disjunction. We assume that a conflict profile is a multi-set, a subset of *Clause* (*L*). We define the following integration task:

For a given conflict profile of disjunctions

$$X = \{x_i \in Clause(L): i = 1, 2,\ldots,n\}$$

it is needed to determine a disjunction $x^* \in Clause(L)$ *which best represents the given disjunctions.*

This problem has been formulated in [17]. In this paper we propose a new algorithm for its solution.

5 Postulates for Knowledge Integration

In this section we introduce some postulates which are used to integrate knowledge based on consensus technique. In what follows we denote $\Pi(Z)$ as the set of all finite subsets with repetitions of set *Z*.

Definition 4

By a consensus function for profiles of logic formulae we understand a function:

$$C: \Pi(Clause(L)) \rightarrow 2^{Clause\ (L)}$$

which satisfies one or more of the following postulates:

*P*1. *For each disjunction* $x^* \in C(X)$ *there should be:*

$$\bigcap_{x\in X} S_L(x) \subseteq S_L(x^*)$$

*P*2. *For each disjunction* $x^* \in C(X)$ *there should be:*

$$S_L(x^*) \subseteq S_L(\vee_{x\in X}\, x\,)$$

*P*3. *If* $X = \{\, x_1, x_2, \ldots, x_n\}$ *and* $\exists i \in [1..n]$: $Material(x_i) \overset{*}{\bigcap} Material(x_j) = \emptyset$, $1 \le j \le n$, $j \neq i$, *then* x_i *should be in* $C(x)$.

*P*4. *A consensus* $x^* \in C(X)$ *should minimize the sum of distances:*

$$\sum_{x\in X} d\ (x^*,x) = \min_{x'\in Clause(L)} \sum_{x\in X} d\ (x',x).$$

*P*5. *A consensus* $x^* \in C(X)$ *should minimize the sum of distances:*

$$\sum_{x\in X} d^2(x^*,x) = \min_{x'\in Clause(L)} \sum_{x\in X} d^2(x',x).$$

By C_{co} we denote the set of all consensus functions determined by Definition 4. Some commentary for postulates is given as follows:

- The main idea of Postulate *P*1 is the common part of opinions which should be included in the component of consensus. The purport of this postulate is similar to Pareto criterion: If all voters vote for the same candidate then he should be finally chosen.
- Postulate *P*2 means that the consensus should not exceed the profiles. This postulate is also in line with a common principle that each issue in consensus should be mentioned by at least one agent (or expert).
- Postulate *P*3 implies the superiority of knowledge. It states that if there is only one agent (or expert) raising opinions on some issues, these opinions should be preserved in consensus.
- Postulates *P*4 and *P*5 are popular criterions for consensus. These criterions are natural for satisfying the condition of "*the best representation*".

6 Analysis of Postulates and Integration Algorithm

In this section we present several properties of postulates and the relationships among consensus functions satisfying them.

The fact that a consensus function *C* satisfies a postulate *P* for a given profile *X* will be written as:

$$C(X) \mid\!- P.$$

The fact that a consensus function *C* satisfies a postulate *P* (that is the postulate is satisfied for all arguments of the function) will be written as:

$$C \mid\!- P.$$

At last, if a postulate *P* is satisfied for all functions from C_{co} then we will write:

$$C_{co} \mid\!- P.$$

We have the following properties of consensus functions:

Theorem 1
*A consensus function which satisfies postulate P*4 *should also satisfy postulates P*1 *and P*2, *that is*

$$(C \mid\!- P4) \Rightarrow ((C \mid\!- P1) \wedge (C \mid\!- P2)$$

for each $C \in C_{co}$.

Theorem 2

The following dependencies are not true:

a) $(C_{co} \vdash P_i) \Rightarrow (C_{co} \vdash P_j)$ *where* $j \neq i$ *and* $1 \leq i, j \leq 5$,

b) $(\exists C \in C_{co})(\exists X \in \Pi(Clause(L))$:
$((C(X) \vdash P1) \wedge (C(X) \vdash P2) \wedge (C(X) \vdash P3) \wedge (C(X) \vdash P4) \wedge C(X) \vdash P5)$.

Theorem 3

A consensus function which satisfies postulates $P2$ and $P4$ *should also satisfy postulates* $P1$ *and* $P3$, *that is*

$$(C \vdash P2) \wedge (C \vdash P4) \Rightarrow ((C \vdash P1) \wedge (C \vdash P3)$$

for each $C \in C_{co}$.

In the below, we propose an algorithm for knowledge integration using logic disjunctive structures. The idea of this algorithm is based on *Theorem* 1, we implement the following steps. Firstly, we build a disjunction from literals which occurred in all disjunctions. Secondly, the literal is taken from the remaining set of literals occurred in all formulae so that the difference between the number of occurrences in the positive form and the negative form is the largest. Finally, we examine this literal if the condition of postulate P4 is satisfied. If so, we add this literal into the consensus. This step is repeated until all the literals occurred in given formulae are examined.

The detail of algorithm is proposed as presented as follows.

Algorithm 1

Input: Profile $X \in \Pi(Clause(L))$
Output: Consensus x^* satisfying postulates $P1$, $P2$ and $P4$.
Procedure:
BEGIN

1. Set $Z = \bigcup_{x \in X} \bigcup_{V \in S_L(x)} V$;
2. Set $S = \bigcap_{x \in X} \bigcup_{V \in S_L(x)} V$;
3. Set $x^* = \bigvee_{m \in S} m$;
4. Set $found$ = False
5. Calculate $S^* = \sum_{x \in X} d(x^*, x)$;
6. While $Z \neq \emptyset$ and not *found* do
   Begin
   - 6.1. Find $z \in Z$ such that $\sum_{x \in X} d(x^* \vee z, x)$ is minimal;
   - 6.2. $Z = Z \setminus \{z\}$;
   - 6.3. if $\sum_{x \in X} d(x^* \vee z, x) \leq S^*$ then
     - 6.3.1. $x^* = x^* \vee z$
     - 6.3.2. $S^* = \sum_{x \in X} d(x^*, x)$

     else set *found* = True

   End

END.

We can prove easily that the computation complexity of Algorithm 1 is $O(m \cdot n)$ where $m = card(Z)$ and $n = card(X)$.

7 Conclusions

In this paper a model for knowledge integration using disjunctive structure is presented. The authors proposed some postulates for integrating process and some properties of these postulates are mentioned. The presented algorithm is only satisfying one of these properties. The future works should concern the deeper analysis of properties and propose other algorithms satisfying these properties.

References

1. Arrow, K.J.: Social Choice and Individual Values. Wiley, New York (1963)
2. Atkeson, C., Andrew, M., Stefan, S.: Locally weighted learning. Artificial Intelligence Review, 11–73 (1996)
3. Bock, H.H., Day, W.H.E., McMorris, F.R.: Consensus rules for committee elections. Mathematical Social Sciences 35, 219–232 (1998)
4. Day, W.H.E.: The complexity of computing metric distances between partitions. Mathematical Social Science 1, 269–287 (1981)
5. Deja, R.: Using Rough Set Theory in Conflicts Analysis, Ph.D. Thesis (Advisor: A. Skowron), Institute of Computer Science, Polish Academy of Sciences, Warsaw (2000)
6. Diday, E.: Recent Progress in Distance and Similarity Measures in Pattern Recognition. In: Proc. of 2nd International Joint Conference on Pattern Recognition, pp. 534–539 (1974)
7. Domingos, P.: Rule Induction and Instance-Based Learning: A Unified Approach. In: The 1995 International Joint Conference on Artificial Intelligence (IJCAI 1995), pp. 1226–1232 (1995)
8. Grant, J., Hunter, A.: Measuring inconsistency in knowledge bases. Journal of Intelligent Information Systems 27(2), 159–184 (2006)
9. Hunter, A.: Paraconsistent Logics. In: Gabbay, D., Smets, P. (eds.) Handbook of Defensible Reasoning and Uncertain Information, pp. 13–43. Kluwer Academic Publishers, Dordrecht (1998)
10. Jung, J.J., Jo, G.S.: Consensus-based Evaluation Framework for Cooperative Information Retrieval Systems. In: Nguyen, N.T., Grzech, A., Howlett, R.J., Jain, L.C. (eds.) KES-AMSTA 2007. LNCS (LNAI), vol. 4496, pp. 169–178. Springer, Heidelberg (2007)
11. Knight, K.: Measuring Inconsistency. Journal of Philosophical Logic 31, 77–98 (2002)
12. Kohonen, T.: The Self-Organizing Map. Proceedings of the IEEE 78(9), 1464–1480 (1990)
13. Lipski, W., Marek, W.: Combinatorial Analysis. WNT Warsaw (1986)
14. McMorris, F.R., Mulder, H.M., Powers, R.C.: The median function on median graphs and semilattices. Discrete Appl. Math. 101, 221–230 (2002)
15. Nguyen, N.T.: A Method for Integration of Knowledge Using Fuzzy Structure. In: Proc. of the 2007 IEEE/WIC/ACM International Conferences on Web Intelligence and Intelligent Agent Technology, pp. 11–14 (2007)
16. Nguyen, N.T.: Using Distance Functions to Solve Representation Choice Problems. Fundamenta Informaticae 48(4), 295–314 (2001)

17. Nguyen, N.T.: Advanced Methods for Inconsistent Knowledge Management. Springer, London (2007)
18. Pawlak, Z.: An Inquiry into Anatomy of Conflicts. Journal of Information Sciences 109, 65–78 (1998)
19. Tran, T.H., Nguyen, N.T.: An Algorithm for Agent Knowledge Integration Using Conjunctive and Disjunctive Structures. In: Nguyen, N.T., Jo, G.S., Howlett, R.J., Jain, L.C. (eds.) KES-AMSTA 2008. LNCS (LNAI), vol. 4953, pp. 703–712. Springer, Heidelberg (2008)

Three Parameters Refinement for Distributed 3D-Image Rendering

Dariusz Król[1] and Adam Bogdał[2]

[1] Institute of Applied Informatics, Wrocław University of Technology
Wybrzeże S. Wyspiańskiego 27, 50-370 Wrocław, Poland
dariusz.krol@pwr.wroc.pl
[2] adam.bogdal@gmail.com

Abstract. In this study, we present Java-based architecture for 3D-image rendering that supports distributed processing between autonomous nodes. We aim to improve the performance of presented system by decreasing computational time. Our research consists of three major parts: modification of thread priority, multiplication tasks on single machine and dynamic changing the bucket size. Presented results indicate that we succeed on average more than 20 percent compared with the original rendering system.

Keywords: image-based rendering, collaborative visualization, multi-agent systems, grid computing, performance study.

1 Introduction

Modern distributed systems based on desktop computers [9] are constructed by integrating diverse sets of components that are intelligent, heterogeneous and concurrent [2]. Extending the capabilities of grid [6] and agent environments [11] towards more user-centered approaches has been recognized as an area of major interest. In this paper, we deal with technical aspects that must be considered in using an image-based rendering system [4], [7], [10]. Rendering is the process by which an abstract description of a 3D scene is converted to a 2D graphic device. To provide the necessary levels of performance, parallel rendering is used [5]. Rendering of a single frame can take even several hours. In distributed system each frame is divided into tiles which are processed in parallel. There are several examples of such systems including specialized graphics computers, ray-tracing, software based terrain and radiosity renderers. Although it may not be difficult to find out if a given system has been implemented or not, it is hard to assess the quality and maturity of such an implementation.

Experiments were conducted on Helios platform [12]. The rendering engine is the Sunflow open source rendering system, built around a flexible ray tracing core and an extensible object-oriented design. The distributed computations are managed by the JGrid service-oriented grid system that uses the Jini technology as its base. The application will be evaluated in terms of the total time of rendering of 3D images considering three different criteria: the thread priority, the number of parallel started processes and increasing or reducing the bucket size in the single task. On the base of our experience these changes should give good results of the computation time of rendering

I. Lovrek, R.J. Howlett, and L.C. Jain (Eds.): KES 2008, Part I, LNAI 5177, pp. 262–269, 2008.

in every experiment. A positive effect of the alteration of the software on the general productivity of the rendering system will confirm it. To our knowledge there has been no work that applied proposed three parameter refinement to Helios.

In our experimental setup, a computer network consisting of four computers was built (see Fig. 1). Three of them execute the process of the rendering, and the last one is performing the role of the client with additionally started lookup service for working nodes. Three of the computers run under Linux and one under Windows XP. All have Java Virtual Machine 6. 0. They have processors from 1.6 to 1.8 GHz and 512 MB up to 1 GB of memory.

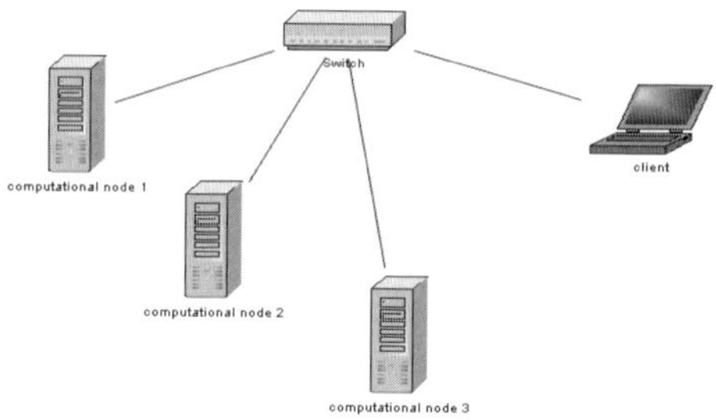

Fig. 1. The architecture of experimental testbed

The configuration of nodes included changes in the transient-compute.config where among others. The number of processors and their speed, as well as priority of working threads was placed. Two files were used in the SC formats which were subjected to the process of the rendering. They were examined before and after entering changes into the Helios system. Studied files have the diverse complexity hence the time of rendering is taking out from a few till several dozen minutes. Two test patterns see Fig. 2, come from the official Helios web site.

The paper is organized as follows. In Section 1, we explain in detail how our experiments have been designed and implemented on Helios platform. In Sections 2, 3, and 4

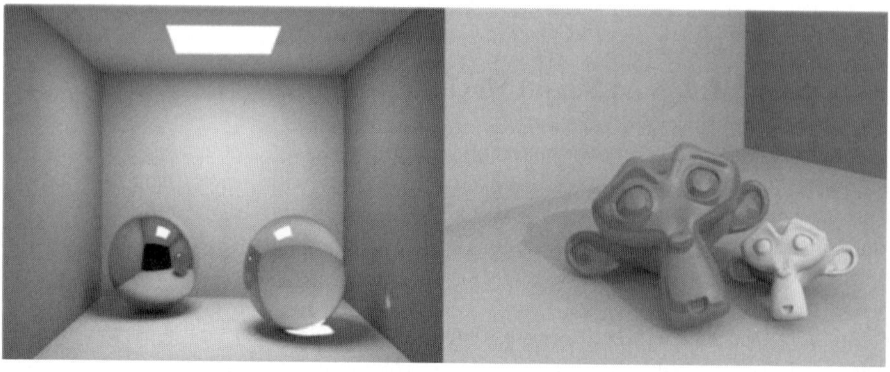

Fig. 2. Images for rendering process

we briefly discuss performance issues by presenting some experimental results. We conclude the paper in Section 5.

2 Modification of Thread Priority

Every thread of the Java program is associated with the priority which is informing how one should treat the given thread in the attitude to others. The priority determines the relative importance of the thread. The thread with the higher priority can supersede the lower one. We first look at the performance in Fig. 3. Every experiment requires less time needed for rendering of the image for a few minutes in the case of increasing the priority of working thread. Increasing the priority of working threads did not influence the general time of the rendering process negatively, all of them perform better than earlier.

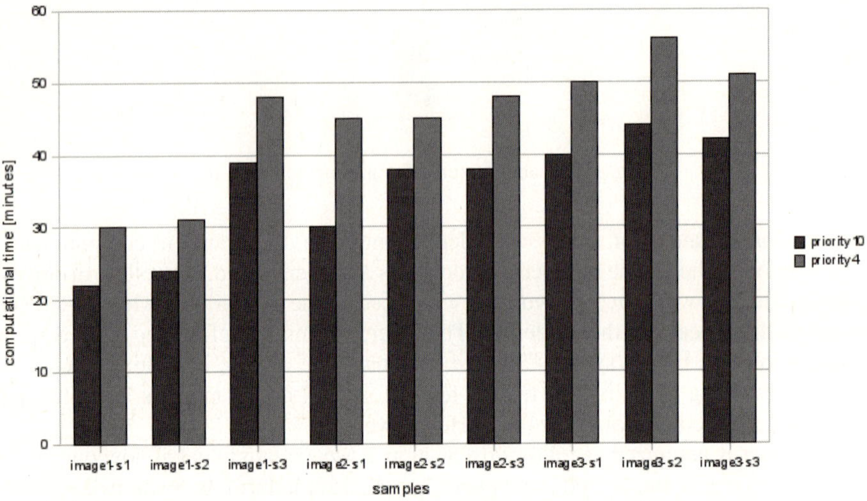

Fig. 3. Performance for images w.r.t. the standard or modified numbers of threads

3 Multiple Tasks on Single Machine

Helios by default starts on every node one task of rendering. The next experiment was supposed to check, whether in the case of computers with a processor over 1 GHz starting next tasks will have a positive effect on the general time of rendering. Original software was modified to that end. Correct discovering such a computer is based initially on appropriate value of the processor speed in the configuration file on the node.

In the measurements were used the following scenes: cornell-_box_jensen.sc – called later image1 and monkey.sc – called image2. They differ markedly from themselves in the complexity and the size. The details for each image are shown in Table 1.

Table 1. Specifications for two images

	Image1	Image2
Primitives	3	125958
Scene diameter	196,98	21,60
Light samples	32	257
Max ray trace depth diffuse	4	80
Max ray trace depth reflection	3	60
Max ray trace depth refraction	2	40
Resolution	800x600	1280x1024

Series of tests depicted in Fig. 4 are pointing at a positive effect of the modification. Each of the tests carried out fell out better than at the original software. It is possible to conclude that tasks allocation into individual computers in the original software did not use all available resources. Starting the second task on the same machine precipitated the entire process of rendering. The alteration of the application allowed for increasing the productivity of the system rendering at the same network architecture (architecture described in Fig. 1).

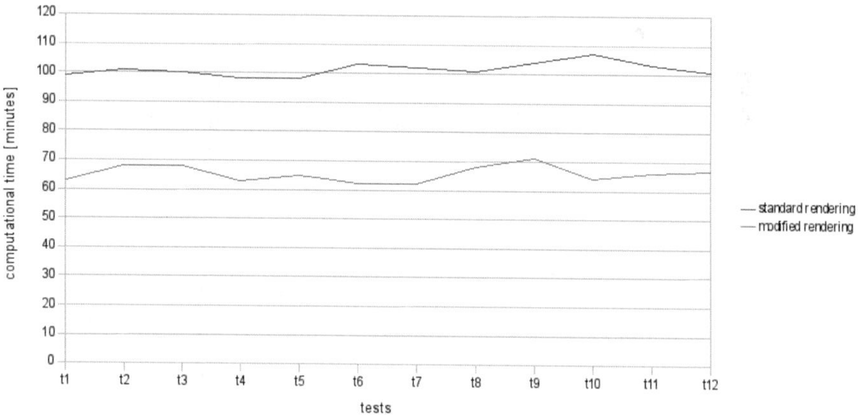

Fig. 4. Performance for image1 w.r.t. the standard or modified numbers of workers (2)

Using more complex image2 in the next experiment also shows in Fig. 5, that changes had a positive effect on the time of the image processing. The mean value of CPU usage was about 93% what at processing such a complicated file is entirely satisfactory and is confirming that it was possible to use better resources of nodes compared to the initial state. Extending the client application for the dynamic allotment of the next tasks can increase the productivity of the rendering system irrespective of the complexity of the processed scene.

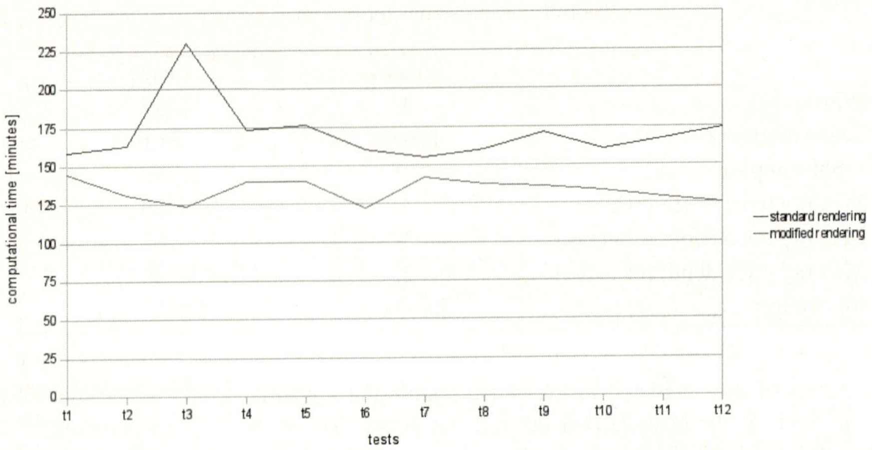

Fig. 5. Performance for image2 w.r.t. the standard or modified numbers of workers (2)

Previous experiments showed that starting two processes on the node had a positive effect on the time of rendering the images. So, that it is possible to recognize that it is optimal number of processes, similar evaluations were conducted for three started at the same time point. Fig. 6 presents the results. The total time of rendering a little bit lowered, what means, that the node received too much data for processing. It is possible to reduce ensuing delays through increasing system resources of individual node; however, such action will make the system less elastic and will impose limitations.

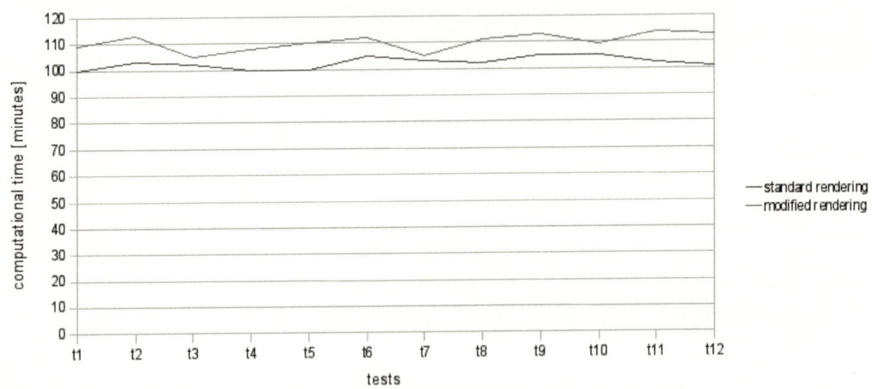

Fig. 6. Performance for image1 w.r.t. the standard or modified numbers of workers (3)

4 Dynamic Change the Bucket Size

The node in the moment of finishing computation is sending the result as well as taking the next task. Next experiment was supposed to prove, whether at the change of the size of the bucket of data sent in the isolated task what is being combined with the rarer data transfer by the network, the time will lower.

In one task a bucket of data of the scene is by default being sent about dimensions 32x32 pixels. After every processing the given fragment of the scene is sent to the client and next one is taken. In the experiment a bucket of data was increased twice, and then they were examining what influence it would have, when data will be sent so often as before implementing changes. The node will have more data for computation in the isolated task.

As can be seen in Fig. 7, the required time for the process of the rendering after the refinement of the application was reduced, it shows, that in the sequence of rendering of the scene, the system wasted time to the data transfer through the network. For nodes does mean more time processing the single bucket of data, however they gained quite a lot of time on account of the low numbers of connections with the client computer what in the end influenced for increasing the efficiency of the rendering system.

Series of experiments carried out on the scene with the greater level of the complexity (see Fig. 8) looks similarly as in the previous examination. The increased

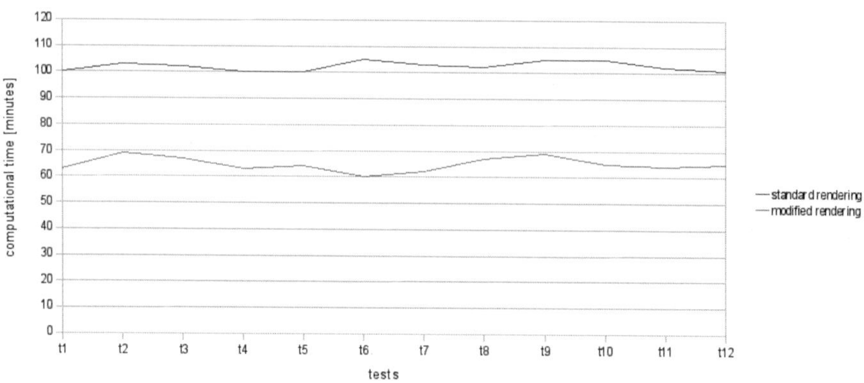

Fig. 7. Performance for image1 w.r.t. the standard or modified bucket size

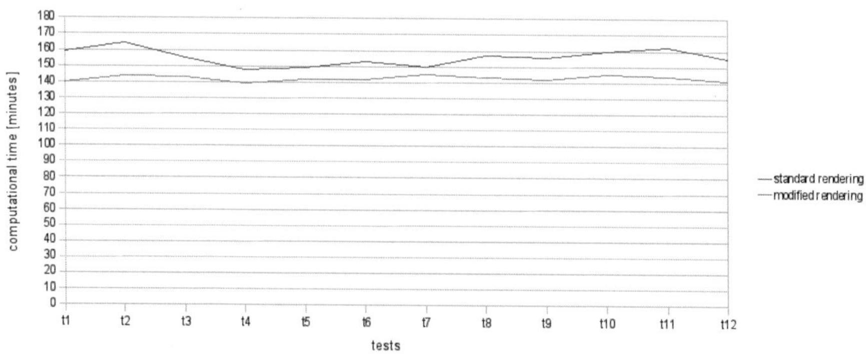

Fig. 8. Performance for image2 w.r.t. the standard or modified bucket size

bucket of data which already required the greater computing than in the previous example, let increase the performance of the rendering system. It is possible and so to conclude, that single bucket of data sent in the one task did not consume in the effective way the available resources.

The results illustrate that the time difference of rendering after making modifications to the application is decreasing according to the complexity of scenes. And so such a solution is putting the upper limit of the details of rendered scenes a little bit lower than in the case of the original application.

5 Conclusion

Working out the scalable method of the distributed 3D-image rendering was a main goal of the paper. The most known network systems of the rendering were analyzed. The research focused on experiments examining the influence of extending the renderer by elements which fully let exploit the accessible network and the hardware resources. The results indicate that we succeed on average more than 20 percent compared with the original rendering system.

Experimental results indicate the following:

- All well-known solutions to the distributed rendering function very well in non-complex computer networks. Only Helios application on the JGrid platform was in the state to accommodate oneself effectively to implemented changes in the case of the necessity of the reconfiguration of the network.
- The environment configuration for the network rendering is a time-consuming and difficult process; badly selected number of nodes and the distance between them can cause reducing the effectiveness of the rendering.
- Helios turned out to be the best system to the network rendering in the case of using heterogeneous resources and dynamic reconfiguration.
- JGrid is a perfect environment for dynamic connecting or disconnecting computers in the any time without the influence on the stability of the rendering process.

Experiments conducted on the basic version and modified version of Helios show, that:

- Modifications are necessary in the configuration files of the environment; they influence on efficient using the system resources.
- The change of threat priorities can change computational time of rendering.
- The number of parallel started processes influences directly for increasing or reducing the bucket size in the single task what allows for the improvement of rendering time.

Moreover, we consider that great Helios potential results from Jini technology, RMI dynamic discovering and registering actually. The framework is based on open source software, so the total cost of ownership is minimal. However, commercial rendering software can also be modified in similar way to meet the needs of users.

We believe that a combination of novel rendering algorithms [1], [8] and intelligent agent architecture [3] allows using more effectively distributed computer resources to approach real-time performance in image rendering. A multi-agent system that includes dynamic propagation mechanisms is under construction.

References

1. Amor, M., Boo, M., Pardon, E.J., Bartz, D.: Hardware Oriented Algorithms for Rendering Order-Independent Transparency. The Computer Journal 49(2), 201–210 (2006)
2. Artail, H., Kahale, E.: MAWS: A platform-independent framework for mobile agents using Web services. J. Parallel Distrib. Comp. 66, 428–443 (2006)
3. Carrascosa, C., Bajo, J., Julian, V., Corchado, J.M., Botti, V.: Hybrid multi-agent architecture as a real-time problem-solving model. Expert Systems with Applications 34, 2–17 (2008)
4. Chong, A., Sourin, A., Levinski, K.: Grid-based Computer Animation Rendering. In: 4th International Conference on Computer Graphics and Interactive Techniques in Australasia and South-East Asia GRAPHITE 2006, pp. 39–48. ACM, New York (2006)
5. Crockett, T.W.: An introduction to parallel rendering. Parallel Computing 23, 819–843 (1997)
6. Doulamis, N.D., Doulamis, A.D., Varvarigos, E.A., Varvarigou, T.: Fair Scheduling Algorithms in Grids. IEEE Transactions on Parallel and Distributed Systems 18(11), 1630–1648 (2007)
7. Litke, A., Tserpes, K., Varvarigou, T.: Computational workload prediction for grid oriented industrial applications: the case of 3D-image rendering. In: 2005 IEEE International Symposium on Cluster Computing and the Grid, pp. 962–969. IEEE, Los Alamitos (2005)
8. Luo, Y.B., Chen, D.F., Xiao, T.Y.: A distributed image-based virtual prototyping with novel rendering tactics. Int. J. Adv. Manuf. Technol. 26, 236–242 (2005)
9. Ryu, S.H., Kim, H.J., Park, J.S., Kwon, Y.W., Jeong, C.S.: Collaborative object-oriented visualization environment. Multimed Tools Appl. 32, 209–234 (2007)
10. Schlechtweg, S., Germer, T., Strothotte, T.: renderBots – Multi Agent Systems for Direct Image Generation. Computer Graphics Forum 24(2), 137–148 (2005)
11. Yau, S.M.T., Leong, H.V., Si, A.: Distributed agent environment: application and performance. Information Sciences 154, 5–21 (2003)
12. Helios. Sunflow Distributed Rendering [01.04.2008], `http://sfgrid.geneome.net/`

CBR to Explain Medical Model Exceptions

Olga Vorobieva[1,2] and Rainer Schmidt[1]

[1] Institute for Medical Informatics and Biometry, University of Rostock, Germany
rainer.schmidt@medizin.uni-rostock.de
[2] Sechenov Institute of Evolutionary Physiology and Biochemistry,
St.Petersburg, Russia

Abstract. In medicine many exceptions occur. In medical practise and in knowledge-based systems too, it is necessary to consider them and to deal with them appropriately. In medical studies and in research exceptions shall be explained. We present a system that helps to explain cases that do not fit into a theoretical hypothesis. Our starting points are situations where neither a well-developed theory nor reliable knowledge nor a proper case base is available. So, instead of reliable theoretical knowledge and intelligent experience, we have just some theoretical hypothesis and a set of measurements.

In this paper, we propose to combine CBR with a statistical model. We use CBR to explain those cases that do not fit the model. The case base has to be set up incrementally, it contains the exceptional cases, and their explanations are the solutions, which can be used to help to explain further exceptional cases.

1 Introduction

In medicine many exceptions occur. In medical practise and in knowledge-based systems too, these exceptions have to be considered and have to be dealt with appropriately. In ISOR-1, we demonstrated advantages of CBR in situations where a theoretically approved medical decision does not produce the desired and usually expected results [1].

In medical studies and in research exceptions shall be explained. The present research is a logical continuation of our previous work. It is still the same system and the same structure of dialogues, but now ISOR-2 deals with situations where neither a well-developed theory nor reliable knowledge nor a proper case base is available. So, instead of reliable theoretical knowledge and intelligent experience, we now have just some theoretical hypothesis and a set of measurements. In such situations the usual question is, how do measured data fit to theoretical hypotheses. To statistically confirm a hypothesis it is necessary, that the majority of cases fit the hypothesis. Mathematical statistics determines the exact quantity of necessary confirmation [2]. However, usually a few cases do not satisfy the hypothesis. We examine these cases to find out why they deviate from the hypothesis. This approach is justified by a certain mistrust of statistical models by doctors, because modelling results are usually unspecific and “average oriented” [3], which means a lack of attention to individual "imperceptible" features of concrete patients.

I. Lovrek, R.J. Howlett, and L.C. Jain (Eds.): KES 2008, Part I, LNAI 5177, pp. 270–277, 2008.

Our approach starts in a situation where no case-base with complete solutions is available but has to be set up incrementally. So, we must

1. Construct a model,
2. Point out the exceptions,
3. Find causes why the exceptional cases do not fit the model, and
4. Develop a case-base.

So, we combine CBR with a statistical model. The idea to combine CBR with other methods is not new. For example Care-Partner resorts to a multi-modal reasoning framework for the co-operation of CBR and Rule-based Reasoning (RBR) [4]. Another way of combining hybrid rule bases with CBR is discussed by Prentzas and Hatzilgeroudis [5]. The combination of CBR and model-based reasoning is discussed in [6]. Statistical methods are used within CBR mainly for retrieval and retention (e.g. [7, 8]). Arshadi proposes a method that combines CBR with statistical methods like clustering and logistic regression [9]).

ISOR-2 is domain-independent and it can be applied on various problems. Here we present it's first application, namely on dialyse patients who where offered to participate in a specific fitness program.

Dialyse and fitness. Hemodialyse means stress for a patient's organism and has significant adverse effects. Fitness is the most available and a relative cheap way of support. It is meant to improve a physiological condition of a patient and to compensate negative dialyse effects. One of the intended goals of this research is to convince the patients of the positive effects of fitness and to encourage them to make efforts and to go in for sports actively. This is important because dialyse patients usually feel sick, they are physically weak, and they do not want any additional physical load [10]. At our University clinic in St. Petersburg, a specially developed complex of physiotherapy exercises including simulators, walking, swimming etc. was offered to all dialyse patients but only some of them actively participated, whereas some others participated but were not really active. The purpose of this fitness offer was to improve the physical conditions of the patients and to increase the quality of their lives.

2 Incremental Development of an Explanation Model for Exceptional Dialyse Patients

For each patient a set of physiological parameters is measured. These parameters contain information about burned calories, maximal power achieved by the patient, oxygen uptake, oxygen pulse (volume of oxygen consumption per heart beat), lung ventilation and others. There are also biochemical parameters like haemoglobin and other laboratory measurements. More than 100 parameters were planned for every patient. But not all of them were really measured. Parameters are supposed to be measured four times during the first year of participating the fitness program. There is an initial measurement followed by a next one after three months, then after six months and finally after a year. Unfortunately, since some measurements did not happen, many data are missing. Therefore the records of the patients often contain different sets of measured parameters. It is necessary to note that parameter values of dialyse patients essentially differ from those of non-dialysis patients, especially of healthy people, because dialyse interferes with the natural, physiological processes in an organism. For statistics, this means difficulties in applying statistical methods based on correlation and it limits the usage of a knowledge base developed for normal people. Inhomogeneity of observed data, many missing values, many parameters for a relatively small sample size, all this makes our data set practically impossible for usual statistical analysis.

2.1 Setting Up a Model

We start with a medical problem that has to be solved based on given data. In our example it is: "Does special fitness improve the physiological condition of dialyse patients?" More formal, we have to compare physical conditions of active and non-active patients. Patients are divided into two groups, depending on their activity, active patients and non-active ones. According to our assumption active patients should feel better after some months of fitness, whereas non-active ones should feel rather worse. We have to define the meaning of "feeling better" and "feeling worse" in our context. A medical expert selects appropriate factors from ISOR's menu. It contains the list of field names from the observed data base. The expert selects the following main factors

- F1: O2PT - Oxygen pulse by training
- F2: MUO2T - Maximal Uptake of Oxygen by training
- F3: WorkJ – performed Work (Joules) during control training.

Subsequently the "research time period" has to be determined. Initially, this period was planned to be twelve months, but after a while the patients tend to give up the fitness program. This means, the longer the time period, the more data are missing. Therefore, we had to make a compromise between time period and sample size. A period of six months was chosen.

The next question is whether the model shall be quantitative or qualitative? The observed data are mostly quantitative measurements. The selected factors are of quantitative nature too. On the other side, the goal of our research is to find out whether physical training improves or worsens the physical condition of the dialyse patients. We have to compare each patient with his/her own situation just before the start of the fitness program. The success shall not be measured in absolute values, because the health statuses of patients are very different. Thus, even a modest improvement for one patient may be as important as a great improvement of another. Therefore, we simply classify the development in two categories: "better" and "worse". The changes are assessed depending on the number of improved factors:

- Weak version of the model: at least one factor has improved
- Medium version of the model: at least two factors have improved
- Strong version of the model: all three factors have improved.

The final step means to define the type of model. Popular statistical programs offer a large variety of statistical models. Some of them deal with categorical data. The easiest model is a 2x2 frequency table. Our "Better/ Worse" concept fits this simple model very well. So the 2x2 frequency table is accepted. The results are presented in table 1.

Table 1. Results of Fisher's Exact Test, performed with an interactive Web-program: http://www.matforsk.noIola/fisher.htm

Improve-ment mode	Patient's physical condi-tion	Active	Non-active	Fisher Exact p
Strong	Better	28	2	< 0.0001
	Worse	**22**	21	
Medium	Better	40	10	< 0.005
	Worse	**10**	12	
Weak	Better	47	16	< 0.02
	Worse	**3**	6	

According to our assumption after six months of active fitness the conditions of the patients should be better. Statistical analysis shows a significant dependence between the patient's activity and improvement of their physical condition. Unfortunately, the most popular Pearson Chi-square test is not applicable here because of the small values "2" and "3" in table 1. But Fisher's exact test [3] can be used. In the three versions shown in table 1 a very strong significance can be observed. The smaller the value of p is, the more significant the dependency.

Exceptions. Though the performed Fisher test confirms the hypothesis, there are exceptions, namely active patients whose health conditions did not improve.

These exceptions should be explained. Explained exceptions build the case base. According to table 1, the stronger the model, the more exceptions can be observed and have to be explained. Every exception is associated with at least two problems. The first one is "Why did the patient's condition get worse?" Of course, "worse" is meant in terms of the chosen model. Since there may be some factors that are not included in the model but have changed positively, the second problem is "What has improved in the patient's condition?" To solve this problem we look for significant factors where the values improved.

2.2 Setting Up a Case Base

We intend to solve both problems (mentioned above) by means of CBR. So we begin to set up the case-base up sequentially. That means, as soon as an exception is explained, it is incorporated into the case-base and can be used to help explaining further exceptional cases. We chose a randomly order for the exceptional cases. In fact, we took them in alphabetical order.The retrieval of already explained cases is performed by keywords. The main ones are "problem code", "diagnosis", and "therapy". In the situation of explaining exceptions for dialyse patients the instantiations of these keywords are "adverse effects of dialysis" (diagnosis), "fitness" (therapy), and two problem specific codes. Besides the main ISOR-2 keywords additional problem specific ones are used. Here the additional key is the number of worsened factors. Further keywords are optional. They are just used when the case-base becomes much bigger than the 22 cases of the strong model (table 1) and retrieval is not simple any longer.

However, ISOR-2 does not only use the case-base as knowledge source but further sources are involved, namely the patient's individual base (his medical history) and observed data (partly gained by dialogue with medical experts). Since in the domain of kidney disease and dialyse the medical knowledge is very detailed and much investigated but still incomplete, it is unreasonable to attempt to create an adequate knowledge base. Therefore, a medical expert, observed data, and just a few rules serve as medical knowledge sources.

Expert knowledge and artificial cases. Expert's knowledge can be used in many different ways. Firstly we use it to acquire rules, secondly it can be used to select appropriate items from the list of retrieved solutions, to propose new solutions and last but not least – to create artificial cases. Initially artificial cases are created by an expert, afterwards they can be used in the same way as real cases. They are created in the following situation. An expert points out a factor F as a possible solution for a query patient. Since many values are missing, it can happen that just for the query patient values of factor F are missing. The doctor's knowledge in this case can not be applied, but it is sensible to save it anyway. Principally there are two different ways to do this. The first one means to generate a correspondent rule and to insert it into ISOR-2's algorithms. Unfortunately, this is very complicated, especially to find an appropriate way for inserting such a rule. The alternative is to create an artificial case. Instead of a patient's name an artificial case number is generated. The other attributes are either inherited from the query case or declared as missing. The retrieval attributes are inherited. This can be done by a short dialogue (figure1) and ISOR-2's algorithms remain intact. Artificial cases can be treated in the same way as real cases, they can be revised, deleted, generalised and so on. During the explanation of the 22 cases of the strong model just four artificial cases were generated.

The problem: Why did some patients conditions became worse? As results we obtain a set of solutions of different origin and different nature. There are three categories of solutions: additional factor, model failure, and wrong data.

The most important and most frequent solution is the influence of an additional factor. Only three main factors are obviously not enough to describe all medical cases. Unfortunately, for different patients different additional factors are important. When ISOR-2 has discovered an additional factor as explanation for an exceptional case, the factor has to be confirmed by a medical expert before it can be accepted as a solution. One of these factors is Parathyroid Hormone (PTH). An increased PTH level sometimes can explain a worsened condition of a patient [11]. PTH is a significant factor, but unfortunately it was measured only for some patients. Some exceptions can be explained by indirect indications. One of them is a very long time of dialyse (more than 60 months) before a patient began with the training program. Another solution is a phosphorus blood level. We used the principle of artificial cases to introduce the factor phosphorus as a new solution. One patient's record contained many missing data. The retrieved solution meant high PTH, but since PTH data in the query patient's record was missing too, an artificial case was created, who inherited all retrieval attributes of the query case while the other attributes were recorded as missing. According to the expert high phosphorus can explain the solution. Therefore it is accepted as an artificial solution or a solution of an artificial case. We regard two types of model failures. One of them is neglected data. In fact, three of the patients did not show an improvement in the considered six month but in the following, neglected six months. So, they were wrongly classified and should really belong to the "better" category. The second type of model failure is based on the fact that the two-category model was not precise enough. Some exceptions could be explained by a tiny and not really significant change in one of the main factors.

The problem: What in the patient's condition became better? There are at least two criteria to select factors for the model. Firstly, a factor has to be significant, and secondly there must be enough patients for which this factor was measured at least for six months. So, some principally important factors were initially not taken into account because of missing data. The list of solutions includes these factors (figure 1): haemoglobin, maximal power (watt) achieved during control training. Oxygen pulse and oxygen uptake were measured in two different situations, namely during the training under loading and before training in a state of relax. Therefore we have two pairs of factors: oxygen pulse in state of relax (O2PR) and during training (O2PT); maximal oxygen uptake in state of relax (MUO2R) and during training (MUO2T). Measurements made in a state of relax are more indicative and significant than those made during training. Unfortunately, most measurements were made during training. Only for some patients correspondent measurements in relax state exist. Therefore O2PT and MUO2T were accepted as main factors and were taken into the model. On the other side, O2PR and MUO2R serve as solutions for the current problem.

2.3 Illustration of ISOR-2's Program Flow

Figure 1 shows the main dialogue of ISOR-2 where the user at first sets up a model (steps one to four), subsequently gets the result and an analysis of the model (steps five to eight), and then attempts to find explanations for the exceptions (steps nine and ten). Finally the case base is updated (steps eleven and twelve). Now we explain the steps in detail.

At first the user has to set up a model. To do this he has to select a grouping variable. In this example CODACT was chosen. It stands for "activity code" and means that active and none active patients are to be compared. Provided alternatives are the sex and the beginning with the fitness program (within the first year of dialyse or later). In another menu the user can define further alternatives. Furthermore, the user has to select a model type (alternatives are "strong", "medium", and "weak"), the length of time that should be considered (3, 6 or 12 months), and main factors have to be selected. In the example three factors are chosen: O2PT (oxygen pulse by training), MUO2T (maximal oxygen uptake by training), and WorkJ (work in joules during the test training).

When the user has selected these items, the program calculates the table. ISOR-2 does not only calculate the table but additionally extracts the exceptional patients from the observed database. In the menu, the list of exceptions shows the code names of the patients. In the example patient "D5" is selected" and all further data belong to this patient. The goal is to find an explanation for the exceptional case "D5".

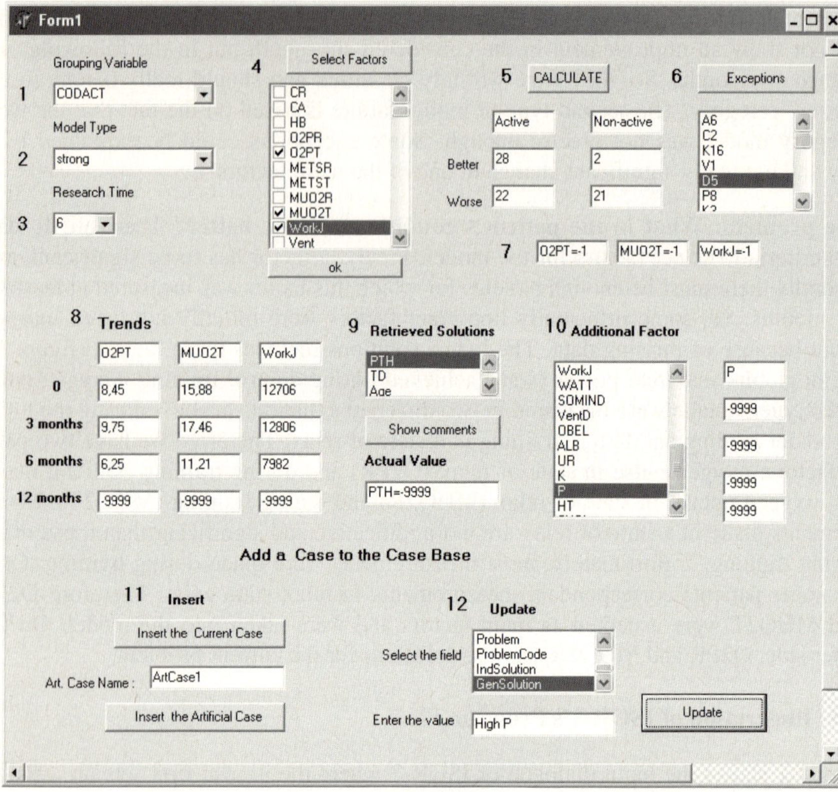

Fig. 1. ISOR-2's program flow

In point seven of the menu it is shown that all selected factors worsened (-1), and in point eight the factor values according to different time intervals are depicted. All data for twelve months are missing (-9999).

The next step means creating an explanation for the selected patient "D5". From the case base ISOR-2 retrieves general solutions. The first retrieved one in this example, the PTH factor, denotes that the increased Parathyroid hormone blood level may explain the failure. Further theoretical information (e.g. normal values) about a selected item can be received by pressing the button "show comments". The PTH value of patient "D5" is missing (-9999). From menu point ten the expert user can select further probable solutions. In the example an increased phosphorus level (P) is suggested. Unfortunately, phosphorus data are missing too. However, the idea of an increased phosphorus level as a possible solution shall not be lost. So, an artificial case has to be generated.

The final step means inserting new cases into the case base, query cases and artificial cases. Query cases are records of real patients from the observed database. The records contain a lot of data but they are not structured.

Artificial cases inherit the key attributes from the query cases (step seven). Other data may be declared as missing, by the update function data can be inserted. In the example of the menu, the generalised solution "High P" is inherited, it may be retrieved as a possible solution (step nine) for future cases.

3 Conclusion

In this paper, we have proposed to use CBR to explain cases that do not fit a statistical model. Here we used one of the simplest models. However, it is relatively effective, because it demonstrates statistically significant dependencies, in our example between fitness activity and health improvement of dialyse patients, where the model covers about two thirds of the patients, whereas the other third can be explained by applying CBR.

The presented method makes use of different sources of knowledge and information, inclusive medical experts. It seems to be a very promising method to deal with a poorly structured database, with many missing data, and with situations where cases contain different sets of attributes.

References

1. Schmidt, R., Vorobieva, O.: ISOR: A Case-Based System for Investigations of Therapy Inefficacy. In: Gabrys, B., Howlett, R.J., Jain, L.C. (eds.) KES 2006. LNCS (LNAI), vol. 4251, pp. 334–341. Springer, Heidelberg (2006)
2. Kendall, M.G., Stuart, A.: The advanced theory of statistics, 4th edn. Macmillan publishing, New York (1979)
3. Hai, G.A.: Logic of diagnostic and decision making in clinical medicine. Politheknica publishing, St. Petersburg (2002)
4. Bichindaritz, I., Kansu, E., Sullivan, K.M.: Case-based Reasoning in Care-Partner. In: Smyth, B., Cunningham, P. (eds.) EWCBR 1998. LNCS (LNAI), vol. 1488, pp. 334–345. Springer, Heidelberg (1998)
5. Prentzas, J., Hatzilgeroudis, I.: Integrating Hybrid Rule-Based with Case-Based Reasoning. In: Craw, S., Preece, A.D. (eds.) ECCBR 2002. LNCS (LNAI), vol. 2416, pp. 336–349. Springer, Heidelberg (2002)
6. Shuguang, L., Qing, J., George, C.: Combining case-based and model-based reasoning: a formal specification. In: Proc APSEC 2000, p. 416 (2000)
7. Corchado, J.M., Corchado, E.S., Aiken, J., et al.: Maximum likelihood Hebbian learning based retrieval method for CBR systems. In: Ashley, K.D., Bridge, D.G. (eds.) ICCBR 2003. LNCS (LNAI), vol. 2689, pp. 107–121. Springer, Heidelberg (2003)
8. Rezvani, S., Prasad, G.: A hybrid system with multivariate data validation and Case-based Reasoning for an efficient and realistic product formulation. In: Ashley, K.D., Bridge, D.G. (eds.) ICCBR 2003. LNCS (LNAI), vol. 2689, pp. 465–478. Springer, Heidelberg (2003)
9. Arshadi, N., Jurisica, I.: Data Mining for Case-based Reasoning in high-dimensional biological domains. IEEE Transactions on Knowledge and Data Engineering 17(8), 1127–1137 (2005)
10. Davidson, A.M., Cameron, J.S., Grünfeld, J.-P., et al. (eds.): Oxford Textbook of Nephrology, vol. 3. Oxford University Press, Oxford (2005)

Reasoning for Incomplete Authorizations

Yun Bai

School of Computing and Mathematics
University of Western Sydney
Locked Bag 1797, Penrith South DC
NSW 1797, Australia
ybai@scm.uws.edu.au

Abstract. Authorization plays an important role to control access to the system resources. It enforces security mechanism in compliance with the polices and rules specified by the security strategies. However, the security rules may not be always complete. In certain situations, we need to evaluate and reason about an incomplete security domain. In this paper, we propose an approach to reason under incomplete security domain by extended logic programs, discuss properties of unknown and conflict queries, and solve these problems by defining a procedure of evaluating the logic programs.

Keywords: authorization, formal specification, incomplete domain, logic program.

1 Introduction

Ensuring system security is an essential issue of a wide variety of computing and IT systems such as data management systems, e-trading systems, database transaction systems, etc. Research in system security has long been an important area in computer security. Authorization or access control is a fundamental mechanism to prevent malicious or unauthorized attempt to the system resource. It only allows the authorized users performing authorized operations on the shared data resource of the system. Study on the formal specification of authorization has become a major challenge in the current development of secure computing and IT systems. In [11], a logic language for expressing authorization rules is proposed. It uses predicates and rules to specify the authorizations. This work mainly emphasizes the representation and evaluation of authorizations. The work of [3] describes an authorization mechanism based on a logic formalism. It mainly investigates the access control rules and their derivations. In [4], a formal approach based on C-Datalog language is presented for reasoning about access control models. [12] develops a logical language called *delegation logic* to represent authorization policies, credentials in large-scale, distributed systems. The work emphasizes the delegation depth and a variety of complex delegation principals. In [7] security policy management using logic program approach is discussed. [13] proposes a formal approach using default logic to represent and

I. Lovrek, R.J. Howlett, and L.C. Jain (Eds.): KES 2008, Part I, LNAI 5177, pp. 278–285, 2008.

evaluate authorizations. [2] discusses a formal approach for specifying XML document security.

So far, certain research has been done using logic formalism in authorizations as mentioned above. These works either focus on authorization representation, evaluation or delegation. Little has been done in reasoning about incomplete authorization domains. It is a new area needs to be investigated to make the security system more robust by working in an even incomplete authorization domain. This paper is to address high level authorization specification and resolution for incomplete authorizations by using extended logic programs. We first propose a logic language by using logic programs to specify authorization rules, and investigate the properties of the problem of unknown and conflict queries, then evaluate it under incomplete domain by using the concept and techniques of the extended logic programs.

The paper is organized as follows. Section 2 describes authorization rules, its specification and evaluation or query answering. Section 3 investigates incomplete authorization issue and analyzes the properties of the problems. Section 4 proposes an extended logic programming approach to solve the conflict queries. Section 5 investigates the unknown queries, adjusts them to fit into the class of conflict queries and uses the same approach to solve them. Finally, section 6 concludes the paper.

2 Policy Base Representation

We define that all the authorization rules form a *policy base or PB*, it is specified by an *authorization domain.* The PB is represented by a logic language $\mathcal{L}$ with predicates, functions symbols and logic connectives.

In $\mathcal{L}$, the fact that a subject S has read right for object O is represented using a ground atom $read(S,O)$. A ground atom is an atom with no variables. S is a senior member is represented as $senior(S)$. A member S belongs to a group G is represented by $S \in G$. Similarly, we represent inclusion relationships between a group and its subgroup as $G_1 \subseteq G_2$. In general, we define a *literal* which represents a *fact* F to be an atomic formula of $\mathcal{L}$ or its negation, while a *ground fact* is a fact without variable occurrence.

Generally, we use lower case letters for variables and upper case letters for constants. $\neg\neg F$ is viewed as F.

For instance, If there is a policy base specified as:

$$PB = \{read(A,O), read(B,x) \leftarrow read(A,x)\}.$$

This PB states that currently A can read O; and B can read whatever A reads. The query $read(B,O)$ will generate answer "yes" while the query $read(C,O)$ will generate answer "no".

Let's have a look of another PB:

$$PB = \{senior(A), senior(B), review(x) \leftarrow senior(x)\}.$$

It states that both A and B are senior members; if a member is senior then he can have the review right. The query $review(A)$ will get answer "yes", while the query $review(C)$ will yield "no". General logic programs adopt the closed world assumption. That is, if we cannot deduce $review(C)$, we conclude that $\neg review(C)$ holds.

In our policy base, we apply extended logic program concept and consider the classical strong negation $\neg$ as well as the weak negation: negation as failure *not*.

A *rule* of a policy base is an expression of the form:

$$F_0 \leftarrow F_1, \cdots, F_m, notF_{m+1}, \cdots, notF_n. \tag{1}$$

where each F_i $(0 \leq i \leq n)$ is a literal. F_0 is called the *head* of the rule, while $F_1, \cdots, F_m$,*not* $F_{m+1}, \cdots$, *not* F_n are called the *body* of the rule. Obviously, the body of a rule could be empty. In this case, it represents an authorization fact. A rule is *ground* if no variable occurs in it.

An *extended logic program* is a collection of rules of form (1). In a rule, the set $\{F_1, \cdots, F_m\}$ is the literals without weak negation; the set $\{notF_{m+1}, \cdots, notF_n\}$ is the literals with weak negation.

All the rules of an authorization domain form a policy base which specifies the restriction conditions to access system resource of an organization. It is formally defined as:

Definition 1. *A policy base is a finite set $D = \{R_i\}$, (i=1,2, ...k) where R_i is a rule of the form $F_0 \leftarrow$ or $F_0 \leftarrow F_1, \cdots, F_m, notF_{m+1}, \cdots, notF_n$ where m>=0, n>m.*

Example 1. Here is an example of a policy base.

$D = \{R_1, R_2, R_3, R_4\}$, where
R_1: $read(S_1, O_1) \leftarrow$
R_2: $\neg read(S_1, O) \leftarrow$
R_3: $O \in O_1 \leftarrow$
R_4: $read(S_1, O) \leftarrow read(S_1, O_1), O \in O_1,\ not\ \neg read(S_1, O)$

This PB states that currently S_1 can read O_1; S_1 cannot read O; O is a member of O_1; if S_1 can read O_1 and O is a member of O_1 and it is not specified that S_1 can not read O, then S_1 has the right to read O.

Example 2. The following is a different policy base.

$D = \{R_1, R_2, R_3, R_4, R_5, R_6\}$, where
$R_1 : junior(Alice) \leftarrow$
$R_2 : experienced(Alice) \leftarrow$
$R_3 : senior(Bob) \leftarrow$
$R_4 : review(x) \leftarrow senior(x)$
$R_5 : review(x) \leftarrow junior(x), experienced(x)$
$R_6 : \neg review(x) \leftarrow \neg experienced(x), \neg senior(x)$

This PB represents the current authorization information about a reviewing system: Alice is a junior but experienced member; Bob is a senior member; a senior member has the review right; a junior but experienced member has the review right as well; an inexperienced, non-senior member does not have the review right.

3 Query Evaluation

Given a policy base and a query, the evaluation process is to decide either to grant or deny such a query in compliance with the system authorization restrictions.

Obviously, in Example 1, rules R_2 and R_4 conflict with each other as their heads are complementary literals. Query $read(S_1, O)$ yields both $read(S_1, O)$ and $\neg read(S_1, O)$. At the moment, we do not have complete information to answer query $read(S_1, O)$.

For Example 2, suppose we also have the rules:

$R_7 : junior(Carl) \leftarrow$

The policy base does not include any information about either Carl is experienced or not, that is, the current PB about Carl's information is incomplete. Hence, query $review(Carl)$ yields neither $review(Carl)$ nor $\neg review(Carl)$.

Again in Example 2, if for another member Peter, we are not quite sure if he is a senior or junior member, so we have:

$R_8 : senior(Peter)$ or $junior(Peter) \leftarrow$

The PB yields two answer sets regarding Peter:

$\{junior(Peter)\}$ and
$\{senior(Peter), review(Peter)\}$

Hence, the query $review(Peter)$ cannot be decided.

The above examples show inconsistent policy bases due to incomplete information. In some situation, the policy base is complete initially, but after certain updates or modification, it becomes inconsistent or incomplete.

These two examples indicate two cases of incomplete authorizations. Example 1 specified a policy base where both the query and the negation of the query are achieved. We call this kind of query conflict query. Example 2 describes a situation where either the query or the negation of the query cannot be achieved. We call such a query unknown query. We will discuss the two situations separately and provide solutions to solve both the incomplete authorization reasoning.

4 Solving Conflict Query

We discuss the conflict query with an example.

Example 3. For instance, we initially have the following policy base:

$D = \{R_1, R_2, R_3\}$, where
R_1: $read(S, O) \leftarrow$
R_2: $S_1 \in S \leftarrow$
R_3: $read(S_1, x) \leftarrow read(S, x), S_1 \in S$

It says that currently S can read O; S_1 is a member of S; if S_1 is a member of S then S_1 can read whatever S entitled to read. The answer set for this PB is:

$\{read(S, O), read(S_1, O), S_1 \in S\}$

Now, the new information R_4: $\neg read(S_1, O)$ is just acquired to the policy base. It conflicts with the existing policy base where it implies $read(S_1, O)$. We need to define a preference order to solve this conflicts. Suppose we prefer the update, that is we set the newly added R_4 higher preference. This order is represented as $R_3 < R_4$. After adding the new information, the new policy base has the following answer set:

$$\{read(S, O), \neg read(S_1, O), S_1 \in S\}$$

We call the logic program with partial ordering $<$ on the rules *ordered logic program* $\mathcal{P}$ [14]. $\mathcal{P}$ is defined as $(D, <)$, where D is an extended logic program defined in Definition 1, $<$ is a strict partial ordering on the rules of D. The partial ordering $<$ in $\mathcal{P}$ plays an essential role in the evaluation of $\mathcal{P}$. We also use $\mathcal{P}(<)$ to denote the set of $<$-relations of $\mathcal{P}$. Intuitively $<$ represents a preference of applying rules during the evaluation of the program. In particular, if $R < R'$ holds in $\mathcal{P}$, rule R' would be preferred to apply over rule R during the evaluation of $\mathcal{P}$.

The evaluation of an ordered logic program or an OLP will be based on its ground form. It is to find the answer set of the policy base. Given an OLP $\mathcal{P} = (D, <)$. We say $\mathcal{P}$ is *well formed* if there does not exist a rule R' that is an instance of two different rules R_1 and R_2 in D and $R_1 < R_2 \in \mathcal{P}(<)$. In the rest of this paper, we will only consider well formed OLPs in our discussions, and consequently, the evaluation for an arbitrary program $\mathcal{P} = (D, <)$ will be based on its ground instantiation $\mathcal{P}' = (D', <')$. Therefore, in our context a ground ordered logic program may contain infinite number of rules. If so, we will assume that this ground program is the ground instantiation of some program that only contains finite number of rules.

Definition 2. *Let D be a ground extended logic program and R a rule with the form $F_0 \leftarrow F_1, \cdots, F_m$, not $F_{m+1}, \cdots$, not F_n (R does not necessarily belong to D). Rule R is* revoked *by Π iff Π has an answer set and for any answer set $Ans(D)$ of D, there exists some $F_i \in Ans(D)$, where $m + 1 \le i \le n$.*

This definition explains that for any rule $F_0 \leftarrow F_1, \cdots, F_m$, not $F_{m+1}, \cdots$, not F_n, if any of not $F_{m+1}, \cdots$, not F_n is in the answer set, this rule will be false, can be removed from the policy base D.

Let us consider example 2 once again. If we choose $R_2 < R_4$ then R_2 is revoked by $\mathcal{D} - \{R_2\}$, rule R_2 should be ignored during the evaluation of $\mathcal{D}$. We will get the unique answer set $\{read(S, O_1), O \in O_1, read(S_1, O)\}$.

To calculate the set of access facts of a policy base of an authorization domain, we need to evaluate its corresponding extended logic program. That is, to find the answer set of ordered logic program $\mathcal{P}$. Now, we present the procedure for finding the answer set. We start from a reduced set or the reduct of $\mathcal{P}$.

Definition 3. *Let $\mathcal{P} = (D, <)$ be an ordered extended logic program. $\mathcal{P}^<$ is a* reduct *of $\mathcal{P}$ with respect to $<$ if and only if there exists a sequence of sets Π_i ($i = 0, 1, \cdots$) such that:*

1. $D_0 = D$;
2. $D_i = D_{i-1} - \{R_1, R_2, \cdots \mid$
   (a) *there exists* $R \in D_{i-1}$ *such that* *for every* j $(j = 1, 2, \cdots)$, $R < R_j \in \mathcal{P}(<)$ *and* $R_1, \cdots,$ *are revoked by* $D_{i-1} - \{R_1, R_2, \cdots\}$, *and*
   (b) *there does not exist a rule* $R' \in D_{i-1}$ *such that* $R_j < R'$ *for some* j $(j = 1, 2, \cdots)$ *and* R' *is revoked by* $D_{i-1} - \{R'\}\}$;
3. $\mathcal{P}^< = \bigcap_{i=0}^{\infty} D_i$.

In Definition 3, $\mathcal{P}^<$ is a ground extended logic program obtained from D by eliminating some *less preferred rules* from D. In particular, if $R < R_1$, $R < R_2$, $\cdots$, and $D_{i-1} - \{R_1, R_2, \cdots\}$ revokes $\{R_1, R_2, \cdots\}$, then rules $R_1, R_2, \cdots$ will be eliminated from D_{i-1} if no less preferred rule can be eliminated (i.e. conditions (a) and (b)). This procedure is continued until a fixed point is reached. It is worth to note that the generation of a reduct of an OLP is based on the ground form of its extended logic program part. Furthermore, if $R_1 < R_2$ holds in an OLP where R_1 or R_2 includes variables, then $R_1 < R_2$ is actually viewed as the set of $<$-relations $R'_1 < R'_2$, where R'_1 and R'_2 are ground instances of R_1 and R_2 respectively.

Using Definitions 3, it is easy to conclude that in example 1, if we assign $R_2 > R_4$, $\mathcal{P}$ has a unique reduct as follows:

$$\mathcal{P}^< = \{read(S_1, O_1) \leftarrow, \neg read(S_1, O) \leftarrow, O \in O_1 \leftarrow\}$$

from which we obtain the following answer set of $\mathcal{P}$:

$$Ans^P(\mathcal{P}_1) = \{read(S_1, O_1), \neg read(S_1, O), O \in O_1)\}$$

If the preference ordering is $R_2 < R_4$, $\mathcal{P}$ has a unique reduct as follows:

$$\mathcal{P}^< = \{read(S_1, O_1) \leftarrow, O \in O_1 \leftarrow, read(S_1, O) \leftarrow read(S_1, O_1), O \in O_1, not\ \neg read(S_1, O)\}$$

from which we obtain the following answer set of $\mathcal{P}$:

$$Ans^P(\mathcal{P}_1) = \{read(S_1, O_1), O \in O_1, read(S_1, O)\}.$$

5 Solving Unknown Query

Section 4 discussed how to evaluate the logic program with conflict query and get the answer sets of a PB. In this section, we investigate the property of the unknown queries, then adjust the policy base to make it fit into the cases discussed in Section 4, then use the same evaluation approach to process the query.

Let's revisit the Example 2 for *Carl*. In this situation, the policy base does not provide enough information in order to answer query *review*(*Carl*). An intuitive solution is to add rules for further checkup for undecided query. For example, we may add one more rule as:

$$R_9 : furthercheckup(x) \leftarrow review(x), \neg review(x)$$

To better process the evaluation, we add two more rules:

$R_{10} : review(x) \leftarrow not \neg review(x)$
$R_{11} : \neg review(x) \leftarrow not review(x)$

For the rule with disjunction *or* in the head such as R_8, we split the PB into multiple PSs with each contains one literal of the head. Then each of the PB is evaluated separately. A query answers "yes" if it evaluated positive in every answer set.

Definition 4. *A query Q of a policy base D is granted iff for each answer set A_i (i=0,1,2,...) of D, $Q \in A_i$.*

Now we check $review(Carl)$ again. Rules $R_4 - R_7$, and $R_9 - R_{11}$ are used to process the query, obviously we reach $furthercheckup(Carl)$ which is reasonable in commonsense.

For member $Peter$, Rules $R_4 - R_6$, $R_9 - R_{11}$ and $senior(Peter)$ forms a new PB, while $R_4 - R_6$, $R_9 - R_{11}$ and $junior(Peter)$ forms the other new PB. It get answer $furthercheckup(Peter)$.

6 Conclusion

In this paper, we proposed an approach to solve incomplete information reasoning in authorization domains. We employed an extended logic program to answer queries about an authorization domain specified by a logic language. The queries are investigated and classified by conflict and unknown queries. For the conflict queries, we assign each rule a preference ordering, using a fixed point semantics to remove those less preferred rules (the rules will not take effect under current state), then using answer set theory to evaluate the authorization domain to get the preferred authorizations. For the unknown queries, we investigate their properties, justify them to fit into the class of conflict queries, then use the same approach to evaluate them.

In our future work, we will consider the implementation issue with authorization evaluation and query processing. A related work using logic programs for conflict resolution in reasoning has been implemented in [15]. It is our future work to use logic programs(stable model semantics) to implement the approach for incomplete authorization reasoning.

References

1. Apt, K.R., Bol, R.N.: Logic programming and negation: A survey. Journal of Logic Programming 19(20), 9–71 (1994)
2. Bai, Y.: On XML Document Security. In: International Conference on Software Engineering and Data Engineering, pp. 39–42 (2007)
3. Bertino, E., Buccafurri, F., Ferrari, E., Rullo, P.: A Logic-based Approach for Enforcing Access Control. Computer Security 8(2-2), 109–140 (2000)

4. Bertino, E., Catania, B., Ferrari, E., Perlasca, P.: A Logical Framework for Reasoning about Access Control Models. ACM Transactions on Information and System Security 6(1), 71–127 (2003)
5. Bettini, C., Jajodia, S., Wang, X.S., Wijesekera, D.: Provisions and Obligations in Policy Management and Security Applications. In: Proceedings of the Very Large Database Conference, pp. 502–513 (2002)
6. Chomicki, J., Lobo, J., Naqvi, S.: A Logical Programming Approach to Conflict Resolution in Policy Management. In: Proceedings of International Conference on Principles of Knowledge Representation and Reasoning, pp. 121–132 (2000)
7. Crescini, V., Zhang, Y.: A logic Based Approach for Dynamic Access Control. In: Proceedings of 17th Australian Joint Conference on Artificial Intelligence (AI 2004), pp. 623–635 (2004)
8. Damiani, E., Vimercati, S., Paraboschi, S., Samarati, P.: A Fine Grained Access Control System for XML Documents. ACM Transactions on Information and System Security, 160–202 (2002)
9. Gelfond, M., Lifschitz, V.: The stable model semantics for logic programming. In: Proceedings of the Fifth Joint International Conference and Symposium, pp. 1070–1080. MIT Press, Cambridge (1988)
10. Gelfond, M., Lifschitz, V.: Classical negation in logic programs and disjunctive databases. New Generation Computing 9, 365–386 (1991)
11. Jajodia, S., Samarati, P., Sapino, M.L., Subrahmanian, V.S.: Flexible Support for Multiple Access Control Policies. ACM Transactions on Database Systems 29(2), 214–260 (2001)
12. Li, N., Grosof, B., Feigenbaum, J.: Delegation Logic: A Logic-based Approach to Distributed Authorization. ACM Transactions on Information and System Security 6(1), 128–171 (2003)
13. Woo, T.Y.C., Lam, S.S.: Authorization in Distributed systems: A Formal Approach. In: Proceedings of IEEE Symposium on Research in Security and Privacy, pp. 33–50 (1992)
14. Zhang, Y., Bai, Y.: The Characterization on the Uniqueness of the solution $\neg holds(S_1, R, O)$ can be derived. Answer Set for Prioritized Logic Programs. In: Proceedings of the International Symposium on methodologies on Intelligent Systems, pp. 349–356 (2003)
15. Zhang, Y., Wu, C.M., Bai, Y.: Implementing Prioritized Logic Programming. AI Communications 14(4), 183–196 (2001)

A Place Judgment System Based on an Association Mechanism for Natural Language Interaction

Seiji Tsuchiya[1], Eriko Yoshimura[2], Ren Fuji[1,3], Hirokazu Watabe[2], and Tsukasa Kawaoka[2]

[1] Institute of Technology and Science, The University of Tokushima
Minami Josanjima, Tokushima, 770-8506, Japan
[2] Dept. of Knowledge Engineering & Computer Sciences, Doshisha University
Kyo-Tanabe, Kyoto, 610-0394, Japan
[3] School of Information Engineering, Beijing University of Posts and Telecommunications
Beijing, 100876, China
{tsuchiya,ren}@is.tokushima-u.ac.jp,
eyoshimura@indy.doshisha.ac.jp,
{hwatabe,tkawaoka}@mail.doshisha.ac.jp

Abstract. We are conducting research aiming to develop new interfaces that follow the mechanism of human communication, particularly focusing on human common sense. In this paper, we propose a technique which is able to associate the person and thing existing in the place and the event done on the place from the word expressing the place based on an association mechanism. F-mesure of the place subject and place object was 88.0%, and 84.3% respectively. The average of F-mesure of both was 86.2%. This result shows that the place judgment system has achieved a judgment similar human's sense.

Keywords: Place, Common Sense, Concept Base, Degree of Association.

1 Introduction

Humans manipulate smooth communications by retrieving, understanding and judging the characteristics of the place from conversations consciously or unconsciously. For example, it is usual that humans talk to when meeting the acquaintance in a hospital, "Do you visit to patient?" and "Is your condition somewhere bad?", etc. Such the question is an objection originated very naturally. The question can be done because the humans know "A hospital is facilities where a sick person is examined and treated, and provides with a hospitalization facilities" and understand "To Visit a hospitalized patient" and "It is necessary to cure a fault of a condition". In a word, a natural conversation does not consist if "Examination" and "Treatment", "Sickness", "Sick person", "Hospitalization", and "Visit", etc. cannot be associated from "Hospital" even if the place "Hospital" can be specified.

We propose a technique which is able to associate the person and thing existing in the place and the event done on the place from the word expressing the place

I. Lovrek, R.J. Howlett, and L.C. Jain (Eds.): KES 2008, Part I, LNAI 5177, pp. 286–293, 2008.

based on an association mechanism. We think that the proposal technique in this paper greatly contributes to a development of a robot which can make smooth conversation with human beings.

2 Place Judgment System

Figure 1 shows the structure of a place judgment system. The place judgment system consists of a knowledge base that contains the relationships between the nouns expressing the place and their relevant persons, things and events (hereinafter referred to as the "place judgment knowledge base") and an unknown word processing method that processes unknown words that do not exist in the place judgment knowledge base as known words by relating the unknown words to the known words that exist in the place judgment knowledge base. The unknown word processing realizes the association of the words, processes the words using the Concept Base [1, 2], which is a large database generated automatically from multiple electronic dictionaries, and the calculates the Degree of Association method [3], which evaluates the relationships between words (hereinafter these two methods are collectively referred to as the "Association Mechanism"). In this paper, nouns expressing the place express concrete place like "Hospital" and "Shrine".

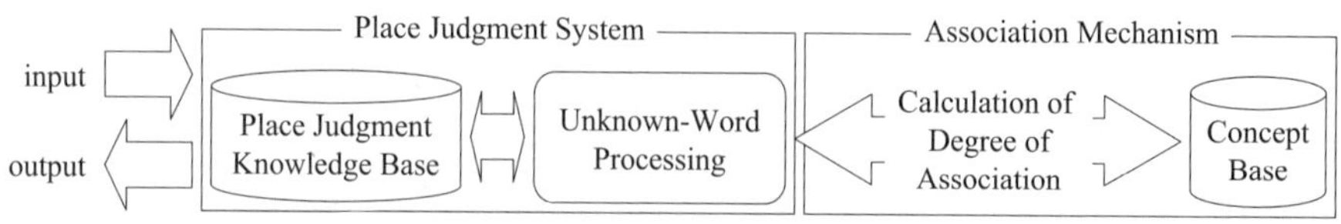

Fig. 1. Structure of a place judgment system

3 Place Judgment Knowledge Base

To realize place judgment, knowledge that relates the noun expressing the place (hereinafter referred to as the "place word") to its relevant persons, things (hereinafter referred to as the "place subject") and events (hereinafter referred to as the "place object") is required. However, it is difficult and inefficient to define and register such relations for all nouns. Therefore, the nouns used often in everyday life were selected as representative words, and these words were related to their retrieved the persons, things and events and registered in the place judgment knowledge base. Table 1 shows an example of the place word, place subject and place object.

The place judgment knowledge base was created based on an existing thesaurus [4] using a tree structure to represent its knowledge efficiently. The tree structure of this thesaurus describes the upper/lower level relationship and the whole/part relationship of the semantic attributes (nodes) of 2710 words that represent semantic usages of general nouns. The above-mentioned representative words are the leaves or nodes of the place judgment knowledge base (thesaurus). Especially, the representative words which are the nodes of the thesaurus are

Table 1. Example of a place word, place subject and place object

Place word	Place subject	Place object
Hospital	Doctor , Sick person, ...	Examination, Admission, ...
School	Teacher , Student, ...	Education, Study, ...

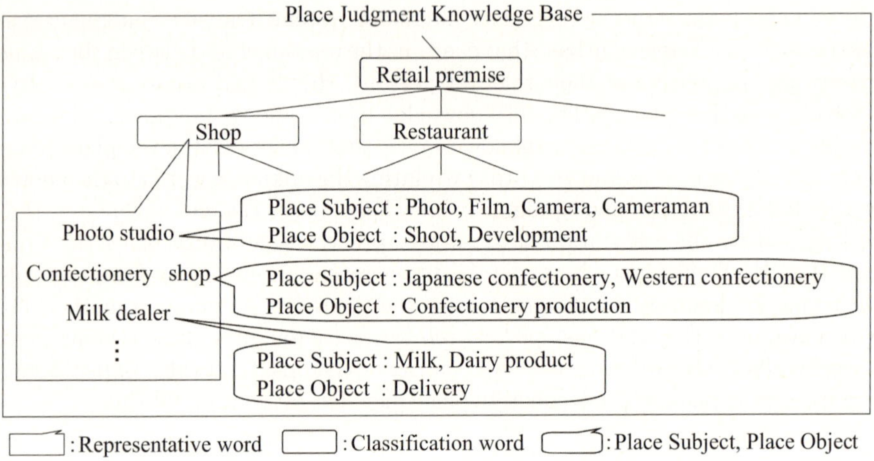

Fig. 2. Example of the place judgment knowledge base

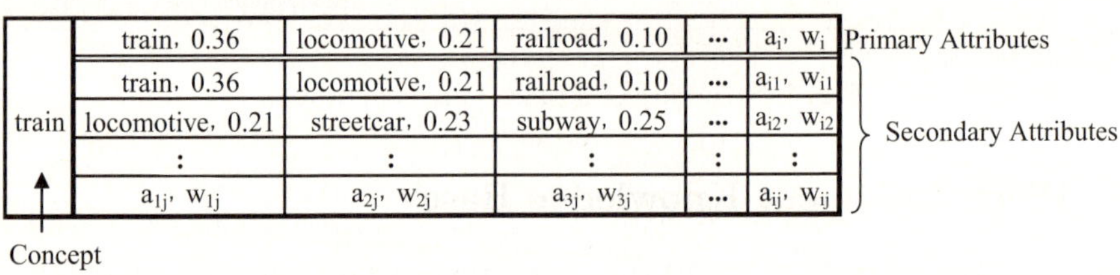

Fig. 3. Example of the concept gtrainh expanded as far as Secondary Attributes

called classification words. The representative words of 443 words (the classification words are 120 words) are registered in the place judgment knowledge base. Figure 3 shows an example of the place judgment knowledge base.

4 Concept Base and Degree of Association Algorithm

4.1 Concept Base

The Concept Base is a large-scale database that is constructed both manually and automatically using words from multiple electronic dictionaries as concepts and independent words in the explanations under the entry words as concept attributes. In the present research, a Concept Base containing approximately 90,000 concepts was used, in which auto-refining processing was carried out after the base had been manually constructed. In this processing, attributes considered inappropriate from the standpoint of human sensibility were deleted and necessary attributes were added.

In the Concept Base, Concept A is expressed by Attributes a_i indicating the features and meaning of the concept in relation to a Weight w_i denoting how important an Attribute a_i is in expressing the meaning of Concept A. Assuming that the number of attributes of Concept A is N, Concept A is expressed as indicated below. Here, the Attributes a_i are called Primary Attributes of Concept A.

$$A = \{(a_1, w_1), (a_2, w_2), \cdots, (a_N, w_N)\}$$

Because Primary Attributes a_i of Concept A are taken as the concepts defined in the Concept Base, attributes can be similarly elucidated from a_i. The Attributes a_{ij} of a_i are called Secondary Attributes of Concept A. Figure 2 shows the elements of the Concept "train" expanded as far as Secondary Attributes.

4.2 Degree of Assosiation Algorithm

For Concepts A and B with Primary Attributes a_i and b_i and Weights u_i and v_j, if the numbers of attributes are L and M, respectively $(L \leq M)$, the concepts can be expressed as follows:

$$A = ((a_1, u_1), (a_2, u_2), \cdots, (a_L, u_L))$$
$$B = ((b_1, v_1), (b_2, v_2), \cdots, (b_M, v_M))$$

The Degree of Identity $I(A, B)$ between Concepts A and B is defined as follows (the sum of the weights of the various concepts is normalized to 1):

$$I(A, B) = \sum_{a_i = b_j} \min(u_i, v_j)$$

The Degree of Association is calculated by calculating the Degree of Identity for all of the targeted Primary Attribute combinations and then determining the correspondence between Primary Attributes. Specifically, priority is given to determining the correspondence between matching Primary Attributes. For Primary Attributes that do not match, the correspondence between Primary Attributes is determined so as to maximize the total degree of matching. Using the degree of matching, it is possible to give consideration to the Degree of Association even for Primary Attributes that do not match perfectly.

When the correspondences are thus determined, the Degree of Association $R(A, B)$ between Concepts A and B is as follows:

$$R(A, B) = \sum_{i=1}^{L} I(a_i, b_{xi})(u_i + v_{xi}) \times \{\min(u_i, v_{xi}) / \max(u_i, v_{xi})\}/2$$

In other words, the Degree of Association is proportional to the Degree of Identity of the corresponding Primary Attributes, and the average of the weights of those attributes and the weight ratios.

5 Unknown Word Processing

It is easy to judge the relating place subjects and place objects from the known word of the place word. Because, place subjects and place objects which relate to a known word are registered in the place judgment knowledge base.

When the place word is an unknown word, the classification word with a strong relation is judged by a method described in paragraph 5.1. As a result, place subjects and place objects related to a classification word can be used. However, because place subjects and place objects related to the classification word are inherited to the node, those ranges of an expression are very large. Therefore, specific place subjects and place objects relevant to an unknown word cannot be correctly judged only from the use of them.

Then, specific place subjects and place objects to an unknown word are judged by a method which is described in paragraph 5.2 using attributes of the Concept Base.

5.1 Judgment of the Classification Words

In the unknown word processing, the classification word with the extremely strong semantic relation is decided using the Degree of Association. However, it cannot be judged that the classification word with a high Degree of Association is the place word related with an unknown word. For example, when the relations between an unknown word "Doctor" and classification words are calculated, the Degree of Association of a classification word "Hospital" is highest. Consequently, unknown word "Doctor" is judged to be a place word by mistake.

Then, the Degree of Association between all nodes of the thesaurus used when the place judgment knowledge base is constructed (2710 words) and an unknown word is calculated. And, it can be judged that an unknown word is a place word when the word with an extremely strong semantic relation is a classification word registered in the place judgment knowledge base. As a result, in the above-mentioned example, the Degree of Association of a node "Doctor" is higher than a classification word "Hospital" when the relations between all nodes of the thesaurus and an unknown word is calculated.

In addition, the threshold of Degree of Association used for the judgment is 0.23. This threshold is a value experimentally calculated.

5.2 Judgment of the Place Subject and the Place Object

Specific place subjects and place objects relevant to an unknown word are judged by using place subjects and place objects relevant to a classification word and an unknown word's attributes expanded from the Concept Base.

Specific place subjects relevant to an unknown word are decided by the following conditions.

(1) The word is a place subject relevant to a classification word.
(2) The word is the same as a first attribute of an unknown word.

(3) The first attribute has the Degree of Association of 0.18 or more. The Degree of Association is value between the first attribute and the unknown word.

In addition, specific place subjects relevant to an unknown word are decided by the following conditions, too.

(1) The first attribute of an unknown word has the Degree of Association of 0.62 or more. The Degree of Association is value between the first attribute and the specific place subject.

Specific place objects relevant to an unknown word are decided by the same method as the above-mentioned. However, the threshold of Degree of Association is 0.13. In addition, these thresholds are values experimentally calculated.

6 Evaluation of the Place Judgment System

6.1 Method of Evaluation

Specific place subjects and place objects of an unknown word are judged by "degree of common sense" and "share of mind". These are evaluation indexes which judge proximity of humans' sense. The degree of common sense and accuracy rate are the same meanings. And, The share of mind and recall rate are the same meanings. Therefore, a pseudo F-measure is introduced as an overall evaluation index. F-measure is defined by the following expressions.

$$\text{F-measure} = (2 * \text{degree of common sense} * \text{share of mind}) / (\text{degree of common sense} + \text{share of mind})$$

The degree of common sense. The evaluation to place subjects and place objects judged to be related to an unknown word is done at the following three levels.

(A) "common sense" (correct answer)
(B) "not out-of-common-sense"
(C) "out of common sense" (an error)

For example, when an unknown word is "Police", place subjects and place objects are judged as follows: "common sense" is place subject "Policemen" and place object "Investigation"; "not out-of-common-sense" is place subject "Instructor" and place object "Education"; "out of common sense" is place subject "Doctor" and place object "Camping".

The above-mentioned evaluation is done by three test subjects' decision by majority. If the judgments of three test subjects are different, the evaluation is judged "not out-of-common-sense". The degree of common sense is calculated from the evaluation result as follows.

$$\text{Degree of common sense} = (\text{No. of (A)} + \text{No. of (B)}) / (\text{No. of (A)} + \text{No. of (B)} + \text{No. of (C)})$$

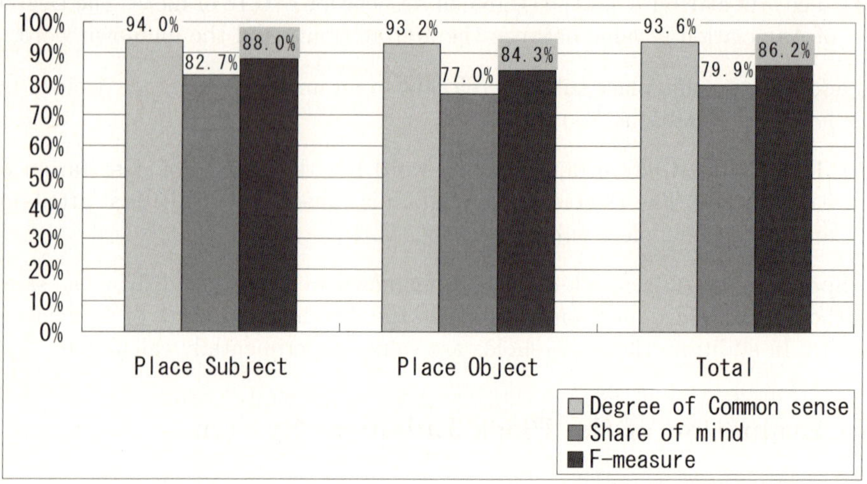

Fig. 4. Result of the place judgment system

Share of mind. The share of mind is calculated based on the number of place subjects and place objects which human can judge from a place word. The 150 words used in everyday life which are unknown words were used for the investigation of the number which human was able to judge. On the average of the number of words which three test subjects were able to judge, place subjects were 11.4 words, and place objects were 5.8 words. The share of mind is calculated based on the above-mentioned result as follows.

$$\text{Share of mind} = \text{Ave. of (No. of (A) + No. of (B))} / \\ \text{Ave. of No. of words which human can judge.}$$

6.2 Evaluation Result

Place words of 300 words collected by the questionnaire were used as evaluation data to evaluate the effectiveness of the place judgment system. In the breakdown, the representative words were 75 words, and the unknown words were 225 words. Figure 4 shows the result of the place judgment system.

F-mesure of the place subject and place object is 88.0%, and 84.3% respectively.In the breakdown, the degree of common sense is 94.0%, and 93.2% respectively. And, the share of mind is 82.7%, and 77.0% respectively. This result shows that the place judgment system has achieved a judgment similar human's sense.

7 Conclusion

In this paper, we proposed the technique which was able to associate the person and thing existing in the place and the event done on the place from the word expressing the place based on an association mechanism for natural language interaction.

F-mesure of the place subject and place object was 88.0%, and 84.3% respectively. The average of F-mesure of both was 86.2%. This result shows that the place judgment system has achieved a judgment similar human's sense.

References

[1] Hirose, T., Watabe, H., Kawaoka, T.: Automatic Refinement Method of Concept-base Considering the Rule between Concepts and Frequency of Appearance as an Attribute. Technical Report of the Institute of Electronics, Information and Communication Engineers. NLC 2001-93 109–116 (2002)

[2] Kojima, K., Watabe, H., Kawaoka, T.: A Method of a Concept-base Construction for an Association System: Deciding Attribute Weights Based on the Degree of Attribute Reliability. Journal of Natural Language Processing 9(5), 93–110 (2002)

[3] Watabe, H., Kawaoka, T.: Measuring Degree of Association between Concepts for Commonsense judgments. Journal of Natural Language Processing 8(2), 39–54 (2001)

[4] NTT Communication Science Laboratory: NIHONGOGOITAIKEI. Iwanami Shoten (1997)

Knowledge Maturity Based on Processing and Creation of Information

Sabah S. Al-Fedaghi

Computer Engineering Department
Kuwait University
P.O. Box 5969 Safat 13050
Kuwait
sabah@alfedaghi.com

Abstract. This paper investigates the problem of how to model knowledge maturity (KM). It utilizes a flow model that includes five information flow stages: receiving, processing, creation, transfer, and communication. Knowledge is carried in the stream of information flow with continuing improved "maturity." It is proposed that knowledge is the result of cycling between the processing and creation stages. Hence, KM is defined in terms of the frequency of cycling between these two stages. Theoretical applications of this approach are also discussed.

1 Introduction

Recently, a new flow model has been introduced and applied in several fields such as information security and supply chains [1], [2]. It includes five information flow stages: receiving, processing, creation, transfer, and communication. The new contribution in this paper is to adopt such a model in the area of information maturity.

Accordingly, knowledge is defined as new information generated from current information. KM is measured by the frequency of cycling between the processing and creation stages, where new information participates in the creation of newer information. In the next section we review the flow model and some of its features.

2 The Flow Model

A flow model is a uniform method to represent things that "flow," i.e., are exchanged, processed, created, transferred, and communicated. "Things that flow" include information, materials (e.g., manufacturing), money, etc. The notion of flow is a widely used concept in many fields of study. In economics, the goods circular flow model is well known; in management science there is the supply chain flow, etc. In computer science, the classical model of flow is the 1949 Shannon-Weaver communication model representing electrical signal transfer from sender to receiver. It reflects the concept of "flow" in terms of three stages: information being transmitted, information in the channel, and information being received. Flow of information means the movement from one information sphere (the sender) to another information sphere.

I. Lovrek, R.J. Howlett, and L.C. Jain (Eds.): KES 2008, Part I, LNAI 5177, pp. 294–303, 2008.

In the flow model (FM) approach, the concept of flow is understood in a more comprehensive way. An FM has a number of different components with a spatial assembly of these components relative to each other and to time, and links between the components that indicate flow of items. To simplify the review of FM, we introduce flow in terms of *information* flow.

The lifecycle of information is a sequence of states as information moves through stages of its lifecycle, as follows:

1. Information is received (i.e., it has arrived at a new sphere, analogous to passengers arriving at an airport).
2. Information is processed (i.e., it is subjected to some type of process, e.g., compressed, translated, mined).
3. Information is disclosed/released (i.e., it is designated as released information, ready to move outside the current sphere, analogous to passengers ready to depart an airport).
4. Information is transmitted (e.g., from a customer's sphere to a retailer's sphere).
5. Information is created (i.e., it is generated as a new piece of information using different methods such as data mining).
6. Information is used (i.e., it is utilized in some action, analogous to police rushing to a criminal's hideout after receiving an informant's tip).
7. Information is stored. Thus, it remains in a stable state without change until it is brought back to the stream of flow again.
8. Information is destroyed.

The first five states of information form the main stages of the stream of flow, as illustrated in Figure 1. For example, *storing* information is a *sub-state* because it occurs while it is created (stored created information), processed (stored processed information), and received (stored received information). The five stages scheme can be applied to humans and organizations. It is reusable, because a copy of it is assigned to each agent. Suppose that a small organization has a manager and a secretary. Additionally, it has two departments with two employees in each department. It then comprises nine information schemes, one for the organization at large, one for the manager, one for the secretary, one for department 1, and so forth (Figure 2; dotted lines denote communication).

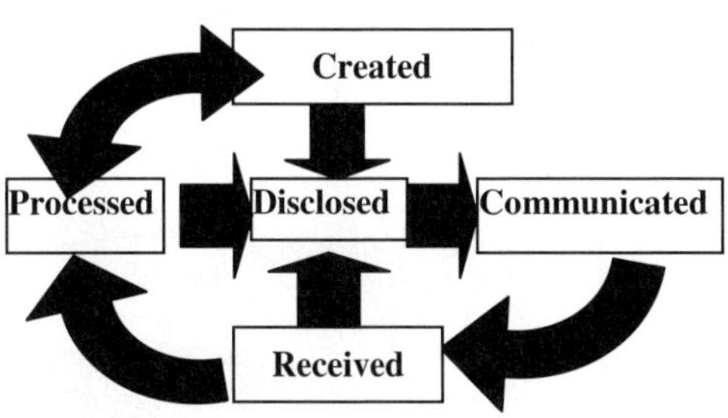

Fig. 1. Stages of flow of information

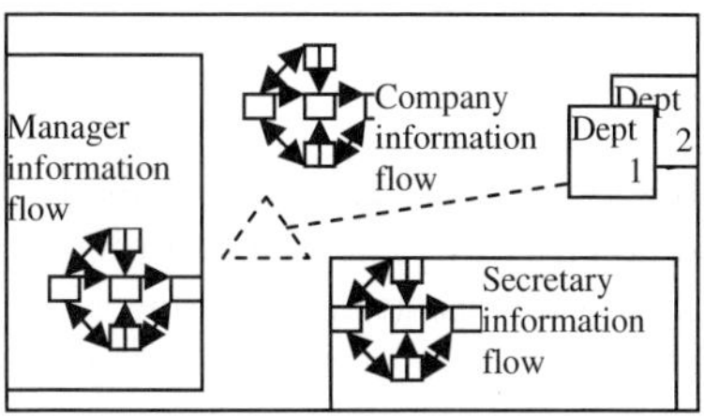

Fig. 2. Each sphere has its own copy of the five stages scheme

The five information states are the only possible "existence" patterns in the stream of information. We can start at any point in the stream and follow the flow of information in different flow paths. Suppose that information enters the processing stage, where it is subjected to some process. The following are ultimate possibilities:

1. It is stored.
2. It is destroyed.
3. It is disclosed and transmitted to another sphere.
4. It is processed in such a way that it generates implied information (e.g., *a is the father of b and b is the father of c* generates the information that *a is the grandfather of c*).
5. It is processed in such a way that it generates new information (e.g., comparing certain statistics generates the new information that *Smith is a risk*).
6. It is *used* to generate some action (e.g., upon decoding or processing the information, the FBI sends its agents to arrest the spy who wrote the encoded message). In the *used* sub-stage, information is not a patient. The *patient* is a term that refers to the thing that receives the action.

The storage and uses/actions sub-stages can be found in any of the five stages. However, in the release and transfer stages, information is not usually subject to these sub-stages, so we apply these sub-stages only to the receiving, processing, and creation stages, without loss of generality. Figure 3 shows the interiors of these three stages.

The "storage" in each stage represents a static state of information. Thus, when information is received, it may be stored in its received condition for a later time when it is activated by being brought into the flow stream.

The "black holes" in Figure 3 represent the destruction of information in the corresponding stage. Implicit in Figure 3 is the fact that information may be duplicated through copying. In addition, the "Thinking" rectangle in the processing stage represents processes that generate implied (curved arrow) and new information (arrow to the creation stage).

Figure 3 is a detailed version of Figure 1, in which the receiving stage leads to the processing stage, which in turn leads to the creation stage. The creation

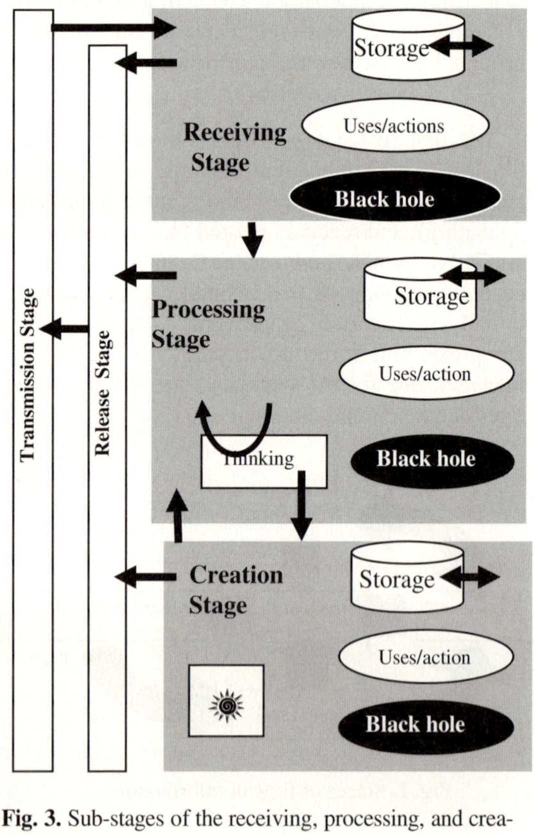

Fig. 3. Sub-stages of the receiving, processing, and creation stages

stage may lead back to the processing stage. These three stages may lead directly to the disclosure/release stage, then to the transmission stage, which in turn leads to the receiving stage of *another sphere*.

The "sun" in the creation stage denotes information that has not been generated without using other information. This internal information is produced within. The outside information has not participated directly in the creation of this information.

3 Motivating Examples

The so-called *knowledge maturing process* provides a conceptual framework for the design of the required integrating processes in organizations [4]. Maier and Schmidt (2007) proposed "a systematic characterization of the knowledge maturing process, … for explaining integration barriers between the different disciplines concerned with learning in organizations" to the "quality" of knowledge and "consolidating knowledge," or "putting it into the context of a bigger whole."

Maier and Schmidt's structuring of the knowledge maturity process has five phases that can be explained in the context of the FM as follows.

Emergence of Ideas. This emergence occurs in the processing and creation stages of FM. As we will see later, the "informal discussion" among "individuals" mentioned by Maier and Schmidt (2007) is communicating information among several FMs, each representing an individual.

Distribution in Communities. The development of common shared terminology also emerges in the processing or creation stages of FM after the information carrying knowledge is circulated in the flow of FM.

Formalization. According to Maier and Schmidt (2007), "Artefacts [are] *created* in the preceding two phases." Also, "in this phase, purpose-driven structured documents are created, e.g., project reports or design documents." Such creation of artifacts and documents is performed in the creation stage of FM. However, FM distinguishes between "pure creation" of information (knowledge), which has never existed before, and mere processing that changes the form of information or generates implied information. So, it seems that what Maier and Schmidt (2007) are describing as "project reports" belong to the processing stage, not to the creation stage of FM.

Ad-Hoc-Training. This phase seems to be performed in the processing stage of the FM, where information is formulated in "a pedagogically sound way, enabling broader dissemination" [4].

Formal Training. This phase also seems to be performed in the processing stage of the FM; however, more information from other resources is brought into the formulation of materials that have become "teachable to novices" [4].

We observe that this "knowledge maturing process" is in fact a sequence of phases of information creation and processing. It does not distinguish between processing and creation of information (knowledge); in addition, the phases are cut off from the knowledge/information flow that includes other FM stages.

The Maier and Schmidt model does not reflect the evolution of knowledge maturity; rather, it describes steps of evolving knowledge to become teachable materials.

This is of course understandable in the context of learning that Maier and Schmidt (2007) are trying to integrate. However, the problem being considered by this paper is to develop a more general flow model for information/knowledge maturity, which will be defined later. Knowledge matures not only for the purpose of teachability, but also to be applied to all information processing activities in organizations.

In the context of maturing information/knowledge for teachability, FM, with its circulation of streams of information, provides a better foundation than Maier and Schmidt's model. As we have shown, the phases of that model can be interpreted in FM. Producing "mature" teaching materials is the result of information/knowledge circulation in FM, which refines the knowledge with more processing that produces new information from current information.

Maier and Schmidt (2007) also adopt a model of "organizational knowledge processing" [5]. Because of space limitation, we highlight only the model elements where numbers and descriptions refer to knowledge: (1) acquisition, (2) identification, (3) individual learning, (4) sharing, (5) institutionalization (6) knowledge (theories) in use, (7) feedback, (8) refining and repackaging, (9) dissemination to the environment, and (10) services offered to employees.

The general phases of the model can be scrutinized as before. In general, while these models present useful tools for analyzing knowledge maturity, we propose to formalize them as FM models. In this paper, instead of attempting such a venture, directed at specific proposed models, we target the general problem of analyzing the notion of knowledge maturity in terms of the flow model.

4 Knowledge Maturity Definition

The notion of Knowledge Maturity was derived from the basics of management strategies aimed at incorporating awareness of quality in all organizational processes (e.g., Total Quality Management). It has been applied in many fields; for example, in e-learning, KM refers to quality improvement based on several ideas such as software process improvement and capability methodologies. KM models are usually introduced in terms of processing phases, as described in the previous section. "A knowledge maturity model defines stages of maturity that an organization can expect to pass through in its road to improve its overall knowledge-centric practices and processes and ultimately business performance" [6].

One of the uses of KM is in the area of decision making and assessment of process. Decision makers can verify the quality of development of their knowledge base in order to move forward to the next step.

We approach KM in a stricter way, as follows:

- Knowledge is defined as new information generated from available information. For example, in any deduction process, the conclusion based on premises is knowledge.
- Knowledge is produced from knowledge. Thus, current knowledge serves as information to generate newer knowledge.
- Maturity of knowledge is a *measure* of the amount of processing that produces knowledge from information.

Figure 4 illustrates this relationship between processing and creation of information. In FM, producing knowledge from information is the product of the processing stage followed by the creation stage. Notice the cycle between these two stages in Figures 1 and 2. It can be repeated over and over again, where previously produced knowledge becomes the fuel for generating new knowledge.

The definition of KM as a *measure* of the amount of processing that produces knowledge from information has an application-neutral definition based only on the notions of knowledge and information. It can provide a foundation for applying KM in any field such as management and e-learning.

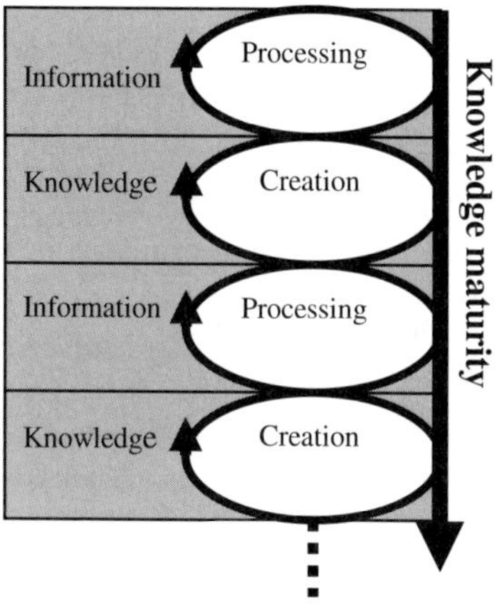

Fig. 4. KM increases with the recurring generation of knowledge from information

5 Knowledge Maturing Process

"Information flow" interconnects information processing spheres of information-processing entities, including persons, organizations, and sub-organizations, and carries with it processed and created knowledge/information. Knowledge in such flow can metaphorically be compared with the *cream or butter* produced from the flow of milk. Knowledge is carried in the stream of information flow with continuing improved "quality" and "maturity." Hence, we can speak of flow of knowledge as a feature of circulating flow of information among different FM stages and FM schemes of informational spheres.

Knowledge is carried by the information flow. Pieces of knowledge are represented by the corresponding new pieces of information. The FM processing and creation stages are the main sources of producing knowledge. The first signs of (new) knowledge emergence from "knowledge springs" appear in these two stages. The new knowledge then pours into the stream of information flow as "valuable information" (metaphorically, *cream of the flow*). The creation and processing stages are the hubs of discovering and developing new knowledge. This is applied to informational spheres of human beings as well as to organizations.

With this framework of flow of information/knowledge, our problem becomes developing a better understanding of how and when to create knowledge. To accomplish this, we concentrate on a measure of "how much processing," f, has generated knowledge. In Figure 4, this measure reflects how far we have progressed in the recurring Processing (information) and Creation (knowledge) Cycle (PCC). The PCC frequency, denoted as f, can be measured in terms of the number of circulations between the processing that resulted in the creation of a particular piece of knowledge.

Let X be a set of medical data. Suppose that a doctor examined (processed) the data twice to reach (create) the medical judgment x′. x′ (e.g., *the patient has a particular disease*) is knowledge or new information. In this case $f = 2$, as illustrated in 5. The PCS with $f = 2$ can be written as: $X \rightarrow \{X \cup x'\} \rightarrow x'$. That is, the doctor concluded $x' \notin X$, then he or she re-examined his or her judgment and reached the same conclusion, x'. Notice that f reflects the deduction ($\rightarrow$) that produces new information.

Alternatively, ($f = 2$) can be achieved through the following: $X \rightarrow \{X \cup x''\} \rightarrow x'$. That is, the doctor concluded x'', then he or she re-examined his or her judgment x'' and changed it to the final judgment x'.

In both cases the PCC frequency f is 2. This process is analogous to a person who analyzes information to reach a conclusion, then returns to the analysis stage to analyze information that now includes the conclusion, to reach a new conclusion, and so forth. Kurzweil (2007) characterizes this process in the phrase, "knowledge builds on previous knowledge."

There are two ways to implement $f > 1$.

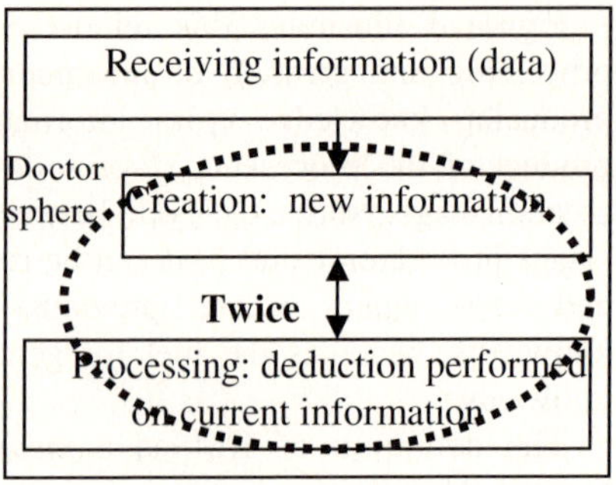

Fig. 5. Internal $f = 2$, where PCC (dotted ellipse) is performed twice

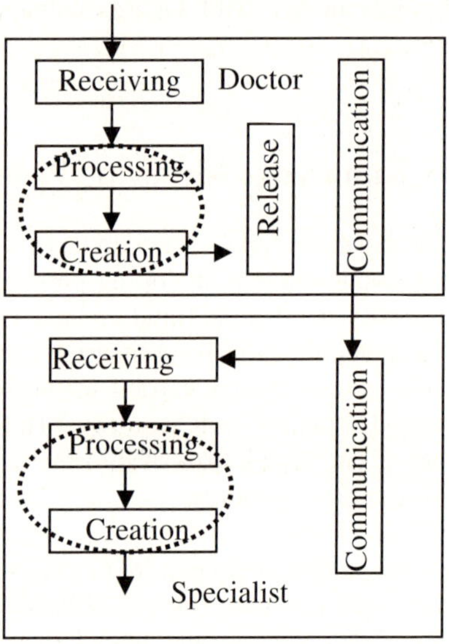

Fig. 6. Example of PCC of $f = 2$

1. The method illustrated in Figure 5 shows that repeated PCC is applied twice. It is possible to increase the number of iterations between the processing and creation stages by the same agent. Accountants usually repeat the accounting process several times, starting from the beginning, to be certain about the results.
2. Achieving $f > 1$ can be accomplished through use of several agents that perform the PCC using the same initial information. Figure 6 illustrates this case for the doctor example of Figure 5. Instead of examining the medical data twice, the doctor sends it to a specialist who processes it and makes a judgment. Assuming the doctor and the specialist examined the data once, we have in this case $f = 2$.

6 Application to Decision Making

Kurzweil (2007) proposed a model for "a relationship between knowledge (amount of knowledge and how 'mature' or 'truthful' it is), uncertainty (how much is not known and how variable the known knowledge is) and time," as shown in Figure 7. "The more time spent learning about the future, the more knowledge is acquired and the more certain it becomes." In terms of decision making, Kurzweil continues, "The relationship between the amount of knowledge, certainty and time is not linear since knowledge builds on previous knowledge, but the 'knowledge maturity surface' gives a general idea of the decision and action space that is inherent to the front-end of the innovation process and the strategy creation process" [3].

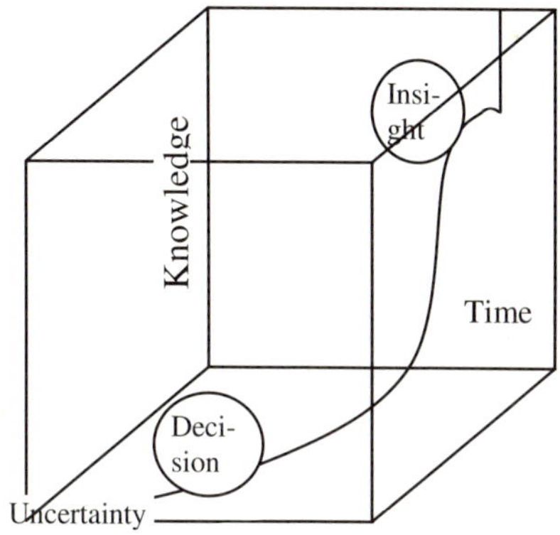

Fig. 7. Kurzweil's model

We propose a similar model, where time is replaced by a scale of the minimum time to finish one PCC, denoted as MinTime(PCC). Kurzweil's time is not an accurate indicator of the maturity of knowledge. Information may not evolve into knowledge because it is stored in one of the FM stages. Even when we have time, we may not increase f. Factor f is an indicator of the amount of effort to "maturitize" knowledge, and its value is a multiple of MinTime(PCC).

Suppose X is the initial set of information received to produce a decision. MinTime(PCC) is the minimum time necessary to perform one processing/creation cycle to reach (create) a decision (knowledge). The horizontal axis in Kurzweil's model (Figure 7) represents MinTime(PCC) units.

Example: Imagine the organizational structure of a company that has several departments that in turn include sections. A decision about a project is first made at the section level, then the department reviews the project and the section's decision. At the organization level, a final decision is made after reviewing the department decision. Suppose that M1 = MinTime(PCC)_1, M2 = MinTime(PCC)_2, and M3 = MinTime(PCC)_3 are the minimum times of the section, department, and organization, respectively. Ignoring communication time, the required minimum time is Total = M1 + M2 + M3.

The deadline for making a decision (D) must equal to or grater than Total, otherwise one of the organizational units would not be able to complete one PCC. Notice that $f > 1$ does not mean that $f >$ MinTime(PCC). For example, if the section assigns analysis of the proposed project to two different teams, then $f = 2$, while the time required for making a decision is (one) MinTime(PCC), because the teams work parallel to each other. If Total is less than D, then f may be increased in any of the organizational units without using parallel PCCs. Since we surmise that the maturity (and certainty) of knowledge increases with an increase of the number of PCC, then we have to increase the units of MinTime(PCC).

Let M1 = 1, M2 = 1, M3 = 1, D = 5, where each number denotes a week. For example, M1 = 1 means that the section needs one week to complete one processing/creation sequence to reach a decision about the project under consideration. The difference between Total and D (the deadline to reach a final decision) is 2. Strategies to make a decision in this particular project involve the "distribution" of the extra two PCCs (or weeks) among the organizational units. For example, it may be decided to adopt the strategy shown in Figure 8.

The first row in the figure is the horizontal axis in Kurzweil's model. It replaces "Time" in Figure 7. Each slot is the minimum time to complete one PCC. If the organization needs two weeks to complete a PCC, then slot 5 is equal to two weeks, while the slots of the section (1, 2) and department (3, 4) each equal one week.

PCC	1	2	3	4	5
Section	x	x/y			
Department			x	x	
Organization					x/y

Fig. 8. Assigning PCCs to units. x, and y indicate that the PCC is performed by two different agents

What is the advantage of such an approach? It is a first model for using knowledge (e.g., strategies for decisions) based on the maturity of such knowledge (e.g., *f*). The model is uniformly applied to different levels of organization, including humans. Knowledge maturity is now a well-understood notion. If we want to increase the maturity of the knowledge, increase *f*s, the frequency of PCC. This is a commonsense notion: rethinking a decision makes it a more mature judgment.

7 General Application

FM can be applied to the general situation of development of a project such as production of materials for education. Instead (maybe alongside) of Maier and Schmidt's structuring of the knowledge maturity process into the five phases: emergence of ideas, distribution in communities, formalization, ad-hoc-training, and formal training, we use the five FM stages. The FM stages of receiving, processing, creation, transfer, and communication are generic stages that can be applied to represent the maturity of progress in any project-building operation.

Suppose six people are involved in the project. Figure 9(a) assigns an FM five-stages assembly to each of them. The dotted lines represent the informal communication among them. Knowledge maturity is attached to each of them in terms of the thinking and rethinking of the project. This phase includes Maier and Schmidt's (2007) two phases of emergence of ideas and

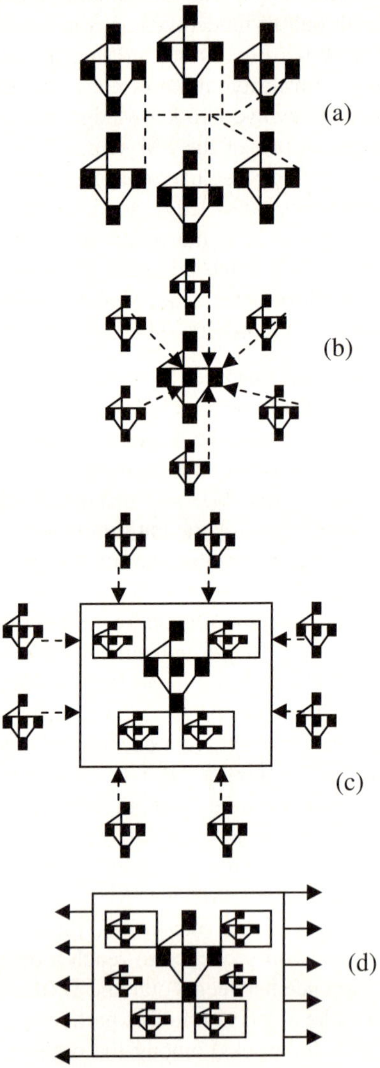

Fig. 9. Project development stages

distribution in communities. Figure 9(b) reflects the project emergence (the FM five stages assembly in the middle) as an independent informational sphere. The arrows indicate the six project builders "inputting" their maturing knowledge into the project. The project now has a formal representation, where "it" can receive, process, create, transfer, and communicate knowledge and information. In Figure 9(c) the project has matured enough to have its own sub-spheres, as shown in the box. The six original project builders continue pouring knowledge into the project, possibly joined by others (thus eight FM assemblies shown in the figure). The maturity of completion of the project and knowledge is reflected in progress from Figure 9(a) through 9(d). At (e) the project is complete and serves (outward arrows) its purpose.

This description of the maturity of development of a project using FM is very general. It can be used as a blueprint to formalize a specific application such as knowledge maturity in e-learning. Space limitation does not allow further discussion of this topic.

8 Conclusion

Applying a flow model to the notion of information maturity introduces a promising approach that clarifies several concepts in the field of KM. Information flow forms a basis for defining KM as new information generated from available information.

References

1. Al-Fedaghi, S., Al-Saqabi, K., Thalheim, B.: Information Stream Based Model for Organizing Security. In: Symposium on Requirements Engineering for Information Security, Barcelona, Spain, March 4th-7th (2008)
2. Al-Fedaghi, S.: Some Aspects of Personal Information Theory. In: 7th Annual IEEE Information Assurance Workshop, United States Military Academy, West Point, N. Y (2006)
3. Kurzweil, R.: Time Horizons, Knowledge Maturity, and Strategy, Inovo: The process of Innovation (December 2007), http://inovo-innovation.wetpaint.com/page/Time+Horizons,+Knowledge+Maturity,+and+Strategy?t=anon
4. Maier, R., Schmidt, A.: Characterizing Knowledge Maturing: A Conceptual Process Model for Integrating E-Learning and Knowledge Management. In: 4th Conference Professional Knowledge Management, Potsdam, Germany (2007)
5. Maier, R.: Knowledge Management Systems. Information and Communication Technologies, 2nd edn., Berlin (2004)
6. The Knowledge Company, Knowledge Optimization Services: KM Maturity Model Services – KMmm (2006), Accessed January 2008, http://www.knowledgecompanyinc.com/images/TKCI_-_KM_Maturity_Model_-_0307c.pdf

On Attaining Higher Level of Certainty in Evaluation Processes

Sylvia Encheva[1] and Sharil Tumin[2]

[1] Stord/Haugesund University College, Bjørnsonsg. 45, 5528 Haugesund, Norway
sbe@hsh.no
[2] University of Bergen, IT-Dept., P. O. Box 7800, 5020 Bergen, Norway
edpst@it.uib.no

Abstract. Reuse of electronic learning materials has been of interest since 1970s. Most of the work was focused on how to define, develop and make reusable learning objects accessible by different learning systems. However, very little has been done with respect to automated evaluation of quality and appropriativeness of such objects. To attain a higher level of certainty in the evaluation process we propose application of many-valued logic.

Keywords: many-valued logic, reusable learning objects.

1 Introduction

The reuse of electronic learning materials has been of interest since 1970s. In [7] a learning object is defined as any entity, digital or non-digital, that may be used for learning, education or training. Somewhat different definitions can be found in [11] and [15]. A variety of learning technology systems and interoperability standards providing reuse of learning objects (LOs) and interoperability of content across delivery have been developed ([8], [10], and [14]). A formal model for selecting LOs based on student's individual learning preferences was presented in [3].

Consider a federated learning system providing a large number of LOs upon a lecturer request. Both the quality and effectivity of services can be considerably improved if each suggested LO has been first evaluated by experts. In this paper we focus on a decision making process facilitating automated evaluation of LOs done by several experts. A large number of experts will certainly contribute to the quality of the decision making process. However in practice it is quite difficult to find many experts who are willing to devote time and efforts to such evaluations. Another difficulty is related to the rapidly increasing number of possible responses when more experts are involved.

We first present a model based on responces from three experts. Afterwards we show how the responses of a larger number of experts can be handled by the same system.

I. Lovrek, R.J. Howlett, and L.C. Jain (Eds.): KES 2008, Part I, LNAI 5177, pp. 304–311, 2008.

The rest of the paper is organized as follows. Related work, basic terms and concepts are presented in Section 2. The model is described in Section 3. The paper ends with a description of the system in Section 4 and a conclusion in Section 5.

2 Background

The five-valued logic introduced in [4] is based on the following truth values: *uu* - unknown or undefined, *kk* - possibly known but consistent, *ff* - false, *tt* - true, *ii* - inconsistent.

A truth table for the ontological operation ∨ is presented in Table 1.

Table 1. Truth table for the ontological operation ∨ in five-valued logic

∨	uu	kk	ff	tt	ii
uu	uu	kk	uu	tt	ii
kk	kk	kk	kk	tt	kk
ff	uu	kk	ff	tt	ii
tt	uu	tt	tt	tt	tt
ii	ii	kk	ii	tt	ii

The seven-valued logic presented in [5] is based on the following truth values: *uu* - unknown or undefined, *kk* - possibly known but consistent, *ff* - false, *tt* - true, *ii* - inconsistent, *it* - non-false, and *if* - non-true. A truth table for the ontological operation ∨ in seven-valued logic is presented in Table 2.

Table 2. Truth table for the ontological operation ∨ in seven-valued logic

∨	uu	kk	fi	ff	ii	tt	it
uu	uu	uu	fi	ff	fi	uu	uu
kk	uu	kk	fi	ff	fi	kk	uu
fi	fi	fi	fi	ff	fi	fi	fi
ff	ff	ff	ff	ff	ff	ff	ff
ii	fi	fi	fi	ff	ii	ii	ii
tt	uu	kk	fi	ff	ii	tt	it
it	uu	uu	fi	ff	ii	it	it

Let P be a non-empty ordered set. If $sup\{x, y\}$ and $inf\{x, y\}$ exist for all $x, y \in P$, then P is called a *lattice*, [2]. In a lattice illustrating partial ordering of knowledge values, the logical conjunction is identified with the meet operation and the logical disjunction with the join operation.

Nested line diagrams are used for visualizing large concept lattices, emphasizing sub-structures and regularities, and combining conceptual scales, [16]. A nested line diagram consists of an outer line diagram, which contains in each node inner diagrams.

An approach for integrating intelligent agents, user models, and automatic content categorization in a virtual environment is presented in [13]. Federated systems are discussed in [1], [6], and [9].

3 Appropriativeness of Learning Objects

Suppose a course builder requests a LO, that has a certain property, from a database in a federated learning system. The system suggests a LO according to the LO's level of appropriativeness with respect to that property. Every LO is first evaluated by three experts. Each expert is sending his/her opinion on the LO's appropriativeness with respect to that property using a Web-based form. The expert can choose among the following options:

c - this LO is totally appropriate
i - this LO is totally inappropriate
n - no response is provided
p - this LO is partially appropriate, i.e. usefull but something is still missing
b - this LO is both appropriate and inappropriate. This means that the particular LO provides useful information and at the same time conveys ideas that are misleading with respect to the property required by the course builder.

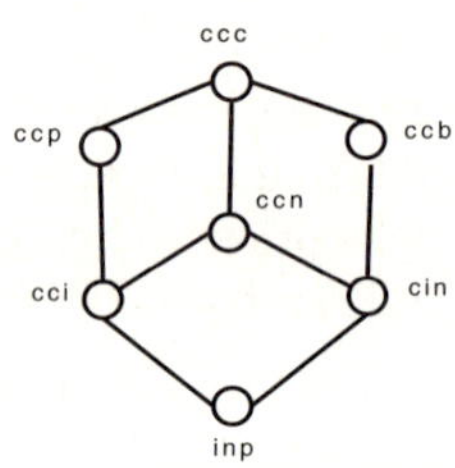

Fig. 1. Responses related to the truth value *tt*

The three experts' responses result in thirty five combinations. These responses are divided in five sets, corresponding to truth values in the five-valued logic (see f. ex. Fig. 1).

Each set contains seven combinations. Seven-valued logic is used to navigate within a set [5] and five-valued logic [4] for combining combinations of responses of several groups of experts.

The ontological operation '$\vee$' is commutative and associative for five-valued logic and for the seven-valued logic. This allows working with non-ordered responses of single experts and non-ordered combinations of responses of groups of experts. The opinions of all experts have equal effect on the final recommendation, the order in which opinions arrive in the database does not effect the final recommendation, and the system can apply a greedy algorithm while accommodating responses of several groups of experts.

A lattice illustrating relations among corresponding truth values in the nested lattice is presented in Fig. 2.

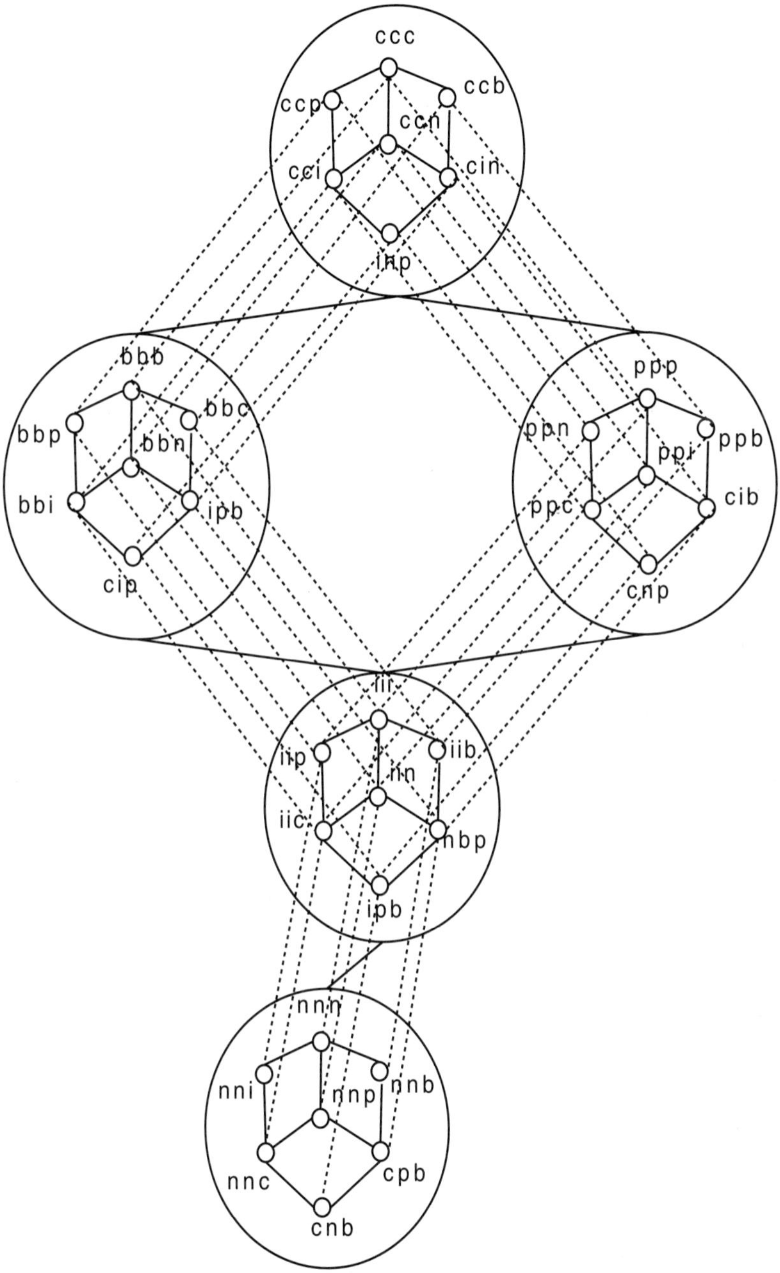

Fig. 2. Nested lattice

3.1 Experience with the System

The system has been used by one university and two university colleges. LOs initially developed for courses in logic and informatics-1 on Bachelor and Master level were suggested for inclusion in these subjects given by the educational institutions.

Among the most popular ones are LOs containing Java applets and practical examples. In informal interviews lecturers expressed satisfaction and were interested in expending the number of educational institutions being able to use and contribute to the system. Some students suggested that students' evaluation of LO's could be also a part of the recommendation process.

4 System Implementation

Each LO is defined to be atomic, that is to say, it can not be further divided into simpler components. Each LO is a self-contained Web object attached to its meta-data that describes its content as accurately as possible. Any LO can be viewed together with its meta-data by expert tutors, course builders and administrators at any time. Only authorized expert tutors can change meta-data attached to a particular LO.

The expert tutors can review and change a particular LO appropriativeness options for a particular course usage. An expert tutor can submit a usage statement on a particular LO and then encourage other experts to give their reviews.

Meta-data for each LO contain, among other properties, the following important information

1. level
   - introduction
   - bachelor
   - master
   - PhD

2. type
   - theory
   - example
   - exercise
   - quiz
   - test

A usage statement can have the following form.

Unit **LO-1X34** of type **theory** with level **introduction** can be used for topic **logic** in subject **informatics-1**.

The expert that introduces the usage statement will be the first to give an appropriativeness ranking. She will then invite at least two other experts defined

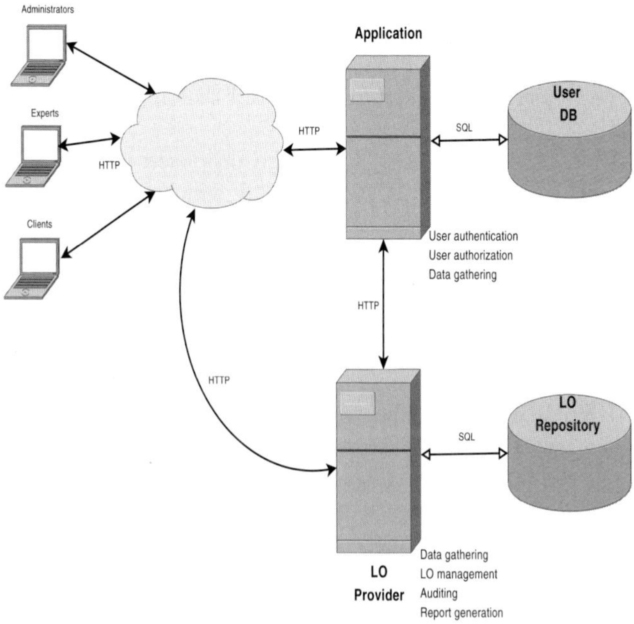

Fig. 3. The system

in the system to rank this particular usage statement. Each expert will receive an email with a Web link with appropriate HTTP get parameters related to a particular usage statement and the expert identifier.

In the process of making a course, a client (a course builder, a teacher or a tutor) can consult the system for a list of usage statements containing the appropriate particulars for the course. The presented usage statements are those that are already reviewed by at least three experts. The ranking is a pair of recommendation following the structure presented in Fig. 2, for example

$$\{\text{tt, ccp}\} \rightarrow \{\text{'highly recommended', '2 total and 1 partial recommendations}\}$$

A skeletal prototype system (Fig. 3) for decision support for course building from federated LO was implemented using a three-tiers Web application server architecture. The presentation layer is handled by an Apache Web server. The logic layer is written in Python. The data layer is implemented using SQLite database engine.

Apache is a modular Web server that can incorporate a high level scripting language as a module such as Python using mod_python. Python provides a programming environment for implementing handler for dynamic content, data integration and intelligent software agents. Python is a scriptable language that promotes modular system development paradigm. A base system implementing the system basic architecture can be built quickly by writing basic modules.

Step-wise improvements on these modules can later be done to fill up the functional details. The back end SQLite databases are used to store both static and dynamic data. SQLite is a small footprint and administration-free database engine. SQLite can easily implement parallel and distributed databases without much overhead.

Having Python as a programing layer, the proposed system can implement Web services quit easily using REST method which we call HTTP-RPC. A client application makes a call for a service by opening a Web link and sends parameters to the service in the request string. The server responds by sending an XML reply page back to the client. There is no need to implement a special communication protocol as every request/response message is handled using the underlaying HTTP.

A long running program is implemented by a two phase technique. A client requests a particular long running service by opening a Web link and request parameters. The server replays back to the client via a unique request identifier together with a temporary Web link and starts that particular long running program in the background. Results of the long running program will be stored in a database index by the uniquely client related request identifier. After an appropriate waiting time period for a particulate service, the client can then open the given Web link and the request identifier as a request parameter. If the service has been completed, the server then sends the resulting data back to the client in XML, otherwise a 'not completed' status message will be sent back. It is the job of the client to resubmit a request for result and eventually delete the result on the server database. The technique of caching results can also be useful for sharing results of similar requests from different clients at different time.

5 Conclusion

This paper is devoted to automated evaluation of learning objects with respect to their appropriativness for being enclosed in a pre-determined subject. The evaluation process is based on application of many-valued logic which contributes to a better decision making process.

At this stage we involved three educational institutions and two subjects and have used only opinions of experts in the decision making process. Some of the issues related to expending the number of participating educational institutions are more political than technical.

We are collecting all comments and suggestions and plan to further develop the presented prototype system where we are going to incorporate students' feedback.

References

1. Busse, S., Kutsche, R.-D., Leser, U.: Strategies for conceptual design of federated information systems. LNCS, vol. 1626, pp. 255–269. Springer, Heidelberg (1999)
2. Davey, B.A., Priestley, H.A.: Introduction to lattices and order. Cambridge University Press, Cambridge (2005)

3. Encheva, S., Tumin, S.: Cooperative Shared Learning Objects in an Intelligent Web-Based Tutoring System Environment. In: Luo, Y. (ed.) CDVE 2005. LNCS, vol. 3675, pp. 227–234. Springer, Heidelberg (2005)
4. Ferreira, U.: A Five-Valued Logic and a System. Journal of Computer Science and Technology 4(3), 134–140 (2004)
5. Ferreira, U.: Uncertainty and a 7-Valued Logic. In: Proceedings of The 2nd International Conference on Computer Science and its Applications (ICCSA 2004), San Diego CA, USA (June 2004)
6. James, A., Conrad, S., Hasselbring, W. (eds.): Engineering Federated Information Systems, Proceedings of the 5th Workshop EFIS 2003, Coventry (UK), Akad. Verlagsgem. / IOS Press (infix) (2003)
7. IEEE Learning Technology Standards Committee (LTSC), Draft Standard for Learning Object Metadata Version 6.1 (2001), http://ltsc.ieee.org/doc/
8. http://www.harvestroad.com/
9. Kutsche, R.-D., Conrad, S., Hasselbring, W. (eds.): Engineering Federated Information Systems, Proceedings of the 4th Workshop EFIS, Berlin (2001)
10. http://kmr.nada.kth.se/el/ims/md-lomrdf.html
11. Longmire, W.: Content and Context: Designing and Developing Learning Objects. Learning Without Limits, Informania 3 (2000)
12. Rehak, D., Dodds, P., Lannom, L.: A Model and Infrastructure for Federated Learning Content Repositories. In: WWW 2005, Chiba, Japan (2005)
13. Santos, C.T., Osòrio, F.S.: Integrating intelligent agents, user models, and automatic content categorization in virtual environment. In: Lester, J.C., Vicari, R.M., Paraguaçu, F. (eds.) ITS 2004. LNCS, vol. 3220, pp. 128–139. Springer, Heidelberg (2004)
14. http://www.adlnet.org/index.cfm?fuseaction=scormabt
15. Wiley, D.A.: "Connecting Learning Objects to Instructional Design Theory: A Definition, a Metaphor, and a Taxonomy." The Instructional Use of Learning Objects, Bloomington, In: Agency for Instructional Technology (2002)
16. Wille, R.: Concept lattices and conceptual knowledge systems. Computers and Mathematics with Applications 23(6-9), 493–515 (1992)

A User Interface for the User-Centred Knowledge Model, t-UCK

Anne Håkansson

Department of Information Science, Computer Science,
Uppsala University, Uppsala, Sweden
Anne.Hakansson@dis.uu.se

Abstract. This paper presents a user interface to the User-Centred Knowledge Model (t-UCK). T-UCK is a knowledge modelling tool for designing knowledge-intensive systems. The model centres round the various users, i.e., both the design users and the end users, and facilitates the use of a conceptual model for handling different types of knowledge, the reasoning strategy and other functionality. For the design users, the conceptual model is presented through a modelling view of the contents used for developing the system. For the end users, the conceptual model has a parallel consulting view used for sessions with the system. Both these views are directly modelled into the system through a graphical modelling language, the Unified Modelling Language (UML). UML is a general-purpose modelling language, which in a modified form it can be used for development of knowledge-based systems.

Keywords: Knowledge-based systems, knowledge modelling, visualisation, graphic modelling, user centred modelling, Unified modeling language.

1 Introduction

A user-centred knowledge model, called t-UCK, is used to model a knowledge-intensive system from the users point of view. The model is developed to *transfer knowledge* from the design users and the end users, which involves transferring domain specific knowledge [15] from a source through a system so that it eventually reaches a receiver. This process is usually divided into two sub-processes, knowledge acquisition and knowledge elicitation. *Knowledge acquisition* is a wider term and includes eliciting, modelling and representing the domain knowledge [9]; knowledge elicitation constitutes the communication between the domain expert and the knowledge engineer and is the process of extracting, interpreting and analysing domain knowledge. In t-UCK, knowledge acquisition implicitly includes elicitation.

The usefulness of a system often depends on the extent of the users' contribution during the development process. It also depends on the system's ability to match its users. For matching, *conceptual design* has applied to bridge the gap between the conceptual model and the different users' mental models [13; 10]. The essence of the design is to correspond the users' mental models to the actual artefact. The closer the users' models match the way in which the application actually works, the more successful the operation of the application by the user will be. In the model, t-UCK, the conceptual

I. Lovrek, R.J. Howlett, and L.C. Jain (Eds.): KES 2008, Part I, LNAI 5177, pp. 312–321, 2008.

design minimise the distinction between the users' intentions and the execution of these intentions, realised by taking users' previous knowledge and experience into account. The *design model* is held by the designer and is used to bridge the gap between the conceptual model and the *users' models*, see Figure 1.

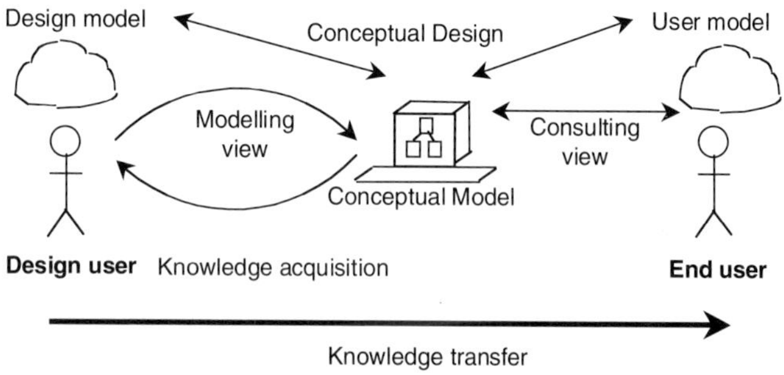

Fig. 1. The User-Centred Knowledge Model

Graphic modelling of the knowledge is achieved by incorporating the conceptual model [12] in t-UCK. In a sketch of the system as seen from the perspective of the experts, the *conceptual model* is usually placed as a mediator between a domain expert's expertise and the creation of code within the system [1]. Moreover, the conceptual model can function as an intermediate link and thereby simplify the task of knowledge engineering [3]. In t-UCK, the domain expert's expertise forms an integral part of the system when the conceptual model is visualised at the user interface. Thus, this conceptual model can be beneficial for the domain experts, the knowledge engineer and the end users. The model represents the domain experts' view of the domain and relevant domain knowledge as well as the knowledge engineer's and the end users' views of the domain. Since these different users' interests in the system are likely to differ, they will have different views of the conceptual model, i.e., we have to cater for a *modelling view* and a *consulting view*. The modelling view supports the domain expert and the knowledge engineer first and foremost, whereas the consultation view supports the end users. Using several different graphical views may bring the user closer to a system and vice versa and, thus, provide a more user-centred modelling of the domain, but also a user-centred consultation with the system.

The combination of the views can provide a means for making the system more transparent by viewing both the domain knowledge and the reasoning strategies. This dual view should support changing the knowledge to eliminate errors. As a user interface, we use graphical modelling language, the Unified Modelling Language (UML) [2; 11]. UML is a general-purpose modelling language and can be used for developing knowledge-based systems with rules or frames representation [5; 6; 4]. The different diagrams show the content of the knowledge base but also the reasoning strategy of the system. Moreover, the diagrams illustrate the system's functionality.

2 Related Work

UML is usually used for modelling object-oriented systems. UML diagrams are used in CommonKADS to build knowledge-based systems in object-oriented fashion. Diagrams are used to model the state of the system as it changes over time and to model the dynamic behaviour of the system and provide an image of the sequence of events and decision-making [15]. Diagrams are also used to describe the context of the information for the task analysis and the structure of objects handled in a task. The diagrams are also incorporated to present the actors and the services (or use-cases) and, include additional chunks of information that are difficult to model, e.g., large or complex systems [15].

UML can also be used for modelling of other types of systems, such as frame-based [4] and rule-based systems. UML is a visual language for object-oriented systems and frames have many similarities to object-oriented language, e.g., encapsulation, inheritance, polymorphism and messages passing. Therefore, UML can be straight forward applied to frames. The UML diagrams applied onto frames are class diagrams, object diagrams, use cases, activity diagrams such as sequence diagrams and collaboration diagrams [4]. We apply UML diagrams to rule-based knowledge management systems that have been developed in a declarative fashion. This affects the UML's diagrams, since they cannot be used in their original form as they are used in CommonKADS. In its current form, UML is not directly applicable for modelling knowledge in systems that are rule-based. However, UML can easily be adapted to knowledge management systems utilizing rules in the knowledge base.

3 Applying Different Diagrams

A visual conceptual model can be used to present an already implemented system, thereby, making the system more transparent and the contents of the system more accessible [5; 6; 7; 8]. Visualisation of the conceptual model supports modelling and consulting by visually presenting the contents of the system and the operation of the interpretation mechanism. The visualisation can support modelling the knowledge base, altering the functionality and changing the reasoning strategy. It can also support consulting the system for specific purposes. In addition, it should be able to provide visual explanations for the design users and the end user.

The domain knowledge and the reasoning strategy is presented in diagrams from the graphical modelling language, UML. Many types of diagrams encompassed by UML are utilised, with the exception of the component diagram and the deployment diagram. Domain knowledge and reasoning strategies are presented in use-case diagrams, class diagrams, object diagrams, interaction diagrams, i.e., sequences and collaboration diagrams, activity and state-chart diagrams, and packages. The diagrams are used for different purposes, but together they will constitute a unit describing the contents of the system. This implies putting the contents of the diagrams together to illustrate the domain knowledge, reasoning strategy and functionality of the current system.

3.1 Use-Case Diagram

In t-UCK, use-case diagrams are used to describe the tasks a system can perform from both the design users' and the end users' point of views. This is illustrated in Figure 2, where the use-cases are presented in the ellipses. In designing this diagram, the requirements of the knowledge-based systems' performances are modelled.

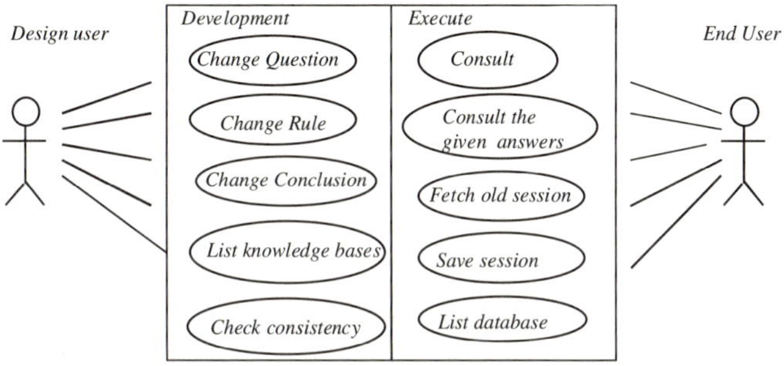

Fig. 2. Example of use-cases for design users and end users

As illustrated in the example, the design user can develop the knowledge-based systems by adding, deleting or editing rules, questions or conclusions. Moreover, the knowledge bases can be listed and the consistency of the rules checked. This example also demonstrates how the end users interact with the system by consulting, consulting the given answers, fetching previous sessions, saving sessions and listing the contents of the database. Within each alternative, there are more detailed tasks, which also need to be illustrated by use cases at a detailed level.

3.2 Class Diagram and Object Diagram

Class diagrams are used as templates for the parts to be inserted into the system. From the templates or the classes, objects are created, as demonstrated in Figure 3 below.

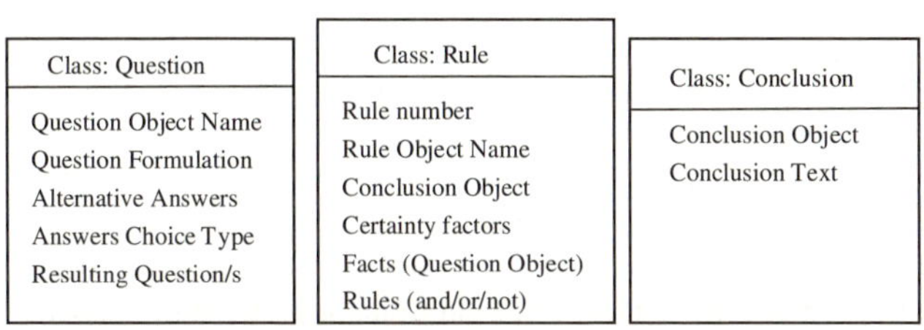

Fig. 3. Templates for questions, rules and conclusions

The classes consist of questions, rules and conclusions. These classes have relationships that connect them. For example, the class: rule is related to both questions and conclusions by the relationships facts (the answers from a consultation) and conclusion objects (used by the interpreter). The rules are also related to other rules, which is illustrated by comprising them as rules with *and*, *or*, or *not*. The objects, that are instances of classes and created from classes, are stored in the knowledge bases. The domain expert can use the template to insert new domain knowledge, but the end users can only use the objects, e.g., by answering questions correctly using the single or multiple answer alternatives.

Object diagrams model instances of class diagrams. The object diagram is used to show a set of objects and their relationship. It is also used to categorise the objects. A graphic illustration of these different objects is presented in Figure 4.

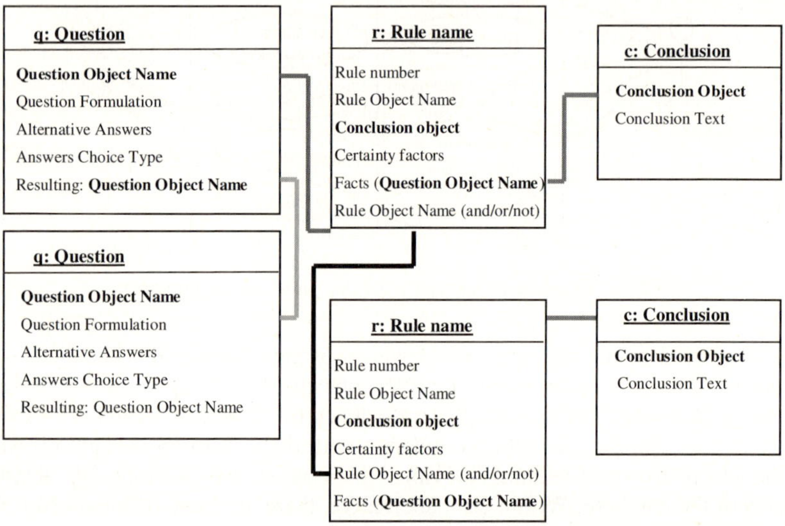

Fig. 4. Object diagram with questions, rules and conclusions

The lower-case character in the header of the object denotes that the object is of the class rule (r), question (q) or conclusion (c). Following this character, the name of the category is specified, i.e., the category that the object belongs to. One purpose of using categories is that a hierarchical diagram can be established from them.

3.3 Interaction Diagrams with Sequence and Collaboration Diagrams

A consultation, or an interpretation, of the contents of the knowledge-based system involves call sequences, i.e., the sequences in which different parts (rules) interact with each other. There is a difference between static and dynamic relations, so one of the challenges is to illustrate how different rules are related, statically and dynamically. A static relation is a constant relationship between rules in a knowledge base, e.g., one single rule might be related to several others and is dependent on them. A dynamic relation is comprised of the connections that are important during execution since the interpretation depends on them. For example, a rule is dependent on the end user's input, which becomes the fact when a specific question has been answered. The static relations in the knowledge base are presented in sequence diagram and the dynamic relations are presented in collaboration diagram.

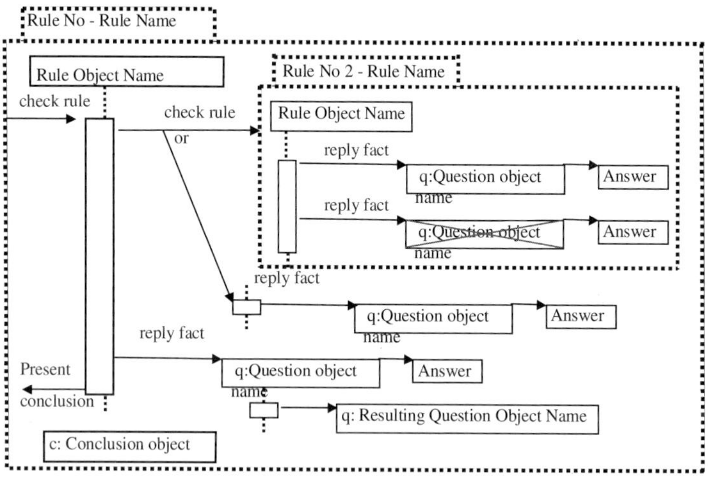

Fig. 5. A sequence diagram including rules, questions and conclusions

The sequence diagram illustrates the static relationships between questions, rules and facts. The connections between these parts are illustrated in Figure 5.

The rule "Rule No – Rule Name" contains one rule (check rule in the Figure) and two facts (reply fact). The fact is an answer that has a value connected to a question. Usual, the time-line is "and-clause" relation. In this case, the relation also includes an "or-clause", which means that either the rule or the fact can be used during the consultation. Moreover, one of the facts "Question object" is marked with a cross, which means that the fact are not allowed to be satisfied.

The collaboration diagram illustrates the stepwise exposition of the parts involved in making an interpretation to draw a specific conclusion. These diagrams can make it easier to get an overview of the entities in the sets. Since the collaboration is dynamic it is possible to check the result of using certain input data. The interrelation between these parts shows how they are interlinked, as demonstrated in Figure 6.

For example, to reach the conclusion "Conclusion Name", the inputs, "Input 1" and "Input 2", have been inserted into the diagram and the rules and the facts must be

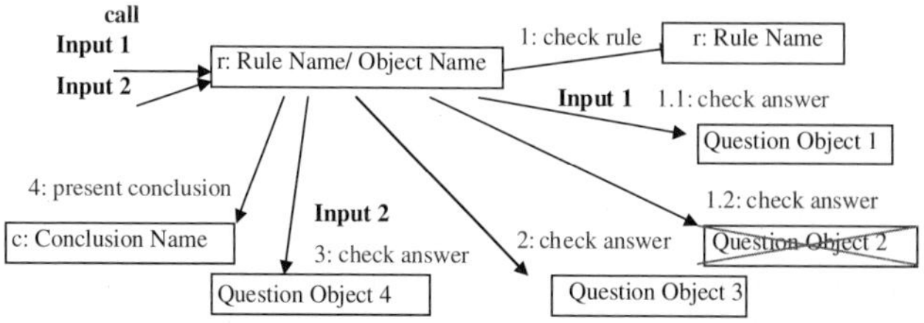

Fig. 6. A dynamic presentation of rules for a conclusion

satisfied. Some of the facts use the input data as answers; other facts have to fetch answers from the database. The relationships between rules and facts show the order of the invoking of rules and facts. The cross symbolizes "the answer cannot be true" which corresponds to negation as failure found in the programming language Prolog. The diagram may also clarify the system's reasoning process. For instance, the design user must be aware of the resulting questions and whether one question is more specific than another. Besides, the end user could be aware that there is a follow-up question (subsequent question), which means there is dependence between questions.

3.4 State-Chart Diagram and Activity Diagram

The state-chart diagram presents the dynamic view of a system and specifies the life cycle of the use-case instances, as exemplified in Figure 2. The activity diagram is a special kind of state-chart diagram that illustrates the flow from activity to activity and clarifies the sequence of actions.

In state-chart diagrams, all states and their content have a description, e.g., explaining how they behave, what constraints there are, and the dependence on others. For instance, the chart can present the objects' states or the interpreter's states and the information flow between them. At a detailed level, state diagrams can be utilised to insert production rules with facts and other rules. When several rules have been inserted it is possible to produce meta-rules. The state diagrams can also present the procedural execution of the programming code. For example, the state-chart diagram views the execution of the use-cases, as shown in Figure 7a.

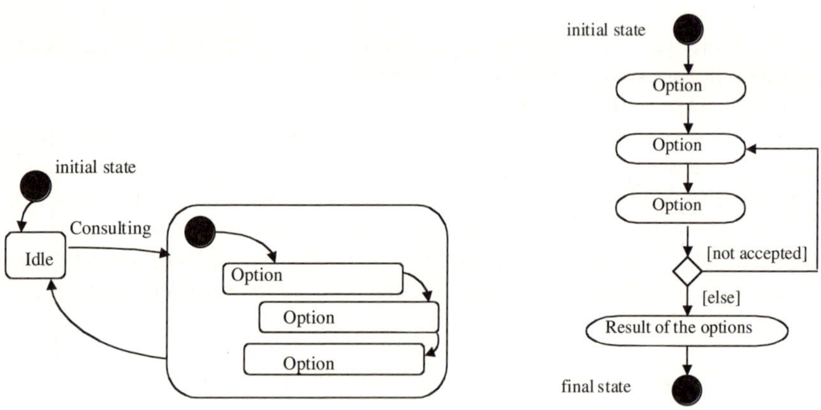

Fig. 7. a) A state-chart diagram for, developing the knowledge base **b)** An activity diagram for presenting the use-cases

The activity diagram can show procedural behaviour in a declarative representation, giving an understanding of how use-cases work in the system, see Figure 7b. At a detailed level, activity diagrams can be used to view the consultation and even to follow the interpretation. Compared to the interaction diagrams shown in Figures 5 and 6, this activity diagram demonstrates how to display the execution at a level of abstraction. The design user can follow the process stepwise to judge if it is acceptable. Even the end user can follow this process.

During the consultation a kind of state-chart may be generated where each part involved should be presented to the end user. A specific part, a question, can be studied to get information about when and in what context it is used. This may also clarify the functionality of the system.

3.5 Packages

When one needs to comprehend large systems, a package can be used to encapsulate data to decrease the complexity of the system's content. Hence, the design user will not need to deal with all the information simultaneously and can more easily comprehend the content as a whole. For example, rules that are connected with some other rules can all be packed into one package, see Figure 8.

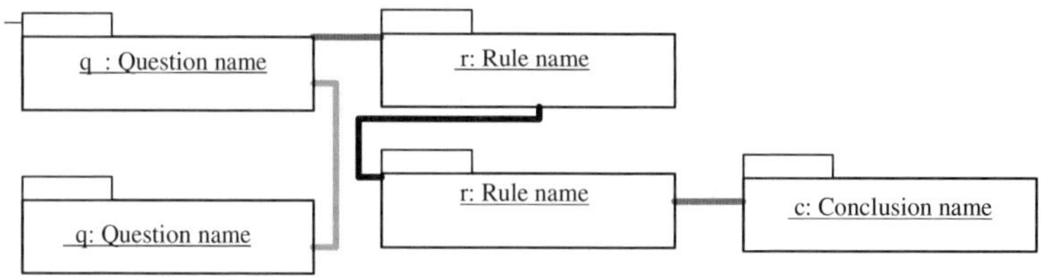

Fig. 8. Package of the rules that handle childhood diseases at a lower level

The problem with a package is that somehow rules shared by other rules have to be identifiable to inform the design user about the sharing. However, by unfolding this package, all rules can be presented in a sequence or in a collaboration diagram. This can support the design user when a rule is edited since the dependence rules can be rapidly identified.

The package can present the knowledge in a hierarchy structure with the most abstract knowledge being found on the highest level. By using packages, the end user only has to deal with one piece of knowledge and with just one level of detail at a time. Nevertheless, the packages can be utilised at different levels of details. This can be benefit the end users, with different expertise, since they may need various kinds at various depth of knowledge.

4 Conclusions and Further Work

We have presented a user interface to a user-centred knowledge model, so-called t-UCK. The model is based on the users' mental models, which is reflected in the modelling view and the consultation view. The views contain several different diagrams, which shows the content of the knowledge base but also the reasoning strategy of the system and functionality. To support the different users, we use modified diagrams of UML that can be combined for purposes.

Parts of t-UCK have been used in 40-50 different knowledge system projects and works well for these projects. Almost all diagrams have been utilised except for object-diagrams, state-chart diagrams and activity diagrams since they have only been at the sketch phase earlier. Nonetheless, during the development, the diagrams have been changed to better suit their purposes. The major changes concern the relationships between rules and facts in the sequential and collaboration diagrams and the negation as failure cross. The altered diagrams are presented in this paper.

Some of the diagrams, sequential and collaboration diagrams have been implemented in Prolog with graphic interface but also in Java with Tcl/Tk. Next step is to implement all diagrams and provide a automatically translation of the diagrams to programming code. Moreover, investigations are needed to check the advantages of the user interface.

References

1. Alonso, F., Fuertes, J.L., Martinez, L., Montes, C.: An incremental solution for developing knowledge-based software: its application to an expert system for isokinetics interpretation. Expert Systems with Applications 18, 165–184 (2000); In: Darlington, K.: The Essence of Expert Systems. Prentice Hall, England
2. Booch, G., Rumbaugh, J., Jacobson, I.: The Unified Modeling Language User Guide. Addison Wesley Longman, Inc., Amsterdam (1999)
3. Durkin, J.: Expert System Design and Development. Prentice Hall International edn. Macmillian Publishing Company, New Jersey (1994)
4. Helenius, E.: UML and Knowledge Engineering of Frame Based Knowledge Systems (UML och systemutveckling av framebaserade kunskapssystem). Master Thesis, Department of information science, Computer Science Division, Uppsala University (2001)
5. Håkansson, A.: UML as an approach to Modelling Knowledge in Rule-based Systems. In: Bramer, M.A., et al. (eds.) Proceedings of ES2001, The Twenty-first SGES International Conference on Knowledge Based Systems and Applied Artificial Intelligence, pp. 187–200. Springer, Heidelberg (2001)
6. Håkansson, A.: Supporting Illustration and Modification of the Reasoning Strategy by Visualisation. In: Tessem, B., et al. (eds.) Proceedings of The Eighth Scandinavian Conference on Artificial Intelligence, SCAI 2003. Frontiers in Artificial Intelligence and Applications, vol. 103. IOS Press, Amsterdam (2003)
7. Håkansson, A.: Transferring Problem Solving Strategies from the Expert to the End Users - Supporting understanding. In: Proceedings of 7th International Conference on Enterprise Information Systems, ICEIS-2005, INSTICC, Portugal, vol. II, pp. 3–10 (2005) ISBN: 972-8865-19-8
8. Håkansson, A.: Modelling from Knowledge versus Modelling from Rules using UML. In: Khosla, R., Howlett, R.J., Jain, L.C. (eds.) KES 2005. LNCS (LNAI), vol. 3681, pp. 393–402. Springer, Heidelberg (2005)
9. Jackson, P.: Introduction to Expert Systems. Addison Wesley, England (1999)
10. Lantz, A.: Computer Mediated Communication in a Work Context: an Interdisciplinary Approach. Department of Psychology. Stockholm University, Sweden (1996)
11. Larman, C.: Applying UML and Patterns: An Introduction to Object-Oriented Analysis and Design and Iterative Development, 3rd edn. Prentice Hall, Englewood Cliffs (2004)

12. Luger, G., Stubblefield, W., Cummings, B.: Artificial Intelligence: Structures and Strategies for Complex Problem Solving. The Benjamin/Cummings Pub. Company, Inc. (1993)
13. Norman, D.A.: Cognitive engineering. In: Norman, D.A., Draper, S.W. (eds.) User Centred System Design: New Perspectives on Human-Computer Interaction, Lawrence Erlbaum Associates, Hillsdale (1986)
14. Sandahl, K.: Developing Knowledge Management Systems with an Active Expert Methodology. Dissertation No. 277, Linköping University, Sweden (1992)
15. Schreiber, G., Akkermans, H., Anjewierden, A., de Hoog, R., Shadbolt, N., Van de Velde, W., Wielinga, B.: Knowledge Engineering and Management – the CommonKADS Methodology. The MIT Press, Cambridge (2001)

Focus Support Interface Based on Collaborative Learning Activity

Yuki Hayashi, Tomoko Kojiri, and Toyohide Watanabe

Department of Systems and Social Informatics,
Graduate School of Information Science, Nagoya University
{yhayashi,kojiri,watanabe}@watanabe.ss.is.nagoya-u.ac.jp

Abstract. With the development of information and communication technologies, learners can study easily with others in the distributed environment. However, it is still hard for them to share collaborative interaction with others efficiently because of the limited communication means. In order for learners to study collaboratively with others and immerse in learning, it is important for them to grasp directly the actions occurred in the learning environment, such as making utterances, facing to other learners, writing memos, etc. Moreover, they should observe the collaborative learning environment appropriately according to their focusing intentions. In this paper, we analyze the activities occurred in the collaborative learning environment, and propose a method of detecting learner's focusing intention. Then, we address the effective view change based on the focusing intention in the collaborative learning system.

Keywords: Round-table interface, CSCL, Awareness, Learning activity, Focusing target.

1 Introduction

As the drastic development of information and communication technologies, learners can study with others without sharing the same physical space [1, 2]. Computer supported collaborative learning (CSCL) is one of the interesting learning styles in which plural learners study collaboratively in the shared virtual space. However, as the communication band of the network is restricted, learners cannot acquire much information. In order to collaborate with others smoothly/effectively and immerse in learning, it is very important for learners to understand other learners' actions and situations appropriately in a distributed learning environment. *Awareness* is a key concept to concentrate on this subject [3, 4, 5]. By reflecting awareness information such as understanding of others' facial expressions, the situation, phenomenon, etc., the learning environment enables learners to discuss more collaboratively.

To concentrate on the learning process and communicate with others effectively, we proposed a round-table interface where learners can grasp the learning situation from their own views according to their focusing intentions [6, 7]. In this interface, other learners were represented as camera images and were arranged around the round-table in the 3-dimentional learning environment. The learner's focusing target was situated in the center of learner view. In addition, the learner view was changed

I. Lovrek, R.J. Howlett, and L.C. Jain (Eds.): KES 2008, Part I, LNAI 5177, pp. 322–329, 2008.

automatically according to the focusing degree for the learner's target. In order to estimate learners' focusing target, a method of calculating the focusing degrees for other learners proposed by Kojiri, et al. [8] was proposed. In this method, the focusing degree was determined by type, target, and contents of utterance. However, this method did not consider the learners' actions other than making utterances. In the real world, learners usually change their focusing targets based on not only utterances but also actions of learners such as looking to the other learners and their private works. Thus, learning activities consisted of the learners' actions effect the focusing intention of the learners. By considering learning activities in determining the focusing degree, learners' focusing intentions can be grasped more correctly. Thus, the detection method of focusing degree should be modified in order to consider the various activities during the learning. In addition, the learner view in the interface must be displayed based on the estimated target.

This paper describes a method of calculating focusing degrees and detecting the focusing target based on the collaborative learning activities. According to the calculated focusing degrees, the learner view in the round-table interface is changed effectively. By reflecting the learners' focusing intentions, they can grasp the learning situation from the interface more easily/naturally.

2 Approach

2.1 Focusing Target Learner Based on Learning Activities

In the real world, learners make various actions while discussing with others so as to accomplish a common exercise, such as focusing on others, exchanging textbook, writing memo, and so on. If learners are aware of such actions which are performed in the virtual collaborative learning environment, the interaction among learners becomes more smooth and effective. Therefore, learner's focusing intention should be calculated according to the learning activities and reflected to the learner view through the interface.

In the collaborative learning environment, there are some objects which relate to the learning. Currently, we focus on the virtual learning environment where learners can make an utterance using the text-chat, and write down their solutions/ideas to the memo-sheets. In this learning environment, learners and their memo-sheets are existing objects. During the learning, learners make some actions such as making utterances to other learner/learners, writing down ideas to their own memo-sheets, observing memo-sheet of other learner, looking to the other learners, and so on. In all actions, a learner is the active object. On the other hand, the memo-sheet is observed and activated by its owner or other learners, so the memo-sheet is a passive object. If a learner makes action to other learner, he/she may be interested in the target. On the other hand, since memo-sheets represent the owners' answer/idea, the action to the memo-sheets may represent the interest to its owner. If the learner makes action to his own memo-sheet, he/she can be regarded to have the interest to him/her.

According to the focusing intention, the focusing target of the learner is changed through the actions. We defined the learners' actions into two types; *private actions* and *public actions*. Private actions is occurred by the learner to his/her own possession. In this case, the action of writing/observing his/her memo-sheet is private actions. On the other hand, public actions correspond to the action from learners to other learner/learners or their possessions. The action transitions among learning objects can be classified into seven patterns. Figures 1 and 2 show actions occurred in our learning environment. In these figures, nodes named *learner X*, *learner A* and *learner B* correspond to individual learners and their memo-sheets. The source of the arrow indicates the learner who makes an action and the direction of the arrow shows the target of the action. Based on these actions, we summarize the various cases of changing focusing intention of *learner X* in Table 1. Table 1 describes *learner X*'s target learner *(Target learner based on action according to learner X's individual actions*). As a result of the private action, *learner X* himself is the target in action 1. On the contrary, the target becomes the other learner based on the public actions shown as actions 2-7. The timing of changing target is also different from types of actions. Utterances are recognized by all learners, so the target learner is changed every time when the utterances are occurred in the learning. On the other hand, other actions, such as observing memo-sheet of other learner, are only perceived if they are seen in the learner view. Thus, the targets learner based on action is changed when these actions are observed from the learner.

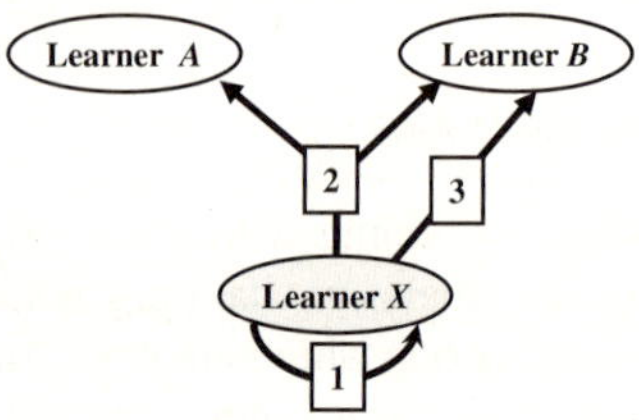

Fig. 1. Actions from learner *X*

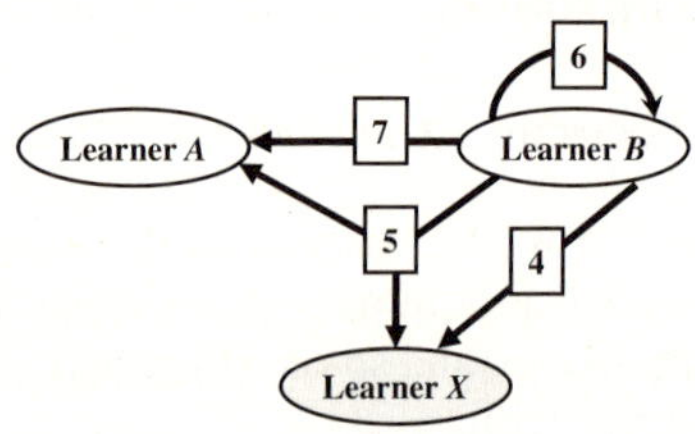

Fig. 2. Actions from other learner

Table 1. Target learner of learner *X* based on action

	Action	From	To	Target learner based on action
1	Writing down to learner *X*'s memo-sheet Observing learner *X*'s memo-sheet	Learner *X*	Learner *X*'s memo-sheet	Learner *X*
2	Making utterance to all learner	Learner *X*	All learners	Not change
3	Making utterance to learner *B* Observing learner *B*'s memo-sheet	Learner *X*	Learner *B* and *B*'s memo-sheet	Learner *B*
4	Making utterance to learner *X* Looking learner *X* Observing learner *X*'s memo-sheet	Learner *B*	Learner *X* and *X*'s memo-sheet	Learner *B*
5	Making utterance to all learner	Learner *B*	All learners	Learner *B*
6	Writing down into learner *B*'s memo-sheet Observing learner *B*'s memo-sheet	Learner *B*	Learner *B* and *B*'s memo-sheet	Learner *B*
7	Making utterance to learner *A* Looking learner *A* Observing learner *A*'s memo-sheet	Learner *B*	Learner *A* and *A*'s memo-sheet	Learner *A*

2.2 Learner's View Change According to Focusing Target

In the real world, the learner commonly observes various phenomena from his/her view according to focusing intention. The learner puts his/her focusing target in the center of his/her view and observe him/her carefully. Moreover, the size of the focusing target in learner view is changed according to his/her focusing degrees. Namely, when the learner focuses on the target learner carefully, the focusing target becomes large in his/her view. On the other hand, the size of the focusing target decreases in learner view if he/she does not pay attention to the current focusing target. To attain the smooth/effective learning in the virtual learning environment, the learner view through the interface should be displayed appropriately according to the focusing degree for focusing target.

As private actions, the learner writes down or observes his/her memo-sheet. If the learner concentrates on his/her private actions, he/she looks down the direction of his/her memo-sheet. If the learner focuses on the public actions, he/she looks up to see the focusing target. Hence, the learner view moves between other learners and his/her memo-sheet according to the focusing degree for himself/herself. In Figure 3, the learner's direction of the view changes from learner *B* to his/her own memo-sheet. On the contrary, the learner view among other learners is also changed based on the public actions. When the learner focuses on the other learner/memo-sheet, the direction of the learner view turns to the focusing target so as to observe the detailed information. In addition, if the focusing target is changed to other learner, the learner swings his/her own view to the focusing target. In Figure 4, the focusing target of the learner is changed from learner *B* to learner *A* based on the public actions, so the learner's direction of the view is turned to the learner *A*.

In our interface, the object displayed in the learner view is changed automatically according to the learner's focusing target. In order to reflect the learner's focusing intention, focusing degrees for all learners including learner himself are calculated based on private/public actions. Then, the learner who has the largest focusing degree is determined as a focusing target. The focusing target is appeared in the center of our interface window. When, the learner himself is the focusing target, the learner's direction of the view turns to his/her memo-sheet. Moreover, the size of the focusing target is changed according to the focusing degree for the target. In order to change the size of the target, the distance between the learner and the focusing target is changed

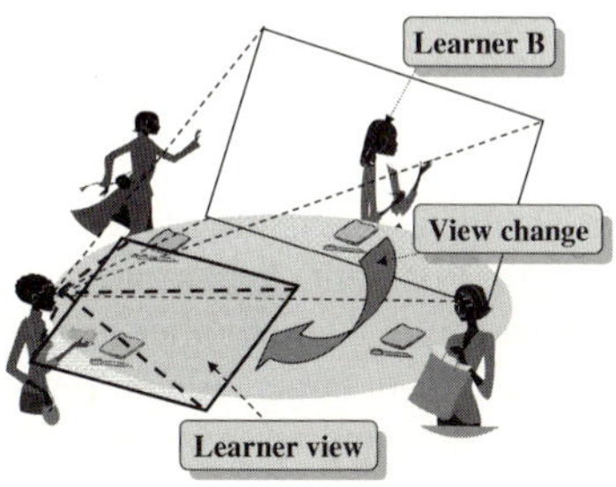

Fig. 3. Learner view changed by private action

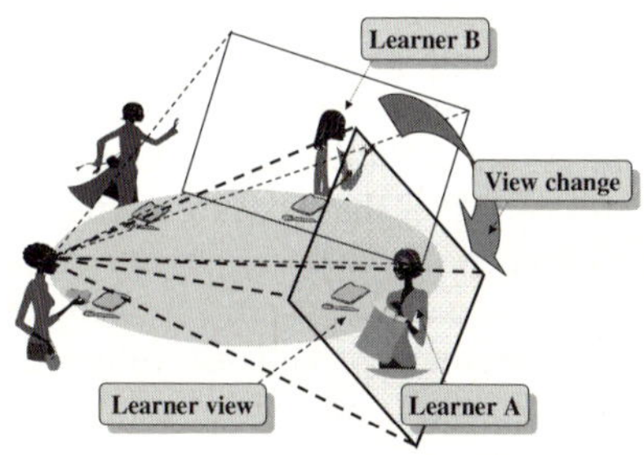

Fig. 4. Learner view changed by public action

according to the focusing degree. From the position based on the focusing degree, the direction of learner view changes up-and-down and right-and-left toward the focusing target.

3 Detection of Focusing Target

The target learner based on action is changed according to private/public actions shown in Table 1. In order to determine the focusing target, the focusing degrees for individual learners are calculated by Expression (1). Expression (1) represents the method of calculating the focusing degree *F(n, t)* of learner *n* at time *t*. *F(n, t)* corresponds to focusing degree of learner *n* at the time *t, N* is the number of learners participated in the learning environment, α_i is the constant number which represents the change of the focusing degree on a certain action *i*. α_i is defined for each action.

$$F(n,t+1)=\begin{cases} \dfrac{F(n,t)+\alpha_i}{\sum_{\forall n' \in N} F(n',t)+\alpha_i} & \text{(if learner } n \text{ is the target learner based on action)} \\ \dfrac{F(n,t)}{\sum_{\forall n' \in N} F(n',t)+\alpha_i} & \text{(otherwise)} \end{cases} \qquad (1)$$

$$(0 \le F(n,t+1) \le 1)$$

When the action *i* is occurred, α_i is added to the current focusing degree of the target learner based on action. Then, the focusing degrees of all learners are normalized to set the total focusing degree to 1. The focusing target is determined as a learner whose focusing degree is the largest.

4 Learner View

In our collaborative learning environment, the other learners' positions are assigned around the round-table. Learners are represented using polygon objects attached with their camera images. Their memo-sheets are also arranged on the round-table in front of their camera images.

Learner view for each learner's interface is set by the location and direction of his/her view in the learning environment determined by his/her focusing degree for the target. When the focusing target is another learner, the distance from the center of the round-table is determined according to the focusing degree. Expression (2) is a method of calculating distance *d(t)* from the center of the round-table interface at time *t* and Figure 5 shows its illustration. In Expression (2), *F(n, t)* calculated based on Expression (1) is the focusing degree of learner *n* at time *t*. The learner takes the position between d_{min} and d_{max} in the virtual learning environment. When the learner is situated at d_{min}, he is eagerly focusing on the focusing target. On the other hand, when the learner is at d_{max}, he is not fully focusing on the focusing target. According to this expression, the learner gets nearer to the center of the round-table as the focusing degree becomes larger.

$$d(t) = (1 - F(n,t)) \times d_{\max} + F(n,t) \times d_{\min} \qquad (0 \le F(n,t) \le 1) \qquad (2)$$

On the other hand, the focusing intention of learner his/herself is represented as the view angle between the other learner's camera image and learner's own memo-sheet, measured from the learner location *d(t)* calculated by Expression (2). That is, the focusing intention of learner is represented as up-and-down direction. Figure 6 represents the illustration of changing angle. The angle θ is calculated according to the focusing degree of learner his/herself based on Expression (1). As the learner's own focusing degree becomes larger, the view direction of the learner goes down to the memo-sheet. If the focusing target is the learner himself, the horizontal direction of the view and the location of the learner in our interface are determined by the target learner who has the second-largest focusing degree. Based on the calculated position *d(t)*, angle θ, and focusing target, the learner view is displayed in our interface window.

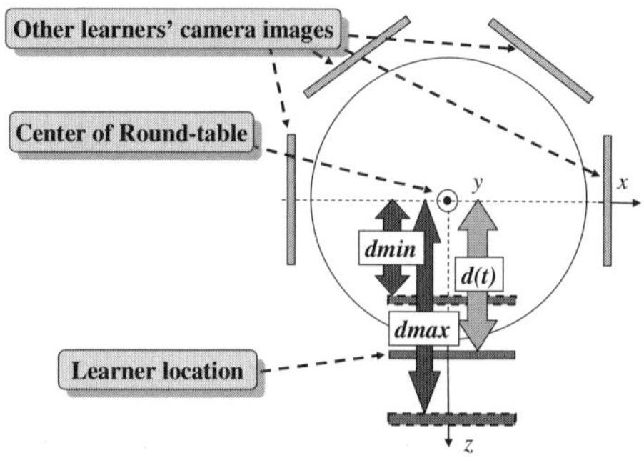

Fig. 5. Distance between learner and center of round-table

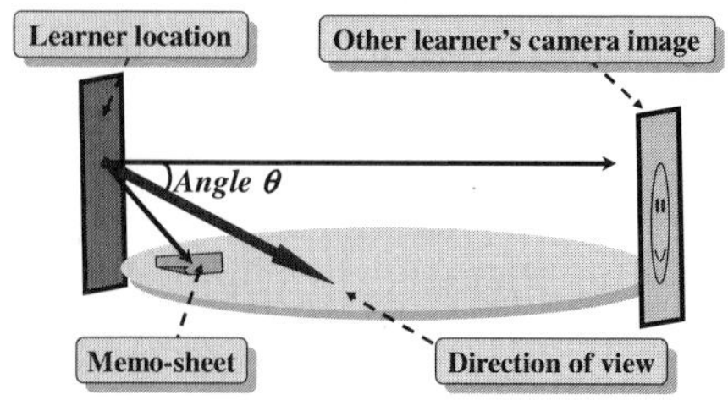

Fig. 6. Angle between other learner and learner's own memo

5 Prototype System

We embedded our interface which shows the learner view based on our detection method of focusing target in the collaborative learning system HARMONY, which has been previously developed in our laboratory [9].

Figure 7 shows windows in our interface. Learners can make their utterances with target learner information attached to the text-chat window and observe the learning environment through the round-table window. The camera images of other learners which exist in the learner view are situated around the round-table. Camera images of others are faced to their focusing targets. The learner situated in the center of the round-table window is the learner's focusing target. The learner can observe or write down his/her answer by clicking his/her own memo-sheet in the round-table window. When the focusing degree of learner himself becomes largest, his/her own memo-sheet window appears to the side of round-table window. This window is disappeared if the focusing degree for the learner himself is decreased and becomes smaller than a certain value. On the other hand, other learners' memo-sheets can be seen from the learner only if the memo-sheet in the round-table window is clicked and its possessor becomes the focusing target. Focusing learner's memo-sheet window is not editable. It also disappears when the focusing target is changed to other learner.

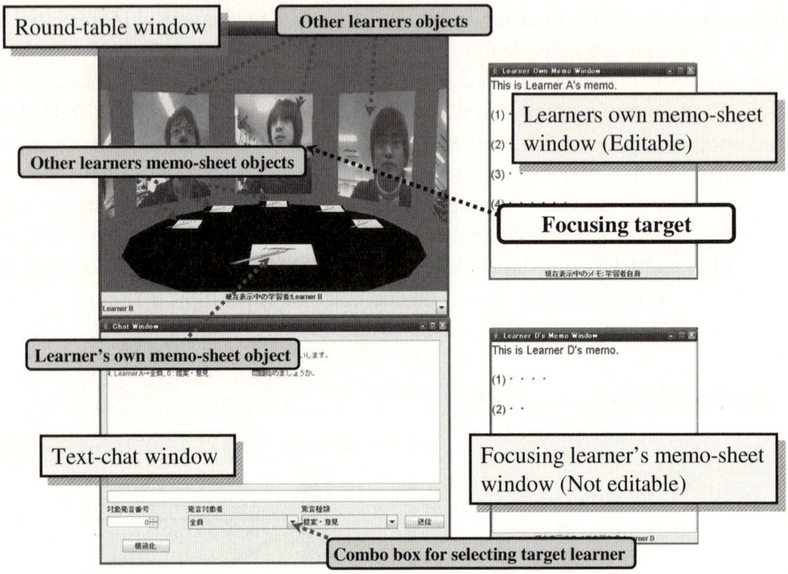

Fig. 7. Windows in our interface

The learner view is changed based on other learners' focusing degrees. Figurer 8 is an example of view change in the round-table window according to the focusing degree. In this example, six learners (learner *A* to *F*) participate in the collaborative learning, and the *A*'s round-table window is displayed. In Figure 8(a), *A* focuses on *D* whose focusing degree is 0.757. Since the *A*'s own focusing degree is 0.001, *A* does not face to *A*'s memo-sheet. If the focusing degree of *E* is increased to 0.942 and *A*'s focusing target is changed to *E* as shown in Figure 8(b), the camera image of *E* moves to the center of the interface. On the other hand, when the *A*'s own focusing degree is increased to 0.159 as shown in Figure 8(c), *A*'s view is inclined to the memo-sheet so as to observe memo-sheet easily.

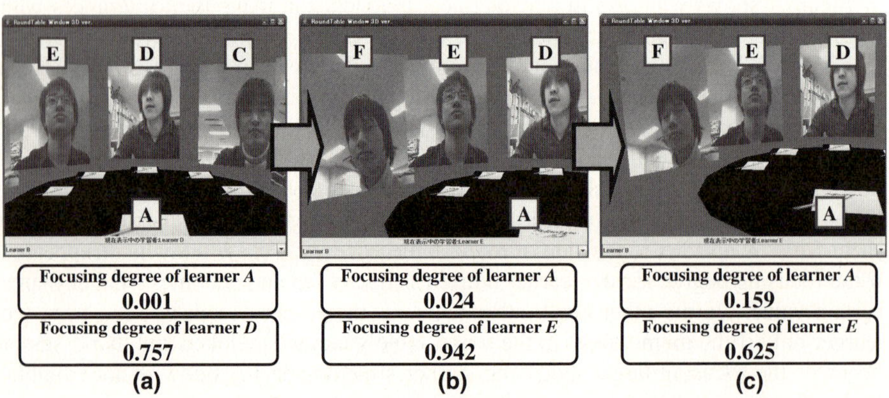

Fig. 8. Example of view changes according to focusing degrees

6 Conclusion

In this paper, we analyzed the learning activities in the collaborative learning environment, and proposed a method of detecting the focusing target according to the actions. Then, we developed the round-table interface which reflects the focusing intention of the learner. In this interface, the object displayed in the learner view is changed automatically according to the learner's focusing target and focusing degrees toward other learners. For our future, we should evaluate the correctness of our detection method and the effectiveness of the view change in the interface based on learners' actions.

Currently, α_i which indicates the change of the focusing degree on a certain action *i,* is prepared for each type of action. Interests for actions are different for individual learners. Therefore, the mechanism for changing α_i during the learning according to the individual learner's behavior must be considered.

References

1. Adelsberger, H.H., Collis, B., Pawlowski, J.M.: Handbook on Information Technologies for Education and Training. Springer, Heidelberg (2002)
2. Andriessen, J.H.E.: Working with Groupware. Springer, Heidelberg (2003)
3. Gutwin, C., Stark, G., Greenberg, S.: Support for Workspace Awareness in Educational Groupware. In: Proc. of ACM-CSCL 1995, pp. 147–156. ACM, New York (1995)
4. Prasolova-Førland, E., Divitini, M.: Supporting Social Awareness: Requirements for Educational CVE. In: Proc. of ICALT 2003, pp. 366–367 (2003)
5. Nakanishi, H., Yoshida, C., Nishimura, T., Ishida, T.: Free Walk: Supporting Casual Meetings in a Network. In: Proc. of CSCW 1996, pp. 308–314 (1996)
6. Hayashi, Y., Kojiri, T., Watanabe, T.: Immersive Round-Table Interface in Collaborative Learning. In: Apolloni, B., Howlett, R.J., Jain, L. (eds.) KES 2007, Part III. LNCS (LNAI), vol. 4694, pp. 769–776. Springer, Heidelberg (2007)
7. Hayashi, Y., Kojiri, T., Watanabe, T.: Computer-Supported Focusing Interface for Collaborative Learning. In: Proc. of ICCE 2007, pp. 123–130 (2007)
8. Kojiri, T., Ito, Y., Watanabe, T.: User-oriented Interface for Collaborative Learning Environment. In: Proc. of ICCE 2002, vol. 1, pp. 213–214 (2002)
9. Kojiri, T., Ogawa, Y., Watanabe, T.: Agent-oriented Support Environment in Web-based Collaborative Learning. Int'l Journal of Universal Computer Science 7(3), 226–239 (2001)

Grammar-Based Argument Construction

Jörn Sprado and Björn Gottfried

Centre for Computing Technologies, University of Bremen, Germany

Abstract. The science of history motivates the investigation on how arguments are constructed, which are later to be used in argumentation frameworks. This is difficult as much arguments in history are frequently built upon sources which are mutually inconsistent. In this paper it is shown how formal grammars can be employed in order to construct arguments. Thereby, it is investigated how the interplay of expressiveness and parsing effciency motivates in particular to use mildly context sensitive grammars. These systems are more expressive than context free grammars, but can still be parsed within polynomial time.

Keywords: Formal grammars, argument construction, cross-serial dependencies, historical theories, historical information sources.

1 Introduction

The science of history poses a challenging problem in the context of argumentation. Usually different views exist on how things have been like in the past, different views being frequently mutually inconsistent. The historian however does not aim at finding the ultimate truth. Rather the goal among historians is to investigate how different views relate, that is to say arguments are to be made to justify views and it is of interest how those views can be defended.

The problem can be stated more clearly by employing Dung's definition of an argumentation framework [9]. It consists of a set of arguments and of a binary relation called *attack*, telling us that one argument represents an attack against another argument. In our context this means that those arguments are mutually contradictory, and the reason is that either a single source is inconsistent in itself or that a number of different historical sources support different theories. Complementing the notion of an argumentation framework we are here in particular interested in generating arguments given historical sources. Then, however, a new aspect emerges: it shows that arguments can be composed of sub-arguments, entailing sets of arguments that are found at different levels of a hierarchy of arguments. While Modgil [15] has formalised hierarchical argumentation frameworks as an extension to Dung's system, it is the aim of the present work to focus on the generation of the hierarchical structure given a number of facts in a specific temporal order, these facts also being referred to as the object or domain level in [17] and in our case semantic tokens which already refer to conceptual objects as opposed to their natural language description found in a source. Making explicit this structure of how facts relate to sub-arguments and arguments enables the historian to compare

I. Lovrek, R.J. Howlett, and L.C. Jain (Eds.): KES 2008, Part I, LNAI 5177, pp. 330–340, 2008.

and to communicate such relations. Concerning the five steps an argumentation process consists of according to [3], we concentrate on the first step, namely the construction of arguments.

1.1 Example

As a running example we consider profit-and-loss accounts concerning people living in the past. These are of great interest for the historian, because possessions and the change of possessions tell us a lot about the socioeconomic structure in the past, and as a consequence, we can learn much about the biography of individuals or even about political developments in specific areas, for example.

Imagine a historian tries to determine if a concrete person participated in a revolt or not. For this purpose, he analyses different sources like court records, land registers and correspondences being interested in detailed information about the person over a specific period in the past. How do valuable possessions change over this time period (e. g. to lose a forest or to lose a meadow)? Is there any other evidence given for participating in the revolt or an abscondence (cf. Section 2.2 and 2.3)? Do these sources form a consistent model? Are these sources compatible with assumptions the historian makes? Being faced with a wealth of information sources it is an exhausting endeavour to sort out dependencies among historical characters, to test the consistency of sources, or to check the validity of hypotheses.

Information extracted from a source are events in a temporal order, and arguments are supported by these events. Specific chains of events reflect the historian's theories which he tries to align with the source under study. As already mentioned the examples we will use throughout this paper concern an individual for whom we want to show how events influence his wealthiness and change in wealthiness. That is, arguments can be found when compiling events in the temporal order they occur, such sequence of events justifying, for example why someone might have been active in a revolt uprising; a sequence in this case might simply consists of three semantic tokens:

< *end-revolt* >

with “<" and “>" denoting a decrease and increase in wealth, respectively. In other words, this sequence reads: before the revolt comes to an end a decrease of wealth is observed (<). However, after the end of the revolt things change and an increase of the wealth of our subject can be observed (>). Such a chain of events would generate an argument that speaks for our peasant to have participated in the revolt and who succeeded in overcoming his poverty (according to a theory[1] to be defended). According to the temporal order of such events, it is shown how chains of events develop and this is what the historian uses in justifying specific arguments (which are part of his theories).

1.2 Challenges

The inconsistency of historical sources, however, poses a large problem, let alone their vagueness. It is therefore the purpose of many historical investigations to expound

[1] By the term *theory* we refer to the historians' assumptions which are to be confirmed or refuted.

which kinds of arguments can be made independent on whether they form together a consistent set of arguments. Thereby, it is not about proving them to be true but of finding arguments which can be persuasively defended. While Dung and Modgil provide means for doing this on the basis of sets of abstract arguments which support or attack each other, the historian needs also to align specific arguments (his theories) to given sources. This entails several problems we shall focus on in the present investigation:

- The temporal order of events, given in sources, is to be considered.
- Dependencies among events occur, e. g. one event has to precede another one.
- Dependencies between arguments and events occur.
- Despite of these dependencies, chains of events given in a source are to be parsed efficiently, i. e. with polynomial time, to be of practical use.
- The structure of alignment is to be made explicit in order for the historian to comprehend and communicate the generated argument.

In particular dependencies among events are a challenge in that they define a specific context which is to be maintained when collecting additional facts from sources. This is the reason why we will essentially focus on this problem of *cross-serial dependencies* which, if correctly dealt with, enables the construction of argumentation chains that can even deal with overlapping events: it makes a difference when an increase or decrease in wealth is observed regarding specific events, and it makes therefore a difference whether event B starts after the end of event A or before A ends, for example. This must be taken into account in the formalism we are looking for.

1.3 Structure

The paper is structured as follows. The next section shows how formal grammars are employed in order to assemble events to form an argumentation. While the first step consists in considering simply the temporal order of events, dependencies are taken into account in the following sections. Thereby, it is shown that the introduced formalism satisfies both constraints, to parse structures in polynomial time and to make explicit the parsed structures by means of argumentation trees. The outlook shows further steps in particular towards the integration of the present approach into an established ontology of the historical domain.

2 Argumentation Trees

This section elaborates on difficulties and possibilities arising when aligning theories of the historian with chains of events. For this purpose grammars will be used to parse sequences of events in the form of semantic tokens into possible explanations (arguments). Thereby we follow the line of argument of Kiefer et al. [12], concerning the employment of formal grammars for dealing with the analysis of sequences of events.

2.1 Temporal Order of Events (Regular Grammars)

A grammar can generally be defined as a quadruple $G = (N, \Sigma, P, S)$ where

- N is an alphabet of non-terminals,
- Σ is an alphabet of terminals with $N \cap \Sigma = \emptyset$,

- P is a set of production rules $P \subseteq (N \cup \Sigma)^* \times (N \cup \Sigma)^*$, and
- $S \in N$ is a start symbol.

When considering the event sequence of the running example we would like to define a simple *regular grammar* $G_{reg} = (N_{reg}, \Sigma, P_{reg}, S)$ which consists of a set of non-terminals $N_{reg} = \{$ARGUMENTATION, EVENT, GAIN, LOSE$\}$, a set of terminals $\Sigma = \{$<, >, *start-revolt, end-revolt*$\}$, a set of regular production rules $P_{reg} = \{p_1, ..., p_4\}$:

p_1: ARGUMENTATION	$\rightarrow$	EVENT | GAIN | LOSE
p_2: LOSE	$\rightarrow$	< EVENT
p_3: EVENT	$\rightarrow$	*end-revolt* GAIN
p_4: GAIN	$\rightarrow$	> | ϵ

and a start symbol $S =$ ARGUMENTATION. Regular production rules might also have a non-terminal on the right hand side of a rule; but in order to be regular they must be expressible in either left- or right-regular manner. Depending on the position of the terminal symbol a rule is called either left- or right-regular; e. g. LOSE $\rightarrow$ < EVENT is a left-regular production rule, since the terminal < precedes the non-terminal EVENT. One possible sequence satisfied by grammar G_{reg} is the example given in Section 1.1. Note that rules having a number of alternatives such as p_4 make sense in as much later it should be possible to include different kinds of reasons for profit-and-loss accounts.

2.2 Dependencies among Adjacent Events (Context Free Grammars)

Being interested in allowing a higher expressiveness and a more natural description of the rules, we extend production rules by allowing for adjacent non-terminals, for example:

REVOLT	$\rightarrow$	LOSE EVENT GAIN
EVENT	$\rightarrow$	*start-revolt* | *end-revolt*
LOSE	$\rightarrow$	<
GAIN	$\rightarrow$	>

These rules are telling us that the argument REVOLT can be supported when either a decrease of wealth is observed before the start of the revolt and an increase in wealth after the revolt started, or if a decrease of wealth is observed before the end of the revolt while there is an increase after the revolt. Unfortunately, these kinds of production rules are not allowed in regular languages. However, *Context Free Grammars* (CFGs) are able to handle those rules.

A CFG is a grammar with a set of production rules P that follow the notation $n \rightarrow \omega$ where $n \in N$ and $\omega \in (N \cup \Sigma)^*$. Thus, let $G_{cfg} = (N_{cfg}, \Sigma, P_{cfg}, S)$ be a grammar by extending the set of non-terminals of the regular grammar with $N_{cfg} = N_{reg} \cup \{$ABSCOND$\}$ and by considering a number of context free rules $P_{cfg} = \{p_5, ..., p_{12}\}$ which are defined as follows:

p_5: ARGUMENTATION	$\rightarrow$	GAIN
	|	LOSE
	|	REVOLT
	|	ABSCOND

p_6: EVENT → START-REVOLT ∣ END-REVOLT
p_7: START-REVOLT → *start-revolt*
p_8: END-REVOLT → *end-revolt*
p_9: GAIN → >
p_{10}: LOSE → <
p_{11}: REVOLT → LOSE EVENT GAIN
∣ GAIN EVENT GAIN
∣ LOSE START-REVOLT GAIN END-REVOLT LOSE
p_{12}: ABSCOND → GAIN EVENT LOSE
∣ GAIN END-REVOLT GAIN
∣ GAIN START-REVOLT LOSE END-REVOLT GAIN

CFGs in comparison to regular grammars, which could be parsed by a finite state machine, have less restrictions. Nevertheless, as a type-2 grammar of the well-known *Chomsky Hierarchy* [7] they can still be parsed in polynomial time.

Parse trees are obtained when parsing the given event sequence using a grammar such as G_{cfg}. Two parse trees are shown in Figure 1. Each tree represents an argument, which itself might consist of a number of hierarchical-ordered arguments, for a defined time slice. The hierarchical structure splits arguments into sub-arguments and facts. As a well-formed structure, such a parse tree makes argumentation explicit and easy to comprehend for the non-expert.

In general, different parse trees do not form conflict-free sets of arguments. This means that if there are two mutually ambiguous (cf. Figure 1) or even contradictory parse trees then there will be more than one possible theory in argumentation. Note how this contrasts to describing a programming language where different parse trees yield different programs and thus, unambiguity is necessary [2].

2.3 Dependencies among Non-adjacent Events (Tree Adjoining Grammars)

So far we argued at a high level of abstraction, i. e. profit-and-loss accounts have been considered without reference to specific objects; until now we do not respect concrete properties, e. g. to lose a forest or to lose a meadow, or we do not distinguish different events, such as two different revolts. Since, however, a specific event cannot end before it started, there is a dependency between events which we need to represent.

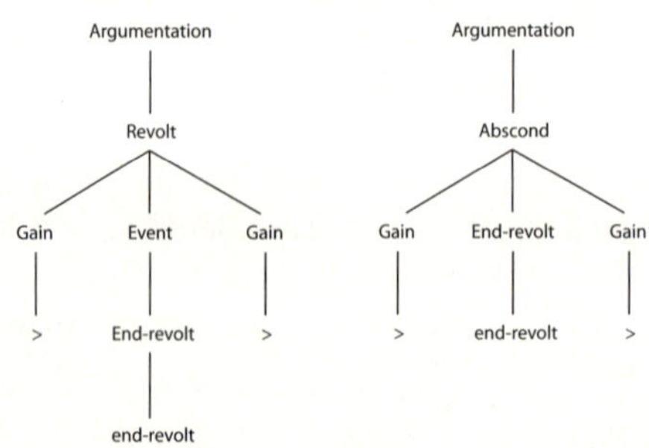

Fig. 1. An ambiguous sequence having two different parse trees created by the context free grammar G_{cfg}

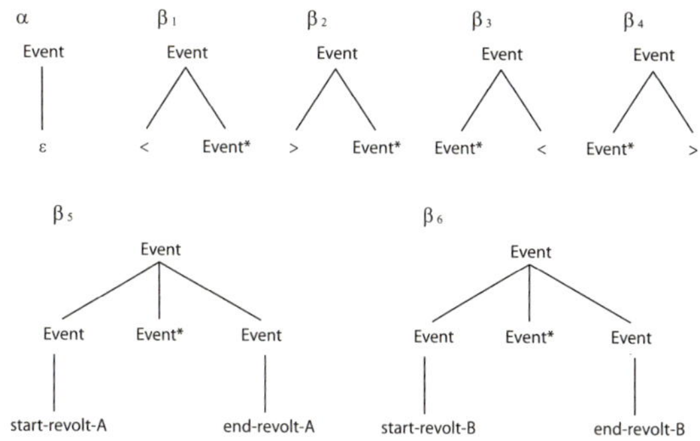

Fig. 2. An initial tree α and a set of auxiliary trees $\beta_1 - \beta_6$ of a Tree Adjoining Grammar G_{tag}

From the grammatical point of view such cross-serial dependencies cannot be represented by CFGs [12]. Therefore, we need a more powerful grammar than a type-2 grammar. In the Chomsky Hierarchy this would usually mean to resort to a type-1-grammar. Such a grammar is context sensitive. That is, the word problem is decidable but only within exponential time by a linear-bounded automata. There is a trade-off between computational and linguistic complexity; restricting the linguistic complexity could enable more efficient parsing algorithms as in the case of *Mildly Context Sensitive Grammars* (MCSG). The main idea of this category of grammars is to restrict cross-serial constraints, and thus, to allow grammars to be parsed in polynomial time. In particular, the class of *Tree Adjoining Grammars* (TAGs[2], cf. [11]) belongs to MCSGs; they allow dependencies to be maintained.

By contrast to context free grammars TAGs generate a language based on trees instead of rules as elementary structures. A TAG is defined as a tuple $G = (I, A)$ where I is a set of initial trees and A is a set of auxiliary trees. An *initial tree* $\alpha \in I$ is defined as a tree with a start symbol as a root node and a set of non-terminal and terminal symbols as leaf nodes. An *auxiliary tree* $\beta \in A$ is defined as a tree with a set of terminal symbols and exactly one non-terminal leaf node. The latter is referred to as a foot node which is marked with an "*". Note that the foot node and root node have to refer to the same non-terminal in an auxiliary tree. In addition, operations called *substitution* and *adjunction* are defined to build *derived trees* from the elementary structures, i. e. from the initial and auxiliary trees. Substituting the root node of a tree α with the foot node of another tree β a new tree γ is derived. However, the more interesting operation will be the adjoining operation, which allows dependencies between objects of the set of terminals and non-terminals to be maintained (such as between the start and end of a revolt).

Adjoining an auxiliary tree β (having the root node N) with a tree γ (containing a node also labelled with N) yields a tree γ'. In the latter the subtree t of γ dominated by N is deleted and the node N will be associated with tree β which will be finally connected to the root node of subtree t. An example illustrates that TAGs are a reasonable formalism to tackle the issue of cross-serial dependencies. Considering the intersection

[2] See [1] for a detailed introduction to TAGs.

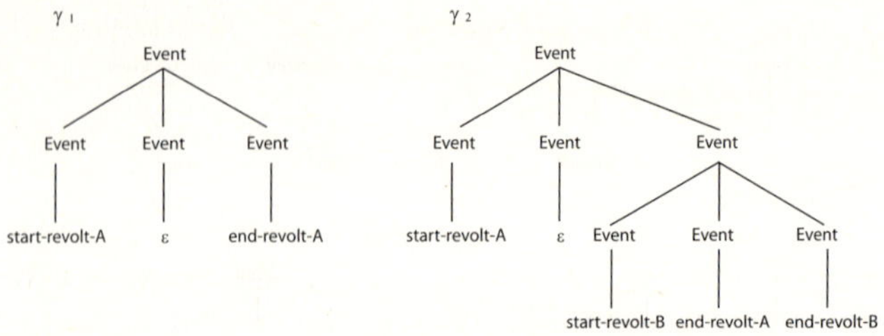

Fig. 3. The derived trees γ_1 and γ_2 of G_{tag}

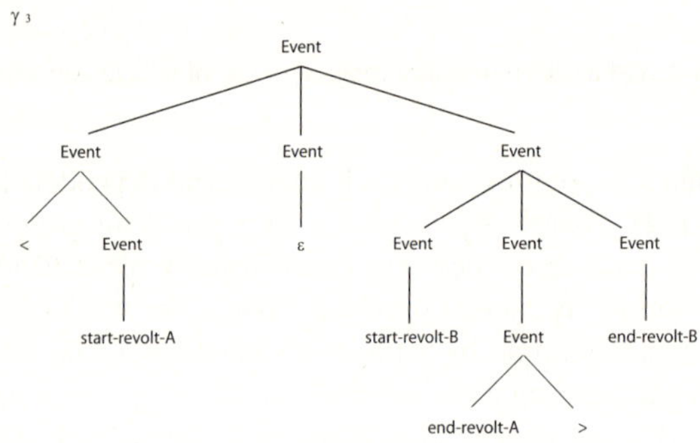

Fig. 4. A derived tree γ_3 of G_{tag}

of events in the running example we define a grammar $G_{tag} = (I, A)$ with $I = \{\alpha\}$ and $A = \{\beta_1, ..., \beta_6\}$ shown in Figure 2. Substituting the root node of the initial tree α with the non-terminal foot node of auxiliary tree β_5 we get a derived tree γ_1. In Figure 3 we also show the adjunction of two trees representing dependencies between a revolt A and another revolt B resulting in a derivation tree γ_2. After two adjoining operations the tree γ_3 consists of additional arguments considering profit-and-loss-accounts (see Figure 4). The operations are:

$$\gamma_1 = \text{SUBSTITUTION}(\alpha, \beta_5)$$
$$\gamma_2 = \text{ADJUNCTION}(\gamma_1, \beta_6)$$
$$\gamma_3 = \text{ADJUNCTION}((\gamma_2, \beta_1), \beta_4)$$

The example shows how to handle cross-serial dependencies within a TAG. The adjoining of trees guarantees that dependencies between events are not destroyed when enriching a given argumentation by further facts; e. g. when considering yet another revolt, its begin and end being nested in the first revolt in a specific way (revolt B overlaps revolt A). Eventually, arguments could be constructed by the historian's theory of G_{tag} stating a decrease in wealth of a peasant before the revolt A starts and an increase in wealth after the corresponding revolt ends. That revolt B starts before A ends might have been extracted from another source. This second revolt can however be embedded in the current argumentation of how the wealthiness of the peasant under investigation changes. In a similar way much more complex arguments can be stated by establishing further initial- and auxiliary trees, those arguments probably consisting of a number of nested dependencies.

3 Discussion

The proposed approach is purely generative, i. e. it allows arguments to be constructed given inconsistent sources. This is possible since the generation process only aims at constructing persuasive arguments based on appropriate chains of events, as opposed to requiring a given knowledge base to be formally consistent; something which is hardly given regarding different, vague historical sources. Clearly, the approach can be generalised to other domains. The historical domain however helps in clarifying problems occurring in the process of constructing arguments.

3.1 Related Work

A considerable amount of literature has been published on argumentation in AI. A survey is given by Bench-Capon et al. [5] which also summarises a number of trends and concerns. While one trend is the continuing enrichment of the formal theory of argumentation we focus on employing formal grammars for constructing arguments. In addition to this work which can be seen as an exploitation of Natural Language Processing (NLP) and argumentation in the historical domain, there has been an increasing interest in applying argumentation-based methodologies to multiagent system (MAS) applications [14]. Moreover, there will be effort on merging work in the area of argumentation and Machine Learning [16]. While MASs are irrelevant for our purpose, ML could be of interest when it comes to the automatic analysis of sources and the automatic generation of grammars.

In contrast to most of the work based on the abstract argumentation framework of Dung [9], we do not determine sets of "acceptable" arguments or examine whether a statement can be regarded as justified (cf. [6, 4]). The aim of the current work is rather to identify those chains of events which support possible theories of historians, these chains of events being kinds of argumentation chains that aid in justifying the historian's theories.

More closely related to our approach are those which also consider the problem of dependencies in temporal ordered lists of objects, in particular when the consideration of a context is relevant. Then, in arguing about cross-serial dependencies some authors have made attempts to employ grammars, especially in the area of plan recognition [10, 12]. Geib et al. have primarily investigated the relationship between grammatical formalisms from NLP and representations for plan recognition, whereas Kiefer et al. have applied MCSGs for the purpose of intention recognition; similarly as [13] have proposed to represent cross-serial dependencies by TAGs.

3.2 Challenges

In the introduction a number of problems have been identified. These problems occur when it comes to the alignment of arguments (the historian's theories he tries to defend) to given sources (events described by semantic tokens). This paper has outlined a solution to these problems:

– The temporal order of events at the domain level is considered by the order of terminals in a grammatical formalism.
– Dependencies among events are dealt with in several ways, depending on how complex these interrelationships are:
  - by the appropriate order of terminals and non-terminals, it is sometimes sufficient to take just regular grammars;
  - by two or more successive non-terminals, allowing more sub-arguments to be followed one another to be represented in a single rule (context free grammars);
  - dependencies between events that do not necessarily follow directly each other, such as a terminal in between two non-terminals which correlate (tree adjoining grammars).
– Dependencies between arguments and events are represented by parse trees.
– Chains of given events are parsed within polynomial time, since the most expressive formalisms we need to deploy are TAGs.
– The structure of alignment between theory and facts is made explicit by parse trees of the grammars employed; these *argumentation trees* easing the comprehension, communicating, and persuasive defending of the historian's theory.

3.3 Outlook

Concerning the five steps an argumentation process consists of according to [3], we concentrated on the first step, that is the construction of arguments. The next step concerns the definition of interactions between arguments. In our case this would concern relations between non-terminals which represent specific arguments. Interactions between arguments are either possible by the grammar itself or have to be specified explicitly by relations, involving probably both attacks and supports, as in bipolar argumentation frameworks [3]. On the other hand, before proceeding with the second step in the argumentation process, the construction of arguments should be further investigated. The reason is that argument construction is much more complex than what could be presented in this work.

Additionally, it should be investigated how to connect the process of argument construction by grammars to given argumentation frameworks. That is to say that argumentation frameworks (such as Dung's) require to determine which arguments and which kinds of (attack and support) relations do exist. While the construction process aids in determining this (based on specific sources), the result of the construction process must however fit the formal specification of the argumentation frameworks under consideration.

Then, quite another issue concerns the combination of grammars which represent theories with established ontologies, namely in the historical domain the CIDOC upper level ontology [8]. Concepts used in the grammar for both terminals and non-terminals should be used in the sense of what CIDOC defines, hence, CIDOC concepts define our semantic tokens. This enables to better compare theories of different historians and information extracted from different sources.

4 Summary

The construction of arguments is investigated by employing formal grammars. They have the advantage of making explicit how arguments compose into sub-arguments and facts. This allows easy comprehension by the non-expert and forms the basis for argumentation processes. Tree Adjoining Grammars have been identified to form appropriate formalisms which are expressive enough in order to allow for cross-serial dependencies. Additionally, these formalisms can still be parsed within polynomial time, making the approach manageable in applications.

References

[1] Abeillé, A., Rambow, O.: Tree Adjoining Grammar: An Overview. In: Abeillé, A., Rambow, O. (eds.) Tree Adjoining Grammars: Formalisms, Linguistic Analysis, and Processing. CSLI Publications, Stanford (2000)
[2] Aho, A.V., Ullman, J.D.: Foundations of computer science. Computer Science Press, Inc., New York (1992)
[3] Amgoud, L., Cayrol, C., Lagasquie, M.-C., Livet, P.: On bipolarity in argumentation frameworks. International Journal of Intelligent Systems (2008)
[4] Baroni, P., Giacomin, M.: On principle-based evaluation of extension-based argumentation semantics. Artif. Intell. 171(10-15), 675–700 (2007)
[5] Bench-Capon, T.J.M., Dunne, P.E.: Argumentation in artificial intelligence. Artif. Intell. 171(10–15), 619–641 (2007)
[6] Caminada, M., Amgoud, L.: On the evaluation of argumentation formalisms. Artificial Intelligence 171(5-6), 286–310 (2007)
[7] Chomsky, N.: Three models for the description of language. IEEE Transactions on Information Theory 2(3), 113–124 (1956)
[8] Doerr, M.: The CIDOC Conceptual Reference Module: An Ontological Approach to Semantic Interoperability of Metadata. AI Magazine 24(3), 75–92 (2003)
[9] Dung, P.M.: On the Acceptability of Arguments and its Fundamental Role in Nonmonotonic Reasoning, Logic Programming and n-Person Games. Artificial Intelligence 77(2), 321–358 (1995)
[10] Geib, C.W., Steedman, M.: On natural language processing and plan recognition. In: International Joint Conference on Artificial Intelligence (2007), pp. 1612–1617 (2007)
[11] Joshi, A.K.: Tree Adjoining Grammars: How Much Context-Sensitivity is Required to Provide Reasonable Structural Descriptions?. In: Dowty, D.R., Karttunen, L., Zwicky, A.M. (eds.) Natural Language Parsing: Psychological, Computational, and Theoretical Perspectives, pp. 206–250. Cambridge University Press, Cambridge (1985)
[12] Kiefer, P., Schlieder, C.: Exploring Context-Sensitivity in Spatial Intention Recognition. In: Gottfried, B. (ed.) 1st Workshop on Behaviour Monitoring and Interpretation (BMI 2007). CEUR Workshop Proceedings 296, pp. 102–116 (2007)
[13] Kuhlmann, M., Möhl, M.: Extended cross-serial dependencies in Tree Adjoining Grammars. In: Eighth International Workshop on Tree Adjoining Grammars and Related Formalisms (TAG+8), Sydney, Australia (2006)
[14] Maudet, N., Parsons, S., Rahwan, I. (eds.): ArgMAS 2006. LNCS (LNAI), vol. 4766. Springer, Heidelberg (2007)

[15] Modgil, S.: Hierarchical Argumentation. In: Fisher, M., van der Hoek, W., Konev, B., Lisitsa, A. (eds.) JELIA 2006. LNCS (LNAI), vol. 4160, pp. 319–332. Springer, Heidelberg (2006)
[16] Možina, M., Žabkar, J., Bratko, I.: Argument based machine learning. Artif. Intell. 171(10-15), 922–937 (2007)
[17] Wooldridge, M., McBurney, P., Parsons, S.: On the meta-logic of arguments. In: Dignum, F., Dignum, V., Koenig, S., Kraus, S., Singh, M.P., Wooldridge, M. (eds.) AAMAS, pp. 560–567. ACM, New York (2005)

Towards the Knowledge Capitalisation and Organisation in the Model of Conformity-Checking Process in Construction

Anastasiya Yurchyshyna[1,2], Catherine Faron-Zucker[2], Nhan Le Thanh[2], and Alain Zarli[1]

[1] CSTB, Centre Scintifique et Technique du Bâtiment, 290 route des Lucioles, BP 209, 06904 Sophia Antipolis, France
{anastasiya.yurchyshyna,alain.zarli}@cstb.fr

[2] I3S, Université de Nice Sophia-Antipolis, CNRS 930 route des Colles, BP 145, 06903 Sophia Antipolis, France
{Catherine.Faron-Zucker,Nhan.Le-Thanh}@unice.fr

Abstract. This paper presents an ontological method aimed at the capitalization of expert knowledge in the context of semi-automatic checking model of the conformity of construction projects against a set of construction norms. The efficiency of our ontology-based reasoning model relies on two keystones. First, it is based on the matching of construction projects represented by RDF graphs to technical conformity queries formalized as SPARQL queries. Second, our reasoning model integrates the meta-knowledge on conformity checking process. Our approach of capitalizing such knowledge is based on the development of semantic annotations of conformity queries and organizing them into a query base. This helps to formalize the expert knowledge in form of expert rules scheduling the matching operations of the checking process. Semantic annotations of conformity queries also help to generate a structured conformity report, which interprets the results of reasoning in terms of conformity checking in construction.

Keywords: Conformity checking, capitalization of construction knowledge, organization of the base of conformity queries, Semantic Web in Construction.

1 Introduction

Nowadays, the construction industry is managed and regulated by complex technical rules that define the execution of construction products and their components. However, their current representations are still mostly paper-based (e.g. texts with diagrams, tables) and require a human interpretation to be practically applied.

Construction projects (e.g. public buildings) are commonly represented in the Industry Foundation Classes (IFC) model. Developed by the International Alliance for Interoperability, the IFC model is an object oriented data model for Building Information Modelling (BIM) that contains information about all aspects of a building throughout its lifecycle. The model is provided with XML syntax for information

I. Lovrek, R.J. Howlett, and L.C. Jain (Eds.): KES 2008, Part I, LNAI 5177, pp. 341–348, 2008.

exchange: ifcXML[1]. However, the ifcXML representation is sometimes insufficient to describe the complexity of the building information flow: the IFC model is semantically richer than any XML language (in ifcXML, it is impossible to model the constraints on data types values, e.g. non negative *length,* which can be represented, however, by the IFC data model). Moreover, the IFC model is not particularly oriented at conformity-checking problem.

Our work intentionally dwells within the scope of a double trend. First, we propose an expressive model for checking a construction project against technical norms. It is based on the matching of norm representations with the representations of construction projects. Second, we develop an approach of semantic annotation and organisation of these norms aimed to optimise our conformity-checking model. In this research, we particularly focus on the method of capitalisation of expert knowledge and its integration into our checking model. Its efficiency is explained by: (i) ontological representation of regulation knowledge and ontology-based reasoning; (ii) semantic annotation and organisation of conformity queries; (iii) integration of meta-knowledge characterizing the checking process in construction, in form of expert rules that help to conduct the validation process and explain its results.

In next section, we analyse the conformity-checking problem in construction and discuss some aspects of the related research. In section 3, we present our approach for formal representation of the knowledge taking part in the checking process. Section 4 is devoted to the organisation of conformity queries aimed to optimise the reasoning model represented in section 5. Section 6 describes the prototype C3R developed to validate the model. To conclude, we discuss the ongoing works and the perspectives.

2 Analysis of Conformity Checking Problem in Construction

The complexity of the conformity-checking problem in construction is explained by the large amount of its multidisciplinary components and the interdependencies among them. In this paper, we focus on the following aspects.

By analysing a standard ifcXML *representation of a construction project* in the context of conformity checking, we notice that it (i) is redundant; (ii) contains voluminous data; at the same time, (iii) is semantically insufficient. The first axis of our research concerns, therefore, the problem of the development of an *intelligent* representation of a project oriented the specific problem of conformity checking. Our research is based on works on the development of a general construction ontology buildingSMART [1], the construction of an application-oriented ontology [11] as well as the projects aiming at the development of the IFC-to-OWL conversion tool [9].

The current representations of technical construction norms (paper format, non formalised, non interoperable, interpretation required) cannot be used *as-is* for modelling the conformity reasoning. Several ongoing works, as for example, SMART-codes™[2], the CICA[3] project and [8] are held today to develop a formal representation of the conformity requirements for compliance checking. We analyse these works to

[1] http://www.iai-international.org/IFCXML
[2] http://www.iccsafe.org/SMARTcodes
[3] http://www.cicacenter.org

define a base of constraints representing technical norms, which we use to validate out model. However, the knowledge extraction from texts is out of scope of our reasoning-oriented modelling, and we don't particularly focus on it: the formalisation of the queries is done manually with the help of CSTB experts.

The reasoning on the checking is based on the *graph homomorphism approach* (graph projection). We have adapted the approach of [5] on the *validation of knowledge bases* [10] constructed for our problematic of conformity checking.

Our research is also largely based on the problem of (semi)formalisation of expert knowledge to be integrated into the conformity-checking model. It is based on [4] and multiple interactions with domain experts – mainly from CSTB. The particular interest represents the capitalisation of this tacit knowledge and the organisation of different types of expert knowledge for further reasoning that, to our knowledge, is innovative for the conformity checking modelling in the construction sector.

3 Knowledge Representation

We adopt the ontological approach and the semantic web technologies [2] to develop the knowledge acquisition method (Fig. 1) of complex and multidisciplinary knowledge characterising the conformity-checking process in construction. Here, we present a synthesized description of our knowledge acquisition method. A more detailed explanation and corresponding examples could be found in [6].

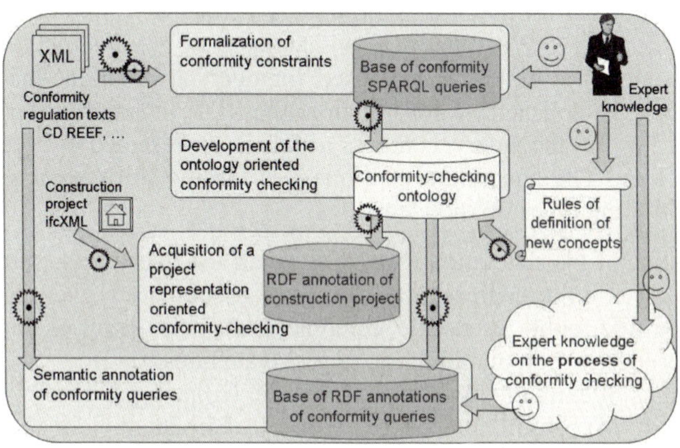

Fig. 1. Knowledge acquisition method

The first phase of our method aims to *acquire the formal representations of technical construction norms*. We have developed a base of accessibility queries by extracting them from the CD REEF, the electronic encyclopaedia of construction texts and regulations, edited by CSTB, and by formalising them as SPARQL queries in collaboration with construction experts from the CSTB.

The second phase aims at the *automatic development of an ontology oriented conformity checking* on the basis of the concepts from the acquired SPARQL queries. These concepts are organized as hierarchies and described in the RDFS language. The acquired ontology is then enriched by non-IFC concepts from formalized conformity queries. The intervention of domain experts is required in this case to define new non-IFC concepts in terms of the checking ontology (e.g. GroundFloor class is defined by a resource of type IfcBuildingStorey situated on the level of entering into a building).

The third phase is dedicated to the *acquisition of a construction project representation oriented conformity checking*. This representation is based on its initial ifcXML representation and is guided by the acquired conformity-checking ontology. We develop an XSLT stylesheet that filters this ifcXML to extract the data relative to the conformity checking ontology and organizes them as RDF triples. The acquired RDF is then enriched with non-IFC concepts extracted from conformity queries (e.g. a project representation is enriched by GroundFloor concept calculated on the basis of its initial IFC-based data (e.g. IfcDoor, IfcStair, etc.)

The acquired queries, however, contain only conformity constraints, but have no supplementary information, guiding the checking process: e.g. the scheduling of queries. The forth phase of our method aims thus at *the development of semantic annotation of conformity queries.* We propose a special RDF annotation of a query, developed according to its *tag-based context*: possible values for certain tags are concepts/properties of the conformity-checking ontology. To do it, we combine two main methods of document annotation: annotation by content of the document and annotation by its external sources [7]. First, we annotate a query by its content by modifying the approach [7] that annotates a conformity document by *element to check*. Instead, we propose to annotate a query by *set of its key concepts*. In other words, we define *keyConcept* tag in the RDF annotation of a query, which value is a list of primitive concepts from the conformity-checking ontology extracted from the SPARQL representation of this query. We remark also that there is a *semantic correspondence* between different types of knowledge used for query annotation. In our case, this correspondence is established by construction: the RDF annotation of a query is constructed according to the query content, which is represented by the conformity-checking ontology. The annotation of a query according to external sources allows representing different types of knowledge:

1. Characteristics of the regulation text from which the query was extracted (i) thematic (e.g. accessibility); (ii) regulation type (e.g. circular); (iii) complex title composed of the title, publication date, references, etc.; (iv) level of application (e.g. national), (v) destination of a building (e.g. private house).
2. Characteristics of extraction process: (i) article, (ii) paragraph from which a query was extracted, (iii) current number (e.g. 3 query of 1 paragraph of *Door* article).
3. Formalised expert knowledge. It is tacit « common knowledge » on the process of conformity-checking that is commonly applied by domain experts: (i) knowledge on domain and sub domain of the application of a query (e.g. Stairs); (ii) knowledge on checking practice (e.g. if a room is *adapted,* it is always *accessible*)
4. Application context of a query. This group specifies the aspects of query application for certain use cases. For example, the requirements on the maximal height of stairs handrail vary from 96 cm (for adults) to 76 cm (for kids). In this case, it is important to know the destination of a building (e.g. school).

Characteristics and possible values of the first two groups are automatically extracted from the CD REEF. The knowledge described by the last two groups is defined partially and/or has to be explicitly formalised by domain experts.

4 Organisation of the Base of Conformity Queries

The organization of the base of conformity queries defines the optimal scheduling of matching procedures, which captures the tacit knowledge and experience of the experts of the construction industry (mostly from CSTB). It is based on the semantic classification of conformity queries: the identification of groups of queries for which the reasoning is similar. We have defined 3 types of such semantic classification:

1. *Classification by construction* corresponding to the criteria that were used for generating the semantic annotations of these queries (e.g. thematic). The classes are classified by the criterion possible values (e.g. accessibility, acoustic, etc.).
2. *Classification by key concepts* corresponding to primitive concepts of the conformity-checking ontology. This classification is, in fact, the classification by *specialisation/generalisation relations* existing between the graph patterns of key concepts. For example, (i) the class concerning a door (door, entrance door, main door) is defined by the primitive concept IfcDoor; (ii) the class concerning a building (public building, three-floor house, school) is defined by IfcBuilding.
3. *Classification by application condition* for queries that should be checked only under certain conditions (e.g. for a building with multiple entrances). It is a classification by *specialisation/generalisation relations* existing between the graph patterns representing the condition of query application. For example, the application condition of a query *in school, all doors are ...* is a specialisation of the application condition of a query *in public building, all doors are ...*as the graph representing school is the specialisation of the one representing *public building*.

5 Conformity Checking Model

Our conformity-checking model [6] is based on the matching of norm representations with representations of the construction project. The effectiveness is gained by the scheduling of the conformity queries on the basis of their RDF annotations aiming at reducing the number of matching procedures and generalizing their validation results. The matching operations results are interpreted in terms of (non)conformity of the construction project and the reasons of the eventual non-validation are identified.

5.1 Validation of the Project According to the Norm

The elementary reasoning mechanism of our conformity-checking model is the matching of a construction project representation with representations of construction norms. Practically, in the context of conformity checking in construction, we check the *non-conformity* condition: a construction project is conform to a query, if there is no projection of the SPARQL representation of this query into the RDF of the project. If such projection is found for some elements, it means that these elements cause the *non-conformity* of the project against this query. Otherwise, the projection can not be established if the RDF of the project does not contain enough information which is "asked" by the query.

5.2 Formalization of Expert Reasoning

In collaboration with CSTB experts, we have identified a set of so called expert rules guiding the process of the conformity checking in construction. These rules are applied to classes of conformity queries and define the optimal scheduling of their checking according to the priorities between the whole query classes. They minimise, therefore, the general time-cost characteristics of the checking process. The groups of expert rules correspond to the types of classification of conformity queries:

1. Classification *by construction.* Each type of such classification and possible values are defined externally: in regulation texts and/or by checking practices. The scheduling of the treatment of these queries corresponds to the order/hierarchy of their classes. We defined 3 types of expert rules illustrated by simplified examples:
   - Classification type: according to the type of regulation text. Identified classes: (i) decrees, (ii) circulars. The scheduling corresponds to the explicit hierarchy of regulation texts: (1) decrees, (2) circulars.
   - Classification type: according to the application domain. Identified classes: (i) vertical circulation, (ii) stairs, (iii) elevator. The scheduling is defined and validated by a domain expert: e.g. (1) vertical circulation, (2) stairs and elevator with the same priority.
   - Classification type: according to the regulation text. Identified classes: different regulations in our base. The scheduling is defined by a user during the checking process: e.g. all queries extracted from Circular 82-81 of 4/10/1982.
2. Classification by *key concepts* (by *specialisation/generalisation relations* between their graph patterns). Queries representing more specialised knowledge are treated in priority. For example, an *entrance door* query is prior to a *door* query (*entrance door* is a specialisation of *door*), because if a construction project is non-conform to the first one, it will be automatically non-conform to the second one.
3. Classification by *application condition* (for queries checked only under certain conditions). The priority is also reserved for queries, which application condition is more strict knowledge in comparison to application conditions of other queries. For example, if we interest on the accessibility of a *school*, we should start by checking queries applied to *public building receiving sitting public* and continue by checking more general queries applied to *public building receiving public*

Our current work is focused on the identification of hybrid expert rules: the rules taking into account different types of query classification. To do this, we propose to define a *context* of checking process that corresponds to *typical* user scenarios, i.e. how the checking process is held in practice (e.g. (1) choose thematic, (2) choose element to check, (3) schedule queries according to *key concepts* set of expert rules).

5.3 Analysis and Interpretation of Results in Terms of Conformity Checking

The results of the checking process (validation/non-validation, elements causing the non-validation and its possible reasons) are used to generate a conformity report, which interprets them in the terms of conformity checking. First, it lists the *failed* queries: (i) queries that fail in the checking process because of the non-compatibility of the construction project; (ii) queries, which graph pattern is more general in comparison to the ones previously failed, (iii) queries, which annotation representing the condition of its application is more general in comparison to the annotation of another failed query. The other possible reason of validation failure is the case when the representation of the construction project does not contain enough information for matching. In the case of such incomplete representations, it is useful to precise the lacking elements (the pattern sub graphs of the query which can not be matched), so that the user could know the reason of non-verifiability and/or complete the project representation (e.g. define the functionality of a room: kitchen). The conformity report grouping conformity queries by classes is automatically generated on the basis of query annotations (e.g. type of the regulation text, its application level, etc.). For each query, it indicates its success or failure, and provides a detailed description of the non-conformity and/or non-verifiability reasons.

6 The C3R Prototype

To validate our conformity-checking approach, we develop the C3R[4] system (Fig.2) that implements the algorithms of reasoning by expert rules according to organized conformity queries. For the checking operation, C3R relies on the semantic search engine CORESE [3], which answers SPARQL queries asked against an RDF/S knowledge base.

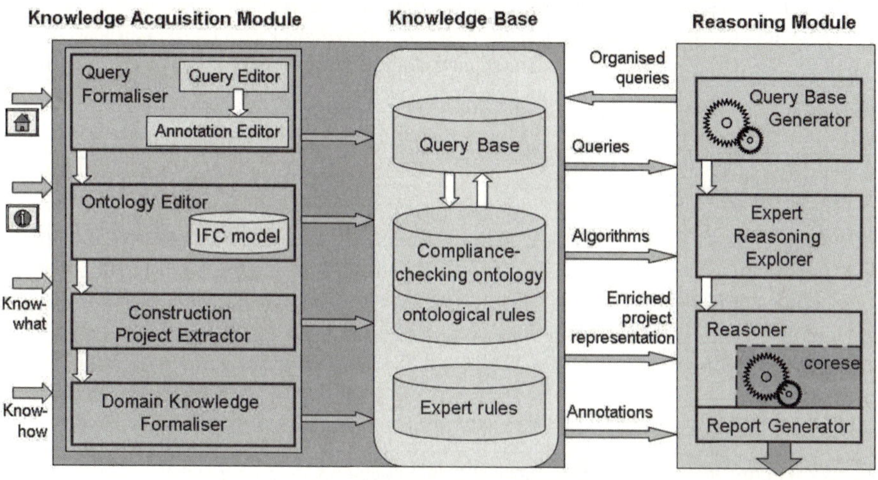

Fig. 2. C3R architecture

The results of this analysis are interpreted in terms of conformity checking in construction. They are then used to generate a structured conformity report, which groups conformity queries by classes and explains the possible reasons of non-conformity.

[4] Conformity Checking in Construction with the help of Reasoning.

7 Conclusions and Perspectives

We have presented an ontological method aimed at semi-automatic checking the conformity of construction projects represented by RDF graphs against a set of construction norms formalized as SPARQL queries. Our reasoning model is based on the matching of RDF project representations to SPARQL conformity queries. It also integrates the meta-knowledge on conformity checking process. Our approach of capitalizing such knowledge is based on the development of special semantic annotations of conformity queries that are organized into a query base, according to these annotations. This allows formalizing the expert knowledge in form of expert rules that schedule matching operations of the checking process. Semantic annotations of conformity queries also help to generate a structured conformity report, which interprets the results of reasoning in terms of conformity checking in construction.

The ongoing works focus on the incremental development of the conformity-checking ontology and the C3R prototype, its evaluation by experts, and the formalization of hybrid expert rules.

References

1. Bell, H., Bjorkhaug, L.: A buildingSMART Ontology. In: Proc. of the European Conference on Product and Process Modeling (ECPPM 2006), Valencia, Spain, pp. 185–190 (2006)
2. Berners-Lee T.: Reflections on Web Architecture. Conceptual Graphs and the Semantic (2001), http://www.w3.org/DesignIssues/CG.html
3. Corby, O., Dieng, R., Faron-Zucker, C.: Querying the Semantic Web with Corese Search Engine. In: Proc. Prestigious Applications of Intelligent Systems PAIS, ECAI, Valencia (2004)
4. Dai, W., Drogemuller, R.: Computer modelling of buildings for compliance checking. In: Proc. Building Control Commission Int. Convention, Melbourne, Australia, pp. 176–186 (1999)
5. Dibie-Barthélemy, J., Haemmerlé, O., Loiseau, S.: Refinement of Conceptual Graphs. In: Delugach, H.S., Stumme, G. (eds.) ICCS 2001. LNCS (LNAI), vol. 2120, pp. 216–230. Springer, Heidelberg (2001)
6. Faron-Zucker, C., Yurchyshyna, A., Le Thanh, N., Lima, C.: Une approche ontologique pour automatiser le contrôle de conformité dans le domaine du bâtiment. In: Actes des 8èmes journées Extraction et Gestion des Connaissances, EGC 2008, RNTI-E11, Cépaduès, Sophia Antipolis, France, pp. 115–120 (2008)
7. Mokhtari, N., Dieng-Kuntz, R.: Extraction et exploitation des annotations contextuelles. In: Actes des 8èmes journées Extraction et Gestion des Connaissances, EGC 2008, RNTI-E11, Cépaduès, Sophia Antipolis, France, pp. 7–18 (2008)
8. Nguyen, T.-H., Asa, E.: Integrating building code compliance checking into a 3D CAD system. In: Proc. of Joint International Conference on Computing and Decision Making in Civil and Building Engineering, Montreal, Canada, June 14-15 (2006)
9. Schevers, H., Drogemuller, R.: Converting IFC data to the web ontology language. In: 1st International Conference on Semantics, Knowledge and Grid (SKG 2005), Beijing, China, November 2005, pp. 28–29 (presented, 2005)
10. Sowa, J.F.: Conceptual Structures: Information Processing in Mind and Machine. Addison-Wesley, Reading (1984)
11. Yang, Q., Zhang, Y.: Semantic interoperability in building design: Methods and tools. Computer-Aided Design 38(10), 1099–1112 (2006)

A Web-Based Customer Support Knowledge Base System

S. Wood and R.J. Howlett

Intelligent Systems
Engineering Research Centre, University of Brighton
Moulsecoomb, Brighton, BN2 4GJ, UK
stuart.wood200@gmail.com, r.j.howlett@brighton.ac.uk

Abstract. This paper discusses the design and implementation of a web-based customer support system. The support facility operates using an adaptable knowledge base capable of diagnosing user problems. The system was developed for use by a software reseller in order to provide 24 hour support to its customer base accessible via the World Wide Web (WWW). Through integration with the company's in-house systems the support facility is able to extract expert knowledge from company staff and tailor the information for customer use.

1 Introduction

Developments in how information is searched for and expectations on how information should be presented are changing rapidly. Customer support is a major department within any successful company, but with the emergence of the Internet can customers only be expected to receive week day, working hours support in a world no longer constrained by time and location? This paper looks at the implementation of a web-based customer support knowledge base system to diagnose customer problems 24 hours a day via the WWW.

The strength of a knowledge base system in its concentrated knowledge within a specific subject domain allows any potential user access to its valuable 'expert' information at all times. For such a system, the incorporation of concentrated information gathered from a single, or, multiple experts allows the system to outperform many human experts, with the added attribute of continual accessibility.

The design and implementation of the web support system was gradually implemented into the host company over a 12 month period. The company requesting the system provide accounting software to their clients throughout the UK and were in need of a diagnostic web facility to enable them to provide an interactive out-of-hours service.

Prior to the implementation of the system, large amounts the company's field expert's time would be spent answering incoming telephone calls, from company customers, regarding issues encountered with the supported software. A direct benefit of the web support system would be to limit these telephone calls by creating the system as the optimum choice for a customer that has a problem. With many of the telephone calls duplicated from customer to customer the web support system would provide a central depositary for all previous issues created by the interaction between the customer and field expert. The system would provide the company with not only an

I. Lovrek, R.J. Howlett, and L.C. Jain (Eds.): KES 2008, Part I, LNAI 5177, pp. 349–361, 2008.

attribute in the pursuit of new custom, but also propel the company above its sector competitors who were not offering this service.

2 Web-Based Customer Support Systems

Online diagnostic support is an area that continues to grow as the need for access to support information at all times of the day becomes a necessity [1]. The advent of the ever accessible 24 hour knowledge factory [2] will become a necessity for any business with a customer-base, not just an advantage over competitors.

Giant companies with huge online presence, such as Microsoft or Apple for example, have built online customer support facilities in order to provide relief to their call centre staff and offer a form of support to their customers at all times. Indeed even local authorities, funded by UK government, continue to encourage customers to use the WWW to access information, pay taxes and even diagnose problems.

The National Health Service has a simplified version of their decision support telephone system available via their website. The diagnostic tool enables a user to help diagnose common health problems through a variety of multiple choice questions. In it's first year, 1998, NHS Direct took more than 100,000 calls from patients seeking advice on medical problems with that number increasing to in excess of six million during 2003 and 2004 [3]. This growth in awareness coincides with NHS Directs figures for their interactive web service activated during 2001.

Table 1. NHS Direct Online Activity

Year	Visits (1,000s)
2001-2002	2027
2002-2003	3972
2003-2004	6541
2004-2005	9319
2005-2006	14000*
2006-2007	19000*
2007-2008	25000*

Note: * Indicates the projected activity of NHS Direct Online based on the upward trajectory of 2005 [4].

The success of the NHS Direct web service is evident through the increase in visitors shown in the table above. NHS Direct predicts that for the year 2007-2008 it will accommodate 25million visitors [4].

As awareness in a service grows this should be reflected in a growth of service use. The success of the online support system could be determined by both its use and the decline in calls to the respective company's telephone support centre. This success will depend on the accuracy and efficiency of the design of the system in providing a user friendly application where users feel both empowered and informed.

The experience of the user, during the interaction process, will decide on whether they are likely to return to the system or if, in future, they will bypass the web support and opt for a time consuming telephone call. Potential reward for a user in the form of an accurate result heightens their experience and should ensure that they will return to the support area and use the facility again. Interactivity of a web application is a highly desirable feature of any website [5] and by encouraging the interaction between user and system the process, over time, will develop the basic user into the informed user. The informed user would then be able to help improve and grow the system.

3 Expert System and Knowledge-Based System Techniques

In designing a system of this type all possible techniques of implementation must be reviewed in order to find a solution that is appropriate. Expert, knowledge-based and case-based reasoning techniques are three ways in which the system could be implemented within the timescale allotted to the web support project.

Expert systems are computer programs that emulate human thinking processes in a problem solving situation. Their ultimate goal is to arrive at the same results that a specific human mental process would produce [6]. The expected output of an expert system would be a conclusion, recommendation or decision [7]. This process of reason allows the expert system to act humanlike. It is this attribute combined with the speed of reasoning and the accuracy of the decision ultimately decides how successful the system is.

The basic components of an expert system are a knowledge base and an inference engine. The knowledge held within the knowledge base is usually obtained by interviewing the specific experts within the field. After the knowledge elicitation process the knowledge engineer is able to organise the data into a collection of inference rules. The rules are developed using an If-Then structure, stored within the inference engine, which are then accessible by the expert system via an array of reasoning techniques.

3.1 Reasoning Techniques and Case Compare

Forward chaining, backward chaining and case-based reasoning (CBR) are three reasoning techniques used during inference. CBR provides a more complex reasoning technique than forward and backward chaining using partial matching to compare solutions to each other. Forward chaining and backward chaining provide reasoning often used within simplified rule based systems, providing adequate solutions to non-complex problems.

CBR does not need to rely on general knowledge within a knowledge base in the same way that forward & backward chaining do. CBR is capable of using specific knowledge of previous experiences, or cases, in order to find a viable solution to the problem. This different approach allows CBR to find a similar past case, adapt the case if necessary, and reuse it in the new problem situation. CBR is then able to save the new case and add it to its knowledge base for use in the future.

This reuse of previous cases is much like how a human would revisit past experiences when faced with a problem. CBR systems have intuitive appeal because much of human problem solving capability is experience based, that is, humans draw on past experience when solving problems and can readily solve problems that are similar to ones encountered in the past.

In order to find an appropriate case to solve the current problem CBR uses a variety of techniques to compare the current case with an existing case. An example of how CBR can compare new cases with existing cases is by breaking down the case into a selection of keywords. The keywords can then be compared to previous cases to see how closely they are matched. The solutions are then ranked dependent on how closely they are matched.

The simplified concept of keyword matching has been continually used as part of a search engine's algorithm on the World Wide Web (WWW). A search engine takes the input from the user via a user interface provided by an Internet browser. The input then proceeds through several processes of cleansing in order to extract the keywords. Often the algorithms, employed by the search engines, will delete frequently occurring words such as 'a', 'the', 'an', 'and', 'off', 'at', and 'to' for example. These frequently occurring words, known as stop-words, will almost certainly appear in all cases, new and existing. If a search engine was to search the WWW, with stop-words included, then there is an increased likelihood of returning far more results than necessary.

With the stop-words deleted from the input, the algorithms employed by the search engines can then compare the most important keywords of the input against the most important keywords within the page of a website. The reasoning of 'most important' again depends on the factors engineered by the search engine's algorithms. Different levels of importance are decided by the algorithm which is then able to rank the web pages dependant on their optimisation toward the keyword.

The same process of keyword matching can be employed through CBR. For example, each previous case within a knowledge base can have a series of keyword attributes associated with it. These keywords can then be compared with the keywords extracted from the new case to determine if it is an appropriate match and contains a viable solution to the new case or user problem.

The following example shows the same question asked in different ways. This scenario could occur if different users are attempting to describe the same problem and therefore trying to find the same solution.

Q1: Not able to complete year end report
Q2: How can I print a year end report?
Q3: printing year end reports how can I do this?
Q4: Reports not working

By firstly extracting the keywords from the text and extracting the less important stop-words the questions then become:-

Q1: year, end, report
Q2: print, year, end, report
Q3: printing, year, end, reports
Q4: reports

Once the keywords are identified the system can then morph inflectional forms of words into their stem form. Highlighted by a 2001 study [8] this method of morphing will allow each keyword to be compared to a list of stem words. In the example above 'printing' could be considered the inflectional form of the stem word 'print' and 'reports' the inflectional form of its stem 'report'. This level of compare would then leave the following results.

Q1: year, end, report
Q2: print, year, end, report
Q3: print, year, end, report
Q4: report

This process of stopping and stemming allows the keywords from the same question posed in a variety of ways to be likened and, more importantly, take the form of the keywords relating to the existing cases that they can then be compared to.

4 Web Support System Design

The system is designed as a knowledge base system capable of growing through its use by user and field engineer input. Ensuring the satisfaction of each user is of fundamental importance within the development of the system. The identifiable users of the system are the customer, field expert, knowledge engineer and the company management. To ensure satisfaction the system is developed to encourage use by the customer while requiring minimal intervention from both the field expert and knowledge engineer

The system was implemented incrementally to allow the knowledge base to be available at the beginning of the developmental life-cycle. This ensures that as much

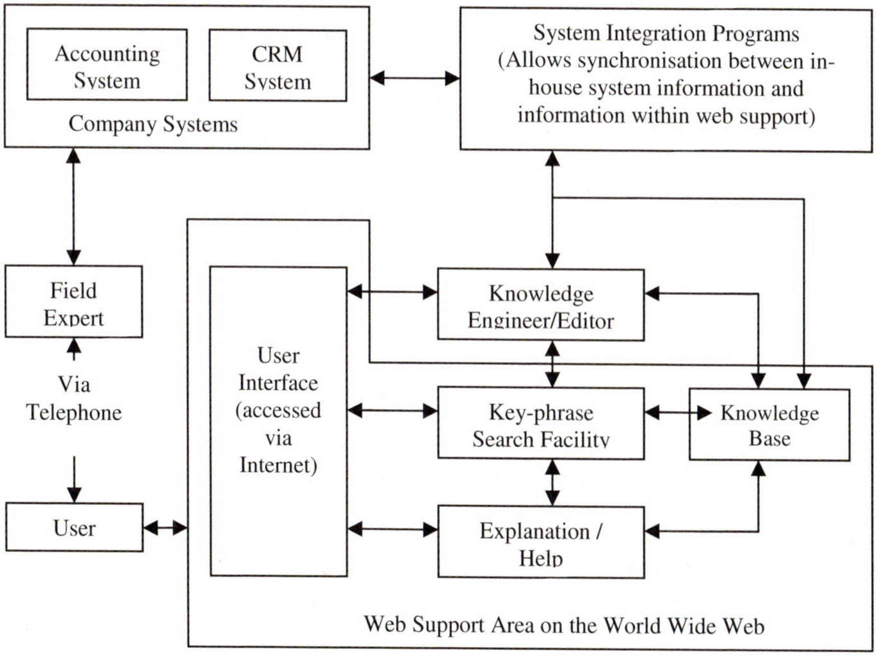

Fig. 1. Web Support System Design

data as possible is available to customer users when accessed. The design of the system, shown within the diagram below, is modelled on a classical knowledge base system and could, in effect, be used as a stand alone system functioning within any sector and not just within the constraints of this project.

The diagram shows how the customer is able to interact with the web support system via the user interface over the WWW. It also illustrates how both the user and the field expert are able to provide information for the web support system each time they interact via the telephone. The system integration programs providing the synchronicity execute a number of tasks allowing company information to be accessible to customers via the web.

Users are provided with a web support login which provides access to areas of the system. Once they are logged in the user is presented with customised information relevant to their requirements. As the system is only available to contracted clients of the host company, each user will have to be connected to a company that is contracted with the host. This contractual information is uploaded from the host company accounting system to the web server where an access profile is created for each individual user. Having logged in to the system the user will be recognised along with their contractual details. User support and the problem diagnosis can only be carried out for the specific contracts held. Once a user has logged onto the system, they, along with details on their product contracts with the host, will be identified. The host is only accountable for the products in the contract and thus support and problem diagnosis will only be provided for these products.

During the process of the telephone conversation a field expert will create a related issue within the host company's in-house Customer Relationship Management (CRM) system. The issue is populated with relevant information regarding the enquiry which is then uploaded automatically to the web support area. The customer user is able to view the status of the issue in the web support area, including the solution once the issue has been resolved. This method of knowledge elicitation allows the field expert to populate the web support system with information without the requirement of extra effort. The field expert is also able to populate the knowledge base with generic issues which are not customer specific and could relate to any customer.

4.1 Knowledge Base

It is most important that the knowledge base has a simplistic, robust design to promote ease of growth and maintainability. The knowledge base can be thought of as a series of database tables within the online database alongside associate retrieval pages within the web support area that store, retrieve and display the information.

At the beginning of this section, Figure 1 illustrates the Web Support area structure. This structure can be compared with the structure of a classical knowledge base system with the search facility acting like an Inference engine within an expert knowledge base system.

The Web Support knowledge base takes the form of questions, answers and service categories within tables on the online database. A knowledge base entry can be described as a question with its solution relating to a particular service category. By simplifying the process of defining a knowledge base entry with an associated powerful search tool, it is expected that the knowledge base can grow quickly and accurately, being both easily maintainable and understandable to all users.

4.2 Advanced Search

Since the modern evolution of the Search Engine, the introduction of Google and its main goal of improving the quality of web search engines [9], the necessity of accuracy has remained. The need for accuracy of search queries, or problem diagnosis, is at the forefront of all users conclusions when introduced to the idea of a new system. The accuracy of the system defines both its lifespan and function as an effective tool for problem diagnosis.

The role of the perfect search engine is to find the exact results for the terms that are being searched. Lin et al [10] discuss the role of context within question answering systems combined with the role the interface has to play in the presentation of answers to search queries. A requirement of the web support system is to obtain as much information from the user as possible. The accuracy of the presented search results can be enhanced by encouraging the user to enter a phrase to be searched, as opposed to a word.

Minimal input by the user would be to the detriment of the accuracy of the results. Customer surveys carried out by the host company highlighted that accuracy is the most important element in finding a solution, but surveyed users continue to define keyword search as an effective medium, when a 'key phrase' would be more appropriate. The process of extracting the correct solution from a customers input specification must be as efficient and accurate as possible.

Maximum knowledge extracted from the user regarding their problem allows the proposed system to better diagnose the problem, while satisfying the users needs of questions answered in a single step. Acquiring the knowledge from the user into the system is achieved via an interface encouraging natural language and trust on the part of the user. Trust in the system can be developed in the user over repeat usage during repeat visits.

Within the search facility the user is required to select a service that the question relates to. This will eliminate all non viable solutions, before the search takes place, highlighted by Minsky [11]:

> *'For any problem worthy of the name, the search through all possibilities will be too inefficient for practical use'*

Since the knowledge base holds entries relating to several different products, by allowing the user to select the product to which the problem is related, large numbers of non-viable solutions can be disregarded allowing the remainder of the search process to concentrate on viable solutions only.

Next the user is required to enter either a question which their problem relates to or an error number. The web support search facility encourages the usage of natural language inviting the user to ask a question as if they were addressing another human being. The input search phrase then goes through a cleansing process which strips the string of its stop-words as explained within section 3.1. Once cleansed, the string will contain only crucial words relating to the symptom of the problem. These symptoms are then morphed into their inflectional form, which can then be matched against the potential solutions, which have been morphed using the same process. This process of cleansing on both sides of the match allows the greater potential for a solution to the problem.

The user can expect to receive more accurate results by incorporating more key symptoms into a question or phrase entered into the text box within the search interface. This method of diagnostics allows users to decide the level of accuracy they require.

4.3 Knowledge Elicitation

A knowledge base should allow its content to evolve both in terms of volume and accuracy. If a knowledge base system is unable to update the information held within its database in keeping with its environment then it becomes redundant as soon as it is invoked. Lederberg [12] states that databases:

> *'... should not be thought of as static, final repositories but as bulletin boards, subjected to dynamic critical attention by the entire knowledgeable community'*

The necessity of keeping the knowledge base system dynamic within its environment is paramount to the systems success or failure as a useful application. This necessity is mirrored within the systems knowledge elicitation process. The semi-automated knowledge elicitation process allows the field expert to easily upload knowledge base entries for display on the web with no distraction of effort.

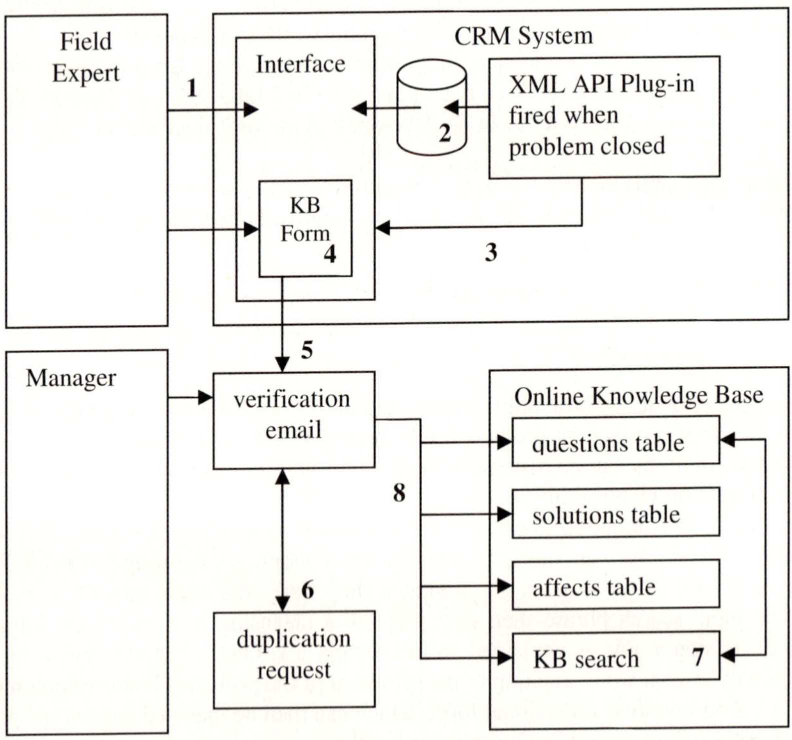

Fig. 2. Knowledge base entry lifecycle

The importance of the elicitation process has not been underestimated within the development of the web support system. Although it is a challenging process [13], concerns regarding the change in work pattern to the field experts were reduced with the automation of the elicitation process from CRM system to the Web support system. This method allows ease of data capture without the need for the field expert to change their working practice. Each query initiated by a telephone call by the customer is documented within the CRM system by the field expert and the elicitation plug-in duplicates the information once the query is marked as a solution. On close of the query the field expert is prompted by the elicitation plug-in to decide as to whether the data should or should not be uploaded to the knowledge base.

The diagram above shows the lifecycle of a knowledge base entry from its entry into the CRM system to its verified state within the knowledge base.

1. The field expert accesses the CRM system via the interface to close a problem which has been solved.
2. When the status of a problem changes to 'closed' it triggers the upload plug-in.
3. The upload plug-in extracts all data from within the problem and populates the knowledge base upload form.
4. The knowledge base upload form is presented to the user via the interface. The user can edit the form if necessary and then select the upload entry button to create the knowledge base entry.
5. Once the upload entry button has been selected the entry is parsed into a string variable. This is then passed within the Uniform Resource Locator (URL) to a transmitting page on the website. When the transmitting page receives the request, the URL is decoded. The solution, question and service variables are extracted from the URL and inserted into the online database via Structured Query Language (SQL) statements. The entry is not yet verified at this stage and will not be included within the web support knowledge base until the verification and duplication processes have been performed.
6. The verification email is sent to the manager of the expert whom selected the problem for upload. The manager is then able to authorise the content and make sure it is acceptable for generic use within the knowledge base. The email also includes a hyperlink allowing the manager to check the existing knowledge base for duplicates to the new entry.
7. On selecting the hyperlink the manager is taken to the knowledge base search screen where results of the nearest matches to the potential knowledge base entry are displayed. The manager then decides if the new knowledge base entry should be verified, deleted or modified via three further hyperlink options within the verification email.

The simplicity of this process allows the real time duplication of queries within the CRM system to be available within the knowledge base. The process compliments the way the field experts operate by encouraging participation without unnecessary distraction or effort.

4.4 System Intelligence - Issue Tracking

The ability of the user to track the progress of a query via the web support system encourages self-sufficient usage. Instead of depending on consultation with time-restricted field experts the user can independently check the progress of their queries within the support area. Further functionality allows users to track issues within their user history. This function provides solutions to closed issues that have been diagnosed by the field experts if the knowledge base can not provide the answer. If the question and solution of this issue does not contain private information specific to the company then the knowledge base is updated to include this entry for use by all users.

4.5 Issue Logging

Issue logging is the process enabling the user to log an issue which can then be tracked by use of the issue tracking outlined in section 4.4. Issue logging takes place if the knowledge base has been unsuccessful in diagnosing the problem posed by the user. Once the problem has been logged the details are emailed to the relevant field expert. Once the email is received diagnosis is carried out by the field expert and on completion both the problem and knowledge base can be updated to include this new entry.

4.6 Financial Information

Integrating with in-house financial systems, users' outstanding invoices, credit notes and current contracts are uploaded each day and visible by an approved account holder within web support. This feature allows the user to access outstanding account transactions held between the host company and the users company. It gives the user access to an account balance and to each of the outstanding invoices. Any of the invoices to which a user has access can be emailed or saved by the user for administrative purposes.

5 System Performance

As the support system is in the public domain and accessible to all customers free of charge at all times via the WWW, its performance can be measured in terms of its use as shown in Table 2 below.

Table 2. Web Support Statistics 2007

Month	Unique Logins	Page Views	Issues Created	KB Entries (Cumulative)
1	43	153	-	513
2	24	117	-	588
3	39	448	-	610
4	51	331	-	662
5	112	705	1018	720
6	125	658	784	772
7	110	614	711	820
8	134	687	590	850

Table 2 shows the number of system logins (Unique Logins), pages viewed, telephone issues created by support staff (Issues Created) and the knowledge base growth within the first eight months of system deployment. These results show the growth within usability of the system. A greater volume of pages were visited, along with the increased number of logins to the system over the same period. For the system to be deemed a success, increased logins and pages viewed must be maintained and users must continue to use the system over an extended period of time.

The increased access could be attributed to the increase in knowledge base entries over the same period. The KB entry column within the table shows the increase in entries over the eight month period. Each entry corresponds to the individual upload by the field expert and entry verification by management. In the same period over 3500 individual issues were uploaded to the system. These 3500 issues relate specifically to an individual user's company queries and are not open to access by other users. Only generic issues can be duplicated and entered into the public domain of the web support system to become a knowledge base entry accessible to all users.

An encouraging aspect of system performance is the deduction in support calls received by the support team over the same period. The issues created by the support team from each telephone call, has reduced sharply during the last four months of the system's lifespan with a 42% drop recorded in created issues from Month 5 through to Month 8. Unfortunately the tracking software was not in place prior to Month 5 so it is impossible to predict the numbers of issues created during Month 1 to Month 4. This reduction could be attributed to the fact that users are bypassing the telephone support option, a primary objective, and diagnosing their problems via web support.

5.1 System Scalability, Evolution and Classical Comparison

The system is designed to be easily accessible and maintainable to all of its potential users. It is able to grow through the simplicity of its connection with its field experts, within knowledge elicitation, and through customers' use of issue logging, tracking and knowledge base searching. The system encourages participation and learning by all users maintaining its scalability and relevance via the WWW. By encouraging the user to learn through interaction, the system provides the user with real-time web learning, therefore enhancing the users web experience [1], without the requirement of sourcing through documents and training manuals [14].

The system differs from conventional expert systems as it uses specific methods relating to its surroundings to extract information, develop and sustain itself. A classical expert system gives a recommendation to a given input, the web support system provides an empowered user different methods to diagnose their problem offering historical solutions as well as a search facility.

No rules are created by an inference engine within the system, thus the information is not restricted to the rules of the engine and each solution can be reviewed within the process of a query. Accuracy and redundancy of data are quantified within the verification process of each individual entry. If the system contains matches to a proposed upload then the verification process highlights this allowing the verification manager to make changes if appropriate.

6 Conclusions

The web support system provides a real-time solution to customer software support. From a first time user through to an informed user, the system provides each of its members with a fast, scalable and accessible platform for knowledge. The primary goal is to provide results to user queries over the WWW. Web support maintains this process with the added functionality of real time issue tracking, issue logging and financial information view.

Through the systems seamless integration with financial and CRM applications it is able to capture knowledge without the requirement of cultural change for the company experts. This can be expected to continue allowing the system to increase in size. Problems caused by an expected increase in the size of the system have been anticipated with the incorporation of a duplication check as an entry is uploaded with the opportunity of amending or rejecting the entry.

The system could be easily adapted for a range of support uses on the WWW. Each of the components of the system could be deployed individually, for instance if a company required the capacity to provide its customers with the ability to see outstanding financial information or to facilitate customer access to current issues. The real strength of the system is within the range of services offered to help the user and encourage re-use and interaction.

Future work encompasses increased interaction between the host company's in-house systems and the user through Web support, and the upgrade of the system's user interface to accommodate system growth. Further interaction between user and system will be continually assessed through feedback requests and user suggestion. System awareness will be developed to ensure all new contract holders are granted access immediately, with consultants demonstrating the system during the sales presentations and implementation process.

References

1. Rodgers, W., Negash, S.: The effects of web-based technologies on knowledge transfer. Communications of the ACM 50(7), 117–122 (2007)
2. Gupta, A., Seshasai, S.: 24-hour knowledge factory: Using Internet technology to leverage spatial and temporal separations. ACM Transactions on Internet Technology (TOIT) 7(3) (2007)
3. NHS Direct. Telephone Usage Statistics, `http://www.statistics.gov.uk/cci/nugget.asp?id=1335`
4. NHS Direct. New Media. Multi-Channel Strategy 2005-2008 (2004)
5. Ghose, S., Dou, W.: Interactive functions and their impacts on the appeal of Internet presence sites. Journal of Advertising Research 38, 29–43 (1998)
6. Englard, B., et al.: Expert systems in accounting. The CPA journal, 58–62 (April 1989)
7. Eppinette, M., Inman, R.A.: Expert systems and the implementation of quality customer service. Industrial Management & Data Systems 97(2), 63–68 (1997)
8. Hui, S.C., et al.: A web-based intelligent fault diagnosis system for customer service support. Engineering Applications of Artificial Intelligent 14, 537–548 (2001)
9. Brin, S., Page, L.: The Anatomy of a Large-Scale Hypertextual Web Search Engine, Computer Science Department, Stanford (1998)

10. Lin, J., et al.: The Role of Context in Question Answering Systems. In: CHI 2003 (Extended Abstracts), pp. 1006–1007 (2003)
11. Minsky, M.: Steps toward artificial intelligence. Proc. Inst. Radio Eng. 49, 8–30 (1961)
12. Lederberg, J.: How DENDRAL was conceived & born, ACM Symposium on the history of medical informatics. National Library of Medicine (1987)
13. Sagheb-Tehrani, M.: The design process of expert systems development: some concerns. Expert Systems 23(2), 116–125 (2006)
14. Khalifa, M., Lam, R.: Web-Based Learning: effects on learning process and outcome. IEEE Transactions on Education 46(4), 350–356 (2002)

Incremental Determinization of Finite Automata in Model-Based Diagnosis of Active Systems

Gianfranco Lamperti, Marina Zanella, Giovanni Chiodi, and Lorenzo Chiodi

Dipartimento di Elettronica per l'Automazione, Via Branze 38, 25123 Brescia, Italy

Abstract. Generating a deterministic finite automaton (DFA) equivalent to a nondeterministic one (NFA) is traditionally accomplished by *subset-construction* (*SC*). This is the right choice in case a single transformation is needed. If, instead, the NFA is repeatedly extended, one transition each time, and the DFA corresponding to each extension is needed in real-time, *SC* is bound to poor performances. In order to cope with these difficulties, an algorithm called *incremental subset-construction* (*ISC*) is proposed, which makes up the new DFA as an extension of the previous DFA, avoiding to start from scratch each time, thereby pursuing computational reuse. Although conceived within the application domain of model-based diagnosis of active systems, the algorithm is general in nature, hence it can be exploited for incremental determinization of any NFA. Massive experimentation indicates that, while comparable in space complexity, incremental determinization of finite automata is, in time, far more efficient than traditional determinization by *SC*.

1 Introduction

Automata and language theory was an active field till the end of the millennium, while today there is far more research on applications of automata rather than on their foundations. This work is no exception, since it stems from the domain of model-based diagnosis (MBD) of active systems [1], monitoring-based diagnosis [2,3,4] in particular, where each considered uncertain system–observation is typically represented as a finite automaton which is progressively growing over time. However, the authors' proposal of an incremental algorithm for NFA determinization is not application-specific, on the contrary it achieves a generality which has led to the present paper. The notion of incrementality is not univocally defined in automata theory, rather its meaning varies from a contribution to another. So it is worth clearly stating which is the meaning given here to this concept. In our approach, if the NFA includes t transitions, a sequence $\langle \mathcal{A}_{\mathrm{n}_0}, \mathcal{A}_{\mathrm{n}_1}, \ldots, \mathcal{A}_{\mathrm{n}_t} \rangle$ of $t+1$ NFAs is considered, with $\mathcal{A}_{\mathrm{n}_0}$ being the NFA composed of the initial state only, and each $\mathcal{A}_{\mathrm{n}_i}$ being an extension of $\mathcal{A}_{\mathrm{n}_{i-1}}$ by one extra transition. The i-th call to the incremental algorithm generates the i-th DFA $\mathcal{A}_{\mathrm{d}_i}$ equivalent to $\mathcal{A}_{\mathrm{n}_i}$ by taking as input $\mathcal{A}_{\mathrm{n}_{i-1}}$, $\mathcal{A}_{\mathrm{d}_{i-1}}$, and the new transition that extends $\mathcal{A}_{\mathrm{n}_{i-1}}$ to $\mathcal{A}_{\mathrm{n}_i}$. This transition is bound to exit an existing state of $\mathcal{A}_{\mathrm{n}_{i-1}}$. No other algorithm for NFA determinization is incremental in this way in the literature. The incremental construction of automata was faced also by [5,6] but in a perspective quite different from the present paper, both in the addressed task (and, consequently, in the input/output) and in the adopted notion of incrementality. The quoted contributions, in fact, do not perform

I. Lovrek, R.J. Howlett, and L.C. Jain (Eds.): KES 2008, Part I, LNAI 5177, pp. 362–374, 2008.

the task of determinizing an NFA, as in the present paper, they instead take as input an acyclic DFA (*dictionary*) and update it by adding a new entry. Such a new entry is not a transition, as in the present paper, but a whole word, that is a sequence of symbols (a string). Moreover, both contributions aim at achieving the *minimality* of the generated DFA. The two quoted contributions differ with each other for the former operates in two steps, construction of a DFA, and minimization, while the latter (and subsequent work [7,8]) produces the minimal dictionary in one step. The algorithm proposed in the present paper, being an incremental version of *SC*, which does not achieve minimality, does not produce a minimal DFA itself. More recently, also cyclic DFAs have been considered for progressively adding new strings to them. In [9], an algorithm is proposed, which modifies any (existing) minimal DFA, be it cyclic or not, so that a string is added to (or removed from) the accepted language. Note that it is not possible to introduce new cycles by just adding simple strings while the transitions that are progressively added to the NFA taken as input by the algorithm proposed in the present paper can create new cycles at any time. In [10], the addition of a string to a cyclic DFA is performed in a *semi-incremental* way, that is, according to a two-step process. A further notion of incrementality is defined in [11], where the considered task is the minimization of a DFA and the proposed algorithm is incremental in the sense that it may be halted at any time, yielding a partially-minimized automaton. An interesting proposal inherent to NFA determination comes from the context of XML stream processing [12]. The considered task is filtering a stream of XML packets based on a given so-called query tree, a structure which can be mapped into an NFA. The quoted paper is aimed to show that the DFA equivalent to such an NFA can be used effectively for this purpose. The idea is to generate the DFA in a lazy way and exploit it (for the filtering intent) while generating it. The DFA is lazy in that it is constructed at run-time on demand. Initially it has just the initial state, then, whenever an attempt to make a transition into a missing state is accomplished, such a state is computed. Now, leaving the application domain apart and focusing on the generation of the DFA, this is somewhat incremental in [12] as it is not performed before-hand in a single step, instead it is performed iteratively at run time. However, this situation is quite different from that coped with in the present paper: the former takes into account a given single NFA, which never changes at run time, whereas, in the latter, the NFA is growing at run time, transition by transition. In other words, the approach in [12] translates, transition by transition, a given unchanging NFA into the equivalent DFA, while our work repeatedly translates an NFA that is growing transition by transition into the equivalent DFA. Still more, [12] deals with a single translation, which is performed fragment by fragment and starts from scratch, while our work deals with several translations, one for each transition of the NFA, and each translation, the first one excluded, does not start from scratch, since it actually updates the result of the previous one.

2 Motivating Application-Domain

The need for the incremental generation of a DFA based on an incremental specification of an NFA stems from the domain of model-based diagnosis (MBD) of active systems [1], specifically, monitoring-based diagnosis [2,13,4]. MBD [14] aims to diagnose a physical system based on the model of the system and relevant observations. The discrepancy between the normal behavior of the system and the observation allows the diagnostic engine to generate candidate diagnoses, where each candidate is a set of faulty components (or, more accurately, a set of faults ascribed to components). MBD can be applied either to *static* or *dynamic* systems. Roughly, modeling a dynamic system requires managing time, as the system behavior is time-varying [15,16,17]. Among such systems are discrete-event systems (DESs) [18], whose behavior is typically modeled as networks of components, with each component being a communicating automaton [19]. A DES may be either synchronous or asynchronous. If synchronous, a system transition of the DES involves several parallel (synchronized) transitions by different components. If asynchronous, a system transition involves just one component transition. Active systems are a special class of asynchronous DESs, where components may exchange events to one another by means of *links*. During operation, the active system reacts to external events by performing system transitions, which possibly trigger new transitions by generating events towards neighboring components through links. Thus, the active system evolves according to its model, which incorporates both normal and faulty behavior, by performing a sequence of component transitions within its *behavioral space*. The latter is an automaton that specifies the global behavior of the system.[1] As such, the evolution of the system is a sequence of transitions, that is, a path within the behavior space, and is called the *history* of the system. The problem lies in the ambiguity of the mode in which the system is evolving, because only a (possibly small) subset of component transitions are visible by an external observer (the diagnostic engine). If the transition is visible, it generates an *observable label*.[2]. Consequently, the history is perceived by the observer as a sequence of observable labels, called the *signature*. The diagnostic engine performs consistent reasoning and eventually provides the candidate diagnoses, where each candidate is a set of faulty transitions, and corresponds to one or several *candidate histories*, each one equally possible. In large, distributed systems, the problem is complicated by the way observable labels are conveyed to the observer, which may involve multiple (possibly noisy) channels. This causes a distortion of the signature, called a *relaxation*, where each label is perceived as a set of *candidate labels*, while the total temporal ordering among labels is relaxed to *partial temporal ordering*. The result is an *uncertain temporal observation* [21], which is represented by a DAG, where nodes are marked by candidate labels, while edges define partial temporal ordering among nodes. However, the observation graph, namely $\mathcal{O}$, is inconvenient for processing as is. A surrogate of it, namely $Isp(\mathcal{O})$, the *index space* of $\mathcal{O}$, is used instead. The index space is a DFA whose regular language is the whole set of candidate signatures of the relevant observation. The point is, $Isp(\mathcal{O})$ is derived via

[1] The behavioral space can be derived from the description of the system in terms of component-automata and links, but this may be impractical for large systems. Thus, a strong requirement for MBD of active systems is to perform diagnosis without the explicit generation of the behavior space [20].

[2] However, such a label is not necessarily an identifier of the transition, as different transitions may generate the same label.

subset-construction by an NFA called *prefix space*, denoted $Psp(\mathcal{O})$, which is directly derived from $\mathcal{O}$. Thus, three transformations occur for a signature Σ:

$$\Sigma \rightsquigarrow \mathcal{O} \Longrightarrow Psp(\mathcal{O}) \Longrightarrow Isp(\mathcal{O})$$

where the former ($\rightsquigarrow$) depends on the nature of both the communication channels and the observer, and, as such, is beyond the scope of the diagnostic engine, while the others ($\Longrightarrow$) are artificially performed by the diagnostic engine for computational purposes. In monitoring-based diagnosis, candidate diagnoses must be generated each time a piece of observation is received. Typically, the observation graph is received as a sequence of *fragments*, with each fragment carrying information on one node and the arcs coming from its parents. These are called *fragmented observations*. At the reception of each fragment, the index space is to be updated based on the extension of the prefix space. The point is, *Generating the sequence of index spaces via subset-construction may become computationally prohibitive in real applications, as each index space is generated from scratch at each new fragment*. A better solution is to make subset-construction incremental, so that each index space in the sequence is generated as an update of the previous one, thereby pursuing computational reuse.

3 Determinization of Finite Automata

Determinizing an NFA amounts to generating a DFA equivalent to the NFA, thereby sharing the same regular language. Shown in Fig. 1 is an NFA (left) along with the equivalent DFA generated by *SC* (right).[3] Accordingly, each state in the DFA is identified by a (proper) subset of the states of the NFA. For example, the initial state is marked by 01, which is a shorthand for the subset $\{0, 1\}$. *SC* yields the DFA starting from the ϵ-closure of the initial state of the NFA, where such an ϵ-closure becomes the initial state of the DFA, and by progressively generating the successor subsets of a given subset as the ϵ-closure of the set of NFA states reached by a specific symbol from the NFA states in the subset. *ISC*, the incremental determinization algorithm we propose (whose detailed specification is postponed to Section 4), produces the same results as *SC* but it acts in a different way. In order to give an intuitive idea of how *ISC* works, we consider three scenarios, each of which extends the same NFA (displayed on the left of Fig. 1) by a new transition.

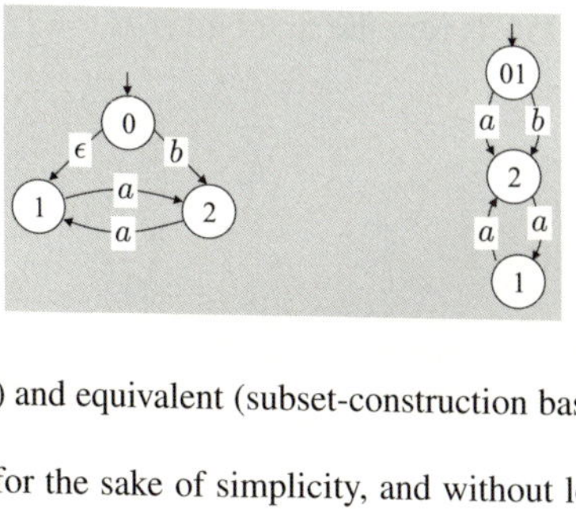

Fig. 1. NFA (left) and equivalent (subset-construction based) DFA (right)

[3] Throughout this section, for the sake of simplicity, and without loss of generality, we do not consider final states.

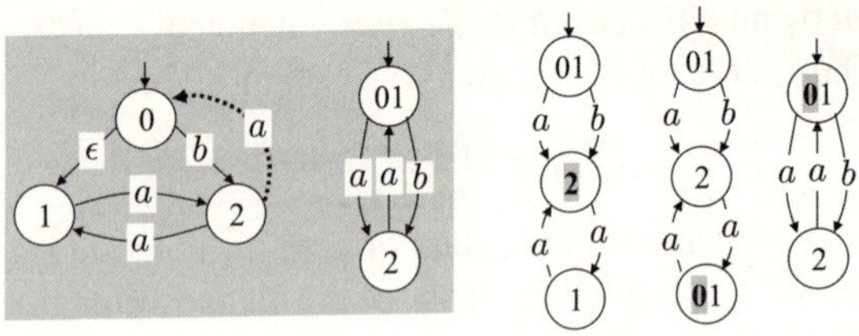

Fig. 2. Determinization, first scenario: NFA extension by $2 \xrightarrow{a} 0$

In the first scenario (Fig. 2), the NFA is extended by transition $2 \xrightarrow{a} 0$ (dotted arrow). The corresponding DFA, as computed by *SC*, is outlined on the right, within the box (including two states). We now show how the same DFA can be generated by *ISC*, starting from the DFA on the right of Fig. 1. The progressive evolution of such a DFA to the new one is outlined in Fig. 2, on the right of the box. Incremental determinization is based on a data structure called *bud set*, with each *bud* being a triple $(S_\mathrm{d}, \ell, \mathbb{S})$, where S_d is the object-identifier of a DFA state, ℓ a symbol of the alphabet, and $\mathbb{S}$ a subset of NFA states. Roughly, the bud set parallels the accumulator of DFA states in *SC*. Just as new DFA states are inserted into the *SC* accumulator and thereafter processed, so the new buds are accumulated in the bud set and processed one by one. In *SC*, the first state inserted into the accumulator is the DFA initial state. In incremental determinization, the bud set is initialized by one or several buds relevant to the source state of the new transition in the NFA. Considering Fig. 2, where the new transition is $2 \xrightarrow{a} 0$, only one bud is generated, namely $B_1 = (2, a, \{0, 1\})$, where $\mathbb{S} = \{0, 1\}$ is the ϵ-closure of the target state (0) of the new transition. The algorithm loops, by picking up a bud at each iteration, until all buds are processed. While processing each bud, new buds are possibly inserted into the bud set. The nature of processing of each bud depends both on the bud and the current configuration of the DFA. In our example, processing (the only bud) B_1 means inserting state 0 into the DFA state 1, thereby extending the latter to 01. This causes the generation of the new bud $B_2 = (01, b, \{2\})$. Note how such an extension determines a duplication of state 01, which must be eliminated by merging the two DFA states (see the DFA on the right of Fig. 2). Finally, bud B_2 is processed without any further action, as 2 is already included in the target state of the transition exiting 01 and marked by b. As expected, the final DFA equals the one obtained by *SC*.

The second scenario is displayed in Fig. 3: the NFA (of Fig. 1) is extended by transition $2 \xrightarrow{\epsilon} 0$. In this case, if *ISC* is run, the first bud is $B_1 = (2, \epsilon, \{0, 1\})$, the processing

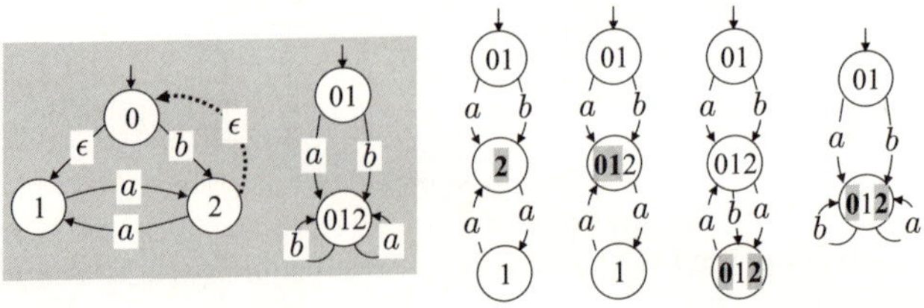

Fig. 3. Determinization, second scenario: NFA extension by $2 \xrightarrow{\epsilon} 0$

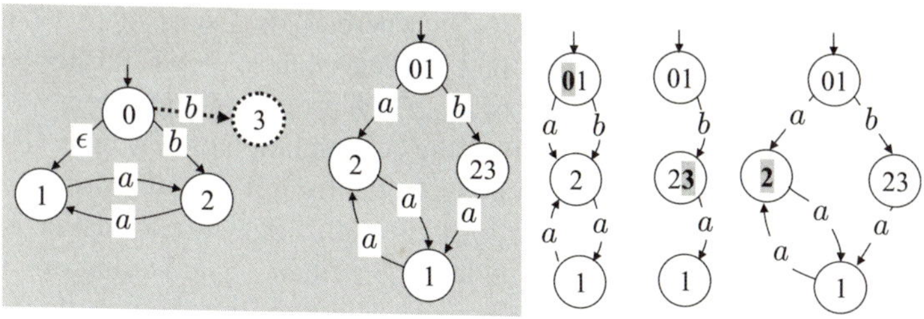

Fig. 4. Determinization, third scenario: NFA extension by $0 \xrightarrow{b} 3$

of which causes the extension of the DFA state 2 (of Fig. 1) by $\mathbb{S} = \{0, 1\}$, thereby obtaining the new subset 012. Based on this addition, two new buds are generated, namely $B_2 = (012, a, \{0, 1, 2\})$ and $B_3 = (012, b, \{0, 1, 2\})$. Processing B_3 causes the extension of the DFA state 1 by $\mathbb{S} = \{0, 2\}$, and the creation of a transition marked by b from 012 to the extended state. The state extension, in its turn, yields two new buds, B_4 and B_5. Moreover, a new merge is required for the duplicated state 012, which moves the DFA to its final configuration, as outlined on the right of Fig. 3 (the processing of B_2, B_4 and B_5 leaves such a configuration unchanged).

In the third scenario (Fig. 4), the DFA of Fig. 1 is extended by transition $0 \xrightarrow{b} 3$, which involves the new target state 3. If *ISC* is adopted, bud $B_1 = (01, b, \{3\})$ is generated. However, processing B_1 requires some attention to the current configuration of the DFA. In fact, 01 is already exited by a transition marked by b, with target state 2. The point is, state 2 is also entered by two other transitions (both marked by a).Thus, it would be misleading simply extending state 2 by $\mathbb{S} = \{3\}$, as this would be inconsistent with the other transitions. Instead, on the one hand, we remove the two entering transitions, on the other, buds $B_2 = (01, a, \{2\})$ and $B_3 = (1, a, \{2\})$ are generated as surrogates of the missing (removed) transitions. Then, the DFA state 2 is extended by $\mathbb{S} = \{3\}$.[4] The subsequent processing of B_2 causes the generation of state 2 (and the creation of bud $B_4 = (2, a, \{1\})$, while processing B_3 causes the creation of transition $1 \xrightarrow{a} 2$. Finally, the processing of bud B_4 leaves the DFA in its final configuration.

4 Incremental Subset-Construction

This section provides the pseudo-coded algorithm for incremental determinization of finite automata we propose. This algorithm, called *Incremental Subset-Construction*, namely *ISC*, takes as input an NFA $\mathcal{A}_{\mathrm{n}}$, the equivalent DFA $\mathcal{A}_{\mathrm{d}}$, and a new transition T_{n} for $\mathcal{A}_{\mathrm{n}}$, exiting a state already in $\mathcal{A}_{\mathrm{n}}$. *ISC* updates $\mathcal{A}_{\mathrm{d}}$ so that the new DFA is equivalent to the extension of $\mathcal{A}_{\mathrm{n}}$ by T_{n}. The algorithm is supported by the accumulator $\mathbb{B}$, the

[4] One may argue that removing all the other transitions and transforming them into buds is uselessly complicated. A better solution would be just generating an extended copy of the node and redirecting the transition towards such a node. Considering Fig. 4, this amounts to skipping the intermediate step. Although this is a correct solution for acyclic automata, when cyclic automata are considered, this shortcut is bound to generate spurious states in the final DFA, which are unreachable from the initial state.

bud set. Each bud in $\mathbb{B}$ is a triple $(S_{\mathrm{d}}, \ell, \mathbb{S})$, where S_{d} is a state of $\mathcal{A}_{\mathrm{d}}$, ℓ is a label, and $\mathbb{S}$ is a subset of nodes in $\mathcal{A}_{\mathrm{n}}$. Each bud indicates that S_{d} needs further processing, which is bound to change the topology of $\mathcal{A}_{\mathrm{d}}$. Each processed bud is removed from $\mathbb{B}$. However, processing a bud possibly causes the accumulation of new buds. Throughout the pseudo-code, we keep a distinction between the (object) identifier of a state in $\mathcal{A}_{\mathrm{d}}$ and its content, where the former is a symbol (e.g. S_{d}), while the latter is a set of nodes in $\mathcal{A}_{\mathrm{n}}$ (e.g. $\mathbb{S}$). The content of a node S_{d} is written $\|S_{\mathrm{d}}\|$. During execution, the content may change, while the identifier is fixed. The algorithm exploits three auxiliary subroutines. Auxiliary function *New* (Lines 5–10) generates a new state S'_{d} in $\mathcal{A}_{\mathrm{d}}$, with content $\mathbb{S}$. Auxiliary procedure *Extend* (Lines 11–22) takes as input a state S_{d} in $\mathcal{A}_{\mathrm{d}}$ and adds to its content the subset $\mathbb{S}$ of states in $\mathcal{A}_{\mathrm{n}}$. Two extra operations are needed. First (Line 16), the bud set $\mathbb{B}$ is updated by the buds relevant to S_{d} and the labels exiting nodes in $(\mathbb{S} - \|S_{\mathrm{d}}\|)$ in $\mathcal{A}_{\mathrm{n}}$. Second, if the content of the extended node S_{d} equals the content of a node S'_{d} already in $\mathcal{A}_{\mathrm{d}}$ (Line 18), then the two nodes must be merged into a single node (Line 19). Such a fusion is carried out by the auxiliary procedure *Merge* (Lines 23–31) as follows. All transitions entering (exiting) S_{d} are redirected to (from) S'_{d} (Lines 26–27). If S_{d} is the initial state $S_{0\mathrm{d}}$ of $\mathcal{A}_{\mathrm{d}}$, then S'_{d} becomes the new initial state (Line 28). After the removal of S_{d} from $\mathcal{A}_{\mathrm{d}}$ (Line 29), all buds relevant to S_{d} are converted to S'_{d} (Line 30). Note how the redirection of transitions exiting S_{d} may cause nondeterminism exiting S'_{d}, that is, pairs of transitions exiting S'_{d} that are marked by the same label ℓ. Such a nondeterminism is maintained throughout the processing and, eventually, eliminated without any algorithmic coercion, based on the fact that, in the end of bud processing, the states reached by nondeterministic transitions are necessarily equal.[5] The body of *ISC* is in Lines 32–60. After the extension of $\mathcal{A}_{\mathrm{n}}$ by the new transition T_{n} (Line 33), the bud set $\mathbb{B}$ is initialized with buds of the form $(S_{\mathrm{d}}, \ell', \mathbb{S}')$, where the content of S_{d} includes the state S_{n} exited by T_{n}, ℓ' is the label marking T_{n}, and $\mathbb{S}'$ is the ϵ-closure of the state S'_{n} reached by T_{n}. Then, a loop is iterated until $\mathbb{B}$ becomes empty (Lines 35–58). At each iteration, a bud $(S_{\mathrm{d}}, \ell, \mathbb{S})$ is considered (Line 36). Four main rules are defined, $\mathcal{R}_1 \mathinner{..} \mathcal{R}_4$, depending on the nature of the elements of the bud. Additionally, three sub-rules of $\mathcal{R}_4$ are given, $\mathcal{R}_4^1 \mathinner{..} \mathcal{R}_4^3$. Each rule is defined based on the pattern [*Condition*] $\Rightarrow$ *Action*. On the termination of the loop, the set of final states of $\mathcal{A}_{\mathrm{d}}$ is updated (Line 59).

1. **Algorithm** *ISC*$(\mathcal{A}_{\mathrm{n}}, \mathcal{A}_{\mathrm{d}}, T_{\mathrm{n}})$
2. $\mathcal{A}_{\mathrm{n}} = (\mathbb{S}_{\mathrm{n}}, \Sigma, \mathbb{T}_{\mathrm{n}}, S_{0\mathrm{n}}, \mathbb{S}_{f\mathrm{n}})$: an NFA,
3. $\mathcal{A}_{\mathrm{d}} = (\mathbb{S}_{\mathrm{d}}, \Sigma, \mathbb{T}_{\mathrm{d}}, S_{0\mathrm{d}}, \mathbb{S}_{f\mathrm{d}})$: a DFA equivalent to $\mathcal{A}_{\mathrm{n}}$,
4. $T_{\mathrm{n}} = S_{\mathrm{n}} \xrightarrow{\ell'} S'_{\mathrm{n}}$: a new transition for $\mathcal{A}_{\mathrm{n}}$, where $S_{\mathrm{n}} \in \mathbb{S}_{\mathrm{n}}$;
5.   **auxiliary function** *New*$(\mathbb{S})$: a new state in $\mathcal{A}_{\mathrm{d}}$
6. $\mathbb{S}$: a subset of states in $\mathcal{A}_{\mathrm{n}}$;
7.   **begin** {New}
8.   Create a new state S'_{d} in $\mathcal{A}_{\mathrm{d}}$, where $\|S'_{\mathrm{d}}\| = \mathbb{S}$;
9.   **return** S'_{d}
10.   **end** {New};

[5] This comes from the fact that nondeterministic transitions in the intermediate $\mathcal{A}_{\mathrm{d}}$ are generated by the merging of states with identical content.

auxiliary procedure $\mathit{Extend}(S_d, \mathbb{S})$
S_d: a state in $\mathcal{A}_d$,
$\mathbb{S}$: a subset of states in $\mathcal{A}_n$;
begin {Extend}
if $\mathbb{S} \not\subseteq \|S_d\|$ **then**
Update $\mathbb{B}$ with the buds relevant to S_d and $(\mathbb{S} - \|S_d\|)$;
Insert $\mathbb{S}$ into $\|S_d\|$;
if $\mathcal{A}_d$ includes a state S'_d such that $\|S'_d\| = \|S_d\|$ **then**
$\mathit{Merge}(S_d, S'_d)$
end-if
end-if
end {Extend};
auxiliary procedure $\mathit{Merge}(S_d, S'_d)$
S_d, S'_d: the states in $\mathcal{A}_d$ to be merged;
begin {Merge}
Redirect to S'_d all transitions entering S_d and remove duplicates;
Redirect from S'_d all transitions exiting S_d and remove duplicates;
if S_d is the initial state S_{0d} **then** $S_{0d} := S'_d$ **end-if**;
Remove S_d from $\mathcal{A}_d$;
Convert to S'_d the buds in $\mathbb{B}$ relevant to S_d
end {Merge};
begin {ISC}
Insert into $\mathcal{A}_n$ the new transition $T_n = S_n \xrightarrow{\ell'} S'_n$;
$\mathbb{B} := \{(S_d, \ell', \mathbb{S}') \mid S_d \in \mathcal{A}_d, S_n \in \|S_d\|, \mathbb{S}' = \epsilon\text{-}\mathit{closure}(S'_n)\}$;
loop
Remove a bud $B = (S_d, \ell, \mathbb{S})$ from $\mathbb{B}$;
Based on B and $\mathcal{A}_d$, if applicable, select one rule in $\mathcal{R}_1 \mathinner{..} \mathcal{R}_4$:
$(\mathcal{R}_1)$ $[\ell = \epsilon] \Rightarrow \mathit{Extend}(S_d, \mathbb{S})$;
$(\mathcal{R}_2)$ $[\ell \neq \epsilon, \nexists$ a transition exiting S_d and marked by ℓ, $\mathcal{A}_d$ includes a state
S'_d such that $\|S'_d\| = \mathbb{S}] \Rightarrow$
Insert a new transition $S_d \xrightarrow{\ell} S'_d$ in $\mathcal{A}_d$;
$(\mathcal{R}_3)$ $[\ell \neq \epsilon, \nexists$ a transition exiting S_d and marked by ℓ, $\mathcal{A}_d$ does not include
a state S'_d such that $\|S'_d\| = \mathbb{S}] \Rightarrow$
$S'_d := \mathit{New}(\emptyset)$,
Insert a new transition $S_d \xrightarrow{\ell} S'_d$ in $\mathcal{A}_d$,
$\mathit{Extend}(S'_d, \mathbb{S})$;
$(\mathcal{R}_4)$ $[\ell \neq \epsilon, \exists$ a transition exiting S_d and marked by $\ell] \Rightarrow$
$\forall T_d = S_d \xrightarrow{\ell} S'_d$ where $\mathbb{S} \not\subseteq \|S'_d\|$: select one rule in $\mathcal{R}_4^1 \mathinner{..} \mathcal{R}_4^3$;
$(\mathcal{R}_4^1)$ $[S'_d = S_{0d}] \Rightarrow$
$S''_d = \mathit{New}(\emptyset)$,
Redirect T_d towards S''_d,
$\mathit{Extend}(S''_d, \|S'_d\| \cup \mathbb{S})$;
$(\mathcal{R}_4^2)$ $[S'_d \neq S_{0d}, \nexists$ another transition entering $S'_d] \Rightarrow \mathit{Extend}(S'_d, \mathbb{S})$;
$(\mathcal{R}_4^3)$ $[S'_d \neq S_{0d}, \exists$ another transition entering $S'_d] \Rightarrow$

55. $\forall \bar{T}_{\text{d}} = \bar{S}_{\text{d}} \xrightarrow{\bar{\ell}} S'_{\text{d}}$ where $\bar{S}_{\text{d}} \neq S_{\text{d}}$ **or** $\bar{\ell} \neq \ell$:
56. Remove $\bar{T}_{\text{d}}$ and update $\mathbb{B}$ with the buds relevant to $\bar{S}_{\text{d}}$ and $\bar{\ell}$,
57. *Extend*$(S'_{\text{d}}, \mathbb{S})$;
58. **while** $\mathbb{B} \neq \emptyset$;
59. Update the set of final states of $\mathcal{A}_{\text{d}}$
60. **end** $\{$ISC$\}$.

As an example of the application of *ISC*, consider the NFA, called $\mathcal{A}_{\text{n}}$, displayed on the top-left of Fig. 5, without (dotted) transition $T_{\text{n}} = N_0 \xrightarrow{\epsilon} N_2$. The corresponding DFA, namely $\mathcal{A}_{\text{d}}$, is displayed on the bottom-left of the same figure. Based on $\mathcal{A}_{\text{n}}$, $\mathcal{A}_{\text{d}}$, and the new transition T_{n}, we now show the execution of the *ISC* algorithm step by step, with the help of Fig. 5 and Table 1. Table 1 outlines, for each iteration of the main loop in *ISC*, the chosen bud and the bud set at the end of the iteration (before next iteration). In particular, after the execution of Line 34, $\mathbb{B}$ will contain two buds, $(D_0, \epsilon, \{N_1, N_2, N_3\})$ and $(D_2, \epsilon, \{N_1, N_2, N_3\})$, as N_2 is both included in D_0 and D_2, and the ϵ-closure of N_2 is $\{N_1, N_2, N_3\}$. Then, the loop iterates five times:[6]

(1) Bud $(D_2, \epsilon, \{N_1, N_2, N_3\})$ is chosen, corresponding to rule $\mathcal{R}_1$. This is expected to cause the extension of $\|D_2\|$ by $\mathbb{S} = \{N_1, N_2, N_3\}$ (Line 38), which, however, has no effect, since $\mathbb{S} \subset \|D_2\|$ (see Line 15).
(2) Bud $(D_0, \epsilon, \{N_1, N_2, N_3\})$ is chosen, corresponding to rule $\mathcal{R}_1$. This causes the extension of $\|D_0\|$ by $\{N_1, N_2, N_3\}$ (see Fig. 5, *Iteration 2*, top), and the generation of the new buds $(D_0, a, \{N_1, N_2, N_3\})$ and $(D_0, b, \{N_1, N_2, N_3\})$. Since such an extension causes $\|D_0\| = \|D_2\|$, a merge of D_0 and D_2 is required (Line 19). The effect of the merge is displayed in Fig. 5, *Iteration 2*, right, where all transitions exiting D_0 are redirected from D_2 (Line 27). Then, D_2 becomes the new initial state (Line 28), while D_0 is removed (Line 29).[7] Finally, the two buds, previously generated by *Extend* and relevant to D_0, are renamed to D_2.
(3) Bud $(D_2, a, \{N_1, N_2, N_3\})$ is chosen, corresponding to rule $\mathcal{R}_4$. Of the two transitions exiting D_2 and marked by a, namely $D_2 \xrightarrow{a} D_1$ and $D_2 \xrightarrow{a} D_3$, only the former is relevant to the processing, as the latter is such that $\mathbb{S} = \{N_1, N_2, N_3\} = \|S'_{\text{d}}\|$. Specifically, sub-rule $\mathcal{R}_4^2$ applies, which causes the extension of $\|D_1\|$ to $\{N_1, N_2, N_3\}$, with the generation of the new bud $(D_1, a, \{N_1, N_2, N_3\})$. The resulting automaton is displayed in Fig. 5, *Iteration 3*, bottom. As before, the extension of D_1 makes $\|D_1\| = \|D_3\|$, thereby requiring a merge of D_1 and D_3, whose effect is displayed in Fig. 5, *Iteration 3*, right. Also, note how the merging causes bud $(D_1, a, \{N_1, N_2, N_3\})$ to be renamed to $(D_3, a, \{N_1, N_2, N_3\})$.
(4) Bud $(D_2, b, \{N_1, N_2, N_3\})$ is chosen, corresponding to rule $\mathcal{R}_4$. However, since $\mathbb{S} = \{N_1, N_2, N_3\} \subseteq \|S'_{\text{d}}\|$ (false condition at Line 48), no action is performed.
(5) The last bud, $(D_3, a, \{N_1, N_2, N_3\})$, is considered, corresponding to rule $\mathcal{R}_4$. As before, since condition at Line 48 does not hold, no action is performed.

[6] In each iteration of the loop, a bud is chosen. *ISC* does not constraint the order with which buds are picked up from the bud set, as this does not affect the final result.

[7] Note how the effect of the merge causes the nondeterminism of the intermediate $\mathcal{A}_{\text{d}}$, owing to two transitions exiting D_2 and marked by the same label a.

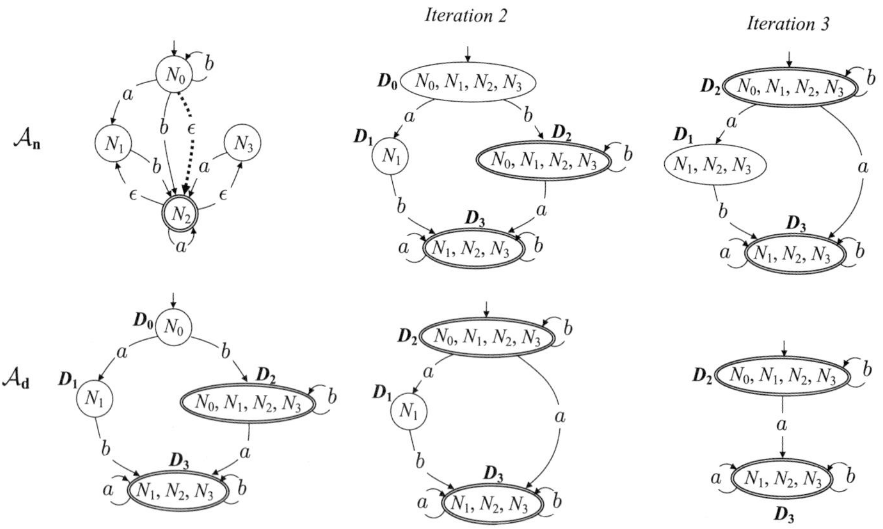

Fig. 5. Tracing of *ISC* execution

Since $\mathbb{B}$ is empty, the loop terminates. As both states in $\mathcal{A}_d$ are final already, this concludes the run of *ISC*, which correctly led to the generation of the DFA equivalent to $\mathcal{A}_n$ augmented by T_n.

5 Experimental Results

In order to assess the practical value of incremental determinization of finite automata, we implemented *ISC* in Java language, under Linux operating system, and ran a number of experiments. The developed software environment allows for both performance analysis and symbolic tracing of executions. For comparison purposes, besides *ISC*, we also implemented both *SC* and a generator of NFAs. The task of NFA determinization was never faced before in the literature in an incremental way, therefore experimental results obtained by applying the proposed method can only be compared with the results inherent to monolithic algorithms accomplishing the same task, namely *SC*. Both

Table 1. Details of *ISC* steps outlined in Fig. 5

Iteration	*Chosen bud*	*Bud-set before the next iteration*
		$\{(D_0, \epsilon, \{N_1, N_2, N_3\}), (D_2, \epsilon, \{N_1, N_2, N_3\})\}$
1	$(D_2, \epsilon, \{N_1, N_2, N_3\})$	$\{(D_0, \epsilon, \{N_1, N_2, N_3\})\}$
2	$(D_0, \epsilon, \{N_1, N_2, N_3\})$	$\{(D_2, a, \{N_1, N_2, N_3\}), (D_2, b, \{N_1, N_2, N_3\})\}$
3	$(D_2, a, \{N_1, N_2, N_3\})$	$\{(D_2, b, \{N_1, N_2, N_3\}), (D_3, a, \{N_1, N_2, N_3\})\}$
4	$(D_2, b, \{N_1, N_2, N_3\})$	$\{(D_3, a, \{N_1, N_2, N_3\})\}$
5	$(D_3, a, \{N_1, N_2, N_3\})$	$\emptyset$

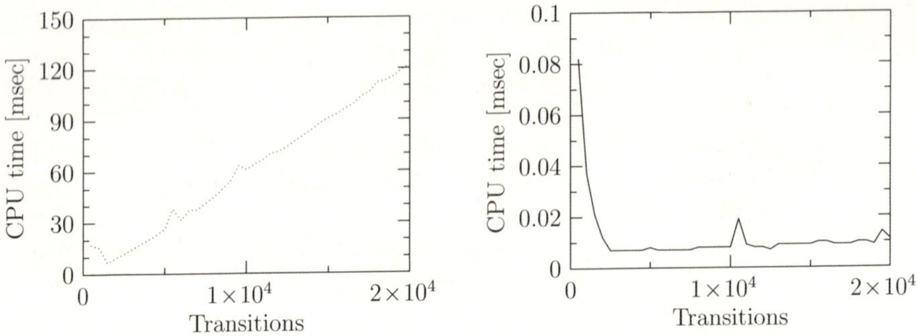

Fig. 6. CPU time for experiment $E : (S = 10000, T = 20000, \sigma = 50)$

CPU time and memory allocation were considered. The experiment presented in this paper is defined by a triple $E = (S, T, \sigma)$. A corresponding NFA, namely $\mathcal{N}$, was randomly generated, having number of states, transitions, and symbols (the alphabet marking the transitions) equal to parameters S, T, and σ, respectively. A sequence of increasingly growing NFAs was considered, with the last NFA being $\mathcal{N}$. Then, both *SC* and *ISC* were run on the sequence of NFAs, the former starting from scratch at each new NFA-to-DFA transformation, the latter updating the previous DFA incrementally. The parameter-values of the experiment are $S = 10000$, $T = 20000$, and $\sigma = 50$. Therefore, 20000 runs were performed of both *SC* and *ISC*. The CPU time for *SC*, represented on the left of Fig. 6, grows linearly. The curve for *ISC*, which is close to zero in all the range of transitions, is outlined (on a different scale) on the right of Fig. 6. Interestingly enough, after a certain point (about 2500 transitions), the response time for *ISC* keeps being constant. So, at the last (200000th) transition, *SC* needs 121.973 msec, while the CPU time for *ISC* is only 0.011 msec, about four orders of magnitude less than *SC*. As to the memory allocation, experiments indicate that *ISC* needs slightly more memory than *SC* does.

6 Conclusion

This paper deals with incremental subset-construction (*ISC*), an algorithm for the incremental construction of a DFA equivalent to a given NFA, where the NFA grows transition by transition. Each run of the algorithm processes a new transition and updates the previous DFA. *ISC* is a variant of the subset-construction algorithm (*SC*). If *ISC* were not available, *SC* should be used instead, that is, each time a new transition is added to the NFA, this (updated) NFA would be transformed into the equivalent DFA by *SC*, starting from scratch every time, that is, without exploiting in any way the previous DFA. The idea of the new algorithm is to exploit the previous DFA in order to obtain the current DFA. In fact, the previous DFA is a compiled version of the previous NFA, that is, it already includes most of the knowledge needed to obtain the current DFA. *ISC* is a general algorithm that can be used in a broad variety of applications since it can perform the transformation of any NFA, whether cyclic or acyclic, and does not generate any DFA state which is unreachable from the initial DFA state, which is an asset of the proposed approach. The only limiting assumption of *ISC* is that the transition extending the NFA at each iteration is bound to exit from an existing state of the NFA itself, that is, it cannot exit from a new state to be added to the NFA. New states are possibly added to the NFA only as final states of new transitions. A challenge for future work is generalizing the notion of the portion of NFA to be considered at each iteration, so as to relax this constraint. A question still needs investigation: how the CPU-time is affected by the order according to which buds are processed. This would be the starting point for optimizing *ISC* based on specific heuristics. A formal proof of the soundness and completeness of *ISC* is a final engagement for future work.

References

1. Lamperti, G., Zanella, M.: Diagnosis of Active Systems – Principles and Techniques. The Kluwer International Series in Engineering and Computer Science, vol. 741. Kluwer Academic Publisher, Dordrecht (2003)
2. Lamperti, G., Zanella, M.: A bridged diagnostic method for the monitoring of polymorphic discrete-event systems. IEEE Transactions on Systems, Man, and Cybernetics – Part B: Cybernetics 34(5), 2222–2244 (2004)
3. Cerutti, S., Lamperti, G., Scaroni, M., Zanella, M., Zanni, D.: A diagnostic environment for automaton networks. Software – Practice and Experience 37(4), 365–415 (2007)
4. Lamperti, G., Zanella, M., Zanni, D.: Incremental processing of temporal observations in model-based reasoning. AI Communications 20(1), 27–37 (2007)
5. Revuz, D.: Dictionnaires et lexiques: méthodes et algorithmes. PhD thesis, Institut Blaise Pascal, Paris, France (1991)
6. Daciuk, J.: Incremental construction of finite-state automata and transducers, and their use in the natural language processing. PhD thesis, University of Gdansk, Poland (1998)
7. Daciuk, J., Watson, B., Watson, R.: Incremental construction of minimal acyclic finite state automata and transducers. In: International Workshop on Finite State Methods in Natural Language Processing, Ankara, Turkey, pp. 48–56 (1998)
8. Daciuk, J., Mihov, S., Watson, B., Watson, R.: Incremental construction of minimal acyclic finite state automata. Computational Linguistics 26(1), 3–16 (2000)
9. Carrasco, R., Forcada, M.: Incremental construction and maintenance of minimal finite-state automata. Computational Linguistics 28(2), 207–216 (2002)
10. Daciuk, J.: Semi-incremental addition of strings to a cyclic finite automaton. In: Klopotek, M., Wierzchon, S., Trojanowski, K. (eds.) Advances in Soft Computing, pp. 201–207. Springer, Heidelberg (2004)
11. Watson, B., Daciuk, J.: An efficient incremental dfa minimization algorithm. Natural Language Engineering 9(1), 49–64 (2003)
12. Green, T., Gupta, A., Miklau, G., Onizuka, M., Suciu, D.: Processing xml streams with deterministic automata and stream indexes. ACM Transactions on Database Systems 29(4), 752–788 (2004)
13. Lamperti, G., Zanella, M.: Monitoring and diagnosis of discrete-event systems with uncertain symptoms. In: Sixteenth International Workshop on Principles of Diagnosis – DX 2005, Monterey, CA, pp. 145–105 (2005)
14. Hamscher, W., Console, L., de Kleer, J. (eds.): Readings in Model-Based Diagnosis. Morgan Kaufmann, San Mateo (1992)
15. Dvorak, D., Kuipers, B.: Model-based monitoring of dynamic systems. In: Eleventh International Joint Conference on Artificial Intelligence – IJCAI 1989, Detroit, MI, pp. 1238–1243 (1989)

16. Lackinger, F., Nejdl, W.: Integrating model-based monitoring and diagnosis of complex dynamic systems. In: Twelfth International Joint Conference on Artificial Intelligence – IJCAI 1991, Sydney, Australia, pp. 2893–2898 (1991)
17. Struss, P.: Fundamentals of model-based diagnosis of dynamic systems. In: Fifteenth International Joint Conference on Artificial Intelligence – IJCAI 1997, Nagoya, Japan, pp. 480–485. Morgan Kaufmann, S. Francisco (1997)
18. Cassandras, C., Lafortune, S.: Introduction to Discrete Event Systems. The Kluwer International Series in Discrete Event Dynamic Systems, vol. 11. Kluwer Academic Publisher, Boston (1999)
19. Brand, D., Zafiropulo, P.: On communicating finite-state machines. Journal of ACM 30(2), 323–342 (1983)
20. Baroni, P., Lamperti, G., Pogliano, P., Zanella, M.: Diagnosis of large active systems. Artificial Intelligence 110(1), 135–183 (1999)
21. Lamperti, G., Zanella, M.: Diagnosis of discrete-event systems from uncertain temporal observations. Artificial Intelligence 137(1–2), 91–163 (2002)

A Knowledge Based Approach for Capturing Rich Semantic Representations from Text

Peter Z. Yeh, Daniel R. Farina, and Alex Kass

Accenture Technology Labs, San Jose CA 95113, USA
{peter.z.yeh,daniel.r.farina,alex.kass}@accenture.com

Abstract. In this paper, we present a knowledge based approach to capture semantic representations from natural language for a class of applications where the representations of interest are known in advance. Our approach performs this task by generating phrases from these representations and matching these phrases against text using a set of syntactic and semantic transformations. The representation that best matches a piece of text is selected as its meaning. We evaluate our approach on a corpus of news articles collected from over 150 online news sources, and show how our approach performs well on capturing semantic representations from text.

1 Introduction

One of the goals of AI is to build Natural Language Understanding (NLU) systems that can produce rich semantic representations for tasks such as question answering, information extraction, and information retrieval.

For example, to produce the semantic representation for the sentence *"The man received a loan from the bank"*, most NLU systems first produce a syntactic representation that captures the syntactic relationships (i.e. subject, direct object, etc.) between the atomic constituents (i.e. nouns, verbs, adjectives, and adverbs) in the sentence. These systems then select from an ontology the most appropriate concept and semantic relation for each constituent and syntactic relationship respectively. Given the context, *Financial-Institution* may be the most appropriate concept for "bank", and *recipient* may be the most appropriate semantic relation for "subject" (see Figure 1).

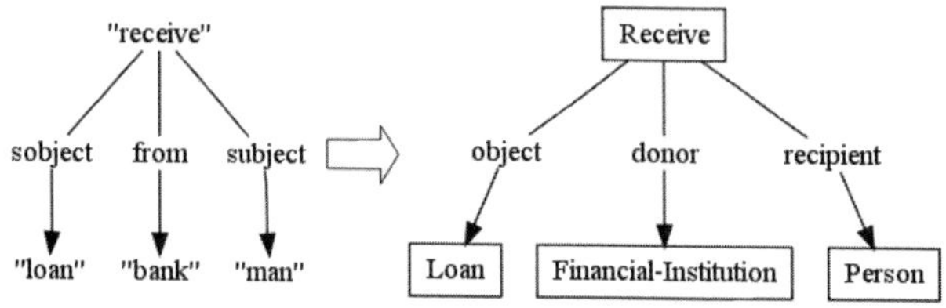

Fig. 1. Left: The syntactic representation of our example sentence which captures the syntactic relationships between the constituents. **Right:** The semantic representation of the sentence.

I. Lovrek, R.J. Howlett, and L.C. Jain (Eds.): KES 2008, Part I, LNAI 5177, pp. 375–383, 2008.

Several promising approaches have been reported that produce rich semantic representations from text [1,2], but they are limited in scope. These approaches only produce representations for controlled languages [3] or for restricted domains. These limitations are due to the difficulty of understanding unrestricted natural language.

Other solutions that do not have these limitations only address part of the problem. Solutions like [4,5,6] focus on determining just the meaning of a word within a context, and solutions like [7,8,9] focus on determining just the semantic relations between a verb and its arguments. These solutions do not produce a complete semantic representation, and previous research has shown the value of combining these separate tasks into a unified approach [1].

In this paper, we report a knowledge based solution that addresses these limitations for a class of applications where the semantic representations of interest are known in advance. This requirement is satisfied by applications that monitor sources like news, blogs, etc. for pre-determined information of interest to generate actionable insights ranging from technology maturity assessments to detecting business threats and opportunities. For example, an application to track technology maturity would know in advance events that affect technology maturation (e.g. product deployments, adoption by a Fortune 500 company, etc) [10]. Similarly, a business intelligence application would know in advance events that could signal potential threats (e.g. a supplier in financial trouble) and opportunities (e.g. a recall by a competitor) for an organization [11].

Our solution captures semantic representations from text by generating phrases from representations of interest and matching these phrases against text using a set of syntactic and semantic transformations. The representation that best matches a piece of text is selected as its meaning. We evaluate our approach in the context of an application to track technology maturity. We use a corpus of news stories collected from over 150 online news sources and show how our approach performs well on capturing semantic representations from unrestricted text.

2 Knowledge Requirements

Our approach requires an ontology to capture semantic representations from text. This ontology needs to provide rich representations that are linguistically motivated. Hence, we chose the Component Library (CLib) built by Barker et al. [12] over other resources like WordNet [13] and FrameNet [14]. The semantics of concepts in WordNet are limited mostly to hypernyms, meronyms, and synonyms while FrameNet focuses primarily on the semantic roles played by the syntactic arguments of verbs. The CLib, on the other hand, provides a rich domain independent upper ontology with about 80 semantic relations and about 500 generic concepts that can be composed and extended to build domain specific ones.

2.1 Semantic Relations

One type of knowledge in the CLib is semantic relations. These relations fall into three general categories: 1) Relations between an event and an entity such as *agent*, *instrument*, *object*, and *donor*. These relations are in the spirit of case roles proposed by

Fillmore [15] and others [14]. 2) Relations between entities such as *has-part*, *possesses*, and *material*. 3) Relations between events such as *caused-by*, *prevents*, and *enables*.

Each semantic relation also has information about its syntactic realization [16] – i.e. how the relation surfaces in a sentence. For example, *agent* can surface as the subject (e.g. "A man hit a ball") or as a prepositional phrase marked by the preposition "by" (e.g. "A ball was hit by a man"). From this information, our approach can look up the syntactic realizations for each semantic relation to match against text.

2.2 Events and Entities

The second type of knowledge in the CLib covers events and entities. Each event is similar to the frames in FrameNet and encodes knowledge about the participants in the event, where (and when) the event occurred, and other events that are caused (or prevented). For example, a *Buy* event encodes knowledge about the object bought, the donor, and the recipient (see Figure 2 left).

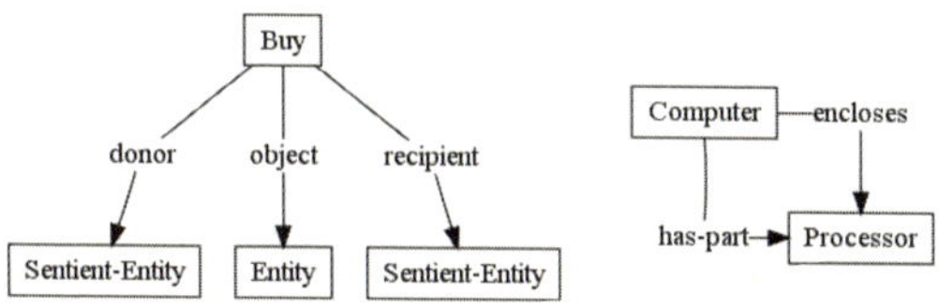

Fig. 2. Left: The encoding for the *Buy* event drawn as a conceptual graph [17]. **Right:** The encoding for the concept of *Computer*.

Each entity encodes knowledge about its parts, its spatial relationship to other entities, and the roles it can play. For example, the representation of a computer says: a computer has a processor part and also encloses the processor part (see Figure 2 right).

Each concept in the CLib is annotated with appropriate senses from WordNet to provide information about the concept's lexical realization. For example, the *Buy* event is annotated with the WordNet senses of buy#1, purchase#1, etc. From this information, our approach can look up the lexical realizations for each concept to match against text.

3 Our Approach

Our approach captures semantic representations from text by generating phrases from representations of interest and matching these phrases against text using a set of syntactic and semantic transformations. The representation that best matches a piece of text is selected as its meaning.

3.1 Generate Phrase

Our approach takes a semantic representation, represented as a conceptual graph, and generates a set of phrases from it. For example, to generate phrases from the representation R shown in Figure 3, our approach first selects a concept in R to serve as the root. This root concept is treated as the main verb or noun – depending on whether this concept is an *Event* or *Entity* respectively – when matching against text.

Our approach prefers events over entities because events typically lead to better matches, so it selects *Award* as the root. If there are multiple events (or entities) to select from, then our approach randomly selects one.

Once a root is selected, our approach randomly selects one of its lexical realizations to match against text. In this example, our approach selects "grant" for *Award*.

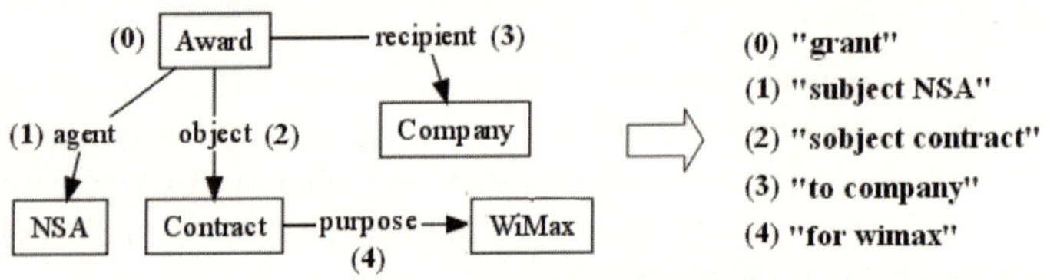

Fig. 3. Left: The semantic representation of interest R. **Right:** The phrases generated from R. The numeric labels in R (see left) correspond to the phrases generated.

Our approach then generates a phrase for each semantic relation (i.e. edge) and its value (i.e. the concept pointed to) in R. This phrase is generated by randomly selecting one of the syntactic realizations for the semantic relation and one of the lexical realizations for its value. For example, our approach would generate the phrase "to company" for the semantic relation *recipient* and its value *Company* because "to" is a syntactic realization for *recipient* and "company" is a lexical realization for *Company*. Figure 3 shows all phrases generated from R.

3.2 Match Phrase

Our approach first matches the lexical realization generated from the root concept in R to text. For example, given the sentence *"Wimax contract awarded by NSA to Nokia"* which we will refer to as S, our approach matches "grant" (the lexical realization generated from *Award*) to this sentence. We say a lexical realization matches a sentence if the realization's stem is identical to the stem of one of the constituents in the sentence.[1]

There is no match between "grant" and S, so our approach tries to apply one of the following transformations to improve the match:

- **Lexical Realization Replacement:** Replace one lexical realization with another for the same concept. For example, replace "purchase" with "buy" for the concept *Buy*.
- **Syntactic Realization Replacement:** Replace one syntactic realization with another for the same semantic relation. For example, replace "subject" with "by" for the semantic relation *agent*.
- **Active to Passive Voice:** Change the syntactic realization for a semantic relation that surfaces as the syntactic object to the syntactic subject. For example, change "sobject" (the syntactic object) for *recipient* to "subject".

[1] We use the Porter stemmer [18] to stem each word.

- **Modifier Conversion:** Turn the lexical realization for a concept C into a modifier of the lexical realization for another concept that C is an attribute of. One concept is an attribute of another if they are directly related through a semantic relation. For example, in Figure 3 *WiMax* is related to *Contract* through the *purpose* relation. Hence, the lexical realization for *WiMax* can be turned into a modifier of *Contract* resulting in "wimax contract".
- **Lexical Realization Specialization:** Replace the lexical realization for a concept with the lexical realization for one of its subclasses or instances. For example, replace "move" for the concept *Move* with "walk" for the concept *Walk*, a subclass of *Move*.

Our approach can apply the *Lexical Realization Replacement* transformation to replace "grant" with "award", another lexical realization for the root, and this transform results in a match with S. If the root cannot be matched, then our approach disregards R as the meaning of S and moves on to matching the next sentence.

Once the root is matched, our approach matches the remaining phrases generated from R to the subject, syntactic object, and prepositional phrases in S. We say a phrase p generated from R matches a phrase q in S if they begin with the same syntactic marker (e.g. "subject", "sobject", "by", etc.) and the body of p matches q under the same criterion used for matching the root concept (see above). This match process is repeated until all phrases generated from R are matched or the match cannot be improved further through the use of transformations. Table 1 shows how the phrases generated from R match the phrases in S.

Table 1. Matches between the phrases generated from R (see first column) and the phrases in S (see second column). Transformations used to enable a match are given in the third column, and an explanation for how the transformation was used is given in the last column.

Phrases Generated from R	Phrases in S	Transformation Applied	Explanation
sobject contract	subject wimax contract	Active to Passive Voice	*sobject* changed to *subject*
subject NSA	by NSA	Syntactic Realization Replacement	*subject* replaced with *by*, another syntactic realization for the *agent* relation.
to company	to Nokia	Lexical Realization Specialization	*company* specialized to *nokia*, the lexical realization for the concept *Nokia* which is an instance of the concept *Company*.
for wimax	subject wimax contract	Modifier Conversion	*wimax* turned into a modifier of *contract*, *WiMax* is an attribute of *Contract* in R

Once the match is complete, our approach computes a score for the match to determine how well a representation captures the meaning of a piece of text. This score is computed using the following equation:

$$\frac{\sum_{p \epsilon R} w(p) match(p, S)}{|R|} \tag{1}$$

where $|R|$ is the number of phrases generated from R; $match(p, S)$ is whether a phrase generated from R matches the sentence S (1 if they match, 0 otherwise); and $w(p)$ is the weight of p on a scale from 0 to 1. This weight can be set to the statistical frequency of the phrase, the ontological depth of the relation and concept that the phrase was generated from, etc. In our implementation, we set $w(p)$ to always return 1.

Returning to our example, the score between R and our example sentence S is 1 because all the phrases generated from R could be matched to S. Hence, the meaning of S is most likely captured by R.

4 Evaluation

4.1 Performance Task

Our approach was used in a system – the Technology Investment Radar[2] – that tracks the maturity of technologies of interest to inform investment decisions. This is an important problem for business executives as they often have difficulty determining when a technology's potential will be realized. For example, many executives in the wireless industry recognize WiMax as a technology that may have a significant impact on their industry, but they are less certain about whether (and especially when) that impact will be realized.

These executives, therefore, need a way to accurately determine the maturity of technologies that relate to their business in order to inform investment decisions such as when to develop in-house expertise on the technology, when to start offering products based on the technology, and so forth.

The Technology Investment Radar offers this capability, and our approach was used in this tool to continuously scan a variety of online news sources for information affecting technology maturity. This information is provided by a semantic model of technology maturation that was built by extending the Component Library (see Section 2). This model encodes the stages that a technology advances through as it matures and the events (and entities) that determine the technology's placement within these stages. For example, a company announcing plans to invest in a technology suggests that the technology is just beginning to emerge while an adoption of the technology by a Fortune 500 company suggests that the technology is more mature. Hence, this maturation model satisfies our approach's requirement that the representations of interest are known in advance.

4.2 Corpus

We evaluate our approach using a corpus of news articles collected from 158 online sources over an eight month period. These sources range from technology oriented venues (e.g. Techworld, VoIP Forum, RCR News, etc.) to mainstream media (e.g. BBC, NY Times, Yahoo! News, etc.).

---

[2] The Technology Investment Radar is being piloted with Accenture's Wireless Community of Practice: an organization within Accenture that performs strategy and technology consulting within the wireless technology space.

To construct a gold standard for evaluation, we employed two human graders to produce semantic representations from these news articles. However, due to constraints on both time and resources our graders were not able to encode all these articles, which numbered over 250,000. Hence, we reduced this corpus using the following methodology, but we want to point out that our approach processed all 250,000 articles in actual usage within the Technology Investment Radar system.

We first kept only articles that referenced one of the technologies being tracked by the Technology Investment Radar. This resulted in over 3,200 articles from which we randomly selected 1,000. To further reduce the encoding demand on our graders, we used only the titles of these 1,000 articles for evaluation. Hence, the resulting corpus contained 1,000 sentences with the average length of each sentence being 8.83 words.

Once a corpus for evaluation was constructed, each human grader was instructed to produce a semantic representation for each sentence from the corpus using concepts and relations from both the Component Library and the technology maturation model. The first grader produced 346 representations, and the second grader produced 337 representations. The graders did not produce representations for sentences containing concepts not covered by the ontologies they had access to. The agreement between the two graders is 0.617 for the Cohen kappa, which suggests substantial agreement.

4.3 Experiment Setup and Result

We compared our approach to a baseline that uses only the lexical realizations for the concepts in a representation (and the realizations for their subclasses) to match against text. A comparison with the state of the art [1,2] could not be made for two reasons. 1) These approaches produce semantic representations for controlled English only, so they are not able to process the unrestricted English found in our corpus. 2) These approaches produce a semantic representation as the output. Our approach, on the other hand, starts with a representation of interest and matches the phrases generated from this representation to a piece of text to capture its meaning. Hence, it is difficult to establish a direct, meaningful metric for comparison.

Both our approach and the baseline were given the representations generated by the human graders to match against the corpus of 1,000 sentences. The representation that best matches a sentence is selected as its meaning. We say a match is correct if the grader produced the same representation for the sentence.

The output of both approaches were graded using precision (i.e. the number of correct answers over the total number of answers given by an approach) and recall (i.e. the number of correct answers over the total number of answers given by the human grader). Table 2 shows the results of this evaluation.

Our approach performed significantly better than the baseline on precision across both graders ($p < 0.01$ for the $\mathcal{X}^2$ test in each case). Our approach performed better because it matched the relationships in a representation against each sentence. This requirement, for example, allowed our approach to differentiate the semantic role played by O2 (a mobile communications company) in the sentences *"iPhone contract awarded to O2"* and *"O2 awards 2G/3G network contract"*, which the baseline could not.

This same requirement, however, hurt our approach on recall when the syntactic realization for a semantic relation failed to surface in a sentence. For example, one of

Table 2. The performance of each approach on the task of capturing semantic representations from text, given as percentages. Grader 1 produced 346 representations from our corpus of 1,000 sentences and grader 2 produced 337 representations.

	Grader1		**Grader2**	
	Precision	Recall	Precision	Recall
Our Approach	89.61	79.77	82.58	75.96
Baseline	64.89	81.21	61.33	81.89

the graders said that "Latin America" is the *site* of the *Deploy* event in the sentence "Telmex launches Latin America wimax", but the syntactic realizations for *site* – which include "in", "at", etc. – failed to surface in this sentence. This allowed the baseline to perform better than our approach across both graders, but the difference was not statistically significant in either case. Hence, our approach was able to significantly improve precision without any significant loss to recall.

5 Future Work

The results we observed are encouraging, but several issues remain to be addressed. We found that euphemisms and metaphors are common in unrestricted English, and they have a negative effect on recall. For example, the human graders recognized that "green light" means "approval" in the sentence *"EarthLink gets green light from Philadelphia for Wi-Fi network"*, but our approach could not capture this meaning. It can only capture the literal meaning of a sentence. We are currently investigating whether euphemisms and metaphors can be handled in a scalable manner through the use of additional transformations during the match phase of our approach.

We also observed that the improper handling of negation by our approach had an adverse effect on precision. For example, our approach would incorrectly match a representation of wimax purchase to the sentence *"Intel backs off plans to purchase wimax"*. To address this problem, we are currently exploring ways to automatically insert constraints into a representation of interest. These constraints can capture information such as negation, and if matched to a piece of text, then the representation of interest would be discarded as its meaning.

Finally, we are exploring the use of our approach to capture the meaning of text in other domains like military intelligence (e.g. detecting reports of insurgent activities) and business risk assessment (e.g. detecting reports of natural disasters that can disrupt an organization's operations).

6 Conclusion

In this paper, we presented a knowledge based approach to capture semantic representations from unrestricted natural language for a class of applications where the representations of interest are known in advance. Our approach performs this task by generating phrases from these representations and matching these phrases against text using a set of syntactic and semantic transformations. The representation that best matches a piece of text is selected as its meaning.

We also evaluated the performance of our approach. We used a corpus of 1,000 sentences constructed from the titles of news articles collected from 158 online news sources over an eight month period. The results showed that our approach can significantly improve precision without any significant loss in recall. We gave reasons for why our approach performed well and proposed directions for future work.

References

1. Yeh, P., Porter, B., Barker, K.: A unified knowledge based approach for sense disambiguation and semantic role labeling. In: AAAI (2006)
2. Barker, K., Agashe, B., Chaw, S., Fan, J., Glass, M., Hobbs, J., Hovy, E., Israel, D., Kim, D., Mulkar, R., Patwardhan, S., Porter, B., Tecuci, D., Yeh, P.: Learning by reading: A prototype system, performance baseline and lessons learned. In: AAAI (2007)
3. Clark, P., Harrison, P., Jenkins, T., Thompson, J., Wojcik, R.: Acquiring and Using World Knowledge Using a Restricted Subset of English. In: FLAIRS (2005)
4. Patwardhan, S., Banerjee, S., Pedersen, T.: Using Measures of Semantic Relatedness for Word Sense Disambiguation. In: Gelbukh, A. (ed.) CICLing 2003. LNCS, vol. 2588. Springer, Heidelberg (2003)
5. Mihalcea, R., Tarau, P., Figa, E.: PageRank on Semantic Networks, with Application to Word Sense Disambiguation. In: COLING (2004)
6. Vasilescu, F., Langlais, P., Lapalme, G.: Evaluating Variants of the Lesk Approach for Disambiguating Words. In: LREC (2004)
7. Gildea, D., Jurafsky, D.: Automatic Labeling of Semantic Roles. Computational Linguistics 28(3) (2002)
8. Hacioglu, K.: Semantic Role Labeling Using Dependency Trees. In: COLING (2004)
9. Pradhan, S., Ward, W., Hacioglu, K., Martin, J., Jurafsky, D.: Semantic Role Labeling Using Different Syntactic Views. In: ACL (2005)
10. Yeh, P., Farina, D., Kass, A.: Semantic interpretation of the web without the semantic web: Toward business-aware web processors. In: IEEE ICSC (2007)
11. Kass, A., Cowell-Shah, C.: Using lightweight nlp and semantic modeling to realize the internet's potential as a corporate radar. In: AAAI Fall Symposium (2006)
12. Barker, K., Porter, B., Clark, P.: A Library of Generic Concepts for Composing Knowledge Bases. In: KCAP (2001)
13. Fellbaum, C.: WordNet: An Electronic Lexical Database. MIT Press, Cambridge (1998)
14. Baker, C., Fillmore, C., Lowe, J.: The Berkeley FrameNet Project. In: COLING (1998)
15. Fillmore, C.: Some Problems for Case Grammar. In: 22nd Annual Round Table Meeting on Linguistics and Language Studies (1971)
16. Barker, K.: Semi-Automatic Recognition of Semantic Relationships in English Technical Texts. PhD thesis, University of Ottawa (1998)
17. Sowa, J.F.: Conceptual Structures: Information Processing in Mind and Machine. Addison-Wesley, Reading (1984)
18. Porter, M.: An algorithm for suffix stripping. Program 14(3) (1980)

Ontology-Based Design Pattern Recognition

Damir Kirasić[1] and Danko Basch[2]

[1] Information Support Center
[2] Department of Control and Computer Engineering in Automation
University of Zagreb
Faculty of Electrical Engineering and Computing

Abstract. This paper presents ontology-based architecture for pattern recognition in the context of static source code analysis. The proposed system has three subsystems: parser, OWL ontologies and analyser. The parser subsystem translates the input code to AST that is constructed as an XML tree. The OWL ontologies define code patterns and general programming concepts. The analyser subsystem constructs instances of the input code as ontology individuals and asks the reasoner to classify them. The experience gained in the implementation of the proposed system and some practical issues are discussed. The recognition system successfully integrates the knowledge representation field and static code analysis, resulting in greater flexibility of the recognition system.

Keywords: knowledge-based system, ontology-based system, static code analysis, description logics, OWL application, formal pattern definition.

1 Introduction

Design pattern recognition is a part of the code analysis field. As design patterns [1] are descriptions of the *design level*, they may be, and inevitably will be implemented in a number of different ways. Once implemented, design patterns become concrete structures of interacting programming components, and thus difficult to detect.

Discovery of design patterns, and other types of program features, can be used as a basis for diverse objectives like: bug finding [2]; security vulnerabilities discovery [3]; program model checking [4,5]; program design recovery and reverse engineering [6]; code optimisation [7]; parallelism discovery [8]; software documentation management [9].

A number of design pattern detection techniques have been proposed [14]. Different techniques vary vastly at code and pattern representations, algorithms applied, and overall system architecture. Most of the proposed methodologies are based on static code analysis, but some authors advocate dynamic analysis techniques [10], or combined approach [11]. There is a great variety of approaches to input code representation as well to pattern representations. For example, [12] uses graph-based descriptions that are translated to matrices for both input code and patterns. The PINOT tool [13] uses an abstract syntax tree (AST),

I. Lovrek, R.J. Howlett, and L.C. Jain (Eds.): KES 2008, Part I, LNAI 5177, pp. 384–393, 2008.

a control-flow graph (CFG) and a data-flow graph (DFG) for code and pattern representations. For a more complete survey and comparison of 26 different techniques see [14].

The purpose of this paper is to describe the ontology-based pattern recognition system. The attribute "ontology-based" refers to the fact that the OWL ontology [15] [16, Chapter 14] forms the core of the system. This ontology precisely and formally defines various code patterns.

The main objectives of the recognition system can be summarised as follows:

- Create an expandable framework for the recognition of various program features;
- Allow for precise (possibly formal) definition of program features that are searched for;
- Separate code pattern description from the rest of the system;
- Create an expandable and easily maintainable pattern ontology;
- Use existing libraries, APIs and proven technologies as much as possible;

The above objectives, actually, list the requirements for the proposed software system. This paper presents the architecture and the design of such a system and describes some experience gained during its implementation.

The recognition system is envisioned as a framework that can be used as a stand-alone utility, or as a subsystem for various larger systems, such as a compiler front end or IDE plug-ins.

The proposed system brings new methodology to the field of static code analysis. It introduces architecture that uses knowledge base as a building block for the recogniser system. Another contribution is in the field of the pattern definition area. To the best of the author's knowledge, there is no pattern detection methodology that uses OWL ontology as a pattern definition medium.

The research project (currently a work in progress) is being done as a part our Ph.D. research at the University of Zagreb, Faculty of Electrical Engineering and Computing.

The paper is organised as follows. Section 2 presents an overall overview of the proposed system. The following sections describe main modules of the system: parsing and AST generation (section 3), ontologies (section 4), code patterns and recognition procedure (section 5). Some practical problems encountered during the implementation and possible further lines of work are discussed in section 6. The final section reaches a conclusion.

2 Architecture Overview

The overall architecture of the system is shown in figure 1. The main modules of the system are: parser, analyser and ontologies. Viewed as a black box, the whole system takes some source code as input and creates an augmented abstract syntax tree (AST) and analyser reports as output.

Currently, the parser takes the C programming language as input and generates AST of the input program in the form of the XML document [17]. (Section 3 brings more details about parsing and AST generation).

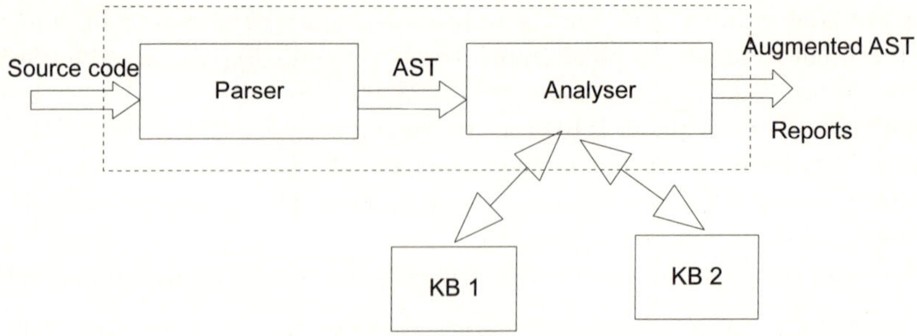

Fig. 1. Architecture overview

The analyser module takes previously generated AST as input and tries to find code patterns defined in the code pattern ontology. In order to find the specific pattern, the analyser has to follow a sequence of steps that, for complicated patterns, cannot be fully described in the ontology. Therefore, those patterns have to be "known" not only to the ontology but to the analyser code as well. Finally, AST is augmented with additional information and emitted as output. In addition, various reports can be generated. For example, the AST node that represents some program loop can be augmented with information that the loop can be parallelised, etc.

Two ontologies are created for the purpose of the analyser: (i) the programming language ontology and (ii) the code pattern ontology. Both ontologies have been created manually using the Protégé ontology editor [18].

The intent of the architecture presented is to separate the code pattern descriptions from the rest of the system as much as possible for the reasons mentioned in the introduction. Ideally, even the task of the analysis itself should be defined in an ontology.

3 Parsing and AST Representation

The parser module is generated by the ANTLR [20], a well known and widely used language recognition tool. ANTLR is a framework tool for constructing parsers, interpreters and similar programs from language grammars. Along with grammatical rules, the language grammars usually contain actions - code blocks that are executed at the appropriate phase of the parsing process. The ANTLR supports various "target languages". The same grammar can be used for the generation of different parsers - parsers written in a different "target" programming language. C# has been chosen as a target language for the reasons briefly explained later in this section.

The parser generated by the ANTLR is a full-fledged parser that can be used as a compiler front end. It supports full ANSI C grammar and is easily extended and configured in various ways. For the purpose of this project, C# and Java parsers have been written, but currently are not fully integrated into the rest of the system.

The output of parsing is, as usual, an abstract syntax tree (AST). The generated parsers can create predefined, default AST node types but can be configured to create custom made nodes. By specifying "XML element", as a type for AST node, parser generates a tree of XML elements - actually, a complete XML document. It is not sufficient to specify only the "XML element" as the AST node type but one has to write a custom made class that implements prescribed ANTLR interface.

The parser does full translation. This means that all programming constructs found in the input are preserved and emitted in a different form at the output. For example, for the input code segment:

```
int x = 0;
x = x + 1;
```

the parser will create the following XML document:

```
<?xml version="1.0" encoding="utf-8"?>
<CML version="0.7">
  <!--Created by CToXml 0.7 from 'intx_plus.c'-->
  . . .
  <Declaration>
    <Type>int</Type>
    <DirectDeclarator>x</DirectDeclarator>
    <Initializer>
      <Operator>=</Operator>
      <DecimalLiteral>0</DecimalLiteral>
    </Initializer>
  </Declaration>
  <Statement>
    <AssignmentExpression>
      <Operator>=</Operator>
      <Lvalue>
        <Id>x</Id>
      </Lvalue>
      <Plus>
        <Id>x</Id>
        <DecimalLiteral>1</DecimalLiteral>
      </Plus>
    </AssignmentExpression>
  </Statement>
</CML>
```

It is obvious that the AST (XML document) created is significantly larger then the plain text input code (the actual AST is even larger). But the AST now belongs to the well known realm of XML documents. It is a world with hundreds of tested and optimized supporting tools, APIs, parsers, checkers, translators, query engines, etc.

The reason for choosing C# as a target language is that it has powerful XML processing support. Among other things, version 3.5 of C# supports a new

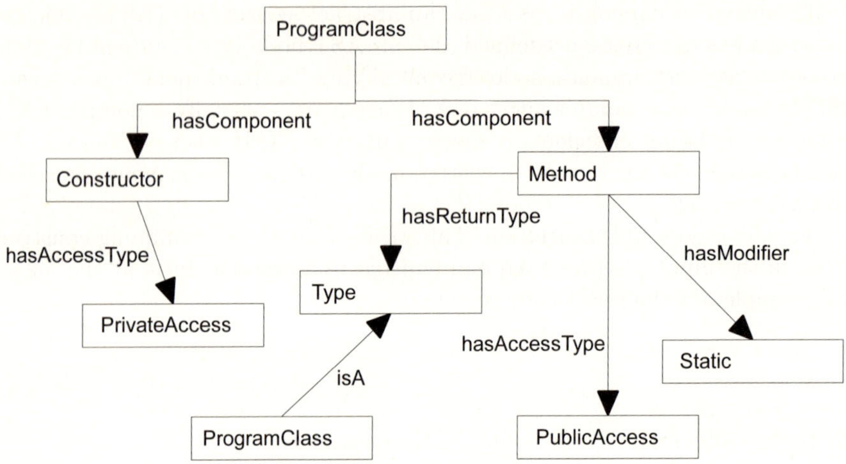

Fig. 2. Simplified programming ontology consists of concepts (boxes) and properties (arrows). A property defines the binary relation between two concepts. Note that `ProgramClass` is a subconcept of a more general `Type`. That is the only concept- subconcept relation shown in figure.

lightweight XML document model and the "LINQ to XML" querying constructs [19] that are part of the C# language (not another XML library).

4 Programming and Pattern Ontologies

The purpose of introducing ontologies in this project is manifold: (i) ontology precisely defines *what* has to be found; (ii) ontology brings clear separation of *knowledge* about patterns from the *procedures* of finding them; (iii) ontology brings expandability to the whole system; (iv) ontologies can be easily reused and integrated with other ontologies and other software systems.

For the purpose of the project, two OWL ontologies have been created. The first, called the programming ontology, defines taxonomy of general programming concepts like: Variable, Statement, ForLoop, Type, etc. This ontology contains the OO programming terms as well: ProgramClass, Constructor, Method, AccessType, etc. Figure 2 shows a simplified version of the programming ontology.

The second ontology, called the pattern ontology, defines specific patterns that have to be found.

An example will clarify usage of both ontologies. First, we can assume that the input language is an OO programing language like Java or C#. Second, we suppose that the recogniser has to find all the instances of the singleton classes [1,21]. The simplest definition of a singleton is: *"A singleton is a program class that can be instantiated exactly once"*. When this description is translated to code, it inevitably leads to certain program structure (code pattern), that the analyser will try to recognise. Implementing (one version of) a singleton as a Java/C# class, following conditions must be meet: (i) Only private constructors are allowed; (ii) there must be at least one public method (or public static final field) that has the same return type as the class it is defined in; (iii) that public method must return the same instance of the class, no matter how many times called.

The above conditions actually define *restrictions* that have to be applied on a programming class to get the singleton class. Once the restrictions on the program class are clearly articulated, it is not difficult to write a description of the singleton pattern:

```
Class(Singleton partial
   intersectionOf(
   (restriction(hasConstructor someValuesFrom PrivateConstructor))
   (restriction(hasConstructor allValuesFrom PrivateConstructor)))
   ProgramClass)
```

The above description is confined to the first restriction only and defines the `Singleton` as a subclass of the `ProgramClass`. A singleton must have at least one `PrivateConstructor` (in order to avoid automatic constructor generation by compiler) and all of its constructors have to be private. The above code is written in the OWL abstract syntax [15] which is more compact than the official XML OWL syntax.

A second and third restrictions are more complicated and require much more space. Moreover, the second restriction cannot be defined only with OWL but needs SWRL [22] - the rule language that extends the OWL capabilities. Therefore, the reasoner (Pellet [23]), used in this project, must understand both the OWL and the SWRL.

The second restriction has a non-tree structure. We have to state that "the return type of the public static method is the same as the type introduced by the class definition". That kind of non-tree situation is well known problem in the realm of the OWL. (Usually, it is described in terms of `hasUncle` property [22]).

The informal definition (with intuitive meaning) of the rule needed for the second restriction could be written as:

$$ProgramClass(?c) \wedge hasPublicMethod(?c, ?m) \wedge$$

$$hasReturnType(?m, ?c) \rightarrow Singleton(?c)$$

Two approaches are possible now. Either we can try to formulate *all* the restrictions using SWRL rules only, or we can use the OWL to describe as much as we can and then fill the gaps with rules. Adopting the second approach (and assuming that class `SingletonCandidate` complies to the first restriction) the second restriction has to be reformulated to:

$$SingletonCandidate(?c) \wedge hasPublicMethod(?c, ?m) \wedge$$

$$hasReturnType(?m, ?c) \rightarrow Singleton(?c)$$

In order to test the validity of the pattern definition we can define, in the ontology editor, the test individual as an instance of the ProgramClass and ask the reasoner to classify [16,24,25] all the individuals present in the ontology. If the reasoner finds that all the restrictions are satisfied, it can deduce that the test individual is a singleton. The same logic is followed in the analyser module that is presented in the next section.

5 Finding Code Patterns

The main module of the whole system is the "Analyser" box shown in Fig. 1 section 2. It takes an XML document (AST) as input and augments some of the input nodes according to the patterns defined in the pattern ontology. The analyser can produce additional reports as well. Note that the XML document constructed by the parser does not have to be saved in a file and then be re-parsed again by the analyser. It is held as in-memory tree that is passed to the analyser.

Depending on the type of the pattern, the recogniser will inspect the input tree and try to match input nodes with a specific pattern. In case of the singleton pattern, described in the previous section, it will construct an instance of the `ProgramClass`. The properties of the constructed individual will correspond to the actual properties found in the input tree. Once the construction is finished, the newly constructed instance is added to the pattern ontology. Now the reasoner is asked if the added individual is an instance of the `Singleton` class. If the reasoner answers positively, we have a match.

Other types of patterns can require different scenarios. One of the main objectives of this project is to explore possible ways of pattern description and recogniser-ontology interaction.

The interaction between the analyser and the ontologies is accomplished by the OWL API [26] - open source Java library. The .jar libraries were previously translated from .jar to .dll.

6 Practical Issues and Further Work

Three situations encountered during the implementation phase seem to be important and worth elaborating: (i) non-tree ontology structures; (ii) analyser code most "know" about ontologies; (iii) ontology and reasoner APIs are defined for Java but the project uses C#.

The non-tree structure of the ontology is already shown in Fig. 2 in section 4. The `ProgramClass` has public static method that returns the same type as the class the method is defined in. Therefore, the same relations should be more accurately depicted as in Fig. 3.

The non-tree property of the structure is more obvious in Fig. 3 than in Fig. 2. The presence of non-tree structures were the rationale for the introduction of the SWRL rules. This approach implies the usage of the SWRL-supporting reasoner as well. It would be much cleaner to have pure OWL only.

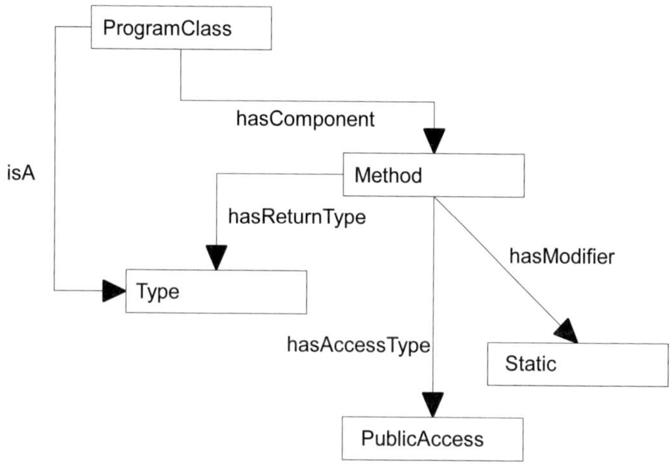

Fig. 3. The non-tree structure from Fig. 2

The analyser must "know" about patterns. For example, the reasoner must know that for the singleton pattern it has to construct a new instance, and that it has to add a newly created instance to ontology, and that it has to ask the reasoner about inherited types etc. It would be very desirable to have these steps defined in ontology as well. Another possibility is to have a uniform pattern-independent algorithm for *all* pattern descriptions.

Almost all libraries and APIs for ontology processing are written in Java but the project is written in C#, for the reasons mentioned before. Currently, the project uses Java bytecode to MS CIL translator: IKVMC (see http://www.ikvm.net). This tool was used for the translation of huge .jar libraries to .dll libraries. Although no problems were encountered, it would be much more desirable to have native C# libraries for OWL processing

In addition, a more detailed evaluation of this work has to be done. Primarily, comparison with other similar approaches and verification of accuracy (soundness and completeness) has yet to be performed.

7 Conclusion

The proposed system uses an ontology as a basis for pattern definitions. The main points that describe this system can be summarized as follows:

- The parser generates AST as an in-memory lightweight XML tree.
- The AST can be easily and elegantly processed by many tools and APIs (such as MS LINQ to XML).
- The code patterns searched for are formally described in the separate standalone ontology.
- The pattern recognition system can be easily expanded, modified and integrated with other systems.

– The system successfully integrates formal knowledge bases and static code analysis.

Although the pattern recogniser works well, there are many features which can be improved. The most important area of possible improvement seems to be the area of knowledge-algorithm integration.

References

1. Gamma, E., Helm, R., Johnson, R., Vlissides, J.: Design patterns: elements of reusable object-oriented software. Addison-Wesley Professional, Reading (1995)
2. Rutar, N., Almazan, C., Foster, J.: A Comparison of Bug Finding Tools for Java. In: Proceedings of the 15th IEEE International Symposium on Software Reliability Engineering, St. Malo, France (November 2004)
3. Pistoia, M., Chandra, S., Fink, S.J., Yahav, E.: A survey of static analysis methods for identifying security vulnerabilities in software systems. IBM System Journal 46(2) (2007)
4. Engler, D., Musuvathi, M.: Static Analysis versus Model Checking for Bug Finding. In Verification. In: Steffen, B., Levi, G. (eds.) VMCAI 2004. LNCS, vol. 2937, pp. 191–210. Springer, Heidelberg (2004)
5. Gulavani, B.S., Henzinger, T.A., Kannan, Y., Nori, A.V., Rajamani, S.K.: Synergy: A New Algorithm for Property Checking. In: FSE 2006: 14th Annual Symposium on Foundations of Software Engineering (November 2006)
6. Niere, J., Schafer, W., Wadsack, J., Wendehals, L., Welsh, J.: Towards pattern-based design recovery. In: Proceedings of the 24rd International Conference on Software Engineering, pp. 338–348. ACM, New York (2002)
7. Lattner, C., Adve, V.: LLVM: A Compilation Framework for Lifelong Program Analysis & Transformation. In: Proceedings of the 2004 International Symposium on Code Generation and Optimization (CGO 2004), Palo Alto, California (March 2004)
8. Tarditi, D., Puri, S., Oglesby, J.: Accelerator: Using Data Parallelism to Program GPUs for General-Purpose Uses. In: Proceedings of the 12th international conference on Architectural support for programming languages and operating systems 2006, San Jose, CA, USA, October 21 - 25 (2006)
9. Zhang, Y., Witte, R., Rilling, J., Haarslev, V.: An Ontology-based Approach for Traceability Recovery. In: 3rd International Workshop on Metamodels, Schemas, Grammars, and Ontologies for Reverse Engineering (ATEM 2006), Genoa (October 2006)
10. Ng, J.K.-Y., Guhneuc, Y.-G.: Identification of Behavioral and Creational Design Patterns through Dynamic Analysis. In: Zaidman, A., Hamou-Lhadj, A., Greevy, O. (eds.) Proceedings of the 3rd International Workshop on Program Comprehension through Dynamic Analysis, October 2007, pp. 34–42 (2007)
11. Pattersson, N.: Measuring precision for static and dynamic design pattern recognition as a function of coverage. In: International Conference on Software Engineering, St. Louis, Missouri (2005)
12. Tsantalis, N., Chatzigeorgiou, A., Stephanides, G., Halkidis, S.: Design Pattern Detection Using Similarity Scoring. IEEE transaction on software engineering 32(11) (November 2006)

13. Shi, N., Olsson, R.A.: Reverse engineering of design patterns from Java source code. In: 21st IEEE/ACM International Conference on Automated Software Engineering (2006)
14. Dong, J., Zhao, Y., Peng, T.: Architecture and Design Pattern Discovery Techniques - A Review. In: Arabnia, H.R., Reza, H. (eds.) Proceedings of the 2007 International Conference on Software Engineering Research & Practice, SERP 2007, Las Vegas Nevada, June 25-28, 2007, vol. II, pp. 621–627 (2007)
15. Dean, M., Schreiber, G. (eds.): OWL Web Ontology Language Reference, W3C Recommendation (February 10, 2004), `http://www.w3.org/TR/owl-ref/`
16. Baader, F., Calvanese, D., McGuinness, D., Nardi, D., Patel-Schneider, P.F. (eds.): The Description Logic Handbook, 2nd edn. The theory, Implementation and Applications. Cambridge University Press, Cambridge (2007)
17. Bray, T., Paoli, J., Sperberg-McQueen, C.M., Maler, E., Yergeau, F. (eds.): Extensible Markup Language (XML) 1.0, 4th edn. W3C Recommendation (August 16, 2006), `http://www.w3.org/TR/2006/REC-xml-20060816/`
18. Knublauch, H., Fergerson, R.W., Noy, N.F., Musen, M.A.: The Protégé OWL Plugin: An Open Development Environment for Semantic Web Applications. In: McIlraith, S.A., Plexousakis, D., van Harmelen, F. (eds.) ISWC 2004. LNCS, vol. 3298, Springer, Heidelberg (2004)
19. Rattz Jr., J.C.: Pro LINQ: Language Integrated Query in C# 2008. Apress (2007)
20. Parr, T.: The Definitive ANTLR Reference: Building Domain-Specific Languages. The Pragmatic Programmers (2007)
21. Bloch, J.: Effective Java Programming Language guide. Addison-Wesley Professional, Reading (2001)
22. Horrocks, I., Patel-Schneider, P.F., Boley, H., Tabet, S., Grosof, B., Dean, M.: SWRL: A Semantic Web Rule Language Combining OWL and RuleML. W3C Member Submission 21 May 2004 (2004), `http://www.w3.org/Submission/SWRL/`
23. Sirin, E., Parsia, B., Grau, B.C., Kalyanpur, A., Katz, Y.: Pellet: A practical OWL-DL reasoner. Journal of Web Semantics 5(2) (2007)
24. Motik, B., Shearer, R., Horrocks, I.: Optimized Reasoning in Description Logics using Hypertableaux. In: Pfenning, F. (ed.) CADE 2007. LNCS (LNAI), vol. 4603, pp. 67–83. Springer, Heidelberg (2007)
25. Horrocks, I., Sattler, U.: A Tableau Decision Procedure for SHOIQ. J. of Automated Reasoning 39(3), 249–276 (2007)
26. Horridge, M., Bechhofer, S., Noppens, O.: Igniting the OWL 1.1 Touch Paper: The OWL API. In: OWLED 2007, 3rd OWL Experienced and Directions Workshop, Innsbruck, Austria (June 2007)

Cognitive Systems for Medical Pattern Understanding and Diagnosis

Lidia Ogiela

AGH University of Science and Technology
Al. Mickiewicza 30, PL-30-059 Krakow, Poland
logiela@agh.edu.pl

Abstract. In the paper will be presented a new way of intelligent cognitive systems and pattern analysis using cognitive categorization. Such an understanding will be based on the linguistic and cognitive mechanisms of pattern recognition and classification. The goal is making computer analysis of the meaning for some selected classes of medical patterns. The approach presented will show the possibilities of automatic and intelligent disease detection and its classification based on cognitive resonance processes. Cognitive categorisation systems operate by executing a particular type of thought, cognitive and reasoning processes which take place in the human mind and which ultimately lead to making an in-depth description of the analysis and reasoning process.

1 Introduction

The process of computer data analysis that has been developing incessantly for a number of years is now moving to jobs of not just simply interpreting analysed data, but it concentrates mainly on deeper reasoning and an attempt at the computer understanding of that data. It is precisely for activities aimed at understanding the analysed data that a special class of intelligent information and decision-support systems called cognitive categorisation systems has been developed. Such systems do not just conduct simple analyses, but mainly strive to reveal the semantic information contained in image data, and then run procedures leading to its machine understanding based on previously defined semantic information. Such a process was possible due to applying formalisms of linguistic perception and understanding of data to automatic reasoning processes combined with the purely human process of interpreting, analysing, understanding and reasoning which occurs in the human mind.

The most important element in this analysis and reasoning process is that it occurs both in the human cognitive/thinking process and in the system's information/reasoning process that conducts the in-depth interpretation and analysis of data. It should be added that this process is based on cognitive resonance (Fig. 1) which occurs during the examination process, and which forms the starting point for the process of data understanding consisting in extracting the semantic information and the meaning contained in the analysed type of data that makes reasoning possible.

The applications of Computational Intelligence (CI) methods in the area of biomedical engineering problems, and creation of intelligent information systems, include some

I. Lovrek, R.J. Howlett, and L.C. Jain (Eds.): KES 2008, Part I, LNAI 5177, pp. 394–400, 2008.

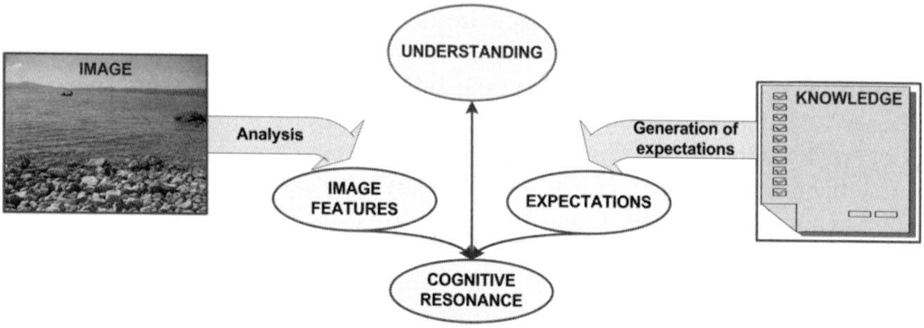

Fig. 1. Cognitive resonance in automatic understanding process

classical algorithms like data processing and analysis procedures, pattern classification, neural modeling, and genetic computation [1, 4]. Such algorithms allow to make description of the analyzed objects, its classification, creation of behavioral models, or solve optimization problems connected with particular task. However in many biomedical, economical, or engineering problems such traditional techniques of analysis may occur completely insufficient. This is especially visible when solving problems leads to the necessity of merit content understanding. Automatic understanding is something more than signal processing – it needs also some knowledge and it demands special type of data processing. Details of natural understanding are very complicated therefore, we can talk about understanding in terms of cognitive science.

Nevertheless in this paper I propose methods of artificial imitation of understanding processes. I describe the general methodology of the machine understanding procedures and show how to use this methodology for solving selected biomedical, economical and also engineering problems [2, 7].

2 Automatic and Machine Analysis and Understanding

Trying to explain what automatic understanding is, and how we can force the computer to understand the image content we must demonstrate the fundamental difference between a formal description of an image and the content meaning of the image, which can be discovered by an intelligent entity, capable of understanding the profound sense of the image in question.

The most important difference between traditional methods of image processing and the new concept of image understanding is that there are two-directional interactions between features extracted from the image and expectations resulting from the knowledge of image content. Applying automatic understanding the input data stream must be compared with the stream of demands generated by a dedicated source of knowledge. Such demands are always connected with a specific hypothesis of the image content semantic interpretation. As a result, we can emphasise that the proposed 'demands' are a kind of postulates, describing (basing on the knowledge about the image contents) the desired values of some (selected) features of the image. The selected parameters of the image must have desired values when some assumption about semantic interpretation of the image content is to be validated as true. The fact that the parameters of the input image are different can be interpreted as a partial falsification of one of possible hypotheses about the meaning of the image content, however, it still cannot be considered the final solution. Such a specific model of inference we can name the 'cognitive resonance' (fig. 1).

Cognitive resonance and cognitive categorisation have been developed by combining intelligent information systems with cognitive systems in which cognitive resonance and cognitive analysis occur.

Our method of image understanding is based on the same processes connected with cognitive resonance.

3 Example of Understanding Medical Image

The connection between proposed methodology and mathematical linguistics, especially a linguistic description of images, is a very important aspect of the automatic image understanding method. There are two reasons for the selection of linguistic methods as a fundamental tool for understanding patterns.

The first one results from the fact, that during the understanding process no classes or templates are known a priori. In fact, the possible number of potential classes goes to infinity. So it must be a tool that offers us the possibilities to describe a potentially infinite number of categories.

The second reason owns to the fact that in the linguistic approach, after processing, we obtain a description of the image content without the use of any classification known *a priori*. This is possible because of a very strong generalisation mechanism within the grammar parsing process.

The only problem consists in a correct adjustment of the terms and methods of formal grammars and artificial languages when applying them in the field of images.

Cognitive categorization approach may be applied to medical visualization. Author has a great experience in application of such a way of semantic analysis for interpretation of various medical images [5, 6, 8]. Below, an example of interpretation of food bones will be presented.

The cognitive analysis of images showing foot bones has been conducted using formalisms for the linguistic description, analysis and interpretation of data, which include such formalisms as graph grammars and to identify and intelligently understand the analysed X-ray images of bones of the foot.

In order to perform a cognitive analysis aimed at understanding the analysed data showing foot bone lesions, a linguistic formalism was proposed in the form of an image grammar whose purpose is to define a language describing the possible layouts of foot bones which are within physiological norms and the possible lesions of foot bones.

The analysis of foot bones in the for example dorsoplanar projection formed the basis for defining a graph used to make a model description of the foot bone skeleton (Fig. 3) which employs the known anatomical rules of this part of the lower extremity (Fig. 2).

Topographic relationships were introduced for the thus defined, spanned graph describing the foot bone skeleton in the dorsoplanar projection. These relationships describe the location of particular structures in relation to one another, as well as the possible pathological changes within the foot (Fig. 4).

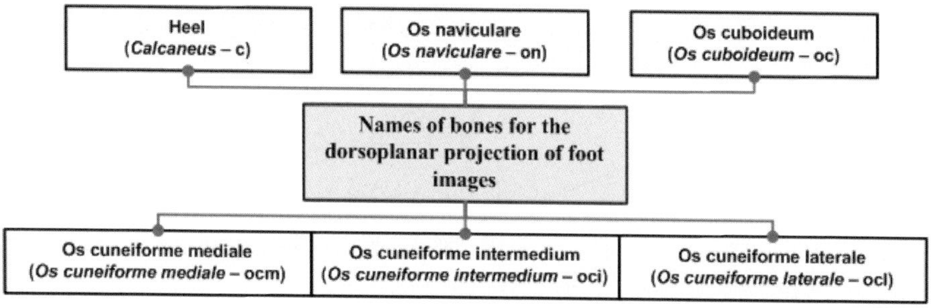

Fig. 2. Names of bones for the dorsoplanar projection of foot images

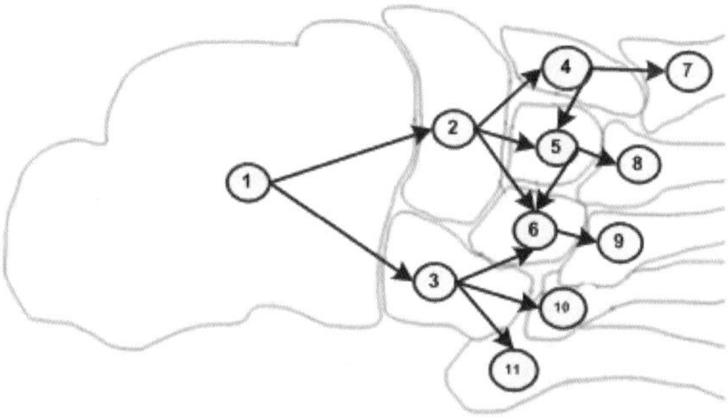

Fig. 3. A graph describing the foot bone skeleton in the dorsoplanar projection

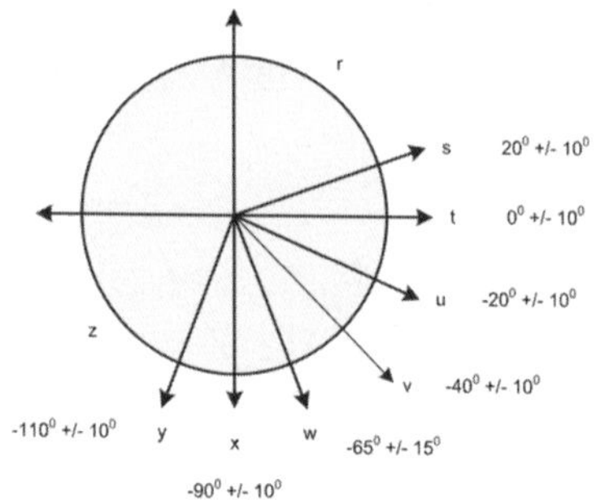

Fig. 4. A relation graph for the dorsoplanar projection of the foot

The introduction of such spatial relationships (Fig. 4) and the representation in the form of a graph spanned on the skeleton of foot bones were used to define the graph proper, in which all the adjacent foot bones were labelled as appropriate for the analysed dorsoplanar projection (Fig. 5). This graph shows bones that are already numbered and which have been assigned labels in line with searching the graph across. (bfs/wfs-wide first serach). Such a representation creates a description of foot bones using the so-called IE graph. This is an ordered and oriented graph for which the syntactic analysis will start from the distinguished apex number 1 (Fig. 5).

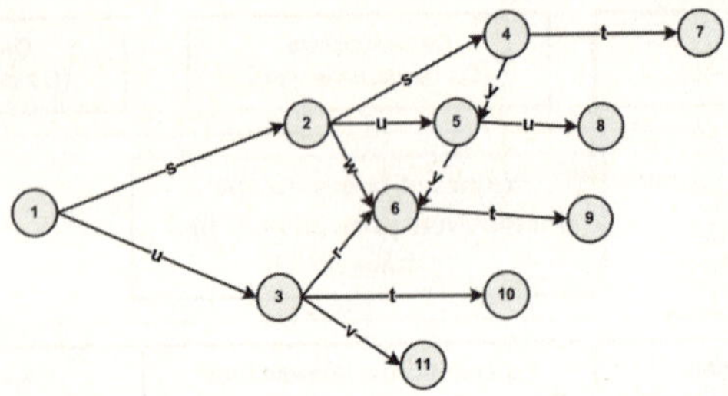

Fig. 5. A graph with numbers of adjacent bones marked based on the relation graph for the dorsoplanar foot projection

For the purposes of the analysis conducted, a formal definition of the graph grammar was introduced, which takes into account the developed linguistic description of correct connections between foot bones:

$$G = (N, \Sigma, \Gamma, ST, P)$$

where:

The set of non-terminal labels of apexes:

N={ST, CALCANEUS, OS NAVICULARE, OS CUBOIDEUM, OS CUNEIFORME MEDIALE, OS CUNEIFORME INTERMEDIUM, OS CUNEIFORME LATERALE, M1, M2, M3, M4, M5}

The set of terminal labels of apexes:

Σ ={s, t, u, v, w, x, y, c, on, oc, ocm, oci, ocl, m1, m2, m3, m4, m5},

Γ – the graph shown in Fig. 5, The start symbol S = ST,

P – a finite set of productions shown in Fig. 6.

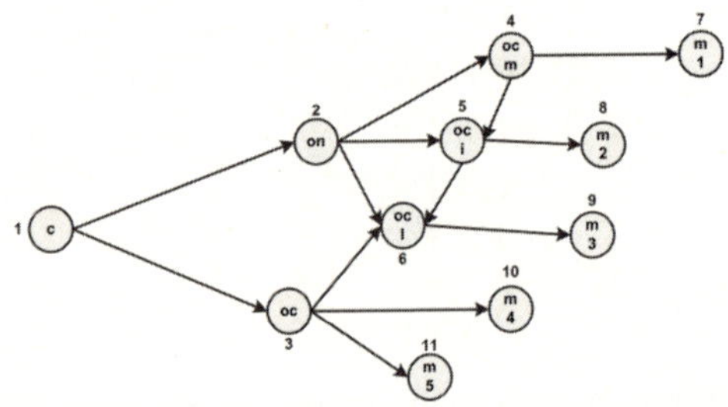

Fig. 6. A set of productions defining the interrelations between particular elements of the structure of foot bones for the dorsoplanar projection

Our analysis of image-type data understanding was aimed at an in-depth understanding of the images analysed, in this case also of specific lesions. Figure 7 shows the possibilities for describing various disease cases by expanding the set of linguistic rules to include additional grammatical rules.

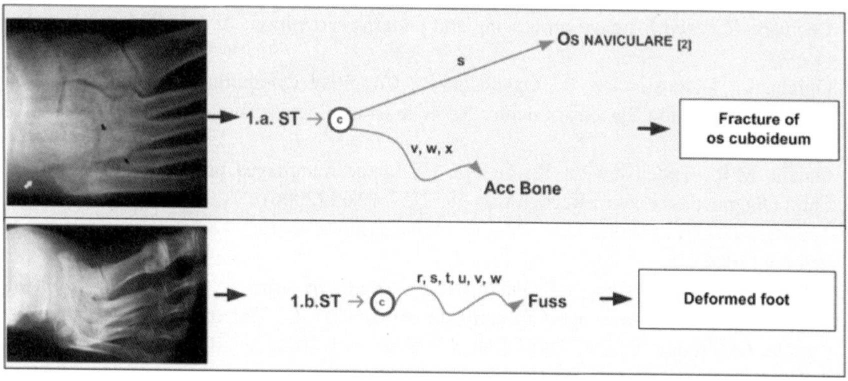

Fig. 7. Examples of using the automatic understanding of foot bone lesions detected by the UBIAS system in the dorsoplanar projection

The presented examples of the cognitive analysis and interpretation of data, describing the lesions appearing in foot bones, show possible cases, namely: fractures and deformations foot.

The types of foot bone lesions shown above and detected by an intelligent and cognitive system have been presented using a selected type of projection for foot bone imaging. Obviously, similar solutions can be proposed for the remaining projection types, that is the lateral projection (external and internal).

4 Conclusion

In the paper we present a new approach based on cognitive categorization to data analysis and pattern understanding. We have described the general concept of machine cognitive inference which allows extracting the semantic information from the analyzed patterns. Applying such a methodology we successfully attempted to develop an experimental implementation of the IT systems relevant to many decision support problems including the intelligent and cognitive systems.

Acknowledgement

This work has been supported by the Ministry of Science and Higher Education, Republic of Poland, under project number N516 025 32/2881.

References

1. Bankman, I. (ed.): Handbook of Medical Imaging: Processing and Analysis. Academic Press, London (2002)
2. Davis, L.S. (ed.): Foundations of image understanding. Kluwer Academic Publishers, Norwell (2001)
3. Khan, M.G.: Heart Disease Diagnosis and Therapy. Williams & Wilkins, Baltimore (1996)

4. Leondes, C.T. (ed.): Image processing and pattern recognition. Academic Press, San Diego (1998)
5. Ogiela, L., Tadeusiewicz, R., Ogiela, M.R.: Cognitive Computing in Intelligent Medical Pattern Recognition Systems. Lecture Notes in Control and Information Sciences 344, 851–856 (2006)
6. Ogiela, M.R., Tadeusiewicz, R., Ogiela, L.: Image Languages in Intelligent Radiological Palm Diagnostics. Pattern Recognition 39, 2157–2165 (2006)
7. Tadeusiewicz, R., Ogiela, M.R.: Medical Image Understanding Technology. Springer, Heidelberg (2004)
8. Tadeusiewicz, R., Ogiela, L., Ogiela, M.R.: Cognitive Analysis Techniques in Business Planning and Decision Support Systems. In: Rutkowski, L., Tadeusiewicz, R., Zadeh, L.A., Żurada, J.M. (eds.) ICAISC 2006. LNCS (LNAI), vol. 4029, pp. 1027–1039. Springer, Heidelberg (2006)

Detection and Analysis of Cell Nuclear Phases

Donggang Yu[1], Tuan D. Pham[1], and Xiaobo Zhou[2]

[1] Bioinformatics Applications Research Centre
James Cook University Townsville, QLD 4811, Australia
[2] HCNR Centre for Bioinformatics
Harvard Medical School, Boston, MA 02215, USA

Abstract. Automated analysis of molecular images has increasingly become an important research in computational life science. In this paper some new and efficient algorithms for detecting and analyzing cell phases of high-content screening are presented. The conceptual frameworks are based on the morphological features of cell nuclei. Furthermore, the novel detecting and analyzing strategies of feed-forward and feed-back of cell phases are proposed based on grey feature, cell shape, geometrical features and difference information of corresponding neighbor frames. Experiment results tested the efficiency of the new method.

Keywords: Nuclear phases, cell screening, feature extraction, morphological feature, feed-forward detection, feed-back detection, shape recognition, normal cellular cycle.

1 Introduction

The tracing and recognition of cell phases using fluorescence microscopy images play an important role for any automated high-content screening that helps scientists to better understand the complex process of cell division or mitosis [1]-[5]. High content screening concerns with the tracking of cell cycle progression (interphase, prophase, metaphase, and telophase), which can be identified by measuring nuclear changes. The most difficult task of such analysis is finding different stages during cell mitosis. For example, four image frames taken from a database of high-content screening are enhanced by contrast stretching [6] and shown in Fig. 1.

A typical nuclear migration during cell division is shown in Fig. 2. We develop in this paper new and efficient algorithms for detecting and analyzing cell phases based on grey feature, shape recognition, geometrical features and prior information of normal cellular cycle.

The rest of this paper is organized as follows. Section 2 describes the preprocessing of cell images. Section 3 describe the method of detecting and analyzing nuclear phases. Finally, Section 4 concludes the findings of our work.

I. Lovrek, R.J. Howlett, and L.C. Jain (Eds.): KES 2008, Part I, LNAI 5177, pp. 401–408, 2008.

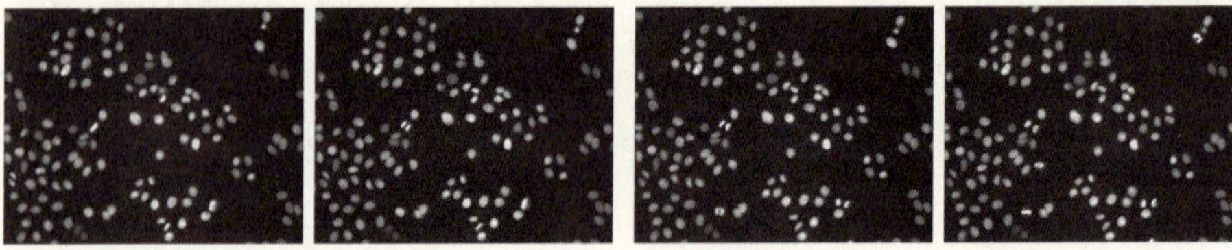

(1) Frame 9. (2) Frame 10. (3) Frame 11. (4) Frame 12.

Fig. 1. Nuclear images of a series of frames in one cell-cycle screening

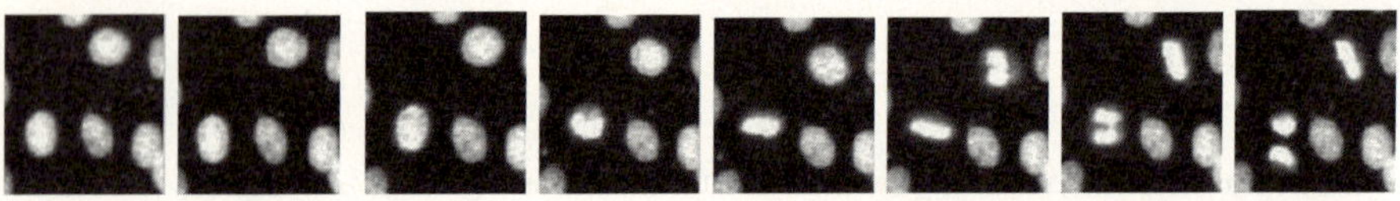

(1) (2) (3) (4) (5) (6) (7) (8)

Fig. 2. Part of sample images of one cell-cycle screening (frame time:15 minutes)

2 Preprocessing

The Ostu's method [7] is used to segment images in Fig. 2, and the binary results are shown in Fig. 3. The chain code set of contour k is represented as:

$$C_k = \{c_0, c_1 ... c_i, ... c_{n-1}, c_n\} \tag{1}$$

where i is the index of the contour pixels. The difference code, d_i, is defined as:

$$d_i = c_{i+1} - c_i. \tag{2}$$

Smooth following
In smooth followed contours, $|d_i|$ equals 0 or 1 [8].

Linearization
Suppose that the direction chain code set of the smoothed contour is

$$\{c_l^{ln}[i] \quad (i = 0, ...(n_l^{ln} - 1))\}, \tag{3}$$

where ln is the ln-th line of a smoothed contour and n_l^{ln} is the number of points of the ln-th line. A linearized line has the following property: [8]
if

$$d_{ij} = c_l^{ln}[i] - c_l^{ln}[j] \quad (i = 0, ...k - 1), (j = 0, ...k - 1), \tag{4}$$

then

$$| d_{ij} | \leq 1 \quad (i = 0, ...k - 1), (j = 0, ...k - 1). \tag{5}$$

Therefore, a linearized line contains only two elements (being represented by $cdir1$ and $cdir2$) whose chain codes meet the above equation [8].

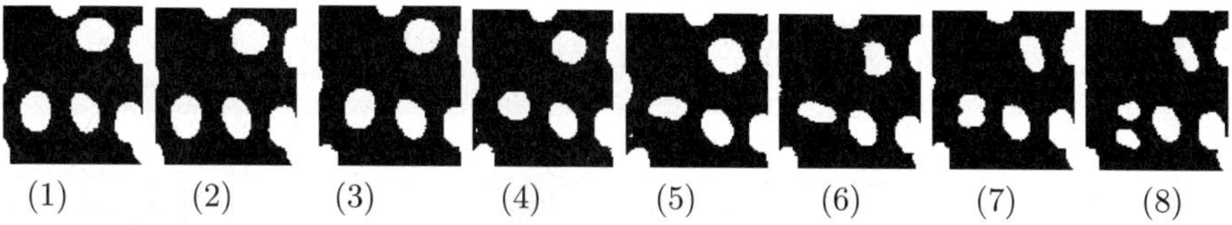

Fig. 3. Part of binary sample images of one cell-cycle screening (frame time:15 minutes)

The structural points

The structural points are defined based on the structure patterns of element codes of two lines. Assume that $line[ln]$ is the current line and that $line[ln-1]$ is the previous line.

Definition 1. The convex point in the direction of code 4 (represented with the character "∧"): If the element codes 3, 4 and 5 occur successively as a group of neighborhood linearized lines, then one convex point can be found as follows; if $cdir1$ of $line[ln]$ is code 4, $cdir2$ is code 5 and the direction chain code of the last pixel of $line[ln-1]$ is code 3, then the first pixel of the current line $line[ln]$ is a convex point which is represented with "∧".

Similar to Definitions 1, other structural points can be defined and found. These points are convex points "v", "[', ")", "F", "o", "T", "s", and concave points "m", "$", "]", "(", "f", "O", "t" and "S" which are shown in Fig. 4(1) respectively [8].

The preprocessing results of images in Fig. 3 can be found and shown in Fig. 4(2) based on the above algorithms.

Separation and reconstruction of touching cell nuclei

Another problem for identifying the size and shape of the cell nuclei is that they are touching [9]. If two or more cells are touched, there are is one concave

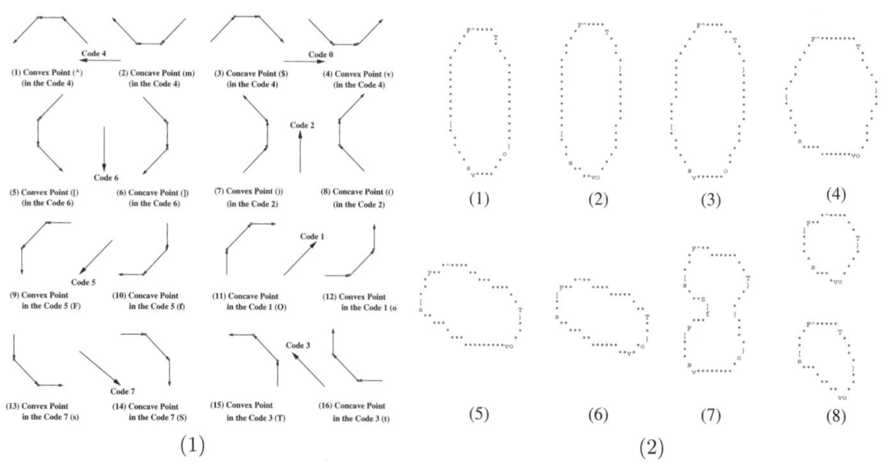

Fig. 4. (1) Structural patterns of structural points. (2) The preprocessing results (smooth following, linearization, extraction of structural points) of nuclear contour and cell morphological structures of different phases (refer to Fig. 3).

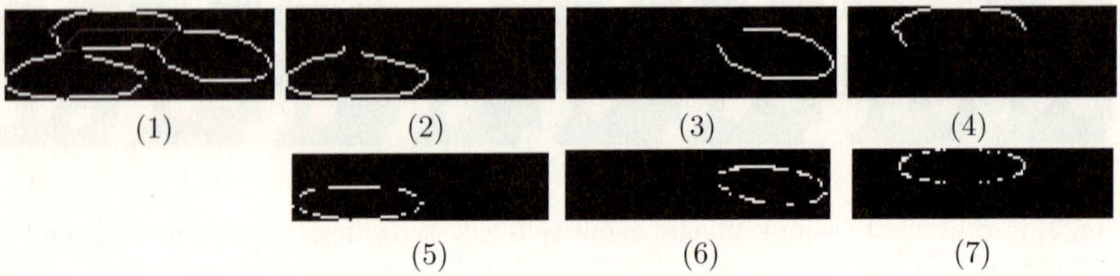

Fig. 5. The contour, separated arcs and reconstructed ellipses of one touching cell nuclei

structural point at least on its outer contour. Also, its size is larger than that of one cell image as touching cell image consists of two or more cells. One example is shown in Fig.5 where all separation points and separated arcs can be found, and touching cell images can be reconstructed by these data based on the methods developed in [9] [10].

Shape recognition of cell nuclei
In order to trace progresses of cells, it is necessary to recognize the cell shapes. The ellipse shapes can be three types, skew, horizontal and vertical.

Morphological model 1: Ellipse shapes $e_{(5,1,2,6)}$ and $e_{(6,2,1,5)}$.
For these shapes, there are no concave structural points on the cell contour.

Let $c_{5,6,1,2}$ be the total number of codes 5, 6, 1 and 2, c_t be the total number of all codes, $c_{5,1}$ be the total number of codes 5 and 1, and $c_{6,2}$ be the total number of codes 6 and 2, on the cell contour respectively. If (1) its outer contour mainly consists of chain codes 5, 6, 1 and 2 ($c_{5,6,1,2} \geq \frac{1}{2}c_t$); (2) the number of chain codes 5 and 1 is more than that of chain codes 6 and 2 ($c_{5,1} \geq c_{6,2}$), then the cell image shape is recognized as the shape $e_{(5,1,2,6)}$, otherwise ($c_{5,1} < c_{6,2}$) the cell image is recognized as the shape $e_{(6,2,1,5)}$.

The shape, $e_{(5,1,2,6)}$, is a skew ellipse in the direction of code 5 or 1, and the shape, $e_{(6,2,1,5)}$, is a vertical ellipse.

Based on the above recognition model, the cell images in Figs. 4(2)(2-3) are recognized as shape $e_{(6,2,1,5)}$.

Morphological model 2: Barbell shapes.
If (1) there are two concave structural changes; (2) there is one pair of corresponding concave structural points, "$\wedge$" and "\$" (horizontal), "]" and "(" (vertical), "f" and "O" (skew), or "t" and "S" (skew), then cell image contour can be recognized as the barbell shape. The cell image in Fig. 4(2)(7) can be recognized as a barbell shape.

Also, other morphological structures of cell images can be described.

Extracting geometrical features of cell nuclei of each frame
Extracting geometrical features of each frame is as follows: binarizing the frame image; labeling the frame image; extract the area of each object; extract the centroid of each object; extract the lengths of the major and minor axes of each object.

3 Tracing Detection and Analysis of Cell Nuclear Phases

About 90 percent of a cells time in the normal cellular cycle is spent in interphase (http://biology.about.com/od/mitosis/ss/mitosisstep.htm). Therefore, if we can determine how many nuclei are in metaphase, anaphase, telophase and prophase in a frame, then the rest of nuclei are in interphase. Tracing detection and analysis of cell nuclear phases are as follows.

Metaphase

In metaphase, the nuclear membrane disappears completely, and a spindle with bright grey is formed, which is distinguished from its other phases. Its morphological features as follows: spindle shape (ellipse shape); larger rate between major and minor axes of the ellipse; there are some pixels of nuclei which are in high grey value.

Let the lengths of major and minor axis a cell be represented as L_{maa} and L_{mia}(the length being the distance between two corresponding pixels) respectively, and L_t be the length threshold of L_{maa}, then $L_{maa} > L_t = 22$ and $(L_{maa}/L_{mia})>1.5$ if the cell is in metaphase based on the statistic results of nuclear samples in our used database.

Let G_i be the grey value of ith pixel of a cell, G_n be the number of pixels whose grey value equals 225 or is larger than 225 ($G_i \geq 225$), G_a be the total number of pixels in the cell and G_m be the average number of pixels in the cell samples. Based on the statistic results of nuclear samples, G_m=320, and $(G_n/G_a)>(1/3)$ for most of nuclei which are in metaphase. However, the minimum G_n is 35 for a very few nuclei which are in metaphase. Therefore, the threshold of G_n, G_{nt}, is set as 35, and the threshold of G_a, G_{at}, as $G_m \times 1.5 = 480$. That means $G_a \leq 480$ and $G_n \geq 35$ if the cell is in metaphase.

Based on the prior knowledge the normal cellular cycle, the metaphase of cell nuclei have all the above morphological features, and other phases of nuclei not. Therefore, the metaphase of a cell can firstly be detected in a frame.

For example, let 55th cell in Frame 9 is represented as 9_55. For the object 9_55, G_n=127, G_a=322, L_{maa}=28, L_{mia}=14, and his shape is $e_{(5,1,2,6)}$ based on the method described in preprocessing. Therefore, the object 9_55 in Frame 9 meets the requirement of the above morphological features of metaphase, and the cell is in metaphase.

Feed-forward detection of phases (Metaphase, Anaphase, Telophase and Prophase)

If the phase of nuclei is detected as metaphase in a frame, the phase of the cell should be metaphase or anaphase in the next frame. If the morphological features of the cell meet the conditions of metaphase, then the phase of the cell is in metaphase in the next frame. Otherwise, the phase of the cell is in anaphase, and two spindles are formed in parallel. The morphological features of nuclear anaphase in the next frame is: (1) the previous phase of the cell is metaphase; (2) the cell shape is a barbell shape or two parallel small ellipse shapes (spindle); (3) the centroid of the cell approximate that of the cell in the previous frame; (4) some pixels of the cell are in high grey value (>225).

Furthermore, after detecting the anaphase of the cell, the phase of the cell should be telophase in the following frames. In telophase, the cell has been divided into two small ellipses (spindle) whose centroid is close in the following frames. The number of the following frames is 2 based on the prior information of the normal cellular cycle. (http://biology.about.com/od/mitosis/ss/mitosisstep.htm). The above detecting procedure call feed-forward detection.

Therefore, metaphase, anaphase and telophase of cell can be detected and analyzed between the neighboring frames based on feed-forward detection method.

One example is shown in Fig. 6(1). The cell 10_55 in Frame 10 is detected as being in anaphase because it meet following conditions: there two parallel spindles with high grey values; its centroid coordinate in Frame 10 is near that of the cell 9_55 in Frame 9. Furthermore, two spindles begin to be formed at opposite poles, and move away. Therefore, the nuclei 11_55, 11_58, 12_54, and 12_57 are detected as telophase based on their morphological features, the centroid coordinates of being in corresponding neighboring frames.

Feed-back detection of phases

If the phase of a cell is detected as metaphase in a frame, the phase of the cell should be metaphase or prophase in the previous frame. The centroid of the cell in the previous frame approximates that of the cell in the current frame. If the morphological features of the cell meet the conditions of metaphase, then the phase of the cell is in metaphase in the previous frame. Otherwise, the phase of the cell is in prophase in the previous frame based on the prior information of the normal cellular cycle. After detecting the first prophase of cell, the phase of the cell in six back neighboring frames is determined as prophase based on the prior information of normal cellular cycle. The above procedure calls feed-back detection.

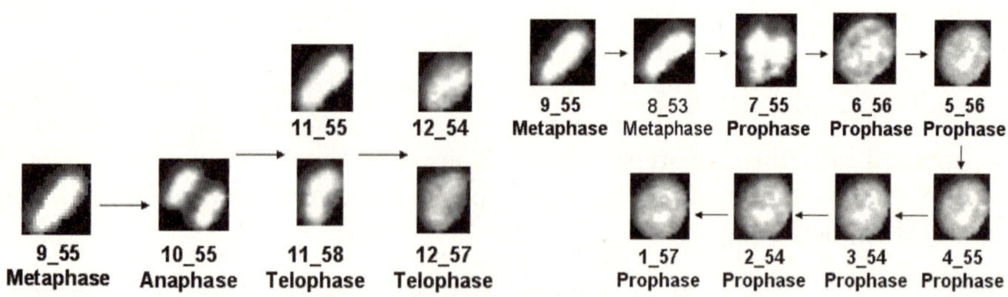

(1) One example of feed-forward detection. (2) One example of feed-back detection.

Fig. 6. Feed-forward detection and feed-back detection of phases for the 55th object in Frame 9

One example is shown in Fig. 6(2). Cell 9_55 is in metaphase. Based on its centroid coordinates and ones in its back neighboring frame (Frame 8), the cell is 8_53 in Frame 8. Because the cell 8_53 meet the morphological conditions of metaphase (spindle with high grey value), cell 8_53 is in metaphase. Furthermore, the cell is the cell 7_55 in back neighboring frame (Frame 7). Because it does not meet the conditions of morphological features of metaphase, and the cell 7_55 is in prophase. After the cell's being firstly in prophase, the cells of six back neighboring frames (Frames 6-1) are detected to be in prophase based on the prior information of normal cellular cycle. Therefore, The cells 6_56-1_57 in Frames 6-1 are determined in prophase respectively.

Based on the above analysis method, the phase of cells (9_7, 9_8, 9_53 and 9_99) can be detected and shown in Fig. 7 if one cell is not in interphase in Frame 9.

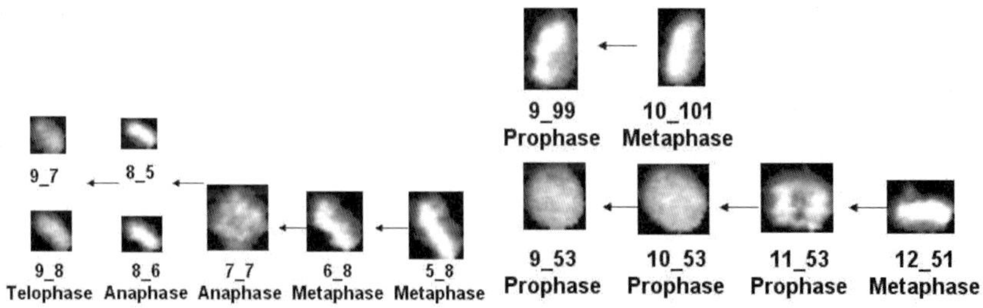

(1) Phase detection of the cells 9_7 and 9_8. (2) Phase detection of the cells 9_53 and 9_99.

Fig. 7. Phase detection of cells 9_7, 9_8, 9_53 and 9_99 in Frame 9 by feed-forward and feed-back detection methods

Interphase: If the phase of a cell cannot be detected as anyone of prophase, metaphase, anaphase or telophase, the cell is in interphase except it is dead.

Dead: Average division time is 1920 minutes (series of 128 frames). If a cell is not detected as metaphase for 140 frames, the cell is dead.

Detecting metaphase of the touching nuclei
If some nuclei are touching and a spindle shape with high grey value is contained in the region, Otsus multi-threshold method need to be used to segment corresponding binary cell image. Two examples are shown in Fig. 8.

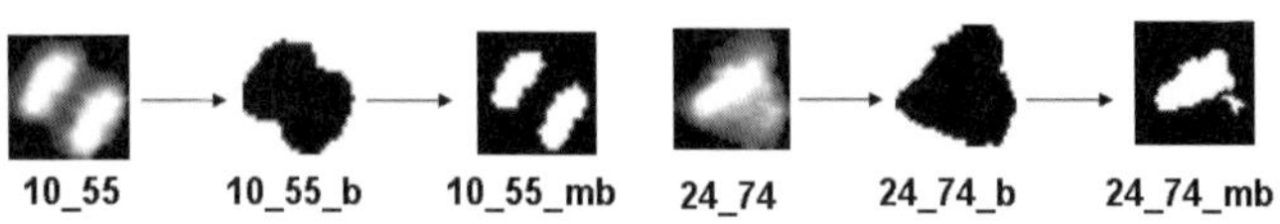

Fig. 8. Two examples of extracting spindles of nuclei based on multi-threshold method and prior information of spindle

4 Experiments and Conclusion

An efficient and new method has developed to recognize, trace and analyze the phases of nuclei in frames of a high-contents cell cycle based on shape recognition, morphological and geometrical features, grey feature and prior information of normal cellular cycle. This method simulates artificial intelligence. For example, there are 134 nuclei in Frame 9 based on labeling and segmented result. Twelve cells are in prophase, one cell is in metaphase, no cell is in anaphase, six nuclei are in telophase, and 115 nuclei are in interphase based on the method. Based on the new method, all nuclei in all frames of high-contents cell cycle can be traced to determined which phase the cell is. One database which consists of 240 frames is used to test the efficiency of the method. These frames are taken from time-lapse fluorescence microscopy with time interval of 15 minutes. The identification rate of cell phases is 94.68 percent for all nuclei. The identification errors of cell phases are caused by wrong segmentation of touching cells especially when the frames are the late one and there are more complicated touching nuclei in frames.

Acknowledgement. This work was supported by the Australia Research Council ARC-DP grant (DP0665598) to T. D. Pham. The cell images were provided by Dr. Randy King of the Department of Cell Biology, Harvard Medical School.

References

1. Fox, S.: Accommodating cells in HTS. Drug Discovery World 5, 21–30 (2003)
2. Feng, Y.: Practicing cell morphology based screen. European Pharmaceutical Review 7, 7–11 (2002)
3. Yarrow, J.C., et al.: Phenotypic screening of small molecule libraries by high throughput cell imaging. Comb. Chem. High Throughput Screen 6, 279–286 (2003)
4. Pham, T.D., Tran, D., Zhou, X., Wong, S.T.C.: An automated procedure for cell-phase imaging identification. In: Proc. AI 2005 Workshop on Learning Algorithms for Pattern Recognition, pp. 52–29 (2005)
5. Pham, T.D., Tran, D.T., Zhou, X., Wong, S.T.C.: Classification of cell phases in time-lapse images by vector quantization and Markov models. In: Greer, E.V. (ed.) Neural Stem Cel l Research, Nova Science, New York (2006)
6. Davies, E.: Machine Vision: Theory, Algorithms and Practicalities, pp. 26–27, 79–99. Academic Press, London (1990)
7. Ostu, N.A.: Thresholding selection method from graylevel histogram. IEEE Trans. Systems Man Cybernet SMC8, 62–66 (1978)
8. Yu, D., Yan, H.: An efficient algorithm for smoothing, linearization and detection of structure feature points of binary image contours. Patt. Recog. 30(1), 57–69 (1997)
9. Yu, D., Pham, T.D., Zhou, X.: Analysis and Recognition of Touching Cell Images Based on Morphological Structures, Computational and Information Science. In: Apolloni, B., Howlett, R.J., Jain, L. (eds.) KES 2007, Part I. LNCS (LNAI), vol. 4692, pp. 439–446. Springer, Heidelberg (2007)
10. Fitzgibbon, A., Pilu, M., Fisher, R.B.: Direct Least Square Fitting of Ellipse. IEEE Trans. Pattern Analysis and Machine Intelligence 21, 476–480 (1999)

Simple Perceptually-Inspired Methods for Blob Extraction

Paolo Falcoz

Università degli Studi di Milano – Dipartimento di Tecnologie dell'Informazione
falcoz@dti.unimi.it

Abstract. Studies in Visual Attention (VA) and eye movements have shown that humans generally only attend a few regions of interest (ROIs) in an image. The idea is to perform image analysis only on a small neighborhood of each ROI. This regions can be thought of as the most informative parts of the image, and as such can be analyzed with respect to colors, textures, and shapes. In this paper we will focus on color-driven blob extraction. Inspiration from the human retina guides the definition of *neighborhood* of a ROI, while psychological factors in human color perception are used to drive color selection.

Keywords: blob extraction, regions of interest, color perception.

1 Introduction

Studies in Visual Attention (VA) and eye movements [5],[10] have shown that humans generally only attend a few regions of interest (ROIs) in an image, which are determined in part by their information content. Ten different methods to find ROIs are listed and compared in [7]; we will make use of the Michaelson Contrast, a simple way to estimate contrast variations. Even though this is specifically suited for terrain analysis, it is also considered to be an important feature in human vision.

Since only a small set of ROIs are usually required by the brain to recognize a complex visual input, the idea is to extract a set of ROIs using Michaelson Contrast, and to perform image analysis only on a small neighborhood of each ROI: this should decrease processing costs, without impacting on important image features. The word *should* is used because our exploration is still at the beginning and we don't have performance figures yet.

The process can easily be extended from one ROI to many, therefore here we will discuss the simpler case only. In particular, the most prominent region of interest will be considered; multiple ROIs can be obtained considering the second most prominent region, the third, and so on. Related to the concept of ROI there is the problem of finding a good definition for *neighborhood*: given a region of interest – from now on referred to as *focus point* – how many pixels should be considered as important for effective image processing? How should this neighborhood be shaped?

I. Lovrek, R.J. Howlett, and L.C. Jain (Eds.): KES 2008, Part I, LNAI 5177, pp. 409–416, 2008.

There are of course many answers, but we wanted to push a step further the analogy with human perception system. Inspiration comes from the human retina, the light-sensitive layer at the back of the eye that covers about 65 percent of its interior surface. Photosensitive cells called *rods* and *cones* in the retina convert incident light energy into signals that are carried to the brain by the optic nerve. In the middle of the retina is a small dimple called the *fovea* or fovea centralis. It is the center of the eye's sharpest vision and the location of most color perception. In fact, while cones are concentrated in the fovea, rods are absent there but dense elsewhere. Measured density curves for the rods and cones on the retina show an enormous density of cones in the fovea; to them is attributed both color vision and the highest visual acuity [3].

We will try to formalize the idea of rods and cones different distributions in section 2.3. Section 3 deals with blob extraction, and specifically with color flattening (section 3.1)and histogram processing (section 3.3).

2 Region of Interest Extraction

2.1 Michaelson Contrast

Michaelson contrast $\mathcal{C}$ is most useful in identifying high contrast elements, generally considered to be an important choice feature for human vision. Given L_M the overall mean luminance of the image I, and $\mathcal{L}_m$ the mean luminance within a 7×7 surrounding of the center pixel I_{ij},

$$\mathcal{C}_{ij} = \|(\mathcal{L}_m - L_M)/(\mathcal{L}_m + L_M)\|$$

The result of Michaelson contrast is a matrix which associates to each pixel of the original color image a measure of how much luminance changes; greater the value, bigger the variation. Examples are given in figure 1.

2.2 Focus Point

The focus point could be chosen as the pixel with the greatest Michaelson contrast, but this in general could lead to bad choises due to noise; to overcome this limitation, a mean filter is used. In the mean filter the value of each pixel is the sum of the values of all its neighbors divided by the neighborhood's cardinality. This can easily be done by convolving $\mathcal{C}$ with a square mask $m^{\mathcal{C}}$ of size k such that

$$m^{\mathcal{C}}_{ij} = \frac{1}{k^2}, \forall i, j = 1 \dots k$$

and then choose the focus point to be the maximum of the resulting matrix. Focus point fp will be

$$fp = \max(\mathcal{C} \otimes m^{\mathcal{C}})$$

where $\otimes$ is the convolution operator.

Fig. 1. Original images (**a**, **b**), corresponding Michaelson contrast (**c**, **d**), and focus point (red dot in **e**, **f**)

2.3 Mask Construction

The next step is to build the focus matrix. The idea is to weight the blobs based on their distance from the focus point; the further they are from the focus point the less they weight. By *weight* of a blob we mean the mass of a blob, that is the number of pixels belonging to it. Usually each pixel is given a unitary weight, but in this case the weight depends on the position of the pixel relative to the focus point. If the focus point changes, the same pixel will likely have a different weight.

A bidimensional Gaussian distribution $\mathcal{G}(x, y)$ was chosen as the distance function. The weight W_{ij} of a pixel I_{ij} is the value of $\mathcal{G}(i, j)$ scaled between $[0 \ldots 1]$.

$$W_{ij} = \mathcal{G}(i, j) = \frac{1}{2\pi\sigma^2} e^{-\frac{1}{2}\frac{(i^2+j^2)}{\sigma^2}} \tag{1}$$

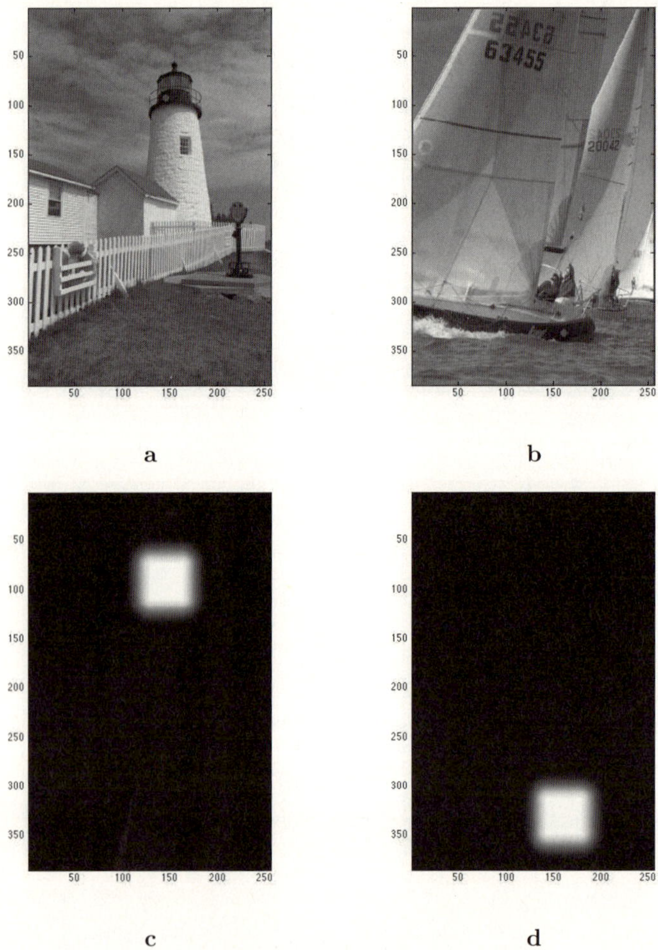

Fig. 2. Focus point (**a**, **b**) and resulting focus matrix (**c**, **d**)

Scaling is done via the rule of three. Note that the focus mask is first entirely generated, and then centered in the focus point.

$\mathcal{G}$ parameters are chosen under the assumptions that I coincides with the observer's field of view, that is what we see is *exactly* what we can see. We define I_w and I_h respectively to be input image I's *width* and *height*; given $k = \max(I_w, I_h)$, and the fact that points at distance more than 3σ can be considered effectively 0 [9], then

$$\sigma = \frac{(k-m)}{6} \tag{2}$$

where m is the fovea size relative to the current field of view. Since human field of view is about 180°, with the fovea covering about 15°, then

$$m = \frac{15}{180}k \tag{3}$$

The reason why we subtract m from k in (2) is that the region representing the fovea in the focus mask has a perfect color vision, so its weight is 1.

The process of focus matrix generation can be summarized in the following few points:

1. calculate macula size m;
2. generate a $(k-m) \times (k-m)$ weight matrix W;
3. uniformly scale W in the closed interval $[0 \dots 1]$;
4. expand W by adding a square $m \times m$ matrix of ones at its center.

The focus matrix is then centered in the focus point, as shown in Figure 2.

3 Blob Processing

Now that the focus mask has been created, we are able to use it for blob selection. This involves some additional steps like color flattening, color quantization, and histogram processing.

3.1 Color Flattening

Of all the huge number of digital images created each day, only a very limited amount is shot with professional cameras; the great majority is taken with low cost, low quality equipment, meaning non uniform colors and evident noise.

To cope with this the first step is to flatten colors by performing several steps of bilateral filtering [8]. The effectiveness of this approach is to combine a low-pass filter with a range filter

$$h(x) = k^{-1}(x) \int_{-\inf}^{\inf} \int_{-\inf}^{\inf} f(\xi)c(\xi, x)s(f(\xi), f(x))\mathrm{d}\xi$$

where

$$k(x) = \int_{-\inf}^{\inf} \int_{-\inf}^{\inf} c(\xi, x) s(f(\xi), f(x)) \mathrm{d}\xi$$

The simple Gaussian filtering has been used, in which both the closeness function $c(\xi, x)$ and the similarity function $s(f(\xi), f(x))$ are Gaussian functions of the Euclidean distance between their arguments. Closeness then becomes

$$c(\xi, x) = e^{-\frac{1}{2}\left(\frac{d(\xi, x)}{\sigma_d}\right)^2}$$

where

$$d(\xi, x) = d(\xi - x) = |\xi - x|$$

while similarity becomes

$$s(\xi, x) = e^{-\frac{1}{2}\left(\frac{\delta(f(\xi), f(x))}{\sigma_r}\right)^2}$$

where

$$\delta(\phi, f) = \delta(\phi - f) = |\phi - f|$$

To achieve perceptually uniform colors – colors where mathematical distance equals perceptual distance – other color spaces should be used instead of RGB and the strictly correlated HSV and HSL, such as CIE-Lab or CIE-Luv. Anyway, for simplicity HSV has been chosen here, and all the color processing has been done in this space.

3.2 Color Quantization

After color flattening a quantization step is performed: using a uniform quantization approach, the total number of colors is reduced to (at most) 512 – or 3 bits per color plane.

3.3 Histogram Processing

Color histogram processing is a key step, and can be divided in two stages: histogram construction and histogram enhancement.

Histogram construction means counting the number of pixels of each (quantized) color, giving each pixel an equal weight of 1. Usually the most represented colors are chosen for blob extraction. Here we will use the focus mask built in section 2.3 to give each pixel a different weight depending on the distance from the focus point. This means that a color globally poorly represented but locally close to the focus point has a chance to be selected over a globally well represented but locally distant (from the focus point) color. Typical examples are skies, seas, meadows et cetera.

Histogram enhancement takes the move from some general agreements on physical and emotional understanding of certain colors [1] [4] [11]. The color red is associated with fire and is considered as an aggressive color, whereas blue is associated with water and coolness. Warm colors in general call for action [6], and as such we tend to give them priority over cool colors; this idea is captured in the histogram enhancement step by multiplying warm colors by a fixed factor (a sort of *bonus*).

We empirically define warm colors those colors having the following HSV values

$$\begin{cases} 0 \leq x \leq 60 \quad x \in H,\ H = \{h \in \mathcal{R}, 0 \leq h \leq 360\} \\ 0 \leq y \leq 0.12\ y \in S,\ \ S = \{s \in \mathcal{R}, 0 \leq s \leq 1\} \\ 80 \leq z \leq 255\ \ z \in V,\ V = \{v \in \mathcal{N}, 0 \leq v \leq 255\} \end{cases}$$

Of course other ranges can be defined; an example could be skin detection.

After histogram enhancement the *nbest* most represented colors are selected for blob extraction.

3.4 Blob Extraction

Blob extraction step takes the *nbest* selected colors and scans input image to find the biggest *nblobs* blobs. For each *nbest* color, a binary mask representing all the pixels with that color is first computed, and all 8-connected objects are labeled. Labeling is achieved using the techinique outlined in [2]. Each blob's mass is then calculated, and focus matrix weights are used. The effect is to make bigger but distant blobs less important than smaller but closer blobs (Figure 3). Blobs are ordered according to their mass, and only the biggest *nblobs* are taken.

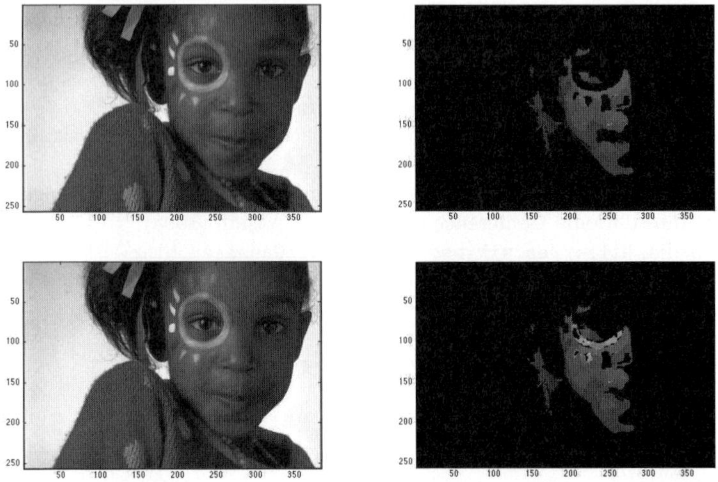

Fig. 3. Blob detection without (top figure) histogram enhancement and with (bottom figure) histogram enhancement. Yellow blobs take precedence over dark brown blobs. Focus point corresponds to image's center. Settings: $nbest = 8$, $nblobs = 4$.

4 Conclusions

The goal of this paper – and the ultimate sense of our exploration – is to show that making use of simple facts and observations from the fields of visual perception, visual attention, and human eye's anatomy can help discriminating important information from "background" information. We applied few simple concepts to blob processing, and showed how these could be used to select blobs with specific *good* characteristics against less desirable blobs. We are not able to give performance figures yet, so we cannot state how advantageous our approach is compared to full blob processing. We are aware of the many limitations of our work, but we are also convinced that if we want to successfully accomplish such a difficult task as reproducing human vision, then also simple and marginal facts should be considered.

References

1. Churchland, P.: The Engine of Reason: the Seat of the Soul. MIT Press, Cambridge (1995)
2. Haralick, R.M., Shapiro, L.G.: Computer and Robot Vision, vol. I, pp. 28–48. Addison Wesley, Reading (1992)
3. Hecht, E.: Optics, 2nd edn. Addison-Wesley, Reading (1987)
4. Hundert, E.M.: Lessons from an optical illusion. Harvard University Press, Cambridge (1995)
5. Norton, D., Stark, L.W.: Eye movements and visual perception. Sci. Am. 224, 34–43 (1971)
6. Panchanathan, S., et al.: The Role of Color in Content–Based Image Retrieval. In: Proc. IEEE International Conference on Image Processing, vol. 1, pp. 517–520 (2000)
7. Privitera, C.M., Stark, L.W.: Algorithms for Defining Visual Regions-of-Interest: Comparison with Eye Fixations. IEEE Trans. on Pattern Analysis and Machine Intelligence 22(9), 970–982 (2000)
8. Tomasi, C., Manduchi, R.: Bilateral Filtering for Gray and Color Images. In: Proc. IEEE International Conference on Computer Vision (1998)
9. Wikipedia, `http://en.wikipedia.org/wiki/Gaussian_blur`
10. Yarbus, A.L.: Eye movements and vision. Plenum, New York (1967)
11. Zeki, S.: A Vision of the Brain. Blackwell, Oxford (1994)

A Study on Self Practice Navigation System for the "Kana" Strings

Kenta Kouyama[1], Naoto Hara[1], Toshiaki Kuroiwa[2], Seiu Yamashita[3], Akiko Ono[1], Isao J. Ohsugi[1], Junichi Yoshino[1], and Hiroshi Ichimura[1]

[1] Salesian Polytechnic 4-6-8 Oyamagaoka Machida-shi Tokyo 194-0215 Japan
{s07603,s07610,ono,ohsugi,yoshino,ichimura}@salesio-sp.ac.jp
[2] SymbolicTechnology Co., Ltd. 3-15 Kandanishiki-Mahchi Chiyoda-ku Tokyo 101-0054 Japan
kuroiwa@symbolic-technology.co.jp
[3] Penmanship R&D Center 3-5-17 Kirigaoka Yokohama-shi Kanagawa 226-0016 Japan
seiu@mono.so-net.ne.jp

Abstract. The "kana" strings are an indigenous character to Japan. Chinese characters were simplified and made into the original phonograms. The culture of the "kana" strings was intended in commuting lessons under the master. Meanwhile, IT devices are greatly improved and they are now applied in various fields (For example: In the art field, preservation of the cultural heritage and e-Leaning etc.). Therefore, it was examined whether an IT device can be applied to practice in the "kana" strings. A method of outputting a beautiful model was already studied [1]. But it was considered whether the practice by the navigation support method is effective [2]. This paper describes the construction method of self "kana" strings practice system that adopts the navigation support method.

1 Introduction

IT devices are greatly improved and they are now applied in various fields. The art field is used for preservation of the cultural heritage and e-Leaning etc. A system is constructed for training "kana" strings of the culture indigenous to Japan by the aid of IT. There was a similar study in which models of "kana" strings were generated. However, in our system we do not generate a copybook and compare a disciple student's writings with models, but we use a navigation system. Dynamism is more important than static beauty of the generated models of the "kana" strings. Therefore, the navigation reflects both dynamism and static beauty. It is originally formulated by us, based on opinions of the master. "Yamatokotoba" was used for the spoken language in ancient Japan. Then, we Japanese started to apply man'yogana to the written language (Fig.1-a). Man'yogana employs the Chinese character. It consists of ideograms and phonograms. This is very difficult. Therefore, it has developed in Japanese original writings characters.

- Katakana
  The katakana used a part of the letterform of man'yogana. Man'yogana allocated Chinese character every one sound, so it is very complication. The katakana used a part of Chinese character which was used man'yogana (Fig.1-b).

I. Lovrek, R.J. Howlett, and L.C. Jain (Eds.): KES 2008, Part I, LNAI 5177, pp. 417–424, 2008.

- Hiragana
  This recognized the character of man'yogana as a continuous body. This has naturally developed and simplified man'yogana around the woman in the heian era. "The Tale of Genji" the oldest novel in the world was written in the hiragana (Fig.1-c). Sentences chiefly spelt by the hiragana are called the "kana" strings.

(a) Man'yogana (b) Katakana (c) Hiragana

Fig. 1. Japanese characters

Today, the hiragana and the katakana are used in Japan. However, this system is intended for the "kana" strings as not simple lines of characters we daily use but a traditional artistic production. The culture of the "kana" strings was handed down in commuting lessons under the master. The system enables disciple students to practice alone without their masters. Whether it is possible to hand down the beauty of "kana" strings by IT are deliberated. The practice by the IT support is operated not by the pattern matching but by navigating the size and the movement of a brush which based on master's knowledge. The knowledge is based on formal knowledge that was analyzed from implicit knowledge the master obtained by the experience. This paper describes the design of this system.

2 "Kana" Strings

The "kana" strings have vivid beauty and the speed writing as their features. The speed writing has "renmen", which is our cursive hand; characters are connected to each other to make feeling some kind of streams. "Renmen" has "keiren" and "iren".

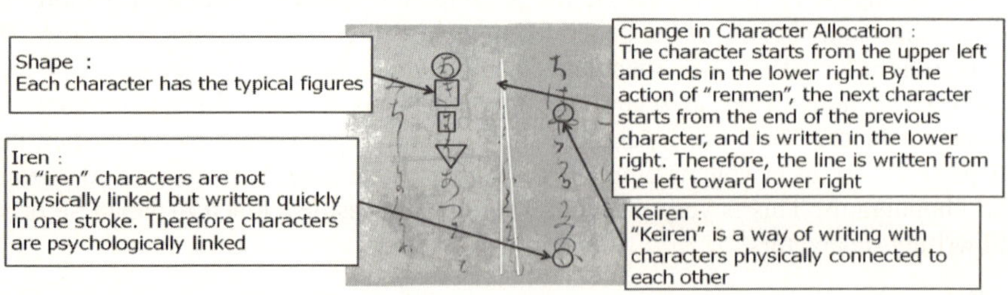

Fig. 2. Features of the "kana" strings

The features of the 'kana' strings in Fig 2 are converted into formal knowledge which is based on the master's opinions for constructing our system. This master targets the writings of Kinoturayuki, who is one of the most famous calligraphers in the Heian era.

3 Self Practice Navigation System

3.1 System Overview

The disciple students are the one in stage of "rinsho"[1] practice. Disciple students in this stage can write "kana" characters but cannot write "kana" strings properly. The purpose of this practice is to enable disciple students to write "kana" strings in the form of "iren" and "keiren".

This system uses a liquid crystal tablet. The liquid crystal tablet is a device with which the users can input their handwriting on real paper, using a pen. Furthermore, the liquid crystal displays the navigation. This system is composed of three units (Fig.3).

- The interface unit displays the navigation for disciple students to practice on the liquid crystal tablet.
- The Knowledge Database possesses implicit knowledge concerning the "kana" strings, which has already been analyzed.
- The navigation information generation unit produces information to display the navigation.

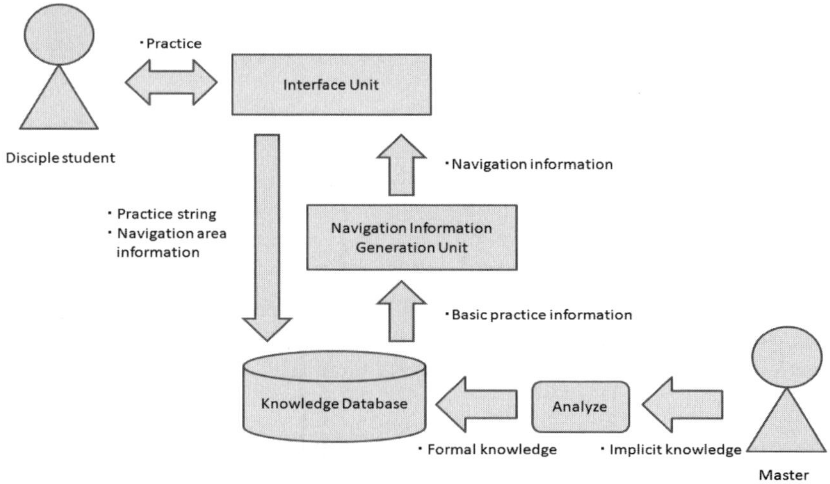

Fig. 3. System overview

[1] "Rinsho" is aimed at learners in the intermediate course. In the practice disciple students write "kana" strings, seeing a model, for example, "*sunsyoannshikishi*" [3] or "koya-gere" [4] .

3.2 Interface Unit

3.2.1 Evaluation of Interface

The master, Seiu Yamashita evaluated the LCD tablet. Table 1 shows the evaluation result and the solutions.

Table 1. Problem and solution

Problem	Solution
The nib slips.	The sheet (example: PVC) is placed on the tablet.
	Felt is used for the point.
The pen top is far from the display.	The drawing method will be changed.

3.2.2 Navigation Method

To write a beautiful line of the "kana" strings, shapes, sizes of each character on the line, and allocation and balance of "renmen" are important.

This interface unit is designed to display the following methods on the screen (Fig.4).

(a) A method to show shape of character
(b) A method to show starting points and endpoints
(c) A method to show an outline
(d) A method to show directions of "renmen"

This navigation system instructs disciple students by indicating one or more method at the same time. This method is designed by us based on master's implicit knowledge.

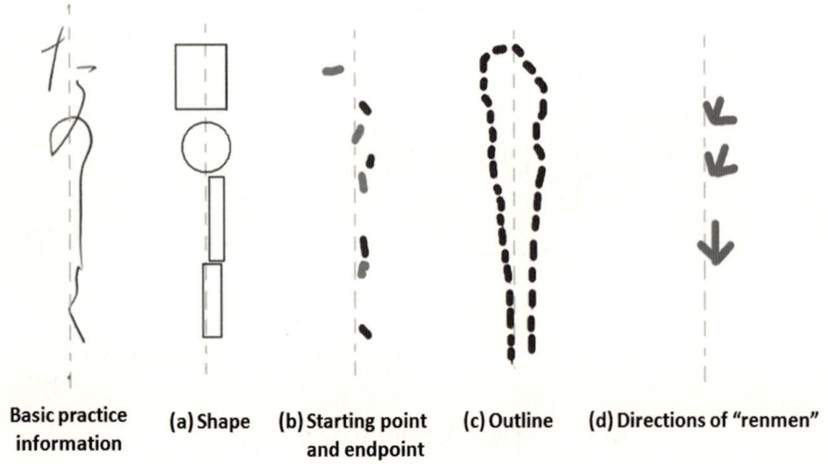

Fig. 4. Navigation examples

3.2.3 Screen Composition

The system screen is composed of the input screen and the practice screen on the LCD tablet (Fig.5). On the input screen, the disciple student types sentences he/she wants to write. After he/she inputs the sentences, the practice screen is displayed. On the practice screen, the disciple student actually practices the input sentences. One line of the "kana" strings in Figure 5(b)-① is displayed, and the navigation is displayed in the screen ②.

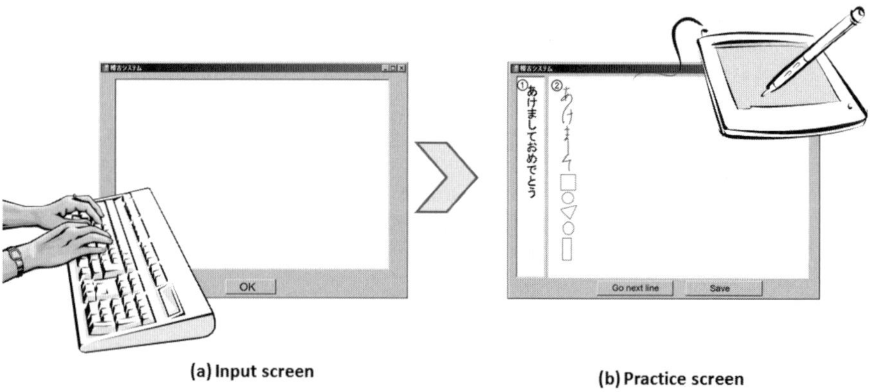

Fig. 5. System screen composition

3.3 Navigation Information Generation Unit

The Navigation Information Generation Unit gets basic information from Knowledge Database and it generates information for navigation from the basic information (Fig.6). Knowledge Database is described at section 3.4.

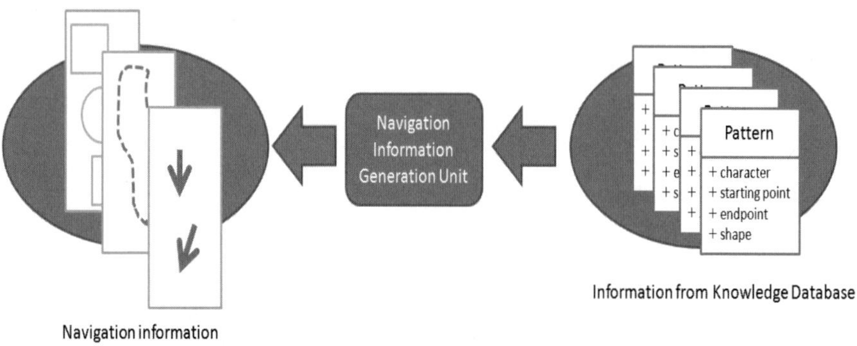

Fig. 6. Navigation Information Generation Unit

A navigation which was generated by this unit is given to the Interface Unit. And the single or the combined navigation is displayed on the screen.

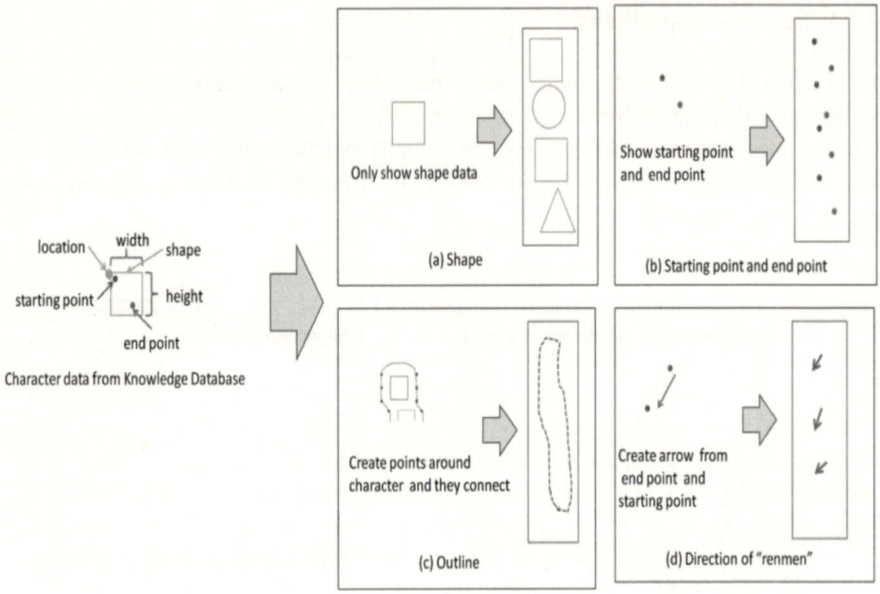

Fig. 7. Generation method of navigation

The method of generating each navigation is illustrated in Fig 7 and below described.

(a) Shape navigation (Fig.4-a)
It shows shape figures from Information of Knowledge Database
(b) Starting point and end point navigation (Fig.4-b)
It shows starting points and end points (Fig.7-b).
(c) Outline navigation (Fig.4-c)
First, it plots six points of each character. The points are top, center and bottom of both sides of a character. Next, the points are connected as outline (Fig.7-c).
(d) Direction of "renmen" navigation (Fig.4-d)
It obtains angle of each line which is drawn from an end point to a starting point and creates arrow from the angle (Fig.7-d).

3.4 Knowledge Database

3.4.1 Construct

Authors designed three classes of "line", "allocation" and "pattern" as information which intended the feature of the "kana" strings. The composition of these classes is the following (Fig.8).

(a) Line class (Information of the practice screen and the strings)
(b) Allocation class (Allocation information of each character)
(c) Pattern class (Basic information of character)

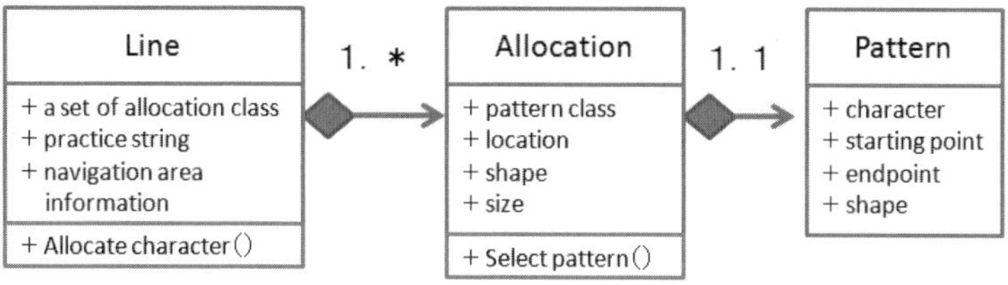

Fig. 8. Class diagram

The line class intends information on one line of the "kana" string. This is a class to generate the allocation class. The line class possesses information on the sentences, and width and height in the navigation area on the screen. In addition, it has the allocation class generated from this information in each character (Fig.9-a).

The allocation class is generated by the line class at each practice, and exists in each character. Information of the allocated character consists of the coordinates on the screen, and the shape and the size of the character (Fig.9-b). This class has the best pattern class selected from the information of this allocation class.

The pattern class has fixed information, which is not generated in every practice. Even if the sound is the same, the pattern class is different, because of differences in the shape, the starting point and the endpoint (Fig.9-c).

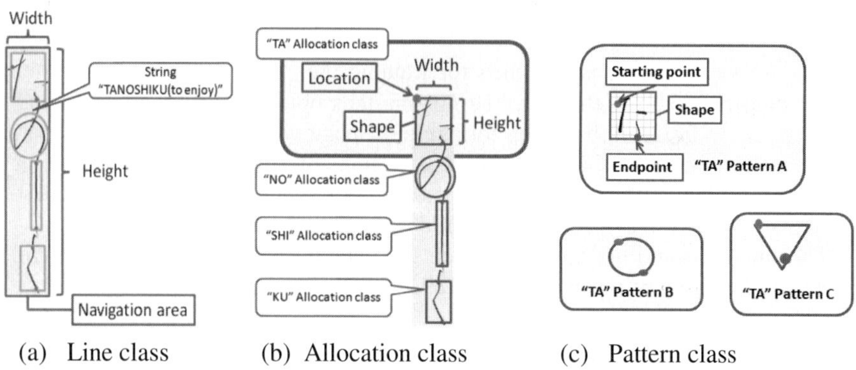

Fig. 9. Each class information

3.4.2 The Process of Basic Practice Information Generation

Figure 10 describes the process of generating information to practice the "kana" strings. First, the disciple student inputs a character string that he/she wants to write on the input screen. The database takes in this string and information of navigation area, and generates the line class. Second, each character taken into the Knowledge Database string is allocated and allocation class is generated. At that time, information of the line class, peripheral information on front and back characters and adjacent lines, formalized implicit knowledge are used for the allocation. Finally, the best pattern is selected, based on the shape and the size in the allocation class from a set of each character pattern. The analyzed implicit knowledge is used for selecting a pattern.

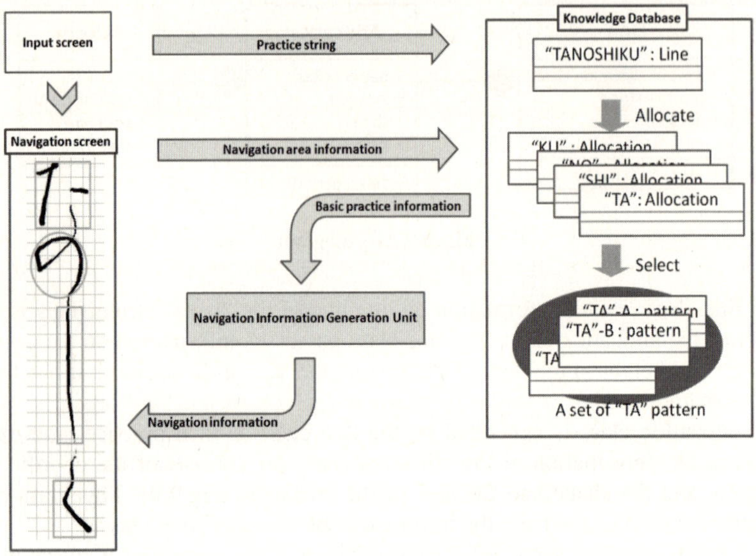

. **Fig. 10.** Basic practice information generation processing

4 Conclusion

In this work, a system is designed for training the "kana" strings. This system is newly constructed from the 'kana' strings master's opinions. The navigation and the database were designed, based on formal knowledge which was generated according to the master's opinions. This system will be implemented and evaluated in near future. The disciple students at targeted practice state 'actual answers' to the questionnaire that is this system useful will be used as an evaluation method.

The authors would like to express their gratitude for receiving the Grant-in-Aid for Scientific Research of Japan No.19300289.

References

1. Takashi, Y., Horita, J., Yamashita, S., Matsumoto, A., Suzuki, M., Ichimura, H.: A Study on the Automated Generation Method of Copybook Based on the Structural Categorization of Hiragana. In: Proceedings of the IEICE General Conference, vol. 154 (2006)
2. Kuroiwa, T., Hara, N., Kouyama, K., Yamashita, S., Yoshino, J., Ohsugi, I.J., Ichimura, H.: A Study on Navigation System of Vivid Handwriting Method for Japanese Traditional Writers Using IT. KEER 2007, E-8 (2007)
3. Kinoturayuki (presumption): Sunsyoannshikishi. NIGENSHA Co. Ltd. (2006)
4. Kinoturayuki (presumption): Kouya-gire. NIGENSHA Co. Ltd. (2003)

A Fuzzy Thresholding Circuit for Image Segmentation

Angel Barriga and Nashaat M. Hussein

Instituto de Microelectronica de Sevilla(CNM-CSIC)/University of Seville,
E.T.S. Ingeniería Informática,
Avda. Reina Mercedes s/n
41012 Sevilla, Spain
barriga@imse.cnm.es

Abstract. The paper describes a technique for image segmentation based on binary thresholding and the hardware implementation of the segmentation circuit. The technique applies a fuzzy logic strategy to calculate the threshold. The hardware design methodology is based on a high level behavioral description of the fuzzy system. The knowledge base is described using the standard hardware description language VHDL. The hardware design criterion is the low cost and, in special way, a high processing speed.

Keywords: Image segmentation, image thresholding, Fuzzy Logic application, VHDL fuzzy system description.

1 Introduction

Image segmentation is one step in many image analysis techniques. By means of the segmentation the image is divided in parts or objects that constitutes it. The image is divided in regions from which there are applied a decision criteria that allow to classify the different regions of the image in the different objects that constitute it. In the case of considering only one region the image is divided in object and background. The level at which this subdivision is made depends on the application. The segmentation will finish when all the objects of interest for the application have been detected. The segmentation is an essential step in diverse processes of image processing. Among others, segmentation is used for measurements on a region, to make three-dimensional reconstructions of a zone of the image, for the classification or automatic diagnosis, or to reduce the information of the images.

The image segmentation algorithms are based generally on two basic properties of the image grey levels: discontinuity and similarity. Inside the first category the techniques tries to divide the image by means of the sharp changes on the grey level. An example inside this category is the edges detection. In the second category there are applied thresholds techniques, growth of regions, and division and fusion techniques.

The simplest segmentation problem appears when the image is formed by only one object that has homogenous light intensity on a background with a different level of luminosity. In this case the image can be segmented in two regions using a technique based on a threshold parameter. Thresholding then becomes a simple but effective tool to separate objects from the background. Most of thresholding algorithms are

I. Lovrek, R.J. Howlett, and L.C. Jain (Eds.): KES 2008, Part I, LNAI 5177, pp. 425–432, 2008.

initially meant for binary thresholding. This binary thresholding procedure may be extended to a multi-level one with the help of multiple thresholds $T_1, T_2, \ldots, T_n$ to segment the image into n+1 regions [1-3]. Multi-level thresholding based on a multi-dimensional histogram resembles the image segmentation algorithms based on pattern clustering.

The thresholding techniques that are based on applying fuzzy logic allow to solve the problem of the vagueness information of the image [4-7]. This paper is focused in the binary thresholding. The proposed technique is based on applying fuzzy logic focused in the hardware implementation of the algorithm. The paper is organized in four sections. Section 2 discusses some thresholding strategies. In section 3 the proposed thresholding method is presented. Finally, the design of the circuit that computes the threshold is described and the realization results are analyzed.

2 Image Thresholding Techniques

The thresholding techniques allow classifying the pixels in two categories (black and white). With this transformation a distinction between the objects of the image and the background is made. In order to obtain this binary image the comparison of the pixels values with a threshold T is performed. Thus if the set of values of pixels of the image is G then

$$G = \{0,1,\ldots,L-1\} \begin{Bmatrix} 0 & black \\ L-1 & white \end{Bmatrix}$$

A threshold $T \in G$ is defined so that the following transformation is made

$$y_{i,j} = \begin{cases} 0 & if \quad x_{i,j} < T \\ L-1 & if \quad x_{i,j} > T \end{cases} \tag{1}$$

where $x_{i,j}$ is a pixel of the original image and $y_{i,j}$ is the pixel corresponding to the binary image. In case of an grey-level image in that the pixels codifies with 8 bits the range of values that the pixels take corresponds to the range between 0 and 255 (L=256). It is usual to express the above mentioned range with normalized values between 0 and 1. Figure 1 shows an example for the case of the image of Lena.

Several different methods for select the threshold value T exists. In [8] there are analyzed 40 methods for threshold selection. These methods are classified in six categories in agreement with the information that is used: histogram shape-based methods, clustering, entropy, object attribute-based methods, spatial and local methods.

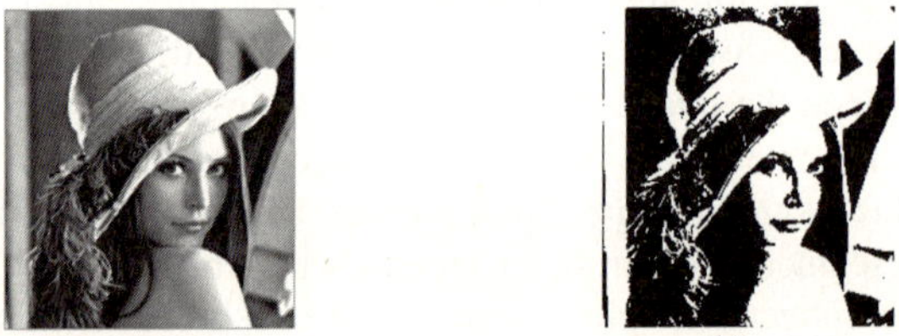

a) b)

Fig. 1. a) Original Lena's image, b) binary image with T=0.5

The most widespread methods for threshold calculation are based on the analysis of the histogram. A basic technique for threshold calculation is based on the frequency of grey level. In this case the threshold T is calculated by means of the following expression:

$$T = \sum_{i=1}^{L} p_i i \tag{2}$$

where i is the grey level that takes values between 1 and 256 for the case of a codification with 8 bits, p_i represents the grey level frequency (also known as the probability of the grey level). For an image with n pixels and n_i pixels with the grey level i:

$$p_i = n_i / n \quad and \quad \sum_{i=1}^{L} p_i = 1 \tag{3}$$

A widely used method is the Otsu's method [9]. This method is based in maximizing the variance between the classes by means of an exhaustive search.

The techniques that apply fuzzy logic for the threshold calculation are based mainly on three types of measures of fuzziness [4]: entropy, Kaufmann`s measure, and Yager's measure.

The technique based on the entropy consists of minimizing the dispersion of the system. This way the pixels of the image are grouped in two classes corresponding to the objects and to the background. Huang and Wang [5] consider that the averages of the data corresponding to each class are μ_0 and μ_1. The membership function of each class is defined as:

$$u_x(x) = \begin{cases} \dfrac{1}{1 + \dfrac{|x - \mu_0|}{x_{\max} - x_{\min}}} & if \quad x < T \\ \dfrac{1}{1 + \dfrac{|x - \mu_1|}{x_{\max} - x_{\min}}} & if \quad x > T \end{cases}$$

The calculation of the threshold T is based on the entropy of a fuzzy set that is calculated using the function of Shannon:

$$H_f(x) = -x \log x - (1 - x) \log(1 - x)$$

The threshold will be that one that minimizes the entropy of the data:

$$E(T) = \frac{1}{M} \sum_i H_f(\mu_x(i)) h(i)$$

The Kaufmann's measure of fuzziness is defined as [6]:

$$D(A) = \left[\sum_{x \in X} |\mu_A(x) - \mu_C(x)|^w \right]^{\frac{1}{w}}$$

This method is based on using the distance metric to the set A. When $w=1$ the Hamming's distance is used whereas if $w=2$ it is the Euclidean distance.

The method of Yager [7] is based on the distance between a fuzzy set and its complementary. This way it is based in minimizing the following function:

$$D_2(T) = \sqrt{\sum_i \left|\mu_x(i) - \mu_{\bar{x}}(i)\right|^2}$$

where $\mu_{\bar{x}}(i) = 1 - \mu_x(i)$.

3 The Fuzzy Threshold Inference Module

The proposed technique consists in applying fuzzy logic in the calculation of the threshold T. Basically, from a formal point of view, this technique is based on the calculation of the average applied to the histogram of the image. An advantageous aspect of this technique is that the calculation mechanism improves the processing time since it only requires processing the image once and allows calculating in a direct way the value of the threshold. From the point of view of hardware realisation there is a low cost architecture of the fuzzy processing module as it will be seen in a next section.

The fuzzy system has an input that receives the pixel that is going to be evaluated and an output that corresponds to the result of the fuzzy inference. Once read the image the output shows the value of threshold T. Basically the operation that makes the fuzzy system corresponds to the calculation of the centre of gravity of the histogram of the image with the following expression:

$$T = \sum_{i=1}^{M} \sum_{j=1}^{R} \alpha_{ij} c_{ij} \Bigg/ \sum_{i=1}^{M} \sum_{j=1}^{R} \alpha_{ij} \qquad (4)$$

where T is the threshold, M is the number of pixels of the image, R is the number of rules of the fuzzy system, c is the consequent of each rule and α is the activation degree of the rule.

The universe of discourse of the histogram is divided in a set of N equally distributed membership functions. Figure 2 shows a partition example for $N=9$. Triangular membership functions have been used since they are easier to implement in terms of hardware. The above mentioned functions have an overlapping degree of 2 what allows to limit the number of active rules.

The membership functions of the consequent are equally distributed singleton functions. The use of singleton-type membership functions allow to apply simplified

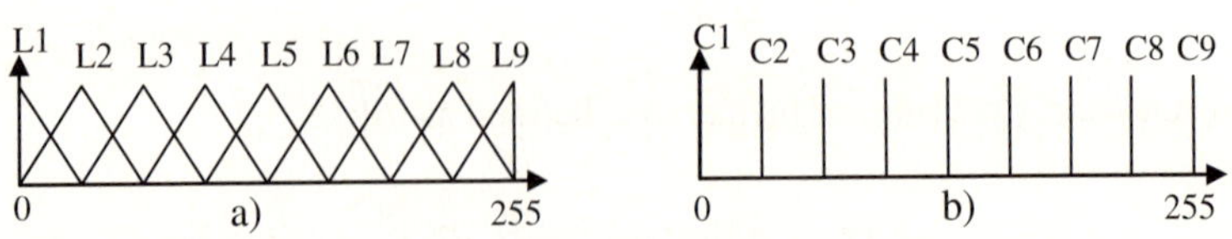

Fig. 2. Membership functions for $N=9$, a) antecedent, b) consequent

defuzzification methods such as the Fuzzy Mean. This defuzzification method can be interpreted as one in which each rule proposes a conclusion with a "strength" defined by its grade of activation. The overall action of several rules is obtained by calculating the average of the different conclusions weighted by their grades of activation. These processing characteristics based on active rules and simplified defuzzification method allows a low cost and high speed hardware implementation.

The rule base shown in figure 3 use the membership functions defined in figure 2. The knowledge base (membership functions and the rules) is common for any images, for that reason the values can store in a ROM memory.

It is possible to optimize the expression shown in equation (4) if the system is normalized. In this case the sum, extending to the rule base, of the consequent activation degree takes value 1 ($\sum_{j=1}^{R} \alpha_{ij} = 1$). Then equation (4) transforms in:

$$T = \frac{1}{M} \sum_{i=1}^{M} \sum_{j=1}^{R} \alpha_{ij} c_{ij} \tag{5}$$

For each pixel the system makes the inference in agreement with the rule base of figure 3. The output of the system accumulates the result corresponding to the numerator of equation (5). The final output is generated with the last pixel of the image that allows to perform the final division.

if x is L1 then c is C1;	if x is L6 then c is C6;
if x is L2 then c is C2;	if x is L7 then c is C7;
if x is L3 then c is C3;	if x is L8 then c is C8;
if x is L4 then c is C4;	if x is L9 then c is C9;
if x is L5 then c is C5;	

Fig. 3. Rulebase for N=9

4 Hardware Implementation

The design methodology used in the implementation of the fuzzy system circuit for threshold generation is described in [10]. This methodology is based on the hardware description language VHDL as a way to describe and model the system at high level. To achieve a behavioural modelling of the fuzzy inference module a VHDL description style have been used. In this style the system structure description (fuzzy sets, rule base) and the operator description (connectives, fuzzy operations) are defined separately. This makes it possible to describe independently both the fuzzy system structure and the processing algorithm. The fuzzy system description must be synthesizable in order to generate the hardware realizations.

Figure 4 shows the VHDL architecture of the fuzzy system. The rule base is described in the architecture body. It is a rule base structure with the 9 rules of figure 3.

Each rule can be divided into two components: the antecedent of the rule and the consequent. The antecedent is an expression of the input variable related to it linguistic values. The consequent set the linguistic value of the rule output.

```
architecture knowledge_base of threshold is
 signal R: consec;
 -- MF for x
 constant L1: triangle:=((0,   0,  31),(0, 32));
 constant L2: triangle:=((0,  31,  63),(32,32));
 constant L3: triangle:=((31, 63,  95),(32,32));
 constant L4: triangle:=((63, 95, 127),(32,32));
 constant L5: triangle:=((95,127, 159),(32,32));
 constant L6: triangle:=((127,159,191),(32,32));
 constant L7: triangle:=((159,191,223),(32,32));
 constant L8: triangle:=((191,223,255),(32,32));
 constant L9: triangle:=((223,255,255),(32,0));
 --MF for z
 constant C1: integer := 0;   constant C2: integer := 31;
 constant C3: integer := 63;  constant C4: integer := 95;
 constant C5: integer := 127; constant C6: integer := 159;
 constant C7: integer := 191; constant C8: integer := 223;
 constant C9: integer := 255;
begin
   R(1) <= rule( (x = L1), C1 );
   R(2) <= rule( (x = L2), C2 );
   R(3) <= rule( (x = L3), C3 );
   R(4) <= rule( (x = L4), C4 );
   R(5) <= rule( (x = L5), C5 );
   R(6) <= rule( (x = L6), C6 );
   R(7) <= rule( (x = L7), C7 );
   R(8) <= rule( (x = L8), C8 );
   R(9) <= rule( (x = L9), C9 );
   Zout<=defuzz(R);
end knowledge_base;
```

Fig. 4. VHDL architecture of the knowledge base

The processing mechanisms of the fuzzy operation '*is*' (=) and the inference '*then*' (*rule*(,)) are not defined in the VHDL description. Only the structure of the rulebase is defined. Such a description is a high level description because it does not assume any specific implementation criteria. It only describes the knowledge base in terms of a behavioral rule base.

Linguistic labels represent a range of values within the universe of discourse of input and output variables. These labels can be described by functions in order to compute the membership degree of a certain input value. Membership functions associated to a linguistic label can be triangular or trapezoidal. The declarative part of the architecture of figure 4 shows the definition of such membership functions.

The data type *triangle* is defined in a VHDL package called 'xfvhdlfunc'. This type contains the definitions of the points that define a triangle as well as the slopes. On the other hand the rule base expresses the knowledge of figure 3. The function *rule* also is defined in the package 'xfvhdlfunc'. This function makes the inference of the rule. The set of rules is evaluated concurrently since the signal assignments in the architecture body are concurrent. The operator "=" also has been redefined taking advantage of the overload properties of VHDL functions. Figure 5 shows the VHDL package with the definitions of the data types and the specific functions.

The functions used in the description of the fuzzy system have been described in agreement with the restrictions of VHDL for synthesis. This has allowed generating the circuit that implements the fuzzy system using conventional tools for circuit synthesis. The fuzzy inference module circuit is a combinational circuit that makes the fuzzy inference. The output of the fuzzy system corresponds to the numerator of the defuzzification stage in agreement with equation (5).

The fuzzy inference module and the divider stage produce the threshold value of the image. The area occupied on a FPGA of the Xilinx Spartan3 family has been of 1,180 slices what is equivalent to 56,845 equivalent gates. The circuit can operate at a frequency of 137 MHz which would allow that the processing of a CIF image (352x288 pixels) will carry out in 0.7 ms. In the case of a VGA image (1024x768) the required time to calculate the threshold is 5.7 ms.

```
PACKAGE xfvhdfunc IS
 type points is array (1 to 3) of integer;
 type slopes is array (1 to 2) of integer;
 type triangle is record
    point: points;
    slope: slopes;
 end record;
. . .
 -- Consequents
 type two_int is array (0 to 1) of integer;
 type consec is array (1 to RULES) of two_int;
 -- FUNCTIONS
 function "=" (x: integer; y: triangle) return integer;
 . . .
 function rule (x: integer; y: integer) return two_int;
 function defuzz(x: consec) return integer;
END xfvhdfunc;
```

Fig. 5. VHDL package with the data types and functions definitions

Figure 6 shows some results for images of 352x288 pixels. The threshold values are shown in table 1. There has been compared Otsu's method, the method based on the grey level frequency, and the proposed method based on the calculation of the threshold applying fuzzy logic. In all the cases Otsu's method gives place to smaller values in the threshold. The fuzzy technique gives very similar results to those who are obtained considering the grey frequency.

Table 1. Thresholds values applying Otsu method, grey frequency, and the proposed method

	Otsu	Frequency	Proposed method
ducks	130	133	133
stork	87	117	115
bridge	150	177	178
objects	114	137	133

Fig. 6. Test images and results based on proposed fuzzy method

5 Conclusions

A technique based on fuzzy logic to calculate the value of a threshold was proposed. The technique allows realizing a binary segmentation of images. The main objective of the proposed thresholding method has been an efficient hardware implementation in terms of cost and processing speed. It facilitates the development of real time analysis and images processing systems where it has his main field of application.

Acknowledgements. This work was supported in part by the Spanish Education and Research Council under grant no. TEC2005-04359/MIC, and by the Andalusia Regional Government under grant no. TIC2006-635.

References

1. Liao, P.-S., Chen, T.-S., Chung, P.C.: A Fast Algorithm for Multilevel Thresholding. Journal of Information Science and Engineering 17, 713–727 (2001)
2. Cao, L., Shi, Z.K., Chenp, E.K.W.: Fast automatic multilevel thresholding method. Electronics Letters 38(16), 868–870 (2002)
3. Oh, J.-T., Kim, W.-H.: EWFCM Algorithm and Region-Based Multi-level Thresholding. In: Wang, L., Jiao, L., Shi, G., Li, X., Liu, J. (eds.) FSKD 2006. LNCS (LNAI), vol. 4223, pp. 864–873. Springer, Heidelberg (2006)
4. Forero-Vargas, M.G., Rojas-Camacho, O.: New formulation in image thresholding using fuzzy logic. In: 11th Portuguese Conf. on Pattern Recognition, pp. 117–124 (2000)
5. Huang, L.K., Wang, M.J.: Image thresholding by minimizing the measure of fuzziness. Pattern Recognition 28, 41–51 (1995)
6. Kaufmann, A.: Introduction to the theory of fuzzy subsets. Academic Press, London (1975)
7. Yager, R.R.: On the measure of fuzziness and negation. Part 1: membership in the unit interval. Int. Journal of Genet. Syst. 5, 221–229 (1979)
8. Sezgin, M., Sankur, B.: Survey over image thresholding techniques and quantitative performance evaluation. Journal of Electronic Imaging 13, 146–165 (2004)
9. Otsu, N.: A Threshold Selection Method from Gray Level Histogram. IEEE Trans. on Systems, Man and Cybernetics SMC-8, 62–66 (1978)
10. Barriga, A., Sánchez-Solano, S., Brox, P., Cabrera, A., Baturone, I.: Modelling and Implementation of Fuzzy Systems based on VHDL. International Journal of Approximate Reasoning 41(2), 164–278 (2006)

A Method for Reading a Resistor by Image Processing Techniques

Yoshihiro Mitani[1], Yuuki Sugimura[1], and Yoshihiko Hamamoto[2]

[1] Ube National College of Technology, Ube, 755-8555, Japan,
[2] Faculty of Engineering, Yamaguchi University, Ube, 755-8611, Japan

Abstract. The resistance of a resistor is defined by colored lines printed on the resistor's body. Normally, people read it by sight. Though, if a computer performs this instead, we can reduce the costs. In this paper, we propose a method for reading this resistance by image processing techniques. We extract colors from a real resistor's picture, and classify it by its colors. The experimental results show the effectiveness of the proposed method.

1 Introduction

Three elements: object, light and sense are mutually related to understand the color of the object. The resistance of a resistor is defined by colors (Fig. 1 shows a resistor's picture). In Fig. 1, the 4 lines (orange, blue, black and gold in this sequence) describe the resistance. In this case $36\Omega \pm 5\%$. If a computer performs this instead, the cost to read a resistor is considered to be reduced. K. L. Chan and H. Wang reported reading resistor values by color image processing [1]. K. L. Chan and H. Wang showed effectiveness of color feature space $L^*a^*b^*$ for reading color bars. Furthermore, K. L. Chan and H. Wang discussed the combination of Bayes classifier and fixed boundary method. However, a method of color bar extraction from a real image is not shown. Therefore, in this paper, we propose a method for reading a resistance of a resistor from a real image by image processing techniques [2]. The method for reading this resistance consists of two steps. First, colors printed on a resistor are extracted by segmentation. In our approach, the color regions segmentation is based on the k-means method [3] by using the combined feature vector of the color feature vector and the position feature vector. Then, we use the 1-NN classifier [4] [5] for classifying the color of the resistor. It's important to notice that we are working since the beginning with real pictures. The real images are photographed under three types of illumination: dark, lighted, and well-lighted. In the color classification experiment, not only the RGB color values but also the $L^*a^*b^*$ and $L^*u^*v^*$ color values [6] are examined. From the results we understand a well illuminated environment is required. Also, the $L^*a^*b^*$ and $L^*u^*v^*$ color feature spaces show to be effective.

I. Lovrek, R.J. Howlett, and L.C. Jain (Eds.): KES 2008, Part I, LNAI 5177, pp. 433–439, 2008.

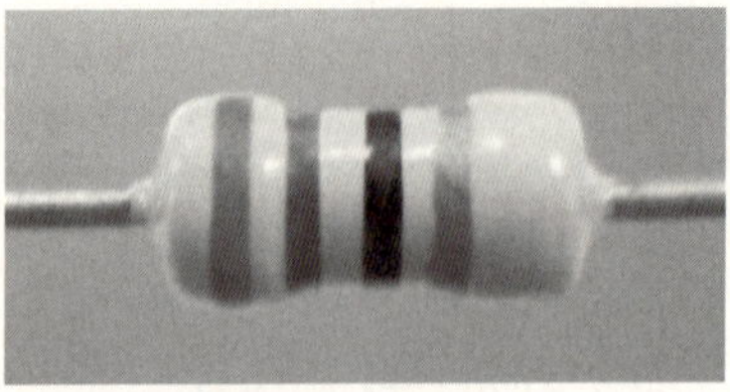

Fig. 1. A resistor

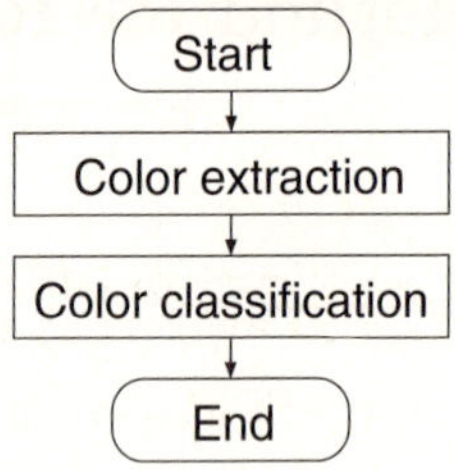

Fig. 2. A flow of the proposed method

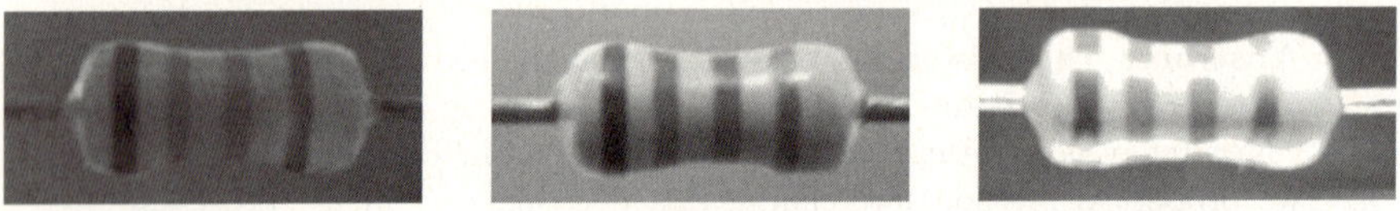

(a) 100 [lx] (b) 3000 [lx] (c) 6000 [lx]

Fig. 3. Examples of resistors at each illumination

2 A Method for Reading a Resistor

We focus on the resistor with 4 color bands, as it's shown in Fig. 1. In this paper, we are going to work with 11 different colors for the lines at the resistor (black, brown, red, orange, yellow, green, blue, purple, gray, white, and gold). We won't work with the ones with silver lines or blank spaces because of its rare use. The real resistor images are taken a picture in different three types of illumination (dark, lighted, and well-lighted). The dark situation is about 100[lx]. While the lighted and well-lighted situations are about 3000[lx] and 6000[lx] respectively. Fig. 3 shows examples of resistors at each illumination. The resistor position is central while the color lines are in the right order to be read (from left to right). We used 30 pieces of the real images at each illumination. The size of the real image is $500 \times 250(width \times height)$ pixels.

The proposed method consists of color extraction and color classification. Fig. 2 shows a flow of the proposed method.

2.1 Color Extraction of the Resistor

It is difficult to extract colors of a resistor directly from a real image. Firstly, we describe the extraction of the main body from the real image. Fig. 4 shows a process of extracting a resistor. The 4 color values at 10×10 pixels from each corner are taken(Fig. 4 (a)). The average color value of the 4 color values is used as a background color value(Fig. 4 (b)). The image background with around zero value is obtained by taking the absolute values of all pixel values of the real image subtracted from the background color value(Fig. 4 (c)). The color image is converted into a monochromatic image(Fig. 4 (d)). The vertical and

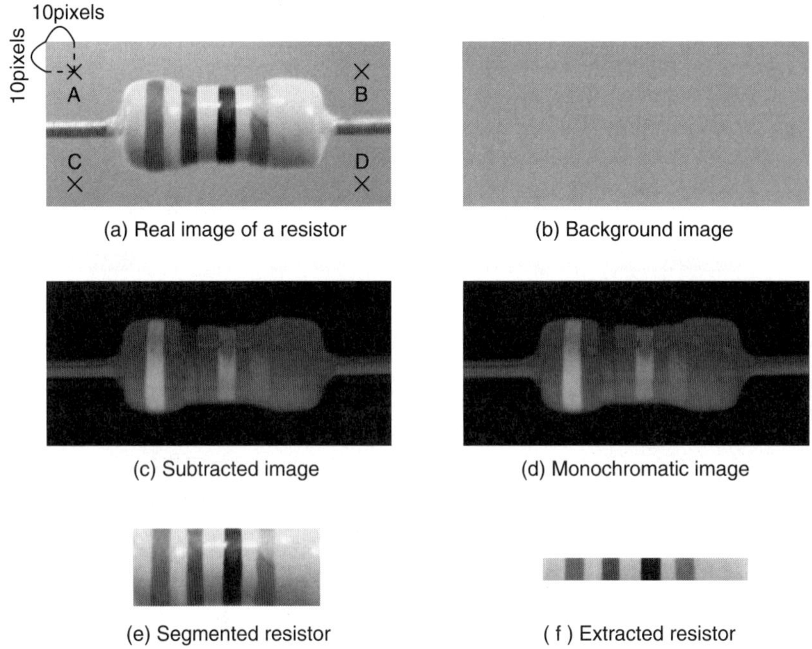

Fig. 4. Extraction of a resistor

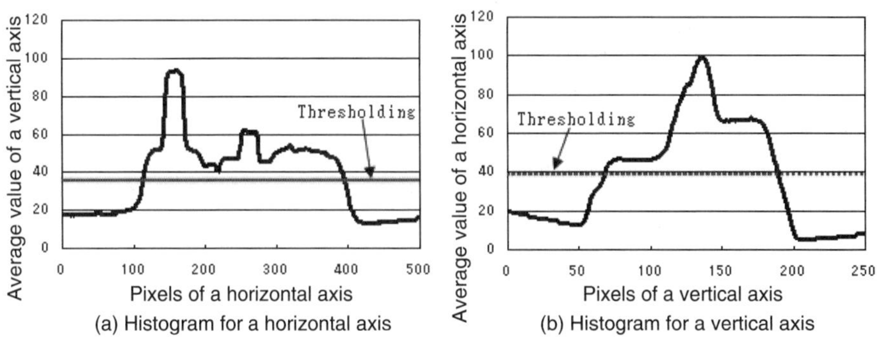

Fig. 5. Histograms of the resistor image

horizontal histograms for the density values are made by averaging ones of the monochromatic image. Fig. 5 shows the density histograms of the resistor image. By thresholding these histograms with a method based on the discriminant analysis [7], the main body of the resistor is obtained from the real image(Fig. 4 (e)). Moreover, a central part of a 15 pixels width is extracted(Fig. 4 (f)).

In the extracted resistor image, each of pixel values differs. Secondly, we explain a method of extracting positions of colors to be read. Fig. 6 shows an

Fig. 6. Segmentation of a resistor

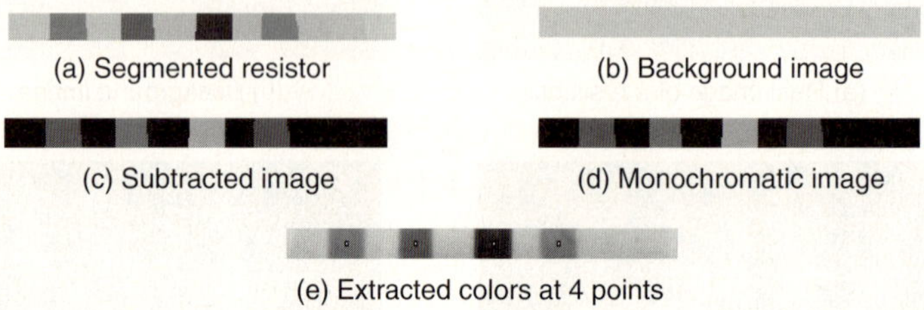

Fig. 7. Extraction of 4 colors

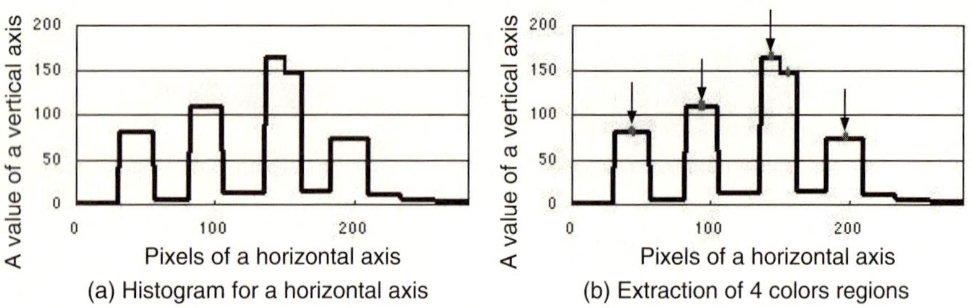

Fig. 8. Process of extracting 4 colors

extracted resistor and a segmented resistor. The 4 color bands and their background are divided by grouping regions between similar colors. For segmentation of the colors, the k-means method is used. Then, a combined feature vector $\boldsymbol{w}$ of a position feature vector $\boldsymbol{u}$ and a color feature vector $\boldsymbol{v}$ is used.

$$\boldsymbol{w} = (tx, (1-t)r, (1-t)g, (1-t)b)^T, \quad 0 \leq t \leq 1, \tag{1}$$

$$\boldsymbol{u} = (x)^T, \tag{2}$$

$$\boldsymbol{v} = (r, g, b)^T, \tag{3}$$

where t denotes a weight for the combined feature vector, x is a position of the pixel from the left. While r, g, and b are the values of R, G, and B, respectively. The combined feature vector is essentially equivalent to a position feature vector when $t = 1$, and a color feature vector when $t = 0$, respectively. In the k-means method, initial clusters are determined as below. We divide the RGB color feature space into $n \times n \times n$ subspaces. In each of the subspace, the number of patterns is counted. And, we select k subspaces in descending order, and their average vectors are computed. Thus, we select k average vectors as initial clusters of the k-means method. In the experiment, we used $n = 12$ and $k = 12$.

In the color regions segmented image, 4 colors to be read are extracted as follows. Fig. 7 shows a process of extracting 4 colors from the segmented resistor. Both the left and right side of color values are taken. The average color value is used as a background color value(Fig.7 (b)). And the image background with around zero value is obtained by taking the absolute values of all pixel values of a real image subtracted from the background color value(Fig.7 (c)). The color image is converted into a monochromatic image(Fig.7 (d)). The histogram for the density values is made by taking 1 pixel width for the vertical direction of the monochromatic image. From the histogram, the density values of the regions in descending order are obtained. Furthermore, we select the top 4 regions, but the low ranked region next to each other is ignored. Fig. 8 illustrates the process of extracting 4 colors. In each region, the color of the central pixel of the selected regions is extracted(Fig.7 (e)). Thus, we obtain 4 colors to be read from the real image.

2.2 Color Classification of the Resistor

For the colors obtained by color extraction of the resistor, we investigate what color they are. In the color classification, 1-NN classifier [4] [5] is used. The 1-NN classifier is a well known method in the pattern recognition field. In the experiment, the test samples are used as colors extracted from the real images. On the other hand, the training samples are used as colors preliminary extracted from the real images different from the test samples. Note that the training samples and test samples are statistically independent. Fig. 9 is the training samples at each illumination. In the experiment, 20 training samples per one color are used. In other words, 220 training samples are used.

In this paper, we use not only the RGB color feature but also the $L^*a^*b^*$ and $L^*u^*v^*$ color features [6]. The $L^*a^*b^*$ and $L^*u^*v^*$ color features are known to be the uniform color space. The $L^*a^*b^*$ and $L^*u^*v^*$ values are made from the

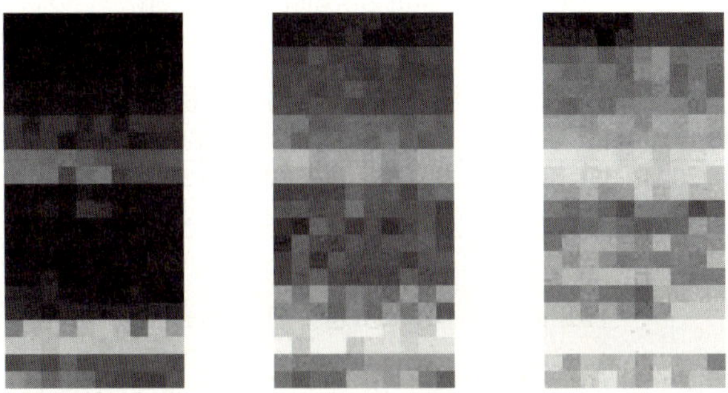

(a) 100 [lx] (b) 3000 [lx] (c) 6000 [lx]

Fig. 9. Training samples at each illumination

RGB values via the XYZ values. The $L^*a^*b^*$ and $L^*u^*v^*$ values are computed as shown in the reference [6].

3 Experimental Results

The effectiveness of the proposed method is examined by using 30 real images at different illuminations. Firstly, the number of images rightly extracted is investigated in terms of cutting the background of the resistor. Table 1 shows the result of extracting the resistor at each illumination. In the table, x / y means that x is the number of images rightly extracted, and that y denotes the number of the real images. Almost the results show a good performance independently by the illumination. Secondly, the number of images rightly extracted is investigated in terms of extracting the colors of the resistor. In the experiment, the value of t in the combined feature vector of a color feature vector and a position feature vector changes from 0.0 to 1.0 increasing 0.1 every time. We regard it as successful when all the 4 colors are extracted rightly from the real image. Table 2 is the result of color extraction at each illumination. From the results, when the value of t is 0.8, the number of images rightly extracted is high at every illumination. And, the higher the illumination is, the better the result shows. When the illumination is about 6000[lx] and $t = 0.7$ and 0.8, 28 of 29 pieces succeed. On the other hand, 6 and 21 pieces succeed when $t = 0.0$ and 1.0 respectively. This means the combined feature vector is working out effectively. Finally, the number of images rightly extracted is investigated in terms of color

Table 1. Result of extracting the resistor

Illumination	No. of the images
100[lx]	30 / 30
3000[lx]	29 / 30
6000[lx]	29 / 30

Table 2. Result of color extraction

Illumination	Values of t										
	0	0.1	0.2	0.3	0.4	0.5	0.6	0.7	0.8	0.9	1.0
100[lx]	10/30	8/30	8/30	3/30	1/30	1/30	3/30	16/30	17/30	15/30	14/30
3000[lx]	9/29	11/29	12/29	15/29	17/29	15/29	10/29	24/29	24/29	16/29	16/29
6000[lx]	6/29	7/29	8/29	10/29	9/29	13/29	15/29	28/29	28/29	21/29	21/29

Table 3. Result of color classification

Illumination	RGB color	$L^*a^*b^*$ color	$L^*u^*v^*$ color
100[lx]	2 / 17	6 / 17	3 / 17
3000[lx]	17 / 24	16 / 24	21 / 24
6000[lx]	22 / 28	24 / 28	23 / 28

classification. In the experiment, we use $t = 0.8$ at every illumination. In the color classification, the $L^*a^*b^*$ and $L^*u^*v^*$ color features as well as the RGB color feature are used. We regard it as success that all the 4 colors from the extracted colors are classified rightly. Table 3 is the result of color classification at each illumination. The results of the $L^*a^*b^*$ and $L^*u^*v^*$ color features are better than that of the RGB color feature. The result is also better when high illumination. By using the RGB color feature, it is difficult to classify the gold color. While on the other hand, the use of the $L^*a^*b^*$ and $L^*u^*v^*$ color features improve the classification of the gold color.

4 Conclusion

In this paper, we have proposed a method to read the resistance of a resistor by image processing techniques. Experimental results support the proposed method is promising. While in classifying colors, we recommend to use the $L^*a^*b^*$ or $L^*u^*v^*$ color feature spaces. In the classification, we use only 1-NN classifier. In order to classify colors accurately, the investigation of other classifiers is being planned as part of future work.

References

1. Chan, K.L., Wang, H.: Reading resistor values by color image processing. In: Ho, A.T., Rao, S., Cheng, L.M. (eds.) Proc. SPIE, Automatic Inspection and Novel Instrumentation, vol. 3185, pp. 157–168 (1997)
2. Russ, J.C.: The Image Processing Handbook, 3rd edn. CRC-Press, Boca Raton (1999)
3. Duda, R.O., Hart, P.E., Stork, D.G.: Pattern Classification, 2nd edn. Wiley-Interscience, Chichester (2001)
4. Cover, T.M., Hart, P.E.: Nearest neighbor pattern classification. IEEE Trans IT-13, 21–27 (1967)
5. Fukunaga, K.: Introduction to statistical pattern recognition, 2nd edn. Academic Press, London (1990)
6. Plataniotis, K.N., Venetsanopoulos, A.N.: Color Image Processing and Applications. Springer, Heidelberg (2000)
7. Otsu, N.: A threshold selection method from grey-level histograms. IEEE Trans. Syst. Man Cybern. 9(1), 62–66 (1979)

Networked Virtual Marionette Theater

Daisuke Ninomiya[1], Kohji Miyazaki[1], and Ryohei Nakatsu[2]

[1] Kwansei Gakuin University, School of Science and Technology
2-1 Gakuen, Sanda, 669-1337 Japan
aaz61232@kwansei.ac.jp, miyazaki@nirvana.ne.jp
[2] National University of Singapore, Interactive & Digital Media Institute
Blk E3A #02-04, 7 Engineering Drive 1, Singapore 117574
idmdir@nus.edusg

Abstract. This paper describes a system that allows users to control virtual marionette characters based on computer graphics (CG marionette characters) with their hand and finger movements and thus perform a marionette theatrical play. The system consists of several subsystems, and each subsystem consists of a web camera and a PC. It can recognize a hand gesture of its user and transform it into a gesture of a CG marionette character. These subsystems are connected through the Internet, so they can exchange the information of the CG marionette character's movements at each subsystem and display the movements of all characters throughout the entire system. Accordingly, multiple users can join the networked virtual marionette theater and enjoy the marionette play together.

Keywords: Marionette, puppet, virtual theater, hand gesture, image recognition.

1 Introduction

The culture of controlling puppets with the hands to perform theatrical play has been common throughout the world from ancient times. In Japan, there is a type of puppet theater called Bunraku, which arose about three hundred years ago [1][2]. In Europe, too, various kinds of puppet play have been performed and enjoyed. The puppet play using a puppet called a "marionette" has been the most popular variety [3]. Marionette play and puppets have become very popular in recent years, largely due to the movie called "Strings [4]" (Fig. 1). This paper describes a networked virtual marionette theater that is basically a distributed system consisting of several subsystems connected through the Internet. Each subsystem can recognize the hand and finger gestures of the person in front of its web camera and then transform them into the motions of a marionette character based on computer graphics (CG marionettes). Each subsystem exchanges the information of actions performed by its marionette character with such information from the other subsystems. The display of each subsystem shows a virtual scene where multiple marionette characters, each controlled by a different user, interact. Thus multiple users, even if they are in separate locations, can gather in a virtual marionette theater and perform a theatrical marionette play.

I. Lovrek, R.J. Howlett, and L.C. Jain (Eds.): KES 2008, Part I, LNAI 5177, pp. 440–447, 2008.

Fig. 1. A scene of "Strings"

2 Related Works

Technologies based on three-dimensional computer graphics have made tremendous progress in recent years. We can see photographically real CG objects and CG characters in movies and games. Furthermore, the technologies based on CG animation have also progressed rapidly. Animations of fluid [5] and the destruction of objects [6] have been studied. Moreover, the movements of a crowd based on an artificial-intelligence approach [7] and movements of humans based on inverse kinematics [8] have been proposed. Motion capture systems have been widely used for the control of CG characters [9]. Although the realization of human-like motions of CG characters has been eagerly pursued, the realization of marionette-like motions has seldom been studied. Since the movements of marionette characters are unique and have been loved by people throughout history, it is worth studying a system by which non-experts of marionettes can easily manipulate their movements and generate marionette-like behaviors using CG characters.

3 System Concept

The following elements typically compose a marionette theater.

(1) Puppets called "marionettes"
(2) Speech of each puppet
(3) Scene settings
(4) Music

In a large performance, various kinds of marionette puppets appear and the scene settings are changed frequently, depending on the story's plot, and even a live orchestra is sometimes used to generate music. Therefore, even if people wanted to enjoy manipulating marionette puppets and creating theatrical play, it could be very difficult. On the other hand, if we introduced virtual marionette characters based on computer graphics instead of using real marionettes with physical bodies, it would become significantly easier for users to generate and change most of the above elements of marionettes, speech, backgrounds, and music. In addition, by developing a networked virtual marionette theater, multiple users, manipulating their own marionette characters, can gather in a virtual theater and let their virtual puppets interact with other puppets, thus creating the performance of a virtual theatrical play.

4 System Structure

4.1 Overview

The entire system is made from a group of subsystems connected through a network. The structure of the whole system is shown in Fig. 2, and the structure of each subsystem is illustrated in Fig. 3. Each subsystem consists of a PC and a web camera. The posture of a user's hand is captured by the web camera, and then hand-gesture recognition is carried out. Then the recognition result of a hand posture is reflected in the gestures of a CG marionette character.

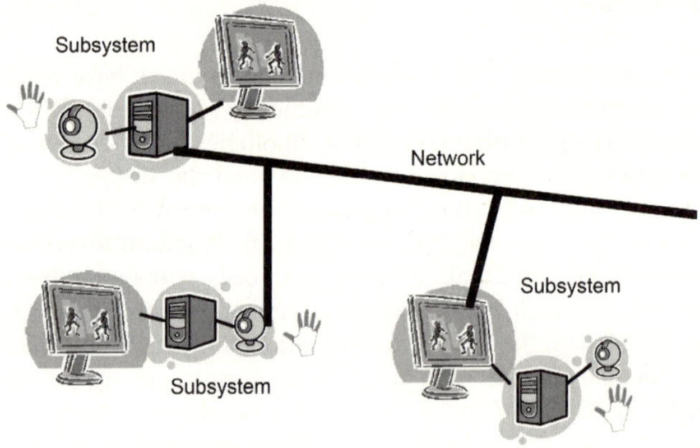

Fig. 2. Structure of entire system

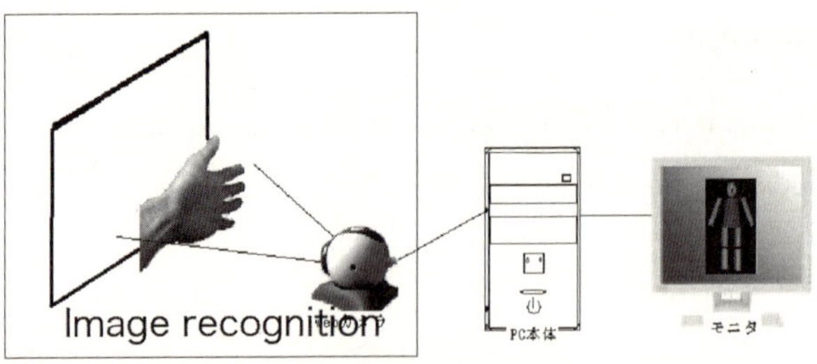

Fig. 3. Structure of subsystem

4.2 Hand-Gesture Recognition

In this section, a real-time hand-gesture recognition method is described for use in the recognition of a user's hand gesture for each subsystem [5]. There have been several research efforts on the real-time recognition of hand gestures [6][7]. Most of them use rather complicated systems such as multiple cameras. On the other hand, we tried to develop a simpler system using a single web camera. The recognition process consists of the following sub-processes.

4.2.1 Extraction of Hand Area (Fig. 4)

Using the color information of a hand, the image area corresponding to a hand is extracted from the background. In this case, HSV information obtained by the transformation of RGB information is used. Then, using a median filter, the noise contained in the extracted image is deleted.

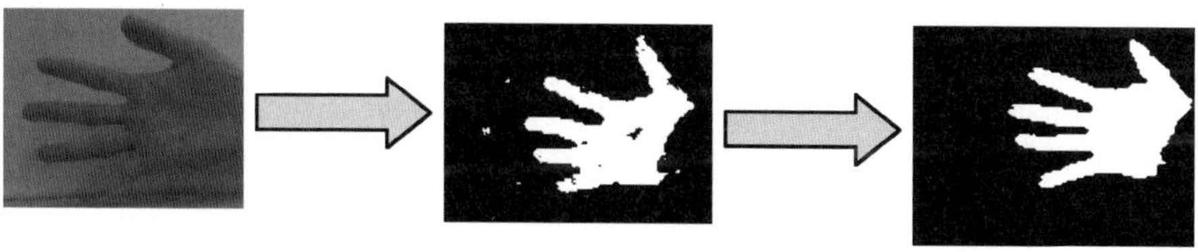

Fig. 4. Extraction of hand area

4.2.2 Extraction of Finger Information Using Histogram

The length of each finger is calculated by using simple histogram information. Figure 5 shows the information of a histogram corresponding to finger length. Depending on the angle of finger bending, the height of the histogram varies. This means that from the height information of the histogram, the bending angle of a finger can be calculated.

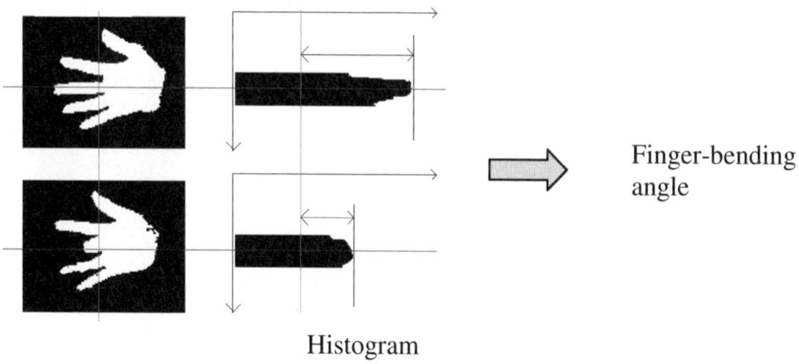

Fig. 5. Extraction of finger-length information

4.2.3 Optimization of Separating Each Finger's Histogram

Depending on the angle of each finger against the x axis (or y axis), it is sometimes difficult to clearly separate a histogram corresponding to each finger. Therefore, for the information extraction of each finger, rotation transformation is carried out to achieve the optimum separation of partial histograms corresponding to each finger.

4.2.4 Bending-Angle Estimation of Each Finger

Figure 5 also shows a comparison between two histograms varying with the bending angle of a finger. By comparing the length of a histogram to the original (longest) histogram when the bending angle is zero, the bending angle of the finger is calculated.

4.3 Control of CG Marionette

Each finger is assumed connected to a certain part of a CG marionette through a virtual string. The relationship between five strings and the part of the marionette to which each string is attached is illustrated in Fig. 6. Here, t1 ~ t5 are virtual stings, and p1 ~ p8 are the parts composing the marionette model, where a1 ~ a7 are joints of these parts. The bending angle of each finger calculated in the above process is reflected directly in the length of each string. In this way, the angle of each joint of the marionette, corresponding to p1, p2, p3, p4, p5, p6, p7, and p8, is determined. Therefore, by bending each of the five fingers appropriately, a user can control the motion and gestures of a virtual CG marionette.

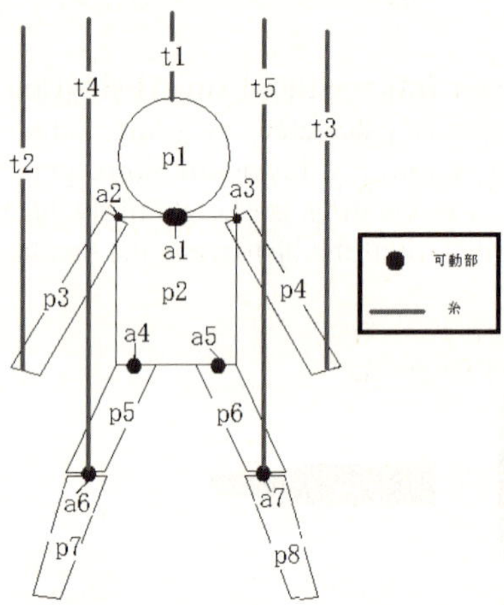

Fig. 6. Model of a virtual marionette

4.4 Background and CG Characters

We are planning a system that allows us to easily change scenes as well as characters, so we have developed various kinds of backgrounds and characters based on computer graphics. We are trying to develop an "Interactive Folktale System [8]" that offer users the ability to generate Japanese folktales as animation and to enjoy the interactions with creators of other characters in the system. Therefore, we have prepared various kinds of backgrounds and characters for our virtual marionette system. Figure 7 shows one of the marionette character in three different backgrounds.

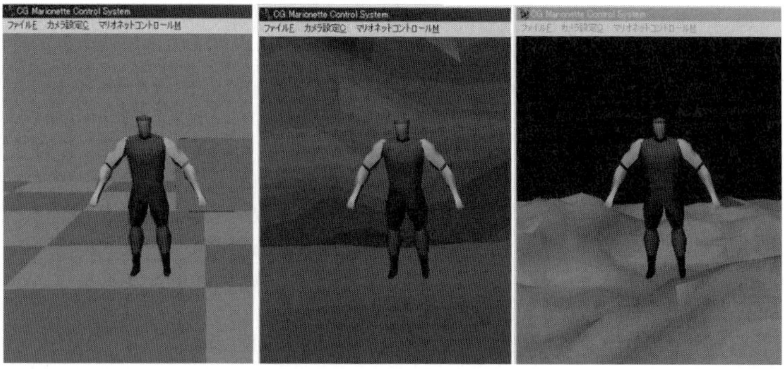

Fig. 7. Examples of virtual marionette characters

4.5 Networked Marionette Theater

The virtual marionette system we have developed as a first prototype toward the networked virtual marionette system consists of a hand-gesture recognition unit and an animation generation unit. This prototype system would work as a sub-system in the distributed system. In each subsystem, the recognition results of the other subsystems are shared. Furthermore, all of the CG characters and backgrounds are shared among these subunits. Using these recognition results as well as the CG characters and backgrounds, each subsystem can simultaneously create the same scene where multiple CG characters, each of which is controlled by its own subsystem, appear and behave in the same way.

5 Evaluation of the System

We have carried out an evaluation of a subsystem, which is the basis of the whole system and the instrument with which a user can control one virtual marionette character. We selected 20 students as subjects for this evaluation's tests. All of them know about marionette puppets but have never manipulated them. We asked them to manipulate both a real marionette puppet and a virtual CG marionette used in this system. After that we asked them several questions. The questions and the answers are summarized as follows.

(1) Is the movement of a virtual marionette "unique" compared with other CG characters?

Definitely Yes (4), Comparatively Yes (12), Neutral (4), Comparatively No (0), Definitely No (0)

(2) Is the movement of a virtual marionette "real"?

Definitely Yes (0), Comparatively Yes (1), Neutral (15), Comparatively No (4), Definitely No (0)

(3) Did you feel that your hand gestures were closely reflected in the movements of a virtual marionette?

Definitely Yes (0), Comparatively Yes (15), Neutral (3), Comparatively No (1), Definitely No (1).

From the first question, it is clear that 80% of the subjects said that there is some unique aspect in the movement of the virtual marionette. This means that the authors succeeded in their intention to develop a system in which the particular movement of a marionette is regenerated. For the second question, the fact that most of the subjects answered "neutral" indicates that the meaning of "real" is somewhat difficult for them to associate with the marionette's movement. For the third question, 75% of the subjects answered that the marionette correctly moved according to their hand gestures. These results show that the recognition method introduced here works very well and gives people the feeling that they are directly manipulating the virtual marionette characters. Moreover, they again expressed the feeling that the system successfully reproduced the particular movement of a marionette.

6 Conclusions

In this paper, we proposed a system in which users can easily manipulate virtual marionette characters with their hand gestures. For the recognition of hand gestures, simple real-time hand-gesture recognition was realized by using histogram information of an extracted hand area. The recognition result is reflected in the movement of the marionette character by connecting each finger movement to a particular part of the virtual marionette by a virtual string. Furthermore, the concept of networked marionette theater was proposed in which several subsystems are connected by a network. Here, multiple users can perform theatrical marionette play by manipulating their own marionette characters. Finally, we carried out an evaluation test to assess the feasibility of a subsystem. By using twenty subjects and letting them manipulate both a physical marionette as well as a virtual one, we obtained evaluation results indicating that by using this virtual marionette system, even a non-expert of marionette manipulation can have the feeling of manipulating marionettes and thus can participate in a theatrical marionette performance.

For our further work, we need to improve the recognition accuracy of the hand-gesture recognition. Moreover, we need to develop adequate contents to refine the entire networked virtual marionette theater, and we also need to carry out an evaluation of the whole system by letting people use the system.

References

1. Keene, D.: No and Bunraku. Columbia University Press (1990)
2. `http://www.lares.dti.ne.jp/bunraku/index.html`
3. Currell, D.: Making and Manipulating Marionettes. The Crowood Press Ltd. (2004)
4. `http://www.futuremovies.co.uk/review.asp?ID=319`
5. Stam, J., Fiume, E.: Depicting Fire and Other Gaseous Phenomena Using Diffusion Process. In: Proceedings of SIGGRAPH 1995 (1995)

6. O' Brien, J.F., Hodgins, J.K.: Graphical modeling and animation of brittle fracture. In: Proceedings of SIGGRAPH 1999 (1999)
7. Courty, N.: Fast Crowd. In: Eurographics 2004 (2004)
8. Boulic, N., Thalmann, M., Thalmann, D.: A Global Human Walking Model With Real-Time Kinematic Personification. The Visual computer, 344–358 (1990)
9. Lee, J., Hoon Lee, K.: Precomputing avatar behavior from human motion data. Graphical Models 68(2), 158–174 (2004)
10. Ninomiya, D., Miyazaki, K., Nakatsu, R.: Study on the CG Marionette Control Based on the Hand Gesture Recognition. Annual Meeting of Game Society of Japan (in Japanese) (2006)
11. Wah Ng, C.: Real-time gesture recognition system and application. Image and Vision Computing 20 (2002)
12. Utsumi, A., Ohya, J., Nakatsu, R.: Multiple-camera-based Multiple-hand-gesture-tracking. Transaction of Information Processing Society of Japan 40(8), 3143–3154 (1999) (in Japanese)
13. Miyazaki, K., Nagai, Y., Wama, T., Nakatsu, R.: Concept and Construction of an Interactive Folktale System. In: Ma, L., Rauterberg, M., Nakatsu, R. (eds.) ICEC 2007. LNCS, vol. 4740, pp. 162–170. Springer, Heidelberg (2007)

A Generic Methodology for Classification of Complex Data Structures in Automotive Industry

Dymitr Ruta

British Telecom (BT) Group* and Bournemouth University**

Abstract. Driving a vehicle is a complex mixture of real-time processes of cognition and control. Recent advances in pattern recognition and machine learning brought automotive industry to the verge of direct AI applications in vehicles. They would continuously sense the external environment, monitor vehicle's internal systems and track drivers actions in an attempt to support driver's decisions, making them better informed or even take over decision process if the reliability and confidence of perceived safety critical situations outperform human performance. This work intends to contribute to the automated real-time classification methodology suitable for applications in the realistic circumstances of driving a vehicle. It proposes a coherent strategy to a fast extraction of simple but robust features out of complex data structures like images, continuous and discrete signals etc. It advocates the use genetic algorithm for feature selection paired with simple classifiers suitable for further combination at the decision level. The presented generic methodology is open for various data transformations, classifiers, features and scales well with the data size. It has been tested in two independent competitions: NISIS Competition 2007 concerned with automated classification of pedestrian images and Ford Classification Challenge 2008 dedicated to symptoms detection from high-frequency signal patterns. The model was announced the winner of the NISIS'2007 Competition achieving pedestrian recognition rate exceeding 95% and is now under evaluation for the second challenge.

1 Introduction

One of the biggest challenges in automotive industry that still hinders further development towards autonomous driving and self-aware vehicle control systems is the ability to sense, understand and recognise the internal and external environment of a driven vehicle. One of the key components of this challenge is the ability to detect characteristic objects like other cars, road signs, traffic lights, pedestrians as well as correctly recognise and diagnose internal systems' signals, faults and symptoms. Let us consider two representative cognition problems: pedestrian detections from 2-dimensional images and symptom detection from 1-dimensional signals and show how these two seemingly distant problems can be solved using the same coherent methodology.

* BT Group, Chief Technology Office, Intelligent Systems Lab, Adastral Park, Orion MLB 1, PP12, Ipswich IP53RE, UK, dymitr.ruta@bt.com

** Bournemouth University, School of Design, Engineering and Computing, Smart Technology Research Centre, Poole House, Talbot Campus, Fern Barrow, Poole, Dorset, BH12 5BB, UK, druta@bournemouth.ac.uk

I. Lovrek, R.J. Howlett, and L.C. Jain (Eds.): KES 2008, Part I, LNAI 5177, pp. 448–455, 2008.

1.1 Pedestrian Detection (NISIS Competition 2007)

While road signs or traffic lights are fairly structured, well defined and rather invariable in in time, detecting moving and deformable people whose shape, pose, clothing and scene background are highly variable is particularly difficult. Pedestrian detection is a complex process that starts from the analysis of the whole scene to extract small patches or areas of attention likely to include pedestrian images [2], [6], [7]. Two-dimensional images are not the inputs that can be consumed by a classification model. A feature generation process is required to extract good, discriminative variables to provide a numerical distinction between the pedestrian and its background.

Although many high performing classifiers are reported in the supervised learning literature, the main effort in related recent work on pedestrian classification is clearly focussed on extracting good features out of the images. Dalal and Trigs [2] proposed a method of using histograms of oriented gradients paired with support vector classifier and achieved high pedestrian recognition rates under both daylight conditions and infrared detection by night. Feed-forward neural networks were used in combination with local receptive fields [9] and with filtered gradient images [11]. Papageorgiou and Poggio [7] used Support Vector Machine in a combination with overcomplete sets of wavelet features. Viola et al. [8] proposed an efficient cascade of single-feature detectors trained by AdaBoost [3]. Other approaches tried to seek an improvement by decomposing the pedestrian recognition task into simpler component tasks [5].

There are striking differences of the classification performance reported in the literature with false positive rate prone to huge variability stretching up to several orders of magnitude [11], [7]. Lack of large representative image datasets and massive inconsistency in composing the images makes it difficult to extract synergic merits from complementary methods to progress further towards objectively better performing models. To avoid such dissonance the presented model uses an established DaimlerChrysler Pedestrian Classification Benchmark Dataset (DCPCBD)[1] used for benchamrking in many recently developed pedestrian recognition systems [6] and is comparatively assessed within the International NISIS Competition 2007 [2].

The objective of this competition was to devise an automated classification system that would be able to detect images of pedestrians against a background or other objects seen along the road. The image patches of 36x18 pixels in 8-bit gray scales were preselected from larger images captured by the camera installed on the front of a vehicle. Some examples are shown in Figure 1.

The competition data comprised 9800 images from DCPCBD with equal pedestrian/non-pedestrian class priors. The dataset has been split into labelled training set (37.5%) and unlabelled testing set (62.5%). The objective of the competition was to built a classification model that having learned on the training set would yield the highest recognition rate on the testing set.

[1] DaimlerChrysler Pedestrian Classification Benchmark Dataset provided by S. Munder and D.M. Gavrila, http://www.science.uva.nl/research/isla/downloads/pedestrians/

[2] Nature Inspired Smart Information Systems Competition 2007. Description and results: http://www.nisis.risk-technologies.com/msc/competition2007.aspx

Fig. 1. Examples of pedestrian images

1.2 Symptom Detection (Ford Classification Challenge 2008)

Vehicle components become increasingly complex while labour cost increasingly expensive. Repairing a vehicle is now a matter of replacement of the smallest subcomponent that can be identified as faulty. Moreover if a faulty component is not spotted early enough it may damage larger linked components and often multiply the cost of a repair or even compromise driver security. A reliable and automated diagnostic tool is therefore a highly useful and desirable asset of any vehicle diagnostic station as well as in-vehicle self-diagnostic systems. The real challenge in this space is to to detect various symptoms based on cheap and fast signal measurements instead of invasive and costly direct inspections. There is a number of ways vehicle systems can be examined for quality but there is a significant class of problems in which high frequency signal can be used to measure vibrations of various parts [10], electric current flow of internal circuits [1] or even sound characteristics of moving elements [4].

The problem of symptom detection based on 1-dimensional high-frequency signal patterns has been addressed within Ford Classification Challenge 2008.[3]. The objective of this challenge was to devise a classification model to distinguish between the presence and the lack of a particular symptom in a vehicle subsystem based on a diagnostic data composed of continuous signal batches of 500 samples per diagnostic session. The data has been provided by Ford Motor Company and was organised in two separate sets FordA and FordB with separate training, validation and testing parts in the order of thousands of signal patterns. While FordA dataset was consistent across its different parts, validation and testing parts of FordB dataset were injected with a significant noise component to examine model's resistance to heavy noise. Validation set labels were provided after the first stage of the challenge that required validation set classification. The competitive models were to be assessed in terms of a combination of classification accuracy and false positive rate measures.

2 Data Preprocessing

Both images and signals as well as other complex data objects have been preprocessed by means of a number of transformations that intend to expose differences between classes of objects in various projections and orientations.

[3] Ford Classification Challenge run in conjunction with IEEE/WCCI 2008.

2.1 Differencing

Let $X^{[N \times M]}$ stand for an object with the object elements: $x_{i,j}, i = 1, .., N, j = 1, .., M$. The directional difference transformations denoted by $T_D^{direction}(X)$ simply calculate the aggregated absolute difference between the current element (pixel or signal value) and the neighbouring elements along the given direction. Due to lack of space only horizontal difference is formally presented:

$$T_D^{-}(X) = X^{-} = x_{i,j}^{-} : \begin{cases} \underset{i \wedge j \neq 1,M}{\forall} x_{i,j}^{-} = |2x_{i,j} - x_{i,j-1} - x_{i,j+1}|/2 \\ \underset{i \wedge j=1}{\forall} x_{i,j}^{-} = |x_{i,j+1} - x_{i,j}| \\ \underset{i \wedge j=M}{\forall} x_{i,j}^{-} = |x_{i,j} - x_{i,j-1}| \end{cases} \tag{1}$$

but by using the same logic, the vertical $T_D^{|}X$ and diagonal differences: $T_D^{\backslash}(X)$ and $T_D^{/}(X)$ have been computed along with the circular average difference $T_D^{\odot}(X) = X^{\odot} = (X^{-} + X^{|} + X^{/} + X^{\backslash})/4$. Note that one-dimensional signal would use only horizontal transformation $T_D^{-}(X)$ while for high dimensional objects the number of directions scales up with the number of different orientations. The second and higher orders n of differencing along a particular orientation r can be achieved recursively by: $T_D^{r^{n+1}}(X) = T_D^{r}(T_D^{r^n}(X))$.

The intention of these transformations is to capture the location and magnitude of significant local gradients that may indicate the presence of the shape contour in case of images or simply measure trends and their changes for univariate signals.

2.2 Transformations to the Frequency Domain

Transforming a data object to the frequency domain makes sense if there is any indication of repetitiveness in the object's data. Sound, vibrations or electric current signals all share a great deal of periodicity and are naturally suited for time-frequency analysis. However, one can argue that regularly spaced elements on images can provide some space-frequency characteristics that can be exploited for learning purposes. Good example of such transformation is Discrete Fourier Transformation decomposing the original signal into a sum of multiple frequency components: $Y_k = \sum_{j=1}^{N} x_j e^{-2\pi i(j-1)(k-1)/N}$. Given the transform X_k the most informative is a characteristic of a signal amplitude $A = |X_k|$ as a function of the frequency k. Assuming 2-dimensional data objects like images Fourier transformation can be calculated along a number of orientations. For simplicity we consider only horizontal and vertical orientations within the frequency range of up to half of the original signal length:

$$T_F^{-}(X) = y_{p,q}^{-} = \sum_{j=1}^{M} x_{p,j} e^{\frac{-2\pi i(j-1)(q-1)}{M}}, \quad T_F^{|}(X) = y_{p,q}^{|} = \sum_{j=1}^{N} x_{j,q} e^{\frac{-2\pi i(j-1)(q-1)}{N}} \tag{2}$$

2.3 Other Transformations

An important global characteristic of the complex data objects like images or signals is a distribution of their component values. To avoid excessively wide distributions histograms were cut off form both sides of the value range collecting all elements with values more distant than the 3 standard deviations from the mean into the boundary bins. For consistency distribution $T_H^r(X) = Y_H^r$ is defined along particular orientation r of an object, but due to small image sizes it was applied to the whole object.

So far we considered transformation from one or two dimensional object to 1 or 2 dimensional objects. Object statistics provide further simplification by transforming an object to a single value reflecting its statistical properties. Mean, variance, min, max, various moments are the typical statistics that can always be used to extract features. In fact such single-valued statistics immediately become object features.

Separate treatment has been applied to FordB dataset of the Ford Classification Challenge whose validation and testing part was significantly contaminated by the noise. In an attempt to remove the noise without harming the signal structure both the clean X and contaminated X^* sets of signals have been transformed to the frequency domain using Fast Fourier transformation T_F and averaged across different patterns to get $\overline{T_F(X)}$ and $\overline{T_F(X^*)}$. From the frequency domain perspective removing the noise means subtracting the average amplitude excess from the contaminated set across the frequency spectrum which formally can be expressed by:

$$Y = T_F^{-1}(T_F(X^*) - \overline{T_F(X^*)} + \overline{T_F(X)}) \quad (3)$$

where T_F^{-1} denotes inverse discrete Fast Fourier transformation. Such denoising treatment was applied to all noise-contaminated parts of FordB dataset prior to feature extraction.

3 Feature Generation

Feature generation is a process of capturing and enumeration of significant object characteristics that would differentiate the classes as much as possible. Despite significant difference between images and signals the same rather simplistic feature generation process has been applied to extract multidimensional feature vectors. It uses the logics of calculating a pair of distances between the object and the averaged class specific object templates in various transformations projections defined in previous section. To generate features the process requires preexisting labelled training set $\mathbf{X} = \{(X_1, \omega_1), .., (X_N, \omega_N)\}$ where $\omega_i \in \{\Omega_1, .., \Omega_C\}$ denotes a class label. First let the class incidence function I_c^k be defined as 1 if $\omega_k = \Omega_c$ or 0 otherwise. Then for each Ω_c the class-specific template can be calculated by: $\overline{X}_c = \sum_{k=1}^N X_k I_c^k / \sum_{k=1}^N I_c^k$.

If the objects are transformed using transformation of certain type t and orientation r, then the c-class template is denoted by $\overline{T_t^r(X)}$. Exploiting the presented transformations the following set of C features can be defined upon the testing image Y: $f_t^c(Y) = |(T_t^r(Y) - \overline{T_t^r(X_c)}|$. These features represent in fact some form of Manhattan distance from the actual objects to different aggregated class templates. Both our problems are in fact two-class problems with the positive $\Omega_1 = 1$ and negative $\Omega_2 = 0$ classes which taking into account all the presented transformations and statistics gives 38 features for pedestrian images data and 18 features for symptom detection signals.

Figure 2 shows some examples of pedestrian and non-pedestrian class templates in various differencing projections, whereas Figure 3 depicts symptom and non-symptom class templates in the distribution projection and frequency spectrum.

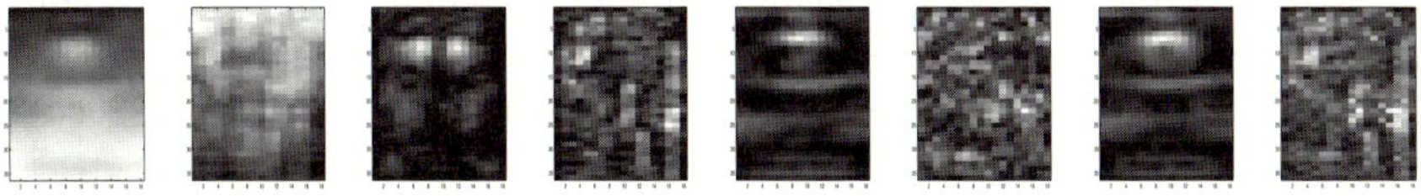

(a) $\overline{x}_1$ (b) $\overline{x}_0$ (c) $\overline{T_D^{-}(X)}_1$ (d) $\overline{T_D^{-}(X)}_0$ (e) $\overline{T_D^{|}(X)}_1$ (f) $\overline{T_D^{|}(X)}_0$ (g) $\overline{T_D^{\odot}(X)}_1$ (h) $\overline{T_D^{\odot}(X)}_0$

Fig. 2. Aggregated pedestrian vs non-pedestrian templates in various difference projections

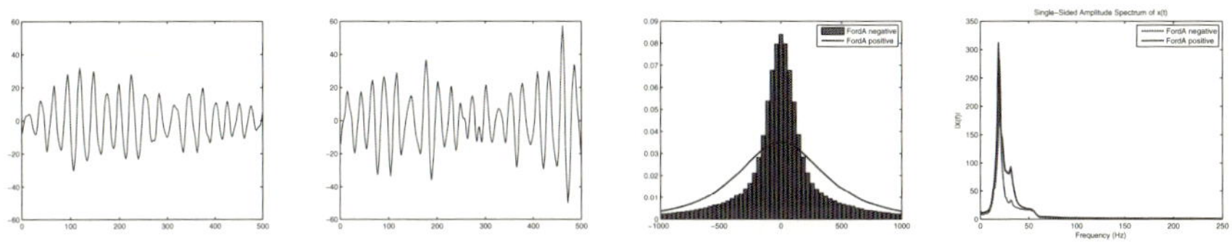

(a) FordA: $\overline{(X)}_{-1}$ (b) FordA: $\overline{(X)}_1$ (c) FordA: $\overline{T_H(X)}$ (d) FordA: $\overline{T_F(X)}$

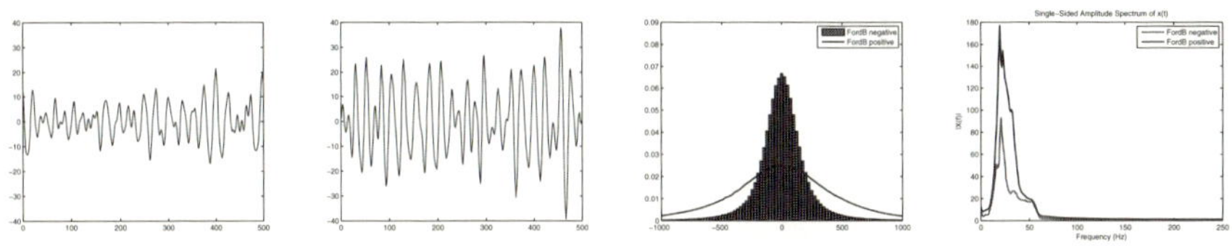

(e) FordB: $\overline{(X)}_{-1}$ (f) FordB: $\overline{(X)}_1$ (g) FordB: $\overline{T_H(X)}$ (h) FordB: $\overline{T_F(X)}$

Fig. 3. Illustration of signal template transformations for both classes of the Ford A/B data sets

4 Feature and Classifier Selection

Given a set of features a taylored genetic algorithm (GA) was applied to find the optimal subset of features. The chromosomes were represented by binary vectors indicating the presence or absence of corresponding features in the model. Using classification accuracy as a fitness function, at each generation a population of 100 parents was doubled by randomly recombined off-springs and then reduced back to its original size retaining the fittest distinct chromosomes. The process continued until for subsequent 10 generations no off-spring showed better fitness than a worst parent.

About 20 different classification methods have been tested in Matlab environment for both considered classification problems. Mixture of Gaussians (MoG) model showed the best classification performance for pedestrian detection problem. Each class was separately modelled by a combination of 4 Gaussians optimised by an expectation maximisation (EM) algorithm. For the symptom detection challenge a k-nearest neighbour (kNN) classifier with Euclidean distance and k=12 yielded the top performance.

5 Experimental Results

For pedestrian detection problem, given the fine-tuned subset of features a full MoG model was built and its performance tested on the validation set yielding low misclassification rate of e=4.86% and real-time capability of delivering about 85 image predictions per second. The next experiment involved the same classifier but applied in multiple versions using top 100 feature subsets found by GA as described in Section 4 and aggregated using mean combiner. The misclassification rate fell down to e=4.37% at an expense of massive increase in computational cost. Following the competition results the model was announced a winner of the NISIS Competition 2007 yielding the expected low misclassification rates of 4.56% and 4.16% for individual and combined versions respectively. The ROC curves comparison shown in Figure 4(a) reveals much greater stability and flexibility in controlling the true and false positive rates for the combined model. It is also worth noting that the presented model outperformed all models presented in [6] that were tested on the same dataset.

For Ford symptom detection challenge the tuned kNN model, was applied to training and validation parts of both FordA and FordB datasets to obtain misclassification rates and ROC curves as shown in Figure 4(b). The figures suggest that the performance is not spectacular but given the unknown nature of a problem and a lack of any benchmark reference it is impossible to assess the presented classification model. The addition of noise for FordB dataset clearly deteriorated classification performance. Preliminary experiments for combined model did not significantly improved the performance hence are not shown here. Given these validation results the model was retrained on all available training and validation sets put together and applied to classify the testing set the results of which will be revealed on IEEE WCCI'2008.

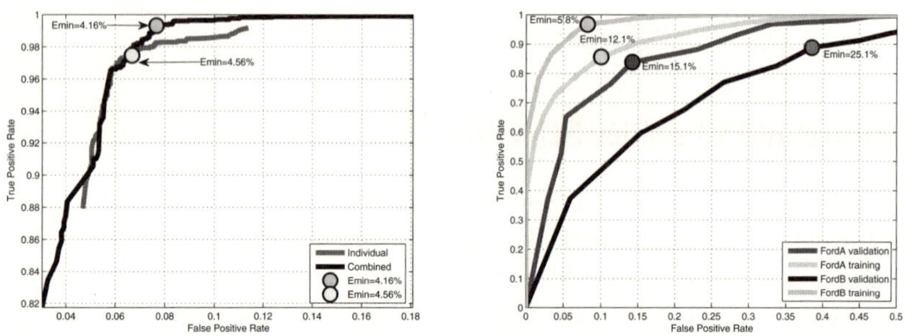

(a) NISIS 2007 (b) Ford Symptoms Detection

Fig. 4. The error rates and ROC curves for pedestrian detection and Ford symptoms detection competitions

6 Conclusions

This work presents a powerful yet simple and generic methodology for classification of images and signals as the examples of complex data objects dealt with in automotive industry. It employs a generic strategy of feature extraction based on measuring Manhattan distance from the objects to class-specific object templates in various transformation projections including 1st and 2nd order differencing, distribution and discrete

Fourier frequency spectrum. The system deals well with missing and noisy and its flexible architecture is open for accommodating different individual and multiple classifiers, feature selection methods or even further combination and fusion at a decision level.

The same classification methodology has been applied to the two quite distinct classification problems of pedestrian images detection and continuous signals symptoms detection. For both cases the final detection system was fine tuned with problem specific subset of parameterised features and classifiers offering various cost-specific model options. With the classification accuracy exceeding 95% and real-time detection capability the presented system was announced the winner of the NISIS Competition 2007 and is still under evaluation within Ford symptoms detection competition.

References

1. Adachi, K.: Failure warning system of electric power unit in vehicle. US Pat.: 510869 (1997)
2. Dalal, N., Triggs, B.: Histograms of oriented gradients for human decision. In: Proc. Int. Conf. on Computer Vision and Pattern Recognition (CVPR), pp. 886–893 (2005)
3. Freund, Y., Schapire, R.: Experiments with a new boosting algorithm. In: Proc. 13th Int. Conf. on Machine Learning, pp. 148–156 (1996)
4. Lerg, G., Devina, A., Roberts, D., Johnson, R.: Vehicle tire leak detection system and method of using the same. US Pat. 588989 (2001)
5. Mohan, A., Papageorgiou, C., Poggio, T.: Example-based object detection in images by components. IEEE Trans. Patt. Analysis and Mach. Intelligence 23(4), 349–361 (2001)
6. Munder, S., Gavrila, D.: An experimental study on pedestrian classification. IEEE Trans. on Pattern Analysis and Machine Intelligence 28(11), 1–6 (2006)
7. Papageorgiou, C., Poggio, T.: A trainable system for object detection. Int. Jrnl. on Comp. Vision 38(1), 15–33 (2000)
8. Viola, P., Jones, M., Snow, D.: Detecting pedestrians using patterns of motion and appearance. In: Proc. Int. Conf. Computer Vision, pp. 734–741 (2003)
9. Wöhler, C., Anlauf, J.: An adaptable time-delay neural-network algorithm for image sequence analysis. IEEE Trans. on Intelligent Transp. Sys. 10(6), 1531–1536 (1999)
10. Yang, C., Adams, D., Ciray, S.: System identification of nonlinear mechanical systems using embedded sensitivity functions. Vibration and Acoustics 127(6), 530–541 (2005)
11. Zhao, L., Thorpe, C.: Stereo- and neural network-based pedestrian detection. IEEE Trans. on Intelligent Transportation Systems 1(3), 148–154 (2000)

Uncovering the Deep Web: Transferring Relational Database Content and Metadata to OWL Ontologies

Damir Jurić, Marko Banek, and Zoran Skočir

Faculty of Electrical Engineering and Computing, University of Zagreb,
Unska 3, HR-10000 Zagreb, Croatia
{damir.juric,marko.banek,zoran.skocir}@fer.hr

Abstract. Organizing the publicly available Web content into highly systematized domain ontologies is a necessary step in the evolvement of the Semantic Web. A large portion of that content called the deep Web is stored in relational databases and it is not accessible to Web search engines. Incorporation of the deep Web data results in domain ontologies richer both in content and in semantic relations. In this paper we introduce a framework for an automatic mapping of relational database metadata and content to domain ontologies written in OWL. Relational constructs: relations, attributes and primary-foreign key associations are translated to OWL classes, datatype properties and object properties. Database tuples become ontology instances. In order to define reference points for integration with other ontologies the constructed ontologies are further enriched with additional semantics from the WordNet lexical database using word sense disambiguation mechanisms. A software implementation of the approach has been developed and evaluated on case study examples.

Keywords: ontology, OWL, relational database, deep Web, WordNet, word sense disambiguation.

1 Introduction

The scientific community has universally recognized the Semantic Web [1] as the prospective evolvement direction of the current Web. The core of the Semantic Web are ontologies written in OWL [2], which formalize the domains by defining classes and their properties and by assigning individuals to the classes.

Web search engines can access only the HTML pages, the "surface" of the Web, while much larger quantity of data, the deep Web [3], remains hidden: the data is available to the users but the pages do not exist until they are created dynamically as the result of a specific search.

The extraction of deep Web data (i.e. the content and the metadata of databases used to generate the pages) is a twofold contribution to creating the circumstances that will lead to the evolvement of the Semantic Web. First, domain ontologies that incorporate the *database content* are much richer than those including only the HTML page content. Second, relational *database metadata* provide additional semantic knowledge that can successfully be transferred to ontologies. This knowledge is lost once the Web pages are created and can thus be obtained only by accessing the databases directly. Uncovering

I. Lovrek, R.J. Howlett, and L.C. Jain (Eds.): KES 2008, Part I, LNAI 5177, pp. 456–463, 2008.

the deep Web causes no security or privacy risks to organizations willing to participate in Semantic Web projects, either by disclosing relational schema metadata or by exposing the already publicly available database content as a whole (i.e. as an ontology), instead in small portions (dynamically created HTML pages).

In this paper we introduce a framework for automatically creating OWL ontologies from extracted relational database metadata and content. We first describe a set of transformations that translate the relational database schema into ontology classes, properties and constraints and then add the database content as ontology instances. One contribution of our work is to apply word sense disambiguation mechanisms to acquire additional semantics from the large spectrum of semantic relations in the WordNet lexical database [4]. Java-based prototype software that evaluates the presented approach on case study examples is another major contribution of the paper.

The paper is structured as follows. Section 2 gives an overview of the related work. Rules for mapping relational metadata to OWL structures are explained in Section 3. Section 4 presents the process of adding additional semantics to the constructed ontologies and gives its evaluation. The architecture of the prototype software tool is illustrated in Section 5. Conclusions are drawn in Section 6.

2 Related Work

While many existing works focus on creating entity-relationship or object-based models from relational databases, only few solutions deal with automatic extraction of database semantics. However, none of them uses a standard ontology language. In [5] ontologies are generated semi-automatically from relational databases using Frame-Logics. The main drawback is the fact that a continuous interaction with the user is required. Similarly, in [6] the primary focus is kept on analyzing key, data and attribute correlations, as well as their combinations. Basic concepts of automatic reverse engineering are considered in [7]. Relations are mapped as classes, their attributes as class attributes and tuples as ontology instances. The primary-foreign key mechanism and specialization are not considered. In [8] we sketched the principles for transforming relational database metadata into OWL structures, but without discussing attribute constrains. Besides, neither a solution for acquiring additional semantics for ontology integration was proposed, nor a software implementation developed. A different approach is presented in [9], where OWL ontologies are created semi-automatically, corresponding to the content of the relational database and based on analyzing the resulting HTML forms.

A detailed overview of the most up-to-date approaches to ontology integration is given in [10]. Providing an unambiguous meaning of ontology components (particularly classes) is regarded as a necessary step to eliminate false matches caused by homonymy of terms. Annotation of the ontology components to be matched with entries from a background ontology with a comprehensive coverage of the domain of interest of the match target ontologies (such as WordNet) is considered necessary for the disambiguation of multiple possible meanings of terms. However, in most of the works annotations are either created manually or supposed to be provided earlier [11, 12]. The ontology matching approach described in [13] provides disambiguation of terms to a certain extent. Disambiguation i.e. choosing the right sense for a word in its occurring context is based on statistical analysis of ontology instances. On the contrary, we apply disambiguation techniques based exclusively on dictionary data.

3 Rules for Mapping Relational Database Schema Components

The process of generating an ontology from a relational database consists of two phases. In the first phase a relational database is translated to an OWL ontology, while in the second phase WordNet is used to enrich the ontology with additional semantics. The first phase is explained in the remainder of this section. The COMPANY database [8] is used as an example to illustrate the mapping procedure (Fig. 1). Primary key attributes are underlined.

Mapping a relational database to an ontology is based on analyzing both the schema metadata (keys and attributes) and the content. The process of mapping a relational schema to OWL consists of several actions performed in the following order: (1) mapping relations, (2) mapping attributes, (3) mapping primary keys, (4) mapping 1:1 and N:1 binary relationships.

```
EMPLOYEE (Fname, Minit, Lname, Ssn, Bdate, Address, Sex,
          Salary, Super_snn, Dnumber)
DEPARTMENT (Dname, Dnumber, Mgr_ssn, Mgr_start_date)
DEPT_LOCATIONS (Dnumber, Dlocation)
PROJECT (Pname, Pnumber, Plocation, Dnumber)
WORKS_ON (Ssn, Pnumber, Hours)
DEPENDENT (Ssn, Dependent_name, Sex, Bdate, Relationship)
```

Fig. 1. Relational schema of the COMPANY database (adopted from [8])

Mapping relations. All entities from an initial entity-relationship diagram exist as relations in the corresponding relational database schema. Relations express a concept similar to ontology classes. Thus, mapping relations into OWL classes is a straightforward process. OWL classes are defined within the *owl:Class* tag.

Mapping attributes. All attributes A_j in a relation R are mapped to corresponding OWL datatype properties P_j. Their domains and ranges are defined within the *owl:DatatypeProperty* tags (the left part of Fig. 2). The domain (*rdfs:domain*) of all those properties is a class C, which corresponds to the relation R. Each property is given the name of the corresponding attribute in addition with the prefix *has* (e.g. the attribute *Lname*, meaning last name, is translated into the datatype property *hasLname*). The range (*rdfs:range*) of each property is the OWL datatype that conforms to the attribute datatype. The database constraint UNIQUE on an attribute results in creating the OWL constraint *maxCardinality=1* on the corresponding property, while NOT NULL implies the cardinality constraint *minCardinality=1*.

Mapping primary keys. A primary key of a relation is an attribute (or a set of attributes) whose value is distinct for each individual tuple. A property created from a primary key attribute should be declared inverse functional. A property P is inverse functional if $P(domainX, rangeZ) = P(domainY, rangeZ)$ implies $domainX = domainY$

i.e. the same range value always denotes a unique instance of the domain [2]. Relation EMPLOYEE has a primary key attribute *Ssn*. The corresponding OWL structure is the inverse functional property *hasSsn* (the right part of Fig. 2). We also state that *minCardinality* for this property is *1* when referring to class *Employee* (the value of a primary key cannot be NULL).

```
<owl:DatatypeProperty rdf:ID = "hasFname">
   <rdfs:domain rdf:resource = "#Employee"/>
   <rdfs:range rdf:resource = "&xsd;string"/>
</owl:DatatypeProperty>

<owl:DatatypeProperty rdf:ID = "hasLname">
   <rdfs:domain rdf:resource = "#Employee"/>
   <rdfs:range rdf:resource = "&xsd;string"/>
</owl:DatatypeProperty>
```

```
<owl:DatatypeProperty rdf:ID = "hasSsn">
   <rdfs:domain rdf:resource = "#Employee"/>
   <rdfs:range rdf:resource = "&xsd;string"/>
   <rdf:type rdf:resource = "&owl;InverseFunctionalProperty"/>
</owl:DatatypeProperty>
```

Fig. 2. Mapping attributes and primary keys to an OWL ontology

Mapping binary relationships. Database relations are connected by the mechanism of primary and foreign keys. Hence, if a foreign key attribute in a relation *A* points to the primary key of some other relation *B*, a semantic association exists between them. In the original ER diagram of the database those associations are specified explicitly as bidirectional *binary relationships*, their names and cardinality constraints being declared explicitly as well. The fact that no such information exists in a relational database causes a problem of naming the ontological structures created as the result of the mapping process. The unidirectional primary-foreign key association between relations *A* and *B* is translated into two OWL object properties (*owl:ObjectProperty*) between the corresponding OWL classes *A* and *B*. We introduce a generic naming mechanism, which adds the *haveRelationTo* prefix to the range class of each property.

When a relation *A* points to a relation *B* with its foreign key, the related binary relationship can either have the cardinality 1:1 or N:1. The cardinality is determined by analyzing the database content. The cardinality to-one corresponds to a functional property in OWL (*owl:FunctionalProperty*). The same instance of such a property's domain class must always be joined to the same instance of the range class.

The association between DEPARTMENT.mgr_ssn and EMPLOYEE.ssn has cardinality 1:1. The resulting object properties are presented in Fig. 3. Two functional properties are created, each of them representing one direction of the relationship: *haveRelationToDepartment* and *haveRelationToEmployee*. The domain and range tags inside the object property tag denote the direction of that part of the relationship. The functional property *haveRelationToDepartment* connects each instance of *Employee* (the domain class) to a single instance of *Department* (the range class).

The only difference between mapping 1:1 binary relationships and N:1 relationships to OWL ontologies is the fact that in the latter case only one of the two created object properties is functional: the one that represents the cardinality to-one. The primary-foreign key association between DEPARTMENT.dnumber and EMPLOYEE.dnumber states that an employee works in a single department (the resulting

```
<owl:ObjectProperty rdf:ID = "haveRelationToDepartment">
    <rdfs:comment>foreign key pair (Mgr_ssn/Ssn)</rdfs:comment>
    <rdfs:domain rdf:resource = "#Employee"/>
    <rdfs:range rdf:resource = "#Department"/>
    <rdf:type rdf:resource = "&owl;FunctionalProperty"/>
</owl:ObjectProperty>
```

```
<owl:ObjectProperty rdf:ID = "haveRelationToEmployee">
    <rdfs:comment>foreign key pair (Mgr_ssn/Ssn)</rdfs:comment>
    <rdfs:domain rdf:resource = "#Department"/>
    <rdfs:range rdf:resource = "#Employee"/>
    <rdf:type rdf:resource = "&owl;FunctionalProperty"/>
</owl:ObjectProperty>
```

Fig. 3. Mapping a 1:1 binary relationship to an OWL ontology

```
<owl:ObjectProperty rdf:ID = "haveRelationToDepartment2">
    <rdfs:comment>foreign key pair (dnumber/dnumber)</rdfs:comment>
    <rdfs:domain rdf:resource = "#Employee"/>
    <rdfs:range rdf:resource = "#Department"/>
    <rdf:type rdf:resource = "&owl;FunctionalProperty"/>
</owl:ObjectProperty>
```

```
<owl:ObjectProperty rdf:ID = "haveRelationToEmployee2">
    <rdfs:comment>foreign key pair (dnumber/dnumber)</rdfs:comment>
    <rdfs:domain rdf:resource = "#Department"/>
    <rdfs:range rdf:resource = "#Employee"/>
</owl:ObjectProperty>
```

Fig. 4. Mapping a N:1 binary relationship to an OWL ontology

functional object property *haveRelationToDepartment2*, see Fig. 4), but more than one employee can work in a department (the non-functional property *haveRelationToEmployee2*).

4 Acquiring Additional Semantics from WordNet

The primary goal of exporting a database into an ontology is to achieve a reliable knowledge source for a particular narrow domain, able to be easily integrated with other ontologies. The integration is impossible without defining a standard reference ontology or dictionary, whose entries have a determined, unambiguous meaning [10]. Thus, we try to associate every class in the target ontology to a WordNet sense.

WordNet [4] is a large taxonomy of the English language, whose searchable lexicon is divided into four categories: nouns, verbs, adjectives and adverbs. Each input word can have more than one meaning (also called word sense). Each word sense can be described by one or more synonyms and is called a synset. A synset is given a description sentence called gloss and may have antonyms.

WordNet includes a set of semantic relations for each word category. The largest spectrum of relations exists for nouns, which comprise about 80% of all WordNet entries [4]. Hyponymy/hypernymy is a transitive relation between nouns that represents subordination/superordination and exactly conforms to the concept of subclasses in ontologies. The part-whole relation is called meronymy/holonymy. Meronymy/holonymy is only occasionally transitive.

The created OWL classes that stem from a database are given some target URI (the value of the *xml:base* attribute, see Fig. 5), while all the associated WordNet-based classes are assigned to the same global URI (*http://www.fer.hr/dbonto/wordnet#*). WordNet class names contain all the synonym words of a synset as well as numbers determining the particular word sense of each word (e.g. *Hall3* in Fig. 5). Classes originating from a database are not declared equivalent to the corresponding WordNet classes (equivalence assumes that two classes are subclasses of each other and share the same set of instances) but only their subclasses (Fig. 5: the database class *Hall* is associated to the third sense of *hall* in WordNet i.e. class *Hall3*). Such a definition enables equally named classes from other databases (having different properties and thus not equivalent to their namesakes) to be associated to the same WordNet synset. In the ontology integration process those classes will possibly be merged into a single new class whose property set is the union of the two original sets.

```
<rdf:RDF xml:base = "http://www.fer.hr/dbonto/faculty#"
         xmlns:x = "http://www.fer.hr/dbonto/faculty#"
         xmlns:wn = "http://www.fer.hr/dbonto/wordnet#"
    <owl:Class rdf:ID = "Hall">
        <rdfs:subClassOf rdf:resource = "&wn;Hall3" />
    </owl:Class>
</rdf:RDF>
```

Fig. 5. Associating a class resulting from a database to a WordNet synset

Recent word sense disambiguation approaches are based exclusively on dictionary data. They analyze either the semantic similarity between the synsets or their glosses and hyponyms. We use both the similarity calculation technique developed in [14] and the gloss-based technique outlined in [15]. Each table name (future ontology class name) is disambiguated using (1) names of all the table's attributes, (2) names of all tables referenced by the foreign keys in the target table, (3) names of all tables that reference the target table. The disambiguated table names must be WordNet entries (either single-word or multiple-word ones): we can disambiguate table names such as *hall* or *academic_year*, but not *student_course*. On the other hand, names of attributes and other tables may be any combinations of words (in that case the contribution of each word is calculated separately) or abbreviations (the software connects to the Abbreviations.com Web page, www.abbreviations.com, and obtains the meaning).

In [14] WordNet synsets (i.e. word senses) are interpreted as graph vertices, connected by edges representing hyponymy/hypernymy and meronymy/holonymy (each edge is given a weight according to the relation type). All possible paths between the target vertex (corresponding to one sense of the target table name) and all vertices corresponding to different senses of the other word (attribute name, related table name or name particle) are constructed. Weights are multiplied across paths and the highest product becomes the semantic similarity between the target synsets. Since the calculated similarities for different attribute and related table names may point to different word senses, we take the arithmetic mean across all attributes and related tables as the final similarity score. We consider the disambiguation process successful if the highest score is at least 1.2 times bigger than the second highest (the ratio of 1.2 has been obtained by experiments on case study examples).

In [15] word senses are disambiguated by finding the overlap among their sense definitions. For each of the two target synsets the following four sets are extracted: (1) all synonyms, (2) all words of the gloss, (3) synonyms in all hyponyms, (4) all words of all the hyponyms' glosses. The existence of an overlap between any of those four sets belonging to the senses *s* and *s'* of words *w* and *w'*, respectively, suggests a correlation between the senses (i.e. synsets). If no overlap exists for other pairs of senses of *w* and *w'* or if the size of that overlap is smaller (in our case at least 1.2 times), the disambiguation is successful. For example, *w'=computer* disambiguates *w=terminal* since it appears only in the gloss of its third sense.

Experiments are performed for the case study example in Fig. 1 (*company database*) as well as our real-world *faculty database*, which maintains the data about the students and the courses they attend. Disambiguation is needed for about 30% of table names, which conform to multiple-sense WordNet entries; other 40% of names match single-sense entries. The disambiguation technique presented in [15] shows higher recall, precision and F-measure for both databases than the technique presented in [14] or when both techniques are applied in parallel and only unanimous results considered (Table 1).

Table 1. Comparison of word sense disambiguation techniques

	# of tables	Yang &Powers			Liu, Yu & Meng			both in parallel		
		Rec.	Prec.	F-m.	Rec.	Prec.	F-m.	Rec.	Prec.	F-m.
Company	2	**1.000**	0.500	0.667	**1.000**	**1.000**	**1.000**	0.500	**1.000**	0.667
Faculty	8	0.250	0.000	0.000	**0.875**	**0.857**	**0.867**	0.625	0.800	0.702

5 Prototype Software Tool

A Java prototype tool (Fig. 6) has been developed in order to test our approach on different relational databases as well as to determine which of the presented word sense disambiguation techniques is more suitable for our purpose. The software was encoded in Java 1.5. The JDBC API v3.0 [16] provides access to the input relational databases. The JWNL API v1.3 [17] is used as an interface to WordNet files (we use WordNet v2.1 downloadable from the Princeton website [4]). Jena v2.4 [18] is applied to produce the output OWL ontologies.

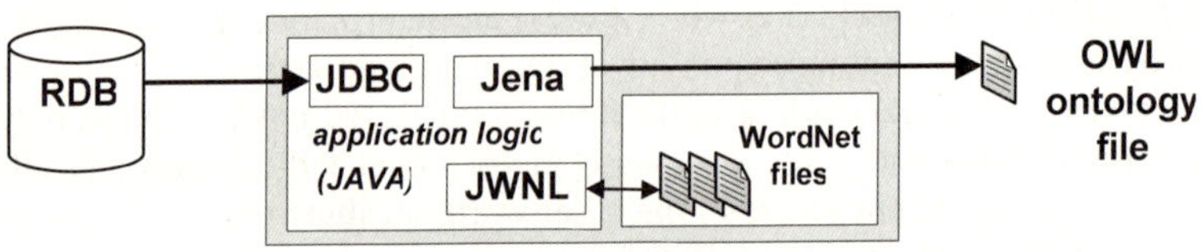

Fig. 6. The architecture of the Java prototype tool

6 Conclusion

This paper presents a framework for an automatic transfer of semantics from relational databases to OWL ontologies in order to exploit the large potential of the "hidden" deep Web relational data in the implementation of the Semantic Web. We use OWL as the target ontology language due to its expressivity and standardization.

In the first phase of the process relational constructs: relations, attributes and primary-foreign key associations between relations, are translated into OWL classes, datatype properties and object properties. The constructed ontologies are enriched in the second phase by acquiring additional semantics from the WordNet lexical database, which defines reference points for integration with other ontologies.

A software implementation of the approach has been developed and tested on two case study examples. The word sense disambiguation mechanism based on analyzing word glosses emerges as the best solution for the second phase of the process.

References

1. World Wide Web Consortium: W3C Semantic Web Activity (last visited: January 16, 2008) (2008), http://www.w3.org/2001/sw/
2. World Wide Web Consortium: OWL Web Ontology Language Guide W3C Recommendation as of February 10 2004 (2004), http://www.w3.org/TR/2004/REC-owl-guide-20040210
3. Bergman, M.K.: The Deep Web: Surfacing Hidden Value. White paper, Deep Content (2001), http://www.brightplanet.com/resources/details/deepweb.html
4. WordNet. A lexical database for the English language (last visited May 5, 2008) (2008), http://wordnet.princeton.edu/
5. Stojanovic, L., Stojanovic, N., Volz, R.: Migrating Data-Intensive Web Sites into the Semantic Web. In: 17th ACM Symposium on Applied Computing, pp. 1100–1107. ACM Press, New York (2002)
6. Astrova, I.: Reverse Engineering of Relational Databases to Ontologies. In: Bussler, C., Davies, J., Fensel, D., Studer, R. (eds.) ESWS 2004. LNCS, vol. 3053, pp. 327–341. Springer, Heidelberg (2004)
7. Dogan, G., Islamaj, R.: Importing Relational Databases into the Semantic Web(last visited: January 16, 2008) (2002), http://www.mindswap.org/webai/2002/fall/Importing_20Relational_20Databases_20into_20the_20Semantic_20Web.html
8. Jurić, D., Skočir, Z.: Building OWL Ontologies by Analyzing Relational Database Schema Concepts and WordNet semantic relations. In: Car, Ž., Kušek, M. (eds.) 9th Int. Conf. on Telecommunications, FER, Zagreb, pp. 235–242 (2001)
9. Benslimane, S.M., Benslimane, D., Malki, M.: Acquiring OWL Ontologies from Data-Intensive Web Sites. In: Wolber, D., Calder, N., Brooks, C.H., Ginige, A. (eds.) Int. Conf. on Web Engineering, pp. 361–368. ACM Press, New York (2006)
10. Euzenat, J., Shvaiko, P.: Ontology Matching. Springer, New York (2007)
11. Giunchiglia, F., Shvaiko, P., Yatskevich, M.: S-Match: an Algorithm and an Implementation of Semantic Matching. In: Bussler, C., Davies, J., Fensel, D., Studer, R. (eds.) ESWS 2004. LNCS, vol. 3053, pp. 327–341. Springer, Heidelberg (2004)
12. Castano, S., Ferrara, A., Montanelli, S.: H-MATCH: an Algorithm for Dynamically Matching Ontologies in Peer-based Systems. In: Cruz, I.F., Kashyap, V., Decker, S., Eckstein, R. (eds.) Proc. Int. Workshop on Semantic Web & Databases, pp. 231–250 (2003)
13. Pan, R., Ding, Z., Yu, Y., Peng, Y.: A Bayesian Network Approach to Ontology Mapping. In: Gil, Y., Motta, E., Benjamins, V.R., Musen, M.A. (eds.) ISWC 2005. LNCS, vol. 3729, pp. 563–577. Springer, Heidelberg (2005)
14. Yang, D., Powers, D.M.W.: Measuring Semantic Similarity in the Taxonomy of WordNet. In: Estivill-Castro, V. (ed.) 28th Australasian Computer Science Conference, pp. 315–322. Australian Computer Society (2005)
15. Liu, S., Yu, C.T., Meng, W.: Word Sense Disambiguation in Queries. In: Herzog, O., et al. (eds.) 2005 ACM International Conference on Information and Knowledge Management, pp. 525–532. ACM Press, New York (2005)
16. Java Database Connectivity - JDBC (last visited May 5, 2008) (2008), http://java.sun.com/products/jdbc/overview.html
17. Java WordNet Library - JWNL (last visited May 5, 2008) (2008), http://jwordnet.sourceforge.net
18. Jena Semantic Web Framework (last visited May 5, 2008) (2008), http://jena.sourceforge.net

Accessing the Distributed Learner Profile in the Semantic Web

Mourad Ouziri

Centre de Recherche en Informatique de Paris 5, Université Paris Descartes, Paris, France
mourad.ouziri@univ-paris5.fr

Abstract. Adaptive e-learning systems need to get complete learner profile to make efficient personalization. However, learner profile is dispersed over multiple heterogeneous e-learning systems. Unfortunately, these e-learning systems are heterogeneous what makes difficult to get complete learner profile. By using semantic web technology, namely Topic Maps, we show how we perform integration of heterogeneous fragments of learner profile to get more complete one. However, this technology does not consider constraints and is not able to make reasoning. So we use, together with Topic Maps, Description Logics to represent constraints and make reasoning over integrated data.

Keywords: Interoperability of learner profile, Topic Maps, Description logics.

1 Introduction

Adaptive learning system aims to make efficient learning activity by considering the learner at the center of the system. Learner is represented in adaptive learning system by the learner profile. Unfortunately, learner profile is dispersed over multiple heterogeneous e-learning systems, which are heterogeneous what makes difficult to get complete learner profile.

To explain this situation, let us consider two adaptive e-learning systems, namely *ELS1* and *ELS2*. To make adaptation of e-learning, each one maintains learner profile for each learner who uses the system. The system *ELS1* uses relational database to store information (see Fig 1. - left). So, some information about learner are maintained as static data in relational tables such as learned courses and goals, and other information are maintained as stored procedures/triggers such as rules classifying learner to a category such as : familiar in databases, unfamiliar in C++ programming, etc. So, e-learning is personalized to each category of learners. For instance, in the *ELS1* we consider that a learner is:

- *unfamiliar* in a given subject if he/she performs less than two subject related courses, and
- *familiar* if he/she performs more than three courses

The e-learning system *ELS2* maintains learner profile in XML. The XML document of Fig. 1 (right) shows that the apprentice *Bob* learned one course about databases that constitutes the only interest.

I. Lovrek, R.J. Howlett, and L.C. Jain (Eds.): KES 2008, Part I, LNAI 5177, pp. 464–472, 2008.

Let us consider that *ELS2* defines three categories of apprentices: *beginner* (if he performs less than two courses), *standard* (if he performs between three and six courses), and *expert* (if he performs more than seven courses).

Learner	Experience
Bob	Databases designing
Bob	SQL Syntax
Bob	C++ in nutshell

Learner	Goal
Bob	Databases
Bob	C++

```
<apprentice id="Bob">
  <studiedCourse>
     <course>
        Introduction to databases
     </course>
  </studiedCourse>
  <interest> Databases </interest>
</apprentice>
```

Fig. 1. An example of a learning profile in relational tables (left) and XML (right)

So, *Bob* profile is incomplete at each e-learning system. Moreover, e-learning systems, *ELS1* and *ELS2*, are heterogeneous in terms of vocabulary, rules, and used technology. That is, *Bob* is not considered to be familiar in *ELS1* nor standard in *ELS2* with regard to only the local profile. However, *Bob* satisfies conditions to be familiar in *ELS1* or medium in *ELS2* if we consider the union of his profiles. The consequence is that *ELS1* (respectively *ELS2*) adapts training to *Bob* as he is unfamiliar (respectively beginner) in databases and should present to him inappropriate learning resources. So incomplete profile should decrease efficiency of e-learning.

Moreover, rules used in both systems are not uniform. Learner who performs five courses is considered to be familiar in *ELS1* and standard in *ELS2*. Connecting these rules is more difficult as they are coded in a different way (PLSQL and XQL). So, many difficulties occur when trying to get complete learner profile. These difficulties are especially due to semantic heterogeneity of existing learner profiles.

Two main solutions are proposed in the literature, namely standardization and interoperability. First, standardization aims to describe learner profile using fixed characteristics called metadata. However, standardization does not deal with constraints. So, we discarded standardization-based solution. Second, interoperability-based solution aims to design a federated structure which allows to access multiple learner profiles as single one. The used formalisms must be able to represent both profile semantics and constraints to deal with semantic and rules heterogeneities.

We propose in this paper a solution for semantic interoperability of distributed learning profile. Our solution aims at making a semantic representation of learner profile fragments and integrate them into a coherent knowledge base. To deal with semantics and rule heterogeneities, we combined two semantic web technologies, namely Topic Maps and Description Logics.

2 Related Works

There are many standards and specifications for learner profile description. Most important standards are PAPI Learner [1] and IMS Learner Information Package [2]. We do not detail here this approach since it is not the closest one to this paper.

To make interoperability of distributed learner profile, learner profiles are represented using an adequate formalism and integrate these representations into a consistent and global one by resolving possible heterogeneities. Many formalisms are used to represent learner profile. Some formalisms such as Bayesian networks [3] and case-based reasoning [4] models are not planned for interoperability purpose.

To make semantic interoperability of distributed learner profile, we think that the formalisms which are suitable to be used should allow to make connection with common ontology. RDF(S) [5] is the most used formalism in the current works. Examples of projects using RDF(S) to represent learner profile are OntoAIMS [6] and ELENA[1]. OntoAIMS [6] uses ontology to represents the aspects of the application semantics, including the user profile. The OntoAIMS user profile is elicited with respect to a OWL [7] domain ontology. User model includes different user perspectives, such as knowledge, personal preferences and interests [8].

Even if the user profile is elicited by means an ontology, this is not enough to make semantic interoperability because user profile defined-attributes are specific to the local ontology of the application. So, relating the defined-attributes of the user profile with those defined by the standards remains to carry out.

A more interesting approach is presented in [9]. It aims at representing learner profile by combining multiple standards in an RDF conceptual model. The proposed model of learner profile is based represented using attributes of multiple standards. That is, in a same RDF graph, profile attributes of a standard may reference other profile attributes of different standard. For instance, PAPI[2] performance may be described by means of IMS RDCEO[3] competency. This approach contributes to exchange learner profiles between learning systems. However, interoperability should deal with heterogeneous existing learner profiles which can use none of the standards.

Description logics [14] is an interesting formalism for learner profile management because they represent semantics of resources and provide powerful reasoning. There is little works that use Descriptions Logics to formalize learner profile. In [15], demander profile and supplier profile are formalized using Description Logics to compute compatibility by matching. In [16], Description Logics are used in mobile environment to manage profiles. In this work, profile is formalized as conjunction of interests and disinterests. That is, `profile ≡ ∃ hasInterest.Interest ⊓ ∀ hasInterest.(¬ DisInterest)`, where `Interest` and `DisInterest` are concepts. Profile is matched to services, also formalized using Description Logics, to determine if service interests user.

These two works consider centralized profile, which is formalized in Descriptions Logics to be matched to other profiles or services. In our work, we are interested mainly to distributed profile and how one can resolve semantic heterogeneity. The only use of Description Logics should not be enough to deal with semantic heterogeneity because Description Logics deal with centralized knowledge.

3 Semantic-Based Integration of Distributed Learner Profile

As motivated in the introduction, learner profile is split up on multiple e-learning systems. Unfortunately, e-learning systems use specific semantics and constraints.

---

[1] ELENA projet, http://www.elena-project.org

[2] IEEE Public and Private Information Specification.

[3] IMS Reusable Definition of Competency and Educational Objectives.

We summarize the problematic in two points. First, e-learning systems use specific metadata and semantics to maintain learner profile. Second, e-learning systems define specific constraints over learner profile. So, we think that use of one formalism should not be sufficient to deal with this problematic. Thus, we use jointly two adapted formalisms to deal with the two problematics above, namely Topic Maps and Description Logics.

That is, Topic Maps are used to resolve semantic heterogeneity of metadata by using shared ontology. Then, Description Logics reasoning is used to deal with constraints and consistency of merged profile. This combination of two formalisms is detailed in the following sections.

3.1 Representing Learner Profile Using Topic Maps

Topic Maps [10] is an expressive formalism. It is able to express any knowledge whatever its complexity [13]. Knowledge expressed using Topic Maps, a topic map[4], is serialized in a XTM [12] document. In Topic Maps, anything is topic. A topic is the formal representation of anything, abstract or real. So, *Bob*, *Goal*, *Experience*, *Learner*, *Has Goal*, *Databases*, *C++* are all represented by topics in Fig. 2(a).

In Fig. 2(a), we represented two main knowledge. First, the goal of the learner *bob-id* consists to learn *databases* and *C++*. Second, the learner *bob-id* has experience (which means that he has learned a course) in *Databases Designing*.

We underline that most important knowledge is expressed in associations. So, the two previous knowledge are expressed by the associations which reify the topics *hasGoal-id* and *hasExperience-id*, respectively.

So, each learner profile is represented by a topic map as seen in Fig. 2(a). Now we integrate the topic maps representing learner profiles into a unique federated topic map. Before that, we enrich topic maps with semantics in order to make semantic-based integration of multiple fragments of distributed learner profile.

3.2 Representing Semantics of Learner Profile

Semantics is shared knowledge which provides common meaning to data/metadata. In our approach, the shared knowledge is given by an ontology, expressed in OWL [7]. OWL is compatible with Topic Maps. We show how OWL and Topic Maps can be jointly used easily to represent semantic knowledge about learner profiles. An example of a OWL ontology is given in Fig. 2(b).

Now, we try to add semantics given by the ontology to the topic map of Fig. 2(a). Technically, content enriching with semantics is very appropriate using Topic Maps and OWL. It is simply done by adding for each topic of a topic map, the element <subjectIdentity> defined in Topic Maps standard [10]. <subjectIdentity> should reference a PSI (Published Subject Indicator [10]). PSIs are a set of unambiguous and well-defined subjects. They are public which means that they are accessed by anyone. In our works, we define ontology as a set of PSIs. That is, each concept of a OWL ontology can be referenced as a PSI.

[4] Topic map (t,m in tiny) references a knowledge base structured with respect to the Topic Maps formalism (T, M in capital letters).

```
<topicMap  xmlns:xlink="http://www.w3.org/1999/xlink">
  <topic id="learner-id"> <topname> a learner entity </topname> </topic>
  <topic id="goal-id"> <topname>a learner goal</topname> </topic>
  <topic id="experience-id"><topname>learner experience</topname></topic>
  <topic id="hasGoal-id"> <topname> learner-goal relationship</topname>
  </topic>
  <topic id="hasExperience-id">
        <topname>learner-experience relationship</topname></topic>
  <topic id="bob-id">
        <instanceOf><topicRef xlink:href="#learner-id"/> </instanceOf>
        <topname> Mr BobCompleteName </topname>
  </topic>
  <topic id="databases-id">
        <instanceOf><topicRef xlink:href="#goal-id"/> </instanceOf>
        <topname> databases </topname>
  </topic>
  <topic id="cpp-id">
    <instanceOf><topicRef xlink:href="#goal-id"/> </instanceOf>
    <topname> C++ </topname>
  </topic>
  <topic id="dbCrs1-id">
    <instanceOf><topicRef xlink:href="#experience-id"/> </instanceOf>
    <topname> Databases Designing </topname>
  </topic>
  <topic id="hasGoal-id"/>
  <topic id="hasExperience-id"/>
   <association>                         /* associations between topics*/
     <instanceOf> <topicRef="#hasGoal-id"/></instanceOf>
         <member> <topicRef xlink:href="#bob-id"/> </member>
         <member> <topicRef xlink:href="#databases-id"/> </member>
         <member> <topicRef xlink:href="#cpp-id"/> </member>
   </association>
   <association>
         <instanceOf> <topicRef="#hasExperience-id"/></instanceOf>
         <member> <topicRef xlink:href="#bob-id"/> </member>
         <member> <topicRef xlink:href="#dbCrs1-id"/> </member>
   </association>
</topicMap>
```

(a)

```
<daml:Class rdf:ID="Learner">
      <rdfs:subClassOf rdf:resource="#Profile"/>
</daml:Class>
<daml:Class rdf:ID="Goal"></daml:Class>
<daml:Class rdf:ID="Course">
      <rdfs:subClassOf rdf:resource="#Resource"/>
</daml:Class>
<daml:ObjectProperty rdf:ID="has-goal">
      <rdfs:range rdf:resource="#Learner"/>
      <rdfs:domain rdf:resource="#Goal"/>
</daml:ObjectProperty>
<daml:ObjectProperty rdf:ID="learnedCourse">
      <rdfs:range rdf:resource="#Learner"/>
      <rdfs:domain rdf:resource="#Course"/>
</daml:ObjectProperty>
```

(b)

```
<topicmap  xmlns:xlink="http://www.w3.org/1999/xlink">
 <topic id="learner-id">
   <subjectIdentity>
    <subjectIndicatorRef link:href="http://onto.org/ontology.daml#Learner"/>
   <subjectIdentity>
 </topic>
 <topic id="goal-id">
   <subjectIdentity>
     <subjectIndicatorRefx link:href="http://onto.org/ontology.daml#Goal"/>
   <subjectIdentity>
 </topic>
 <topic id="experience-id">
   <subjectIdentity>
    <subjectIndicatorRef xlink:href="http://onto.org/ontology.daml#Course"/>
   <subjectIdentity>
 </topic>
 <topic id="hasExperience-id">
  <subjectIdentity>
   <subjectIndicatorRef
           xlink:href="http://onto.org/ontology.daml#learnedCourse"/>
  <subjectIdentity>
 </topic>
</topicmap>
```

(c)

Fig. 2. – **(a)** *Bob* profile in *ELS1* represented in XML Topic Maps – **(b)** Part of a simple OWL ontology for e-learning, supposed at http://onto.org/ontology.daml – **(c)** Learner profile semantics represented in Topic Maps (completes the Fig. a)

So, adding semantics to learner profile consists to make reference from each topic of the profile topic map to its corresponding concept in the ontology using <subjectIdentity> element. In Fig. 2(c), we enrich the learner profile represented in Fig. 2(a) by adding semantics given by the ontology of Fig. 2(b).

In this example, the topic hasExperience-id is interpreted as a learned course. Technically, this is carried out by linking the topic hasExperience-id to its corresponding property learnedCourse in the ontology using the <subjectIdentity> element (see Fig. 2(c)). In Fig. 3, the topic apprentice is defined as a learner and the topic interest as goal.

3.3 Semantic Integration of Distributed Learner Profile

Integrating several fragments of learner profile allows accessing to more complete profile. As shown previously, learner profiles are represented using Topic Maps. Semantis of learner profile is incorporated in the topic maps as references to a shared DAML+OIL ontology. So, semantic integration consists to merge these topic maps into a single one.

Since everything is topic in Topic Maps, this makes Topic Maps a suitable formalism to semantic integration of multiple contents. Associations are too reified by topics. Thus, integration is simply made by merging topics referencing same concept of the shared ontology.

Let us represent just a part of *ELS2* learner profile in the following topic map:

```
<topicmap  xmlns:xlink="http://www.w3.org/1999/xlink">
  <topic id="apprentice-id">
    <subjectIdentity>  <subjectIndicatorRef link:href="http://onto.org/ontology.daml#Learner"/>
    <subjectIdentity>
  </topic>
  <topic id="interest-id">
    <subjectIdentity> <subjectIndicatorRefx link:href="http://onto.org/ontology.daml#Goal"/>
    <subjectIdentity>
  </topic>
  <topic id="studiedCourse-id">
    <subjectIdentity> <subjectIndicatorRef link:href="http://onto.org/ontology.daml#learnedCourse"/>
    <subjectIdentity>
  </topic>
</topicmap>
```

Fig. 3. A part of a topic map representing the learner profile of *ELS2*

Once learner profiles of *ELS1* and *ELS2* are represented in Topic Maps, integrating them into a coherent global learner profile is straightforward. That is, integration of learner profiles aims at identifying the topics of different profiles which reference the same concept of the shared ontology in the <subjectIdentity> sub-element and merge them into a unique topic. So, the topics learner-id and apprentice-id in Fig. 2(c) and Fig. 3, respectively, reference the same concept Learner of the shared ontology. Then, they are merged into the single topic learner-id. In the same way, goal-id and interest-id are merged into the single topic goal-id, and the topics experience-id and course-id are merged into the course-id.

4 Constraints-Based Reasoning in Distributed Learner Profile

Some heterogeneities can not be resolved using only term semantics, as done previously using Topic Maps. In a particular e-learning system, semantics of a given term may not be the one given by the domain ontology. This semantics is usually given by constraints, which are specific to each e-learning system. As example, the term *InterestingLearner* may be defined in system *A* as learners interested in mathematics whereas it may be defined as learners interested in computer science in system *B*. So, system *A* does not consider interesting learners of system *B* as such, and vice versa. In the motivating example, terms *unfamiliar* (of *ELS1*) and *beginner* (of *ELS2*) have different semantics in the domain ontology but they are equivalent with respect to defined constraints.

Constraints are important especially when details on stored data are not accessible. In our example, if the *ELS1* does not share information about studied courses by learners, a *beginner* learner may be integrated into *ELS2* as *unfamiliar* based on constraints defined in both e-learning systems.

The only use of Topic Maps does not allow system to discover that *unfamiliar* and *beginner* are equivalent because Topic Maps does not consider constraints. Then, we use Description Logics [14], which focuses in constraint representation and reasoning.

Constraints in information systems, especially in e-learning systems, are hard coded as shown in the motivation example. Using Description Logics, we show how we make interoperability between these constraints. That is, hard coded constraints are represented in a unique formalism, namely in Description Logics. Then, reasoning is made to deduce semantic relationships among represented concepts.

4.1 A Rapid Overview of Description Logics

Description Logics (DL) [14] are logics developed to represent complex hierarchical structures and make reasoning facilities on these structures. A DL-based knowledge base is composed of two parts: abstract knowledge (TBox) and concrete knowledge (ABox). Concrete knowledge represents a set of facts which are expressed by assertions on individuals. For example, Person (peter) which defines an object *peter* for the concept *Person*. Abstract knowledge is expressed with concepts and roles. Concepts are unary predicates which represent an abstraction of individuals. Roles are binary predicates. They represent relations between concepts. For example,

`Learner ≡ Person ⊓ ∀ learnedCourse.Course ⊓ ≥1 learnedCourses`

means that a Learner is a Person who learned at least one course. Learner, Person, Course are concepts and learnedCourse is a role. DL reasoning deduces the subsumption `Learner ⊑ Person` (which means learners are persons).

4.2 Representing Learner Profile Constraints

Now we represent constraints associated to concepts *familiar*, *unfamiliar* of *ELS1* and *beginner*, *standard*, *expert* of *ELS2* using Description Logics. As done in the related works, we show that using only Description Logics should not enough to make efficient interoperability. So, Topic Maps must be jointly used with DL.

Without using Topic Maps, concepts of *ELS1* and *ELS2* are represented in DL as:

ELS1

`Unfamiliar ≡ learner ⊓ ∀ hasGoal.Goal ⊓ ≤2 hasExperience.Experience`
`Familiar ≡ learner ⊓ ∀ hasGoal.Goal ⊓ >2 hasExperience.Experience`

ELS2

`Beginner ≡ Apprentice ⊓ ∀ hasInterest.Interest ⊓ ∀ studiedCourse.Course ⊓ ≤2 studiedCourse`
`Standard ≡ Apprentice ⊓ ∀ hasInterest.Interest ⊓ ∀ studiedCourse.Course ⊓ ≥2 studiedCourse ⊓ ≤6 studiedCourse`
`Expert ≡ Apprentice ⊓ ∀ hasInterest.Interest ⊓ ∀ studiedCourse.Course ⊓ ≥6 studiedCourse`

DL reasoning should not infer relationships based on previous descriptions of Familiar/Beginner and Unfamiliar/Medium, Unfamiliar/Expert because used terms in descriptions are different. This is a limitation in using only DL as done in the related works.

Now we consider the Topic Maps-based interoperability presented in section 4. That is, `Apprentice` is a `Learner`, `Experience` is a `Course`, `Interest` is a `Goal`, and `hasExperience` and `studiedCourse` are equivalent to `learnedCourse`. Then, preceding descriptions become as follow:

ELS1

Unfamiliar ≡ Learner ⊓ ∀ hasGoal.Goal ⊓ ∀ learnedCourse.Course ⊓ ≤2 learnedCourse

Familiar ≡ Learner ⊓ ∀ hasGoal.Goal ⊓ ∀ learnedCourse.Course ⊓ >2 learnedCourse

ELS2

Beginner ≡ Learner ⊓ ∀ hasGoal.Goal ⊓ ∀ learnedCourse.Course ⊓ ≤2 learnedCourse

Standard ≡ Apprentice ⊓ ∀ hasGoal. Goal ⊓ ∀ learnedCourse.Course ⊓ >2 learnedCourse ⊓ ≤6 learnedCourse

Expert ≡ Apprentice ⊓ ∀ hasGoal. Goal ⊓ ∀ learnedCourse.Course ⊓ >6 learnedCourse

So, DL reasoning infers following knowledge:

Beginner ≡ Familiar, Standard ⊑ Unfamiliar, Expert ⊑ Familiar, Familiar ⊑ Standard ⊔ Expert, Familiar ≡ Standard ⊔ Expert

Inferring this kind of knowledge makes connection between heterogeneous e-learning systems using hard-defined constraints. So, beginners in *ELS2* are taken as unfamiliar learner in *ELS1*. Furthermore, e-learning systems can exchange adaptation rules defined for each category of learners. For example, adaptation rules defined in *ELS1* for unfamiliar learners may be reused in *ELS2* for beginner learners.

5 Conclusion

Use jointly Topic Maps and Description Logics allows to perform reasoning on distributed knowledge. However, translating knowledge from Topic Maps to DL is not always possible. Topic Maps formalism are very expressive. However, Description Logics are based on reasoning algorithms defined on a set of constructors. That is, the expressiveness of Description Logics is restricted by the reasoning algorithms. It represents a paradox between the two formalisms. In our future work, we try to carry out the constraint language specification defined for Topic Maps. Constructors will be built to define constraints on a Topic Maps knowledge base and perform automatic reasoning.

We used in this work single ontology architecture. Multi-ontologies architecture may be studied especially in the peer-to-peer architecture. The use of Distributed Description Logics [11] to make interoperability of distributed learner profile is being studied. The experimentation of the proposed approach is also an important future work.

References

1. IEEE Learning Technology Standards Committee (LTSC), Draft Standard for Learning Technology-Public and Private Information (PAPI) for Learners – PAPI Learner. Core Features, Draft 8 (November 2001), `http://ltsc.ieee.org`
2. IMS LIP. IMS Learner Information Packagin. In: Jisc, C. (ed.). Standards briefings series (2002)

3. Conati, C., Gertner, A., Vanlehn, K.: Using Bayesian networks to manage uncertainty in student modeling. Journal of User Modeling and User-Adapted Interaction 12(4), 371–417 (2002)
4. Gonzálezl, C., Burguillo, J.C., Lamas, M.: A Qualitative Comparison of techniques for Student Modeling in Intelligent Tutoring Systems. In: 36th ASEE/IEEE Frontiers in Education Conference (2006); [10]. Klyne, G., Carroll, J.: Resource Description Framework (RDF): Concepts and Abstract Syntax, W3C Working Draft 23 (January 2003) (accessed october, 2006), http://www.w3.org/TR/rdf-concepts/
5. Brickley, D., Guha, R.V.: RDF Vocabulary Description Language 1.0: RDF Schema. W3C Working Draft (January 23, 2003)
6. Aroyo, L., Dicheva, D.: Aims: Learning and teaching support for www-based education. Int. Journal for Continuing Engineering Education and Life-long Learning (IJCEELL) 11, 152–164 (2001)
7. Smith, M.K., McGuinness, D., Volz, R., Welty, C.: Web Ontology Language (OWL): Guide Version 1.0. W3C Working Draft (November 4, 2002)
8. Denaux, R., Dimitrova, V., Aroyo, L.: Integrating open user modeling and learning content management for the semantic web. In: 10th International Conference on User Modeling (2005)
9. Dolog, P., Schaefer, M.: Learner modeling on the semantic web. In: Proc. of PerSWeb 2005 Workshop, Personalization on the Semantic Web, Edinburgh, UK (July 2005)
10. ISO/IEC 13250 Topic Maps, Dec ISO/IEC FCD (1999)
11. Borgida, A., Serafini, L.: Distributed Description Logics: Directed Domain Correspondences in Federated Information Sources. In: Confederated International Conferences DOA, CoopIS and ODBASE 2002, pp. 36–53. Springer, Heidelberg (2002)
12. Pepper, S., Moore, G.: XML Topic Maps (XTM) 1.0, TopicMaps.Org Authoring Group (2001), http://www.topicmaps.org/xtm/index.html
13. Fresse, E.: Using Topic Maps for the representation, management and discovery of knowledge, XML Europe (2000); Palais des congrès Paris (2000)
14. Baader, F., Calvanese, D., McGuiness, D., Nardi, D., Patel-Schneider, P.: Description Logic Handbook - Theory, Implementation and Applications. Cambridge university press, Cambridge (2003)
15. Cali, A., Calvanese, D., Colucci, S., Di Noia, T., Donini, F.M.: A description logic based approach for matching user profiles. In: Description Logic Workshop (2004)
16. Von Hessling, A., Kleemann, T., Sinner, A.: Semantic User Profiles and their Applications in a Mobile Environment, Fachberichte Informatik 2-, Universität Koblenz-Landau (2005)

A New Travel Time Prediction Method for Intelligent Transportation Systems*

Hyunjo Lee, Nihad Karim Chowdhury, and Jaewoo Chang

Department of Computer Engineering, Chonbuk National University,
Chonju, Chonbuk 561-756, South Korea
{o2near,jwchang}@chonbuk.ac.kr, nihad@dblab.chonbuk.ac.kr

Abstract. Travel time prediction is an indispensable for numerous intelligent transportation systems (ITS) including advanced traveler information systems. The main purpose of this research is to develop a dynamic travel time prediction model for road networks. In this paper we propose a new method to predict travel times using Naïve Bayesian Classification (NBC) model because Naïve Bayesian Classification has exhibited high accuracy and speed when applied to large databases. Our proposed prediction algorithm is also scalable to road networks with arbitrary travel routes. In addition, we compare the proposed method with such prediction methods as link-based prediction model and time-varying coefficient linear regression model. It is shown from our experiment that NBC predictor can reduce mean absolute relative error significantly rather than the other predictors. We illustrate the practicability of applying NBC in travel time prediction and prove that NBC is suitable and performs well for traffic data analysis.

Keywords: Intelligent transportation system (ITS), travel time prediction, Naïve Bayesian Classification, linear regression.

1 Introduction

Numerous intelligent transportation system (ITS) applications, such as trip planning and dynamic route guidance systems, accurate estimation of travel times, are more crucial for traffic data analysis. Effective travel time prediction and dynamic route guidance system can assist travelers to better adjust traveler schedule [1]. Travel time prediction is also becoming increasingly important with the development of advanced travelers information systems (ATIS) [2]. In these applications, travelers want an accurate prediction of travel time from an origin to arrive at a destination. Travel time prediction based on vehicle speed and traffic flow is extremely sensitive to external event like weather condition and traffic incident [2]. Predicting travel time seems to be complex and difficult due to these unavoidable circumstances. On the other hand,

* This work is financially supported by the Ministry of Education and Human Resources Development(MOE), the Ministry of Commerce, Industry and Energy(MOCIE) and the Ministry of Labor(MOLAB) though the fostering project of the Lab of Excellency. It is also supported by the second stage of Brain Korea 21 Project.

I. Lovrek, R.J. Howlett, and L.C. Jain (Eds.): KES 2008, Part I, LNAI 5177, pp. 473–483, 2008.

traffic flow on a designed road depends on daily, weekly and occasional event. Like, daily features distinguish morning and evening rush hour traffic. The time-varying feature to traffic flow is the key concept to estimate accurate travel time.

The main objective of this research is to develop a travel time prediction algorithm that can produce reliable and accurate travel time predictions. For example, travel time prediction algorithm should be applicable on different routes. The reason is that the reliability of a model also depends on the reliability of the real world, and in particular of the people and systems that operate these algorithm. In this paper, we propose a new method for predicting travel time using Naïve Bayesian Classification (NBC). The main idea of NBC method is that based on historical traffic data it will give probable velocity label for any road segment. First, user defines an origin with start time and destination. By using Naïve Bayesian classification we can find high probable velocity label for initial road segment. Then we measure end time for initial road segment and this end time becomes start time of next road segment. Finally using successive operation to each road segment we can measure approximate travel time from origin to destination. The developed algorithm is simple and its performance in terms of prediction accuracy is satisfactory. The result is considered to be superior to forecasting based on linear regression [3] and link based method [1]. The rest of the paper is organized as follows: We introduce some related research in Section 2. Section 3 gives an outline of the characteristics of our proposed method. A concise experimental evaluation is presented in Section 4.Finally, Section 5 concludes with a discussion about future research direction.

2 Related Work

Accurate predictions of travel times on road networks are essential for effective dynamic route guidance system. Currently, travel time predictors have emerged as an active and intense research area. Numerous researches have focused on the accurate prediction of travel time of road networks. The methods employed include artificial neural networks [4] [5] as the non-linear predictors. Broadly used linear models include multivariate linear regression and tree method [6], time-varying coefficient linear regression [7], ARIMA models [8], linear models of system variables [9] and time-varying coefficient linear regression as a component predictor [3]. A linear predictor consisting of a linear combination of the current times and the historical means of the travel times is proposed by Rice et al [9]. The authors compared their linear predictor against several other prediction methods such as historical mean, principle components and nearest neighbors. Kwon et al [6] focused on linear regression method. Their proposed predictor is a linear combination of the current and historical information. They found that their predictor outperforms other predictors like tree method and neural network. Zhang et al [7] proposed a method to predict freeway travel times using a linear model in which the coefficients vary as smooth functions of the departure time. In most exiting research focused on link travel time prediction [1], it is assumed that path travel time is the addition of the travel times on its consisting links. Chen et al [1] focused on two approach path-based and link based. For path based method, probe vehicle's passing is only recorded at the beginning and the end of the path. The average probe travel time is used as the real-time observation of travel time at each time period. In link-based method, record travel times on desired links for those probes entering the links and get the average probe travel time for each link. Final travel time is calculated by adding travel times on all consisting links. Because prior researches focus on implicit characteristics that route must be predefined, their prediction algorithms trained on particular route, regardless of other routes in road networks. Herein little research reflect on predicting travel time for arbitrary travel routes [3]. In their approach, at first they partition the freeway into short segments and observe future travel time on every segment. For a given query route, they predict travel time for all segments that comprise the query route. On the other hand, the approach takes more storage and computation time due to two-step computation.

3 New Travel Time Prediction Method

In this section, we propose a new method for predicting travel time from historical traffic data using naïve Bayesian classification. Bayesian classification is a class of simple probabilistic algorithm which apply Bayes' theorem in order to learn the underlying probability distribution of the data. This can predict class membership probabilities, such as the probability that a given tuple belongs to particular class. A simple Bayesian classifier is known as the Naïve Bayesian classifier [10]. Initially user defines an origin with start time and destination. A route may comprise many segments from origin to destination. First, we introduce our approach for measuring first road segment travel time using naïve Bayesian classification. Then this first road segment's end time becomes start time of next road segment. Finally using successive iteration we can measure approximate travel time from origin to destination.

With the same road network, there can be different road environment for running vehicles on the different time periods of a day. So, the pattern of trajectories of moving objects varies according to the change of time duration of a day. Considering the traffic condition of South Korea, we divide whole day time into nine groups as shown in Table 1.

Table 1. Group definition

Start _Time _Range	Group	Description
06:01 ~ 10:00	1	Morning Rush Hour
10:01 ~ 11:00	2	Morning
11:01 ~ 12:00	3	Early Noon
12:01 ~ 14:00	4	Lunch Time
14:01 ~16:00	5	After Noon
16:01 ~ 18:00	6	Evening
18:01 ~ 22:00	7	Evening Rush Hour
22:01 ~ 00:00	8	Night
00:01 ~ 06:00	9	Late Night

In a traffic data, let $g_1, g_2,..g_i,..g_n$ (where $1 \leq i \leq n$) be the groups in a tuple which are formed by applying timing constraint. A tuple is supported by g_i and it depends on vehicle start time. For example: If a vehicle starts from any road segment during the time interval 10:01 to 11:00, its *Group* will be 2. Let *Velocity _Class* = {*VB, B, F*} be the set of literals called *Velocity_ Class* attribute. The attribute value *VB, B* and *F* means *Very Busy, Busy* and *Free* respectively. Like, if velocity is varying between 0 ~ 1 km/min then *Velocity _ Class* will be *VB* (Very Busy). Complete *Velocity_ Class* on a road segment is shown in Table 2.

Table 2. Velocity class definition

Velocity Range(km/minute)	Velocity _Class
0 ~ 1	VB
1.1~ 2	B
> 2	F

Table 3. Sample traffic data

Vehicle_ID	Road_ID	Group	Start _Time	End _Time	Velocity (km/min)	Velocity _Class
1	1	2	10:01	10:11	1.00	VB
2	1	2	10:05	10:17	0.83	VB
3	1	2	10:05	10:20	0.66	VB
4	1	3	13:00	13:02	5.00	F
5	1	2	10:01	10:15	0.70	VB
6	1	2	10:00	10:04	2.25	F
7	1	3	13:05	13:08	3.33	F
8	1	3	13:06	13:08	5.00	F
9	1	9	6:01	6:07	1.66	B
10	1	2	10:30	10:36	1.66	B

In our traffic data, each tuple has seven attributes. Two attributes *Start _Time* and *End _Time* indicate the period during which a vehicle travels on a particular road segment. It is assumed that, as well as *Start _Time* and *End _Time*, each record has an attribute, *Vehicle _ID*, *Road _ID*, *Group*, *Velocity* and *Velocity _Class*. The value stored in *Group* depends on *Start _Time*. We can calculate *Velocity* by dividing length of road segment and time difference between *Start _Time* and *End _Time.* The value stored in attribute *Velocity _Class* has a relationship with attribute *Velocity.* In our example, we consider all road segment length is 10 km. Table 3 is shown us complete scenario of historical traffic data for any road segment.

3.1 Velocity Class Prior Probability Measure

To find approximate travel time for any road segment, we first introduce *Velocity Class Prior Probability Measure (C)* algorithm. This algorithm computes prior probability for each velocity class. Total number of tuples for any road segment is set to *tt* in line 1. In line 10, *PMc[c]* measure prior probability of each velocity class. Total count of each velocity class (*c. count)* is measured in line 5. *Contain (t, c)* is a procedure which determines whether tuple *t* contains velocity class *c*. According to Algorithm *Velocity Class Prior Probability Measure (C)* and following Table 3, we find (*c.count*) total number of each velocity class. So that we can measure *Total (VB) = 4, Total (B) =* 2 and *Total (F) = 4* for each velocity class like *Very Busy*, *Busy* and *Free* respectively. From Table 3, in our sample traffic data we see that traffic data consist of 10 tuples. Next we evaluate prior probability of each velocity class (*PMc[c]*) in line 10. Following algorithm *Velocity Class Prior Probability Measure (C)* and using Table 3, we find prior probability for each velocity class like *Probability (VB)=0.4, Probability (B)=0.2* and *Probability(F)=0.4*

```
Velocity Class Prior Probability Measure (C){
Input: Velocity Class attributes c∈ C
Output: Velocity Class Prior Probability (PMc) and Velocity Class
        Count(c .count)
        1)   Total Tuple = tt;
        2)   for each class attribute c∈ C do
        3)      for each tuple t in database do
        4)         if Contain(t ,c) then
        5)            c. count ++;
        6)         end if
        7)      end for each tuple
        8)   end for each c
        9)   for each class attribute c∈ C do
        10)  PMc[c] = c. count /tt;
        11)  end for each class
        12)  return PMc; }
Procedure Contain (t, c){
        13)  Let S = c | c∈ C
        14)  If S ⊆t then return 1 else return 0;
        15)  end if }
```

Fig. 1. Velocity Class Prior Probability Measure Algorithm

3.2 Velocity Class Posterior Probability Measure

The algorithm is used to find posterior probability of each velocity class for any road segment with given start time, as shown in Fig. 2. Before this we need to measure conditional probability of each velocity class. This algorithm is combined approach to compute velocity class posterior probability as well as conditional probability and maximize velocity class. User's desired road segment id with start time group is given in line 1.Then we compute posterior probability of each velocity class using sample traffic data as shown in Table 3. Prior to calculate posterior probability, we need to measure velocity class conditional probability for desired road segment with start time group. In algorithm *Velocity Class Posterior Probability Measure(X, C), Mxc* (line 8) is used to compute velocity class conditional probability. Assume that attributes are conditionally independent for a given velocity class. So that in line 17, *CPMXc* measures class conditional probabilities for all attributes that we want to classify. To classify a tuple, *PMXc* (line 19) computes the posterior probability for each velocity class. As for example, a user start travel from *Road _ID 1* at 10:05. According to Table 1, start time group will be 2. So the tuple we wish to classify is X= (*Road _ID =1, Group* =2).

Considering our sample traffic data shown in Table 3 and using line 8 in Fig. 2, the result of each velocity class conditional probability is stored in *Mxc*. Thus, we measure following conditional probabilities like, P (*Road _ID*=1| *Class*=VB) =1, P (*Road _ID*=1| *Class*=B) = 1, P(*Road_ID*=1|*Class*=F)=1, P (*Group*=2| *Class*=VB) =1, P (*Group*=2| *Class*=B) = 0.50, P (*Group*=2| *Class*=F) = 0.25.

```
Velocity Class Posterior Probability Measure(X, C){
Input: Road Segment Ids with Start Time Group and Velocity Class
       attribute
Output: Maximized Posterior Probability of Velocity Class attribute
        1)   X={ x1, x2 | x1 ∈ Road _ Segement _ List  and x2 ∈ Group}
        2)   for each c ∈ C do
        3)      for each x ∈ X do
        4)          Mxc=∅;
        5)          for each t in Database do
        6)              if (c ⊆ t && x ⊆ t ) then
        7)              xc. count ++;
        8)              Mxc = xc. count/c. count;
        9)              end if
        10)         end for each t
        11)     end for each x
        12)  end for each c
        13)  for each c ∈ C do
        14)      PMc = Velocity Class Prior Probability Measure (c);
        15)      CPMXc =1;
        16)      for each x ∈ X do
        17)         CPMXc *=Mxc ;
        18)      end for each x
        19)         PMXc = CPMXc * PMc;
        20)  end for each c
        21)  High _ Velocity _Class = Maximize Velocity Class (PMXc);
        22)  return High _ Velocity _Class; }
```

Fig. 2. Velocity Class Posterior Probability Measure Algorithm

According to line 17 in Fig. 2, now we find complete conditional probability of each velocity class. Using the above conditional probabilities, we obtain following complete conditional probability of each velocity class.

$$P\ (X\ |Class{=}VB) = P\ (Road\ \_ID{=}1|Class{=}VB) \times P\ (Group{=}2|Class{=}VB) = 1$$
$$P\ (X\ |Class{=}B) = P\ (Road\ \_ID{=}1|Class{=}B) \times P\ (Group{=}2|Class{=}B) = 0.50$$
$$P\ (X\ |Class{=}F) = P\ (Road\ \_ID{=}1|Class{=}F) \times P\ (Group{=}2|Class{=}F) = 0.25$$

To find the posterior probability of velocity class attribute, we have measured posterior probability (*PMXc*) in line 19. Before this, *PMc* (line 14) is used to calculate velocity class prior probability. Those are P (X |*Class*=VB) × P (*Class*=VB) = 0.40, P (X |*Class*=B) × P (*Class*=B) = 0.10, P (X |*Class*=F) × P (*Class*=F) = 0.10.

Once posterior probability (*PMXc*) in line 19 is calculated, in addition we need to compute high probable velocity class. *High _Velocity _Class* denotes high probable velocity class is found at this step (line 21). Considering our previous example, user's start road id is 1 with start time 10:05. After calculating posterior probability of each velocity class, we found that *Road _ID 1* will be very busy and its velocity varying 0 ~ 1 km/min at 10:05. The posterior probability of velocity class is used for relative comparison between velocity classes posterior probability.

3.3 End Time Measure

The end time calculation is carried out according to algorithm *End Time Measure (x, c)* is shown in Fig. 3. Once high probable velocity class of any road segment is evaluated, we need to measure average velocity based on high probable velocity class during given start time group. It can be seen that, after measuring average velocity on a road segment during given time group, it is possible to find end time by dividing road segment length and average velocity. The value of average velocity and end time on a road segment is stored in *Velocity* and *End _Time* respectively (lines 4, 5). As we describe earlier, all road segment distance is 10 km. Average velocity and approximate travel time in *Road _ID 1* during *Group 2* will be 0.78 km/min, 13 minutes respectively. User reaches end of *Road _ID 1* at 10:18. This end time becomes start time for next road segment.

```
End Time Measure (x, c){
Input: Road Segment Length and Highest Probable Velocity Class
Output: End time of a Road Segment
    1)  C_k = {c_1, c_2, .... c_k};
    2)  for (i =1; i <= k; i++) do
    3)     if (c == C_i ) then
    4)     Velocity =C_i.average _ velocity ;
    5)     End _Time = x.length / Velocity ;
    6)     end if
    7)  end for
    8)  return End _Time }
```

Fig. 3. End Time Measure Algorithm

3.4 Travel Time Measure

Algorithm *Travel Time Prediction (Start_ Time, Road_ Segment _List, C)* calculates travel time for every road segment according to high probable velocity class obtained from *Velocity Class Posterior Probability Measure (X, C)* procedure. At first user declares desired route with start time. A route may comprise many road segments. We convert start time to corresponding group in line 1. High probable velocity class of any road segment is measured in line 5. Using this velocity class we can find approximate end time of that road segment (line 6). Note that this end time becomes

```
Travel Time Prediction (Start_ Time, Road_ Segment _List, C){
Input: Start_ Time gives start time for initial road segment.
       Road _Segment _List contains all road segments from origin to
       destination. C contains all Velocity Class
Output: Display predicted travel time for every road segment from origin
       to destination
    1)  Group= Convert Start_ Time to Group;
    2)  for (i=0; Road_ Segment _List!=NULL ; i++) do
    3)    X.road _segment _id = Road_ Segment _List .ID;
    4)    X.group = Group;
    5)    High_ Velocity_ Class
    6)          = Velocity Class Posterior Probability Measure (X,C);
    7)    End _ Time = End Time Measure (X, High _ Velocity _Class);
    8)    Group= Convert End _Time to Group;
    9)    Answer = ∪ End _Time;
    10) end for }
```

Fig. 4. Travel Time Prediction Algorithm

start time for next road segment. Finally approximate travel time is shown for any road segment in line 8. After predicting travel time of first road segment, we repeat the steps 2 to 9 for all other road segments that comprise given route.

4 Performance Analysis

4.1 Experimental Environment

We use the mean absolute relative error (MARE) [1] to quantify the performance of different predictor. As we know, MARE is the simplest and well known method for measuring overall error in travel time prediction. MARE measures the magnitude of the relative error over the desired time range. This error measurement is defined as:

$$MARE = \frac{1}{N}\sum_{t}\frac{|x(t) - x^{*}(t)|}{x(t)} \tag{1}$$

where $x(t)$ is the observation value, $x^{*}(t)$ is the predicted value and N is the number of samples.

To assess the performance of different predictors, we conduct our experiment based on synthetic and real data set. The synthetic traffic data is generated using a random trajectory generator that is developed by us. Initially we give trajectory life time and whole travel distance as input. Then the generator choose arbitrary trajectory which traverses given distance and also consist of different number of road segments. Dividing trajectory life time by number of road segments in trajectory, we can calculate average travel time for each segment. As a result, we can get various velocities for every road segment. We generate road network by using California bay data which consists of 170,000 nodes and 220,000 edges. For comparison purpose, we use first 361 days traffic data as the training set and use the last 4 days as our testing set. We apply our algorithm for two time lags, l = 30 and 60 minutes into the future with the current time $x(t)$ ranging from morning 8:00 AM to noon 2:00 PM. As discussed previously, there are *Group* attribute in traffic data that must be specified by user. For creating group list using synthetic data we consider Table 1.

For real data set, we use PNU (Pusan National University) trajectory data generator which provide us real trajectory data. This generator is based on real traffic situation in Pusan City, South Korea. For building PNU generator, they collected real traffic data by using GPS sensor. From this data, traffic pattern of Pusan city was extracted. And according to traffic pattern, generator simulates and generates trajectory data which is almost same as real data. We generate 3000 trajectories using this generator. For our experimental evaluation, we use 3 days traffic data as training set and use 2 days as our testing set. In case of real data, we consider two time lags, 30 and 60 minutes. In addition, we make our group list for real data based on Table 1.

4.2 Performance Result with Synthetic Data

The experimental results of three travel time predictors using synthetic data are shown in Fig.5 and Fig.6. We introduce our prediction algorithm based on Naïve Bayesian Classification method as NBC. Fig.5 (a) and (b) show the prediction performance of different predictors against 30 and 60 min ahead from departure time.

During the time period between 8 AM to 2 PM, NBC predictor outperforms other two predictors. In case of 30 min ahead, NBC predictor exhibits satisfactory performance during morning rush hour. Note that for 60 min ahead, NBC predictor performs very well than other predictors. Although the difference is not so large, link based method performs better than linear regression method. The results in Fig.6 shows the summarized mean absolute relative error (MARE) of different predictors for different time lag. This result shows that NBC predictor reduces MARE 42% in case of prediction 30 min ahead. For lag = 30 min, the reduction in MARE from NBC to Link Based and Linear Regression is 75% and 81% respectively. Almost same trend we observe for lag =60 min. NBC predictor reduces MARE from Link Based and Linear Regression by 58% and 65% respectively.

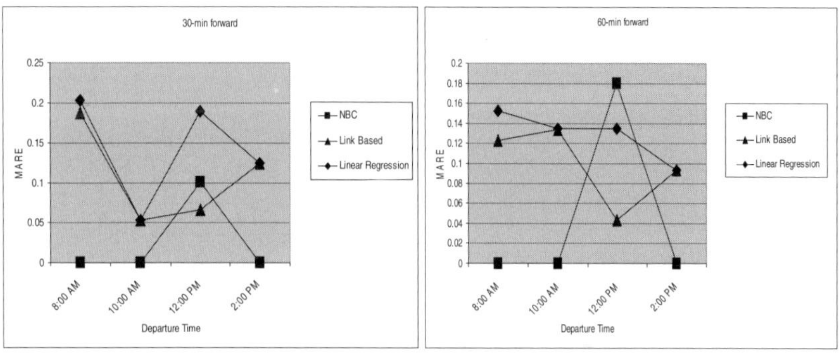

(a) Lag = 30 min (b) Lag = 60 min

Fig. 5. MARE for different travel time predictors (Synthetic Data)

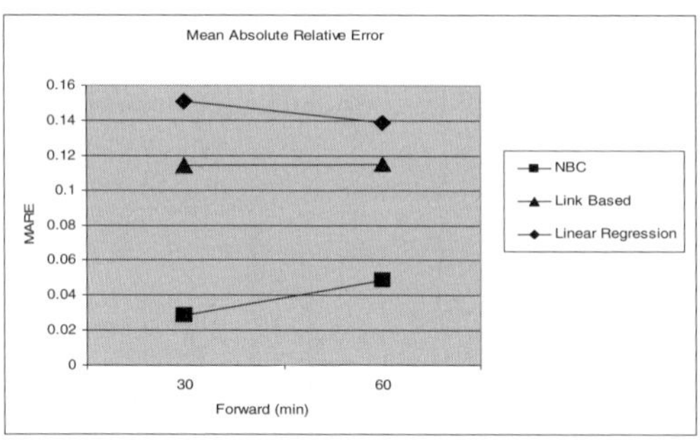

Fig. 6. MARE, averaged over day-hours between 8 AM to 2 PM

4.3 Performance Result with Real Data

We further investigate relative performance between three travel time predictors by examining real data. In this observation, we examine prediction error of three predictors during 8 AM to 6 PM. The prediction performance of three predictors against different time lag = 30 and 60 minutes is shown in Fig.7 (a) and (b) respectively. We can see that, NBC predictor performs much better than link based and linear regression over all two prediction headways. In case of NBC predictor, it is shown that prediction accuracy decreases little as headways time increases. Also we can note that NBC predictor performs well during pick hours of a day. In addition NBC predictor performs worst during non-peak hours which is least interesting. By considering real data, Fig.8 shows the summarized mean absolute relative error (MARE) of different predictors for different time lag. For lag = 30 min, MARE of NBC, Link Based and Linear Regression are 0.00831, 0.00927 and 0.01269 respectively. Thus, NBC reduces MARE from link based by 10%. In case of linear regression, MARE reduction is 35%. A similar trend is also observed for lag =60 min. In addition, NBC predictor reduces MARE from link based and linear regression by 9% and 26% respectively for lag = 60 min. So, finally we can say that NBC model is substantially better than Link Based and Linear Regression model. In addition, NBC model is practically applicable for road networks with arbitrary routes.

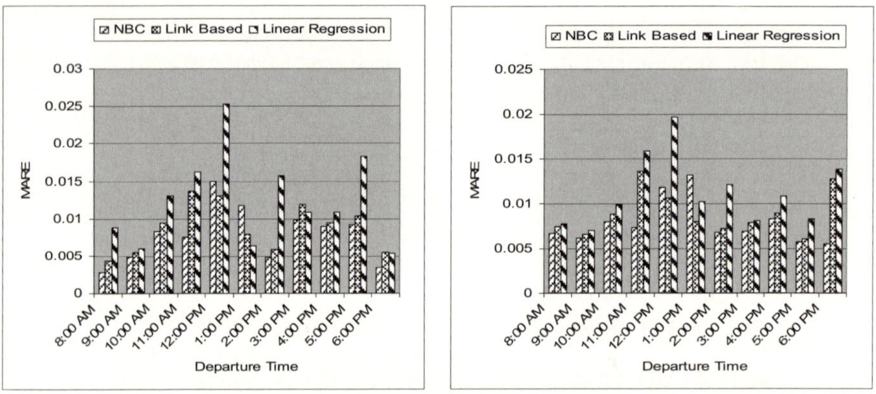

(a) Lag = 30 min (b) Lag = 60min

(a) Lag = 30 min (b) Lag = 60min

Fig. 7. MARE, for different travel time predictors (Real Data)

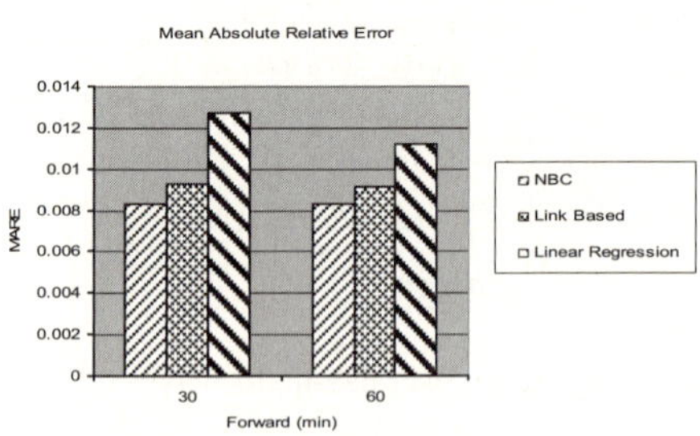

Fig. 8. MARE, averaged over day-hours between 8 AM to 6 PM

5 Conclusions

In this paper we proposed an efficient and scalable method for predicting travel time with arbitrary routes in road network. Though there is much research for predicting travel time in road networks, but there exists little research for travel time prediction considering arbitrary routes. Naïve Bayesian Classifier is a well known statistical classifier. The advantage of naïve Bayesian classification is that it requires a small amount of training data to estimate the parameters necessary for classification. Despite its simplicity, naïve Bayesian classification can often outperform more sophisticated classification methods. Our proposed algorithm also demonstrated feasibility of naïve Bayesian classification in traffic data analysis. Two different kinds of travel time prediction methods were presented and tested using two types of data sets, with impressive results. As our future work, we will extend our system considering not only day time but also week days. Thus a user can ask what will be approximate travel time for a particular route during rush hour on weekly holidays. In addition, we will make a performance evaluation of our system with various methods including ARIMA[8].

Acknowledgments

We would like to thank Prof. Ki-Joune Li for providing us with the PNU (Pusan National University) trajectory data generator.

References

1. Chun-Hsin, W., Chia-Chen, W., Da-Chun, S., Ming-Hua, C., Jan-Ming, H.: Travel Time Prediction with Support Vector Regression. In: Proceedings of IEEE Intelligent Transportation Systems Conference. (2003)
2. Chen, M., Chien, S.: Dynamic freeway travel time prediction using probe vehicle data: Link-based vs. Path-based. J. of Transportation Research Record, TRB Paper N o. 01-2887, Washington, D.C (2001)
3. Kwon, J., Petty, K.: A travel time prediction algorithm scalable to freeway networks with many nodes with arbitrary travel routes. In: Transportation Research Board 84th Annual Meeting, Washington, D.C (2005)
4. Park, D., Rilett, L.: Forecasting multiple-period freeway link travel times using modular neural networks. J. of Transportation Research Record 1617, 163–170 (1998)
5. Park, D., Rilett, L.: Spectral basis neural networks for real-time travel time forecasting. J. of Transport Engineering 125(6), 515–523 (1999)
6. Kwon, J., Coifman, B., Bickel, P.J.: Day-to-day travel time trends and travel time prediction from loop detector data. J. of Transportation Research Record, No. 1717, TRB, National Research Council, Washington, D.C., pp. 120–129 (2000)
7. Zhang, X., Rice, J.: Short-Term Travel Time Prediction. Transportation Research Part C 11, 187–210 (2003)
8. Van der Voort, M., Dougherty, M., Watson, S.: Combining KOHONEN maps with ARIMA time series models to forecast traffic flow. Transportation Research Part C 4, 307–318 (1996)
9. Rice, J., Van Zwet, E.: A simple and effective method for predicting travel times on freeways. IEEE Trans. Intelligent Transport Systems 5(3), 200–207 (2004)
10. Han, J., Kamber, M.: Data Mining: Concepts and techniques, 2nd edn. Morgan Kaufmann, San Francisco (2006)

Measuring Sequence Similarity Trough Many-to-Many Frequent Correlations

Gianluigi Greco and Giorgio Terracina

Dipartimento di Matematica, Università della Calabria,
I-87036 Rende (CS), Italy
{ggreco,terracina}@mat.unical.it

Abstract. Comparing pairs of sequences is a problem emerging in several application areas (ranging from molecular biology, to signal processing, text retrieval, and intrusion detection, just to cite a few) and important results have been achieved through the years. In fact, most of the algorithms in the literature rely on the assumption that matching symbols (or at least a substitution schema among them) are known in advance. This paper opens the way to a more involved mechanism for sequence comparison, where determining the best substitution schema is also part of the matching problem. The basic idea is that any symbol of one sequence can be correlated with many symbols of the other sequence, provided each correlation frequently occurs over the various positions. The approach fits a variety of problems difficult to be handled with classical techniques, particularly where strings to be matched are defined over different alphabets.

1 Introduction

Assessing the similarity between pairs of sequences/strings is a problem that emerges (possibly with different forms and variants related to the distance metric being adopted) in several application areas, ranging from molecular biology, to signal processing, text retrieval, and intrusion detection, just to cite a few. Indeed, defining similarity scores is a pre-requisite for important data mining tasks, in particular for clustering applications.

In some contexts, distance metrics are meant to measure the similarity of pairs of sequences in terms of their mutual (mis)-matches. This is the case, for instance, of the *edit distance*, which evaluates the minimum number of insertions, deletions, and substitutions needed to transform one sequence into the other, and which has found an impressive number of applications in bioinformatics, e.g., in comparing DNA sequences.

In other contexts, especially when the sequences to be compared are defined over different alphabets, the focus is instead on finding correlations between symbols, rather than exact matches. The most noticeable example therein is that of the *parameterized pattern matching* problem, which was firstly considered in [1] to find sections of code in a software system that are the same except for a systematic substitution of parameters. For instance, according to this approach, the sequence `QQQQWWWW` (where `Q` and `W` are parameters) would exactly coincide with `AAAABBBB`, given the possibility of systematically substituting `Q` with `A`, and `W` with `B`.

In the last few years, several techniques for parameterized pattern matching have been proposed in the literature, mainly relying on the assumptions that *(i)* the sequences

I. Lovrek, R.J. Howlett, and L.C. Jain (Eds.): KES 2008, Part I, LNAI 5177, pp. 484–492, 2008.

resulting from the substitutions should exactly match, and that *(ii)* each parameter has to be univocally mapped onto one symbol of the other sequence. As a matter of facts, these limitations are often undesirable and more flexibility in alignment methods is welcome.

For instance, in the context of content-based retrieval of music data, an important problem is to characterize the thematic features of a given music object, which can then be used to answer advanced queries such as "retrieve and cluster all songs similar to" (cf. [5]). However, it is well-known that classical (even approximate) string matching algorithms are not suitable to identify similarities between music samples, in the cases where a given melody is reproduced trough different keys over the song and, hence, where there is no unique matching for the given template melody. In this application context, the ability of discovering frequent patterns over different alphabets might be very beneficial, given its potential ability of singling out approximatively (cf. *(i)* above) repeated melodies over different note sequences (cf. *(ii)*).

The aim of this paper is precisely to move from standard parameterized pattern matching towards more flexible kinds of distance metrics, by introducing a concept of similarity between pair of sequences based on the discovery of *many-to-many frequent* correlations among symbols. The idea is that any symbol of one sequence can be correlated with many symbols of the other sequence, provided each correlation frequently occurs over the various positions.

The similarity score of two given sequences can then be defined as the maximum number of frequent many-to-many correlations occurring over all the possible alignments. Moreover, depending on the application needs, one may desire to prefer alignments with possibly few, but very frequently correlated symbols rather than alignments with many, but less frequently correlated symbols. This paper will investigate these issues. In more detail:

- ▶ This new kind of alignment is formalized in Section 2, and its computational complexity is investigated in Section 3. The result is bad news, since we show that (differently from classical alignment formulations) the problem is NP-hard, except for some trivial cases.
- ▶ Motivated by the above intractability, we then devise a pragmatic approach that is based on a dynamic programming scheme. The approach would exactly solve the problem, in the case where for each pair of symbols a_1 and a_2 over the input strings, the total number of correlations $\#(a_1, a_2)$ in an optimal alignment were known in advance. Since this is not the case in general, a heuristic technique for their estimation is proposed in Section 4.
- ▶ To cope with scenarios where very frequent correlations are more desirable, suitable extensions of the many-to-many alignment problems are discussed.
- ▶ All the algorithms and techniques discussed in the paper have been implemented and a number of tests have been performed in order to assess their efficiency and effectiveness. A report of the results of this experimental activity is illustrated in Section 5.

It is worth pointing out that our approach is not intended to substitute classical alignment approaches in those cases where a scoring schema is clear and/or input strings are defined over the same alphabet. To the contrary, our proposal should be seen as an alternative for those cases where classical approaches are difficult to be applied or they provide unsatisfactory results.

2 Alignments over Different Alphabets

Let Σ_1 and Σ_2 be two possibly overlapping alphabets, and let s_1 and s_2 be two strings (sequences) over Σ_1 and Σ_2, respectively. In the following, the *length* of s_i ($i \in \{1,2\}$), i.e., the number of symbols in it, will be denoted by $len(s_i)$; moreover, for each position $1 \leq j \leq len(s_i)$, the j-th symbol of s_i will be identified by $s_i[j]$, the prefix of s_i composed of its first j symbols will be denoted as s_i^j, whereas the suffix of s_i starting from j will be identified by ${}^j s_i$. Let '$-$' be a symbol not in $\Sigma_1 \cup \Sigma_2$. Then, a string $\bar{s}_i$ over $\Sigma_i \cup \{-\}$ is a *transposition* of s_i, with $i \in \{1,2\}$, if s_i can be obtained by deleting from $\bar{s}_i$ all the occurrences of '$-$'. The set of all the possible transpositions of s_i is denoted by $tr(s_i)$.

An alignment for the strings s_1 and s_2 is a pair $\langle \bar{s}_1, \bar{s}_2 \rangle$ where $\bar{s}_1 \in tr(s_1)$, $\bar{s}_2 \in tr(s_2)$ for which $\nexists\, i$ such that $\bar{s}_1[i] = \bar{s}_2[i] =$'$-$', and where $len(\bar{s}_1) = len(\bar{s}_2)$. For an alignment $\langle \bar{s}_1, \bar{s}_2 \rangle$ and for a pair of symbols $a_1 \in \Sigma_1$ and $a_2 \in \Sigma_2$, let $\#_{\langle \bar{s}_1, \bar{s}_2 \rangle}(a_1, a_2)$—shortly $\#(a_1, a_2)$, when the underlying alignment is clear from the context—denote the cardinality of the set $\{1 \leq i \leq \bar{s}_1 \mid \bar{s}_1[i] = a_1, \bar{s}_2[i] = a_2\}$, i.e., the *correlation* between a_1 and a_2 measured as the number of positions over which $\bar{s}_1$ and $\bar{s}_2$ match with symbols a_1 and a_2, respectively. For any natural number $\kappa \geq 0$, we say that two symbols a_1 and a_2 are *κ-correlated* in the alignment $\langle \bar{s}_1, \bar{s}_2 \rangle$, if $\#_{\langle \bar{s}_1, \bar{s}_2 \rangle}(a_1, a_2) > \kappa$. In fact, we are interested in computing alignments between strings that maximize the total number of κ-correlations, as formalized below.

Definition 1 (Many-to-many κ-alignments). Let $\gamma_\kappa : \mathbb{N} \mapsto \{0,1\}$ be the step function such that $\gamma_\kappa(x) = 1$ if $x > \kappa$, and $\gamma_\kappa(x) = 0$ otherwise. Then, the best *many-to-may κ-alignment* for s_1 and s_2 is the tuple $\langle \bar{s}_1^*, \bar{s}_2^* \rangle$ such that:

$$\langle \bar{s}_1^*, \bar{s}_2^* \rangle = arg \max_{\langle \bar{s}_1, \bar{s}_2 \rangle} \sum_{a_1 \in \Sigma_1, a_2 \in \Sigma_2} \Big(\gamma_\kappa(\#_{\langle \bar{s}_1, \bar{s}_2 \rangle}(a_1, a_2)) \times \#_{\langle \bar{s}_1, \bar{s}_2 \rangle}(a_1, a_2) \Big). \quad \square$$

Note that differently from earlier alignment problems studied in the literature, Definition 1 deals with scenarios where: (1) we take care of many-to-many correlations, i.e., in the scoring we consider the correlations of each symbol $a_1 \in \Sigma_1$ with all the other symbols of Σ_2, and viceversa; and, (2) we look for *frequent* (w.r.t. κ) correlations only, thereby ignoring pairings that might have happened by chance.

Moreover, it is worthwhile noting that it makes sense to deal with $\kappa \geq 1$ only. In fact, the best many-to-many 0-alignment would just depend on the length of the input strings s_1 and s_2, and its score would coincide with $\min\{len(s_1), len(s_2)\}$.

3 Solving the Alignment Problem

In this section, we investigate on the problem of computing the best many-to-many κ-alignment (short: $\texttt{Compute-BA}_\kappa$) for an arbitrary pair of input strings. And, we start with some bad news. Indeed, the number of possible (reasonable) alignments of two strings of lengths m and n is $O\left(\binom{m+n}{n}\right)$ [2], and we next show that computing the best many-to-may κ-alignment precisely requires an exhaustive scan of such a large search space (unless P=NP).

Theorem 1. $\texttt{Compute-BA}_\kappa$ *is* NP-*hard.*[1]

This is rather a surprising result given that intractability in the literature mainly emerges when the alignment is extended to simultaneously hold over an arbitrary number of sequences. Motivated by this result, we shall investigate on pragmatic approaches to face $\texttt{Compute-BA}_\kappa$.

3.1 A Heuristic Solution Scheme

Let s_1 and s_2 be two strings over the alphabets Σ_1 and Σ_2, respectively. Consider the best many-to-many κ-alignment $\langle \bar{s}_1^*, \bar{s}_2^* \rangle$ for s_1 and s_2 and the following function, which is meant to denote the likelihood of a correlation between a_1 and a_2, as for it can be derived from $\langle \bar{s}_1^*, \bar{s}_2^* \rangle$: $\Delta_\kappa(a_1, a_2) = \gamma_\kappa(\#_{\langle \bar{s}_1^*, \bar{s}_2^* \rangle}(a_1, a_2))$. Intuitively, this function indicates whether a_1 and a_2 are κ-correlated. In particular, we may check that the score associated with the best many-to-many κ-alignment precisely coincides with the sum of its associated likelihood scores:

$$\sum_{k \in \{1 \ldots len(\bar{s}_1^*)\}} \Delta_\kappa(\bar{s}_1^*[k], \bar{s}_2^*[k]) = \max_{\langle \bar{s}_1, \bar{s}_2 \rangle} \sum_{a_1 \in \Sigma_1, a_2 \in \Sigma_2} \gamma_\kappa(\#_{\langle \bar{s}_1, \bar{s}_2 \rangle}(a_1, a_2)) \times \#_{\langle \bar{s}_1, \bar{s}_2 \rangle}(a_1, a_2).$$

Armed with the above observation, towards proposing a heuristic approach for computing the best alignment for s_1 and s_2, if the values of Δ_κ were known beforehand (e.g., with an oracle), we may think to solve $\texttt{Compute-BA}_\kappa$ by maximizing the sum of the likelihood scores, with some classical string alignment algorithm using scoring matrices (see, e.g, [3]).

Indeed, it is well-known that this latter problem can efficiently be solved via dynamic programming; in particular, given a pair of indexes i and j ($1 \leq i \leq len(s_1)$, $1 \leq j \leq len(s_2)$), the optimal alignment between the prefix s_1^i of s_1 and s_2^j of s_2, which maximizes the likelihood scores, can be computed just from the optimal alignments previously computed for s_1^{i-1} and s_2^{j-1}, s_1^{i-1} and s_2^j, and s_1^i and s_2^{j-1}. Actually, since Δ_κ requires the knowledge of the best many-to-many κ-alignment, this approach is not constructive. Yet, it suggests that a pragmatic way for facing $\texttt{Compute-BA}_\kappa$ might consists in approximating Δ_κ with a function $\widetilde{\Delta}_\kappa$ intended to estimate the likelihood for the pairs of symbols in $\Sigma_1 \times \Sigma_2$. In a naïve implementation, one may think of fixing beforehand the approximation $\widetilde{\Delta}_\kappa$, thereby disregarding all the knowledge that can be gained after the above dynamic algorithm has completed some intermediate alignments. Our approach is instead more subtle, and is based on an incremental estimation of $\widetilde{\Delta}_\kappa$.

Computing $\widetilde{\Delta}_\kappa$. Consider a generic step of the dynamic programming scheme, where we have to decide whether $s_1[i]$ and $s_2[j]$ have to be paired. In the affirmative case, a position k occurs in the optimal alignment $\langle \bar{s}_1^*, \bar{s}_2^* \rangle$ such that $\bar{s}_1^*[k] = s_1[i]$ and

[1] An on-line appendix with the proof is available at www.mat.unical.it/~ggreco/SM.pdf.

Input: Sequences s_1 and s_2;
Output: Estimation of $\max_{\langle \bar{s}_1, \bar{s}_2 \rangle} \sum_{i \in \{1 \ldots len(\bar{s}_1)\}} \Delta_\kappa(\bar{s}_1[i], \bar{s}_2[i])$;

Var: $C : \{0, \ldots, len(s_1)\} \times \{0, \ldots, len(s_2)\} \times \Sigma_1 \times \Sigma_2 \mapsto \mathbb{N}$, and
$M : \{0, \ldots, len(s_1)\} \times \{0, \ldots, len(s_2)\} \mapsto \mathbb{N}$;

$**\ Initialization\ **$
for each $i \in \{0, \ldots, len(s_1)\}, a_1 \in \Sigma_1, a_2 \in \Sigma_2$ **do** $C[i, 0, a_1, a_2] = 0$;
for each $j \in \{0, \ldots, len(s_2)\}, a_1 \in \Sigma_1, a_2 \in \Sigma_2$ **do** $C[0, j, a_1, a_2] = 0$;
for each $i \in \{0, \ldots, len(s_1)\}$ **do** $M[i, 0] = 0$;
for each $j \in \{0, \ldots, len(s_2)\}$ **do** $M[0, j] = 0$;

$**\ Computation\ **$
for each $i \in \{1, \ldots, len(s_1)\}$ **do**
for each $j \in \{1, \ldots, len(s_2)\}$ **do** {

$$M[i, j] = \max \begin{cases} \text{(A)}\ M[i\text{-}1, j\text{-}1] + \widetilde{\Delta}_\kappa(s_1[i], s_2[j]) \\ \text{(B)}\ M[i\text{-}1, j] \\ \text{(C)}\ M[i, j\text{-}1] \end{cases}$$

case (A): $C[i, j] = C[i\text{-}1, j\text{-}1]$;
$C[i, j, s_1[i], s_2[j]] = C[i, j, s_1[i], s_2[j]] + 1$;
case (B): $C[i, j] = C[i\text{-}1, j]$;
case (C): $C[i, j] = C[i, j\text{-}1]$;
}

Function $\widetilde{\Delta}_\kappa(s_1[i], s_2[j])$: **return** $\gamma_\kappa\ (C[i-1, j-1, s_1[i], s_2[j]]$ + *estimate* $(i, j, s_1, s_2))$;

Fig. 1. COMPUTEALIGNMENT Algorithm

$\bar{s}_2^*[k] = s_2[j]$ and, in fact, we are precisely interested in estimating the value $\Delta_\kappa(\bar{s}_1^*[k], \bar{s}_2^*[k])$. Now, note that computing $\widetilde{\Delta}_\kappa(\bar{s}_1^*[k], \bar{s}_2^*[k])$ requires computing the value $\#_{\langle \bar{s}_1^*, \bar{s}_2^* \rangle}(\bar{s}_1^*[k], \bar{s}_2^*[k]) = \#_{\langle \bar{s}_1^*, \bar{s}_2^* \rangle}(s_1[i], s_2[j])$.

Hence, the basic idea is to estimate the number of the total matches, guided by the fact that the alignment maximizing the likelihood scores for s_1^{i-1} and s_2^{j-1}, say $\langle \bar{s}_1^{i-1}, \bar{s}_2^{j-1} \rangle$, has already been computed. In practice, we may think of estimating $\#_{\langle \bar{s}_1^*, \bar{s}_2^* \rangle}(s_1[i], s_2[j])$ by: *(a)* counting the actual correlations in $\langle \bar{s}_1^{i-1}, \bar{s}_2^{j-1} \rangle$, and by *(b)* estimating the number of correlations that might occur in the remaining portions of s_1 and s_2 that have to be processed, i.e, in the strings ${}^i s_1$ and ${}^j s_2$.

Letting $\widetilde{\#}_{\langle \bar{s}_1^*, \bar{s}_2^* \rangle}(s_1[i], s_2[j])$ denote the estimation thereby computed, we eventually let: $\widetilde{\Delta}_\kappa(a_1, a_2) = \gamma_\kappa(\widetilde{\#}_{\langle \bar{s}_1^*, \bar{s}_2^* \rangle}(a_1, a_2))$.The rationale of the approach is that the more i (resp., j) becomes nearer to $len(s_1)$ (resp., $len(s_2)$), the more the estimate becomes reliable, since the known (optimal) part becomes more relevant than the estimated part.

Figure 1 reports an algorithm for facing Compute-BA$_\kappa$ that is based on the above ideas. As in a standard dynamic programming scheme, the algorithm fills, in a row-wise manner, a support matrix M in which rows correspond to symbols of s_1, columns correspond to symbols of s_2, and where each entry $M[i, j]$ stores the value of the best alignment up to positions i and j over s_1 and s_2, respectively. In more detail, the approach is articulated in two phases:

Initialization: The first row and the first column of M (index 0) are exploited for initialization purposes. Indeed, in the first phase, we set $M[0, j] = 0$ and $M[i, 0] = 0$, i.e. we initialize M in such a way that the first symbol of s_1 can ideally be matched with any symbol of s_2 and vice versa.

Computation: In the second phase, we compute the value of $M[i,j]$ by distinguishing three cases. In fact, the best alignment for s_1^i and s_2^j might be obtained by (A) aligning $s_1[i]$ and $s_2[j]$, (B) deleting $s_1[i]$ (or equivalently aligning $s_1[i]$ to '$-$'), (C) deleting $s_2[j]$ (or equivalently aligning $s_2[j]$ to '$-$').

Note that in case (A) $M[i,j] = M[i-1, j-1] + \widetilde{\triangle}_\kappa(s_1[i], s_2[j])$; in case (B) $M[i,j] = M[i\text{-}1, j]$ and in case (C) $M[i,j] = M[i, j\text{-}1]$.

Eventually, in the computation phase, we need to calculate the expression $\widetilde{\triangle}_\kappa(s_1[i], s_2[j])$ as the sum of two contributions, i.e., $\#_{\langle \bar{s}_1^{i-1}, \bar{s}_2^{j-1}\rangle}(s_1[i], s_2[j]) + \widetilde{\#}_{\langle {}^i s_1, {}^j s_2\rangle}(s_1[i], s_2[j])$.

The function *estimate* is meant to computing $\widetilde{\#}_{\langle {}^i s_1, {}^j s_2\rangle}(s_1[i], s_2[j])$, and a possible implementation for it is discussed in detail in Section 4. Instead, as for the first contribution, we note that $\#_{\langle \bar{s}_1^{i-1}, \bar{s}_2^{j-1}\rangle}(s_1[i], s_2[j])$ can efficiently be computed without scanning the current alignment, but just with the use of another support matrix.

In fact, we use a $len(s_1) \times len(s_2)$ matrix C such that each element (i, j) is associated (in principle) with a $\Sigma_1 \times \Sigma_2$ matrix which summarizes the number of matches of each pair of symbols in the optimal alignment $\langle \bar{s}_1^i, \bar{s}_2^j\rangle$. Then, $\#_{\langle \bar{s}_1^{i-1}, \bar{s}_2^{j-1}\rangle}(s_1[i], s_2[j])$ coincides with $C[i\text{-}1, j\text{-}1, s_1[i], s_2[j]]$. Actually, note that since M is filled in a row-wise manner, only two rows of each of the $\Sigma_1 \times \Sigma_2$ matrices are in fact needed.

We conclude by noticing that the algorithm computes the score associated with the alignment, and that the actual alignment can be reconstructed by backtracking through the decisions that were made by the algorithm, starting from the index (i^*, j^*) storing in M the highest correlation score. We omit details on this phase, since this is standard in string alignments.

Amplification Factor. Now that an approach for solving the basic version of the many-to-many alignment problem has been presented, we may discuss how to cope with some interesting and useful generalizations. As pointed out in the Introduction, in some cases it is desirable to give preference to very frequent alignments. Formally, consider the following function:

$$\gamma_{\alpha,\kappa}(\#(a_1, a_2)) = \begin{cases} \#(a_1, a_2)^\alpha & \text{if } \#(a_1, a_2) > \kappa \\ 0 & \text{otherwise} \end{cases}$$

The parameter α acts now as an amplification factor for the selection of the actual pairings to be done among all the possible frequently correlated symbols. Indeed, for a given value of κ, higher values of α will lead to prefer alignments with possibly few very frequently correlated symbols over many not-very frequently correlated symbols. For instance, with $\alpha = 2$, a pair of symbols correlated 10 times weights as 100 pairs, each one occurring just once in the alignment. Interestingly, the algorithm in Figure 1 remains unchanged to solve this generalized problem, provided $\gamma_\kappa(\#_{\langle \bar{s}_1^*, \bar{s}_2^*\rangle}(a_1, a_2)) \times (\#_{\langle \bar{s}_1^*, \bar{s}_2^*\rangle}(a_1, a_2))^{\alpha-1}$ is now used to estimate $\Delta_\kappa(a_1, a_2)$.

It is worth observing that our scoring scheme precisely aims at maximizing the number of relevant (i.e. frequent w.r.t. κ) matches and, consequently, frequent symbols in a string will be probably matched frequently in the optimal alignment. This is a desirable property in many application contexts, such as the analysis of communication schemas among agents or processes, etc. However, in other application contexts, such as the biological one, frequent symbols should be not overaligned. In order to address this issue, we plan to adopt in future work a different objective function and a scoring schema based on log-odds matrices [4] which positively weighs unexpected matchings and negatively weighs over-expected ones.

4 Heuristics for Counting the Number of Matches

In this section, we complete the discussion of the COMPUTEALIGNMENT Algorithm, by illustrating how $estimate(i, j, s_1, s_2)$ can actually be computed, i.e., how we can estimate the number of correlations between symbols $s_1[i] = a_1$ and $s_2[j] = a_2$ in the suffix of s_1 (resp., s_2) starting from the i-th (resp., j-th) position in the best alignment $\langle \bar{s}_1^*, \bar{s}_2^* \rangle$.

To this end, we propose an heuristic **(H)** that takes into account the amount of edit operations (insertion/deletions) required to correlate all occurrences of a_1 and a_2. In particular, the higher the number of edit operations are needed, the lower is the probability that all of these occurrences will be actually matched in $\langle \bar{s}_1^*, \bar{s}_2^* \rangle$.

The heuristic computes this number by estimating an alignment, under the hypothesis that the only relevant symbols in s_1 and s_2 are indeed a_1 and a_2. To this end, it scans linearly the two strings and just pairs the nearest symbols. After the alignment is determined, **(H)** adjusts the number $\widetilde{\#}(a_1, a_2)$ of matchings between a_1 and a_2 according to the following criteria:

(absolute mean distance- μ) Consider a symbol ${}^i s_1[x] = a_1$ matched with ${}^j s_2[y] = a_2$. We say that the value of $x - y$ is the distance between the occurrence of a_1 in ${}^i s_1$ and of a_2 in ${}^j s_2$. Note that the lower this distance is, the lower number of symbols should be deleted from the original strings in the final alignment. Therefore, the lower the (absolute value) of the mean distance is, the higher the likelihood that the estimated number of matchings is the right one will be.

(variance- σ) For a given mean distance, a low variance would result in a lower number of string edit operations to align all the matched symbols.

From the considerations above, the number of possible correlations $\widetilde{\#}(a_1, a_2)$ is normalized as $\frac{\widetilde{\#}(a_1,a_2)}{\sigma+1}\left(1 - \frac{\mu}{\max\{len(s_1), len(s_2)\}}\right)$. In particular, observe that the value of the mean distance is considered against the lengths of the involved strings; in fact, a mean distance of, say, 10 have a higher impact on strings of length 100 than on strings of length 1000. It is easy to check that values returned by **(H)** belong to the interval $[0, \widetilde{\#}(a_1, a_2)]$.

We conclude by observing that the cost of **(H)** is at most linear in the length of the suffixes under examination. However, using a support array storing for each i (resp., j) the position of the next symbol $s_1[i]$ (resp., $s_2[j]$) in its string, the cost of **(H)** reduces to the maximum number of actual occurrences of $s_1[i]$ and $s_2[j]$ in ${}^i s_1$ and ${}^j s_2$ respectively. This support array can be constructed once for all suffixes in $O(len(s_1))$ (resp., $O(len(s_2))$).

Length	% Noise	α=1	α=2
100	20%	90.79%	98.55%
100	40%	100.00%	100.00%
100	60%	51.28%	80.31%
100	80%	70.00%	89.62%
300	20%	89.17%	100.00%
300	40%	73.03%	99.15%
300	60%	82.50%	97.08%
300	80%	75.00%	100.00%
500	20%	74.75%	100.00%
500	40%	54.33%	100.00%
500	60%	62.00%	98.42%
500	80%	64.00%	100.00%
700	20%	62.14%	100.00%
700	40%	73.10%	100.00%
700	60%	53.71%	99.30%
700	80%	69.29%	100.00%
Average		69.22%	98.21%

(a)

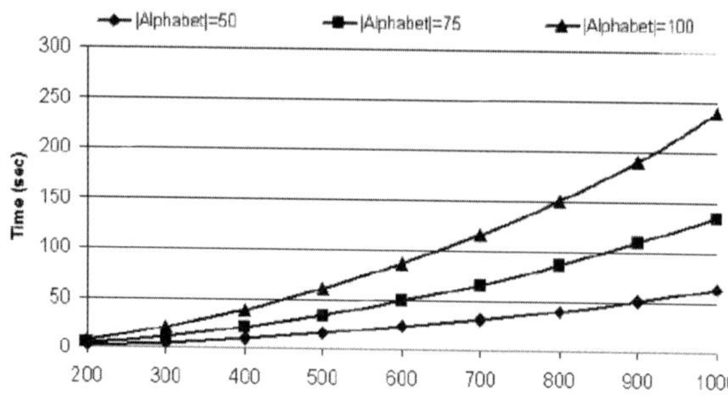

(b)

Fig. 2. Experimental Results

5 Experiments

Accuracy tests. We considered a series of synthetic data sets, generated in such a way that the optimal scores were known in advance. We have then generated various string pairs, characterized by increasing lengths and increasing percentage of noises. String lengths varied from 100 to 700 and the percentages of introduced noise have been 20%, 40%, 60% and 80% of the total string length. Eventually, we ran our system on each string pair, and measured the *precision* of the obtained alignment as the fraction between the score computed by the system and the expected (optimal) one.

Results for this activity are shown in Figure 2.(a). It is easy to observe that the average precision is quite high, and that the heuristic approximates almost exactly the objective function when $\alpha = 2$.

Scalability tests. In order to further analyze the characteristics of our approach on synthetic data, we measured its response times over strings of increasing lengths and generated from alphabets of increasing cardinalities.

Results are reported in Figure 2.(b); here it is possible to observe that the dependency from the length of input strings is actually quadratic (as expected by a dynamic programming approach), but the curve is quite smooth. Moreover, even quite big alphabets can be handled in reasonable time—measured on a Intel Core 2 Duo T7500 with 1Gb Ram.

References

1. Baker, B.S.: Parameterized pattern matching: Algorithms and applications. Journal of Computer and System Sciences 52, 28–42 (1996)
2. Durbin, R., Eddy, S., Krogh, A., Mitchison, G.: Biological Sequence Analysis, Probabilistic models of proteins and nucleic acids. Cambridge University Press, Cambridge (1998)

3. Gusfield, D.: Algorithms on Strings, Trees, and Sequences. Cambridge University Press, Cambridge (1997)
4. Henikoff, S., Henikoff, J.: Amino acid substitution matrices from protein blocks. In: Proceedings of the National Academy of Sciences (PNAS):Biochemistry, pp. 10915–10919 (1992)
5. Hsu, J., Chen, A.L.P., Liu, C.C.: Efficient repeating pattern finding in music databases. In: CIKM 1998: Proceedings of the seventh international conference on Information and knowledge management, pp. 281–288. ACM, New York (1998)

IRPS – An Efficient Test Data Generation Strategy for Pairwise Testing

Mohammed I. Younis, Kamal Zuhairi Zamli, and Nor Ashidi Mat Isa

School of Electrical and Electronic Engineering, Universiti Sains Malaysia,
14300 Nibong Tebal, Penang, Malaysia
{eekamal,ashidi}@eng.usm.my

Abstract. Software testing is an integral part of software engineering. Lack of testing often leads to disastrous consequences including loss of data, fortunes, and even lives. In order to ensure software reliability, many combinations of possible input parameters, hardware/software environments, and system configurations need to be tested and verified against for conformance. Due to costing factors as well as time to market constraints, considering all exhaustive test possibilities would be infeasible (i.e. due to combinatorial explosion problem). Earlier work suggests that pairwise sampling strategy (i.e. based on two-way parameter interaction) can be effective. Building and complementing earlier work, this paper discusses an efficient pairwise test data generation strategy, called IRPS. In doing so, IRPS is compared against existing strategies including AETG and its variations, IPO, SA, GA, ACA, and All Pairs. Empirical results demonstrate that IRPS strategy, in most cases, outperformed other strategies as far as the number of test data generated within reasonable time.

1 Introduction

Software testing is an integral part of software engineering. Lack of testing often leads to disastrous consequences including loss of data, fortunes, and even lives. To ensure acceptable quality and reliability, many combinations of possible input parameters, hardware/software environments, and system configurations need to be considered and verified against for conformance. This consideration often leads to combinatorial explosion problem. Given limited time and resources, it is often impossible to exhaustively consider all of these combinations. Thus, a sampling strategy is needed to select a subset of these combinations in a systematic manner.

Earlier work suggests that pairwise sampling strategy (i.e. based on two-way parameter interaction) can be effective to uncover between 60 to 80 percent of faults [9] [10]. Here, any two combinations of parameter values are to be covered by at least one test [2]. Building and complementing earlier work, this paper proposes and implements an efficient pairwise test data generation strategy, called IRPS. In doing so, IRPS is compared against existing strategies consisting of AETG [2] and its variations [4], IPO [12], SA [15], GA [15], ACA [15], and All Pairs [16]. Empirical results demonstrate that IRPS strategy, in most cases, outperformed other strategies as far as the number of test data generated within reasonable time.

I. Lovrek, R.J. Howlett, and L.C. Jain (Eds.): KES 2008, Part I, LNAI 5177, pp. 493–500, 2008.

2 Related Work

Existing strategies can be categorized into two dominant approaches, that is, algebraic approaches or computational approaches [10].

Algebraic approaches construct test sets using pre-defined rules. Most algebraic approaches compute test sets directly by a mathematical function [10]. Thus, the computations involved in algebraic approaches are typically lightweight, and in some cases, algebraic approaches can produce the most optimal test sets. However, algebraic approaches often impose restrictions on the system configurations to which they can be applied [10] [18]. In a nut shell, algebraic approaches are often based on the extensions of the mathematical methods for constructing orthogonal arrays (OA) [1] [14], and covering arrays (CA) [8] [19]. Some variations of the algebraic approach also exploit recursion in order to permit the construction of larger test sets from smaller ones (see reference [17]).

Unlike algebraic approaches, computational approaches often rely on the generation of the all pair combinations. Based on all pair combinations, the computational approaches iteratively search the combinations space to generate the required test case until all pairs have been covered. Unlike algebraic approaches, the computational approaches can be applied to arbitrary system configurations. Nevertheless, in the case where the number of pairs to be considered is significantly large, adopting computational approaches can be expensive due to the need to consider explicit enumeration from all the combination space.

Adopting the computational approaches as the main basis, an Automatic Efficient Test Generator (or AETG) [2] and its variant (AETG2), employs a greedy algorithm to construct the test case, that is, each test covers as many uncovered combinations as possible. Because AETG uses random search algorithm, the generated test case is highly non-deterministic (i.e. the same input parameter model may lead to different test suites [7]). Other variants to AETG that use stochastic greedy algorithms are: GA (Genetic Algorithm) and ACA (Ant Colony Algorithm) [15]. In some cases, they give optimal solution than original AETG, although they share the common characteristic as far as being non-deterministic in nature.

In Parameter Order (IPO) strategy [11][12], builds a pairwise test set for the first two parameters. Then, IPO strategy extends the test set to cover the first three parameters, and continues to extend the test set until it builds a pairwise test set for all the parameters. In this manner, IPO generates the test case with greedy algorithms similar to AETG. Nevertheless, apart from deterministic in nature, covering one parameter at a time allows the IPO strategy to achieve a lower order of complexity than AETG. All Pairs strategy (i.e. downloadable tool) appears to share the same property as far as producing deterministic test cases is concerned although little is known about the actual strategies employed due to limited availability of references [16][6].

As far as other non-greedy strategies are concerned, some approaches opted to adopt heuristic search techniques such as hill climbing and simulated annealing (SA) [18]. Briefly, hill climbing and simulated annealing strategies start from some known test set. Then, a series of transformations were applied (starting from the known test set) until an optimum set is reached to cover all the pairwise combinations [18]. Unlike AETG and IPO, which builds a test set from scratch, heuristic search techniques can predict the known test set in advance. As such, heuristic search techniques can produce smaller test sets than AETG and IPO, but they typically take longer time to complete [10].

3 The Proposed Strategy

Strategizing to construct minimum test set from the exhaustive test space is a NP-complete problem [11], that is, it is often unlikely that efficient strategy exists that can lways generate optimal test set (i.e. each interaction pair is covered by only one test). Additionally, the size of the minimum pair wise test set also grows logarithmically with the number of parameter and quadratically with the number of values [2]. Motivated by such a challenge, we have opted to develop IRPS as a research vehicle to investigate efficient strategy and data structure implementation to generate optimal pairwise test set that can eventually be generalized for higher order interactions. Adopting the computational approaches as its basis, the IRPS strategy for generating pairwise test data set takes the following steps:

- Step 1: Generates all pairs and store them into compact linked list called Pi.
- Step 2: Search the Pi list and take the desired weight of the candidate case as a test case then delete it from the Pi list.
- Step 3: repeat step 2 until the Pi list is empty.

As indicated above, the generated pairs are stored in compact linked list called Pi, which is a linked list of linked lists. For a test set with N parameters, the Pi list contains (N-1) linked list. Each linked list contains nodes equal to the number of values defined by its parameter as well as an array of linked list that represents the pair of all other variables in the next linked lists.

To understand how the Pi list works, consider a 4 3-valued parameters system, A = {a0,a1,a2}; B = {b0,b1,b2}, C = {c0,c1,c2}, and D {d0,d1,d2}. In this example, we have $\binom{4}{2}3^2 = 54$ possible pairs of combinations.

Table 1. Pi Linked list for storing combination pairs for 4-3 valued parameters

(index) i= 0	i= 1	i= 2
a0	b0	c0
b0 b1 b2	c0 c1 c2	d0 d1 d2
c0 c1 c2	d0 d1 d2	
d0 d1 d2		
a1	b1	c1
b0 b1 b2	c0 c1 c2	d0 d1 d2
c0 c1 c2	d0 d1 d2	
d0 d1 d2		
a2	b2	c2
b0 b1 b2	c0 c1 c2	d0 d1 d2
c0 c1 c2	d0 d1 d2	
d0 d1 d2		

In this case, the complete Pi linked list can be visualized as in Table 1 given earlier. Node a0 with the pairs linked list array contains the following pairs (<a0,b0>, <a0,b1>, <a0,b2>, ……………,<a0,d2>). Here, this list contains only pairs that are based on a0. Similarly, the same observation can be seen with other nodes in the lists. The significant of such arrangement is the fact that less storage unit is required as compared to storing all pairs in clear pairwise combinations. Considering the aforementioned example and assuming each variable takes a unit of storage, then arranging in clear pairwise combinations would require (54*2=108) storage unit. Using similar calculation, adopting our arrangement strategy requires merely 3+(3*9)+3+(3*6)+ 3+(3*3)=33 storage unit.

To describe the IRPS strategy in details, it is necessary to define a number of terminologies. The *weight* of the candidate test case is defined as the number of pairs that are covered by that candidate. For example, the test case combination of a0b0c0d0 covers the pairs (<a0,b0>,<a0,c0>,<a0,d0>,<b0,c0>,<b0,d0>, and <c0,d0>) and the variables b0,c0,d0 in node a0, c0,d0 in node b0, and finally d0 in node c0 , so its weight=6. *The maximum weight*, wmax, for N parameters can be calculated by the following:

$$wmax= N*(N-1)/2$$

Here, if N=4, then wmax=4*3/2=6. *The miss variable* is defined as the difference between the maximum weight and the weight of the candidate test case. *The intersection of node* in the list i with the list (i+1) is defined as the intersection between the node and all nodes given by the first row. IRPS strategy constructs a double linked list that stores the original i node and the intersection with the second node in i+1 list, as well as the rest of the nodes. If the first row in the pairs array is empty, the intersection process will be performed with all values of the nodes in the next list and the miss variable is reduced by one (if miss>0). Otherwise, the intersection process will be terminated and the iteration moves to the next node. *The candidate test* case is obtained by taking the node value in each node in the doubly linked list. For the last node, the candidate test case takes the current value and the first element in the pair array. The candidate test case is taken as a test case only if its weight satisfies the desired weight criteria. If not, the intersection process will continue with the other nodes in the list (by deleting the last node in the doubly linked list and replace it with

```
for (i=0;i<N-;i++) // i is the index of pi list
 begin   //start the search with maximum weight
  w=N(N-1)/2;
  while (list(i) is not empty)
   begin
     if (there exist candidate test case  from the intersection of  a node in ith List
        with the remaining i+1 ,...,N-1 Lists)
       delete the test case from pi list;
     else //not find a test case with the desired weight so :
       w--; //decrease the weight
  end
end
```

Fig. 1. The search algorithm

the intersection with next node in the list, or when there is no next node in the list, the strategy will delete the last two nodes and continue with the iteration). In other words, the intersection process goes horizontally when the target weight is not found and grows vertically in recursive fashion. Finally, the *delete operation* operates by deleting each variable (if they exist) in each node.

Figure 1 depicts the search algorithm for the proposed IRPS strategy. Here, the algorithm is terminated whenever the Pi list is empty in order to guarantee that all pairs are covered and each pair only appears at most once in the final generated test cases (i.e. to achieve optimum solution).

4 Evaluation

Our evaluation has two main goals. Firstly, we want to investigate the growth in the size of the test sets generated by IRPS strategy, as well as the time taken to produce those test sets based on the given number of parameters and values. Secondly, we want to compare the performance of IRPS against existing tools particularly in terms of the size and the time taken to produce the test sets. To perform the evaluation, we have applied IRPS to three series of system configurations. In the first series, the number of parameters (p) and the number of variables (v) are equal to each other, the numbers(n) are (2, 3, 4, 5, 6, 7, 8, 9, 10, 11, 12, 13 and 16) respectively. In the second series, the number of parameters is fixed to be 5, and the number of variables is varied from 2 to 10.

Table 2. Results for n=2 to 11 n n-valued parameters

Case Name	CA1	CA2	CA3	CA4	CA5	CA6	CA7	CA8	CA9	CA10
n=p=v	2	3	4	5	6	7	8	9	10	11
size	4	9	16	25	44	49	64	116	149	121
time	<0.001	<0.001	0.011	0.015	0.087	0.034	0.077	240.2	16.35	0.121

Table 3. Results for 5 parameters with 2 to 10 values

Case Name	CA11	CA12	CA13	CA14	CA15	CA16	CA17	CA18	CA19
value(v)	2	3	4	5	6	7	8	9	10
size	6	12	16	25	44	49	78	96	114
time	0.01	0.015	0.016	0.015	0.077	0.057	0.133	0.178	0.184

Table 4. Results for 2 to 10 parameters with 5 values

Case Name	CA20	CA21	CA22	CA23	CA24	CA25	CA26	CA27	CA28
parameter(p)	2	3	4	5	6	7	8	9	10
size	25	25	25	25	25	37	41	44	45
time	0.053	0.054	0.114	0.015	0.031	0.32	0.78	1.45	1.928

Tables 2, 3 and 4 show the experimental results for the three series of system configurations respectively. The columns in the three tables are self-explanatory. Note that the execution times are shown in seconds, and all the results were collected using a laptop running Windows Vista with 1.6GHZ CPU and 512 MB memory. The entire tool is implemented using Java Development Kit 1.4 (JDK1.4) platforms.

For pairwise interaction, the optimal size can be viewed as the product of the two maximum numbers of variables. This observation can be seen in the case of CA1, CA2, CA3, CA4, CA6, CA7, and CA10 from Table 2. Similar observation can be seen in the case of CA13, CA14, and CA16 from Table 3. The generated test case is also minimal in size, as depicted in CA20, CA21, CA22, CA23 and CA24 from Table 4 respectively. Here, we conclude that the size of generated test case depends linearly on the optimal size of the generated test case.

As far as execution time is concerned, we observe that the execution time is significantly independent on the number of parameters and values when the size is not minimal. This is due to the nature of the algorithm that generates the heavy weighted test case first, deletes them from the Pi list, and then searches again for the uncovered pairs. In this way, the size of the generated test case and the execution time depend on the phenomena of greedy algorithm rather than the number of parameters and values.

We observe that the size and execution time of CA9 (10 10-valued parameters) is greater than CA10 (11 11-valued parameters), according to Table 2, and the size of CA7 (8 8-valued parameters) is greater than CA17 (8 5-valued parameters) according to Tables 2, and 3 respectively. Here, we conclude that the behavior of IRPS is unpredictable in term of the execution time due to the exhaustive search nature when drifting from optimal size, but running the test case generator produces the same test set on every case (thus, IRPS strategy is deterministic).

As for comparison, we have identified the following existing strategies that support pairwise testing: AETG [2] [3], AETG2 [15] [5], IPO [12], SA [15], GA [15], ACA [15], and All Pairs tool [16]. We consider eight systems namely; S1: 3 3-valued parameters, S2: 4 3-valued parameters, S3: 13 3-valued parameters, S4: 10 10-valued parameters, S5: 10 15-valued parameters, S6: 20 10-valued parameters, S7: 10 5-valued parameters, and S8: 1 5-valued parameters, 8 3-valued parameters and 2 2-valued parameters. The system configurations are: AETG2 & SA: C++, Linux, Intel P IV 1.8 GHZ; IPO: Java, Windows 98, Intel P II 450 MHZ; CA, & ACA: C, Windows XP, P IV 2.26 GHZ; AllPairs: Perl, Windows Vista, P IV 1.6 GHZ, 512 MB RAM; and IRPS: Java, Windows Vista, P IV 1.6 GHZ, 512 MB RAM.

Table 5 shows the size of the test set generated by each strategy, and Table 6 shows the execution time for each system. All the problem instances and data for the existing strategies are taken from [12], [15], and [5] except for All Pairs tool (available freely, which we run side by side with our tool). Entries marked with NA are data that are not available in these papers.

Referring to Table 5, IRPS always generate smaller test cases than ALL Pairs and in some cases generates less (i.e. S4, S5, S6, and S7) or equals to that of IPO (i.e.S2, S3). IRPS also generates less the cases compared to AETG2 (except S6), GA and ACA (except S8). While IRPS outperformed AETG in S8, AETG outperformed IRPS in S3, and S6. Finally, SA outperformed IRPS (in S3, S6, and S8). Unlike AETG, AETG2, GA, ACA and IRPS; SA does not have the practical advantage of the

Table 5. Comparison on the size of the test set generated by existing strategies

System	AETG	AETG2	IPO	SA	GA	ACA	All Pairs	IRPS
S1	NA	NA	NA	NA	NA	NA	10	9
S2	9	11	9	9	9	9	10	9
S3	15	17	17	16	17	17	22	17
S4	NA	NA	169	NA	157	159	177	149
S5	NA	NA	361	NA	NA	NA	390	321
S6	180	198	212	183	227	225	230	210
S7	NA	NA	47	NA	NA	NA	49	45
S8	19	20	NA	15	15	16	21	17

Table 6. Comparison on the time taken to generate test set (in seconds) for existing strategies

System	AETG	AETG2	IPO	SA	GA	ACA	All Pairs	IRPS
S1	NA	NA	NA	NA	NA	NA	0.08	<0.001
S2	NA	NA	NA	NA	NA	NA	0.23	0.004
S3	NA	NA	NA	NA	NA	NA	0.45	39.23
S4	NA	NA	0.3	NA	866	1180	5.03	16.35
S5	NA	NA	0.72	NA	NA	NA	10.36	1124
S6	NA	6001	NA	10833	6365	7083	23.3	3213
S7	NA	NA	0.05	NA	NA	NA	1.02	1.928
S8	NA	58	NA	214	22	31	0.35	2.02

greedy algorithm as the implementation is not based on such an algorithm. Here, in the absence of the greedy algorithm, the construction of the test set can not utilize the useful property that the test case created earlier has more significant impact as far as the interaction coverage is concerned [15].

Admittedly, no fair comparison can be made in terms of execution time from existing strategies due to the differences in the computing environments, and the unavailability of the open source code or executable code to run in our platform (with the exception of ALL Pairs tool). Nevertheless, as a general observation; we believe that the execution time for IRPS is still acceptable as compared to other strategies (see Table 6). Not considering the computing differences, IPO outperforms all other strategies. One reason may be that IPO employs deterministic algorithm and needs only one run. Thus, IPO requires much less time to execute than others. SA includes the time taken to find all sized test sets through binary search process, hence, requiring more run time than others. In short, no strategies can clearly be dominant in all.

To conclude, here in this paper, we propose a novel deterministic computational strategy for pairwise testing with efficient data structure for storing and searching pairs. Our initial evaluation results are encouraging particularly in terms of test suite size within acceptable execution time. As part as our future work, we are currently investigating a new parallel search algorithm for IRPS to be implemented under the GRID environment, supported by the USM GRID - Research University Grant.

References

1. Bush, K.A.: Orthogonal Arrays of Index Unity. Annals of Mathematical Statistics 23, 426–434 (1952)
2. Cohen, D.M., Dalal, S.R., Fredman, M.L., Patton, G.C.: The AETG system: An Approach to Testing Based on Combinatorial Design. IEEE Trans. on Software Engineering 23(7), 437–443 (1997)
3. Cohen, D.M., Dalal, S.R., Parelius, J., Patton, G.C.: The Combinatorial Design Approach to Automatic Test Generation. IEEE Software 13(5), 83–88 (1996)
4. Cohen, M.B.: Designing Test Suites for Software Interaction Testing, PhD Thesis, University of Auckland (2004)
5. Cohen, M.B., Gibbons, P.B., Mugridge, W.B., Colbourn, C.J.: Constructing Test Suites for Interaction Testing. In: Proc. of the 25th Intl. Conf. on Software Engineering (ICSE 2003), Dallas USA. IEEE CS Press, Los Alamitos (2003)
6. Copeland, L.: A Practitioner's Guide to Software Test Design. STQE Publishing (2004)
7. Grindal, M., Offutt, J., Andler, S. F.: Combination Testing Strategies: A Survey. In: GMU Technical Report ISE-TR-04-05 (July 2004)
8. Hartman, A., Raskin, L.: Problems and Algorithms for Covering Arrays. Discrete Mathematics 284(1-3), 149–156 (2004)
9. Kuhn, D.R., Wallace, D.R., Gallo, A.M.: Software Fault Interactions and Implications for Software Testing. IEEE Trans. on Software Engineering 30(6), 418–421 (2004)
10. Lei, Y., Kacker, R., Kuhn, D.R., Okun, V., Lawrence, J.: IPOG: A General Strategy for T-Way Software Testing. In: 14th Annual IEEE Intl. Conf. and Workshops on the Engineering of Computer-Based Systems, Tucson, AZ, March 2007, pp. 549–556. IEEE CS Press, Los Alamitos (2007)
11. Lei, Y., Tai, K.C.: In-Parameter-Order: A Test Generating Strategy for Pairwise Testing. In: Proc. 3rd IEEE Intl. Symp. On High Assurance System Engineering, November 1998, pp. 254–261 (1998)
12. Lei, Y., Tai, K.C.: In-Parameter-Order: A Test Generating Strategy for Pairwise Testing. IEEE Transaction on Software Engineering 28(1), 1–3 (2002)
13. Maity, S., Nayak, A., Zaman, M., Bansal, N., Srivastava, A.: An Improved Test Generation Strategy for Pair-Wise Testing. In: Fast Abstract ISSRE 2003 (2003)
14. Mandl, R.: Orthogonal Latin squares: an application of experiment design to compiler testing. Communications of the ACM 28(10), 1054–1058 (1985)
15. Shiba, T., Tsuchiya, T., Kikuno, T.: Using Artificial Life Techniques to Generate Test Cases for Combinatorial Testing. In: 28th Annual Intl. Computer Software and Applications Conference (COMPSAC 2004), Hong Kong, China, September 2004, pp. 72–77 (2004)
16. `http://www.satisfice.com`
17. Williams, A.W., Probert, R.L.: A Practical Strategy for Testing Pair-Wise Coverage of Network Interfaces. In: Proc. of the 7th Intl. Symp. on Software Reliability Engineering (ISSRE), White Plains, New York (1996)
18. Yan, J., Zhang, J.: Backtracking Algorithms and Search Heuristics to Generate Test Suites for Combinatorial Testing. In: Proc. of the 30th Annual Intl. Computer Software and Applications Conference (COMPSAC 2006), Chicago USA, September 2006, vol. 1, pp. 385–394. IEEE CS Press, Los Alamitos (2006)
19. Zekaoui, L.: Mixed Covering Arrays on Graphs and Tabu Search Algorithms. In: MSc Thesis, Ottawa-Carleton Institute for Computer Science, University of Ottawa, Canada (September 2006)

A Procedure Ontology for Advanced Diagnosis of Process Systems

Katalin M. Hangos[1], Erzsébet Németh[1], and Rozália Lakner[2]

[1] Process Control Research Group, Systems and Control Laboratory, Computer and Automation Research Institute, Budapest, Hungary
[2] Department of Computer Science, University of Pannonia, Veszprém, Hungary

Abstract. An ontology for representing operation, safety and control procedures is proposed in this paper that supports diagnosis based on following these procedures and combining observed malfunctions with Failure Mode and Effects Analysis (FMEA) information. The procedure ontology is defined within interconnected components of the process plant, diagnostic analysis (where the FMEA is described) and procedures. The proposed method is illustrated on a simple operating procedure.

Keywords: Knowledge Management, Intelligent Systems, Ontology, Multi-Agent Systems.

1 Introduction

Large-scale complex process plants are safety-critical systems where both the off-line hazard analysis and the real-time diagnosis are of great importance. Because of the great complexity and the large amount of information, a combination of analytical and heuristic methods, called hybrid methods are usually applied [1].

In order to properly manage abnormal conditions, risk management procedures, such as HAZard and OPerability (HAZOP) analysis [2], or Failure Mode and Effects Analysis (FMEA) [3] are applied to most of the process plants in an off-line manner. The results of such an analysis contain a vast amount of diagnostically relevant information that can also be used in a real-time fashion.

The overall aim of our work has been to utilize this risk management information in advanced diagnostic systems [4]. For this purpose, an agent-based diagnostic system has been proposed in [5] that uses HAZOP-FMEA information and bidirectional reasoning to derive a diagnostic conclusion.

The aim of this paper is to further advance the above hybrid diagnostic methodology by including the information contained in the operating, safety and control procedures and use this for diagnosis of complex process plants.

2 The Procedures and the Diagnostic Framework

If one aims at utilizing all diagnostic information and their inter-relationships, then an overall diagnostic framework is needed that is based upon the interconnected components of the process plant, diagnostic analysis and procedures (see [6]).

I. Lovrek, R.J. Howlett, and L.C. Jain (Eds.): KES 2008, Part I, LNAI 5177, pp. 501–508, 2008.

2.1 Procedures and Their Elements

Procedures in a complex process plant have several basic classes:

- *Operating procedures* that are normally written instructions for operational staff that links them to the plant and also the plant to them in terms of their response.
- *Control procedures* that could be continuous control procedures, as well as discrete or sequenced control procedures.
- *Safety procedures* that are frequently executed by dedicated hardware and software components.

Operating, safety and control procedures can be regarded as discrete procedures consisting of sequential and parallel steps. Each step has a precondition, a consequence and at least one action associated with it. Actions are understood in a broad sense, an action can be a manipulation executed by the operating personnel or by a computer through an actuator device (e.g. a valve), a measurement, or even the execution of another procedure.

Formally procedures are usually described using the notions and tools of discrete event systems [9]. The most popular and powerful tools here are the coloured Petri nets (CPNs) [10] and their variants, like timed or hierarchical CPNs.

Procedure steps are the atomic elements of procedures, they correspond to transitions in the describing CPN. Fig. 1. shows a special "procedure step" transition denoted by a rectangle, its preconditions and consequences are logical expressions corresponding to places in the CPN (denoted by ovals). The associated external procedures are denoted by diamonds.

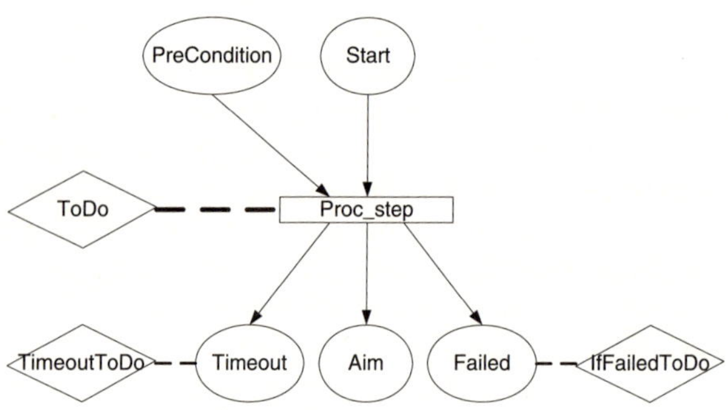

Fig. 1. A simple procedural step

In order to support the diagnosis, three mutually exclusive consequences are associated to each step that are coded in the *Status* attribute of the step:

- the *Aim* (a condition) that is reached if the execution of the step was successful,

- the *Timeout* is detected in case of timing out the action with a possibility to invoke another procedure to handle the situation,
- the *Failed* condition is checked for detecting malfunctions again with the possibility to invoke another procedure.

The preconditions have also been partitioned into two mutually exclusive conditions:

- the *Start* condition serves to connect steps by setting it equal to the successful termination of one or more (previous) step(s),
- the *PreCondition* gives the technologically relevant other conditions expressed in terms of the state variables of the plant.

Note that only the *Start*, *PreCondition* and *ToDo* attributes are specified for the procedure steps in the industrial practice.

2.2 The Plant and Analysis Ontologies

For the sake of modularity, three related ontologies have been developed in Protégé [7] in our diagnostic system: the plant, the analysis and the procedure ontologies. The Plant and Analysis ontologies are components in our earlier agent-based diagnostic system [5], therefore they are only briefly described here. Note that the telling names of the defined classes and attributes are typeset in *italic*.

The Plant ontology captures the attributes, instances of and relationships between the components of the process system, similarly to that in the OntoCape system [8]. Any atomic (i.e. non-divisible) part of the system is regarded as a *SystemComponent*, while the measurable variables are described by *ProcessStates*. The *SystemComponents* are further classified as e.g. *Pipes*, *Valves*, *Tanks* etc, and a *ProcessState* can be a *Temperature*, *Level*, etc.

A class hierarchy is defined among the sub-classes of both *SystemComponents* and *ProcessStates* dictated by the natural hierarchy of the components and variables of the complex process system.

Component	Description	Failure mode	Possible causes	Effects	
				Local	System
VB	TB inflow control valve	Closed	mechanical fail closed operator closed	<NO><Feed to TB>	<NO><Feed to press>
		Opened	mechanical fail opened operator opened	<MORE><Feed to TB>	<MORE><Feed to press>
		Stuck	maintenance failure corrosion	<LESS><Feed to TB>	<LESS><Feed to press>
TA	Bulk tank TA	Broken	corrosion vehicle damage operator damage	<NO><Feed to PA>	<NO><Feed to press>
		Leaked	corrosion	<LESS><Feed to PA>	<LESS><Feed to press>

Fig. 2. The FMEA class

The Analysis ontology describes the results of the blended HAZOP-FMEA analysis [6]. Because of its use in the proposed procedure based diagnostic method, only the FMEA that is produced as a table during the FMEA analysis is described here.

An example of an FMEA table can be seen in Fig. 2. A row in the FMEA table corresponds to a fault entry that is defined for an instance (e.g. "VB" or "TA" in Fig. 2) of a *SystemComponent* giving a *FailureMode* (e.g. "Closed" for "VB") and its possible causes, together with its *LocalEffect* and *SystemEffect.*

3 The Procedure Ontology

Similarly to the Plant and Analysis ontologies, the procedures are also described in a Procedure ontology implemented in Protégé [7] in our diagnostic system. This enables the easy description of the links between the procedures, plant components and diagnostic analysis results.

3.1 The Representation of Procedures

Simple syntactical elements are the ones upon which the syntax of the procedures is built. The most important simple syntactical elements are as follows:

- The *Qualitative state (generalized predicate)* describes a qualitative state for a system component, process state or procedure step state.
- The *Condition* describes a qualitative state for a set of components or variables. It acts as a pre-condition or consequence of a procedural step.
- The *Action* describes an elementary action by specifying the *SystemComponent* or *ProcessState* (*ComponentOrState*) to be manipulated by the *Actor* who performs the manipulation.

A procedure and its steps are represented as classes (*ProcedureStep* and *Procedure*) of Procedure ontology in Protégé [7]. Their attributes and connections are shown in Figure 3. The structure of the procedure step corresponds to the *simple procedure step* introduced in section 2.1.

The *ComponentOrState* specified as the subject of the *ToDo* action in the step gives a pointer to the related entries of the FMEA table: this is the component the possible failure of influences the outcome of the step. The *Failed condition* of a procedure step encoded in the value of the *hasFailed* slot selects the appropriate *FailureMode* in the FMEA table thus pointing to the relevant row. This gives a link to the *LocalEffect* or *SystemEffect* in the FMEA entries that can be and will be used in the diagnostic method based on following the procedures (see later in section 4).

The connection of the procedure steps is represented in an implicit way taking into account that a procedure consists of sequential and parallel steps. The synchronization of the steps, that is, the way they are linked together is described through the attributes of the steps that form the procedure. A procedural step

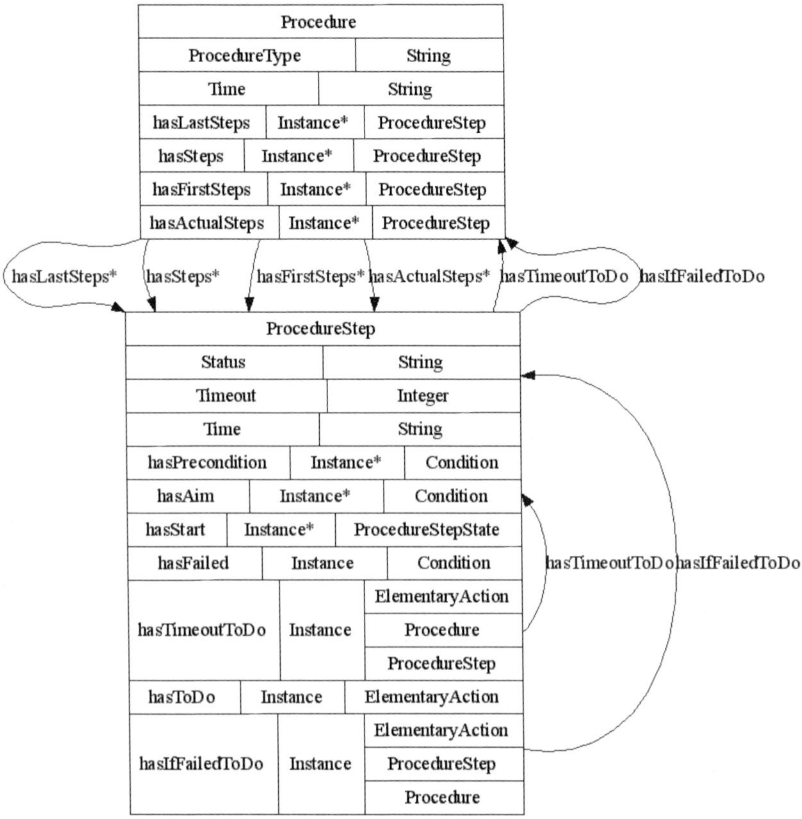

Fig. 3. Structure and connections of procedure and procedure step

in a procedure can be seen as a simple agent that senses its preconditions by monitoring its *Start* and *PreCondition* attributes and performs its *ToDo* action autonomously.

The *Start* attribute of a step is either *START* if it is the first step, or a previous *ProcedureStep* with *Status=OK* for a sequential step, or list of previous *ProcedureSteps* with *Status=OK* for a collector step at the end of a parallel section. The procedure will have its *Status* attribute "failed" or "timeout", if any of its steps fail or have timed out.

Remarks. The procedures encoded in an ontology enable us to

1. link procedures after each other on a conditional basis, e.g. to automatically invoke an emergency procedure when a startup procedure fails,
2. start the diagnostic engine when something has failed or has timed out,
3. link FMEA failure modes, local and system consequences to procedures,
4. construct diagnostic procedures (combined timed checking and a reasoning sequence of simple and parallel steps).

3.2 An Illustrative Simple Example

As a simple illustrative example, let us consider a supply tank with manual feed valve and manual discharge valve can be seen in the top left corner of Fig. 4.

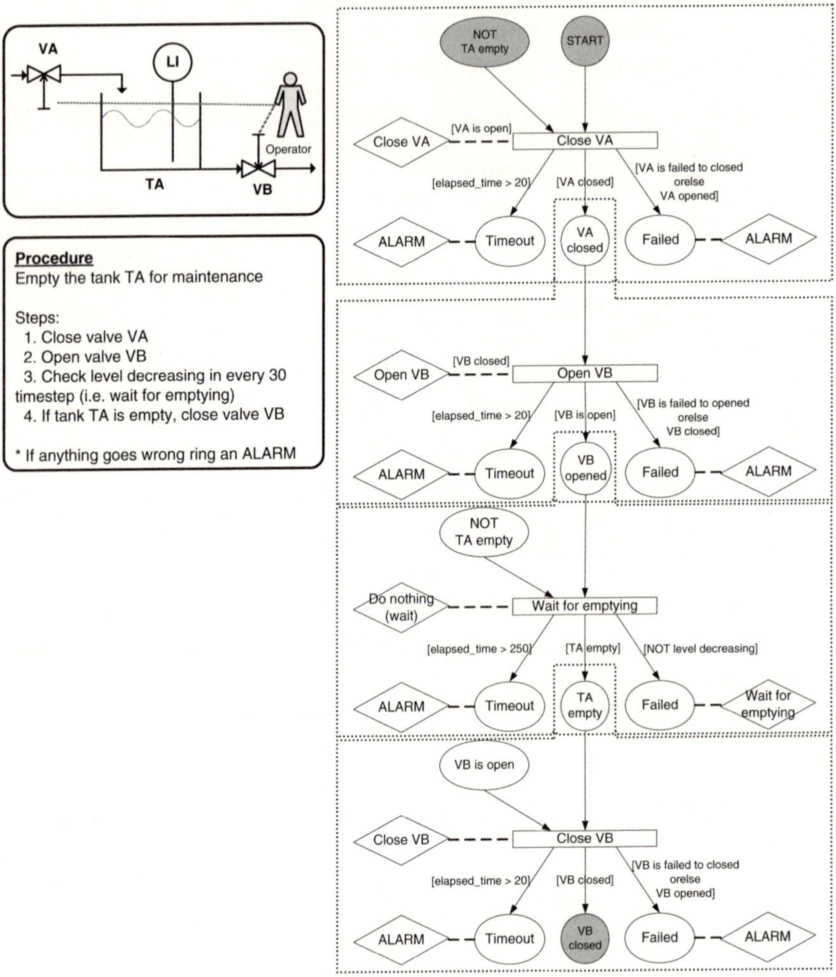

Fig. 4. Supply tank with manual feed valve and manual discharge valve, an illustrative procedure and its steps

An operating procedure that empties and shuts down the tank for maintenance is used for illustrating how a procedure is described by linking together simple procedure steps. The main part of Fig. 4. shows the procedure steps and illustrates the connection of procedure steps.

4 Diagnosis Based on Following the Procedures

4.1 The Diagnostic Principle

The current industrial practice on following the procedures is rather ad-hoc, and done mostly manually by the operating personnel. In most of the cases only the *Start* condition and the most important predicates in the *PreCondition* are specified together with the *ToDo* action. Sometimes (and mostly for safety procedures) a time-out value is given and/or the most important predicates in the *Failed* conditions are also specified.

Therefore, only fault detection (i.e. to detect if something goes wrong) is usually performed when an expected procedure step does not occur, or when a related safety condition is violated. For the fault isolation (i.e. to find out what exactly has gone wrong) to happen, one needs to augment the procedure description with all the possible information in Fig. 1, and connect the results of the fault analysis to the detected fault.

4.2 The Diagnostic Algorithm

The diagnostic algorithm uses the extended procedure description given in section 3.1 implemented in the Procedure ontology. A diagnostic agent called a **watchdog** is then associated to each of the operating, safety or control procedure that checks each executed step for timeout and failure. If any of these occurs then it

- checks if the *ComponentOrState* in the corresponding Action is a *SystemComponent*, and if yes
- locates the FMEA entry related to this *SystemComponent*, and
  - for each of the applicable *FailureMode* of the component locates the *LocalEffect* and *SystemEffect*
  - checks if they are true based on the measurements
  - falsifies those that do not conform with the measurements
  - presents the remaining possible *FailureMode* set to the operator together with any actions related to them.

5 Conclusion and Discussion

An ontology for representing operation, safety and control procedures is proposed in this paper that supports diagnosis based on following these procedures and combining observed malfunctions with fault mode effect analysis (FMEA) information.

Further work includes the development and integration of the People ontology to be able to diagnose human failures, as well.

Acknowledgements

This research has been supported by the Hungarian Scientific Research Fund through grants K67625 and T047198, which is gratefully acknowledged. This work has also been supported by the Australian Research Council under Linkage Grant LP0776636, and by the Control Engineering Research Group of the Budapest University of Technology and Economics.

References

1. Venkatasubramanian, V., Rengaswamy, R., Kavuri, S.N.: A review of process fault detection and diagnosis Part II: Qualitative models and search strategies. Computers and Chemical Engineering 27, 313–326 (2003)
2. Knowlton, R.E.: Hazard and operability studies: the guide word approach. Chematics International Company, Vancouver (1989)
3. Jordan, W.: Failure modes, effects and criticality analyses. In: Proceedings of the Annual Reliability and Maintainability Symposium, pp. 30–37. IEEE Press, Los Alamitos (1972)
4. Németh, E., Lakner, R., Hangos, K.M., Cameron, I.T.: Prediction-based diagnosis and loss prevention using model-based reasoning. In: Ali, M., Esposito, F. (eds.) IEA/AIE 2005. LNCS (LNAI), vol. 3533, pp. 367–369. Springer, Heidelberg (2005)
5. Lakner, R., Németh, E., Hangos, K.M., Cameron, I.T.: Multiagent realization of prediction-based diagnosis and loss prevention. In: Ali, M., Dapoigny, R. (eds.) IEA/AIE 2006. LNCS (LNAI), vol. 4031, pp. 70–80. Springer, Heidelberg (2006)
6. Cameron, I.T., Seligmann, B., Hangos, K.M., Lakner, R., Németh, E.: The P3 Formalism: A basics for improved diagnosis in complex systems. In: Proceedings of the Chemeca Conference, Melbourne, Australia, on CD (2007)
7. The Protégé Ontology Editor and Knowledge Acquisition System (2007), `http://protege.stanford.edu`
8. Yang, A., Marquardt, W., Stalker, I., Fraga, E., Serra, M., Pinol, D.: Principles and informal specification of OntoCAPE, Technical report, COGents project, WP2 (2003)
9. Cassandras, C.G., Lafortune, S.: Introduction to Discrete Event Systems. Kluwer Academic Publishers, Dordrecht (1999)
10. Jensen, K.: Coloured Petri Nets. Basic Concepts, Analysis Methods and Practical Use. Basic Concepts. Monographs in Theoretical Computer Science, vol. 1. Springer, Heidelberg (1997)

Congenital Heart Disease: An Ontology-Based Approach for the Examination of the Cardiovascular System

M. Esposito

Institute for High-Performance Computing and Networking (ICAR)
National Research Council (CNR)
Via Castellino 111, 80131 Naples, Italy
massimo.esposito@na.icar.cnr.it

Abstract. Congenital Heart Disease (CHD) represents the most common group of congenital malformations of the heart and of its blood vessels. In this paper, we present an ontology-based approach to detect abnormalities and malformations due to CHD. In particular, we propose a formal and well-defined model to represent the anatomy of the cardiovascular system, based on the SNOMED vocabulary. The model defines either the anatomy of the cardiovascular system in normal patients or the anatomy characterized by malformations and abnormalities in CHD patients. We have formalized this model in OWL ontologies and SWRL rules and, then, we have used a logic reasoner to identify either CHD patients or the heart abnormalities and malformations they are affected by.

Keywords: Congenital heart disease, ontologies and rules, model checking.

1 Introduction and Related Work

1.1 Introduction

Congenital Heart Disease (CHD) represents the most common group of congenital malformations of the heart and of its blood vessels that affect between 7 and 8 per 1000 live-born infants [1].

In many cases, CHD contributes significantly to infant mortality and morbidity and may only be recognized when the affected infant develops life-threatening symptoms of cardiovascular collapse. The clinical examination of the cardiovascular system at the time of routine clinical newborn examination can enable to identify early those infants who are at risk of adverse or irreversible outcomes as a consequence of congenital heart defects, whilst they are still pre-symptomatic.

The large number of routine clinical investigations on the one hand, and the need of a sound examination of every single case on the other hand have necessarily decreased the cardiologist productivity and the quality of the diagnosis reports.

This highlights the need of an automated approach for the examination of the cardiovascular system that could support the cardiologists, providing outputs which can be used as a 'second opinion' in detecting malformations and abnormalities. Moreover it could also increase the cardiologist productivity and improve the quality of the

I. Lovrek, R.J. Howlett, and L.C. Jain (Eds.): KES 2008, Part I, LNAI 5177, pp. 509–516, 2008.

diagnosis reports. Such an approach should automatically combine the lower-level information, coming from a preliminary segmentation step, with a higher level domain-specific knowledge, closing the existing semantic gap.

We think the Semantic Web languages and technologies, and in particular ontologies and rules, could close this gap for the following reasons:

- Ontologies and rules enable to represent, explicitly and formally, the domain-specific knowledge a cardiologist uses in his clinical investigations. Ontologies and rules can be formalized using semantic representation languages as OWL[1], SWRL [2] and RDF[3]. These languages, characterized by a high degree of expressiveness and modeling power, enable to formalize complex models in an accurate and sound way.
- OWL ontologies and SWRL rules can be processed by logic reasoners. These reasoners perform inference patterns that can be exploited to automatically examine the cardiovascular system and detect abnormalities.

As a result, in this paper, we propose an automated approach, based on the use of ontologies and rules, to examine the cardiovascular system and identify abnormalities and malformations due to CHD.

More precisely, we have realized a formal and well-defined model to represent the anatomy of the cardiovascular system, based on the SNOMED[4] vocabulary. This model defines either the anatomy of the cardiovascular system in normal patients or the anatomy characterized by malformations or abnormalities in CHD patients.
Moreover, we have formalized this model in OWL ontologies and SWRL rules and used a logic reasoner to identify either CHD patients or the heart abnormalities and malformations they are affected by.

The rest of this work is organized as follows. Section 2 describes the semantic approach. Section 3 presents our cardiovascular examination method and overviews some application examples. Finally, section 4 concludes the work.

1.2 Related Work

In the past, many authors have proposed the use of Expert Systems as Clinical Decision Support Systems (CDSS) in order to directly assist physicians with decision making tasks [2], [3],[4],[5].

In [2] the authors investigate the application of artificial neural networks in medical diagnosis, presenting a hybrid fuzzy-neural automatic system and a simple and applied method. In [3] the authors present a rule-based CDSS for the diagnosis of Coronary Artery Disease, based on the development of a fuzzy model. In [4] the authors focus on using decision-theoretic networks for decision-making and discuss, as a typical example, the treatment of patients with aortic coarctation, a kind of CHD. In [5] the authors propose a mechanism for reasoning about the differential diagnosis of cases involving the symptoms of heart failure defining a causal model.

---

[1] http://www.w3.org/TR/owl-semantics/
[2] http://www.w3.org/Submission/2004/SWRL/
[3] http://www.w3.org/RDF/
[4] http://www.snomed.org/

The weakness of all these approaches relies on the poor expressiveness and modeling power of the Expert Systems, that do not enable to formalize a well-defined, unambiguous and structured representation of the domain-specific knowledge. Differently, our approach, exploiting the modeling power of ontologies and rules, enables to describe the knowledge in a structured, organized and more human-understandable manner, facilitating the information formalization and interpretation.

Recently, approaches using ontologies and rules for modeling domain-specific medical knowledge have been proposed [6],[7],[8]. More specifically, in [6],[7] the authors focus on the brain examination and, in particular, they aim at modeling brain knowledge in order to label brain images. These works, similarly to us, use i) ontologies and rules to represent a domain-specific knowledge and ii) logic reasoners to perform reasoning mechanisms.

In [8] the authors use clinical and spatial ontologies representing the human heart to automatically generate a diagram based on a patient's information in cardiology databases. This work, similarly to us, also defines a model for the cardiovascular system, but it is not related to any shared and well-defined collection of medical terminology. Anyway, none of these approaches aim at supporting physicians in the medical diagnosis.

2 The Semantic Approach

2.1 Our Proposal of a Cardiovascular Model

The approach presented in this paper relies on a model that we have defined to provide a unique and uniform representation for the anatomy of the cardiovascular system. This model is fundamentally based on the SNOMED (Systematized Nomenclature of Medicine) vocabulary, that is a systematically organized and computer readable collection of medical terminology covering most areas of clinical information. Our model has been formalized in OWL ontologies and SWRL rules and represents the cardiovascular knowledge by specifying anatomical concepts and relationships between them.

Fundamentally, the cardiovascular system consists of three sub-systems, the heart, the arterial system and the venous system. The heart has four chambers, separated by grooves. Blood is pumped through the chambers, aided by four heart valves, to all parts of the body through a series of vessels, termed arteries. The arteries undergo enormous ramification in their course and end in minute vessels, called arterioles, which in their turn open into a close-meshed network of microscopic vessels, termed capillaries. After the blood has passed through the capillaries, it is collected first into a series of minute vessel, termed venule, and then into a series of larger vessels, called veins, by which it is returned to the heart.

It is worth noting that our model takes into account only the part of the arterial and venous systems that are strictly related to the heart and that can be affected by CHD. First, we have identified the main cardiovascular concepts, that are shown in figure 1, by taking into account the medical terminology specified in SNOMED.

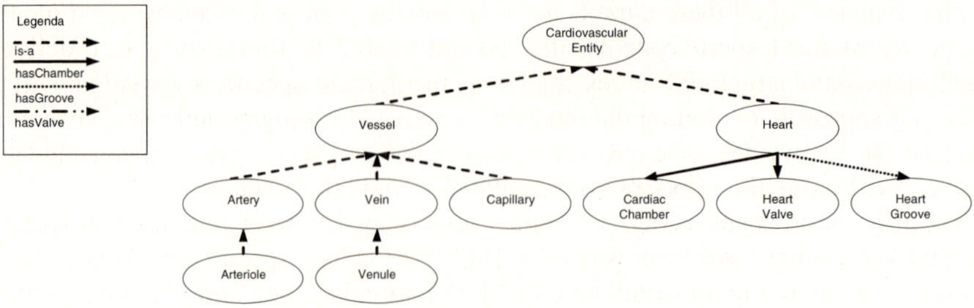

Fig. 1. The main cardiovascular concepts

Table 1. Mereological roles

Role	Domain	Range	Inverse	Trans.	Sym.
HasSegment	Vessel	Vessel	isSegmentOf	yes	no
HasBranch	Vessel	Vessel	isBranchOf	no	no
HasVisceralBranch	Vessel	Vessel	isVisceralBranchOf	no	no
HasParietalBranch	Vessel	Vessel	isParietalBranchOf	no	no
HasTerminalBranch	Vessel	Vessel	isTerminalBranchOf	no	no
HasChamber	Heart	Cardiac Chamber	isChamberOf	no	no
HasGroove	Heart	Heart Groove	isGrooveOf	no	no
HasValve	Heart	Heart Valve	isValveOf	no	no

Table 2. Topological roles

Role	Domain	Range	Inverse	Trans.	Sym.
isConnectedTo	Vessel	Vessel	No	no	Yes
isSeparatedFrom	Cardiac Chamber	Cardiac Chamber	No	no	Yes
isSeparatedFrom VentricleBy	Cardiac Chamber	Heart Groove	Separates VentricleFrom	no	No
isSeparatedFrom AtriumBy	Cardiac Chamber	Heart Groove	Separates AtriumFrom	no	No
emptyIn	Vessel	Cardiac Chamber	isEmptiedBy	no	No
takesOriginFrom	Vessel	Cardiac Chamber	givesOrigin	no	No
isIncludedIn	Heart Valve	Cardiovascular Entity	includes	no	No

Table 3. Relationship between roles

Rule	Antecedent	Consequent
R1	**if**(Cardiac_Chamber_X isSeparatedFromVentricleBy Heart_GrooveY **AND** Cardiac_Chamber_Z isSeparatedFromVentricleBy Heart_GrooveY)	Cardiac_Chamber_X isSeparatedFrom Cardiac_Chamber_Z
R2	**if**(Cardiac_Chamber_X isSeparatedFromAtriumBy Heart_GrooveY **AND** Cardiac_Chamber_Z isSeparatedFromAtriumBy Heart_GrooveY)	Cardiac_Chamber_X isSeparatedFrom Cardiac_Chamber_Z
R3	**if** (Vessel_X HasSegment Vessel_Y **AND** Vessel_Y HasBranch Vessel_Z)	Vessel_X HasBranch Vessel_Z
R4	**if** (Vessel_X HasSegment Vessel_Y **AND** Vessel_Y HasVisceralBranch Vessel_Z)	Vessel_X HasVisceralBranch Vessel_Z
R5	**if** (Vessel_X HasSegment Vessel_Y **AND** Vessel_Y HasParietalBranch Vessel_Z)	Vessel_X HasParietalBranch Vessel_Z
R6	**if** (Vessel_X HasSegment Vessel_Y **AND** Vessel_Y HasTerminalBranch Vessel_Z)	Vessel_X HasTerminalBranch Vessel_Z
R7	**if** (Vessel_T HasSegment Vessel_X **AND** Vessel_T HasSegment Vessel_Y **AND** Vessel_T HasSegment Vessel_Z **AND** Vessel_X isConnectedT Vessel_Y **AND** Vessel_Y isConnectedTo Vessel_Z)	Vessel_X isConnectedTo Vessel_Z

Moreover, we have identified three typologies of relationship: i) the subsumption relationship concerns the generalization/specialization between cardiovascular concepts; ii) the mereological relationship concerns part-whole relations between cardiovascular concepts; iii) the topological relationship concerns neighborhood relations between cardiovascular concepts.

We have modeled these typologies of relationship by a set of roles. Each role has a domain (the set of possible subject concepts) and a range (the set of possible object concepts). Besides, each role can have a corresponding inverse role and it can be transitive (if it remains true across chains of links) or symmetric (if it can be applied in both the directions). Moreover, we have realized a set of SWRL rules, reported in table 3, in order to capture relationships between roles.

2.2 CHD: A Typical Heart Abnormality

The presented model takes into account only the normal anatomy of the cardiovascular system. Nevertheless, the cardiovascular system in a patient affected by CHD is characterized by a different anatomy. In this paper, we consider, as an illustrative example, a typical congenital abnormality, that is the Interrupted Aortic Arch (IAA) defect. We have also taken into account other typical heart abnormalities, but we have not reported their descriptions in this paper for sake of brevity.

Below, we highlight the anatomical differences existing respectively between the cardiovascular systems in a normal patient and in a patient affected by IAA.

In normal patients, the aorta consists of three segments: an ascending segment, a transverse segment or arch, and a descending segment.

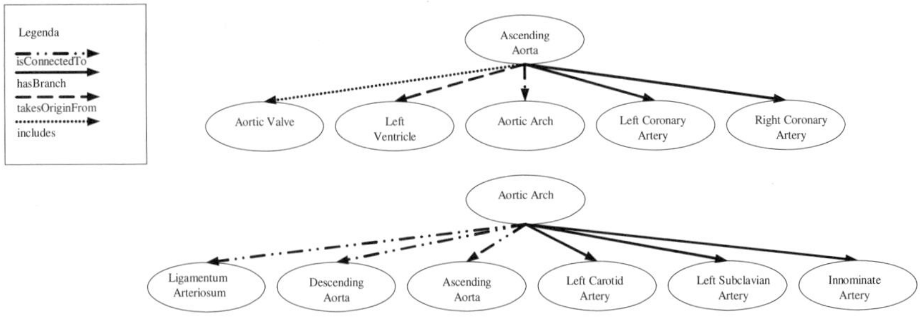

Fig. 2. The Ascending Aorta and the Aortic Arch

The ascending aorta takes its origin from the left ventricle, it gives rise to the left and right coronary arteries and its proximal portion includes the aortic valve.

The aortic arch begins at the innominate artery and ends at the ligamentum or ductus arteriosum. It gives rise to the innominate, left carotid and left subclavian arteries. The descending aorta begins at the ligamentum or ductus and consists of two segments, the thoracic aorta and abdominal aorta.

When a patient is affected by IAA, the aorta does not develop completely in the area of the arch. As a result, the aorta is divided only into two parts, the ascending aorta and descending aorta, that are not connected to each other. IAA can be classified on the basis of the site of interruption. Type A is distal to the left subclavian artery, type B is proximal to the left subclavian artery, and type C occurs between the innominate artery and the left common carotid artery.

As a result, we have modified the presented model, taking into account these differences, in order to describe the anatomy of the cardiovascular system in a patient affected by IAA.

Specifically, the cardiovascular model for patients affected by IAA presents the following changes with respect to the presented one: i) the aorta has two segments, the ascending aorta and the descending aorta (the aortic arch is not present); ii) the ascending aorta is not connected to the descending aorta; iii) IAA Type A: the ascending aorta has the innominate artery, the left subclavian artery and the left carotid artery as its branches; iv) IAA Type B: the ascending aorta has the left subclavian artery and the left carotid artery as its branches; v) IAA Type C: the ascending aorta has the innominate artery as its branch.

3 The Cardiovascular Examination Procedure

3.1 The Method

The method we have used to perform the examination of the cardiovascular system consists of two main steps.

The first step is the segmentation of the cardiac images. We suppose the segmentation procedure is able to identify the anatomical structure of the heart and of its blood vessels and the topological relations existing among them. It is worth noting that, in this paper, we focus our attention only on the next step.

The second step is fundamentally a model checking procedure that performs an automatic formal verification of the outputs of the segmentation step with respect to the defined cardiovascular models. We make use of the logic reasoner presented in [9], that integrates and reasons about ontologies and rules to perform this procedure.

More precisely, we use the presented model and the information produced after the segmentation step to populate the knowledge base of the reasoner. The knowledge base consists of i) the Tbox, populated by using the ontologies formalized in OWL, ii) the Rule-box, populated by using the rules formalized in SWRL, and iii) the Abox, populated with the outputs of the segmentation step, formalized in RDF.

Moreover, we exploit two inference patterns provided by the reasoner, that are the instance checking and the consistency checking. The instance checking determines whether an individual from the Abox is always an instance of a certain concept. The consistency checking determines whether two assertions about a same individual are inconsistent with respect to the TBox. We use these inference patterns to determine whether the current Abox contains a consistent instance of the Tbox.

Our model checking procedure consists of a set of model checking executions that, in turn, perform the instance and consistency checkings on a different cardiovascular model.

We start populating the Tbox with the ontologies and rules that describe a normal cardiovascular system The Abox is populated with the information coming from the segmentation step, that describe the cardiovascular system under exam.

The reasoner performs the instance and consistency checkings to verify whether the cardiovascular system under exam, loaded in the Abox, does not violate the normal cardiovascular model, loaded in the Tbox.

If the reasoner provides a no answer, this means that the cardiovascular system under exam is not coherent and consistent with the normal cardiovascular model. Hence, the patient under exam is affected by a kind of abnormality. Otherwise, the patient is affected by no abnormality. If the patient is affected by a kind of abnormality, we have to change the model loaded in the Tbox with the one describing the anatomy of a cardiovascular system affected, for an example, by IAA Type A and, then, verify whether the individuals in the Abox do not violate it.

We have to change the model in the Tbox and launch a new model checking execution until the reasoner gives a positive answer. When it happens, it is possible to query the reasoner in order to determine which model is currently loaded in the Tbox. This enables to find out the abnormality that affects the patient in exam.

3.2 Some Application Examples

In this subsection we describe two application examples of the cardiovascular examination procedure described above. In particular, suppose that the segmentation has detected the anatomical structure of the heart and of its blood vessels. Moreover, suppose the initial set of information submitted to the reasoner is the following:

I1. *Aorta(aor0)*
I2. *AscendingAorta(asc_aor0)*
I3. *AorticArch(aor_arc0)*
I4. *DescendingAorta(des_aor0)*
I5. *hasSegment(aor0, asc_aor0)*
I6. *hasSegment(aor0, aor_arc0)*
I7. *hasSegment(aor0, des_aor0)*
I8. *InnominateArtery(inn_art0)*
I9. *LeftArtery(lef_art0)*
I10. *LeftSubclavianArtery(lef_sub_art0)*
I11. *isConnectedTo(asc_aor0, aor_arc0)*
I12. *isConnectedTo(aor_arc0, des_aor0)*
I13. *isConnectedTo(asc_aor0, des_aor0)*
I14. *hasBranch (aor_arc0, inn_art0)*
I15. *hasBranch (aor_arc0, lef_art0)*
I16. *hasBranch (aor_arc0, lef_sub_art0)*

This set of information represents a fragment of our current Abox. The information I13 is derived from the application of the rule R7 on the individuals I5, I6, I7, I11, I12. The reasoner performs the instance and consistency checkings and verifies that this information does not violate the normal cardiovascular model loaded in the Tbox. Hence, we can conclude that the patient under exam is normal.

Instead, suppose the initial set of information is the following:

I1. *Aorta(aor0)*
I2. *AscendingAorta(asc_aor0)*
I3. *DescendingAorta(des_aor0)*
I4. *hasSegment(aor0, asc_aor0)*
I5. *hasSegment(aor0, des_aor0)*
I6. *InnominateArtery(inn_art0)*
I7. *LeftArtery(lef_art0)*
I8. *LeftSubclavianArtery(lef_sub_art0)*
I9. *hasBranch (asc_aor0, inn_art0)*
I10. *hasBranch (asc_aor0, lef_art0)*
I11. *hasBranch (asc_aor0, lef_sub_art0)*

Now, this is a fragment of our current Abox. The reasoner performs the instance and consistency checkings and verifies that this information violates the normal cardiovascular model loaded in the Tbox. Hence, we can conclude that the patient under exam is affected by a kind of abnormality.

Now, we load the model describing the IAA Type A in the Tbox. Then, the reasoner performs the instance and consistency checkings and verifies that the individuals in the Abox do not violate the IAA Type A model. Hence, we can conclude that the cardiovascular system of the patient under exam is affected by IAA Type A.

4 Conclusions and Directions for Future Works

In this paper, we have described an ontology-based approach for the examination of the cardiovascular system that aims at detecting abnormalities due to CHD.

The main goal of this work has been to highlight the possible applicability of a novel approach in order to support cardiologists in the medical diagnosis. For this reason, the paper is very descriptive and illustrative and often omits many technical details. Moreover, this work represents only a part of an ongoing research and, in particular, it describes only a first step. As a matter of fact, at the moment, the method has been tested only over a few proof data, and so experimental results and performance evaluations are still missing. Next step of our research will be to investigate the applicability and reliability of the proposed approach in real cases.

Future work will also study i) how to overcome present Semantic Web language and tools limitations and ii) how to extend the approach with uncertainty processing.

References

[1] Knowles, R., et al.: Newborn screening for congenital heart defects: a systematic review and cost-effectiveness analysis. Health Technology Assessment 2005 9(44) (2005)
[2] Moein, S.: A Novel Fuzzy-Neural Based Medical Diagnosis Systems. International Journal of Biological and Medical Sciences 1(3), 146–150 (2008)
[3] Tsipouras, M.G., et al.: A Decision Support System for the Diagnosis of Coronary Artery Disease. In: Proceedings of CBMS 2006, Salt Lake City, Utah, USA, June 22-23 (2006)
[4] Lucas, P.: Knowledge Acquisition for Decision-theoretic Expert Systems. AISB Quarterly 94, 23–33 (1996)
[5] Long, W.: Medical diagnosis using a probabilistic causal network. Applied Artificial Intelligence 3(2-3), 367–383 (1989)
[6] Mechouche, A., et al.: Towards an hybrid system for brain MRI images description. In: Proceedings of OWLED 2006, Athens, Georgia, USA, November 10-11 (2006)
[7] Mechouche, A., et al.: Semantic description of brain MRI images. In: Proceedings of SWAMM 2006 Workshop, Edinburgh, Scotland, May 22 (2006)
[8] Vishwanath, K., et al.: OntoDiagram: Automatic Diagram Generation for Congenital Heart Defects in Pediatric Cardiology. In: Proceedings of AMIA 2005, Washington DC, USA, October 22-26 (2005)
[9] Esposito, M., et al.: An Ontology Service to Support Medical Image Labeling. In: Proceedings of IPCV 2007, Las Vegas Nevada, USA, June 25-28 (2007)

ODDI: Ontology-Driven Data Integration

Paolo Ceravolo[1], Zhan Cui[3], Ernesto Damiani[1], Alex Gusmini[2], and Marcello Leida[1]

[1] Università degli studi di Milano,
Dipartimento di Tecnologie dell'Informazione
via Bramante, 65
26013 Crema (CR), Italy
{ceravolo,damiani,leida}@dti.unimi.it
http://ra.crema.unimi.it/kiwi
[2] agusmini@crema.unimi.it
[3] Intelligent Systems Research Centre, BT Group
Orion Building - Adastral Park - Martlesham Heath
IP5 3RE Ipswich - Suffolk, UK
zhan.cui@bt.com

Abstract. Data Integration systems are used to integrate heterogeneous data sources in a single view. Recent works on Business Intelligence do highlight the need of on-time, trustable and sound data access systems. This require for method based on a semi-automatic procedure that can provide reliable results. A crucial factor for any semi automatic algorithm is based on the matching operators implemented. Different categories of matching operators carry different semantics. For this reason combining them in a single algorithm is a non trivial process that have to take into account a variety of options.

This paper proposes a solution based on a categorization of marching operators that allow to group similar attributes on a semantic rich form. The validation of the system have demonstrate how the aggregation of matching operators is not a trivial problem because traditional aggregators produce a compensation effect on operators that can have very different informative values. For this reason this work is now evolving thought the implementation of aggregators based on logic theories, able to distinguish different properties of matching operators.

Keywords: Data Integration; Mapping Generation; Matching Operators.

1 Introduction

Data Integration is becoming a relevant problem in applications that needs to access, analyse and display data coming from heterogeneous data sources. In the literature different problems related to Data Integration was discussed. A first cluster of issues focus on the generation of G that can be either normative, as in [3], or inductive, as in [15]. A second cluster of issues focus on how to represent the mapping between G and L. Here two main approaches exist. In the *Global as View* approch the mapping is provided on G objects by using a L vocabulary. In the *Local as View* approch the mapping is provided on L objects by using a G

I. Lovrek, R.J. Howlett, and L.C. Jain (Eds.): KES 2008, Part I, LNAI 5177, pp. 517–524, 2008.

vocabulary. In [19] a detailed discussion underlines how these approaches impact on application modeling and data reasoning. The last cluster of issues focus on the problem of query answering, studying the computational complexity related to the different solutions, as in [1] or in [16], and defining effective algorithms for dealing to it, as for instance in [14] or [13].

In a generic data integration framewotk these problems cover nearly the totality of the relevant theoretical aspects to be involved in Data Integration. Moreover, in real-world systems, a factor hardly impacting on the success is related to the execution time. Intelligent Systems Research Center in British Telecom (BTExact) is currently developing a *Real Time Business Intelligence* (*RTBI*) platform: a next generation *Business Intelligence* (*BI*) system that "transforms *data* into *information* into *action*". This requires to relay on semi-automatic mapping procedures. A crucial factor for any semi automatic algorithm is based on the matching operators implemented. Different categories of matching operators carry different semantics. Ad a consequence, an improvement of the mapping procedures can be achieved implementing better aggregation of different matching operators.

This paper Introduces *Ontology Driven Data Integration* (ODDI) a framework developed with the *RTBI*) platform in order to drive experimentations in managing a pallet of matching operators. Because this problem can be of an high complexity, the approach chosen start with a procedure aimed at reducing the dimensions of searching space. First all the available association produced by different operators are combined in a cluster. This cluster collect all the elements that can be associated and express the semantics of the associations. but Mapping Generation is activated only on those set of elements that can be queried without violating any integrity constraints. Obviously developing a data integration system introduces several other problems that need to be handled possibly using the best techniques. Our system is based on an ontology as a tool fo representing the common conceptualization, this brings several benefits but the more relevant is that due to the sound logic basis it is possible to perform reasoning task on the knowledge base such as *Consistency Checking* and *Classification* [4]. The description of the system and of the algorithms building him are out of the scope of this paper, more detail on our system can be found in [6] or [7]. The discussion of the paper mainly focus on matching operators and on their aggregation.

The paper is structuerd as follows: Section 2 briefly discuss related works. In Section 3 we provide an introduction to our system then in Section 4 we define the concept of matching, providing a description of the operators used. Finally in Section 5 we provide a validation of the system that allow to discuss some conclusions, proposed in Section 6.

2 Related Work

Data Integration is a relatively old research topic: the advent of data base systems and the consequent increasing level of interactivity between them have motivated the need of integration.

There are several works that can be considered as milestones for the data integration process. Data Warehousing is one of the first solution presented [11].

Another important step is the use of a mediated schema (*mapping*) for a normative approach to the problem, respect to the traditional procedural approach, which provides loose coupling between the data to integrate and the final representation [19].

Data integration is an extremely complex problem [18] and the solutions proposed so far can be employed successfully only in a restrict application area. Actual trend is to exploit semantic of the mediated schema, in order to provide a semantically rich and machine understandable interface to the integrated data. Initially the mapping is generated manually by a domain expert but nowadays, integration needs to satisfy the request of a dynamically evolving environment, where user assisted mapping generation is not advisable. This consideration leads to another aspect, regarding schema matching, which is the problem of finding correspondences between semantically related entities belonging to different schemas. The most exhaustive survey on matching operators is [20], with the important contribution of [12] in semantic aware matching operators. Where all the solution proposed in literature regarding schema matching problems are presented and evaluated. Recent trends in schema matching are, more than discover new matching operators, to consider more than one matching operator at once and combine the final results. In [10] and [9] the authors present a framework for combining different matching operator and retrieving top-k mappings instead of a single mapping exploiting the statistical monotonicity principle for a measure of quality used to retrieve the best mappings. Another interesting approach is [?] where the authors present a new schema integrating approach capable of considering a certain degree of uncertainty for generating the mapping.

3 Ontology Based Data Integration System

A Data Integration System (DIS) aims at solving the problem of integrating different data sources through a common representation. According to [19], we can define a declarative DIS as the following triple:

$$DIS =< G, L, M >$$

Where G is the global representation, L the set of local representations (the data sources) composed by n single representations $s_1, s_2, ..., s_n$ and M is the map of G over L.

A declarative DIS can be classified in two more categories: *Local As View* and *Global As View* approach ([3], [15], [21]) that refers to the definition of the common representation G. The system that we consider in this paper is based on *Global as View* approach: the mapping between G and L is given by associating each concept in G with the combination of the correspondent set of elements in L.

The use of an ontology as a Global Schema G has been already treated in literature [15].

In [4] the authors describe the added value that using an ontology can provide to the DIS: representing the information using a logic formalism offers the possibility to represent the semantic of the model and to perform reasoning on the data for checking the consistency and the integrity of the global representation, and for discovering additional semantic information about the underlying data.

Due to its logical grounding, the expressive power of an ontology is higher than the local representation L: with an ontology we can represent information derived from logical reasoning on the data, additional knowledge that is not directly accessible from the local representation L but that can be derived from it.

This additional knowledge is extracted by running specific applications, such as reasoner and rules engines, on the ontology. Use of metadata as representation of mapping allows the re-use of this metadata in further steps of the data integration process (query translation using SPARQL and XSLT, data protection) allowing to define a flexible query engine that can consider heterogeneous data sources in a unified and seamless representation. This provide several advantages: form an architectural point of view it allows to relay on a common representation layer; from a data exchange point of view it allows to work on a portable format; but the more important is that all the information produced in the integration process can be seen as metadata carrying information that can be used for performing additional tasks such as for instance: (i) reasoning on the reliability of the results provided by matchers (ii) use the results provided by matcher for ontology enrichment purpose (iii) filtering data according to access control rules. Different languages are available for representing ontologies in a machine readable format but it is out of the scope of this paper to go in detail of the features of the main language. For a comprehensive classification of the most popular ontology languages refer to [8]. The Web Ontology Language standard proposed by W3C [22] is becoming a de-facto standard, supported by different tools for both managing (Proteg, Swoop, Jena API, ...) and processing (Pellet, Racer, Jess, ...) ontologies. For these reasons we adopted OWL, in the OWL-DL version, and its SWRL Horn Rules extension to represent the ontologies. As said, more detail on our system can be found in [6] or [7].

3.1 Representing Mapping

Due to the requirement to develop a semi-automatich algorithm to mapping G and L our system must rely on a fine representation of the data. At first, the system extracts information form the global and local representations through a wrapping process. The wrappers generate the set of elements $e^i_{s_k}, e^j_{s_h}, ...$ of L and G. Notice that the element that the wrappers extracts are correspondent to the information container: columns in case of data bases, attributes in case o ontology. Obviously each wrapper is specific of a certain representation: at the actual state of the system, wrappers are available for several data bases engines (MySQL, Oracle, MSAccess) and for OWL ontologies.

The wrapping process extracts the elements, enriching them with all the information that the wrapper can extract: it is easily possible to extract structural information about elements of the representation. JDBC, in case of data sources and Jena in case of OWL, provide sufficient information about the structure of the element. This way we can collect information about elements' names, eventual relations with other elements not wrapped (attribute/concept, column/table), referential integrity rules, data types end eventual descriptions or annotations.

Moreover wrappers analyses also the instances of the elements: pattern matching techniques such as regular expression are also applied to retrieve information about the nature of the content of the element (web page, email, telephone number, postal code, ...). In some cases, statistical analysis on the data can be evaluated (standard deviation, Gaussian curve). Unfortunately, instance analysis is not always feasible: since G is used as an interface to access data, it does not contain instances. To improve the results, domain experts can annotate the ontology elements with sample values.

The mapping definition according to our approach follows the GAV approach formalizing a mapping as:

$$M =< CM, IC, RC, I_lC >$$

CM (Concept Mapping) is a set of mappings between elements of the local representation L and elements of the global representation G. Intuitively CM represents the mapping between columns or function over a set of columns in to a data type properties of a concept c. IC (Integrity Constraints) is a mapping between elements of L only. Considering relations between objects of the same source schema s_a in the local representation L, such as the typical *primary-key→foreign-key* that canbe easilt extracted using specific wrappers; but also correspondences between elements of different source schemas s_i, s_j of L that are semantically related. These correspondences are generated by a instance-based algorithm that make use of Formal Concept Analysis as searching space [6] discovering related elements by performing queries, builded on the base of the FCA lattice, on L.

By using an ontology as global representation, several relations with different semantic meaning can occur between the same domain and range concepts, so that it is important to consider the semantic of relations in the mapping. RC (Relation Constraints) is a mapping defined like IC, but associated to a relation (Object Property) in the global representation G. In [7] the need of this set has been motivated. Traditionally a DIS do not consider this kind of relations but with the advent of semantic data integration it is crucial to consider this aspect. I_lC (Instance-level Constraints) is a set of constraints that are applied to the queries that retrieve the instances of the elements of the global representation.

4 Matching Operators

This paper proposes a solution based on the aggregation of a variety of matching operators. This allows to support different outcomes supporting different semantics. As discussed in [21] the relations among two elements can be described using set relationships among the sets of instances they represents. In particular we have: equivalence ($\equiv$), inclusion ($\subseteq, \supseteq$), intersection ($\cap$) or disjointedness ($\neq$). This relationship define what we call the semantics of matching operators, due to the different type of mapping they propose. In fig. 1 a description of the matching operators supported in ODDI is presented.

	Description	Semantics	Properties
Name Based Operator	Returns 1 if the label of the objects to match are exactly the same (case insensitive)	≡	Symmetric Transitive
EditDistance Operator	Return a degree of matching close to 1 as much the distance between the label of the objects is small	≡ ⊆ ⊇	Symmetric
Thesaurus Based Operator	Considering the label of two elements (typically the name) returns a relation degree between the two elements	≡ ⊆ ⊇ ≠	Symmetric Transitive
Comment Based Operator	Compares the descriptions of two elements. The descriptions are analyzed using standard machine learning techniques	≡	Symmetric
Instance Based Operator	The operator consider first the datatype and avid to compare frequently similar data (boolean, time stamps, indexes, ...). Then it compare the value of instances.	≡ ⊆ ⊇ ∩	Symmetric

Fig. 1. The matching operators supported in ODDI

5 Validation

Following the approach proposed in Section 4, an important problem, related to the aggregation of the matching operators, is arising. In order to aggregate the values coming from these operators we adopted a Weighted Mean function [2]. As shown in fig. 2 we compared the result obtained by ODDI with the algorithms for ontology alignment presented to the Ontology Alignment Evaluation Initiative 2006 [17]. As it can be noted the ODDI algorithm have acceptable results but

Algo	refalign		edna		automs		coma		DSSim		falcon	
	Prec.	Rec.	Prec.	Rec.	Prec.	Rec.	Prec.	Rec.	Prec.	Rec.	Prec.	Rec.
1xx	1	1	0,96	1	0,94	1	1	1	1	0,98	1	1
2xx	1	1	0,9	0,49	0,94	0,64	0,96	0,82	0,99	0,49	0,91	0,85
3xx	1	1	0,94	0,61	0,91	0,7	0,84	0,69	0,9	0,78	0,89	0,78
	hmatch		jhuapl		OCM		prior		RiMOM		ODDI	
	Prec.	Rec.	Prec.	Rec.	Prec.	Rec.	Prec.	Rec.	Prec.	Rec.	Prec.	Rec.
1xx	0,91	1	1	1	0,95	1	1	1	1	1	1	1
2xx	0,83	0,51	0,2	0,86	0,93	0,51	0,95	0,58	0,97	0,87	0,88	0,67
3xx	0,78	0,57	0,18	0,5	0,89	0,51	0,85	0,8	0,83	0,82	0,72	0,67

Fig. 2. Comparison among different ontology matching algorithms

under the best scores. This partial results is due to the fact that the simple aggregation of the values obtained by the different operators produce a rough compensation, that does not allow to the best results to emerge.

6 Conclusions

This paper addressed the issue of managing a variety of matching operator in a complex Data Integration system executing a semiautomatic process. We stressed the need of supporting multiple matching operators. Also our validation shown that a simple Weighted Mean function does seems the best solution. The idea that operators can be weighted according to they reliability is a promising one. But this cannot be implemented with a simple arithmetic function because the relevance of an operator can depend on the informations detected by other operators. For this reason we are working to a solution based on the exploitation of the properties of marching operators that allow to aggregate outcome according to a semantic rich form. Then Mapping Generation is activated only on those set of elements that can be queried without violating any integrity constraints on data. In order to do that we plan to insert the description of matching operators into a knowledge base. The knowledge base is modeled as a Fuzzy Description Logic theory, which allows to consider the matching value as a degree of membership of a fuzzy concept. In [5] it is demonstrated how Fuzzy Description Logic can allow to implement traditional fuzzy controller or fuzzy rules, enriched by a logical theory in background. Decisions are taken on the base of rules and the successful matching are stored as instances of the mapping ontology with all the additional information discovered.

Acknowledgements

This work was partly funded by the Italian Ministry of Research under `FIRB` contract n. `RBNE05FKZ2_004, TEKNE`. This work is also partially supported by British Telecom (`BT`) research and venturing and is subject of a Patent definition.

References

1. Abiteboul, S., Duschka, O.M.: Complexity of answering queries using materialized views, pp. 254–263 (1998)
2. Aczel, J.: On Weighted synthesis of judgments. Aequationes Math. 27, 288–307 (1984)
3. Braun, P., Lotzbeyer, H., Schatz, B., Slotosch, O.: Consistent integration of formal methods, pp. 48–62 (2000)
4. Cui, Z., Damiani, E., Leida, M.: Benefits of Ontologies in Real Time Data Access. In: Proceedings IEEE/IES Conference on Digital Ecosystems and Technologies (2007)
5. Bobillo, F., Straccia, U.: Fuzzydl: An expressive fuzzy description logic reasoner. In: 2008 International Conference on Fuzzy Systems (FUZZ 2008). IEEE Computer Society, Los Alamitos (2008)

6. Ceravolo, P., Cui, Z., Gusmini, A., Damiani, E., Leida, M.: An fca-based mapping generator. In: 12th IEEE Conference on Emerging Technologies and Factory Automation (2007)
7. Ceravolo, P., Damiani, E., Gusmini, A., Leida, M.: Using ontologies to map concept relations in a data integration system. In: Meersman, R., Tari, Z., Herrero, P. (eds.) OTM-WS 2007, Part II. LNCS, vol. 4806, pp. 1285–1293. Springer, Heidelberg (2007)
8. Corcho, O., Gomez Perez, A.: Evaluating knowledge representation and reasoning capabilities of ontology specification languages. In: Proc. of ECAI 2000 Workshop on Applications of Ontologies and Problem-Solving Methods (2000)
9. Avigdor Gal. Managing uncertainty in schema matching with top-k schema mappings. pp. 90–114 (2006)
10. Gal, A.: Why is schema matching tough and what can we do about it? SIGMOD Rec. 35(4), 2–5 (2006)
11. Inmon, W.H.: Building the data warehouse. QED Information Sciences, Inc., Wellesley (1992)
12. Shvaiko, P., Euzenat, J.: Ontology matching. Springer, Heidelberg (2007)
13. Duschka, O.M., Genesereth, M.R., Levy, A.Y.: Recursive query plans for data integration. Journal of Logic Programming 43(1), 49–73 (2000)
14. Grahne, G., Mendelzon, A.O.: Tableau techniques for querying information sources through global schemas. In: Beeri, C., Bruneman, P. (eds.) ICDT 1999. LNCS, vol. 1540, pp. 332–347. Springer, Heidelberg (1998)
15. Hakimpour, F., Geppert, A.: Global schema generation using formal ontologies (2002)
16. Halevy, A.Y.: Answering queries using views: A survey. VLDB Journal: Very Large Data Bases 10(4), 270–294 (2001)
17. Euzenat, J., et al.: Results of the Ontology Alignment Evaluation Initiative 2006. In: Cruz, I., Decker, S., Allemang, D., Preist, C., Schwabe, D., Mika, P., Uschold, M., Aroyo, L.M. (eds.) ISWC 2006. LNCS, vol. 4273. Springer, Heidelberg (2006)
18. Lenzerini, M.: Data integration is harder than you thought. In: CooplS 2001: Proceedings of the 9th International Conference on Cooperative Information Systems, London, UK, pp. 22–26. Springer, Heidelberg (2001)
19. Lenzerini, M.: Data Integration: A Theoretical Perspective. In: PODS 2002, pp. 233–246 (2002)
20. Rahm, E., Bernstein, P.: A survey of approaches to automatic schema matching. The VLDB Journal 10(4), 334–350 (2001)
21. Parent, C., Spaccapietra, S.: Issues and approaches of database integration. Commun. ACM 41(5es), 166–178 (1998)
22. OWL - Web Ontology Language defintion, http://www.w3.org/TR/owl-features/

Enhancing Recommendations through a Data Mining Algorithm

Alexis Lazanas and Nikos Karacapilidis

IMIS Lab, MEAD, University of Patras, 26504 Rio Patras, Greece
{alexlas,nikos}@mech.upatras.gr

Abstract. This paper reports on the development of a new data mining algorithm that formulates purposeful association rules out of the transactions' database of a transportation management system,. The proposed algorithm is generic and capable to construct such rules by creating a large set of related items. The constructed rules can be used by the system's recommender module, which is responsible for providing recommendations to the associated users. The recommendation process takes into account the constructed rules and techniques that derive from the area of collaborative filtering. Our approach enables users to receive high quality recommendations for their upcoming transactions.

Keywords: Data mining, Knowledge Association Rules, Recommender systems.

1 Introduction

Data mining, also referred to as knowledge discovery in databases, has been defined as a process of non-trivial extraction of implicit, previously unknown and potentially useful information (such as knowledge rules, constraints, regularities, etc.) from databases [12]. Other terms, such as knowledge mining from databases, knowledge extraction, data dredging and data analysis carry a similar or slightly different meaning. By knowledge discovery in databases, interesting knowledge, regularities, or high-level information can be extracted from the relevant sets of data, which can be elaborated from different angles.

In earlier works [16, 18], we presented our approach on the integration of hybrid recommendation techniques into an agent-based transportations transaction management platform, namely *FTMarket*. First of all, we proposed a hybrid approach that combines different recommendation techniques in order to provide the user with more accurate suggestions [15]. The overall process is coordinated by a recommender agent, which is responsible for invoking a correspondent web service which carries out multiple tasks, such as application of knowledge rules, selection of the appropriate recommendation technique, and synthesis of the derived knowledge through the exploitation of collaborative filtering and data mining techniques. The presence of the recommendation agent guarantees that the user will be provided with continuous and dynamically updated recommendations. Extending our previous work, this paper reports on the development of a data mining algorithm for constructing association rules through the exploitation of FTMarket's transactions' database. Our ultimate

I. Lovrek, R.J. Howlett, and L.C. Jain (Eds.): KES 2008, Part I, LNAI 5177, pp. 525–532, 2008.

objective is to make the system provide more accurate recommendations to the users that participate in the recommendation phase [17].

The remainder of the paper is structured as follows: Section 2 refers to related work on data mining algorithms and association rules construction. Section 3 presents the proposed data mining algorithm, while Section 4 discusses its encapsulation into the recommender module of the abovementioned system. Finally, conclusions and future work directions are presented in Section 5.

2 Related Work

The problem of finding association rules falls within the scope of database mining [1, 2, 8, 10, 13], also called knowledge discovery in databases [7, 12]. Related, but not directly applicable, work includes the induction of classification rules [4, 6, 7], discovery of causal rules [5, 11], learning of logical definitions [9], and fitting of functions to data and clustering [1, 17]. In the past few years, the database community studied the problem of rule mining extensively (under the name of association rule mining) [1]. This study was focused on using exhaustive search to find all rules in data that satisfy the user-specified minimum support (minsup) and minimum confidence (minconf) constraints. Although the complete set of rules may not be directly used for accurate classification, effective and efficient classifiers have been built using the rules, e.g. CBA [7], LB [8] and CAEP [4]. The major strength of such systems is that they are able to use the most accurate rules for classification (thus rendering good results). However, they also have some weaknesses, which are inherited from association rule mining. Traditional association rule mining uses only a single minsup in rule generation, which is inadequate for unbalanced class distribution. Classification data often contain a huge number of rules, which may cause combinatorial explosion. For many datasets, the rule generator is unable to generate rules with many conditions, while such rules may be important for classification. An algorithm for mining all association rules, henceforth referred to as the AIS algorithm, was presented in [2] together with the SETM algorithm [8]. Two new algorithms, namely Apriori and AprioriTid [19], use both synthetic and real-life data, and they have been proved to outperform the earlier algorithms. All the above algorithms have a performance gap that is shown to increase with problem size, and ranges from a factor of three for small problems to more than an order of magnitude for large problems [19, 20].

The data mining algorithm we propose in this paper is capable to formulate strong association rules by also taking into account the user's preferences. The constructed rules consist of highly connected elements and are combined with collaborative filtering techniques in the recommendation module of the system, in order to provide a higher level of recommendation to the final user. At this phase, transactions data are gathered through a knowledge construction algorithm. In our case, the data mining process constructs a model from the recommendation module's database that may produce well defined knowledge association rules. After the completion of the above process, the constructed knowledge-based rules participate in the production of knowledge-based recommendation data, which are evaluated and synthesized in the last phase of recommendation. Due to space limitations, issues concerning the knowledge synthesis carried out at this phase are omitted in this paper.

3 The Data Mining Algorithm

3.1 Problem Modeling

To describe the overall problem given the FTMarket's database of transportation transactions, it is desirable to discover the important associations among "items" such that the presence of some items in a transaction will imply the presence of other items in the same transaction. A mathematical model is proposed to address the problem of mining association rules. So, let:

$I = \{i_1, i_2, \ldots, i_m\}$ be a set of literals (items). Let D be a set of transactions, where each transaction T is a set of items such that $T \subseteq I$. Each transaction is associated with an identifier, called TID. Let X be a set of items. A transaction T is said to contain X if and only if $X \subseteq T$. An association rule is an implication of the form $X \Rightarrow Y$, where $X \subset I$, $Y \subset I$ and $X \cap Y = \varnothing$. The rule $X \Rightarrow Y$ holds in the transaction set D with confidence c if $c\%$ of transactions in D that contain X also contain Y. A rule $X \Rightarrow Y$ is defined as having support s in the transaction set D if $s\%$ of transactions in D contain $X \cup Y$. Confidence c denotes the strength of implication (user-defined), and support s indicates the frequencies of the occurring patterns in the rule (also user-defined). It is often desirable to pay attention only to those rules that may have reasonably large support. Such rules with high confidence and strong support are referred to as strong rules in [2, 12].

The task of mining association rules is essentially to discover strong association rules in large databases. The problem of mining association rules is decomposed into the following two steps [3]: (i) discover the large itemsets, i.e., the sets of itemsets that have transaction support above a pre-determined minimum support s, and (ii) use the large itemsets to generate the association rules for the database. It is noted that the overall performance of mining association rules is determined by the first step. After the large itemsets are identified, the corresponding association rules can be derived in a straightforward manner. In the proposed algorithm, candidate itemsets are generated and counted "on-the-fly" as the database is scanned. After processing a transaction, it is determined which of the itemsets that were found to be large in the previous pass are contained in this transaction. New candidate itemsets are generated by extending these large itemsets with other items in the transaction. A large itemset L, is extended with only those items that are large and occur later in the ordering of items than any of the items in L. The candidate items generated from a transaction are added to the set of candidate itemsets maintained for each pass (at the same time, we count the occurrences of the corresponding entries). The following algorithm embodies these ideas:

```
minsup = transportation.get_user_support();
   // the minimum accepted support provided by the user

L1 = {large 1-itemsets}
for (k=2; Lk-1 ≠ Ø; k++) do
{
        Ck = generate_candidates(Lk-1);
               //generate new candidate itemsets
        for each transaction t ∈ D do
        {
        Ct = subset(Ck, t); //Candidates contained in t
        for each candidate c ∈ Ct do
               c.count ++;
        }
        Lk = {c ∈ Ck |c.count ≥ minsup}
   }
Rules = ∪k Lk
```

The `generate_candidates(Lk-1)` function takes as argument L_{k-1}, the set of all large (k-1) itemsets. It returns a superset of the set of all large k-itemsets.

First we join L_{k-1} p with L_{k-1} q :

```
insert into Ck
select p.item1, p.item2, ..., p.itemk-1, q.itemk-1
from Lk-1 p, Lk-1 q
where p.item1 = q.item1, . . ., p.itemk-2 = q.itemk-2, p.itemk-1 <
q.itemk-1;
```

Next, we delete all itemsets `c ∈ Ck` such that some `(k-1)-subset` of `c` is not in `Lk-1`:

```
for each itemset c∈ Ck do
        for each (k-1)-subsets s of c do
        if (s ∉ Lk-1) then delete c from Ck;
```

3.2 An Example

Consider the example transaction set given in Table 1. In each iteration (pass), the algorithm constructs a candidate set of large itemsets, counts the number of occurrences of each candidate itemset, and then determines large itemsets based on a pre-determined minimum support (*minsup*). In the first iteration, the algorithm simply scans all transactions to count the number of occurrences of each item. The set of candidate 1-itemsets, C_1, obtained is shown in Table 2.

Assuming that the minimum transaction support required from the user is 2, the set of large 1-itemsets, L_1, composed of candidate 1-itemsets with the minimum support required, can then be determined. To discover the set of large 2-itemsets, in view of the fact that any subset of a large itemset must also have minimum support, we use $L_1{*}L_1$ to generate a candidate set of itemsets C_2, where "*" stands as the operator for

Table 1. FTMarket's Transactions' table (*D*)

Transaction ID	Items
S00010	{Fuel, Athens, Patras, Zagreb, Unibrokers LTD, Tanker, Economic}
S00011	{Liquid, Athens, Zagreb, Transcargo, Truck, Express}
S00012	{Package, Patras, Athens, Berlin, TransCargo, Truck, Express}
S00013	{Textiles, Athens, Rome, S-Cargo, Airplane, Economic}
S00014	{Liquid, Athens, Patras, Ancona, Rome, Berlin S-Cargo, Railway, Express}

Table 2. The C_1 set of candidate itemsets

C_1

Itemset (C_1)	Support	Add to L_1
{Athens}	5	Yes
{Patras}	3	Yes
{Zagreb}	2	Yes
{Berlin}	2	Yes
{Rome}	2	Yes
{Ancona}	1	No

$\Rightarrow$

L_1

Large Itemset (L_1)	Support
{Athens}	5
{Patras}	3
{Zagreb}	2
{Berlin}	2
{Rome}	2

Table 3. The C_2 set of candidate itemsets

C_2

Itemset (C_2)	Support	Add to L_2
{Athens, Patras}	3	Yes
{Athens, Zagreb}	2	Yes
{Athens, Berlin}	2	Yes
{Athens, Rome}	2	Yes
{Patras, Zagreb}	1	No
{Patras, Berlin}	2	Yes
{Patras, Rome}	1	No
{Zagreb, Berlin}	0	No
{Zagreb, Rome}	0	No
{Berlin, Rome}	1	No

$\Rightarrow$

L_2

Large Itemset (L_2)	Support
{Athens, Patras}	3
{Athens, Zagreb}	2
{Athens, Berlin}	2
{Athens, Rome}	2
{Patras, Berlin}	2

concatenation. C_2 consists of 2-itemsets. Next, the transactions in *D* are scanned and the support of each candidate itemset in C_2 is counted. Table 3 represents the result from such a counting in C_2. The set of large 2-itemsets, L_2, is therefore determined based on the support of each candidate 2-itemset in C_2.

The set of candidate itemsets, C_3, is generated from L_2 as follows: From L_2, two large 2-itemsets with the same first item, such as {*Athens, Patras*} and {*Athens, Berlin*}, are identified first. Then, the algorithm tests whether the 2-itemset {*Patras, Berlin*}, which consists of their second items, constitutes a large 2-itemset or not. Since {*Patras, Berlin*} is a large itemset by itself, we know that all the subsets of {*Athens, Patras, Berlin*} are large and then it becomes a candidate 3-itemset. There is no other candidate 3-itemset from L_3.

Table 4. Generating large itemset L_3

C_3 ⇒ **L_3**

Itemset (C_3)	**Support**	**Add to L_3**
{Athens, Patras, Berlin}	2	Yes

Large Itemset (L_3)	**Support**
{Athens, Patras, Berlin}	2

Then all the transactions are scanned and the large 3-itemsets L_3 are discovered (see Table 4 - L_3). Since there is no candidate 4-itemset to be constituted from L_3, the algorithm ends the process of discovering large itemsets.

4 Recommendation Issues

The association rules formulated using the algorithm discussed in the previous section can be involved in the recommendation phase of FTMarket. At this phase, the constructed rules participate in the knowledge synthesis and are presented to the customer of FTMarket who requests recommendation services concerning an upcoming transaction.

The system's recommender module is responsible of combining collaborative filtering techniques and the association rules (knowledge mining) that have been constructed using the data mining algorithm and is coordinated from the Recommender Agent (*RA*). The software module that implements the algorithm will construct the appropriate association rules following the steps presented in Section 3.1. To establish the communication with the FTMarket's database, we use SQL Server 2005 management studio and the overall procedure is performed through SQL queries sent directly to transactions' tables. The algorithm retains all the appropriate information during its execution cycle, such as: database tables of the candidate rules (C_k) and association rules formulated from large itemsets (L_k). After the completion of the whole process, *RA* performs a synthesis of the recommendation elements (Collaborative Filtering techniques and association rules) and presents the results to the user.

An example of a hypothetical scenario could involve a user who has requested recommendation for a transportation transaction from Athens to Berlin. This transaction can be described as:

Customer ID	*C0002342*
Loading Terminal	*Athens*
Delivery Terminal	*Berlin*
Transportation Plan	*Economic*
Freight Type	*Any*
Carrier	*Any*
Transport Means	*Any*

Taking into consideration the above transaction, the data mining algorithm will try to formulate association rules for the items {*Athens, Berlin*} that participate in the transportation scenario. Assuming that the completed transactions are those contained in Table 1 (Section 3.2), the output will be the rule L_3 = {*Athens, Patras, Berlin*}. The rule L_3 (with a support of 2), suggests that the system should recommend the following transportation scenario:

Customer ID	*C0002342*
Loading Terminal (start)	*Athens*
Trans Loading Terminal	*Patras*
Delivery Terminal (end)	*Berlin*
Transportation Plan	*Express*
Freight Type	*Liquid*
Carrier	*TransCargo, S-cargo*
Transport Means	*Truck, Railway*

Note that the algorithm suggests that the requested transaction should be amended in order to comply with rule L_3. The initial request concerned a direct transportation scenario (from *Athens* to *Berlin*). The L_3 rule though, added an additional transloadings' terminal (*Patras*). Moreover, the suggested transportation plan has been changed, while recommendations are also provided for Freight type, Carrier and Transportation means [15].

5 Discussion and Conclusion

In this paper, we addressed the issue of enhancing the recommendation process of a transportation management system through an innovative data mining algorithm. More specifically, we proposed an efficient algorithm for mining association rules out of the transactions' database tables. These rules are of the form $L_k = \{item_1, item_2, ..., item_k\}$ and represent the frequency of the occurring patterns in the database.

The proposed algorithm enhances the recommendation module of an already implemented system through the knowledge construction process. As a result, the algorithm has a significant contribution to the hybrid character of the overall recommendation approach, due to the fact that the collaborative filtering techniques are strongly supported by the derived association rules. This model of recommendation policy could easily be embedded in a variety of applications that provide recommendations in E-Markets. The main advantage of our approach concerns the high level of "objectivity" it offers. The recommendation is not limited to a score provided by specific users, but it uses rule-based knowledge to construct more accurate - user independent - suggestions.

Future work directions concern the enhancement of the proposed algorithm in a manner of being capable to construct a set of rules (large itemsets) for each item *I* involved in the transaction. Our present implementation considers as candidates items only the loading and destination terminals (cities) and formulates the correspondent rules. A more advanced algorithm should focus on the formulation of association rules that derive from the consideration of additional item categories (fields) of a transaction, such as transportation plans, freight type, transport mean and carrier. The association rules of this type will have the form $Rules = \{L_k , M_k , ..., Z_k\}$, where L_k , M_k , ..., Z_k are the formulated large itemsets for each item category in the transaction's record.

References

1. Anwar, T., Beck, H., Navathe, S.: Knowledge mining by imprecise querying: A classification-based approach. In: Proc. of the IEEE 8th Int'l Conf. on Data Engineering, Phoenix, Arizona (1992)
2. Agrawal, R., Imielinski, T., Swami, A.: Mining association rules between sets of items in large databases. In: Proc. of the ACM SIGMOD Conference on Management of Data, Washington, pp. 207–216 (1993)
3. Anwar, T., Shamkant, B., Beck, H.: Knowledge mining in databases: A unified approach through conceptual clustering. Technical report, Georgia Institute of Technology (1992)
4. Catlett, J.: Mega induction: A test flight. In: Proc. of 8th International Conference on Machine Learning, San Marco, California, pp. 596–599 (1991)
5. Cooper, G., Herskovits, E.: A Bayesian method for the induction of probabilistic networks from data. Machine Learning (1992)
6. Fayyad, U., Weir, N., Djorgovski, S.G.: Skicat: A machine learning system for automated cataloging of large scale sky surveys. In: Proc. of the 10th International Conference on Machine Learning, pp. 749–754 (1993)
7. Han, J., Cai, Y., Cercone, N.: Knowledge discovery in databases: An attribute oriented approach. In: Proc. of the VLDB Conference, Vancouver, Canada, pp. 547–559 (1992)
8. Houtsma, M., Swami, A.: Set-oriented mining of association rules. Research Report RJ 9567, IBM Almaden Research Center, San Jose, California (1993)
9. Muggleton, S., Feng, C.: Efficient induction of logic programs. In: Steve Muggleton, Inductive Logic Programming. Academic Press, London (1992)
10. Michalski, R., Kerschberg, L., Kaufman, K., Ribeiro, J.S.: Mining for knowledge in databases: The INLEN architecture, initial implementation, and first results. Journal of Intelligent Information Systems (113), 1–85 (1992)
11. Pearl, J.: Probabilistic reasoning in intelligent systems: Networks of plausible inference (1992)
12. Piatestsky-Shapiro, G.: Knowledge Discovery in Databases. AAAI/MIT Press (1991)
13. Tsur, S.: Data dredging. IEEE Data Engineering Bulletin 13(4), 58–63 (1990)
14. Barbara, D., Chen, P.: Using Self-Similarity to Cluster Large Data Sets. Data Mining and Knowledge Discovery 7(2), 123–152 (2003)
15. Lazanas, A., Karacapilidis, N., Katsoulis, V.: Applying Hybrid Recommendation Policies through Agent-Invoked Web Services in E-Markets. In: Proc. of the 9th International Conference on Enterprise Information Systems (ICEIS 2007), Madeira, Portugal (2007)
16. Lazanas, A., Karacapilidis, N., Pirovolakis, Y.: Providing Recommendations in an Agent-Based Transportation Transactions Management Platform. In: Proc. of the 8th International Conference on Enterprise Information Systems (ICEIS 2006), Paphos, Cyprus (2006)
17. Sarwar, B.M., Karypis, G., Konstan, J.A., Riedl, J.: Item-Based Collaborative Filtering Recommendation Algorithms. In: Proceedings of the 10th International World Wide Web Conference, Hong Kong. ACM Press, New York (2001)
18. Karacapilidis, N., Lazanas, A., Megalokonomos, G., Moraitis, P.: On the Development of a Web-based System for Transportation Services. Information Sciences 176(13), 1801–1828 (2006)
19. Agrawal, R., Srikant, R.: Fast algorithms for mining association rules. In: Proc. of the VLDB Conference, Santiago, Chile (1994)
20. Srikant, R., Vu, Q., Agrawal, R.: Mining Association Rules with Item Constraints. In: Proc. Third Int'l Conf. Knowledge Discovery and Data Mining, pp. 67–73. AAAI Press, Menlo Park (1997)

Using Sense Recognition to Resolve the Problem of Polysemy in Building a Taxonomic Hierarchy

Lei Liu[1], Sen Zhang[1], Lu Hong Diao[1], Shu Ying Yan[2], and Cun Gen Cao[2]

[1] College of Applied Sciences, Beijing University of Technology
[2] Institute of Computing Technology, Chinese Academy of Sciences
{liuliu_leilei,zhangsen,diaoluhong}@bjut.edu.cn,
{yanshuying,cgcao}@ict.edu.cn

Abstract. Hyponymy is used to build a taxonomic hierarchy. But the terms in hyponymy may have multiple senses. It will cause the problem of polysemy and affect the building of taxonomic hierarchy. In order to solve the problem, we present a method of sense recognition of hyponymy based on vector space model. Firstly we acquire the contexts of hyponymy from Chinese free corpus. Secondly we use Cilin to construct a relation-word vector space. Then we use latent semantic analysis to reduce the dimension of the vector space. In the final phase, we recognize the senses of hyponymy using average-group clustering. Experimental results show that the method can provide adequate discrimination of the different senses.

Keywords: Hyponymy Relation, Latent Semantic Analysis, Sense Clustering, Polysemous Words.

1 Introduction

Automatic acquisition of semantic relations from text has received much attention in the last ten years. Especially, hyponymy relations are important in accuracy verification of ontologies, knowledge bases and lexicons. Hyponymy is a semantic relation between lexical terms.

Given two terms X and Y, there is the hyponymy between X and Y if the sentence "X is a (kind of) Y" is acceptable. X is a hyponym of Y, and Y is a hypernym of X. Hyponymy is also called as subordination, or the "isa" relation. This relation is reflexive, transitive and asymmetrical under the condition of the same word sense. We denote a hyponymy relation as HR(X, Y), as in the following example:

中国是一个发展中国家 ---HR(中国,发展中国家)
(China is a developing country ---HR(China, developing country))

The acquired hyponymy can be used to create a taxonomic hierarchy. But the terms in hyponymy relations may have multiple senses. Term senses are point to multidimensional entities, and can barely be analysed in process of unique assignments to points in taxonomic hierarchy. Polysemy is a widespread and pervasive feature affecting the building of taxonomic hierarchy.

I. Lovrek, R.J. Howlett, and L.C. Jain (Eds.): KES 2008, Part I, LNAI 5177, pp. 533–540, 2008.

There have been many attempts to develop automatic methods to acquire hyponymy from text corpora. One of the first studies was done by Hearst [1]. She proposed a method for retrieving concept relations from unannotated text (Grolier's Encyclopedia) by using predefined lexico-syntactic patterns. Other researchers also developed other ways to obtain hyponymy. Most of these techniques are based on particular linguistic patterns. However, those methods underestimate the influence of the problem of multiple word senses.

Rydin describes an algorithm to create a hypernym-hyponym based lexicon for Swedish texts [2]. Kenji developed a kind of statistical scores method to discover polysemous words [3]. But both of them didn't touch on how to solve the polysemy under the condition where hypernym and hyponym are both polysemous words.

We can also solve the influence of polysemy with drawing lessons from some methods of word sense discrimination in natural language processing [4]. McRoy describe how to combine several information knowledge sources for word sense discrimination. But no quantitative evaluation is given on the relative importance of each knowledge source [5]. Schutze presents an automatic and unsupervised disambiguation algorithm based on clustering. Senses are interpreted as clusters of similar contexts of the ambiguous word. [6].

In this paper, for this problem, we present an algorithm of sense recognition of hyponymy based on vector space model. The rest of the paper is organized as follows. Section 2 elaborates on the principles for this work, section 3 presents the framework of this algorithm, section 4 conducts a performance evaluation of the proposed method, and finally section 5 concludes the paper.

2 Problem Descriptions

In a hyponymy HR(X, Y), both X and Y may be polysemous words. Given a HR(X, Y), let X have i kinds of senses: $M(X)= \{m_1, m_2, \ldots, m_i\}$, and Y have j kinds of senses: $M(Y)=\{m'_1, m'_2, \ldots, m'_j\}$. There is a hyponymy between X and Y if and only if the sense of X is m_a, and the sense of Y is m'_b. We denote this hyponymy relation as $HR(X, Y)|\,(m_a, m'_b)$. For example:

M(病毒)={计算机病毒, 微生物学病毒}
(M(virus)={a computer virus, a virus in virology})
M(程序)={计算机程序, 一系列被执行的步骤}
(M(program)={ a computer program, a series of steps to be carried out })
HR(病毒, 程序)|(计算机病毒, 计算机程序)
(HR(virus, program)|(a computer virus, a computer program))

Definition 1 (hyponymy transitivity of the sense- preserving). Given a set of terms $c_1, c_2, \ldots, c_n$, if $HR(c_1, c_2)|(m_1, m_2)$, $HR(c_2, c_3)|(m_2, m_3)$, …, $HR(c_{n-1}, c_n)|(m_{n\square 1}, m_n)$ then there exists a hyponymy $HR(c_1, c_n)|\,(m_1, m_n)$ between c_1 and c_n.

For describing our problems in detail, we give a set of hyponymy relations implying polysemous in Table 1. Then we use those relations in Table 1 to construct a graph of space structure in Fig 1. In the space structure, we suppose the senses taken by each term in relations should be different.

Table 1. A set of hyponymy relations

HR(病毒, 生物) (HR(virus, biology))
HR(病毒, 程序) (HR(virus, program))
HR(流感, 病毒) (HR(flu, virus))
HR(蠕虫, 病毒) (HR(worm, virus))
HR(蠕虫, 动物) (HR(worm, animal))
HR(审判, 程序) (HR(justice, program))

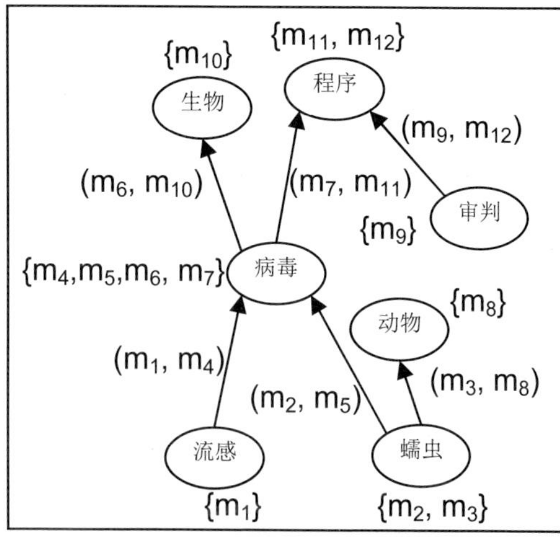

Fig. 1. The space structure of hyponymy

As we can see from the Fig1, all of "病毒"(virus), "蠕虫"(worm) and"程序"(program) are polysemous words. The hyponymy transitivity can't establish because of the polysemy of these words.

For example, given HR(流感, 病毒)|(m_1, m_4) and HR(病毒, 程序)|(m_7, m_{11}), we can't infer whether (流感,程序)|(m_1, m_{11}) establish or not because we don't know whether $m_4 = m_7$. So we firstly need to judge the senses of a term in different hyponymy relations is equal or not. Then we can build a correct taxonomic hierarchy based on hyponymy transitivity of the sense-preserving.

We think the chosen senses of terms be decided mutually in their hyponymy, namely for a HR(c, c')|(m, m'), the sense m of c is decided by c', and the sense m' of c' is decided by c. Here (m, m') is called the sense of the HR(c, c'). According to this hypothesis, the problem of single word sense discrimination can convert into sense discrimination of word pairs in hyponymy. Take the senses m_5 and m_7 of "病毒" in Figure 1 as an example, the problem of judging whether $m_5 = m_7$ can be converted into judging whether (m_2, m_5) = (m_7, m_{11}). We can infer $m_5 = m_7$ if (m_2, m_5) is equal to (m_7, m_{11}), and further we can infer HR(蠕虫,程序)|(m_4,m_7) based on hyponymy transitivity.

Because the sense of a hyponymy is decided by its word pairs, and the senses of a word is generally closely related with the context of its place, we suppose (m, m') is decided by the context feature of the co-occurrence of c and c' in a relation(c, c')|(m, m'). We can think of the feature vector of context as a kind of semantic representation of (m, m'). So we suppose if the contexts of two hyponymy relations are similar, then their senses are similar. The problem of hyponymy sense discrimination is taken as a cluster procedure of hyponymy context.

As an example of "病毒", the related hyponymys include HR(病毒,生物), HR(病毒,程序), HR(流感, 病毒), HR(蠕虫, 病毒). They is called the sense discrimination gather of "病毒", and we denote these relations as **ρ("病毒").** If their contexts are clustered two groups: { (病毒, 生物), (流感, 病毒)} and {(病毒, 程序), (蠕虫, 病毒)}, then we can draw a conclusion of $m_4=m_6$ and $m_5=m_7$. Furthermore, we can infer HR(蠕虫, 程序) and HR(流感, 生物) based on hyponymy transitivity.

3 Method

For a term c and its sense discrimination gather **ρ(c)**, we present a method of hyponymy sense recognition with drawing lessons from some methods of word sense discrimination. Our method consists of four phases as shown in Fig2. In Phase I, we acquire the context feature word. In Phase II, we acquire the contexts of hyponymy in **ρ(c)**, and use these context features to construct a relation-word vector space. In Phase III, we use latent semantic analysis to reduce the dimension of the vector space. In the final phase, we recognize the senses of hyponymy using sense clustering. We also adopt some assistance strategies including heuristic rules and Cilin semantic classes (a Chinese thesaurus).

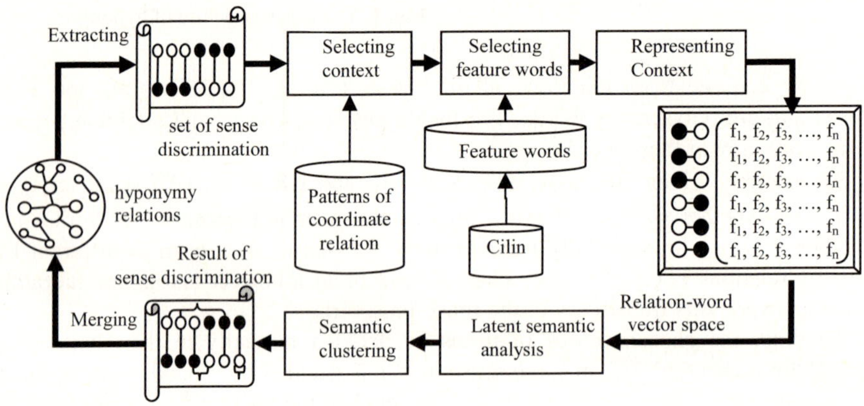

Fig. 2. The framework of sense discrimination of hyponymy

3.1 Phase I: Selecting Feature Words

We take only co-occurrence words as context feature information. The feature word w is a word which satisfies the following conditions: (1) The w is an substantive word, namely the part of speech is a noun, verb, adjective etc; (2) The document number of containing w in the Chinese train corpus $\mathcal{D}$: $DF_w > \alpha$; (3) The inverse document frequency of w: $IDF_w > \beta$; α , $\beta \in$ R are thresholds, IDF_w is defined by the equation (1), and $|\mathcal{D}|$ is the total number of documents in $\mathcal{D}$.

$$IDF_w = \log\left(\frac{|\mathcal{D}|}{DF_w}\right) \tag{1}$$

A set of candidate feature words is selected based on above three conditions from a segmentation lexicon, and we denote FWL=$\{w_1, w_2, w_3, \ldots, w_n\}$. Then for decreasing the influence of synonym feature words, we merge the synonym feature words with

Cilin [7]. Cilin provides the Mandarin synonym sets in a hierarchical structure [8]. If a feature word has multiple Cilin classes, it will be merged multiple times based on every class. After merging synonym feature words, the feature word list changes into FWL={Syn(w_1), Syn(w_2), … , Syn(w_m)}, where Syn(w_i) is a set of synonym feature words of word w_i.

The weight of each feature word is adjusted using three factors: TFIDF score, log-likelihood factor, and distance from the target term pairs of hyponymy.

(1) $TFIDF_w$ is defined by the equation (2), where $|Wef_{w,r}|$ stands for the number of a feature word w appears in the contexts of a hyponymy r, and IDF_w is the inverse document frequency of w in the corpus $\mathcal{D}$.

$$\mathrm{Tfidf}(w, r) = Wef_{w,r} * \mathrm{Idf}_w \tag{2}$$

(2) The log-likelihood factor of a feature word w and a hyponymy r is computed as the equation (3) [9].

$$\mathrm{Logw}(w, r)=\min\{1,|\mathrm{CTU}\ (r)|/10\}*\log(\frac{\Pr(w\mid r)}{\Pr(w)}+1) \tag{3}$$

Where Pr(w) is estimated from the frequency of w in the train corpus, and Pr(w|r) is the frequency of w in the contexts of r. |CTU(r)| is the number of context of r. min{1,|CTU(r)|/10} can avoid poor estimation for hyponymys with sparse contexts.

(3) The equation of distance about feature word w and hyponymy r=HR(c, c') is shown in equation (4). Here dis(w,c) is the number of feature words between w and c.

$$\mathrm{Disw}(w, r)=10/\max\{10, \min\{\mathrm{dis}(w,c), \mathrm{dis}(w,c')\}\}. \tag{4}$$

3.2 Phase II: Constructing Context Feature Vector Space

Hyponymy is different from single term in selecting context, and need to consider the distance problem of the hyponym and hypernym. Given a hyponymy (c_1,c_2), its contexts originate from all document containing the co-occurrence of c_1 and c_2 in corpus, and satisfy the following conditions: (1) the distance of c_1 and c_2 in document $d_1<\alpha$; (2) the distance of radius using c_1 or c_2 as central $d_2<\beta$; α, β are integer thresholds. Here the distance is the number of feature words.

The contexts of hyponymy require containing the co-occurrence of c_1 and c_2, so it may cause the data sparse problem. We present a heuristic method of using a set of coordinate relation patterns for solving this problem. We denote the contexts of (c_1,c_2) as CTU(c_1,c_2)={ct_1, ct_2, …, ct_n}, where ct_i is the i context item of (c_1,c_2).

Given a hyponymy r = HR(c,c'), the context CTU(c,c') = {ct_1, ct_2, ct_3, …, ct_m}, the feature words table FWL = {Syn(w_1), Syn(w_2),…, Syn(w_k)}, then the context feature vector of r can be counted by the equation (5)(6)(7).

$$\mathrm{Unionw}(w_i,r) = \mathrm{Tfidf}(w_i,r)*\mathrm{Logw}(w_i,r)*\mathrm{Disw}(w_i,r) \tag{5}$$

$$\mathrm{Unionw}(\mathrm{Syn}(w_i),r) = \mathrm{Unionw}(w_{i1},r) + \mathrm{Unionw}(w_{i2},r)+...+\mathrm{Unionw}(w_{in},r) \tag{6}$$

$$\mathrm{ConVec}\ (r) = (\ \mathrm{Unionw}(\mathrm{Syn}(w_1),r), \mathrm{Unionw}(\mathrm{Syn}(w_2),r)\ , ...\ , \mathrm{Unionw}(\mathrm{Syn}(w_k),r)\ ) \tag{7}$$

The equation (5) counts the total weight of a feature word in the context of r by the product of the three factors. The weight of synonym feature words is counted by the equation (6). We acquire the context feature vector of r by the equation (7), and each dimension in the context vector is the weight of synonym feature words in the context.

So a high dimension "relation-feature word" matrix $A_{m\times n}$ is constructed, and the row elements are the hyponymy relations in ρ(c) and the column elements are the synonym feature words. If there exists a column that its values all are 0 in $A_{m\times n}$, then throw away this column.

3.3 Phase III: Latent Semantic Analysis

We use LSA (Latent Semantic Analysis) to reduce the noise, redundancy, and ambiguity problem of vector. We construct a k dimensional abstract semantic space that represents origin matrix approximatively and obtain $A'_{m\times n}$. $A'_{m\times n}$ corresponds to a least-squares best approximation to the original matrix $A_{m\times n}$, and ideally represents the important and reliable semantic relations underlying the data in $A_{m\times n}$.

3.4 Phase IV: Sense Clustering

The context vectors in $A'_{m\times n}$ are clustered by bottom-up and group-average agglomerative clustering algorithm. The similarity of between two vectors is counted by the equation (8), where $\vec{F_1}$ and $\vec{F_2}$ is the context vector of hyponymy r_1 and r_2 respectively. The similarity of between two groups is counted by the equation (9), where $CL_1 = \{r_{a1}, r_{a2}, \ldots, r_{am}\}$ and $CL_2 = \{r_{b1}, r_{b2}, \ldots, r_{bn}\}$ are two group hyponymys. Each group in the clustering result has the same sense.

$$\mathrm{Sim}(r_1, r_2) = \sum_{f1\in F1, f2\in F2} f_1 f_2 \Big/ \sqrt{\sum_{f1\in F1} f_1^2 \sum_{f2\in F2} f_2^2} \quad (8)$$

$$\mathrm{SIM}(CL_1, CL_2) = \frac{\sum_{i=1}^{m}\sum_{j=1}^{n} sim(r_{ai}, r_{bj})}{m \times n} \quad (9)$$

4 Evaluation

We adopt three kinds of measures: R (Recall), P (Precision), and F (F-measure). They are typically used in information retrieval.

Let H be the total number of hyponymy relations in the corpus, H_1 be the total number of clustered hyponymy relations, and H_2 be the total number of correct clustered hyponymy relations. We can give the measure of evaluation metrics as follows:

(1) Recall is the ratio of H_2 to H, i.e. $R = H_2/H$
(2) Precision is the ratio of H_2 to H_1, i.e. $P = H_2/H_1$
(3) F-measure is the harmonic mean of precision and recall, i.e. $F = 2RP/(R+P)$

Firstly, a set of candidate feature words (85,816 words) was selected from a segmentation lexicon (116,736 words) under the given condition (1,800,000 web pages as train corpus; $DF_w>1$; Idf>4.0). Then, we acquired a set of feature words (64,661) by merging synonym feature words with Cilin.

We selected three sense discrimination gathers of "病毒"(virus), "蠕虫"(worm), and "程序"(program) from hyponymy relations using the pattern-based method[10].

The algorithm of sense discrimination was processed under the given condition (4G web corpus; the context window $d_1<50$ and $d_2<50$; k=10 in LSA). The result of sense clustering is shown in Table 2, where the first column means the term c and the number of correct relations in ρ(c), and M(F) in the fifth column means the highest F-measure of sense clustering in [0,1] similarity threshold.

We compared our method with the methods of Rydin and Kenji. Rydin used a heuristic rule to resolve the polysemy problem in hyponymy. Kenji used the cluster algorithm automatically to discover the sense of polysemy word.

Table 2. Experimental data and results

ρ(c)	Part of hyponymys	context	non-zero dimension	M(F)	Similarity threshold of M(F)	Senses
病毒 (virus) 67(56)	(病毒,生物) (virus, being) (病毒,程序) (virus, program) (蠕虫,病毒) (worm, virus) (爱虫,病毒) (love bug , virus) (流感,病毒) (flu, virus) (尼姆达,病毒) (nimda, virus)	49 877 670 207 46 101	2926 8176 5103 3372 557 1980	0.831	0.61-0.64	a computer virus a virus in virology
程序 (program) 97(92)	(程序,文件) (pragram, file) (进程,程序) (process, pragram) (审判,程序) (justice, pragram) (触发器,程序)(trigger, program)	3119 274 145 26	12416 5180 3395 327	0.726	0.20-0.37	a computer program a series of steps to be carried out
蠕虫 (worm) 32(27)	(蠕虫,病毒) (worm, virus) (蠕虫,生物) (worm, being) (寄生虫,蠕虫)(helminth, worm) (尼姆达,蠕虫) (nimda, worm)	670 22 23 78	5103 298 173 1758	0.803	0.53-0.62	a software program a softbodied animal

We selected three sense discrimination gathers of "病毒"(virus), "蠕虫"(worm), and "程序"(program) as the example comparing with Rydin's and Kenji's methods. Because Rydin's method only supports the disambiguation of hyponym polysemous word and Kenji's method only supports the disambiguation of hypernym polysemy word, we omitted part of hyponymy relations that their methods can't handle. The average result was acquired in a 4G web corpus as shown in Table 3.

Table 3. The comparison of different methods

Methods	Precision	Recall	F-measure
Rydin	0.953	0.256	0.404
Kenji	0.866	0.535	0.661
Our methods	0.804	0.716	0.755

As we can see from Table 3, Rydin's method has the highest precision in all methods, but its recall is the lowest. This reason is that the merging rules of Rydin's method are very strong, but the number of hyponymy relations matching the rules is too little. Kenji's method need to acquire the hyponyms of hyperym of a polysemous word, so his method has sparse problem of the hyponyms. Our method has the highest

F-measure value though the precision is lower than other methods, and can both support the disambiguation of hypernym polysemy and hyponym polysemy. Furthermore some false hyponymys are recognized. The false hyponymys can't be clustered because their contexts have very low similarity with the correct hyponymys. For example HR(流行, 病毒) (HR (popularity, virus)).

5 Conclusion and Future Work

In this paper we proposed a method of sense recognition of hyponymy based on vector space model for the problem of polysemy in the phase of building taxonomic hierarchy. In our experiment, the average F-measure of sense clustering added to 0.755. Experimental results show that the method can provide adequate discrimination of the different senses. Some problems still exist and need to be resolved further. It may still have a sparse problem though the contexts of hyponymy are added 50% by coordinate relation patterns. We use feature words to represent the contexts of hyponymy, and lost some semantic information. If more features is considered, such as part of speech, even attribute of term, the result may be better.

Acknowledgements. This work is supported by the Natural Science Foundation (grant nos. 60572125, 60773059), and the National 973 Programme (grants nos. 2003CB317008 and G1999032701). The Beijing University of Technology Science Foundation (grant no.97006017200701).

References

1. Hearst, M.A.: Automatic acquisition of hyponyms from large text corpora. In: Proceedings of the 14th International Conference on Computational Linguistics, Nantes, France, pp. 539–545 (1992)
2. Rydin, S.: Building a Hyponymy Lexicon with Hierarchical Structure. In: Proceedings of the SIGLEX Workshop on Unsupervised Lexical Acquisition (2002)
3. Miura, K., Tsuruoka, Y., Tsujii, J.: Automatic acquisition of concept relations from web documents with sense clustering. In: IJCNLP 2004 Interactive Poster/Demo, pp. 37–40 (2004)
4. Purandare, A., Pedersen, T.: SenseClusters - Finding Clusters that Represent Word Senses. In: AAAI 2004, pp. 1030–1031 (2004)
5. McRoy, S.W.: Using multiple knowledge sources for word sense discrimination. Association for Computational Linguistics 18(1), 1–30 (1992)
6. Schutze, H.: Automatic Word Sense Discrimination. Association for Computational Linguistics 24(1), 97–123 (1998)
7. Mei, J.J., Zhu, Y.M., Gao, Y.Q., Yin, H.X.: Tongyici Cilin(Dictionary of Synonymous Words). Shanghai Cishu Publisher, China (1983)
8. Chen, K.-J., You, J.-M.: A Study on Word Similarity using Context Vector Models. Computational Linguistics and Chinese Language Processing 7(2), 37–58 (2002)
9. Karov, Y., Edelman, S.: Similarity- based word sense disambiguation. Computational Linguistics 24(1) (1998)
10. Liu, L., Cao, C., Wang, H.: Acquiring Hyponymy Relations from Large Chinese Corpus. Wseas Transactions on Business and Economics 4(2) (2005)

A Framework for Context-Awareness in Artificial Systems

Fulvio Mastrogiovanni, Antonio Sgorbissa, and Renato Zaccaria

DIST - University of Genova, Via Opera Pia 13, 16139, Genova, Italy
{fulvio,sgorbiss,renato}@dist.unige.it

Abstract. This paper introduces a context model to be used in context-aware cognitive artificial systems. The framework, that is aimed at integrating ontology and logic approaches to context modeling, assumes the availability of both an ontology (i.e., a representation of *what exists*) and a simple inference schema (i.e., *subsumption*). The context model is defined using a formal protocol, which describes *contexts* and *situations* as recursive structures grounded with respect to the ontology. Examples are presented to discuss the proposed model.

1 Introduction

The role of context-awareness in intelligent artificial systems is nowadays well defined: it is commonly accepted that context models are to be found in order to improve the behavior of artificial systems. This ultimately means to semantically label system information, thus providing the system itself with *meta*information suitable to improve the behavior of actual algorithms and artificial reasoning processes. Interestingly enough, the search for contexts is strictly related to cognition modeling: since it has been demonstrated in [14] that human activity does not necessarily follow long-term plans, but is very situation dependent, artificial systems need to use any possible cue about the current context or situation to correctly behave with dealing with humans.

A generic interaction between entities must be grounded to a shared representation of *what exists*. For instance, with respect to a typical Ambient Intelligence (AmI) scenario, the artificial system must be fed with information originating from *ontologies* able to capture the main significance of the data being collected by the underlying sensory system. However, context models must be effectively represented, described, and used in actual systems. With respect to these requirements, a high-level symbolic representation for describing such a model must be made effective. In other words, the context model must be responsible for managing both the acquisition and the interpretation of actual information, so as to be aware of relevant contexts, situations and events.

This paper proposes a framework aimed at integrating the benefits of ontology and logic approaches for context-awareness in artificial systems. Ontologies have been selected since they represent a general purpose approach to model information, whereas logic approaches can be thought of as sound frameworks to

I. Lovrek, R.J. Howlett, and L.C. Jain (Eds.): KES 2008, Part I, LNAI 5177, pp. 541–548, 2008.

perform inferences and to process information. This context model is grounded with respect to an ontology and exhibits formal reasoning capabilities, assuming the availability of a sensory system to provide the ontology with information originating from distributed sources [8]. The paper is organized as follows. Section 2 discusses relevant literature. Next, the context model is introduced in its many facets in Section 3. Finally, Section 4 describes specific examples. Conclusion follows.

2 Related Work

Context models must address a number of basic requirements: (i) *effectiveness* in classifing sensory information; (ii) *reusability* in composing different contexts to generate hierarchical representations; (iii) *expressiveness* in representing both *relational* and *temporal* contexts. The notion of context has been deeply explored in different areas [2][4]. A context has been defined as "any information that can be used to characterize the situation of an entity" [1]: an entity can be a person, a place or even an object which is considered relevant for the scenario at hand.

In a general perspective, it is possible to identify numerical, ontology and logic approaches to context and situation modeling. *Numerical* approaches are aimed at recognizing complex contexts from a limited number of highly expressive information sources, mostly using probabilisitic or bayesian state estimation techniques [9]. Obvious drawbacks associated with this class rely on the high error probabilities characterizing both measurements and data association, and on the difficulty of mapping numerical data to semantic information. *Ontology*-based approaches are characterized by a superior expressiveness in describing concepts and relationships [11]: ontologies have been used to model real-world high-level contexts [7] and specific domains of interests [12][13], or location-based and topological models [10]. Although these systems offer a basic framework for supporting reasoning, they lack in generality (as they focus on specific domains) and extensibility (since context models can not be easily aggregated to build more complex representations). *Logic*-based approaches manage contexts using facts which are stated or inferred from a given set of rules. In particular, [6] identifies two main principles: (i) *locality:* reasoning must occur within a domain defined by a context; (ii) *cross-domains bridging:* relationships can occur between reasoning activities belonging to different contexts. Logic-based architectures have been recently proposed [5]: however, in spite of their powerful reasoning capabilities, they are not suited to grasp the general relationships among entities in a compact way.

3 A Context Model for Artificial Systems

3.1 A Formal Protocol for Context Specification

The process of extracting base symbols from sensory data is carried out within the context model. Specifically, the model represents a variable assignment α under a specific interpretation $\mathcal{I}$, such that a base assertion $\hat{\sigma}$ within the ontology

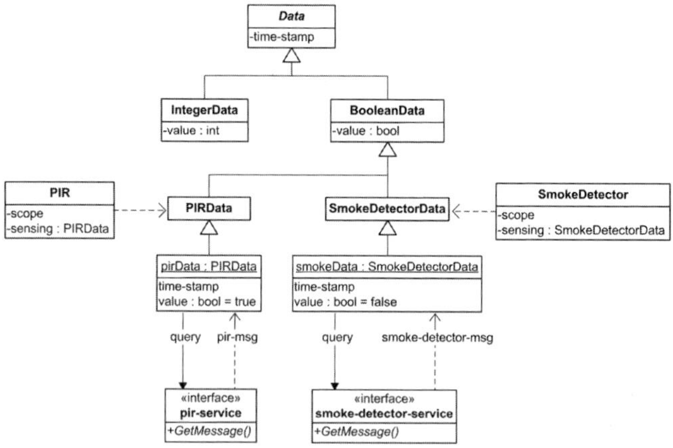

Fig. 1. Representing symbolic information

Σ is satisfied, i.e., $(\mathcal{I}, \alpha) \models \hat{\sigma}$. Put differently, since the context model originates instances of concepts modeling *contexts* and *situations*, it can be viewed also as a computational process taking data acquisition and interpretation into account. Our context model is defined using a protocol composed by four main statements. When $\{\sigma_k\}$ represent *contexts* and *situations*, these four statements define a context model to be used by computational processes to infer events from sensor data. On the basis of this information, predicative structures are then instantiated. Once assessed, predicative structures recursively fill symbolic representations in Σ, thus paving the way for the representation of *contexts* and *situations*. Although an exhaustive specification of the underlying ontology is beyond the scope of the paper, we describe the use of the context model in a working example:

1. Cognitive representations are defined by numerical or ordered values provided by appropriate sources according to predefined symbolic structures $\Sigma = \{\sigma_k\}$, $k = 1, ..., |\sigma_k|$, which are grounded with respect to an ontology.
2. Each symbolic structure belonging to Σ is defined by a vector $\mathbf{r}$ of n individual roles $\{r_i\}$, each one possibly filled by a vector f_i of one or more fillers $\{f_i^j\}$, according to the role definition: $\sigma_k = \wedge\{r_i\} = \{r_1 \wedge r_2 \wedge ... \wedge r_n\}$.
3. An individual $\Sigma \models \sigma_k(\hat{\sigma})$ is given by binding roles and fillers (i.e., a *a variable assignment*) according to the definition of σ_k: $\hat{\sigma} \doteq \wedge\{r_i \bigotimes f_i\} = \{r_1 \otimes f_1 \wedge r_2 \otimes f_2 \wedge ... \wedge r_n \otimes f_n\}$.
4. Symbolic structures $\{\sigma_k\}$ are recursively structured: i.e., each filler f_i^j of a certain role r_i is a symbolic structure.

Example. Let's consider a Smart Home scenario, where an array of distributed sensors collects information about the environment (see Figure 1). Within Σ, data are represented through instances of the concept `Data`. Here, we cosider data originating from PIR and smoke sensors, i.e., $\{\texttt{PIRData}, \texttt{SmokeDetectorData}\} \sqsubseteq \texttt{Data}$.

These are examples of symbolic structures $\{\sigma_k\}$ which can be satisfied in Σ, i.e., $\Sigma \models$ `PIRData` and $\Sigma \models$ `SmokeDetectorData`. In order to create corresponding instances within Σ, we must specify the fillers of the corresponding `value` roles. Furthermore, if we properly define an interpretation $\mathcal{I}$, $\mathtt{PIRData}^{\mathcal{I}}$ describes "the detection of something moving within the sensor range", whereas $\mathtt{SmokeDetectorData}^{\mathcal{I}}$ "the presence of smoke in the nearby".

Actual instances of `Data` are provided by a couple of `services`, namely `pir-service` and `smoke-detector-service`, through the messages `pir-msg` and `smoke-detector-msg`. Let's assume that they are characterized – respectively – by the values *true* and *false* (i.e., according to the interpretation $\mathcal{I}$, "something has been detected" and "there is no smoke in the environment"). The resulting symbolic descriptions of `PIRData` and `SmokeDetectorData` are then given by the bindings $\{\mathtt{value} \otimes \mathit{true}\}$ and $\{\mathtt{value} \otimes \mathit{false}\}$, which have a straightforward correspondence within Σ. In other words, sensor data are mapped onto instances of `Data`, thus updating the `sensing` roles of corresponding `Sensor` instances.

3.2 Developing Representations of Contexts and Situations

Predicates, *contexts* and *situations* are symbolic structures recursively defined on the basis of `Data`. Formally, this requires to identify a proper interpretation $\mathcal{I} =< \Sigma, \cdot^{\mathcal{I}} >$ for concepts in Σ. In particular, we define $\Sigma^{sm} = \{\mathcal{P}_p, \mathcal{C}_c, \mathcal{S}_s\}$ such that $p = 1, ..., |\mathcal{P}|$, $c = 1, ..., |\mathcal{C}|$ and $s = 1, ..., |\mathcal{S}|$. Specifically, $\{\mathcal{P}_p | \Sigma \models \mathtt{Predicate}(\mathcal{P}_p)\}$, $\{\mathcal{C}_c | \Sigma \models \mathtt{Context}(\mathcal{C}_c)\}$, and $\{\mathcal{S}_s | \Sigma \models \mathtt{Situation}(\mathcal{S}_s)\}$. Furthermore, from a structural point of view:

$$\{\Sigma^{sm}\} \sqsubseteq =_1 \mathtt{subject.Entity} \sqcap =_1 \mathtt{begins\text{-}at.INTEGER} \sqcap =_1 \mathtt{ends\text{-}at.INTEGER}$$

where `begins-at` and `ends-at` specify the latest interval when a Σ_l^{sm} definition has been *satisfied*, where $l = 1, ..., |\mathcal{P}| + |\mathcal{C}| + |\mathcal{S}|$. The proposed situation model is then *augmented* by the following operators:

1. *Time.* At the current time instant τ, we say that Σ_l^{sm} is satisfied in τ (and we write $\Sigma_{l,\tau}^{sm}$) if there is an interpretation $\mathcal{I}$ and a variable assignment α_τ such that $(\Sigma, \mathcal{I}, \alpha_\tau) \models \Sigma_{l,\tau}^{sm}$.
2. *Derivative.* A couple of operators, namely *positive derivative* and *negative derivative*, $\{true|false\} \leftarrow \delta_\tau^{t \to f}(\Sigma_{l,\tau}^{sm})$ and $\{true|false\} \leftarrow \delta_\tau^{f \to t}(\Sigma_{l,\tau}^{sm})$ are defined, which are either *true* or *false* when the corresponding $\Sigma_{l,\tau}^{sm}$ is operated by a α_τ such that the truth values changes or remains unaltered.
3. *Length.* An operator, henceforth called *length*, is introduced, *integer* $\leftarrow \lambda_\tau(\Sigma_{l,\tau}^{sm})$. A special predicative symbol $\mathtt{lasts}_\tau(\Sigma_l^{sm}, \mathtt{begins\text{-}at}_l, \hat{\tau})$ can be *hooked* to a Σ_l^{sm} such that `lasts` is *true* as soon as the validity period exceeds $\hat{\tau}$.
4. *Timeline.* A binary operator *order* is introduced, such that $\{true, false\} \prec (\tau_1, \tau_2)$. The operator is *true* iff the time instant τ_1 occurs before τ_2. A special predicative symbol $\mathtt{ordering}_\tau(\{\tau_t\})$, where $t = 1, ..., T$ can be defined, which is true if all the time instants $\{\tau_t\}$ are ordered according to a given constraint.

3.3 Context Recognition and Situation Awareness

In a sense, context-awareness is reduced to many *satisfiability* procedures carried out over instances of $\{\Sigma_l^{sm}\}$. Specifically, given a *symbolic structure* $\mathcal{F}_i$, an interpretation $\mathcal{I}$, a variable assignment at the time instant τ α_τ, satisfied `Predicates`, `Contexts` and `Situations` instances $\hat{\sigma}_k$ are such that $\{\hat{\sigma}_k | (\mathcal{F}_i, \mathcal{I}, \alpha_\tau) \vdash_i \Sigma_{l,\tau}^{sm}(\hat{\sigma}_k)\}$.

We can say that *situations* and *contexts* are encoded in the symbolic structures $\{\mathcal{F}_i\}$ which – on their turn – rely on the definitions of $\{\Sigma_l^{sm}\}$. A symbolic structure is such that $\mathcal{F}_i \doteq \{\sqcap_{k=1}^n \sigma_k | \Sigma \models \sigma_k\}$. In practice, `Predicates` form a set of symbols $\mathcal{P} = \{\mathtt{P}_p : p = 1, ..., |\mathcal{P}|\}$ which are used to build more complex formulas. These formulas are modeled using a set of `Contexts` $\mathcal{C} = \{\mathtt{C}_c : c = 1, ..., |\mathcal{C}|\}$ and a set of `Situations` $\mathcal{S} = \{\mathtt{S}_s : s = 1, ..., |\mathcal{S}|\}$. According to the proposed model, situation awareness is realized aggregating `Predicate` instances in order to satisfy formulas $\mathtt{C}_c$ and $\mathtt{S}_s$ represented within `ontology`. The aggregation assumes the form of the *history* of the n most recent `Predicate` instances $\mathtt{P}_p^i$, $i = 1, ..., n$, which is stored within Σ in a *first-in-first-out* approach.

At each time instant τ, when instances of `Data` are updated, the classification process is carried out over `Predicate` instances, thus – possibly – modifying their truth value. As a consequence, the overall history description $\mathcal{D}_p$ is considered; $\mathcal{D}_p$ is obtained by joining the description of each `Predicate` instance $\mathtt{P}_p^i$, i.e., $\mathcal{D}_p = \mathcal{D}(\mathtt{P}_p^1 \sqcap ... \sqcap \mathtt{P}_p^n), p = 1, ..., |\mathcal{P}|$. In a sense, $\mathcal{D}_p$ represents the system status with respect to both the current situation and the most recent past events. Using $\mathcal{D}_p$ the system can infer what `Contests` $\mathtt{C}_c$ are satisfied at the time instant τ. This is accomplished by checking $\mathtt{subs?[}\mathcal{C}\mathtt{,}\mathcal{D}_p\mathtt{]}$. Specifically, $\mathcal{C}_v(\tau)$ is the collection of `Contexts` $\mathtt{C}_c$ subsuming $\mathcal{D}_p$, i.e., $\mathcal{C}_v(\tau) = \{\mathtt{C}_c \subseteq \mathcal{C} : \mathcal{D}_p \sqsubseteq \mathtt{C}_c\}, c = 1, ..., |\mathcal{C}_v(\tau)|$.

Therefore, all the `Contexts` $\mathtt{C}_c \subseteq \mathcal{C}_v(\tau)$ are occurring in τ. This mechanism is easily iterated for `Situations`. Since $\mathcal{C}_v(\tau)$ varies at each τ, the history description $\mathcal{D}_c$ is considered, obtained by superimposing the description of each `Context` instance $\mathtt{C}_c^j \subseteq \mathcal{C}_v(\tau)$, i.e., $\mathcal{D}_c = \mathcal{D}(\mathtt{C}_c^1 \sqcap ... \sqcap \mathtt{C}_c^{|\mathcal{C}_v(\tau)|}), c = 1, ..., |\mathcal{C}|$. Analogously to `Contexts`, using $\mathcal{D}_c$ the system can infer what `Situation` $\mathtt{S}_s$ are satisfied in τ. Again, this is managed by $\mathtt{subs?[}\mathcal{S}\mathtt{,}\mathcal{D}_c\mathtt{]}$. Current situations are stored in $\mathcal{S}_v(\tau)$, which is the collection of `Situations` $\mathtt{S}_s$ subsuming $\mathcal{D}_c$, where $\mathcal{S}_v(\tau) = \{\mathtt{S}_s \subseteq \mathcal{S} : \mathcal{D}_c \sqsubseteq \mathtt{S}_s\}, s = 1, ..., |\mathcal{S}_v(\tau)|$.

Finally, it is possible to claim that context-awareness is realized by a number of context recognition models $\{\mathcal{F}_i\}$ used to monitor specific patterns of occurrences of events. Furthermore, this model takes *as a whole* data acquisition, symbol grounding and context specification into account.

4 Examples

Several experiments have been performed in a typical Smart Home scenario, the KnowHouse@DIST set-up [8]. Preliminary, two issues must be pointed out: (i) *context grounding*: there is no guarantee that context *semantics* is able to grasp the meaning of *what happens*; (ii) *computational load*: satisfiability procedures require huge amounts of symbols to be considered in subsumption procedures, thereby requiring a trade-off between context complexity and processing capabilities.

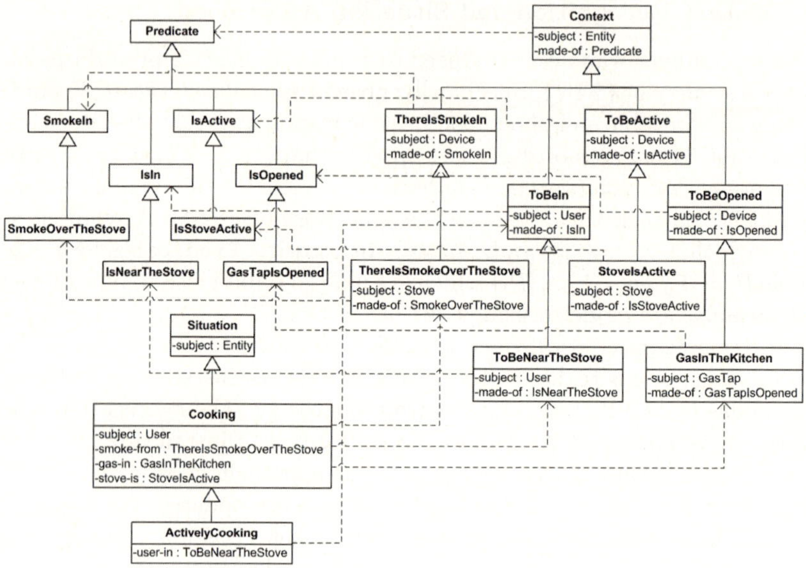

Fig. 2. Hierarchy of `Predicates`, `Contexts` and `Situations` (I)

Cooking. $\sigma_1 \doteq \texttt{Cooking}_\tau \sqsubseteq \texttt{Situation}$ can be abstracted into a *symbolic structure* encompassing all the cases when $\mathcal{U}_1 =$ *doing something*, i.e., $\texttt{Cooking}^\mathcal{I} \sqsubseteq \mathcal{U}_{1,1}^\mathcal{I} \sqsubseteq \mathcal{U}_1^\mathcal{I}$. `Situations` subsumed by $\mathcal{U}_1$ are then characterized by a similar context model used for combining related `Data`: $\texttt{Watching-the-TV}_\tau \sqsubseteq \texttt{Situation}$ is characterized by $\texttt{Watching-the-TV}^\mathcal{I} \sqsubseteq \mathcal{U}_{1,2}^\mathcal{I} \sqsubseteq \mathcal{U}_1^\mathcal{I}$. `Cooking` is defined using the symbols $\{\sigma_k\}$, $k = 2, ..., 5$: $\sigma_2 \doteq \texttt{IsNearTheStove}_\tau\texttt{(User, StoveArea)} \sqsubseteq \texttt{IsIn} \sqsubseteq \texttt{Predicate}$, which is used for describing *location* information; $\sigma_3 \doteq \texttt{SmokeOverTheStove}_\tau\texttt{(StoveArea)} \sqsubseteq \texttt{SmokeIn} \sqsubseteq \texttt{Predicate}$ is used for determining the presence of *smoke* over the stove; $\sigma_4 \doteq \texttt{IsStoveActive}_\tau\texttt{(Stove)} \sqsubseteq \texttt{IsActive} \sqsubseteq \texttt{Predicate}$, exploited for determining the status of the stove in the kitchen; finally, $\sigma_5 \doteq \texttt{GasTapIsOpened}_\tau\texttt{(GasTap)} \sqsubseteq \texttt{IsOpened} \sqsubseteq \texttt{Predicate}$, used to check the gas tap in the kitchen.

Figure 2 depicts `Contexts` and `Situations`. `Situations` are hierarchically organized: i.e., $\sigma_6 \doteq \texttt{ActivelyCooking}_\tau$ relies on $\texttt{Cooking}_\tau$ for its definition. We define $\texttt{Cooking}_\tau$ as follows:

$$\begin{aligned}\texttt{Cooking}_\tau \sqsubseteq\ & \texttt{Situation} \sqcap \\ & =_1 \texttt{smoke-from.ThereIsSmokeOverTheStove}_\tau \sqcap \\ & =_1 \texttt{stove-is.StoveIsActive}_\tau \sqcap =_1 \texttt{gas-in.GasInTheKitchen}_\tau\end{aligned}$$

$\texttt{Cooking}_\tau$ is satisfied when $(\mathcal{F}_{1,1}, \mathcal{I}, \alpha_\tau) \vdash_{1,1} \texttt{ThereIsSmokeOverTheStove}_\tau \sqcap \texttt{StoveIsActive}_\tau \sqcap \texttt{GasInTheKitchen}_\tau$, i.e., α_τ exists which satisfies the constituent `Predicates`: other events can occur as well, but they are not relevant

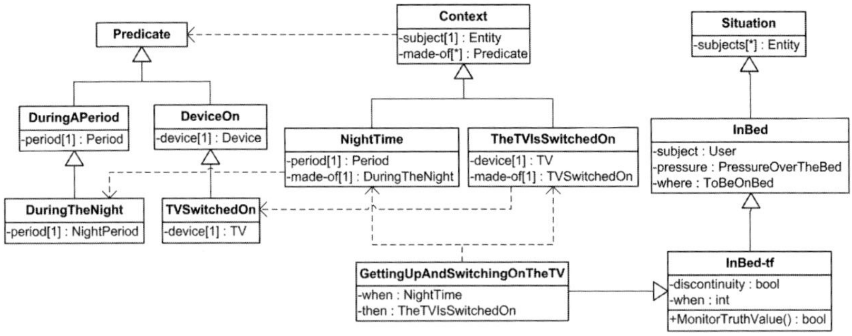

Fig. 3. Hierarchy of Predicates, Contexts and Situations (II)

to satisfy $\mathtt{Cooking}_\tau$: this *minimal* set corresponds to an interpretation $\mathcal{I}$ such that real events coincide with inference procedures.

Wakefulness during the night, getting up and watching the TV instead. This $\sigma_6 \doteq \mathtt{GettingUpAndSwitchingOnTheTV}_\tau \sqsubseteq \mathtt{Situation}$ can be classified by $\mathcal{U}_2$ = *doing something and then something else, which someone is not supposed to*, such that $\mathtt{GettingUpAndSwitchingOnTheTV}^\mathcal{I} \sqsubseteq \mathcal{U}^\mathcal{I}_{2,1} \sqsubseteq \mathcal{U}^\mathcal{I}_2$. This can be very significant as it detects events related to straining, dissatisfaction and other psychological issues. If we assume the availability of a symbolic structure $\mathtt{InBed}_\tau$, its *derivative* $\mathtt{InBed}^{t\rightarrow f}_\tau$ can be defined and used. Therefore, $\mathcal{F}_{2,1}$ (which realizes $\mathcal{U}_{2,1}$) is defined using $\{\sigma_k\}$, $k = 7, ..., 10$. Specifically, $\sigma_7 \doteq \mathtt{IsInBed}_\tau\mathtt{(User,\ BedArea)} \sqsubseteq \mathtt{IsIn} \sqsubseteq \mathtt{Predicate}$ is used for describing *location* information; $\sigma_8 \doteq \mathtt{BedIsPressed}_\tau\mathtt{(Bed)} \sqsubseteq \mathtt{IsPressed} \sqsubseteq \mathtt{Predicate}$ is exploited for determining *pressure* information related to objects located over a bed; $\sigma_9 \doteq \mathtt{DuringTheNight}_\tau \sqsubseteq \mathtt{DuringAPeriod} \sqsubseteq \mathtt{Predicate}$ is satisfied when the system clock is within a given temporal range; finally, $\sigma_{10} \doteq \mathtt{TVSwitchedOn}_\tau \sqsubseteq \mathtt{DeviceOn} \sqsubseteq \mathtt{Predicate}$ is used for monitoring the status of TVs. In Figure 3, the relation $\mathtt{GettingUpAndSwitchingOnTheTV}_\tau \sqsubseteq \mathtt{InBed}^{t\rightarrow f}_\tau \sqsubseteq \mathtt{InBed}_\tau$ holds. $\mathtt{GettingUpAndSwitchingOnTheTV}_\tau$ is defined as follows:

$$\mathtt{GettingUpAndSwitchingOnTheTV}_\tau \doteq \mathtt{InBed}^{t\rightarrow f}_{\tau_1} \sqcap =_1 \mathtt{when.NightTime}_{\tau_2} \sqcap \\ =_1 \mathtt{then.SwitchedOn}_{\tau_3}\mathtt{(TV)} \sqcap \mathtt{ordering}_\tau(\tau_1, \tau_2, \tau_3)$$

which is satisfied in the interpretation $\mathcal{I}$ at the time instant τ if:

$$\mathtt{GettingUpAndSwitchingOnTheTV}^\mathcal{I}_\tau \backsimeq \mathtt{InBed}^{t\rightarrow f, \mathcal{I}}_{\tau_1} \cap \\ \mathtt{when.NightTime}^\mathcal{I}_{\tau_2} \cap \mathtt{then.TVSwitchedOn}^\mathcal{I}_{\tau_3} \sqcap \tau_1 \prec \tau_2 \prec \tau_3$$

In this case, a *temporal constraint* is added: this can be managed by providing $\mathtt{GettingUpAndSwitchingOnTheTV}_\tau$ with an `EvaluateTemporalConstraints()` operator.

5 Conclusion

The presented context model can be used in artificial cognitive systems, managing heterogeneous information at different levels (i.e., numerical as well as symbolic), and "closing the loop" between sensors and actuators. Context-awareness is thus treated as a satisfiability problem over a variable assignment with respect to actual sensor data. The model has been thoroughly described and relevant examples have been presented, detailing how the architecture has been applied to real scenarios and discussing its expressive capabilities.

References

1. Dey, A.K.: Understanding and Using Context. Personal and Ubiquitous Computing 5 (2001)
2. Dourish, P.: What We Talk about when We Talk about Context. Personal and Ubiquitous Computing 8 (2004)
3. Harnad, S.: The Symbol Grounding Problem. Physica D 42 (1990)
4. Loke, S.W.: Representing and Reasoning with Situations for Context-Aware Pervasive Computing: a Logic Programming Perspective. Knowledge Engineering Review 19(3), 213–233 (2005)
5. Augusto, J.C., McCullagh, P., McClelland, V., Walkden, J.A.: Enhanced Healthcare Provision Through Assisted Decision-Making in a Smart Home Environment. In: Proc. of the 2nd Workshop on Art. Intell. Techniques for Ambient Intelligence (AITAmI 2007), Hyderabad, India (January 2007)
6. Giunchiglia, F., Serafini, L.: Multilanguage Hierarchical Logics. Artificial Intelligence 64, 29–70 (1994)
7. Ko, E.J., Lee, H.J., Lee, J.W.: Ontology-Based Context Modeling and Reasoning for U-HealthCare. IEICE Trans. on Information and Systems 8, 1262–1270 (2007)
8. Mastrogiovanni, F., Sgorbissa, A., Zaccaria, R.: An Active Classification System for Context Representation and Acquisition. In: Augusto, J.C., Shapiro, D. (eds.) Advances in Ambient Intelligence, November 2007. FAIA Series (2007)
9. Nguyen, H.T., Qiang, J., Smeulders, A.W.M.: Spatio-Temporal Context for Robust Multitarget Tracking. IEEE Trans. on Pattern Analysis and Machine Intelligence 29(1), 52–64 (2007)
10. Satoh, I.: A Location Model for Smart Environments. Pervasive and Mobile Computing 3(2), 53–73 (2007)
11. Strang, T., Linnhoff-Popien, C.: A Context Modeling Survey. In: Proc. of the 6th Int. Conf. on Ubiquitous Computing (UbiComp 2004), Nottingham, England, September 7-10 (2004)
12. Held, A., Buchholz, S., Schill, A.: Modeling of Context Information for Pervasive Computing Applications. In: Proc. of the 6th World Multiconf. on Syst., Cybern. and Inf (SCI), Orlando, FL (2002)
13. Ranganathan, A., Campbell, R.H.: A Middleware for Context-Aware Agents in Ubiquitous Computing Environments. In: Proc. of the 2003 ACM/IFIP/USENIX Int. Middleware Conference, Rio de Janeiro, Brazil (2003)
14. Waldmann, M.R.: Combining versus Analyzing Multiple Causes: How Domain Assumptions and Task Context Affect Integration Rules. Cognitive Science 31(2), 233–256 (2007)

The Proposal of the Association Response Based on Commonsense in Conversation System

Eriko Yoshimura[1], Seiji Tsuchiya[2], Hirokazu Watabe[1], and Tsukasa Kawaoka[1]

[1] Dept. of Knowledge Engineering & Computer Sciences, Graduate School of Engineering, Doshisha University, Kyo-Tanabe, Kyoto, 610-0394, Japan
{eyoshimura,watabe,kawaoka}@indy.doshisha.ac.jp
[2] Institute of Technology and Science, The University of Tokushima Minamijosanjima-Cho, Tokushima, 770-8506, Japan
tsuchiya@is.tokushima-u.ac.jp

Abstract. This paper proposes one of a response production method on chatting system automatically. Our approach is applied to Japanese language. This is not necessary that it achieve a task. The responses support our conversation flow and let us produce a new topic. The system will get much information for the speaker by performing much conversation with this system. Based on the much information by a conversation with the system, the flexible reply for a speaker will come to be possible. In the present paper, we propose a response by association of the input sentence. For example, we associate "medical treatment" with "hospital". We reply then using the association word. If, like humans, a machine can reply using the association word, it will return various flexible responses.

Keywords: Association response, Natural conversation, Commonsense.

1 Introduction

In recent years, machines have become more and more closely involved in our everyday lives and are becoming increasingly indispensable. For this reason, the ideal machines that we should pursue are machines that coexist with humans.

Many of conversation system tend to use templates. Lots of chatter bots (Eliza[1], Alice[2], jabberwacky[3], etc) have been developed. For example, Eliza which is one of the well-known system acts for counselling by a personification therapist agent. Eliza does not evaluate an answer of a partner for the reply. It memorizes only a part of the content that the partner spoke in the past and replies by using the word. It is prepared for several kinds patterns about the topic.

Like these, as for the natural language processing, task processing type conversation (e.g. automatic systems for tourist information and reservations) becomes the mainstream. However, even under the limited situation, it is known that it is difficult to make a knowledge base of all response case. Moreover, a method using only the prepared template makes monotonous reply and a reply except sentences made by a designer don't appear. So, to make various sentences automatically by machine is important, more than the method to select sentences designer prepared.

I. Lovrek, R.J. Howlett, and L.C. Jain (Eds.): KES 2008, Part I, LNAI 5177, pp. 549–556, 2008.

This paper proposes one of a response production method on chatting system automatically. This is not necessary that it achieve a task. The responses support our conversation flow and let us produce a new topic. The system will get much information for the speaker by performing much conversation with this system. Based on the much information by a conversation with the system, the flexible reply for a speaker will come to be possible.

On human conversation, we return two pattern responses. One is to demand more detail and the other is to speak new topic by association of the sentence. The machine can solve the former by pattern knowledge base of 6W1H(who, what, when, where, whom, when, and how) and some rule. To solve the latter, the machine must solve "association" that is difficult for machine. But, achieving this association response leads to a flexible response. So, this paper proposes a technique for the association response. Our approach is applied to Japanese language. For the analysis of sentences, we used a Japanese parsing system "chasen" was developed by Nara Institute of Science and Technology.

2 Machine Response

The machine cannot understand a true meaning of the conversation like a human. However, by appropriate responses a machine can let us under the impression that a machine can understand. Our purpose is generation of an appropriate association response likes the human by using an association function. (Speaker "I went to the hospital, yesterday": a listener thinks "hospital is a place for treatment" → "there may be sickness on the body of the speaker" → a reply "What were you treated for?")

This section shows classification of the machine response method as follows;

1) Fixed pattern response
   e.g., "Thank you" → "You're welcome"
2) Response demanding more detail [4]
   A) 6W1H response
      e.g., "I bought a new TV."→"Where did you buy it?", "When did you buy it?"
   B) Detailed and concrete response
      e.g., "I bought a book yesterday." → "What kind of book did you buy?"
3) Response to a new topic generated by association with the sentence
   A) Association response from noun and adjective
      e.g., "I ate a cold dessert." → "Did you eat ice cream?"
   B) Association response from noun and verb
      e.g., "I went to a hospital." → "What were you treated for?"

For a fixed pattern response, the response is decided using a database that is prepared beforehand [5]. This response is used for a fixed input sentence. The database in which the pattern is collected by the designer decides the accuracy and the variety of response.

In response to a demand for more detail, the response is decided using simple rules without considering the content [4]. The method of 6W1H response analyzes an input sentence and then generates responses for ask blank information on 6W1H. For example, when one says "I bought a new TV," the sentence has information who bought it and what bought. But the sentence doesn't have information where it bought, why it bought and when it bought. Then, we will ask the blank information to get more detail. In the detail and concrete response technique, if the object in the input sentence is a concrete object (it is judged with thesaurus [6]), more detail is requested using this technique. This response to a demand for more detail using a simple rule is more likely to address the input sentence than a fixed pattern response, but it is easy to fall into a uniform response. It does not introduce new concepts into the discourse.

Therefore, in the present paper, we propose a response by association of the input sentence. For example, we associate "ice cream" with "a cold dessert" and "medical treatment" with "hospital". We reply then using the association word. If, like humans, a machine can reply using the association word, it will return various flexible responses. After the making of various responses, appropriate response can be chosen by a state at the conversation. As a result, the machine can return a different response even in the situation that looks like.

This paper focuses a proposal of association response based on commonsense knowledge. As an example for association response using commonsense knowledge, coverage of this technique confines the commonsense knowledge of the place. We argued about necessity and the thought of the association response. Then, next section explains a part of the association response. Association requires general knowledge about the relations between words and the commonsense of the words. Next, we describe the association judgment mechanism as it relates to general knowledge and commonsense.

3 Association Judgment Mechanism

Humans possess a common knowledge of words and can conduct a discourse based on common knowledge relating to words. For example, humans may possess certain knowledge of airplanes, namely, that airplanes are flying machines that have wings and are related to airports. We also possess the common knowledge that airplanes are larger than people and chairs and are faster than trains and people. Assuming this common knowledge, humans can converse with each other on this subject.

Thus, by modeling human knowledge of discourse and words and making this understandable to machines, we believe it is possible to construct a discourse mechanism similar to the human discourse system. We call the mechanism constructed based on this concept an association judgment mechanism.

The association judgment mechanism is composed of the Concept Association Mechanism and a Commonsense Judgment mechanism. The Concept Association Mechanism defines the common meanings related to words, while the Commonsense Judgment mechanism defines the commonsense related to words. Please refer to references for the detail a concept base, a degree of association and the common sense judgment system. This section explains these outlines.

3.1 Concept Association Mechanism

The Concept Association Mechanism incorporates word-to-word relationships as common knowledge. This mechanism is a structure that includes a mechanism for capturing various word relationships. In this section, we describe the Concept Base [7] and a method of calculating the Degree of Association [7] using this base.

The Concept Base is a knowledge base consisting of words (concepts) and word clusters (attributes) that express the meaning of these words. The Concept Base is automatically constructed from multiple sources, such as Japanese dictionaries. The Concept Base used in the present study contains approximately 90,000 registered words organized in sets of concepts and attributes. These concept and attribute sets are assigned weights to denote their degree of importance. For example, an arbitrary concept, A, is defined as a cluster of paired values, consisting of attribute, a_i, which expresses the meaning and features of the concept, and weight, w_i, which expresses the importance of attribute a_i, in expressing concept A:

$$A = \{(a_1, w_1), (a_2, w_2), ..., (a_N, w_N)\}$$

Attribute a_i is called the first-order attribute of concept A. In turn, an attribute of a_i (taking a_i as a concept) is called a second-order attribute of concept A.

The Degree of Association is a parameter that quantitatively evaluates the strength of the association between one concept and another. The method for calculating the Degree of Association involves developing each concept up to second-order attributes, determining the optimum combination of first-order attributes by a process of calculation using weights, and evaluating the number of these matching attributes. The value of the Degree of Association is a real number between 0 and 1. The higher the number is, the higher the association of the word. Table 1 lists examples of the degree of association.

Table 1. Examples of the degree of association

Concept A	Concept B	Degree of association between A and B
Flower	Cherry blossom	0.208
Flower	Plant	0.027
Flower	Car	0.0008
Car	Bicycle	0.23

3.2 Commonsense Judgment Mechanism

The Commonsense Judgment Mechanism derives commonsense associations from words based on various factors (e.g., quantity, time, location, and percept). Here, "percept" indicates a sense that can be acquired by stimulation through any of the five percepts (vision, hearing, smell, taste, and touch). These associations are constructed using the Concept Association Mechanism. In this section, among the various factors, we focus on location.

The location judgment system [8] obtains information on two points of view with respect to a location. The information is the "object" (What are there?) and the "action" (What do they do?). Using this system, the meaning that a location expresses can be understood as commonsense.

A location judgment knowledge base is made by us include some sets of basic location word, object and action. In addition, using the Concept Base of common knowledge, the system can recollect object and action words that are not contained in the knowledge base (unknown words). Table 2 shows some examples of this system.

Table 2. Examples of the location judgment system

Input	Output		
	Location?	Object	Action
Milk	No	--	--
Book store	Yes	Customer, Book, Magazine,…	Sell, Buy, Display, …
Amusement park	Yes	Roller coaster, Ferris wheel,…	Ride, See, Play, …

4 Association Response

This section explains the association response. In particular, we propose an association response for sentences with location words as an object (verb location association) because location is important point with respect to the conversation topic. This association response is a method of associating a new verb from a "noun of location" and a "location verb". Noun of location and location verb include input sentence. This system can work adequately using the two words. Table 3 shows some examples of this system.

Table 3. Examples of the association response from nouns and verbs

Input	Noun of location	Location verb	Association verb	Output
I went to the cinema.	Cinema	go	see	What movie did you see?
I went to the hospital	Hospital	go	treat	What were you treated for?

Here, "noun of location" means a noun that can be associated with the existence and purpose of a location by itself (e.g., amusement park, post office). The nouns of location have mutual relations with the human being because a speaker is human (e.g., "flower" that receives bees and other insects is not noun of location). This technique used all words that belong to nodes of "location", and "house" meets these conditions in the thesaurus [7]. A thesaurus is a dictionary where words are semantically classified and generally indicated with a tree structure. The thesaurus has two types: 1) a classification thesaurus with words only on leaf nodes and 2) a hierarchical thesaurus with words on root nodes and intermediate nodes besides leaf.

4.1 Construction Database

To judge the input sentence, the association response concerning location by the verb of the input sentence, the location verb database is constructed. The location verb database contains 13 verbs that have object words that are nouns of location in Japanese. This database contains the verbs such as go, visit, see, start, depart, leave, take off, go back, return, and get back.

Then, to construct a sentence from an association word, a return sentence template database is constructed. This template is enlarged in the new association word obtained by the technique for explaining in paragraph 4.2. This database is divided into classes of parts of speech in Japanese. The present paper describes an example when the association word is a verb (Table 4), because other expressions in English are difficult. For a similar reason, we herein omit an explanation of processing that changes the verb for natural sentences.

Table 4. Example of the return sentence template database

What did you [verb]? What were you [verb]? What [noun] did you [verb]? ,......

4.2 Technique of Verb Location Association

The technique of verb location association is as follows:

1) The noun in the input sentence is judged as to whether it is a noun of location by the location judgment system.
2) The basic verb in the input sentence is judged as to whether the location verb database contains the verb.
3) When conditions of steps 1) and 2) are met, step 4) is executed. In other cases, verb location association is not possible.
4) The object and action of the location are obtained from the noun concerning the location of step 1) by using the location judgment system.
5) Verb words are selected from the words of step 4).
6) The word of lowest value in the Concept Base IDF is selected to obtain a frequently used word.
7) A new sentence is generated using the word of step 6) and the sentence is returned to the template database (like table3).

The Concept Base IDF is explained as follows. Simple and comprehensible words are often used for greetings in spoken language. To show the degree of usefulness of a word, we use the Concept Base Inverted Document Frequency (IDF). The Concept Base IDF is the weight related to the frequency of use of a word in the Concept Base. This technique means that low-frequency words in the Concept Base are not used frequently in daily conversation. For example, "therefore" and "but" are same mean word, "but" is used easily in conversation. Such words are excluded by using the Concept Base IDF. The concept IDF can be expressed as follows:

$$idf(t) = log\, N_{ALL} / df(t)$$

t : object word

N_{ALL} : total number of concepts in the Concept Base

$df(t)$: number of t in an attribute.

4.3 Evaluation of Verb Location Association

We evaluated the proposed verb location association technique. We prepared 100 input sentences for verb location association from a junior high school English text. For these sentences, we evaluated the proportion of appropriate association words using the proposed system. The association words are manually (average of 3 people) classified as appropriate or inappropriate. In the present study, we calculate the accuracy of the technique as the proportion of appropriate words. As a result, the accuracy was 74% in this system.

Table 5 shows examples of successes and failures of the proposed system of verb location association.

Table 5. Examples of successes and failures in the verb location association system

	Input	Association verb	Output
Success	I went to the hospital yesterday. I went to the museum by train.	treat for see	What were you treated for? What did you see?
Failure	I went to the station. She visited a historic place in Kyoto.	ride live	What did you ride? What did you live?

As shown in Table 5, for the sentence "I went to the hospital yesterday", the system obtains the natural association words "treat for" from "hospital" and can then generate an appropriate sentence, as follows: "What were you treated for?" For the sentence "I went to the station", although the system obtains the natural association word "ride" from "station", unnatural sentences, such as "What did you ride?" may still be generated. It is unnatural to ask such a question because it is implied that one goes to the station in order to ride a train. When there is only one purpose for an action at a location, the purpose is not asked. Therefore, it is necessary to arrange the object (type and number) for the association verb at the location. Then, if further refinements are made, various association responses will be generated.

5 Conclusion

In the present paper, we proposed a method of association response for conversation growth. By adding the proposed association response to the uniform response generation technique, various natural responses are possible. As a result, this technique is expected to lead to natural conversation response.

Acknowledgments. This work was supported with the Aid of Doshisha University's Research Promotion Fund.

References

1. Weizenbaum, J.: ELIZA - A Computer Program For the Study of Natural Language Communication Between Man and Machine. Communications of the Association For Computing Machinery 9(1), 36–45 (1965)
2. A.L.I.C.E., `http://alicebot.blogspot.com/`
3. Jabberwacky, `http://www.jabberwacky.com/`
4. Nishizawa, Y., Watabe, H., Kawaoka, T.: Intelligent Question Generation for Natural Conversation Using Commonsense Judgment Mechanism, IEICE Technical Report, AI, -99, Vol104, No.728, pp.25–30(2005) (2004)

5. Yoshimura, E., Watabe, H., Kawaoka, T.: An Automatic Enhancing Method of Greeting Sentences Using Association Knowledge Mechanism. Journal of Natural Language Processing 13(1), 117–141 (2006)
6. NTT Communication Science Laboratory, NIHONGOGOITAIKEI Iwanami Shoten book (1997)
7. Kojima, K., Horiguchi, A., Watabe, H., Kawaoka, T.: Concept-Base Construction for a Word Association-System: A Method of Deciding Word Attribute Weights by Logical Relations and Attribute Reliability. In: The 6th World Multiconference on Systemics- Cybernetics and Informatic, vol. 5, pp. 134–139 (2002)
8. Sugimoto, J., Watabe, H., Kawaoka, T.: Constructing a Commonsense Place Judgment System Which Uses a Concept Base, IPSJ SIG Technical Report, 2003-NL-153-11, Vol.2003, No.4, pp.81–88(2003).

Ontological Knowledge Maintenance Methodology*

Rim Djedidi[1] and Marie-Aude Aufaure[1,2]

[1] Department of Computer Science, SUPELEC Campus de Gif
Plateau du moulon, 3 rue Joliot Curie, 91192 Gif sur Yvette Cedex
{Rim.Djedidi,Marie-Aude.Aufaure}@Supelec.fr
[2] INRIA Paris-Rocquencourt / Axis Project
Domaine de Voluceau, 78153 Le Chesnay Cedex
Marie-Aude.Aufaure@Inria.fr

Abstract. Ontologies are used as a key for semantic modelization, offering consensual and formal knowledge specification. They are more and more applied to open and dynamic environments and modeling knowledge that evolve continuously. To take into account all evolving aspects, ontologies have to be adapted to change requirements. In this paper, we propose a methodological approach for ontological knowledge maintenance. The goal of the methodology is to manage ontology evolution while maintaining consistency and evaluating change impact on ontology quality.

Keywords: Ontological knowledge evolution, Ontology representation model, Knowledge consistency, Ontological knowledge quality evaluation.

1 Introduction

Ontologies are used in several areas and applications: semantic web, information retrieval, data and knowledge bases integration, etc. The increasing use of the ontologies has created new interests in knowledge maintenance problems. Ontologies cannot be thought as an achieved conceptualization of well-delimited and static domain. There are many reasons for ontology changes: the continual evolution of the modelled domain, the refinement of the ontology conceptualization, the modification of the application by adding functionalities according to new end-user requirements and the reuse of the ontology for others tasks or applications. To take into account all these evolving aspects, ontologies have to be adapted to change requirements.

In this paper, we propose a methodological approach for ontological knowledge maintenance focusing particularly on OWL ontologies. The goal of the methodology is to guide and validate change application in a systematic and optimized manner, while maintaining consistency and evaluating change impact on ontology quality. The paper is organized as follows: in section 2, we give an overview on what is an ontology, section 3 presents the ontological knowledge maintenance methodology that we defined. The ontology quality evaluation model developed is introduced in section 4. Before we conclude and present our future work, we discuss and compare our approach with related work.

* This work is financed by the French National Research Agency ANR, as a part of the project DAFOE: Differential and Formal Ontology Editor.

I. Lovrek, R.J. Howlett, and L.C. Jain (Eds.): KES 2008, Part I, LNAI 5177, pp. 557–564, 2008.

2 What Is an Ontology?

The word "ontology" stems from philosophy as study of beings and was adapted in artificial intelligence field as "*an explicit specification of a conceptualization*" [1]. Several definitions were given in ontology community literature, covering different aspects of an ontology. Combing these definitions [2] [3] [4], we can say that an ontology is an explicit specification of a shared conceptualization, providing a common terminology and formally describing knowledge of a particular domain in a hierarchical structure of concepts or classes, properties and instances. Ontology can also constraint the interpretation of domain objects by defining a set of rules and axioms, and can be used as a skeleton foundation for a knowledge base.

Concepts represent the classification of domain objects in classes. They are described and related by properties: attributes (literal values), hierarchical properties (is-a relations) and semantic relations. Rules define *antecedent-consequence* form statements describing logical inferences that can be drawn from a particular assertion. Axioms are logical assertions (including rules) explaining and constraining domain interpretation and ontology application. Classes are instantiated by assigning corresponding domain objects (instances or individuals).

3 Ontology Maintenance Methodology

Ontology evolution is defined as the timely adaptation of the ontological knowledge to the arisen changes and the consistent management of these changes [5]. Ontology maintenance is not a trivial task. Changes are captured from a variety of sources and may have many consequences. To capture and handle the impact of a change, we need to formally specify change requirements and to drive a change management process considering change impact at consistency and quality level and automating change validation process.

Change management depends closely on the ontology representation language. We focus on OWL ontology model [6] and particularly on OWL-DL layer as it benefits of Description Logic formalism [7]: well-defined semantic, formal properties comprehensible at complexity and decidability level, several known inference algorithms and existing DL reasoners allowing ontology consistency verification. The process is organized in 4 phases:

1. Inconsistency detection by verifying the consistency of the ontology with respect to ontology consistency definition;
2. Proposition of resolution alternatives for the detected inconsistencies;
3. Evaluation of the impact of inconsistency resolution alternatives on the ontology quality;
4. Change application and final validation.

3.1 Inconsistency Detection

This phase aims to verify consistency and delineate the inconsistent part of the ontology. The formally specified change is applied to a test version of the ontology so that change effects can be analyzed. We start by localizing the minimal inconsistent sub-ontology O' such that: $O' \subseteq O$ and $\forall O'' \subset O'$ O'' is consistent. Two OWL-DL consistency levels are considered: structural consistency and logical consistency.

Structural consistency refers to ontology language constraints. OWL-DL structural consistency defines constraints on constructors (syntax language), elementary axioms and axiom combinations.

Logical consistency refers to the formal semantic of the ontology and to its satisfiability in the meaning that ontology is semantically correct and does not present any logical contradiction. OWL-DL is an axiom-centered language. Concepts and properties have structural description specified through some defined constructors. A semantic is associated to each description based on a domain interpretation [7]. An ontology *O* is consistent if there exists an interpretation *I* that satisfies *O*. Satisfying ontology within an interpretation is constraint by the satisfaction of all the ontology axioms (concept, property and individual axioms). OWL-DL logical constraints cannot be delimited. We choose to focus on a sub-set of logical conditions related to the following axioms: concept and property subsumptions, concept disjointness, cardinality and domain and range.

Consistency is verified by employing Pellet reasoner which supports OWL-DL with both terminological (TBox) and assertional (ABox) levels [8].

3.2 Proposition of Inconsistency Resolution Alternatives

The crucial part of an ontology change management process is not how to determine whether there are contradictions caused by change, but how these contradictions can be resolved. Many resolution strategies can be proposed for a detected inconsistency. A resolution alternative consists in deriving additional changes to apply jointly to a given required change so that the ontology can be maintained in a consistent state.

After consistency verification, axioms causing inconsistency are localized. Rather than opting for deleting solutions, we try to propose alternatives that fusion, divide, generalize concepts and properties and redistribute instances. The idea is to take into account the "ontological continuity" principle stipulating that an existed knowledge cannot be undermined. Indeed, existing knowledge can be referenced by dependant applications or ontologies particularly when the ontology evolves during usage which is more constraining than during construction cycle. So, we tend to minimize axiom removing by proposing substituting solutions. However, if the change itself is a remove operation, we apply it assuming that it's expressed and approved by expert.

We define patterns of resolution alternatives to satisfy the first core of considered logical consistency conditions. The pattern is then adapted to the specific properties of a change described by its formal specification (i.e. its localization on ontology structure).

Example. Let's consider a change *Ch* assigning *I*, an instance of concept *C1* that is a sub-concept of a concept *C*, to a concept *C2* that is also a sub-concept of *C*, defined as disjoint from *C1*. *Ch* causes a disjointness inconsistency. Rather than removing disjointness axiom or instantiation axiom of *I* to *C1*, we propose the following alternatives:

- Instance *I* can be redistributed and assigned to the super-concept *C* and thus the semantic of the instance *I* is preserved (even it's less precise) and disjointness axiom is still true.

- A new concept *C3* can be added as sub-concept of *C* and *I* is assigned to it. *C3* will represent sub-concepts of *C* that are not concerned by disjointness and disjointness axiom between *C1* and *C2* is not removed.
- The same second alternative can be applied while putting *I* as only instance of the new added concept to avoid redundancy for some specific user-defined consistency conditions.

The different proposed alternatives are then evaluated through a qualitative model to guide the choice of the least costly inconsistency resolution.

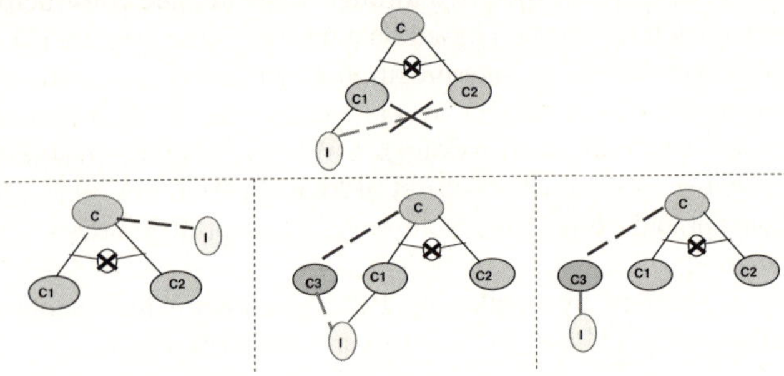

Fig. 1. Example of resolution alternatives

3.3 Evaluation of Inconsistency Resolution Alternatives

Inconsistency resolution can be guided by considering change impact on quality. Alternatives are evaluated through a defined ontology quality evaluation model (cf. section 4), and the alternative that has the less costly impact on quality is chosen. It represents the complementary changes to add to the required change so that ontology consistency can be maintained and ontology quality can be preserved or even improved. Changes can then be applied and validated directly in the next phase.

In the case that all alternatives have negative effect on quality, expert intervention will be solicited. The evaluation results give user complementary information to its expertise and help him deciding about the change and judging if the change cost justifies its relevance.

Quality evaluation techniques are thus, exploited for its full potential to determine the less costly inconsistency resolution and approve useful changes while justifying their cost and minimizing user dependency.

3.4 Change Application

This phase corresponds to the final validation of the required and derived changes while keeping traceability of modifications applied to the ontology. Change analysis and consistency maintenance results are applied on a temporary version of the ontology that could be aborted, if changes are finally rejected and thus, the initial ontology is still preserved. In the other case, a new ontology version is defined, and change historic is kept in a log file.

The architecture of the prototype implementing the methodology is illustrated by the following figure:

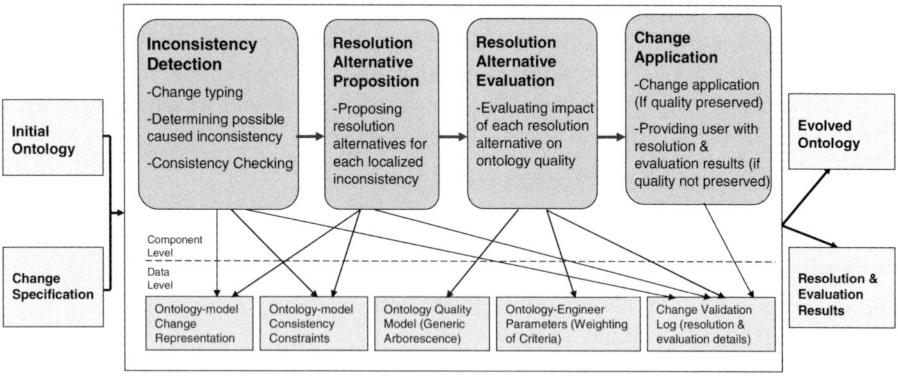

Fig. 2. Ontology change validation system architecture

4 Ontology Quality Evaluation Model

To evaluate change impact on ontology quality, we define a hierarchical model describing and measuring quality through several ontology quality features, organized in an arborescence of criteria and metrics.

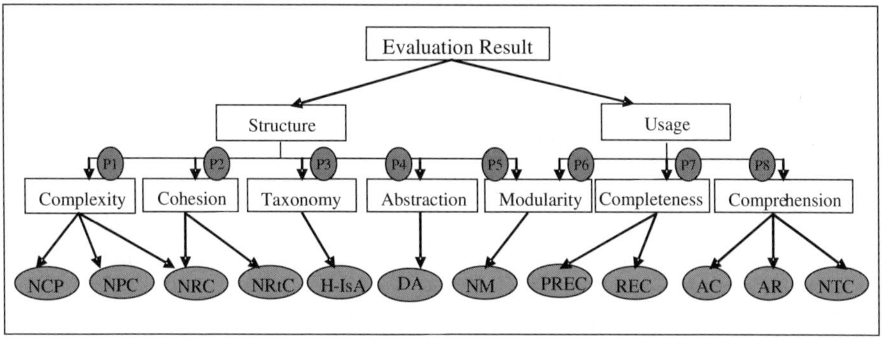

Fig. 3. Ontology quality evaluation model

We consider two principle evaluation aspects: structural aspect and usage aspect. As shown in the arborescence, structural aspect includes complexity, cohesion, modularity, taxonomy and abstraction criteria. Usage aspect is evaluated through ontology comprehension degree, modularity and completeness. Complexity measures concept overlapping defined by concept links. An important overlap between concepts reflects a strong cohesion. Modularity measures the possibility of ontology decomposition to sub-parts (independent modules) that can be reused by other ontologies. Modularity facilitates structure enrichment and maintenance as well as ontology reuse. Taxonomy indicates ontology semantic degree and is inversely proportional to the part of non "is-a" relation. Abstraction measures concept abstraction level (generalization/specification hierarchies). Completeness evaluates if ontology covers relevant properties of the modelled domain. This criterion is based on concept label conformity with keyword domain. Comprehension criterion assesses the facility of understanding ontology throughout the different annotations and concept definitions.

Several quality metrics are defined in literature, we consider quantifiable metrics that redistribute on the different criteria and complement:

1. NCP: Average number of concept path from root.
2. NPC: average number of properties per concept.
3. NRC: Average number of relations per concept.
4. NRtC: Number of direct sub-root concept (number of hierarchy).
5. H-ISA: is-a relation ratio in respect to semantic relations.
6. DA: Depth average of hierarchy.
7. NM: Number of disjoint modules.
8. PREC: Precision.
9. REC: Recall.
10. AC: Percentage of annotated concepts.
11. AR: Percentage of annotated relations.
12. NTC: Average number of terms naming a concept.

Some evaluation criteria can be contradictory: change that leads to a more complex ontology structure for example, can improve cohesion. To face this problem, expert has to weight each evaluation criterion (at the beginning of evolution process) according to its relevance for the modelled domain and the application using the ontology.

5 Related Work and Discussion

Applying a change could spoil consistency and quality of the evolved ontology. However, change impact on consistency and quality has not been considered so far in literature. The most significant approach is presented in [9], author proposes a global evolution process specifying change semantics and maintaining consistency of ontologies represented in KAON formalism. Inconsistency resolution principles were also adapted in [10], to OWL ontologies. But, as we knew, change effects on ontology quality have not been thought-out in evolution researches even so, many studies focused on ontology quality [11] [12] [13] [14].

Inconsistency can be resolved by several ways. Different resolution alternatives can be proposed for a change. In [10], authors present a model for OWL change semantics and introduce resolution strategies based on OWL-Lite (sub-set of OWL-DL) constraints. These strategies are presented to user so that he can decide and validate changes. Nevertheless, we think that ontological knowledge maintenance should also focus on ontology quality. According to this perspective, we propose a complementary approach driving change validation in a more systematic and optimized manner. Rather than requiring expert intervention in resolving inconsistency, we guide him choosing the relevant resolution alternative by evaluating the impact of each alternative on ontology quality. Another difference in comparison to existed researches is that quality evaluation techniques employed to resolve inconsistencies are based on quantifiable criteria. In [10], only some non-quantifiable quality criteria were cited as generic consistency condition defined by user to ensure a well ontology modelization such as meta-properties of OntoClean method including rigidity, identity, unity and dependency [15]. In addition, to minimize axioms removing impact on ontology, authors refer to a set of specific requirements that can be specified by user (ex. fixing the number of axioms that can be removed, defining axiom relevance degree to guide the choice in removing process, etc.). In our approach, these requirements can be considered too. But, we add an evaluation layer upstream to assess change impact on quality by considering quantifiable criteria that are user independent.

To resolve logical inconsistency, Haase and Stojanovic [10] focus on determining and deleting axioms causing inconsistency and propose two alternatives: i) generating the minimal number of changes to apply to obtain the maximal consistent sub-ontology and ii) localizing inconsistency by determining the minimal inconsistent sub-ontology. Axioms that should be removed are presented to user so that he can decide and control evolution process. Taking into account the "ontological continuity" principle stipulating that an existed knowledge cannot be undermined, in our approach, we tend to minimize axiom removing by proposing alternatives that fusion, divide, generalize concepts and properties and redistribute instances.

6 Conclusion and Future Work

In this paper we propose a methodology for ontological knowledge maintenance. The methodology focuses on resolving inconsistencies by proposing several resolution alternatives and guiding the choice of the relevant alternative through an evaluation phase considering alternative impact on ontology quality. We define a hierarchical model describing and measuring ontology quality. The model is employed to guide ontology change validation in a systematic and optimized way, reducing user dependency and justifying change costs.

Currently, we work on enlarging the set of considered OWL ontology changes and analyzing the semantic of consistency resolution of those changes to define more resolution patterns. Besides, we are applying the approach to the enrichment of a medical ontology of Pneumology adding knowledge extracted from medical standards [16] [17] [18].

References

1. Gruber, T.: A translation approach to portable ontology specifications. Knowledge Acquisition 5, 199–220 (1993)
2. Studer, R., Benjamins, V., Fensel, D.: Knowledge Engineering: Principles and Methods. Data and Knowledge Engineering Journal 25, 161–197 (1998)
3. Grüninger, M., Fox, M.: Methodology for the Design and Evaluation of Ontologies. In: Proceedings of IJCAI 1995, Workshop on Basic Ontological Issues in Knowledge Sharing (1995)

4. Swartout, B., Patil, R., Knight, K., Russ, T.: Toward Distributed Use of Large-Scale Ontologies. In: Ontological Engineering, AAAI 1997. Spring Symposium Series, pp. 138–148 (1997)
5. Haase, P., Stojanovic, L.: Consistent Evolution of OWL Ontologies. In: European Conference on Semantic Web ECSW (2005)
6. McGuinnes, D., Van Harmelen, F.: OWL Web Ontology Language Overview, W3C Recommendation (2004)
7. Horrocks, I., Patel-Schneider, P.F.: Reducing OWL Entailment to Description Logic Satisfiability. Journal of Web Semantics 1(4) (2004)
8. Sirin, E., Parsia, B., Cuenca Grau, B., Kalyanpur, A., Katz, Y.: Pellet: A practical OWL-DL reasoner. Journal of Web Semantics 5(2) (2007)
9. Haase, P., Sure, Y.: Incremental Ontology Evolution – Evaluation, Institute AIFB, University of Karlsruhe (2005)
10. Haase, P., Stojanovic, L.: Consistent Evolution of OWL Ontologies. In: European Conference on Semantic Web ECSW (2005)
11. Hartmann, J., Spyns, P., Giboin, A., Maynard, D., Cuel, R., Suarez-Figueroa, M.C., Sure, Y.: Methods for ontology evaluation, KnowledgeWeb D1.2.3 deliverable (2005)
12. Brank, J., Grobelnik, M., Mladenic, D.: A Survey of Ontology Evaluation Techniques. In: Proceedings of the Conference on Data Mining and Data Warehouses SiKDD 2005, Ljubljana, Slovenia (2005)
13. Supekar, K.: A peer-review approach for ontology evaluation. In: Proceedings of the 8th International. Protégé Conference, Madrid, Spain, July 18-21 (2005)
14. Yang, Z., Zhan, D., Ye, C.: Evaluation Metrics for Ontology Complexity and Evaluation Analysis. In: IEEE International Conference on e-Business Engineering, ICEBE 2006 (2006)
15. Guarino, N., Welty, C.: Evaluating Ontological Decisions with OntoClean. Communication of the ACM 45(2), 61–65 (2002)
16. Powell, T.S., Srinivasan, S.J., Nelson, W.T., Hole, L., Roth, V.O.: Tracking Meaning Over Time in the UMLS Metathesaurus. NLM Publication (2002)
17. Rector, A.L., Rogers, J.E.: Ontological and Practical Issues in using a Description Logic to Represent Medical Concepts: Experience from GALEN. In. IMIA WG6 Workshop: Terminology and Natural Language in Medicine. Phoenix Arixona, Manchester University, School of Computer Science Preprints CSPP-35 (2005)
18. Bodenreider, O., Nelson, S.J., Hole, W.T., Chang, H.F.: Beyond Synonymy: Exploiting the UMLS Semantics in Mapping Vocabularies. NLM Publication (2001)

A Hybrid Approach for Data Preprocessing in the QSAR Problem

Adina Cocu, Luminita Dumitriu, Marian Craciun, and Cristina Segal

Department of Computer Science and Engineering,
University "Dunărea de Jos" of Galați, 2 Științei, 800146, Romania
{cadin,lumi,mcraciun,csegal}@ugal.ro

Abstract. One of the approaches in the Knowledge Discovery in Databases (KDD) domain is Predictive Toxicology (PT). Its aim is to discover and represent the relationships between the chemical structure of chemical compounds and biological and toxicological processes. The challenges in real toxicology problems are big amount of the chemical descriptors and imperfect data (means noisy, redundant, incomplete, and irrelevant). The main goals in knowledge discovery field are to detect these undesirable proprieties and to eliminate or correct them. This supposes noise reduction, data cleaning and feature selection because the performance of the applied Machine Learning algorithms is strongly related with the quality of the used data. In this paper, we present some of the issues that can be performed for preparing data before the knowledge discovery process begin.

Keywords: knowledge discovery, prediction, toxicology, knowledge representation.

1 Introduction

One of the directions of Predictive Toxicology domain is Quantitative Structure-Activity Relationship (QSAR) modeling. This approach was introduced in the 1960s by Hansch and co-workers and consists in describing the relationships between the chemical structure of chemical compounds and biological and toxicological processes. Investigating the relationship between the structure and the activity of chemical compounds (SAR) means understanding the toxic activity and allows the prediction of the activity of new compounds based on knowledge about the chemical descriptors. These predictions can be achieved after some operations consisting in pre-processing the data, applying Machine Learning algorithms in order to learn relations and using inference in learned structures.

In order to obtain an explanatory descriptive data model, the QSAR problem needs an inter-disciplinary team that work in a circular process. Thus, results obtained from the knowledge discovery methods must be readable and satisfactory for chemistry specialists. If the results are not acceptable the process is reloaded using other data or other methods. The final results must sustain the integration of the new predicted knowledge within the toxicology domain.

I. Lovrek, R.J. Howlett, and L.C. Jain (Eds.): KES 2008, Part I, LNAI 5177, pp. 565–572, 2008.

Pre-processing is the first step in Knowledge Discovery process and consists of eliminating the irrelevant, noisy, unreliable, incomplete, redundant data, which can have a negative impact on the entire process. The reason is that the process results are sensitive to the quality of data. One of the issues is the dependency between the problem to be solved and the application domain that imposes choosing the right techniques. The majority of these techniques are time and resource consuming. So, the challenge is to choose the most appropriate technique to be applied in sub-stages of the process. In our problem, the steps consist of integrating the data formats from partners, cleaning the data in order to preserve the most important values, selecting the relevant attributes.

Moreover, the data pre-processing has a significant impact on the reduction of the horizontal dimension and, therefore, of the hypothesis space of the data set: less attributes are more comprehensible, the smaller dimension of the input space allows faster training sessions.

The aim of QSAR techniques is to find relations between any biological activity and the properties of a set of molecules (physical and chemical characteristics). However, in its most general form, QSAR is supposed to discover correlations independently of physical and chemical properties.

Among pre-processing techniques there are at least three broad trends in dealing with the problem of features selection: filters, wrappers and embedded methods.

Among the machine learning algorithms the most used ones are based on artificial neural networks [14], decision trees, fuzzy neural network, and Bayesian networks approaches [15]. All these approaches predict the activity of a chemical compound, without being able to explain the predicted value.

2 Techniques Applied to the QSAR Problem

The data mining process is a repetitive one, due to the fact that the mining evaluation is unsatisfactory, or the mining results are meaningless for the user, hence initial conditions/techniques/algorithms may be changed and the whole process restarted.

The knowledge discovery process usually consists of the following stages:

1) achieving a good understanding of the application domain, of the relevant a priory knowledge and of user's final goals;
2) creating or selecting the target dataset, by feature selection and/or data sampling; this step is useful to be done in cooperation with chemistry specialists;
3) pre-processing data (eliminating noise, dealing with missing values, data reduction, data transformation);
4) choosing the data mining technique and the implementation algorithms;
5) the data mining process application itself;
6) the mining result evaluation using automate validating techniques and domain specialist assessment;
7) integrating the obtained knowledge models within the application domain knowledge;

Applying preprocessing techniques on the target dataset is a very important phase in the KDD process. On the dataset there are different types of operations that can be performed, like the following ones:

1. sampling – vertically reducing the initial dataset– is recommended whenever there is a huge amount of instances that is unmanageable when building the data models;
2. noise elimination, is recommended when the initial dataset is obtained in disturbing circumstances that influence the collected data values;
3. data cleaning, checks the initial dataset for inconsistent or missing values and attempts replacing those values with estimated ones;
4. data integration, when several, heterogeneous data sources are used when data is collected;
5. feature selection – horizontally reducing the dataset – is recommended when instances are described through too many attributes;
6. discretization – reducing the size of attribute's domain for computation reasons.

In the feature selection pre-processing field there are three major approaches: filters, wrappers and embedded techniques. In real world application, wrappers and embedded methods seam to select the proper attributes offering superior performance. On the other side, the filter techniques present an insight of the training data set and its statistical properties.

3 Experimental Approach for Preprocessing Techniques

The purpose of our approach is to transform the initial dataset into a complete, consistent, relevant, but manageable data set. During this stage there are many techniques and data transformations that can be applied, but there is no universal procedure, because of the interdependence between the universe of the data and the problem to be solved. It is also a time and resource consuming stage. Our study is based on four of the preprocessing specific activities, namely data integration, data cleaning, feature selection, and discretization.

3.1 Problem Dataset Description and Data Integration

The aim of our study is the assessment of the toxic effects of some organic compounds, possible pollutants in residual waters. The initial dataset was provided by the Center for Applied Biochemistry and Biotechnology (BIOTEHNOL) Bucharest, our partner in a national research project, Topar [13]. It started with 184 chemical substances characterized by about 329 measured toxicological effects, in different contact conditions (ingestion, inhalation, dermal contact, etc.) against almost 20 species of living organisms (birds and mammals). The toxic measures were obtained from the online database ChemID Plus [1]. On the other hand a part of those substances are characterized a set of 266 descriptors supplied by another partner from National Research & Development Institute for Environment Protection (ICIM) Bucharest. The descriptors are grouped in 5 categories: structural, geometrical, topological, electrostatic and quantum-mechanical.

A part of the chemical compounds do not have toxic values for all species and infestation conditions, hence in the pre-processing step were applied Bayesian dependency discovery techniques [16]. Bayesian networks allow computing the probabilities for the dependency relationship between variables. High dependency probabilities lead to the conclusion that some of the data are redundant.

In the first stage, dependency computation has been applied only to toxicological data (without substances descriptors) and it has removed some of the toxicological effects because they showed dependence on others, meaning that they are redundant. The cleaning process conducts to the elimination of 323 toxic measures in two steps. First, we eliminated 96 measures. The reason relied on discovering strong dependencies with others toxicological values regarding others species, meaning that they can not enrich the model. For the next step, we kept 6 toxicological effects, (mouse - oral and intra-peritoneal, rabbit – dermal contact, and rat- oral, intra-peritoneal and dermal contact). The reason for keeping only 6 toxic measures is the lack of coverage in the substance set. The other toxic effects had too few measured values for the interest substance set of the problem.

The training set has also been populated, by the chemists, with 45 of the chemicals and the test set with the rest of 21. Our chemist partners decided to split the dataset in such a way that each toxicity class has representatives on both sets.

3.2 Feature Selection

Feature selection process consists in determining the relevant feature subset, namely features that show a lack of similarity with the target attribute. We used filters and embedded techniques for the feature selection step. Filters evaluate the quality of the attributes and select them independently of the knowledge discovery algorithm that will be used later on. They can be context independent when the contribution of the attribute is individually taken into account (e.g. the Information Theory based measures: Information Gain, Gain Ratio) or context dependent, in connection with others attributes from the training set (e.g. Relief [2], ReliefF [3], RReliefF [4]). The context-independent measures are fast but they completely ignore the other attributes. Context dependent measures have to trade result accuracy with performance. We chose to use context dependent measures because they can detect the information comprised in dependence relations between attributes. In embedded techniques the selection process is included in the knowledge discovery step and machine learning algorithms have the ability to extract the most suited attributes modelling the training data in the same time (e.g. Decision Trees, Bayesian Networks).

In the first step of this study, we have calculated and compared a similarity measure between two or more objects.

We used a combination of the Relief algorithm and the dependence Bayesian network to evaluate the predictive capability of numeric or symbolic attributes. In the end, each attribute is graded according to its calculated predictive capability and with chemistry specialist assessment.

The Relief family methods [4] evaluate the contribution of the values of each attribute in the training data set to distinguish between the most similar instances in the same class, as well as in different class. Using the difference function each attribute is scored being penalized if the values of the attribute are different for the instances in the same class and compensated if the values of the attribute are different for the instances in the opposite class. The partial distance function is set to a neighbor radius of 0.1 and an indiscernability threshold of 0.01. This method is 10-fold cross-validated in order to measure predictive performances.

Bayesian Networks [5] evaluate the importance of one attribute by observing how much the predictive performance of the model drops if we remove the corresponding variable. An important feature makes the prediction accuracy of the model to drop down when it is left out. On the other hand, if removing a feature does not significantly affect the performance of the selected model, then it is less important. The predictive performance of the network is estimated with leave-one-out (LOO) cross validation method [6, 7].

The results of the experiments are obtained with *Weka* [8] benchmark software and *B-Course* [5].

For the 6 chosen toxic measures, the two feature selection methods were performed in order to rank chemical descriptors from the relevance point of view. It was determined that a small set of descriptors frequently appears related with the toxic effects of the chemical, apart from the species.

At the final step, we selected features from dataset after a vote between the results from each selection method. A feature was selected if and only if both methods agreed.

The results obtained in what concerns the feature selection show an interesting connection between the relevant attributes associated with the administration mode (oral, contact, dietary, etc.), regardless of the specie they correspond to. They also reduce the space of 266 descriptors to sets from 30 to 50 relevant descriptors.

In order to validate the results, the 6 datasets and the associated relevant attribute sets have been presented to chemistry specialists. At this moment, the datasets are more readable, due to the relevant attribute sets.

The chemists have selected as target attribute toxicological values for rat LD50 oral administration. They have also decided, according to the ranking of the predictive capability of the attributes obtained from feature selection, to 4 different sets of descriptors for the 66 chemicals: one with 52 descriptors, one with 34, one with 26 and one with 13 descriptors.

3.3 Target Class Discretization

For each chemical, the value of the lethal dose has been discretized by conversion into a toxicity degree according to the Hodge & Sterner scale (see Table 1). The target toxic measure corresponding to rat species in oral administration, take values into an interval between 0.02 and 64000 mg/kg. Another reason for using this scale is the large range of toxic values.

Table 1. Toxicity degree on Hodge & Sterner scale

Degree	Label	LD50 (mg/kg)
1	High	< 1
2	Regular	< 50
3	Moderate	< 500
4	Low	< 5000
5	Very low	< 15000
6	Harmless	> 15000

According to the discovered knowledge type there are two main aspects in knowledge discovery process: the descriptive one and the predictive one. Within the predictive data mining direction there are techniques that mine evolution knowledge, like neural networks, genetic algorithms, pattern recognition and others. Within the descriptive data mining approach there are techniques that discover association rules or frequent sequences of events; make summaries, classes, clusters etc.

The descriptive techniques work with symbolic knowledge and have the capacity to reveal the dataset's structure-related patterns. Most of them have explanatory capabilities. The predictive techniques work with machine learning and inference algorithms and most of them do not have explanatory capabilities.

With the discretized linguistic values of toxicity class we perform classification algorithms using training set with 45 chemicals and test set with 21 substances. The classification methods we used were:

- Artificial neural network implemented in Matlab. The network structure was based on multilayer perceptrons with an input layer containing the same number of neurons like the number of descriptors, an hidden layer with half of the number of inputs and outputs sum, and an output layer with a single neuron. Neuron connection weight was adapted through back propagation algorithm with adaptive learning rate and momentum.
- Neuro-fuzzy interference system ANFIS [9] used for learning the parameters of membership functions, a process similar with neural networks combined with fuzzy inference rules engine.
- Classification and regression tree algorithm CART [10] that divides recursively the data using a hierarchic structure that represents classification rules. Implementation was made with implicit Matlab functions.
- Naïve Bayes classifier [11], a simple probabilistic classifier based on applying Bayes' theorem with naive independence assumptions. The obtained probability model is an independent feature model. The model construction and the assessment of attribute quality were performed with B-Course [5]. B-Course is a web-based tool for causal and Bayesian data analysis. Also, we used Weka [8] for testing reasons, because the b-Course does not support automatic classification accuracy computation with a test dataset.
- NBTree [12], a hybrid decision tree with naïve Bayes classificatory in leafs. This method uses decision tree as the general structure and deploys naive Bayesian for

classification purposes. In the literature is stipulated that naïve Bayesian classifiers work better than decision trees when the sample data set is small.

The obtained results are presented in Table 2 according with dataset split in test set values. The predictive power of the initial set of descriptors is poor. Even if the classification accuracy could not be considered sufficient, a small improvement is observed after the feature selection step. Due to the very large number of descriptors and very low training samples, the artificial neural network and ANFIS models could not be properly trained.

Table 2. The accuracy of prediction for classes according to Hodge & Sterner scale using test dataset

Methods	Initial set	Set no. 1	Set no. 2	Set no. 3	Set no. 4
Neural Net	-	10 / 21 (47.62%)	11 / 21 (52.38%)	9 / 21 (42.86%)	11 / 21 (52.38%)
ANFIS	-	10 / 21 (47.62%)	7 / 21 (33.33%)	8 / 21 (38.10%)	5 / 21 (23.81%)
CART	8/21 (38.1%)	12 / 21 (57.14%)	10 / 21 (47.62%)	10 / 21 (47.62%)	5 / 21 (23.81%)
Naïve Bayes	9/21 (42.86%)	10 / 21 (47.62%)	14 / 21 (66.67%)	14 / 21 (66.67%)	9 / 21 (42.86%)
NBTree	9/21 (42.86%)	15 / 21 (71.43%)	6 / 21 (28.57%)	6 / 21 (28.57%)	5 /21 (23.8%)

Analyzing the confusion matrices on these classification models, it has been observed that the Hodger & Sterner scale used for discretizes of the target attribute is not adequate for this classification purpose.

4 Conclusions and Future Works

Using six toxicity classes, the classification model did not work well for predicting the "very low" and "moderate" toxicity classes. But the model can be used for prediction for a different class definition, by combining domain values for the most relevant descriptors. This prediction can be made with Naïve Bayes classification model obtained from b-Course tool. Also, the classification model is fairly readable, since it is using only a small set of descriptors.

The best results were obtained for the set number three using naïve Bayes classifier, having the highest prediction accuracy. The descriptors used in set number three (less one single descriptor) are also present in sets number two and four, thus confirming their importance.

For future work, beside the pre-processing procedures, redefining toxicity classes is considered in order to improve the prediction capability of the model.

Acknowledgments. This work was partially funded by the TOPAR Romanian project, under the contract CEEX/RELANSIN/535-13/09/2006. We express our thanks to our project partners, especially to Mariana Badea (ICIM) and Florentina Israel (Biotehnol).

References

1. The Specialized Information Services Division of the National Library of Medicine, Information resources and services in toxicology, `http://chem.sis.nlm.nih.gov/chemidplus/`
2. Kira, K., Rendell, L.A.: A practical approach to feature selection. In: International Conference on machine learning, pp. 249–256. Morgan Kaufmann, San Francisco (1992)
3. Kononenko, I.: Estimating attributes: Analysis and Extension of Relief. In: Proc. of ECML, pp. 171–182. Springer, Heidelberg (1994)
4. Kononenko, I.: Evaluating the quality of the attributes. In: Advanced Course on Knowledge Technologies, ACAI 2005, Ljubljana (2005)
5. Myllymäki, P., Silander, T., Tirri, H., Uronen, P.: B-Course: A Web-Based Tool for Bayesian and Causal Data Analysis. International Journal on Artificial Intelligence Tools 11(3), 369–387 (2002), `http://b-course.cs.helsinki.fi/obc/ref.html`
6. Domingos, P., Pazzani, M.: Beyond Independence: Conditions for theoptimality of the simple bayeisan classifier. In: Proceeding of the Thirteenth ICML (1996)
7. Charles, E.: Naïve Bayesian Learning, Technical Report, University of California (1997)
8. Witten, I.H., Frank, E.: Data Mining: Practical Machine Learning Tools and Techniques with Java Implementations. Morgan Kaufmann Publishers, San Francisco (1999)
9. Jang, R.J.S.: ANFIS: Adaptive network-based fuzzy inference system. IEE Trans., Man and Cybernetics 23, 665–685 (1993)
10. Breiman, L., Friedman, J.H., Olshen, R.A., Stone, C.J.: Classification and regression trees. Wadsworth Inc., Belmont (1984)
11. Rish, I.: An empirical study of the naive Bayes classifier. In: IJCAI 2001 Workshop on Empirical Methods in Artificial Intelligence (2001)
12. Liang, H., Yan, Y.: Learning Naive Bayes Tree for Conditional Probability Estimation, CAI (2006)
13. TOPAR Project, `http://www.topar.ro/homeEN.asp`
14. Areej, S., Gongde, G., Daniel, N.: A Study on Applications of Machine Learning Techniques in Data Mining. In: 22nd BNCOD, UK (2005)
15. Neagu, D., Guo, G., Trundle, P., Cronin, M.: A Comparative Study of Machine Learning Algorithms Applied to Predictive Toxicology Data Mining. In: Proc. of SSCT&ETS (2005)
16. Pearl, J.: Bayesian Networks, Causal Inference and Knowledge Discovery, UCLA Cognitive Systems Laboratory, Technical Report (R-281) (March 2001)

Modeling of the Students Scenario on a Learning Course

Yuji Shinoda[1], Kenji Yoshida[2], and Hirotaka Nakayama[3]

[1] Institute of Intelligent Information and Communications Technology, Konan University,
8-9-1, Okamoto, Higashinada, Kobe, Japan
shinoda@center.konan-u.ac.jp
[2] Konan Boys' High School
[3] Dept. of Information Science and Systems Engineering, Faculty of Science and Engineering,
Konan University, Japan

Abstract. Adjusting the content to each student is a major issue in e-Learning. From this viewpoint, a learning course as a series of content also must be adjusted according to the performance of the students. We propose a method that combines clustering and decision tree learning for constructing scenarios of the students' actions. The global statuses of the students are reflected to the clusters, and the local and sequential actions of the students are reflected to the decision trees. The results of e-Learning tests gathered from Japanese junior high school students was processed by our proposed method. We graded the clusters by adaptation to the trees, and selected a set of clusters as a scenario for the students. These scenarios have a possibility of aiding the adjustment, and revision of learning courses.

Keywords: e-Learning, Clustering, k-means, Decision Tree, User Model.

1 Introduction

Adjusting the contents to the student is a major issue in the e-Learning field. e-Learning creates a one-to-one tutoring environment at a low cost, and has the potential to enhance the efficiency of the learning compared to group instruction [1]. Anderson et al constructs cognitive tutors for an advanced learning environment [2]. In this research, the systems not only use information from students actions but also use models of the knowledge of the targeted field. Yoshida et al shows some methods to create human-like action by using the reaction of the students [3]. This system has rules to define the status of the students from their actions on the contents. If a student keeps performs "meaningless" experiments on the interactive contents, the system advises him. These rules are a kind of knowledge. In previous works, the behavior of teachers was reflected to e-Learning systems and produced some remarkable results.

However, teachers' knowledge does not appear only in the contents. In classrooms, they observe the students through a series of lessons and adjust the contents to enhance the effect of their lessons. Teachers somehow define types of students as scenarios of learning behaviors, and find a better approach more suited to the students understanding. These teachers' behavior seems to be based on their experiences and careful observation of their students' circumstances.

I. Lovrek, R.J. Howlett, and L.C. Jain (Eds.): KES 2008, Part I, LNAI 5177, pp. 573–579, 2008.

In e-Learning, constructing student scenarios is difficult because of the following two reasons. Firstly, it is more difficult to gather information compared to actual classrooms. Secondly, there is the possibility that e-Learning courses have different scenarios from actual classrooms. However there are two important merits to finding student scenarios on e-Learning courses. The first merit is that the e-Learning method may provide better information to distribute effective content to its users. This merit is for benefit of the users. The second merit is scenarios make the status of the learning course clearer. This merit is for the benefit of the teachers. Teachers may get useful data for the revision of the course. If the teacher's lessons are combined with an e-Learning course, he may have a chance to give additional lessons to reinforce the students' comprehension.

We propose a method that combines clustering and decision tree learning for constructing these scenarios of student actions. The global statuses of the students are reflected to the clusters, and the local and sequential actions of the students are reflected to the decision trees. The scenarios are shown as a selected clusters based on the rules of decision trees.

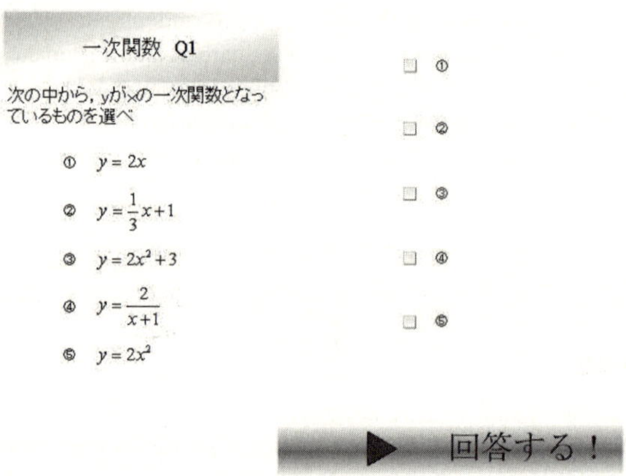

Fig. 1. Image of the Test of Liner Equation (Chapter 2) Question 1

2 Gathering Data

We prepared an on-line test of mathematics for some junior high school students. Our test system was constructed as client-server system. As clients, test contents are constructed by using Flash. Fig. 1 shows an example of the test contents. As a server, we use PostgreSQL to store test result and Java Servlet to process data.

The test consists of four chapters. Chapter 1 is “functions”. Chapter 2 is “proportion”. Chapter 3 is “linear functions”. Chapter 4 is “quadratic functions”. Chapter 1 has 3 questions, chapter 2 has 3 questions, chapter 3 has 5 questions, and chapter 4 has 4 questions. As a result, we have 15 questions through the test.

Our e-Learning site had about 180 students. In the end, we received 154 sets of test results.

3 Algorithm

In this section we propose our method to construct scenarios of student actions. Our method is based on three steps of the processing data. Firstly, we construct clusters by using k-means method as a status of students from global viewpoint. Secondly, we construct decision tree by using C5.0 as a status of students from local and sequential view point. Thirdly, we combine previous two view points, and try to find out our scenarios of students actions. These steps are shown in Fig. 2.

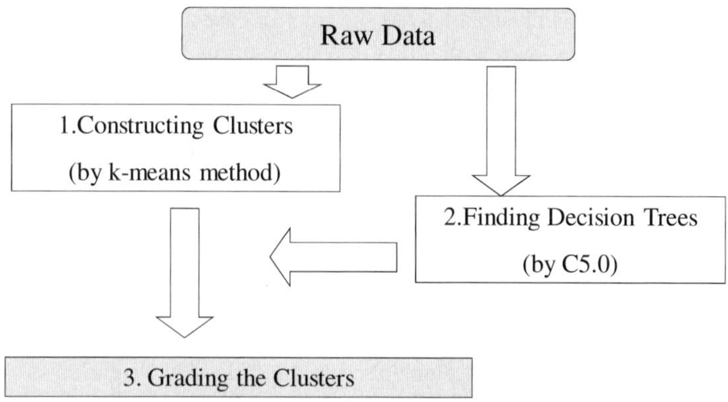

Fig. 2. Outline of Our Algorithm

3.1 Constructing Clusters

As the first step, we construct clusters from data. We construct clusters by using the k-means method [4]. To use this method, we need to decide the number of clusters in advance. However, in general there is no answer about the correct or suitable number of clusters. Finding a scenario of the students provides no information for deciding the number of clusters on the targeted learning course. To avoid this issue, we create some sets of clusters and store them. Two to eight clusters are generated and stored. There is also an issue when we make clusters. A problem will arise if we make clusters with too few elements. However, there is no answer about the best or necessary length. To avoid this issue, we also change the used elements for clustering. We change the length of the element into three to fifteen. We make sets of clusters by changing the number of clusters and the number of elements. As a result, we have ninety-one sets of clusters from seven types of numbers, and thirteen types of length.

3.2 Finding Decision Trees

As the second step, we construct decision trees including the idea that this test is sequential data. In this paper, we use C5.0 [5] [6]. The decision trees reflect the local and sequential actions of the students. Students joined the learning course from chapter 1 to 4 sequentially. This means there are relations among earlier elements and successive ones. If students show a tendency in their test result, there are rules for the later elements based on the former. We use the previous chapters' results to make rules for the next chapter's result.

3.3 Grading the Clusters

As the third step, we compare the clusters and the rule created as a decision tree of the element. By the rule, elements of the clusters are divided into two groups. In one of the groups, the element may be a correct answer. In The other group, the element may be a wrong answer. In this discussion, we use the word to indicate these groups as "right class" and "wrong class", for short. If all elements in one of the clusters are judged as "right class", it means this cluster matches the rule. In case all of the elements in one of the clusters are judged as "wrong class", it means also this cluster matches to the rule. However, if a cluster is divided and half of the elements are "right class" and the other elements are "wrong class", then this cluster does not agree with the rule.

Using this idea, we calculate the score of the clusters by the following equation.

$$S_{ij} = \sum_{k=1}^{j} \left| Er_{ijk} - Ew_{ijk} \right| \tag{1}$$

In this equation, i denotes the number of elements used for clustering. j denotes the number of clusters in a set of clusters. S_{ij} denotes the score of the set of clusters at i and j. Er_{ijk} denotes the number of elements that are judged as “right class” in the k-th cluster in the set of clusters at i and j. Ew_{ijk} denotes the number of elements that are judged as “wrong class” in the k-th cluster in the set of clusters at i and j. From this equation, a set of the clusters S_{ij} is evaluated. As we mentioned in the previous section, we had ninety-one sets of clusters. By using this equation, we grade all of the sets of clusters to find the set of clusters that matches the rules.

4 Data Processing

4.1 Clusters

The summation of the distances between the center of the clusters and the constituent element decreases, if the number of clusters increases. If the number of elements used for clustering increases, the summation of the distance increases. This result is reasonable, and there is no clear information when and how we can generate an effective cluster.

4.2 Decision Trees

We generated rules by C5.0 algorithm. From the result, the rules for element 5, 6, 8, 10, 12, 14, and 15 were found as IF-THEN rules by using the elements in the previous chapters. On the other hand, the rules for element 4, 7, 9, 11, and 13, were found as simple rules that the element is always true. Our grading process is based on how rules divide the elements in the clusters. From this viewpoint, the later simple rules that the element always becomes true have no meaning with regard to our grading process. In this paper, we use these IF-THEN rules for grading.

4.3 Grading Result

Following this procedure, we graded the set of clusters. This time we used the rules for 5, 6, 8, 10, 12, 14 and15 for grading of the 154 students test result. The maximum score for a set of the cluster is 1078 as 7 times 154. The score is shown in Fig. 3. In this figure, x-axis means the number of clusters in a set, and y-axis means the score summation of the score in the set. From Fig. 3, the sets of the clusters generated by using 3,4, and 5, elements do not have a good score. The sets of the clusters generated

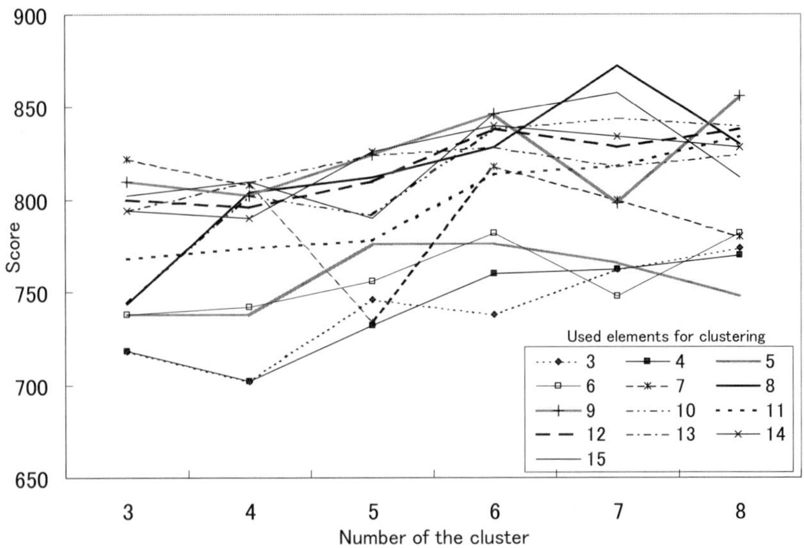

Fig. 3. Grading Result of the Set of Clusters

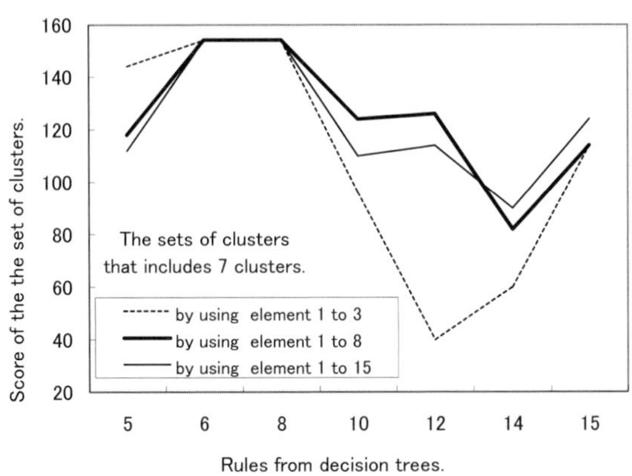

Fig. 4. Score of the Set of Clusters that includes 7 clusters

by using 13, 14, and 15 elements have better scores than other sets. The best set of clusters in this figure is found in generating 7 clusters by using 8 elements.

5 Discussion

However there is a question that this best set of clusters may gain its score only in a few rules.

Fig. 4 shows the score of the set of clusters that include 7 clusters. This figure shows that if we construct a set of clusters in the beginning of the test then the score of the set of clusters decrease on the later rules. On the other hand, the set of clusters using element 1 to 8 scores on many rules.

From this result, we have two suggestions. Firstly, we may have chance to classify students using only half of the test results on our test contents. Secondly, it means there is a possibility that this set of clusters may be suitable for student scenarios.

Fig. 5 shows the detail of the set of clusters. In this figure, the average scores in each chapter of the clusters are shown. From this figure, there is a group that keeps a good score. On the other hand, there is another group that had a bad score in chapter 1, however they recovered in later chapters. This is the scenario of our learning course based on observed data. By using our proposed method, we selected this set of clusters as a scenario for the students. There is possibility to distribute contents based on prediction of students' scenario.

Fig. 5 also shows the status of the test contents. All students' groups have good score in chapter 2. It means questions in this chapter are not suitable to judge the students' skill. From this result, we need to consider two kinds of viewpoint. The first viewpoint is we may adjust the test contents in chapter 2 in the future work. The second viewpoint is we need to focus about normalization of the raw test data.

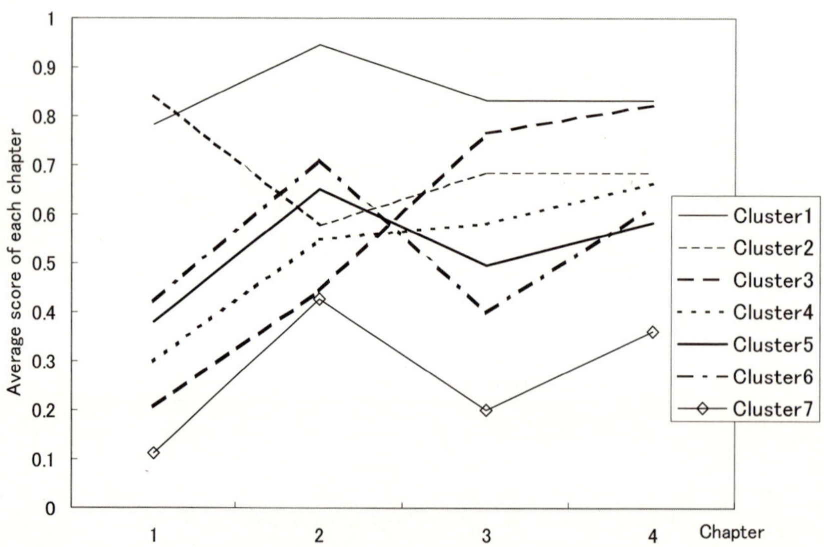

Fig. 5. Detail Status of the Best Set of Clusters

6 Conclusion

In this paper, we proposed a method to find out student scenarios by combining clustering and decision trees. The scenario is shown as a set of clusters. These clusters show what happens to the students on the course. On the other hand, our scenario is one of the candidates. We need discuss carefully our clustering result from stochastic view point.

In future work, we will try to construct a system that distributes content considering these scenarios. We will also try to gather more cases to discuss whether our scenarios have any suggestions for teachers engaged in blended e-Leaning.

Acknowledgement. This research was partially supported by the Ministry of Education, Science, Sports and Culture, Grant-in-Aid for Young Scientists (B), 19700650, 2007.

References

1. Bloom, B.S.: The 2 Sigma Problem: The Search for Methods of Group Instruction as Effective as One-to-One Tutoring. Educational Researcher 13(6), 4–16 (1984)
2. Anderson, J.R., Corbett, A.T., Koedinger, K.R., Pelletier, R.: Cognitive tutors: Lessons learned. Journal of the Learning Sciences 4(2), 167–207 (1995)
3. Yoshida, K., Miyazaki, K., Iwamoto, A., Nakagami, K., Ma, R., Nakayama, H.: Development of a Human-like e-Learning System for Students' Active Learning, IFSR (The First World Congress of the International Federation for System Research). Japan in CD-ROM (2005)
4. McQueen, J.: Some methods for classification and analysis of multivariate observations. In: Proceedings of the Fifth Berkeley Symposium on Mathematical Statistics and Probability, pp. 281–297 (1967)
5. Mitchell, T.: Machine Learning. McGraw-Hill, New York (1997)
6. C5.0, http://www.rulequest.com/

Sensor Network-Based Nonlinear System Identification

Giorgio Biagetti, Paolo Crippa, Francesco Gianfelici, and Claudio Turchetti

DEIT – Dipartimento di Elettronica, Intelligenza Artificiale e Telecomunicazioni, Università Politecnica delle Marche, Via Brecce Bianche 12, I-60131 Ancona, Italy
{g.biagetti,pcrippa,f.gianfelici,turchetti}@deit.univpm.it

Abstract. In this paper, a new algorithm for the identification of distributed systems by large scale collaborative sensor networks is suggested. The algorithm, that uses the distributed Karhunen-Loève transform, extends in a decentralized setting the KLT-based identification approach that have recently been proposed for a centralized setting. The effectiveness of the proposed methodology is directly related to the reduction of total distortion in the compression performed by the single nodes of the sensor network, to the identification accuracy as well as to the low computational complexity of the fusion algorithm performed by the fusion center to regulate the intelligent cooperation of the nodes. The results in the identification of a system whose behavior is described by a partial differential equation in a 2-D domain with random excitation confirms the effectiveness of this technique.

Keywords: KLT, computational intelligence, sensor network, distributed system identification.

1 Introduction

In the last few years collaborative signal processing with distributed sources of data, signals, images and natural phenomena has been gaining importance. Thanks to the recent advances in hardware technologies, today it is possible to realize low-power low-cost wireless devices with limited on-board processing capabilities and sensing units that are able to detect information from the distributed environment. Even though individual sensors can only perform simple local computation and communicate over a short range at low data rate, when deployed in large numbers they can form an intelligent collaborative network interacting with the surrounding environment in a large spatial domain. Sensor networks characterized by low computational-complexity, great learning capability, and efficient collaborative technology are highly desirable to discriminate, regulate and decide actions on real phenomena in many applications such as environmental monitoring, surveillance, factory instrumentation, defence and so on. Based on this premise, the aim of this work is to suggest an innovative framework for the identification of nonlinear non-stationary distributed systems. This approach is based on a centralized intelligent identifier that makes the best identification in a distributed setting on a chosen ensemble of realizations and with no constraints in terms of model kind and/or model order. Methodologically, we define a stochastic setting where

I. Lovrek, R.J. Howlett, and L.C. Jain (Eds.): KES 2008, Part I, LNAI 5177, pp. 580–587, 2008.

the nonlinear system to be identified generates nondeterministic signals, i.e., stochastic processes (SPs), from given initial conditions and random parameters of input signals. In this way, the set of input-output pairs so obtained shows complex but identifiable geometrical relationships in the Hilbert spaces obtained by KLT-transformation of the outputs, because of the intrinsic separability property of the Karhunen-Loève transform (KLT) and the uniformity properties of the systems considered. As a subsequent step we define a computational intelligence technique for approximating the previously mentioned mappings that are able to globally identify the distributed system. The optimization of the global identification accuracy, and the interactions between network elements is performed in a collaborative setting, exploiting and developing the cooperation mechanisms that underpin other related methodologies such as the distributed KLT [1]. This approach is particularly suitable in this context since an iterative algorithm that cooperatively minimizes the overall compression-induced distortion is suggested, and its convergence properties stated and proved.

The proposed technique has been proven on physical models of real phenomena described by partial differential equations (PDEs) with random input signals and simulated in a CAD/CAM simulation environment. The experimental results obtained clearly show the effectiveness of the suggested identification methodology and its excellent compression capability. With no limitations on the type of model, environment geometry and model order, the technique represents an innovative and very powerful framework in a large number of applications.

2 Modeling a Distributed System by the KLT

Let us consider in a domain $D \subset \mathbb{R}^2$ the functional $\mathcal{F}$, depending on the position point $p \in D$, the time t, a generic input signal $u(t)$ and some initial condition c,

$$y(t,p) = \mathcal{F}\left(u(\tau)|_{[t_0,t)}, t, p, c\right), \quad y \in \mathbb{R} \tag{1}$$

represents a scalar field, that is the behavior of a distributed system to be identified.

Assuming the domain D and the time t have been discretized, we can sample the system at S nodal points $p_1, p_2, \ldots, p_S$ yielding the following functionals

$$\begin{aligned} \boldsymbol{y}_\ell(n,\mathbf{x}) = \boldsymbol{y}(n,p_\ell,\mathbf{x}) = \mathcal{F}\left(u(\eta,\boldsymbol{a})|_{[0,n)}, n, p_\ell, c\right) \\ \ell = 1,\ldots,S, \quad n = 0,1,\ldots,L_s-1 \end{aligned} \tag{2}$$

where $\boldsymbol{a}$ is a Random Variable (RV) that parameterizes the process u and $\mathbf{x} = [\boldsymbol{a}\ \mathbf{c}^T]^T$. Let us introduce the vectors $\mathbf{y}_\ell \in \mathbb{R}^{L_s \times 1}$, $\mathbf{y}_\ell \triangleq [\boldsymbol{y}_\ell(0) \cdots \boldsymbol{y}_\ell(L_s-1)]^T$, $\ell = 1,\ldots,S$, and assume $L = S\, L_s$, then the vector $\mathbf{y} \in \mathbb{R}^{L\times 1}$, $\mathbf{y} \triangleq \left[\mathbf{y}_1^T\ \mathbf{y}_2^T\ \cdots\ \mathbf{y}_S^T\right]^T$ is the discrete-space discrete-time representation of the scalar field $\boldsymbol{y}(t,p)$. This representation holds for every system that possesses the properties of uniformity and causality.

It is well-known that, under wide conditions, $\mathbf{y}$, defined as above and with its realizations $\mathrm{y} \in \mathbb{R}^{L\times 1}$, can be represented by the Discrete Karhunen-Loève Transform

(DKLT), also called *canonical representation*. The DKLT and its inverse can be written in matrix form as

$$\begin{cases} \mathbf{y} = \Phi\,\mathrm{k}(\mathbf{x}) \\ \mathrm{k}(\mathbf{x}) = \Phi^T\,\mathbf{y} \end{cases} \tag{3}$$

where $\mathrm{k}(\mathbf{x}) \in \mathbb{R}^{V\times 1}$ is defined as $[\mathrm{k}(\mathbf{x})]_j = k_j(\mathbf{x})$, with $j = 1,\ldots,V$ and $V \leq L$. It is worth noting that the DKLT is the most efficient representation of the SP if the expansion is truncated to use fewer than L orthonormal basis vectors. The matrix $\Phi = [\phi_1 \;\ldots\; \phi_V] \in \mathbb{R}^{L\times V}$ is the reduced orthogonal matrix whose columns ϕ_j, $j = 1,\ldots,V$, are the eigenvectors as obtained from the eigenvalue equation

$$\mathrm{R}_{\mathbf{yy}}\Phi = \Phi\Gamma \tag{4}$$

where $\mathrm{R}_{\mathbf{yy}} = E\{\mathbf{y}\mathbf{y}^T\} \in \mathbb{R}^{L\times L}$ is the autocorrelation matrix of $\mathbf{y}$ and $\Gamma \in \mathbb{R}^{V\times V}$ is the diagonal matrix with (non-null) eigenvalues on the diagonal. The main benefit of this representation is related to the separation property of KLT. On the basis of this property the output of the system can be expressed as a linear combination of products of a function of $\mathbf{x}$ alone and a function of n alone. Since the vectors ϕ_j are determined by means of $\mathrm{R}_{\mathbf{yy}}$, which can be estimated by the realizations of $\mathbf{y}$, the system identification reduces to modeling the functions $k_j(\mathbf{x})$. As $\mathbf{y}$ is a function of $\mathbf{x}$, the terms $k_j(\mathbf{x})$ describe on the space spanned by the columns of Φ curves $C^j_{\mathbf{y}}(\mathbf{x})$, which all together characterize the SP $\mathbf{y}$.

The properties of uniformity and causality determine a smooth behavior of these curves that have then to be reconstructed from an *ensemble* of points extracted by the described approach to perform the identification. Since $\mathrm{k}(\mathbf{x})$ is a no-memory input/output mapping, it can be approximated by a given vector function,

$$\mathrm{k}(\mathbf{x}) \approx \mathcal{G}[\mathbf{x}, \mathrm{W}] \tag{5}$$

where $\mathcal{G}[\cdot]$ is a nonlinear operator and $\mathrm{W} \in \mathbb{R}^{V\times M}$ is a matrix of parameters to be estimated.

3 Identification of a Distributed System Knowing the Output $\mathbf{y}$

With the above considerations in mind, it can be stated that once the structure of the functional $\mathcal{G}[\mathbf{x}, \mathrm{W}]$ has been defined, the identification of the nonlinear system is equivalent to the estimation of W from an *ensemble* of the system's input-output pairs.

In order to derive the identification algorithm [2], it is necessary to relate the stochastic properties of the system (that allowed the development of the general theory) to the available *ensemble* of realizations. Let us then refer to these N realizations of $\mathbf{x}$ as $\mathrm{x}^{(i)} \in \mathbb{R}^{M_{\mathrm{x}}\times 1}$, with $i = 1,\ldots,N$, and to the corresponding realizations of $\mathbf{y}$ as $\mathrm{y}^{(i)} \in \mathbb{R}^{L\times 1}$, with $i = 1,\ldots,N$. Both can be put in matrix form as $\mathrm{X} = [\mathrm{x}^{(1)}\; \mathrm{x}^{(2)} \cdots\; \mathrm{x}^{(N)}]$ and $\mathrm{Y} = [\mathrm{y}^{(1)}\; \mathrm{y}^{(2)} \cdots\; \mathrm{y}^{(N)}]$, where $\mathrm{X} \in \mathbb{R}^{M_{\mathrm{x}}\times N}$ and $\mathrm{Y} \in \mathbb{R}^{L\times N}$. A currently used estimation $\hat{\mathrm{R}}$ of the autocorrelation matrix $\mathrm{R} \in \mathbb{R}^{L\times L}$ is given by

$$\mathrm{R} \approx \hat{\mathrm{R}} = \frac{1}{N}\mathrm{Y}\mathrm{Y}^T\,. \tag{6}$$

The spectral representation of $\hat{\mathrm{R}}$ is

$$\hat{\mathrm{R}}\,\mathrm{U} = \mathrm{U}\,\Lambda \tag{7}$$

where $\mathrm{U} = [\mathrm{u}_1\ \mathrm{u}_2\ \cdots\ \mathrm{u}_V] \in \mathbb{R}^{L\times V}$ is the matrix of eigenvectors and $\Lambda \in \mathbb{R}^{V\times V}$ the matrix of eigenvalues. By projecting all the N realizations onto the basis U we obtain the KLT representation

$$\begin{cases} \mathrm{y}^{(i)} = \mathrm{U}\mathrm{k}^{(i)} \\ \mathrm{k}^{(i)} = \mathrm{U}^T\mathrm{y}^{(i)} \end{cases} \qquad i = 1, \ldots, N \tag{8}$$

and, in matrix form,

$$\mathrm{K} = \mathrm{U}^T\,\mathrm{Y} \tag{9}$$

where $\mathrm{K} = [\mathrm{k}^{(1)}\ \mathrm{k}^{(2)} \cdots\ \mathrm{k}^{(N)}] \in \mathbb{R}^{V\times N}$. Once these projections have been obtained, the problem of approximating $\mathrm{k}(\mathbf{x})$ by a given function $\mathcal{G}[\mathbf{x}, \mathrm{W}]$ corresponds to finding the parameters W that make the following approximation

$$\mathrm{k}^{(i)} \approx \mathcal{G}[\mathrm{x}^{(i)}, \mathrm{W}], \qquad i = 1, \ldots, N \tag{10}$$

hold, so that a model of the system output is

$$\mathbf{y} \approx \mathrm{U}\,\mathcal{G}[\mathbf{x}, \mathrm{W}] \tag{11}$$

In this work we present an identification algorithm that is defined by means of an approximating *mapping* based on neural networks. The proposed nonlinear-in-the-parameter identifier is based on radial basis function networks [3], so that the j-th component of the functional $\mathcal{G}[\mathbf{x}, \mathrm{W}]$ defined in (5) can be put in the following form

$$[\mathcal{G}[\mathbf{x}, \mathrm{W}]]_j = \sum_{l=1}^{M_\mathrm{n}} [\omega_j]_l \exp\left(-[\chi_j]_l \sum_{m=1}^{M_\mathrm{x}} \left([\mathbf{x}]_m - [\Xi_j]_{l,m}\right)^2 \right) \quad , \qquad j = 1, \ldots, V \tag{12}$$

where M_n is the number of neurons in the RBF network, M_x is the dimension of vector $\mathbf{x}$, and $\omega_j \in \mathbb{R}^{M_\mathrm{n}\times 1}$, $\chi_j \in \mathbb{R}^{M_\mathrm{n}\times 1}$, and $\Xi_j = [\xi_j^1\ \xi_j^2\ \cdots\ \xi_j^{M_\mathrm{x}}] \in \mathbb{R}^{M_\mathrm{n}\times M_\mathrm{x}}$ with $\xi_j^m \in \mathbb{R}^{M_\mathrm{n}\times 1}$, for $m = 1, \ldots, M_\mathrm{x}$, are vectors or matrices of weights within $\mathrm{W} = [\mathrm{w}_1\ \mathrm{w}_2\ \cdots\ \mathrm{w}_V]^T \in \mathbb{R}^{V\times M}$, with $M = M_\mathrm{n}\,(M_\mathrm{x} + 2)$, defined so that $\mathrm{w}_j = [\omega_j^T\ \chi_j^T\ (\xi_j^1)^T\ (\xi_j^2)^T\ \cdots\ (\xi_j^{M_\mathrm{x}})^T]^T$. Despite its complexity the neural network-based approximations allow for great flexibility in the choice of the number of free parameters and scale gracefully when M_x increases.

4 Identification of a Distributed System by a Network of Independent Sensors

Let us consider the problem of identifying a distributed system by using a sensor network. In this case the generic variable $y_\ell\,(n)$, $\ell = 1, ..., S$, $n = 0, 1, ..., L_s - 1$ corresponds to the ℓ-th sensor and we assume the sensors are able to transmit only the observed subvector to a *fusion center* and cannot communicate to each other.

By applying a KLT to the observations of each sensor ignoring the dependencies with other terminals we obtain the *marginal KLT*, which is a particular case of (3) and is expressed as

$$\begin{bmatrix} \mathrm{k}_1(\mathbf{x}) \\ \mathrm{k}_2(\mathbf{x}) \\ \vdots \\ \mathrm{k}_S(\mathbf{x}) \end{bmatrix} = \begin{bmatrix} \Phi_{11} & 0 & \cdots & 0 \\ 0 & \Phi_{22} & \cdots & 0 \\ \cdots & \cdots & \cdots & \cdots \\ 0 & \cdots & \cdots & \Phi_{SS} \end{bmatrix}^T \begin{bmatrix} \mathbf{y}_1 \\ \mathbf{y}_2 \\ \vdots \\ \mathbf{y}_S \end{bmatrix} \tag{13}$$

with $\Phi_{ii} \in \mathbb{R}^{L_S \times V_i}$, $i = 1, \ldots, S$. As it is clearly stated by (13) the generic sensor ℓ transmits the subvector k_ℓ (or an approximation of it) so that the output vector $\mathbf{y}$ cannot be reconstructed with a negligible error. This means that the identification approach previously discussed cannot be applied directly to this case due to the lack of a complete knowledge of the output $\mathbf{y}$. It is easy to verify that the marginal KLT will lead to a suboptimal solution to this problem. In general we can search for a solution of the kind

$$\begin{bmatrix} \mathrm{h}_1(\mathbf{x}) \\ \mathrm{h}_2(\mathbf{x}) \\ \vdots \\ \mathrm{h}_S(\mathbf{x}) \end{bmatrix} = \begin{bmatrix} \Psi_1 & 0 & \cdots & 0 \\ 0 & \Psi_2 & \cdots & 0 \\ \cdots & \cdots & \cdots & \cdots \\ 0 & \cdots & \cdots & \Psi_S \end{bmatrix}^T \begin{bmatrix} \mathbf{y}_1 \\ \mathbf{y}_2 \\ \vdots \\ \mathbf{y}_S \end{bmatrix} \tag{14}$$

with $\Psi_i \in \mathbb{R}^{L_S \times V_i}$, $\mathrm{h}_i \in \mathbb{R}^{V_i \times 1}$, $i = 1, \ldots, S$ or in a more compact form,

$$\mathrm{h}(\mathbf{x}) = \Psi^T \mathbf{y} \tag{15}$$

with $\Psi = \mathrm{diag}\left[\Psi_1, \ldots, \Psi_S\right] \in \mathbb{R}^{L \times V}$ and $\mathrm{h} \in \mathbb{R}^{V \times 1}$, $V = \sum_{i=1}^{S} V_i$. In this case the accuracy of the distributed identification is related both to the approximation of the mapping $\mathbf{k}(\mathbf{x})$ and to the minimization of the error $E\left\{\|\mathbf{y} - \hat{\mathbf{y}}\|^2\right\}$ between the real system output $\mathbf{y}$ and its estimation $\hat{\mathbf{y}}$, based on the sensors' observations, and given by

$$\hat{\mathbf{y}} = \mathrm{R}\Psi\left(\Psi^T \mathrm{R}\Psi\right)^{-1} \mathrm{h}(\mathbf{x})\,. \tag{16}$$

However in this case, to the best knowledge of authors, and as it is also pointed out in [1], it is not known a closed-form solution to this problem.

The algorithm developed by Gastpar *et al.*, also known as *distributed Karhunen-Loève transform* is an iterative procedure that aims at finding the matrix Ψ that achieves the MSE best estimate of $\hat{\mathbf{y}}$ in (16).

Without loss of generality, instead of using the general formulation (13), we consider the simple case of two sensors only, corresponding to the variables $\mathbf{y}_1$ e $\mathbf{y}_2$. Assuming the representation $\mathbf{y}_2$ given by the sensor S_2 to be fixed we would determine the representation of $\mathbf{y}_1$ such that $E\left\{\|\mathbf{y} - \hat{\mathbf{y}}\|^2\right\}$ is minimum. The approximation provided by the second sensor can be expressed by $h_2(\mathbf{x}) = \Psi_2^T \mathbf{y}_2 + \mathbf{z}_2$ where $\mathbf{z}_2$ are jointly Gaussian random variables independent of $\mathbf{y}_2$, with zero mean and covariance matrix R_z. Then we can partition the covariance matrix of the entire vector $\mathbf{y}$ into four parts, according to

$$\mathrm{R} = \begin{bmatrix} \mathrm{R}_{11} & \mathrm{R}_{12} \\ \mathrm{R}_{21} & \mathrm{R}_{22} \end{bmatrix} \tag{17}$$

where $\mathrm{R}_{ij} = E\left\{\mathbf{y}_i\mathbf{y}_j^T\right\}$ with $i, j = 1, 2$. Now to find the best estimate of $\mathbf{y}_1$ we define the matrix Ξ, as follows

$$\Xi = \left[\mathrm{R}_{12}^T\ \mathrm{R}_{22}^T\,\Psi_2\right]\begin{bmatrix} \mathrm{R}_{11} & \mathrm{R}_{12}\Psi_2 \\ \Psi_2^T\mathrm{R}_{12}^T & \Psi_2^T\mathrm{R}_{22}\Psi_2 + \mathrm{R}_z \end{bmatrix}^{-1} \tag{18}$$

Let Ξ^* consist the first L_S column of Ξ, thus, we obtain a new matrix

$$\mathrm{R}_w = \begin{bmatrix} \mathrm{I}_{L_S} \\ \Xi^* \end{bmatrix}\left(\mathrm{R}_{11} - \mathrm{R}_{12}\Psi_2\left(\Psi_2^T R_{22}\Psi_2 + \mathrm{R}_z\right)^{-1}\Psi_2^T\mathrm{R}_{12}^T\right)\left[\mathrm{I}_{L_S}\quad \Xi^{*T}\right] \tag{19}$$

with $\mathrm{R}_w \in \mathbb{R}^{L\times L}$, that can be reduced to the diagonal form

$$\mathrm{R}_w = \mathrm{D}_w \mathrm{diag}\left(\lambda_w^{(1)}, \ldots, \lambda_w^{(N)}\right)\mathrm{D}_w^T \tag{20}$$

with $\lambda_w^{(1)} \geq \lambda_w^{(2)} \geq \ldots \geq \lambda_w^{(N)}$, thus we can obtain the basis for representing y_1

$$\Psi_1^T = \left[\mathrm{D}_w^{(V_1)}\right]^T \begin{bmatrix} \mathrm{I}_{L_S} \\ \Xi^* \end{bmatrix} \tag{21}$$

where $\mathrm{D}_w^{(V_1)}$ denotes the matrix consisting of the first $V_1 \leq N$ columns of the matrix D_w and I_{L_S} is the L_S-dimensional identity matrix. In this way by truncating the representation to the first V_1 eigenvectors of $\mathrm{D}_w^{(V_1)}$ we obtain the parameters

$$\mathrm{h}_1(\mathbf{x}) = {\Psi_1}^T\mathbf{y}_1\,. \tag{22}$$

4.1 Best Estimate of the Matrix Ψ Based on the Distributed KLT Algorithm

On the basis of the previous formulation we can now establish a collaborative algorithm that progressively estimates the best approximation for a sensor as a function of the given representation for the other sensors. As it has been demonstrated in [1], by iteratively applying in turn for each sensor formulas (18)–(21), the MSE decreases monotonically, i.e., at the n-th iteration, $E\{\|\mathbf{y} - \hat{\mathbf{y}}^{(n)}\|^2\} \geq E\{\|\mathbf{y} - \hat{\mathbf{y}}^{(n+1)}\|^2\}$. This implies that the algorithm will converge to a stable point, a saddle point or a local minimum, but clearly the convergence to a global optimum cannot be guaranteed.

Having obtained the best estimate of the matrices $\Psi_\ell^{(n)} \approx \Psi_\ell$, $\ell = 1, \ldots, S$, the identification of the distributed system, that is the model of the system, is represented by (16), where the function $\mathrm{h}(\mathbf{x})$ is approximated as

$$\mathrm{h}(\mathbf{x}) = \mathit{\Psi}^T\mathbf{y} \approx \mathcal{G}[\mathbf{x}, \mathrm{W}]\,. \tag{23}$$

5 Experimental Results

To validate the identification technique suggested, the collaborative algorithm was applied to the identification of several distributed systems whose behavior can be described by the solution of a PDE on an elliptical domain. The solutions of such equations have been achieved by using the Matlab PDE Toolbox. In this framework the equations are solved by the finite-element method with non uniform meshes.

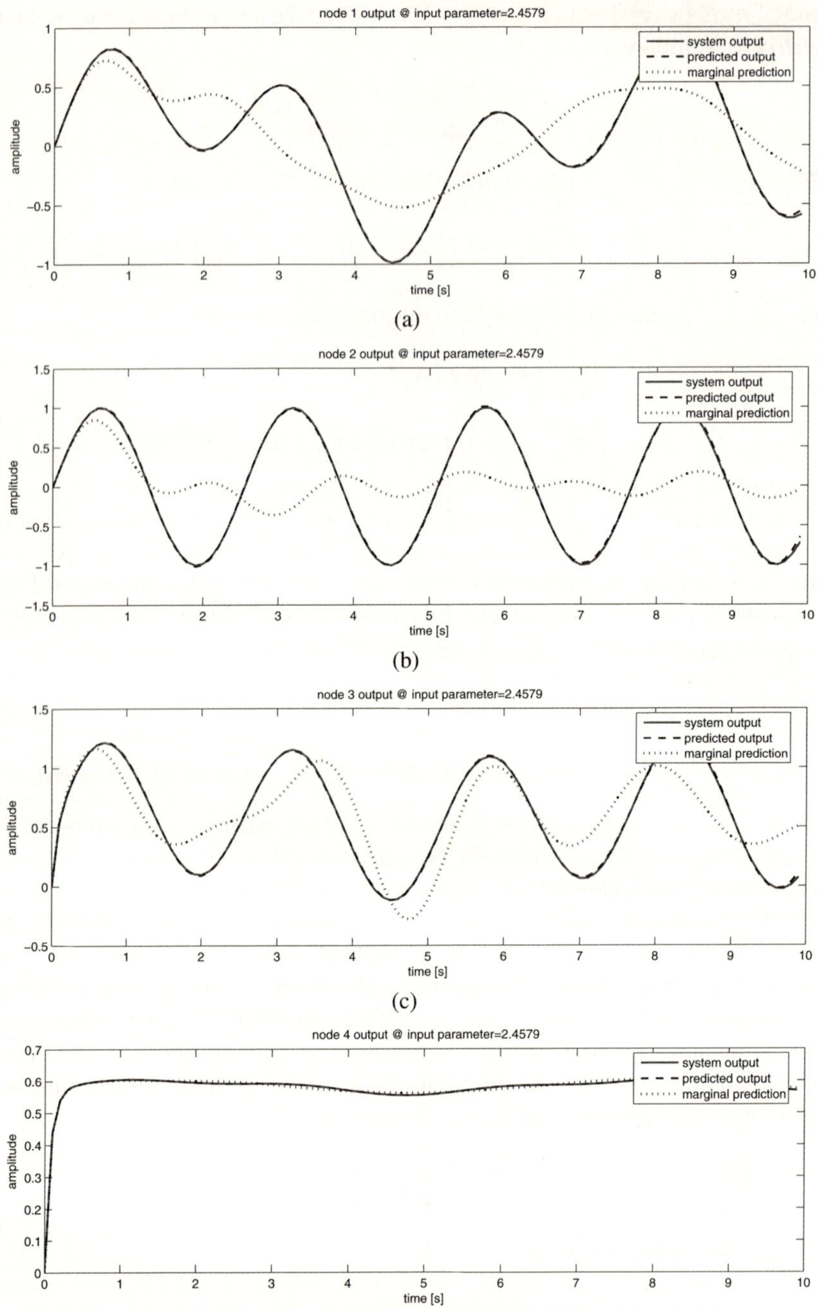

(d)

Fig. 1. Inputs to the sensor network (red solid line) and estimated system output (blue dashed line). The dotted black line represents the approximation that would have been achieved by a simple marginal KLT-based encoding of the system output at the same rate.

One of the experiments carried out made use of a scalar field obtained by the discretization of the following parabolic PDE,

$$\dot{y} - \nabla^2 y + 5\,y = 10\,. \tag{24}$$

The excitation was given as a boundary condition $y = \sin(\omega\, t)$ applied to an arc of the boundary, with ω being proportional to the input parameter. We placed four sensors, labeled 1–4, on randomly chosen knots and selected the best rate allocation for a fixed rate $V = 10$, which resulted to be $V_1 = 1$, $V_2 = 2$, $V_3 = 5$, $V_4 = 2$, i.e. the sensor 1 has one output, the sensor 2 has two outputs, the sensor 3 has five outputs and finally the sensor 4 has two outputs.

The experimental data showed a perfect matching between the sensor outputs and the curve fitting performed by the trained network thus demonstrating the very good performance of the RBF network based algorithm.

The results of the overall identification process for the system generated by the PDE (24) and the comparison between the distributed and the marginal KLT-based techniques are reported in Fig. 1. Here the inputs to the sensor network (red solid line), the estimated system output (blue dashed line) along with the approximation that would have been achieved by a simple marginal KLT-based (dotted black line) encoding of the system output at the same rate have been displayed. It can be easily seen that a huge improvement of this methodology over the marginal KLT can be achieved for the same rate V.

6 Conclusions

In this work an innovative framework for the collaborative identification of distributed systems has been presented. This approach is based on a centralized intelligent identifier that makes the best identification in a distributed setting on a chosen ensemble of realizations. We defined a stochastic setting where the system to be identified generates nondeterministic signals, i.e. SPs from given initial conditions and random parameters of input signals. As a subsequent step we defined a computational intelligence technique for approximating the previously mentioned mappings that is able to globally identify the distributed system. The global optimization of the identification performance was performed in a collaborative setting, exploiting and developing the cooperation mechanisms that underpin other related methodologies such as the distributed KLT. The effectiveness of the proposed collaborative algorithm has been demonstrated in the identification of a distributed system whose behavior is described as the solution of a partial differential equation (PDE).

References

1. Gastpar, M., Dragotti, P.L., Vetterli, M.: The distributed Karhunen-Loève transform. IEEE Trans. Inf. Theory 52(12), 5177–5196 (2006)
2. Turchetti, C., Gianfelici, F., Biagetti, G., Crippa, P.: A computational intelligence technique for the identification of non-linear non-stationary systems. In: Zurada, J.M., Yen, G.G., Wang, J. (eds.) Computational Intelligence: Research Frontiers. LNCS, vol. 5050. Springer, Heidelberg (2008)
3. Haykin, S.: Neural Networks: A Comprehensive Foundation, 2nd edn. Prentice Hall, Upper Saddle River (1999)

Automatic Generation of Computer Animation Conveying Impressions of News Articles

Tadahiko Kumamoto[1], Akiyo Nadamoto[2], and Katsumi Tanaka[3]

[1] Chiba Institute of Technology,
2-17-1, Tsudanuma, Narashino, Chiba 275-0016, Japan
kumamoto@net.it-chiba.ac.jp
[2] Konan University,
8-9-1, Okamoto, Higashinada-ku, Kobe, Hyogo 658-8501, Japan
[3] Kyoto University,
Yoshida-Honmachi, Sakyo-ku, Kyoto, Kyoto 606-8501, Japan

Abstract. This paper proposes a passive viewing interface, called News Reader with Emotional Expressions (wEE), which automatically generates TV program-like animations with synthesized emotional speech and background music from news articles in a user-specified Web news site. A distinctive point of our interface is that it explicitly conveys the impressions of news articles to users by determining appropriate background music and tone of voice for an animated newscaster according to the impressions of each article. Since conventional passive viewing interfaces just read news articles in a monotone, impressions of the articles were not conveyed to users. This caused the bad effect that users did not properly understand news articles in terms of their impressions. We also show that wEE can control (emphasize or soften) impressions of news articles through an on-line experiment with 600 participants.

1 Introduction

Owing to the spread and popularization of the World Wide Web, there is a huge volume of necessary, useful, and interesting information on the Web. Consequently, occasions for accessing and browsing the Web have been sharply increasing even in everyday life. Many individual users, however, do not want to read web pages day after day because reading them is a time-consuming and they have limited time and restricted places in which to read. A passive viewing environment [10,11,13] is therefore needed where users can view and/or listen to web pages as if they were watching television. Such an environment can be realized with passive viewing interfaces that generate TV program-like animations from web pages and present them to users. Passive viewing interfaces enable users to do activities such as housekeeping, bathing, and eating while viewing and/or listening to web pages. Users must also be able to view and/or listen to web pages in a variety of places such as in vehicles, parks, and lavatories.

The basic idea of presenting web pages in a TV program-like fashion was proposed as a TV-program metaphor by Nonogaki et al. in the FRIEND21 Project

I. Lovrek, R.J. Howlett, and L.C. Jain (Eds.): KES 2008, Part I, LNAI 5177, pp. 588–597, 2008.

[10]. Since then, several methods based on this idea have been proposed: segmenting a web page and presenting each segment using synthesized speech and animated agents [13], and transforming web pages into TV information program-like animations by synchronizing the text and image parts of web pages [11]. However, these methods cannot alter the generated animations according to the moods or impressions of the original web pages. They only convert one media into another (web content to animation). This means that the parameters of the animation generation (i.e., direction conditions) such as background music (BGM) and tones of voice of the animated agents are fixed regardless of the impressions. This causes the problem that impressions of web pages are not conveyed to users since hearing is much easier than reading and does not require very much concentration. In other words, the impressions which users gain from a web page by viewing and/or listening to it may be different from those that the users gain by reading it.

In this paper, we take up news articles in a news site on the Web as information sources to be animated, since many news articles are issued and delivered to users everyday and there is a great demand for passive viewing interfaces, and we propose a passive viewing interface, called News Reader with Emotional Expressions (wEE), which determines BGM and tones of voice for each news article collected from a user-specified Web news site, generates a news program-like animation from these news articles, and presents it to users. We also show results of two on-line experiments, and verify the effectiveness of wEE.

2 Related Work

There are numerous studies in which computers explicitly express their computed affective state to users as dialogue participants or convey the affective state of a user to other users as intermediates, where the embodied conversational agents are equipped with synthesized emotional facial expressions to play their roles [8]. Although our system, "wEE," can alter the facial expression of an animated newscaster, in this paper, we focus on sound, which plays a more important role in passive viewing of web pages.

In the field of computer human communication, research on persuasive interfaces has been very active. Hareli et al., based on the idea that one mean by which persuasion can be achieved is by using emotional communication, proposed a theory for displaying emotion, called the "Fitting Principle" [2,3]. According to this theory, in the context of persuasion, communications that include emotional expressions are more persuasive to the extent that the emotion that accompanies a communication fits the content of the message, and communications that lack any accompanying emotional expression, or communications that are accompanied by emotions that do not fit the content of the message, are perceived as less reliable and hence are less persuasive. Our experimental results would complement their work because they used the facial expression and head movement of an embodied conversational agent and kinetic typography to express mood and rhetorical relations, whereas we use the voice of an animated agent and BGM.

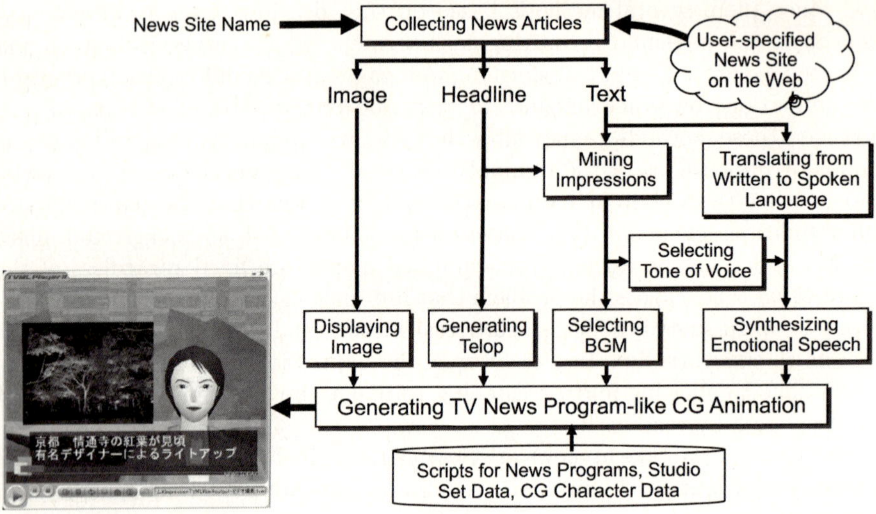

Fig. 1. System architecture of wEE and snapshot of generated animation

It is self-evident in cognitive psychology research on music that music influences human emotions and changes the listener's affective state [12]. However, we have not found any experimental study investigating how people's impressions of the same news article change when the BGM and tone of voice change.

3 News Reader with Emotional Expressions

We developed a passive viewing interface, called the News Reader with Emotional Expressions (wEE). This section presents the system architecture, process flow, and direction method of wEE.

The system architecture of wEE is shown in Figure 1 with a snapshot of a news program generated by wEE, where we replaced the image and telop in the snapshot with those we prepared in order to avoid any copyright problems.

Next, the process flow of wEE is presented: (1) wEE collects news articles in a user-specified genre or category on a user-specified news site on the Web. (2) The headline, main text, and images are extracted from each of the news articles. (3) When an image is extracted, it is adjusted in size and set in a fixed position of the wEE screen as shown in Figure 1. When two or more images are extracted, each image is adjusted in size, and the images are displayed in the fixed position in order, one by one, synchronized with the reading of the corresponding sentences or paragraphs using the method proposed in Ref. [11]. (4) The headline is displayed as a one- or two-line telop according to the number of characters in it. (5) The impression value of the news article is calculated from the headline and the first paragraph of the main text as a real number between 0 and 1 on the impression scale "Bright—Dark." wEE then determines the BGM and the tone of voice for the newscaster according to the impression value. (6) Because the main text of the news articles are in written Japanese, they are modified into spoken Japanese to make the newscaster's speech sound natural. That is, the "DEARU" form of the end of sentence or clause is modified into "DESU-MASU" forms, and article-proper expressions such as "tekijida" and "W-hai" are transformed into expressions for the reading ("taimurii hitto" (run-scoring single) and "waarudo kappu" (World Cup) in the above example). (7) The speech to read is synthesized from the modified main text using a text-to-emotional speech engine [1]. (8) A news program-like animation is generated by embedding information on the studio set and animated agents and the images, telop, BGM, and speech determined in steps (3) to (7) into a TVML-format of the news scenario. (9) Users can watch the animated news program using the TVML player [4], which is free software developed by the NHK Science and Technical Research Laboratories of Japan, and is available at `http://www.nhk.or.jp/strl/tvml/`.

The following paragraphs focus on the main module of wEE, i.e., step (5), which calculates the impression value of a news article, and determines direction of wEE based on the impression value.

First, the input text, $TEXT$, is segmented into words by using a Japanese morphological analysis system called "Juman" [7], and the nouns, verbs, adjectives, and adverbs are extracted from the text.

Next, the impression value $S(w)$ and weight $M(w)$ of each word w in the impression scale "Bright—Dark" are obtained by consulting an impression dictionary, and the impression value $O(TEXT)$ of the input text is calculated using the following equation:

$$O = \sum^{TEXT} (S \times |2S - 1| \times M) \Big/ \sum^{TEXT} (|2S - 1| \times M), \qquad (1)$$

where the $|2S - 1|$ term denotes an inclined distribution depending on the impression value S of each word. Many of the words that appear in a text seem to be independent of the impressions of the text. This inclined distribution is used to remove the adverse effects that such general words have on calculating impression values based on the assumption that the impression values (S) of such general words are approximately 0.5.

The impression dictionary used in this paper was automatically constructed by analyzing the Nikkei Newspaper Full Text Database [9] from the 1990 to 2001 editions[1] using an extended version of the method mentioned in Ref. [6]. The original method created an impression scale from a pair of impression words, but our extended version created an impression scale from two groups of impression words. That is, the extended method calculated which of the two groups would co-occur more frequently with each of words extracted from the newspaper database, and obtained $S(w)$ and $M(w)$ of it.

[1] Each edition contains about 170,000 articles (200 MB), and the newspaper database contains more than two million articles in total.

Table 1. Examples of entries in our impression dictionary

POS	Entry word	S	M	POS	Entry word	S	M
noun	chef	0.894	1.032	noun	debt	0.075	1.183
verb	enjoy	0.842	1.106	verb	oppress	0.080	1.139
adjective	impatient	0.833	1.022	adjective	inactive	0.125	1.048
adverb	easy/leisurely	0.820	1.148	adverb	still	0.176	1.314

Table 2. Impression words composing impression scale "Bright — Dark"

Scale	Impression words
bright	bright, happy, look forward to, like
dark	dark, hard and painful, sad, dislike

A portion of the impression dictionary is shown in Table 1, and the impression words used in constructing the impression dictionary are listed in Table 2, where we translated original Japanese words into English words or phrases. Note that one of the words with the heavier weights (M) than 1.0 were selected out of the words whose impression values (S) are larger than 0.75 or less than 0.25 for each part of speech (POS) (Table 1), and that each entry in the impression dictionary corresponds to a word and has a value and a weight on the impression scale, where negative forms of nouns, verbs, and adjectives such as "mu-teikou (no resistance)," "kesa-nai (do not erase)," and "tanoshiku-nai (be not happy)" are treated as single words.

wEE determines the topic type of a news article based on the impression value O calculated in the above way and selects a musical piece for BGM and the tone of voice for the animated newscaster by using Table 3. The topic type of a news article is "bright" if O is larger than H_1, "dark" if O is less than H_2, and "neutral" otherwise, where H_1 and H_2 are threshold values and H_1 is not less than H_2. Note that the specifications for directing wEE shown in Table 3 were determined based on the results of preparatory experiments in which 600 women and 600 men participated[2]. In these experiments, each participant listened to only the BGM and voice from a news program-like animation[3] and rated each of the news items comprising the animation using the three scales of intelligibility, positive feeling level, and friendliness level.

wEE uses our impression-based music-retrieval system [5] to select a musical piece as BGM from a music database. This system calculates the distance between impressions of every musical piece in a user-specified music database and an input impression, and then presents the musical piece with the impression

[2] 268 (22.3%) were in their twenties, 571 (47.6%) were in their thirties, 247 (20.6%) were in their forties, 88 (7.3%) were in their fifties, and 26 (2.2%) were under 20 or over 60. They all were Internet users, and replied to our questionnaire using their web browsing environment.

[3] All the stimuli presented to the participants were ones recorded on tape to remove the effects of the animations and images.

Table 3. Specifications for directing wEE

Topic	BGM	Tone of voice
bright	bright	bright
neutral	bright	neutral
dark	none	neutral

Table 4. Ten impression scales for input in our impression-based music-retrieval system

Scale No.	Impression scale	Scale No.	Impression scale
1	Quiet — Noisy	6	Leisurely — Restricted
2	Calm — Agitated	7	Pretty — Unattractive
3	Refreshing — Depressing	8	Happy — Sad
4	Bright — Dark	9	Relax — Arouse
5	Solemn — Flippant	10	The mind is restored — The mind is vulnerable

most similar to the input impression. An impression is input by selecting one or more impression scales from the ten (see Table 4) presented by the system and rating each of them on a seven-step scale between 1 and 7, while the impressions of each musical piece are represented by a ten-dimensional vector, each component of which has a real number between 0 and 8 corresponding to the seven-step scale. Because the impression scales designed for wEE are composed of the impression words listed in Table 2, we introduced the following equation to map the value of O onto a real number between 5 (a little bright/happy) and 8 (the brightest/happiest) in each of the impression scale Nos. 4 and 8 when the value of O corresponds to a "bright" or "neutral" topic, so that a bright and happy musical piece can be retrieved.

$$v = \frac{3(O - H_2)}{1 - H_2} + 5 \qquad (\textit{In case that } O \geq H_2)$$

The value of v is entered as input to the impression scales Nos. 4 and 8 of the music-retrieval system.

4 On-Line Experiments on Impression Conveyance

In this section, we analyze the effect that background music (BGM) and the tone of voice for an animated newscaster have on listeners' impressions about news articles based on the results of on-line experiments in which 300 women and 300 men participated, where the 600 participants did not overlap with the 1,200 people who participated in the preparatory experiments described in the previous section. Of the 600 participants, 90 (15.0%) were in their twenties, 273 (45.5%) were in their thirties, 168 (28.0%) were in their forties, 56 (9.3%) were in their fifties, and 13 (2.2%) were over 60. These 600 participants also were Internet users, and replied to our questionnaire using their web browsing environment.

First, we classified the 600 participants into groups A and B, each consisting of 150 women and 150 men and having almost the same age structures. The participants in group A were asked to listen to the BGM and voice from animations which were changed according to the impressions of the original news articles in each, while the participants in group B were asked to listen to the BGM and voice from animations in which the same text was read up but the BGM and the tone of voice for the animated newscaster were fixed to *neutral.*

The BGM and tone of voice for group A were determined according to Table 3. That is, a bright musical piece or the first movement of Beethoven's violin sonata "Spring" was used as BGM when a news article had a bright or neutral topic, and no BGM was played when a news article had a dark topic. A neutral tone was used as a tone of voice for the newscaster when a news article had a neutral or dark topic, and a bright tone of voice was used when a news article had a bright topic. On the other hand, for the participants in group B, a neutral musical piece or Koji Kusanagi's "Next Season" and a neutral tone of voice were used since the neutral BGM and tone of voice were more highly scored on an average in both positive feeling and friendliness levels when the BGM and tone of voice were fixed regardless of topic types. Note that, since the text-to-emotional speech engine we used can synthesize happy, sad, angry, cold, and neutral tones of voice and can change the speed, pitch, and loudness of speech [1], we adjusted these parameters of the engine adequately to synthesize bright and neutral and dark tones of voice. You will find that the dark tone of voice will be used later.

The participants listened to the sound from animated news programs, following instructions on their screens, and rated each of the news programs on a scale from 0 to 10 points in terms of strength or degree of the impressions they felt from the sound. That is, the following five impression scales were used to assess the impressions of the news programs: "Bright—Dark," "Relaxed—Tense," "Angry—Not angry," "Scared—Not scared," and "Like—Dislike." For example, the participants were asked to score ten points when they very strongly agreed with the word on the left side of each scale and zero points when they very strongly agreed with the word on the right side. We prepared three animated news programs, each consisting of one news item. The first news program had the neutral topic that rare gold coins owned by the Ministry of Finance of Japan were put on auction, the second news program had the bright topic that a Nobel prize winner was pro-Japanese and loved Akihabara (the most famous otaku's town in Japan), and the third news program had the dark topic that rabbits were killed in kindergartens and elementary schools. The results of the experiments are shown in Table 5, together with the results of a two-sample z-test between groups A and B. In this table, the significance level is given if the difference is statistically significant at the 1% or 5% level, and the symbol "—" is given otherwise. The symbols μ and σ denote the averages and standard deviations of the participants' scores, and the symbol N denotes the number of participants who rated a news item using the corresponding impression scale. Note that, if

Table 5. Effect of BGM and tone of voice for animated newscaster on impressions of news articles

	BGM and voice tone changed according to impressions				BGM and voice tone fixed regardless of impressions		
	Bright topic						
	Bright BGM & bright tone			Test between	Neutral BGM & neutral tone		
Impression Scale	μ	σ	N	averages	μ	σ	N
Bright — Dark	5.51	2.00	300	1%	4.68	2.14	300
Relaxed — Tense	5.25	1.72	300	—	5.10	2.02	300
Angry — Not angry	4.74	1.84	300	—	4.63	2.24	300
Scared — Not scared	4.16	2.04	300	—	4.48	2.32	300
Like — Dislike	4.38	1.94	300	—	4.34	2.11	300
	Neutral topic						
	Bright BGM & neutral tone			Test between	Neutral BGM & neutral tone		
Impression Scale	μ	σ	N	averages	μ	σ	N
Bright — Dark	4.35	2.32	300	—	4.21	2.25	300
Relaxed — Tense	5.17	2.02	300	—	5.07	2.07	300
Angry — Not angry	4.92	1.99	300	—	4.75	2.18	300
Scared — Not scared	4.43	2.30	300	5%	4.83	2.39	300
Like — Dislike	4.16	2.04	300	—	4.02	2.24	300
	Dark topic						
	No BGM & neutral tone			Test between	Neutral BGM & neutral tone		
Impression Scale	μ	σ	N	averages	μ	σ	N
Bright — Dark	4.23	2.00	300	1%	3.50	2.29	300
Relaxed — Tense	4.66	1.72	300	5%	4.31	2.03	300
Angry — Not angry	5.34	1.91	300	—	5.48	2.46	300
Scared — Not scared	5.11	2.06	300	5%	5.51	2.33	300
Like — Dislike	4.26	2.02	300	1%	3.68	2.16	300

the average value, μ, of a news item in an impression scale is larger than 5.0, the news item has the impression expressed by the word on the left side of the impression scale. On the contrary, if the average value, μ, is smaller than 5.0, the news item has the impression expressed by the word on the right side.

The results of the two-sample z-test shown in Table 5 proved that people's impressions on a news item in an animated news program could be changed by selecting the BGM and tone of voice for the news item. That is, the bright impression of a bright news article was conveyed to the participants in group A by using the bright BGM and tone of voice, while the participants in group B had a little dark impression even on the bright article. And this difference was statistically significant at the 1% level as shown in Table 5. On the contrary, the fear and tension of a dark news article were softened and conveyed to the participants in group A as a little dark impression. This is considered because we determined the specifications for directing wEE so that wEE could become as user-friendly as possible.

Table 6. Effect of dark BGM and dark tone of voice for animated newscaster on impressions of dark news articles

	Dark BGM & dark tone			Test between	Neutral BGM & neutral tone		
Impression Scale	μ	σ	N	averages	μ	σ	N
Bright — Dark	2.98	2.34	300	1%	3.98	2.12	300
Relaxed — Tense	4.62	2.13	300	—	4.49	2.14	300
Angry — Not angry	5.14	1.70	300	—	4.99	2.12	300
Scared — Not scared	5.83	2.45	300	5%	5.43	2.24	300
Like — Dislike	3.67	2.21	300	5%	4.07	2.04	300

The fact that the fear of a neutral news article was softened and conveyed to the participants in group A results from the same reason.

Here, we performed additional experiments in which a dark news article was read in a dark tone of voice and a dark musical piece or Erik Satie's "Gnossienne No.1" was played as BGM. The dark topic was an article that said that the Japanese death toll due to a sovereign remedy for influenza, "Tamiflu," had reached 24 people. In these experiments, the participants in group A were asked to listen to the recording with a dark musical piece and a dark tone of voice, and the participants in group B were asked to listen to the recording with the neutral musical piece and tone of voice. The results of the experiments are shown in Table 6, together with the results of the two-sample z-test between groups A and B. We can see from Table 6 that the impressions of the dark news article became darker and scarier by directing it in a dark manner.

5 Conclusions

We conducted two on-line experiments with a total of 1,800 participants and investigated how impressions of news articles can be conveyed to users in a news program-like computer animation generated from the news articles. As a result, we found that impressions of news articles can be controlled (emphasized or softened) by changing the background music and tone of voice for an animated newscaster. We also presented a passive viewing interface, called News Reader with Emotional Expressions (wEE), which we designed based on this finding. wEE automatically generates news program-like animations with synthesized emotional speech and background music. A distinctive point of wEE is that it explicitly conveys the moods or impressions of news articles to users by determining appropriate BGM and an appropriate tone of voice for an animated newscaster in the animation generation.

We are planning to design the impression scales suited to direct news programs, make our impression mining method more accurate, and develop a method for directing a news program taking the reading history or context of news articles into consideration. And we want to analyze the influences of the different speech and animation algorithm.

References

1. Animo Limited, FineSpeech Ver.2 Emotional Option, http://www.animo.co.jp/products/tts/fs/
2. Hareli, S., Harush, R.: The role of a sender's emotional expression in persuasion. In: The annual meeting of the international society for research of emotions (2004)
3. Hareli, S., Tzafrir, S., Ben-Ze'ev, A.: Experimental issues: emotional expressions, humor and credibility in persuasion. In: Report on basic cues and open research topics in communication and emotions (October 2004)
4. Hayashi, M., Ueda, H., Kurihara, T.: TVML (TV program making language): Automatic TV program generation from text-based script. In: ACM Multimedia 1997 State of the Art Demos, Seattle, USA (October 1997)
5. Kumamoto, T., Ohta, K.: A query by musical impression system using n-gram based features. In: Proc. of IEEE Conference on Cybernetics and Intelligent Systems, Singapore, pp. 992–997 (December 2004)
6. Kumamoto, T., Tanaka, K.: Proposal of impression mining from news articles. In: Khosla, R., Howlett, R.J., Jain, L.C. (eds.) KES 2005. LNCS (LNAI), vol. 3681, pp. 901–910. Springer, Heidelberg (2005)
7. Kurohashi, S., Kawahara, D.: Japanese Morphological Analysis System Juman Version 4.0 (2003)
8. Nakanishi, T., Kitagawa, T., Kiyoki, Y.: An implementation method of automatic composition of the facial expression by any impression words. IPSJ Transactions on Databases 44(SIG8 (TOD18)), 21–36 (2003)
9. Nihon Keizai Shimbun Inc., Nikkei Newspaper Full Text Database DVD-ROM, 1990 to 1995 editions, 1996 to 2000 editions, 2001 edition (2004)
10. Nonogaki, H., Ueda, H.: FRIEND21 project: A construction of 21st century human interface. In: Proc. of the ACM Conference on Human Factors in Computing Systems, pp. 407–414 (April 1991)
11. Tanaka, K., Nadamoto, A., Kusahara, M., Hattori, T., Kondo, H., Sumiya, K.: Back to the TV: Information visualization interfaces based on TV-program metaphors. In: Proc. of the IEEE International Conference on Multimedia & Expo., pp. 1229–1232 (August 2000)
12. Taniguchi, T.: Music and Emotion. KitaOoji Shobo Publisher, Kyoto (1998)
13. Yamaguchi, T., Hosomi, I., Miyashita, T.: WebStage: An active media enhanced World Wide Web browser. In: Proc. of the ACM Conference on Human Factors in Computing Systems, Atlanta, USA, pp. 391–398 (March 1997)

Data Mining for Navigation Generating System with Unorganized Web Resources

Diana Purwitasari, Yasuhisa Okazaki, and Kenzi Watanabe

Saga University, Saga, Japan
{diana,okaz,watanabe}@ai.is.saga-u.ac.jp

Abstract. Users prefer to navigate subjects from organized topics in an abundance resources than to list pages retrieved from search engines. We propose a framework to cluster frequent itemsets (sets of common words) into topics, produce a hierarchical list, and then generate topics sequence from a collection of documents. The framework will regenerate a next sequence when users click a topic. Consider browsing to any topic as a kind of searching for that topic, the framework makes an inquiry using feature terms within the document representation of selected topic as query keywords. Our ranking method in searching process considers content analysis that still retaining spatial information of search keywords and link analysis of documents. Utilizing implementation of navigation generating system the experiments show that a navigation list from clustering results can be settled with regard to variance ratio of *between and within distances*. Agglomerative clustering is used in restructuring the extracted topics in order to produce a hierarchical navigation list.

1 Introduction

With rapid growth of the Internet all information in the unorganized structure Web could cause disorientation for users while navigating. Some users might prefer to navigate particular subjects from already focused and organized topics within abundance resources than to list relevant pages retrieved from search engines. One of ways for structuring when dealing with such amount resources to help the users, who may not yet have enough structured idea about existing topics in the collection and can not decide which should be read first, is by creating a navigation of topics automatically.

We propose a framework to generate hierarchically structured representation of topics to help users to cope with the existent but hidden structure (Fig. 1). *Topics Detecting Module* works to extract feature terms into topics while *Navigation Creating Module* creates a hierarchical model of topics and does labelling for generated navigation of topics sequence. Consider browsing to any topic as a kind of searching for that topic, *Query Searching Module* calculates documents' similarity for inquiry results that act as a new collection in which processes cycle of generating the next topics sequence will take place in more focused collection to learn further topic in detail.

I. Lovrek, R.J. Howlett, and L.C. Jain (Eds.): KES 2008, Part I, LNAI 5177, pp. 598–605, 2008.

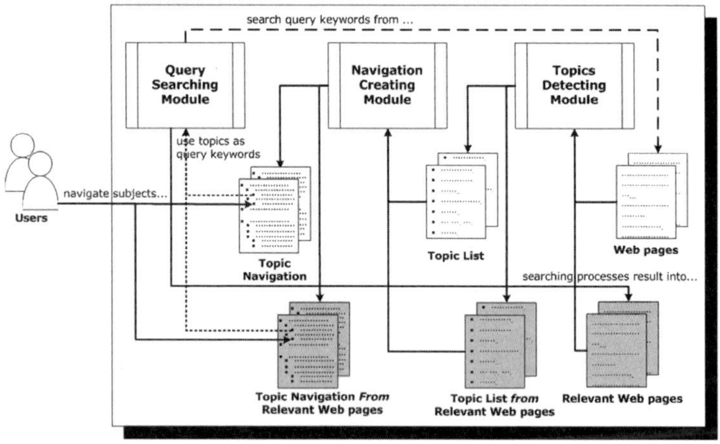

Fig. 1. Framework of navigation generating system to produce topics sequences from Web pages collection

Many research efforts have been engaged to bring structured representation to a large collection of documents in unorganized structure of the Web. WTMS crawls URLs of Web pages for analyzing the links to map subject domain of Web pages based on directory structure or physical domains of Websites [1]. PageCluster not only utilizes Web link structures but also does mining subjects from extracted texts of anchor links [2]. THESUS takes a step further by examining the semantic of Web documents with the help of manually created hierarchical concepts [3].

Our proposed system do not visualize the subjects into a graph of interlinked Web pages [1,2,3] but list a topics sequence as guidance path [4] to let users know which subject should be read first. We produce the sequence without prior action in defining subjects' domain model and analyze documents content instead of relying on preceding users comprehension unlike THESUS [3] or WikiTrails [4].

2 Topics Detecting Module

Topics Detecting Module recognizes topics in the collection for generating a topics sequence. The topics are defined as frequent itemsets (sets of common words) that occurred together. Our adaptation of topic identification with data mining techniques [5] extract terms from the collection, look for frequent itemsets, and then cluster them as sets of common words correspond to the current topics.

2.1 Index Significant Terms for Frequent Itemsets

We use information retrieval techniques (indexing, removing stop words, stemming, and weighting) to create an inverted index [6] and eliminate the indexed

Table 1. List of dependence measures for calculating confidence c

Yule's Q	$\frac{p(x,y)p(\overline{x},\overline{y})-p(x,\overline{y})p(\overline{x},y)}{p(x,y)p(\overline{x},\overline{y})+p(x,\overline{y})p(\overline{x},y)}$
Yule's Y	$\frac{\sqrt{p(x,y)p(\overline{x},\overline{y})}-\sqrt{p(x,\overline{y})p(\overline{x},y)}}{\sqrt{p(x,y)p(\overline{x},\overline{y})}+\sqrt{p(x,\overline{y})p(\overline{x},y)}}$
Two-Way Support	$p(x,y)\log_2\frac{p(x,y)}{p(x)p(y)}$
Linear Correlation Coefficient	$\frac{p(x,y)-p(x)p(y)}{\sqrt{p(x)(1-p(x))p(y)(1-p(y))}}$
Piatetsky-Shapiro	$p(x,y)-p(x)p(y)$
Information Gain	$\log_2\frac{p(x,y)}{p(x)p(y)}$

terms which have weight values less than a threshold. We also apply another filter called entropy which usually exists in decision tree problem [7]. Entropy value measures uncertainty state. For each document, we assume that the terms inside will become test attributes to determine whether those terms are important to the current document. We think entropy value of a term reflects the effectiveness of the term in identifying certain document to others

2.2 List Candidates of Frequent Itemsets

We use data mining parameters of support s and confidence c to list frequent itemsets. Let $T = \{t_1, t_2, \ldots t_k\}$ be a set of k terms. Assume a rule on frequent itemsets of t_i and t_j implies that occurrences of t_i usually followed by occurrences of t_j or the other way round. Support s shows a percentage value of n documents that must contain t_i and t_j; $s_{i,j} = p(t_i, t_j)$. While confidence c measures dependence or correlation between occurrences of t_i and t_j. Note, $p(x)$ is defined as a probability function of x, while $p(x, y)$ is a joint probability function of x and y. In this paper we consider a selection of some dependence measures applied to our domain problem [8]: Yule's Q, Yule's Y, Two-Way Support, Linear Correlation Coefficient, Piatetsky-Shapiro, Information Gain or other works said it as Mutual Information(Table 1).

We choose those measures based on key properties a good dependence measure should satisfy which carefully selected in a way that the properties are appropriate to our problem domain. Our selection apply Piatetsky-Shapiro [9] proposed properties which should obey the following conditions. Since our frequent itemsets are words within documents the occurrence of t_i makes it neither more nor less probable that t_j also occurs. The greater the support for join occurrences of t_i and t_j will give more positive correlation with occurrences of t_i only or t_j only in a collection of documents are equal. When some documents which covering t_i contains t_j (or vice versa), the dependence measure should attain its maximum. Our selection also apply one of Tan et al. proposed properties [10] which states that measure function should unrelate to the number of documents without frequent itemsets.

We apply support s filter and confidence c filter to list important frequent itemsets [5]. We do not consider frequent itemsets with support and confidence value less than some thresholds. We also eliminate a frequent itemsets if its condition falls under these constraints: (a) its support less than average support and its confidence less than average confidence; (b) its support less than a summation of average and standard deviation of support; (c) its confidence less than a summation of average and standard deviation of confidence.

2.3 Partition Frequent Itemsets Based on Hypergraph Relation between Words

A hypergraph is a generalization of a graph in which an edge can connect more than two vertices. This kind of edge is called hyperedge. Each hyperedge describes existing but hidden topic as complex relationships among its corresponding words. To find the hyperedges is a problem to partition the hypergraph.

We do hypergraph partitioning by executing *shmetis* library in hMetis tool [11] with a listing file that stores the hypergraph as the input contains information of vertices, edges and assigns confidence values c between terms t_i, t_j to its corresponding edges. We set the number of desired partitions. Since *shmetis* does not allow to produce perfectly balanced partitions, we specify an upper bound value which letting the number of vertices assigned to each one of the two partitions will be between 45%-55% and each partition at least should contain 5% of total document number in the collection. The output file shows list of vertices and their assigning partition index. We develop *encoding* module to represent frequent itemsets extracted from the collection of documents as hypergraph into the input file. We also create *decoding* module to translate the output file into list of topics and common words that recognizing each topic.

3 Navigation Creating Module

Navigation Creating Module checks relation structure between the topics and finds a document representation for each topic found. Combine structure of subjects and document representations of topics to produce a hierarchical list.

3.1 Create Hierarchical Clusters of Topics

Document contents can have multiple topics. If there are some documents being referenced to a same set of topics, it means that the topics are overlapped and then those topics can be merged into broader topic. Our problem is to organize topics in a nested sequence of groups of merge topics which can be displayed as a tree form. We make appropriate adaptation of agglomerative approach for building a hierarchy from bottom-up.

Our assumptions in initialization are that each document can only belong to exactly one cluster of topics and any cluster without document members will be removed. We use extracted topics as cluster seeds in initialization. Because of those assumptions some initial clusters that do not have any document members will be left out during tree construction.

Before merging topics, we recalculate similarity between any document to soon-to-be-merged topics based on their common words. The similarity between document and topic is derived from TFIDF formula [6] in term weighting process (Eq. 1) [5]. Accumulation of document similarities value with every topics pair is applied in a mutual information style to update proximity matrix of topic clusters *in each iteration time* clustering (Eq. 2) [5]. In the first iteration initial clusters contain single topic which is cluster seed taken from the extracted topics by Topics Detecting Module. For next iterations initial cluster of single topic called tpc_a and other initial cluster of topic called tpc_b could become topics cluster as a result of merging topics (Eq. 2).

$$sim(d_i, tpc_t) = \sum_{k \in t} \frac{tf_{ik} \times (\log(\frac{N}{n_k}))^2}{\sqrt{\sum_{j \in t} (\log(\frac{N}{n_k}))^2} \times \sqrt{\sum_{j \in t} (tf_{ij})^2 (\log(\frac{N}{n_k}))^2}} \tag{1}$$

where tf_{ik} is term frequency of term t_k in document d_i, N is the size of the collection, and n_k is the number of documents with term t_k.

$$sim_{ab} = \frac{\frac{\sum_{i \in docs} sim(d_i, tpc_a) \times sim(d_i, tpc_b)}{N}}{\frac{\sum_{j \in docs} sim(d_j, tpc_a)}{N} \times \frac{\sum_{k \in docs} sim(d_k, tpc_b)}{N}} \tag{2}$$

We show in the next two following sub sections that the similarity function between document and a cluster of single topic or merged topics (Eq. 1) is not only for merging but also used for structuring and representing navigation.

3.2 Create Structured Hierarchical Clusters of Topics

Though agglomerative succeeds to form a tree of hierarchical topics and subtopics, but topics sequence on the same depth level is still unknown. It causes users can not decide which topic should be read first among other topics in certain depth level. A topic here is a node in the tree of topics sequence. A topic will have a document representation where in our problem domain the document is a Web page. We use PageRank score of document to rank nodes. PageRank shows that document which is *linked-to* by many documents will receive higher rank [12].

We calculate rank score of a node as an average value of PageRank scores from a number of documents that similar to all existing topics in a subtree with current node as its subroot. Let a subtree has subroot represented by a cluster of single topic or merged topics called tpc_t with n_t shows total number of nodes in the subtree including the leaf nodes. Note, leaf nodes are initial clusters containing single topic. To calculate rank score of a node tpc_t, we will only consider n_t number of documents which has high similarities to topic tpc_t.

3.3 Create Structured Hierarchical Navigation of Topics

Let node nd_t represents a cluster of single topic or merged topics called tpc_t. To find document representation of a cluster begins with listing of documents similar to tpc_t arranged in a descending order. Select a document d_i from the ordered list which has not been selected as a representation of other nodes. Then calculate the similarity between document d_i and topic tpc_t (Eq. 1). Before assigning document d_i to represent node nd_t, check similarity of document d_i with child nodes of nd_t. Repeat that traversing process until the similarity in child nodes is smaller than in the parent node. We check this to avoid any misrepresent because it is possible that the document is more suitable as sub cluster representation.

4 Query Searching Module

When users click any topic, the system makes an inquiry using feature terms within current document as search keywords of selected topic. Our searching process applies combination ranking factors of link analysis, PageRank Scoring (PRS) [12], and content analysis that still retaining spatial information of search keywords, Fourier Domain Scoring (FDS) [13,14]. Afterward, inquiry results act as a new collection in which the system will produce the next suitable navigation of topics sequence for a collection of more focused subjects.

In searching process, how to interpret the results such as determining ranking method is an important issue. We have studied FDS usage as ranking method [14]. FDS gives higher rank to a document if its content has more terms similar to query terms and occur close together than document where its query terms inside occur far apart. The terms within document should have high occurrence numbers and similar positions. For a better result in ranking method, score of a Web page is a combination of content score and link score [12]. FDS is a method to calculate content score while PRS is a method for computing link score in a Web page. Scoring schema of our searching is defined as a multiplication value of normalized FDS and PRS scores.

5 Experiments

We have experiments using collection of 34 documents as snapshot of English Wikipedia articles without images crawled from main page of category Statistic[1]. We remove duplicate of Wikipedia articles because Wikipedia often makes redirection links from old articles to better coverage articles. We use Oracle Text[2] in extracting terms, Porter Stemmer algorithm in stemming, and TFIDF (term frequency / inverse document frequency) in weighting processes. List of stop words contains words provided by Oracle Text, HTML tags and some common words in Wikipedia articles.

In the implementation we adjust some threshold values based on data experiments. First, weight filter will eliminate all terms with less than 0.01. Second, use terms with a positive entropy value at least 0.06. Third is to set minimum

[1] `http://en.wikipedia.org/wiki/Statistic`

[2] Oracle Text in Oracle Database, `http://www.oracle.com/technology/products/text`

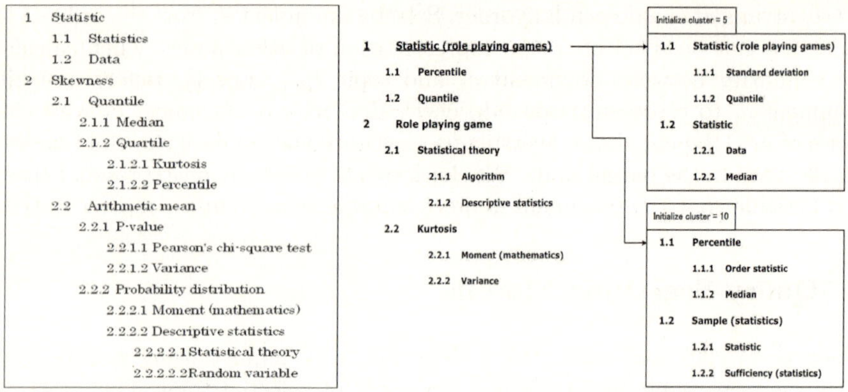

(a) using Piatetsky-Shapiro (b) using Information Gain

Fig. 2. Generated navigation for the first time and after clicking a subject

support s to 0.05, and the fourth is to set minimum confidence c with 1.0 for Information Gain and the other dependence measures with 0.01. Note, in our collection the document frequency of a term ranges from 15 to 20 documents while term frequency averagely exists from 10 to 15 term occurrences.

We choose the appropriate depth for representation of navigation list by considering a performance measure with variance ratio of *between and within distances* from clustering results in our experiments. After observing each generated navigation, we choose to list the subjects retrieved after certain iterating level when its variance ratio value tends to become smaller compare with previous iterating level. We take subject label of topics from title of Web page representation.

Variance ratio for Piatetsky-Shapiro shows more stable inclination of smaller rate. It shows that well-divided clustering results have been formed even four iterations before final iteration and makes its topics sequence has more detailed subjects until five level depths (Fig. 2(a)). Piatetsky-Shapiro also succeeds in filtering lesser number of frequent itemsets. On the other hand Yule's Q shows the least expected results with much shorter list of topics sequence. In fact Yule's Q and Yule's Y results almost two times number of frequent itemsets compare to Piatetsky-Shapiro which mining the least but giving more detailed navigation.

We test regeneration process after users click a topic in navigation with *moderate* depths created using Information Gain as dependence measure. Given that the semantic correctness is based on subjective opinions, the second level navigation not quite matches structurally, yet, with the first one (Fig. 2(b)). There are two versions of regenerated navigation for *Statistic (Role playing game)* subject: the one with 5 seed clusters and 10 seed clusters. The one with 10 seed clusters has better result because document representation of subtopic 1.1 in the second navigation is the same with the first navigation. Variance ratio of *between and within distances* for 10 seed clusters also shows faster rate to become smaller.

6 Conclusions and Future Works

We have implemented our proposed framework and generated topics sequences through experiments. Given that semantic correctness is based on subjective opinions, we have succeeded to extract topics sequence on the first level. However in regenerating next sequence the second navigation does not match structurally with the first one, yet. Next step we will find solutions to overcome that problem.

References

1. Mukherjea, S.: Organizing Topic-specific Web Information. In: HYPERTEXT 2000: Proc. of the Eleventh ACM Conf. on Hypertext and Hypermedia, pp. 133–141 (2000)
2. Zhu, J., Hong, J., Hughes, J.G.: PageCluster: Mining Conceptual Link Hierarchies from Web Log Files for Adaptive Web site Navigation. ACM Trans. Inter. Tech. 4(2), 185–208 (2004)
3. Halkidi, M., Nguyen, B., Varlamis, I., Vazirgiannis, M.: THESUS: Organizing Web Document Collections based on Link Semantics. The VLDB Journal 12(4), 320–332 (2003)
4. Reinhold, S.: WikiTrails: Augmenting Wiki Structure for Collaborative, Interdisciplinary Learning. In: WikiSym 2006: Proc. of the 2006 Intl. Symp. on Wikis, pp. 47–58 (2006)
5. Clifton, C., Cooley, R., Rennie, J.: TopCat: Data Mining for Topic Identification in a Text Corpus. IEEE Trans. on Knowledge and Data Engineering 16(8), 949–964 (2004)
6. Yates, R.B., Ribeiro-Neto, B.: Modern Information Retrieval. Addison-Wesley, Reading (1999)
7. Quinlan, J.R.: Induction of Decision Trees. Machine Learning 1, 81–106 (1986)
8. Geng, L., Hamilton, H.J.: Interestingness Measures for Data Mining: A Survey. ACM Comput. Surv. 38(3) (2006)
9. Piatetsky-Shapiro, G., Frawley, W.J.: Discovery, Analysis, and Presentation of Strong Rules. In: Knowledge Discovery in Databases, pp. 229–248. MIT Press, Cambridge (1991)
10. Tan, P., Kumar, V.: Interestingness Measures for Association Patterns: A Perspective. Technical Report TR00-036, Department of Computer Science, University of Minnesota (2000)
11. Karypis, G.: Multilevel Hypergraph Partitioning. Technical Report 02-25, Comput. Sci. and Eng. Dept., Univ. Minnesota, Minneapolis (2002)
12. Page, L., Brin, S., Motwani, R., Winograd, T.: The PageRank Citation Ranking: Bringing Order to the Web. Technical report, Stanford Digital Library Technologies Project (1998)
13. Park, L.A., Ramamohanarao, K., Palaniswami, M.: Fourier Domain Scoring: A Novel Document Ranking Method. IEEE Trans. on Knowledge and Data Engineering 16(5), 529–539 (2004)
14. Purwitasari, D., Okazaki, Y., Watanabe, K.: A Study on Web Resources' Navigation for e-Learning: Usage of Fourier Domain Scoring on Web Pages Ranking Method. In: ICICIC 2007: Proc. of the Second Intl. Conf. on Innovative Computing, Information and Control (2007)

Exploring Robustness Enhancements for Logic-Based Passage Filtering

Ingo Glöckner[1] and Björn Pelzer[2]

[1] Intelligent Information and Communication Systems Group (IICS), University of Hagen, 59084 Hagen, Germany
ingo.gloeckner@fernuni-hagen.de
[2] Department of Computer Science, Artificial Intelligence Research Group University of Koblenz-Landau, Universitätsstr. 1, 56070 Koblenz
bpelzer@uni-koblenz.de

Abstract. The use of logic in question answering (QA) promises better accuracy of results, better utilization of the document collection, and a straightforward solution for integrating background knowledge. However, the brittleness of the logical approach still hinders its breakthrough into applications. Several proposals exist for making logic-based QA more robust against erroneous results of linguistic analysis and against gaps in the background knowledge: Extracting useful information from failed proofs, embedding the prover in a relaxation loop, and fusion of logic-based and shallow features using machine learning (ML). In the paper, we explore the effectiveness of these techniques for logic-based passage filtering in the LogAnswer question answering system. An evaluation on factual question of QA@CLEF07 reveals a precision of 54.8% and recall of 44.9% when relaxation results for two distinct provers are combined.[1]

1 Introduction

The necessity to incorporate reasoning capabilities into QA systems was recognized as early as 2000 in the NIST roadmap for question answering [1]. Today, there is also experimental evidence for the claim that question answering can profit from the use of logical subsystems: A comparison of answer validators in the Answer Validation Exercise 2006 found that systems reported to use logic generally outperformed those without logical reasoning [2]. There is also evidence from the related task of Recognizing Textual Entailment (RTE) that a high accuracy of entailment recognition can only be achieved by structure-sensitive methods like logic or graph matching [3]. And the best system in the TREC 2006 evaluation of QA systems, PowerAnswer [4], made use of a theorem prover. However, the logical approach suffers from brittleness: A proof of the question from a document of interest and the background knowledge succeeds only if the question and document are analyzed correctly and if every piece of knowledge required to prove the question is actually available in the knowledge base.

[1] Funding of this work by the DFG (Deutsche Forschungsgemeinschaft) under contracts FU 263/12-1 and HE 2847/10-1 (LogAnswer) is gratefully acknowledged.

I. Lovrek, R.J. Howlett, and L.C. Jain (Eds.): KES 2008, Part I, LNAI 5177, pp. 606–614, 2008.

Therefore the successful applications of logic mentioned above resort to robust methods for logic-based knowledge processing, which show a graceful degradation of results when there are errors of linguistic analysis or knowledge gaps. In this paper, we are interested in robustness-enhancing techniques which can be added to existing theorem provers with minor internal changes. Therefore we do not try to build an approximate inference engine by adopting an approximate graph matcher like [5] and adding support for logical rules – this would almost amount to building a prover from scratch. More suitable solutions are the extraction of useful information from failed proofs [6], combining logic-based and shallow features using machine learning [7,8], and finally relaxation techniques [4,6] which reduce the query by subsequently dropping literals until a proof of the simplified query succeeds. But relaxation does not necessarily find the largest provable query fragment, since it only inspects a single sequence of simplification steps. Moreover the choice of skipped literals usually depends on factors like internal literal order of the prover which are arbitrary to some degree. We therefore propose to abstract from such idiosyncratic aspects by combining relaxation results of two different provers. The effectiveness of the relaxation approach is explored in a logical passage filtering task. We also study the effectiveness of robustness-enhancing techniques like extracting useful information from failed proofs and combining logical features with shallow ones by machine learning.

The remainder of the paper is organized as follows. Section 2 introduces the LogAnswer QA system which provides the testbed for the experiments. Section 3 describes the two provers used by LogAnswer and discusses some possible combinations of the relaxation results of both provers. Section 4 presents the results of passage filtering experiments for several robustness-enhancing techniques. The main results of the paper are summarized in Sect. 5.

2 Overview of the LogAnswer System

This section sketches the architecture of the LogAnswer QA system which forms the basis for the experiment described in the paper. The main innovation of LogAnswer is its use of logic for filtering retrieved passages and for answer extraction. By avoiding the inefficiency of answer validation, LogAnswer answers questions in only a few seconds [8]. The system comprises the following processing stages.

Question Analysis. WOCADI [9], a robust parser for German, is used for a deep linguistic analysis of the given question. The syntactic-semantic analysis results in a semantic representation expressed in the MultiNet formalism [10].

Passage Retrieval. The IRSAW QA framework [11] is used for finding passages with a standard IR approach. Using WOCADI, all texts are analyzed prior to indexing, so that no documents must ever be parsed at query time.

Query Construction. The semantic network for the question is finally turned into a conjunctive list of query literals. This step involves a synonym normalization by replacing all lexical concepts with canonical synset representatives. For example, the question *Wie hieß der Sänger von Nirvana?*[2] translates into the logical query

$$\mathrm{val}(X_1, \mathit{nirvana.0}), \mathrm{sub}(X_1, \mathit{name.1.1}), \mathrm{attr}(X_2, X_1), \mathrm{attch}(X_2, X_3),$$
$$\mathrm{sub}(X_3, \mathit{gesangsolist.1.1}), \mathrm{subs}(X_4, \mathit{heißen.1.1}), \mathrm{arg1}(X_4, X_3), \mathrm{arg2}(X_4, \mathit{FOCUS})$$

based on the lexical concepts (word senses) *nirvana.0* (Nirvana), *name.1.1* (name), *gesangssolist.1.1* (singer soloist), and *heißen.1.1* (be named). Here synonym normalization has replaced *sänger.1.1* (singer) with the canonical *gesangssolist.1.1* (singer soloist). The *FOCUS* variable represents the queried information.

Robust Entailment Test. The basic idea of the logic-based passage filtering is that the passage contains a correct answer to the question if there is a proof of the question from the passage representation and the background knowledge.[3] A relaxation loop is used to gain more robustness. Suitable provers must be able to return the proven portion of a query in the case of a failed proof, and also identify the literal which caused a complete proof to fail. The relaxation process then skips the failed literal and tries to prove the resulting query fragment. The process can be repeated until a proof of the remaining query succeeds, and the skip count is a useful feature for recognizing (non-)entailment. Consider the question for the Nirvana lead singer, and this passage (translated from German): *Fans and friends of the lead singer of US-American rock band Nirvana reacted with grief and dismay to the suicide of Kurt Cobain this weekend.* Here the parser misses a coreference and produces two distinct entities for Cobain and the Nirvana lead singer. Thus the system finds only a relaxation proof with skipped literals $\mathrm{sub}(X_3, \mathit{gesangsolist.1.1})$, $\mathrm{attch}(X_2, X_3)$ and the *FOCUS* variable bound to Kurt Cobain. In practice, relaxation is stopped before all literals are proved or skipped. One can then only state upper/lower bounds on the provable literal count, assuming that all (or none) of the remaining literals are provable.

Feature Extraction. The classification of passages rests on the following logic-based features which depend on the chosen limit on relaxation cycles:

- *skippedLitsLb* Number of literals skipped in the relaxation proof.
- *skippedLitsUb* Number of skipped literals, plus literals with unknown status.
- *litRatioLb* Relative proportion of actually proved literals compared to the total number of query literals, i.e. 1 − *skippedLitsUb*/*allLits*.
- *litRatioUb* Relative proportion of potentially provable literals (not yet skipped) vs. all query literals, i.e. 1 − *skippedLitsLb*/*allLits*.
- *boundFocus* Fires if a relaxation proof binds the queried variable.
- *npFocus* Indicates that the queried variable was bound to a constant which corresponds to a nominal phrase (NP) in the text.
- *phraseFocus* Signals that extraction of an answer for the binding of the queried variable was successful.

[2] *What was the name of the singer of Nirvana?*

[3] LogAnswer uses the same background knowledge as the MAVE answer validator [6].

A simplistic solution for answer extraction is used for computing the *phraseFocus* feature. The method leverages information on word alignment, as provided by the parser for nominal phrases (NPs), in order to find answer strings for the bindings of the queried variable. This is done by cutting verbatim text from the original text passage. The basic idea behind the *boundFocus*, *npFocus* and *phraseFocus* features is that the ability of the answer extraction stage to verbalize the answer binding might also have something to say about the relevance of the passage.

In addition to the logic-based features, five 'shallow' features are used which do not require a deep parse and can be computed without the help of the prover:

- *failedMatch* Number of lexical concepts and numerals in the question which cannot be matched with the candidate document.
- *matchRatio* Relative proportion of lexical concepts and numerals in the question which find a match in the candidate document.
- *failedNames* Proper names mentioned in the question, but not in the passage.
- *irScore* Original passage score of the underlying passage retrieval system.
- *containsBrackets* Indicates that the passage contains a pair of parentheses.

The matching technique used to obtain the values of these shallow features also takes into account lexical-semantic relations (e.g. synonyms and nominalizations), see [6]. The *containsBrackets* feature is motivated as follows: Often the queried information is contained in parentheses, e.g. 'Kurt Cobain (Nirvana)', but the linguistic parser has difficulty finding the relationship between the basic entity and the information given in brackets. Thus containment of a pair of brackets in the passage is significant to the relevance judgement.

ML-based Passage Classification. The Weka machine learning toolbench [12] is used for learning the mapping from features of retrieved passages to yes/no decisions concerning containment of a correct answer in the considered passage. The low precision of the original passage retrieval step means a strong disbalance between positive and negative examples in the data sets. In order to emphasize the results of interest (i.e. positive results) and achieve sufficient recall, *cost-sensitive learning* is applied. In the experiments, false positives were weighted by 0.3 while a full weight of 1 was given to lost positives (i.e. false negatives).The Weka Bagging learner with default settings is used as the classifier. It is wrapped in a Weka CostSensitiveClassifier to implement the cost-sensitive learning.

3 Combining Provers for Increased Robustness

This section introduces the two provers used by LogAnswer for logic-based passage filtering and answer extraction. It then discusses how these provers can be combined for improving robustness of knowledge processing.

3.1 The Regular MultiNet Prover

LogAnswer is equipped with a dedicated prover for MultiNet representations, which is part of the MWR+ toolbench.[4] The MultiNet prover is, in principle,

[4] See `http://pi7.fernuni-hagen.de/research/mwrplus`

a regular prover for Horn logic based on SLD resolution. To be precise, the supported logic is even more restricted since the additional assumption is made that all facts are variable-free and that (after skolemization), all variables which occur in the conclusion of a rule also occur in its premise. On the other hand, the prover offers builtin support for special MultiNet constructions (e.g. efficient access to so-called layer features of conceptual nodes and a builtin subsumption check for MultiNet sorts). The prover also offers special support for rules with complex (conjunctive) conclusions which are useful for modelling natural language-related inferences. The translation into Horn logic splits such rules into several clauses, which is inefficient because typically all literals in the conclusion are needed when the rule is applied. The MultiNet prover solves this problem by keeping track of complex conclusions. When applying such a rule, the complex conclusion is cached as a lemma in order to shortcut proofs of other literals from the derived conclusion. The knowledge base can be split into several partitions which can be flexibly combined or exchanged as needed. Iterative deepening is used to control the search. While very limited in expressive power, the MultiNet prover is extremely fast, and proving a question from a passage usually takes less than 20ms [8]. The prover has been systematically optimized both for speed and scalability by utilizing term indexing, caching, lazy computation (index structures are only built on demand), by optimizing the order in which literals are proved, and by removing performance bottlenecks with the help of profiling tools.

3.2 Description of the E-KRHyper Prover

E-KRHyper is an automated theorem proving and model generation system for first-order logic with equality [13]. It is an implementation of the E-*hyper tableau calculus* [14], which integrates a superposition-based handling of equality into the hyper tableau calculus [15]. E-KRHyper is the latest version in the KRHyper-series of theorem provers, developed at the University Koblenz-Landau. Designed for use as an embedded knowledge-processing engine, it has been employed in a number of knowledge-representation applications. E-KRHyper is capable of handling large sets of uniformly structured input facts. The system can provide proof output for models and refutations, and it is able to rapidly switch and retract input clause sets for an efficient usage as a reasoning server. E-KRHyper accepts input first-order input in the common TPTP syntax [16].

The principal data structure used in the operation of E-KRHyper is the E-*hyper tableau*, a tree labeled with clauses and built up by the application of the inference rules of the E-hyper tableau calculus. The tableau is generated depth-first, with E-KRHyper always working on a single branch. Refutational completeness and a fair search are ensured by iterative deepening with a limit on the maximum term weight of generated clauses.

Embedded in the LogAnswer system, E-KRHyper is supplied with the MultiNet axioms transformed into first-order TPTP syntax. The inference process then operates on the axioms and the negated query literals, with a refutation result indicating a successful answer and providing the binding for the queried variable. If the reasoning is interrupted due to exceeding the time limit, then partial results can be retrieved that can guide in the relaxation process.

3.3 Methods for Combining Prover Results

Due to the use of two provers, a pair of results is obtained for each logic-based feature. The most basic approach for incorporating these features into the machine learning approach is juxtaposition (JXT), i.e. both results for each feature are kept and directly passed to the ML method. This method leaves the interpretation of the data entirely to machine learning.

Another approach (OPT) rests on the observation that the relaxation method is generally pessimistic, since it does not necessarily find the largest provable query fragment. This suggests an optimistic combination of relaxation features:

$$\begin{aligned} \mathit{skippedLitsLb} &= \min(\mathit{skippedLitsLb}_1, \mathit{skippedLitsLb}_2), \\ \mathit{skippedLitsUb} &= \min(\mathit{skippedLitsUb}_1, \mathit{skippedLitsUb}_2), \\ \mathit{litRatioLb} &= \max(\mathit{litRatioLb}_1, \mathit{litRatioLb}_2), \\ \mathit{litRatioUb} &= \max(\mathit{litRatioUb}_1, \mathit{litRatioUb}_2). \end{aligned}$$

The other logical features are chosen according to the best value of *skippedLitsLb*. Thus, if $\mathit{skippedLitsLb}_1 < \mathit{skippedLitsLb}_2$, then $\mathit{boundFocus} = \mathit{boundFocus}_1$, $\mathit{npFocus} = \mathit{npFocus}_1$ and $\mathit{phraseFocus} = \mathit{phraseFocus}_1$ using the results of the regular MultiNet prover. Otherwise the values of these features are provided by E-KRHyper. Notice that E-KRHyper sees slightly different queries compared to the MultiNet prover, which is due to the transformation of the queries from the MultiNet format into TPTP formulas. A scaling by query length is necessary so that both prover results refer to the same number of query literals.

4 Evaluation

The questions of the CLEF07 QA track for German served as the starting point for the evaluation, since they target at the corpora currently supported by the IRSAW IR module (CLEF News and Wikipedia).[5] From the full set of 200 questions, all follow-up questions were eliminated since discourse processing is not of relevance here. Definition questions were omitted as well since knowing the logical correctness of an answer is not sufficient for deciding if it is suitable as a definition. The remaining 96 factual questions were checked for parsing quality and two questions for which construction of a logical query failed and one outlier question with an unusually high number 37 supporting passages were discarded. For each of the remaining 93 questions in the test set, IRSAW retrieved up to 200 one-sentence snippets from the pre-analyzed corpora, resulting in a total of 18,500 candidate passages. The snippets with an incomplete parse were eliminated since they cannot be handled by logical processing. The 12,377 passages

[5] See `http://www.clef-campaign.org/2007.html`

Table 1. Quality of passage filtering as a function of allowable relaxation steps n. Abbreviations: RMP (regular MultiNet prover), KRH (E-KRHyper), JTX (juxtaposition), OPT (optimistic combination), IRB (information retrieval baseline, using *irScore* only), SHB (shallow baseline, using all shallow features). The 0ℓ runs use strict proofs and logical features. The 0s runs use strict proofs and both logical and shallow features.

model	n	precision	recall	F-score
RMP	0ℓ	0.786	0.043	0.082
RMP	0s	0.46	0.457	0.458
RMP	0	0.471	0.421	0.445
RMP	1	0.458	0.386	0.419
RMP	2	0.502	0.398	0.444
RMP	3	0.505	0.409	0.452
RMP	4	0.453	0.382	0.415
KRH	0ℓ	0.702	0.13	0.219
KRH	0s	0.449	0.449	0.449
KRH	0	0.462	0.429	0.445
KRH	1	0.490	0.390	0.434
KRH	2	0.565	0.378	0.453
KRH	3	0.533	0.386	0.447
KRH	4	0.518	0.402	0.452
IRB	–	0.291	0.098	0.147

model	n	precision	recall	F-score
JXT	0ℓ	0.702	0.13	0.219
JXT	0s	0.443	0.441	0.442
JXT	0	0.513	0.461	0.485
JXT	1	0.464	0.402	0.430
JXT	2	0.518	0.390	0.445
JXT	3	0.488	0.398	0.438
JXT	4	0.463	0.394	0.426
OPT	0ℓ	0.688	0.043	0.081
OPT	0s	0.462	0.449	0.455
OPT	0	0.496	0.441	0.467
OPT	1	0.435	0.394	0.413
OPT	2	0.493	0.413	0.450
OPT	3	0.548	0.449	0.494
OPT	4	0.514	0.433	0.470
SHB	–	0.433	0.421	0.427

with a full parse (133 per query) were annotated for containment of a correct answer to the question, starting from CLEF07 annotations. The annotation revealed that only 254 of the 12,377 passages really contain an answer.

Table 1 shows the precision, recall, and F-measure for the individual provers and for the best combinations that were tried, along with baseline results for exact proofs and for shallow features. These results were obtained by the cost-sensitive learning approach described in Section 2, using 10-fold cross validation. Compared to the IBS run only based on the *irScore* of the passage retrieval system, all regular runs show a dramatic improvement. The shallow processing baseline (SHB) demonstrates the potential of suitably chosen shallow features: Adding the *irScore* and *containsBrackets* features now increased the F-score of the SHB to 42.7%, compared to 36% in earlier work on the same dataset [8]. The brittleness of the logical approach reflects in very poor retrieval scores when using only logic-based features and exact proofs, see runs labeled '0ℓ'. However, as shown by the '0s' runs, combining the logical features based on exact proofs with the shallow features through ML eliminates the brittleness. The results obtained in this way (in particular the RMP-0s run with an F-score of 45.8%) also clearly outperform the shallow feature baseline. As witnessed by the RMP-n and KRH-n results for $n \in \{0, \ldots, 4\}$, the methods for extracting plausible answer bindings for failed proofs and relaxation show no clear positive effect compared to combining exact proofs with shallow features when a single prover is used. An improvement only occurs when results of two provers are combined. The JXT method which juxtaposes the features determined by the two provers shows good filtering results for $n = 0$ relaxation steps (with an F-score of 48.5%)

but does not appear to work very well otherwise. The OPT method shows more stable performance. For $n = 3$, it achieves the best result in this experiment, with a relative improvement of 7.9% over the best F-score for a single prover.

5 Conclusions

We have discussed robustness-enhancing techniques developed for logical passage filtering in the LogAnswer QA system. The evaluation on factual questions of CLEF07 revealed that combining logic with shallow features using an ML classifier is the most effective technique for increasing robustness. For relaxation and extracting answer bindings for failed proofs, an improvement over the use of shallow features was only achieved when results of two provers were combined.

References

1. Burger, J., Cardie, C., Chaudhri, V., Gaizauskas, R., Harabagiu, S., Israel, D., Jacquemin, C., Lin, C., Maiorano, S., Miller, G., Moldovan, D., Ogden, B., Prager, J., Riloff, E., Singhal, A., Shrihari, R., Strzalkowski, T., Voorhees, E., Weishedel, R.: Issues, tasks, and program structures to roadmap research in question & answering (Q&A). NIST (2000)
2. Peñas, A., Rodrigo, Á., Sama, V., Verdejo, F.: Overview of the answer validation exercise 2006. In: Working Notes for the CLEF 2006 Workshop (2006)
3. Giampiccolo, D., Magnini, B., Dagan, I., Dolan, B.: The third PASCAL recognizing textual entailment challenge. In: Proc. of the Workshop on Textual Entailment and Paraphrasing, Prague, June 2007, pp. 1–9. ACL (2007)
4. Moldovan, D., Bowden, M., Tatu, M.: A temporally-enhanced PowerAnswer in TREC 2006. In: Proc. of TREC-2006, Gaithersburg, MD (2006)
5. Haghighi, A.D., Ng, A.Y., Manning, C.D.: Robust textual inference via graph matching. In: Proc. of HLT/EMNLP 2005, Vancouver, BC, pp. 387–394 (2005)
6. Glöckner, I.: University of Hagen at QA@CLEF 2007: Answer validation exercise. In: Working Notes for the CLEF 2007 Workshop, Budapest (2007)
7. Bos, J., Markert, K.: When logical inference helps determining textual entailment (and when it doesn't). In: Proc. of 2nd PASCAL RTE Challenge Workshop (2006)
8. Glöckner, I.: Towards Logic-Based Question Answering under Time Constraints. In: Proc. of ICAIA 2008, Hong Kong, pp. 13–18 (2008)
9. Hartrumpf, S.: Hybrid Disambiguation in Natural Language Analysis. Der Andere Verlag, Osnabrück (2003)
10. Helbig, H.: Knowledge Representation and the Semantics of Natural Language. Springer, Heidelberg (2006)
11. Leveling, J.: IRSAW – towards semantic annotation of documents for question answering. In: CNI Spring 2007 Task Force Meeting, Phoenix, Arizona (2007)
12. Witten, I.H., Frank, E.: Data Mining. Practical Machine Learning Tools and Techniques. Morgan Kaufmann, San Francisco (2005)
13. Pelzer, B., Wernhard, C.: System Description: E-KRHyper. In: Pfenning, F. (ed.) CADE 2007. LNCS (LNAI), vol. 4603, pp. 508–513. Springer, Heidelberg (2007)

14. Baumgartner, P., Furbach, U., Pelzer, B.: Hyper Tableaux with Equality. In: Pfenning, F. (ed.) CADE 2007. LNCS (LNAI), vol. 4603. Springer, Heidelberg (2007)
15. Baumgartner, P., Furbach, U., Niemelä, I.: Hyper Tableaux. In: Orłowska, E., Alferes, J.J., Moniz Pereira, L. (eds.) JELIA 1996. LNCS, vol. 1126, pp. 1–17. Springer, Heidelberg (1996)
16. Sutcliffe, G., Suttner, C.: The TPTP Problem Library: CNF Release v1.2.1. Journal of Automated Reasoning 21(2), 177–203 (1998)

Comparing Document Classification Schemes Using K-Means Clustering

Artur Šilić[1], Marie-Francine Moens[2], Lovro Žmak[1], and Bojana Dalbelo Bašić[1]

[1] University of Zagreb, Faculty of Electrical Engineering and Computing,
Unska 3, 10000 Zagreb, Croatia
{artur.silic,lovro.zmak,bojana.dalbelo}@fer.hr

[2] Katholieke Universiteit Leuven, Department of Computer Science,
Celestijnenlaan 200A, 3001 Heverlee, Belgium
sien.moens@cs.kuleuven.be

Abstract. In this work, we jointly apply several text mining methods to a corpus of legal documents in order to compare the separation quality of two inherently different document classification schemes. The classification schemes are compared with the clusters produced by the K-means algorithm. In the future, we believe that our comparison method will be coupled with semi-supervised and active learning techniques. Also, this paper presents the idea of combining K-means and Principal Component Analysis for cluster visualization. The described idea allows calculations to be performed in reasonable amount of CPU time.

1 Introduction

Over the past decade, the digitalization of textual data has greatly increased the accessibility of text documents in all areas of human society. This has introduced a strong need for efficient storage, processing, and retrieval of texts, since the number of available documents has become quite large. The first step in solving such large-scale problems involves structuring the data, which can be done by introducing a classification scheme. Since many classification schemes can be used, the question of how to choose the best one among many for a designated task remains.

In this work, we design a method for finding the classification scheme that best fits data given a certain representation. More specifically, we try to quantify how much separation quality is gained through use of a better but more expensive classification scheme. Our work is related to that by Rosell et al. [12], where the objective was to evaluate clustering using two classification schemes. We visualize the data in order to inspect both the separation of clusters and their relationship to classes. The introduced visualization method is similar but distinct from that in the work of Dhillon et al. [4], because we use the Principal Component Analysis (PCA).

The work is structured as follows. We define the problem in Section 2, and we describe the data utilized in Section 3. The methodology of the experiments performed is covered by Section 4, and the results are presented in Section 5. Section 6 presents a broader scope and possible applications of the proposed method. We conclude the work with Section 7.

I. Lovrek, R.J. Howlett, and L.C. Jain (Eds.): KES 2008, Part I, LNAI 5177, pp. 615–624, 2008.

2 Problem Definition

We want to compare classification schemes in order to see how well they separate the space of text documents. Since that there is no "perfect" scheme, we have to find some intrinsic measure of space separation. Therefore, we compare the schemes against clustering. It is clear that introducing clustering introduces a strong bias stemming from the selection of the clustering method and its parameters. For a specific classification problem and its data representation, an appropriate clustering method might be well known from experience; in some cases, the clustering parameters can even be found theoretically. In these cases, the aforementioned bias is justified.

Additionally, the task of selecting the clustering method and parameters based on a predefined data representation and classification scheme remains a very interesting problem. The classification scheme divides the given space in a certain way. This division has a number of geometrical properties, such as balanced/unbalanced, convex/non-convex, and spherical/non-spherical. We want to generalize these geometric properties of a classification scheme into clustering parameters. Using the new clustering parameters, we will be able to reproduce the geometric properties of clusters on a new set of points or new part of the existing space. This investigation has not yet been performed, as the experiments in the following sections deal only with the answer to the previous paragraph.

3 Data

3.1 Document Collection

Our document collection NN9225 consists of 9225 legislative documents from the Republic of Croatia. The documents have been collected by the governmental agency HIDRA [3]. The collection covers dates from 1990 to 2006. Each document is labeled with labels from the two classification schemes described in the following subsection.

3.2 Classification Schemes

We seek to compare two classification schemes in order to understand which one is better suited for the division of a huge collection of documents. Both schemes seek to represent the semantic content of the document. The difference between the two schemes is as follows. The first classification scheme (*Issuer*) is based solely on the issuer of the document, whereas the second classification scheme (*Eurovoc*) is based on the actual content of the document and was manually assigned by legal experts.

***Issuer* scheme.** The *Issuer* classification scheme has 25 classes (issuing institutions), and was developed in an *ad hoc* manner by HIDRA to improve existing online retrieval of official documents. Note that the very same institution maintains the described document collection. If the document was issued by the Ministry of Defense, it is assigned the label "Defense, interior affairs, and national

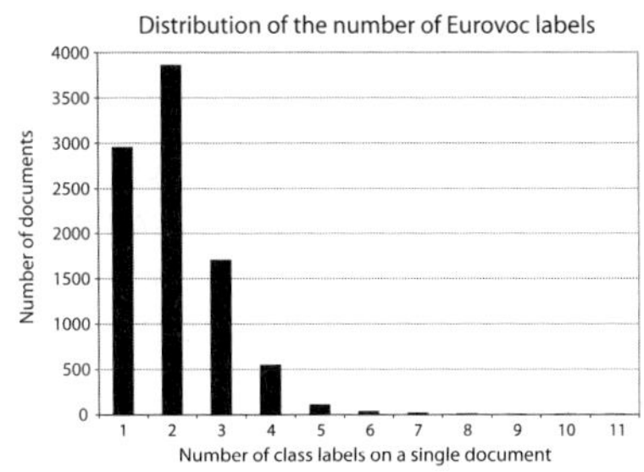

Fig. 1. Distribution of the number of Eurovoc labels over the NN9225 corpus

security" under the *Issuer* scheme. This scheme assumes that issuing institutions have narrow document topics. This classification scheme has obvious flaws. For example, the Ministry of Finance issues legal documents covering much broader topics than just "Finance," because this ministry deals with numerous aspects of the state. After reading such documents, human indexers might assign them multiple semantic labels like "live stock," "agricultural subsidy," or "environmental protection."

In spite of the obvious impairment of the *Issuer* scheme, there was no need for automatic or manual content-based classifiers at the time of document labeling. There was solely a need for information about the issuing institution. Since the document labels were quite easily assigned, the separation of the data set was cheap.

***Eurovoc* scheme.** Eurovoc is a parallel multilingual thesaurus maintained by the Office for Official Publications of the European Communities and officially used by many governmental bodies of the European Union [7]. The Eurovoc thesaurus has a hierarchy with up to eight levels of depth.

Since we wanted to compare the flat *Issuer* scheme with the hierarchical Eurovoc thesaurus classifications, we simplified the Eurovoc by flattening it to its first level of depth. Documents are assigned new labels, which are first-level predecessors of the actual labels. Links in the hierarchy are assumed to be *is-a* relations between classes. For example, a document labeled with the class *"administrative science"* is assigned the grand predecessor class *"science."* The flattened version of the Eurovoc thesaurus is called the *Eurovoc* scheme. Most of the documents in the NN9225 collection have from one to three first level-class Eurovoc labels (see Fig. 1).

Expected tradeoff. Since information about the issuer of legal documents is almost always available, the deployment of the *Issuer* scheme is very cost-effective. The cost of the *Eurovoc* scheme is determined either by the cost of maintaining personnel to index the documents or by the cost of an automatic classifier construction. The two presented schemes are obviously different. Table 1 summarizes the expected tradeoff between the cost and the partition (class separation) quality.

Table 1. Expected tradeoff between the classification schemes

	Issuer	*Eurovoc*
Separation quality	Low	**High**
Cost	**Very low**	High

4 Methodology

4.1 Preprocessing

During the experiments presented in the following subsections, each document from the text collection was represented as a point in the vector space. This vector space was constructed using the standard *bag-of-words* approach, in which each word was linguistically normalized to its lemma [16]. Stop words were removed from the feature space. Our corpus of 9225 documents contains about $1.9 * 10^5$ features, so we used information gain and χ^2 measures for feature selection. We chose to select only 3% of the features, since the early tests have shown that this reduction does not diminish the quality of subsequent processing. At the end of preprocessing, the well-known TF-IDF normalization process was performed [14].

4.2 K-Means Clustering

Clustering is the division of data into groups of similar objects. Among many different clustering algorithms, K-means is one of the simplest and most popular [6], [11], [13]. K-means is not capable of dealing with non-convex shapes, as [8] notes and [13] shows by experiment. Formally, the goal of K-means is to find the minimum of the following potential function:

$$F = \sum_{i=1}^{K} \sum_{x \in C_i} d(c_i, x)^2, \quad c_i = \frac{1}{|C_i|} \sum_{x \in C_i} x, \tag{1}$$

where C_i is the i-th cluster, c_i its centroid, and d a distance function.

Since finding an optimal clustering requires non-polynomial execution time, a heuristic algorithm is used [10]. Different distance measures, such as Euclidian, Mahalonobis, and cosinus-based, are used. For the experiments performed, the cosinus distance is used because this method generates hyperspherical clusters [5]. Additionally, a non-trivial seeding variant is used to enhance the quality and speed of the clustering [1]. Basically, the initial clusters are positioned in such a way that more clusters are present where the data are dense.

4.3 Comparison Approach

The classification schemes are compared with hyperspherical clusters produced by the K-means algorithm. The experiments presented here did not compare many different clustering methods and their parameters. The bias introduced via

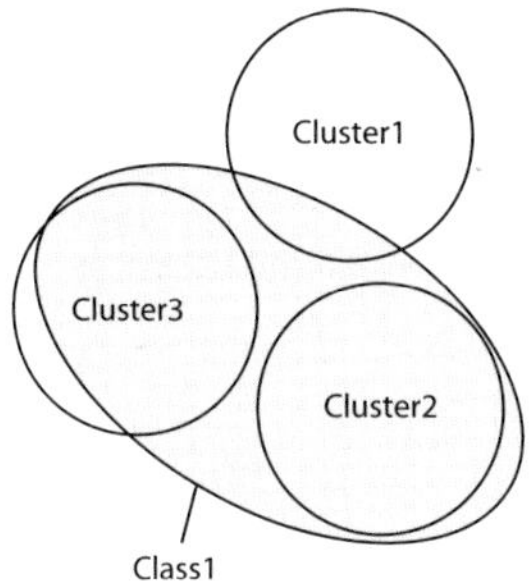

Fig. 2. Clusters 2 and 3 have high precision with respect to class 1

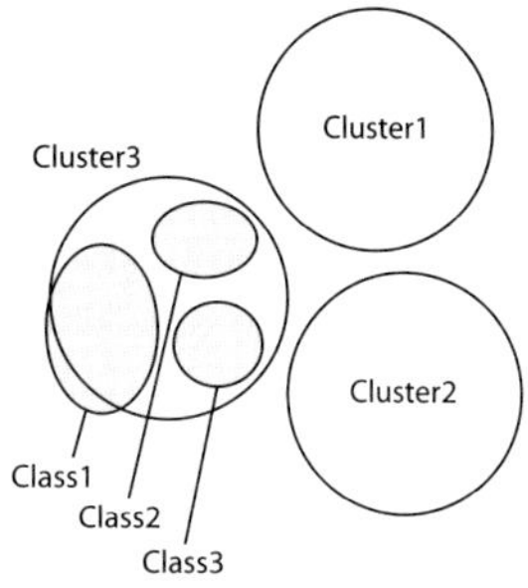

Fig. 3. Cluster 3 have high recall with respect to classes 1, 2, and 3

the selection of a well-known clustering method is tolerated, because we expect the *Issuer* scheme to be inferior to the *Eurovoc* scheme in separating the document collection since it is not based on actual content. The goal of the experiment was to show that our method can detect such separation impairments.

When dealing with texts in the *bag-of-words* space, we expect semantic classes to be separable by linear hyperplanes. Text categorization problems are usually linearly separable, as noted in [9]. If the classes are linearly separable, then they are convex as well. This justifies the use of K-means clustering as a simple baseline, because it generates hyperspherical clusters that are convex, able to cover the whole vector space of presented points, and relatively balanced [15]. The more similar the shape of a class is to the union of disjunctive and evenly distributed hyperspheres, the better the separation. A union of hyperspheres can of course be very complex. In our case, however, the number of clusters is small. Further, we observed after the completion of the experiments that the majority of documents from a particular class is found in just a few clusters.

High recall for a cluster with respect to some class means that one cluster covers a great majority of that class. High recall tells us not about the shapes of classes but rather about the shapes of clusters [see Fig. 3]. On the other hand, higher precision for clusters with respect to classes means that the shapes of the classes in the vector space are spanned by the shapes of clusters [see Fig. 2].

Although we do not generalize this idea for any number of classes, we establish a hypothetical link between the purity of hyperspherical clusters and the separation quality of classes in a classification scheme. Of course, this link has yet to be generally shown and theoretically explained. While our corpus is comprised of legal documents, we expect that the comparison method may easily be extended to corpora in other subject domains and genres since the preprocessing and clustering methods employed do not utilize any domain-specific features.

4.4 Comparison Measures

In the light of this work, a comparison measure between two sets of classes (or clusters) in a set of examples should indicate their closeness. Like other authors

[12], we compare a classification scheme to clustering by utilizing the standard evaluation measures of information retrieval quality. Furthermore, we use the information-theoretic measure of entropy as well.

Precision. Precision, the standard information retrieval measure is used to show how much the cluster i conforms to the class j:

$$p_{ij} = \frac{n_{ij}}{n_i}, \tag{2}$$

where n_{ij} is the number of documents with membership in both the cluster i and the class j, and n_i is the number of documents with membership in the cluster i.

Purity. The purity of a cluster is defined as the maximum precision over all classes:

$$\varrho_i = \max_j(p_{ij}). \tag{3}$$

Entropy. Since precision p_{ij} is the probability that a text drawn at random from the cluster i belongs to the class j, the entropy of a cluster is calculated as follows:

$$E_i = -\sum_j (p_{ij} \log p_{ij}). \tag{4}$$

The entropy of the whole clustering is defined as the weighted average over all of the clusters:

$$E = \sum_{i=1}^{K} \frac{n_i}{n} E_i, \tag{5}$$

where n is the number of documents in the collection.

4.5 Visualization Approach

It is useful to visualize the data, because this provides direct insights and inferences from the illustrations generated. Among many approaches to data visualization, a very popular one involves projecting the data onto a pair of components that will demonstrate the relationships present in the data. In our case, we want to inspect how well the clusters generated by the K-means algorithm separate the vector space. While Linear Discriminant Analysis (LDA) is well suited for this task [6], we employ a slightly different approach using PCA.

For each cluster, we find the concept vector (normalized centroid). Then, we calculate the principal components of the concept vector space and project the whole corpus onto an arbitrary pair of principal components. Since PCA maximizes the variance of concept projections, we expect the projected clusters to be well separated on the visualization plots. This method ignores the within-class scatter of points in the vector space. On one hand, this simplification is a disadvantage because losing some part of the information can lead to suboptimal visualization in comparison to LDA. On the other hand, it is an advantage because

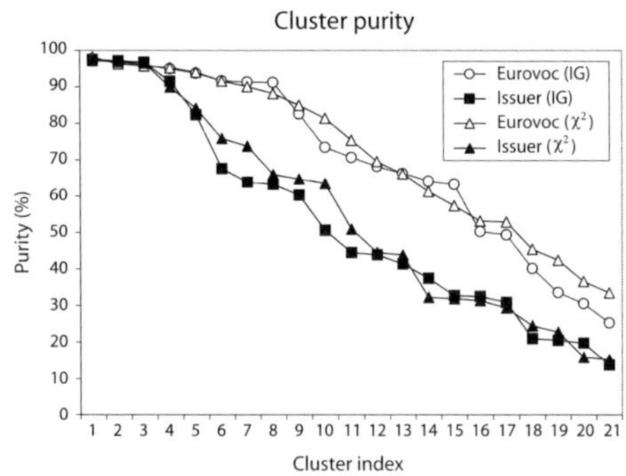

Fig. 4. Purity of clusters with respect to the contained categories (sorted descending)

the computational complexity of the proposed method is drastically lower than that of LDA. This lower complexity arises since the matrix computations are performed on a concept-feature matrix rather than on a document-feature matrix; it is clear that the number of concepts is much smaller than the number of documents (observations). Recall that if we choose to have only 3% of the original features[1], the document-feature matrix yielded is still rather large for matrix calculations performed by ordinary personal computers. Since the visualization method is intended for interactive use on large data sets, its speed is important.

By using the concept vectors, we ignore the within-class scatter. Our approach is thus similar to the work of Dhillon et al. [4]. The difference arises from the fact that we use PCA instead of other matrix calculations to find the projections.

5 Results

5.1 Comparison of the Classification Schemes

The comparison of classification schemes can be seen in Fig. 4, which shows a plot of cluster purities. The clusters are sorted by their respective purities. The *Eurovoc* scheme is closer to the clustering than the *Issuer* scheme because clusters are more pure with respect to the *Eurovoc* scheme.

The graphical data presented in Fig. 4 is numerically summarized in Table 2, which calculates the entropy measures. To eliminate the randomness of the K-means algorithm, the experiment was run five times. The entropy values presented in Table 2 are thus averages of five runs. Both entropy and purity support our expected hypothesis, which stated that the *Eurovoc* scheme would be closer to the K-means clustering.

5.2 Visualization of Clustering

The visualization method was manually evaluated within the research team. Depending on the principle component pair, we could see that the projections from

[1] 5799 features.

Table 2. Entropy of clusters respective to the classification schemes

	Issuer	*Eurovoc*
Entropy	$18,7 * 10^{-3}$	$0,86 * 10^{-3}$
St.dev.	$1,4 * 10^{-3}$	$0,11 * 10^{-3}$

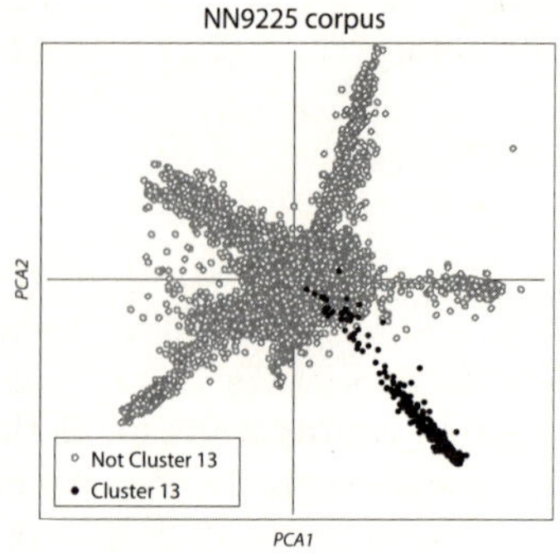

Fig. 5. Separation of a cluster from the rest of the collection

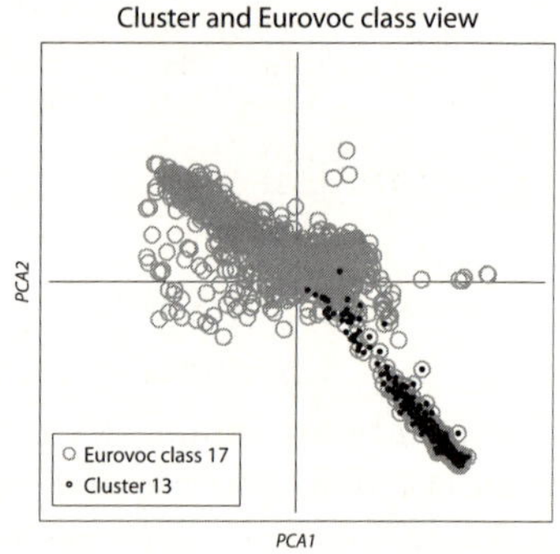

Fig. 6. A cluster spans a part of the *Eurovoc* class Transportation (No. 17)

most clusters were well separated from the other points. One such separation is shown in Fig. 5. In this figure, the NN9225 corpus and one of its clusters are projected onto a pair of principal components. If a cluster is separable in the projected space, then it is separable in the original space as well because the projection dimensions (principal components) are linear combinations of the original dimensions.

Fig. 6 shows the relationship between a class and a generated cluster using the same principal components used in Fig. 5. Cluster 13 has high precision with respect to Eurovoc class 17. This means that Eurovoc class 17 is partly spanned by cluster 13. Since cluster 13 is easily discriminated from the rest of the documents, the spanned part of Eurovoc class 17 is easily discriminated from the rest of the documents as well. Formal evaluation of the visualization is currently being conducted, and manual evaluation with an expert team is in initial stages.

6 Applications and Impact

First, we can compare different directories offered by online services (e.g. Dmoz, Yahoo! Directory, and VLIB) to see how they fit certain text representations and cluster models. This is one of many examples in which the documents are already classified with two or more schemes.

Second, our comparison approach is useful for validating additions, deletions, and other alterations of classification schemes suggested by human experts. Given a document set, its class labels, and a text representation, one could find the clustering method and parameters that produce the clusters closest to the classification scheme. We assume that this clustering will divide the space of texts in the same manner (i.e. using shapes with similar geometric properties) as the given classification. Therefore, a system could validate the suggested scheme changes and verify that they are in line with the rest of the scheme.

Finally, our approach can be useful in a semi-supervised setting. For example, during self- or co-training [2], the clustering can define additional constraints for deciding that an unlabeled example belongs to a certain class. Additionally, the geometric space can be explored in a much more principled manner for active learning [17] when the system chooses examples that are to be manually annotated.

7 Conclusion

This work presents a methodology for comparing different classification schemes using clustering methods. When given a motivated text representation and distance measure, one can choose between several classification schemes according to their fits. The quantitative comparison measures are explained, and their usage is justified. The experiment has confirmed our expectations by showing that the difference in the separation quality of the two presented classification schemes can be detected using the proposed comparison method.

We believe that the ideas introduced are very important and relevant for a much broader scope than the present experiment. Possible future work could explore the effects of using different clustering methods. After that, we could generalize the presented methodology in order to compare hierarchical classification schemes instead of flat ones. Additionally, the inversion of the task, in which we find the clustering method or data representation that fits to some classification scheme, remains to be explored. Finally, the method could be coupled with semi-supervised techniques as discussed in the previous section.

We propose a computationally efficient visualization method combining K-means clustering and PCA that ignores within-class scatter. Possible future work could compare our approach to the work of [4]. The method was useful in discriminating clusters, so visualization helped us to obtain direct insight into the relationships between classes and generated clusters in our collection of text documents.

Acknowledgement

This work has been jointly supported by the Ministry of Science, Education and Sports of the Republic of Croatia and the Government of Flanders under the grants No. 036-1300646-1986 and KRO/009/06 (CADIAL).

References

1. Arthur, D., Vassilvitskii, S.: k-means++: The Advantages of Careful Seeding. In: Proceedings of the 18th Annual ACM-SIAM Symposium on Discrete Algorithms, pp. 1027–1035 (2007)
2. Blum, A., Mitchell, T.: Combining labeled and unlabeled data with co-training. In: Proceedings of the 11th Annual Conference on Computational Learning Theory (COLT 1998), pp. 92–100 (1998)

3. Croatian Information Documentation Referral Agency, http://www.hidra.hr/
4. Dhillon, I.S., Modha, D.S., Spangler, W.S.: Class visualization of high-dimensional data with applications. Computational Statistics and Data Analysis 41(1), 59–90 (2002)
5. Dhillon, I.S., Modha, D.S.: Concept Decompositions for Large Sparse Text Data Using Clustering. Journal of Machine Learning 42, 143–175 (2001)
6. Duda, R.O., Hart, P.E., Stork, D.G.: Pattern Classification, 2nd edn. Wiley, New York (2000)
7. EUROVOC thesaurus, European Union publications office, http://europa.eu.int/celex/eurovoc/
8. Halkidi, M., Batistatis, Y., Vazirgiannis, M.: On Clustering Validation Techniques. Journal of Intelligent Information Systems 17, 107–145 (2001)
9. Joachims, T.: Learning to Classify Text Using Support Vector Machines. Kluwer Academic Publishers, Dordrecht (2002)
10. Lloyd, S.P.: Least squares quantization in PCM. IEEE Transactions on Information Theory 28, 129–136 (1982)
11. Moens, M.-F.: Note on Clustering Large Document Collections. Technical Report, CADIAL, Katholieke Universiteit Leuven (July 2007)
12. Rosell, M., Kann, V., Litton, J.-E.: Comparing Comparisons: Document Clustering Evaluation Using Two Manual Classifications. In: Proceedings of the International Conference on Natural Language Processing (ICON 2004), Hyderabad, India (2004)
13. Satchidanandan, D., Chinmay, M., Ashish, G., Rajib, M.: A Comparative Study of Clustering Algorithms. Information Technology Journal 5, 551–559 (2006)
14. Sebastiani, F.: Machine Learning in Automated Text Categorization. ACM Computing Surveys 34, 1–47 (2002)
15. Su, M.C., Chou, C.H.: A K-means Algorithm with a Novel Non-Metric Distance. IEEE Transactions on Pattern Analysis and Machine Intelligence 23, 674–680 (2001)
16. Šnajder, J., Dalbelo Bašić, B., Tadić, M.: Automatic Acquisition of Inflectional Lexica for Morphological Normalisation. Information Processing & Management (2008) (accepted, to be published) doi:10.1016/j.ipm.2008.03.006
17. Tong, S., Koller, D.: Support Vector Machine Active Learning with Applications to Text Classification. In: Proceedings of the Seventeenth International Conference on Machine Learning (ICML 2000), Stanford, California, pp. 287–295 (2000)

Describing and Generating Web-Based Affective Human-Agent Interaction

Xia Mao, Zheng Li, and Haiyan Bao

P.O. BOX 206 Beihang University, Beijing 100083, P.R. China
{moukyoucn,buaa_david}@yahoo.com.cn,
vivid96912@sohu.com

Abstract. In this paper, we discuss our research on the multi-modal interaction markup language (MIML) which serves as a core for integrating various components, including lifelike agent ActiveX controller, facial expression recognition ActiveX controller and speech emotion recognition ActiveX controller, to describe and generate web-based affective human-agent interaction. Emotional control on lifelike agents provided by MIML makes the human-agent interaction even more intelligent. With the MIML and components we designed, web-based affective interaction can be described and generated easily.

Keywords: Affective Interaction, Web Intelligence, Markup Language, Lifelike Agents.

1 Introduction

Lifelike agents have been used as the middle layer between user and computer. They have shown their potential to allow user to interact with computer in a natural and intuitive manner through human communicative means. Meanwhile, as Nass's researches on human-human and human-computer interactions suggest, people most naturally interact with computers in a social and affectively meaningful way, just like with other people[1]. Researchers like Picard have recognized the potential and importance of emotion to human-computer interaction, dubbing work in this field as "affective computing[2]. Therefore, in order to realize natural, harmonious and believable intelligent human-agent interaction, we must endow lifelike agents with affect, in other words, they ought to be capable of expressing their own emotion and recognizing user's emotional states.

At the same time, there is an emerging number of scripting and representation languages to describe the behavior of the agents, but which have taken different approaches to specify their objectives. The affective presentation markup language (APML) is an XML-based language that represents communicative functions and thus facilitates the scripting of dialogues with agents[3]. The virtual human markup language (VHML) allows interactive Talking Heads to be directed to accommodate various aspects of human-computer interaction. It provides tags for facial and bodily animation, gesture and covers different abstraction levels[4]. The multi-modal presentation markup language (MPML) , which is also XML-based, has been developed with the aim of enabling authors of web

I. Lovrek, R.J. Howlett, and L.C. Jain (Eds.): KES 2008, Part I, LNAI 5177, pp. 625–632, 2008.

pages to add agents for improving human-computer interaction[5][6]. The languages that do not adopt an XML-based approach include scripting technology for embodied personal language (STEP) and parameterized action representation system (PAR). The STEP uses distributed logic programming and action composition operators to specify motions of VRML-based agents[7]. The PAR allows for fine-grained specification of parameters to modify the execution of agent actions[8]. However, all the languages mentioned above can not describe and generate emotion detection behavior.

In this paper, we introduce the multi-modal interaction markup language, a language specifically designed for non-expert users allowing them to describe and generate emotion detection behavior of lifelike agents when creating web-based intelligent interaction system. Three components, including lifelike agent ActiveX controller, facial expression recognition ActiveX controller and speech emotion recognition ActiveX controller, are attached to MIML.

The remaining paper is organized as follows. Section 2 is dedicated to introducing the architecture of the MIML system. In Section 3, the usefulness of the MIML is illustrated by discussing one actual implementations of web-based intelligent interaction system. Finally, we summarize and conclude the paper.

2 The MIML System

The MIML is a markup language compliant with standard XML. The way we employ to describe and generate affective interaction can be divided into three parts: (1)definition of the tags (2)component module (3)MIML compiler. An overview of the MIML system architecture is shown in Fig. 1. The detailed description will be given in the following subsections.

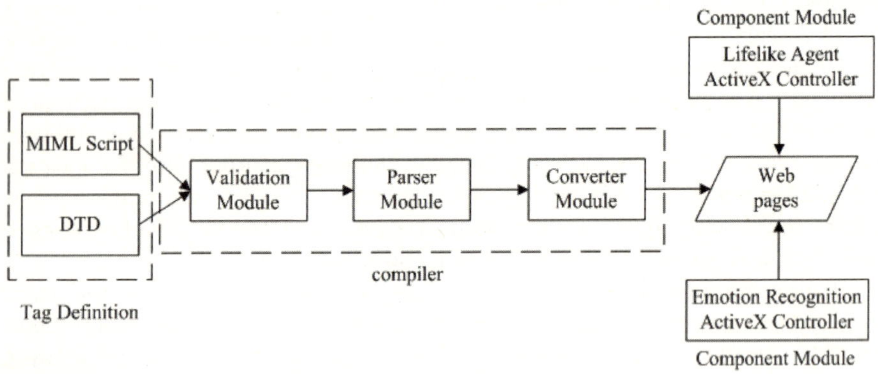

Fig. 1. Architecture of MIML

2.1 Tag Structure

Here we briefly discuss the tags defined in MIML. Fig. 2 illustrates the tag structure. We inherit some tags from MPML and all the tags are easy to learn

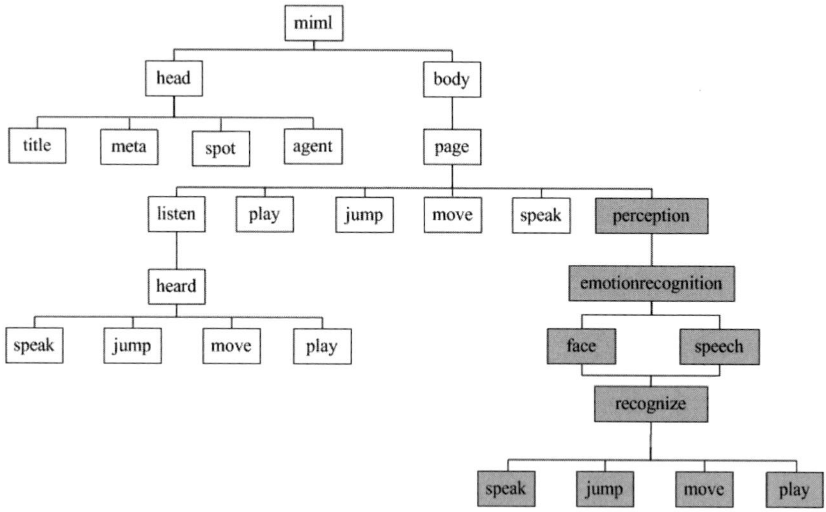

Fig. 2. Tag Structure of MIML

and remember, as they follow the conventions of HTML. The root tag pair of an MIML script document is <miml> and </miml>, which contains all other tag pairs. The tag <head> specifies general information and the tag <body> refers to the sequence of events comprising the actions of lifelike agents. Table 1 summarizes the tags in Fig. 2.

The tags in white boxes in Fig. 2 are used to control the agents' basic behavior, which are extensively discussed in paper [5] and [6]. We will give a detailed description of the tags we proposed to create web-based affective interaction, which are placed in colored boxes. Document Type Definition(DTD) for the tags is summarized in Fig. 3. It defines the grammar and allows the author to make the lifelike agents recognize the user's emotional state through facial expression and speech in the web-based interaction system. The lifelike agents can make various responses (verbal and non-verbal behavior) according to the detective result.

The root tag pair to describe the interaction is <perception> which includes one sub-tag <emotionrecognition>. The <emotionrecognition> involves two sub-tags now: <face> and <speech>, which are designed to control the facial expression recognition and speech emotion recognition respectively. The "align" attribute in tag <face> specifies the destination spot where the controller is located in web pages. The tag <recognize> and "result" attribute attached to it are equivalent to the C "switch" and "case" instruction respectively. It compares the return value of the controller with the value of the "result" attribute, and executes the script included in the <recognize> tag if they are identical. The value of the "result" attribute related to facial expression and speech emotion must be one of the anger, happiness, surprise and neutral.

Table 1. Instruction of Tags

Tag	Description
title	name of the miml script document
meta	information of the author
spot	position of the lifelike agents
agent	name of the agent
page	name of the background webpage
listen	root tag of the speech input module
heard	accept speech input from users
play	play the pre-defined actions
speak	speak sentences through Text-to-Speech (TTS) system
move	move on the screen to the destination spot
jump	jump to the other page
perception	root tag of the perception module
emotionrecognition	root tag of the emotion recognition module
face	control the facial expression recognition
speech	control the speech emotion recognition
recognize	represent the different cases of the recognition result

```
<!ELEMENT perception(emotionrecognition)>
<!ELEMENT emotionrecognition(face|speech)?>
<!ELEMENT face(recognize)>
<!ATTLIST face align CDATA #REQUIRED>
<!ELEMENT recognize (speak|play|move|jump)+>
<!ENTITY  %BASIC-FACEEMOTIONS
      "(anger|happiness|surprise|neutral)+">
<!ATTLIST recognize  result %BASIC-FACEEMOTIONS #REQUIRED>
<!ELEMENT speech(recognize)>
<!ELEMENT recognize (speak|play|move|jump)+>
<!ENTITY  %BASIC-SPEECHEMOTIONS
      "(anger|happiness|surprise|neutral)+">
<!ATTLIST recognize  result %BASIC-SPEECHEMOTIONS #REQUIRED>
```

Fig. 3. DTD for Perception Tags

2.2 Component Module

The lifelike agents ActiveX controller and emotion recognition ActiveX controller constitute the component module. These components should be installed before creating web-based affective interaction system. The MsAgent package is used to provide basic functions. With these functions, the agent can move freely within the computer display, speak aloud (by displaying text on the screen), and listen for spoken voice commands. The agents also can express emotions by performing different actions and changing speech parameters, such as speech rate and pitch changes. Other agent systems can be used with appropriate diver programs. Due to the need of speech dialogue feature, it has to incorporate voice commands and TTS (Text-To-Speech) engines. As for the emotion recognition ActiveX controller, we have designed the algorithms to realize the facial expression recognition and speech emotion recognition, which are extensively discussed in our complementary paper [9] and [10] respectively. In short, these ActiveX controllers provide interface to the tags and can be called in the web pages easily.

2.3 MIML Compiler

The MIML compiler is composed of validation module, parser module and converter module. Firstly, the validation module invokes the DTD to check the MIML script text file for syntactical errors. Then the parser module calls SAX (Simple APIS for XML) in MSXML.DLL to parse the MIML script. Finally the converter module generates Vbscript code that is executable in web browser to perform an interaction. The backgrounds of the MIML interaction are constituted of HTML pages. The Vbscript code will be embedded into the appointed HTML pages automatically by the converter. Currently, MIML assumes Microsoft Internet Explorer 6.0 (or higher) to run the interaction. Some of the rules for the converter are listed in Fig. 4.

```
<perception>
   <emotionrecognition>
      <face align="right">
        <recognize result="happiness">
            <speak emotion="happiness">
               Oh,you are smiling!
            </speak>
        </recognize>
        <recognize result="anger">
            <speak emotion="sadness">
               Oh,you are not happy!
            </speak>
               ...
        </recognize>
       </face>
   </emotionrecognition>
</perception>
```

```
Sub FaceReFunction
     a = Face.Camload()
     Set FaceReRequest1 = agent.Play("RestPose")
End Sub

Sub AlertUser(a)
     If a = "happy" Then
          agent.Play "Pleased"
          agent.Speak("Oh,you are smiling! ")
     ElseIf a = "anger" Then
          agent.Play "Sad"
          agent.Speak("Oh,you are not happy! ")
     ...
     End If
End Sub

Sub Agent_RequestComplete (ByVal Request)
     Select Case Request
       Case FaceReRequest0
         ID = SetTimeout("FaceReFunction", 100)
       Case FaceReRequest1
         ID = SetTimeout("AlertUser(a)", 100)
      End Select
 End Sub
```

Fig. 4. Rules for Converter

3 Illustration

In this section, we will describe a web-based scenario that instantiates the affective human-agent interaction described and generated by MIML. The scenario is web-based E-commerce recommendation system, and the lifelike agent plays as a virtual recommender. The script fragment of the E-commerce recommendation system is demonstrated in Fig. 5.

Fig. 6 is one of the interaction system results after compiling of the MIML script. In Fig. 5, line 7 means the background web page is "main.html". In line 8-11, the agent "genie" expresses his welcome to the user with happiness emotion and is enabled to accept speech command from the user to decide what kind of commodity the user are interested in (line 12-17, the value attribute in the

```
1  <miml>
2   <head>
3      ...
4      <agent id="genie" character="genie"/>
5   </head>
6   <body>
7      <page id="1" ref="main.html">
8           <speak emotion="happiness">
9              Welcome, nice to meet you! What kide of commodity do you want?
10             Tell me through the microphone or click the icon.
11          </speak>
12        <listen>
13          <heard word="wine">
14            <jump des="wine.html"/>
15          </heard>
16        ...
17        </listen>
18     </page>
19     <page id="2" ref="wine.html">
20          <speak emotion="happiness">
21             Hello, I will introduce the wine to you!
22             ...
23          </speak>
24          ...
25        <perception>
26           <emotionrecognition>
27            <face align="right">
28             <recognize result="happiness">
29                <speak emotion="happiness">
30                 Oh,you are smiling. you must be interested in the wine.
31                </speak>
32                <jump des="wine1.html"/>
33             </recognize>
34             <recognize result="anger">
35                <speak emotion="sadness">
36                  Oh,you are not happy. I will introduce the other wine to you!
37                </speak>
38                <jump des="wine2.html"/>
39             </recognize>
40             ...
41            </face>
42           </emotionrecognition>
43        </perception>
44        ...
45     </page>
46     ...
47   </body>
48 </miml>
```

Fig. 5. Script Fragment of E-commerce Recommendation System

<heard> tag can be any word according to the author's requirement, if a speech input text recognized by the voice commands engines, then the actions described inside <heard> tag are executed). When user says "wine", the web page will jump to "wine.html" in which the agent will introduce wine to the user with happiness emotion (line 20-23). Then the agent can detect the user's emotional state by facial expression recognition controller to judge whether the user is satisfied with the wine or not, followed by a branching edge of multiple alternatives (line25-43) where part of the first branch is shown (line 28-33,Fig. 6). This branch is selected when the recognition result is "happiness", then the agent will ask the user to order the wine. If not, the agent will introduce another wine to the user (line 34-39). In this example, we only use the facial expression recognition controller. The authors also can employ the speech emotion recognition controller if they need when describing the interaction with lifelike agents.

A 60-person user study was conducted to quantitatively measure the "performance" of the system. Before the usage of our system, the normal facial expression and emotional speech in our database are played to assist the users to express their own emotions better. The average real-time recognition rate for

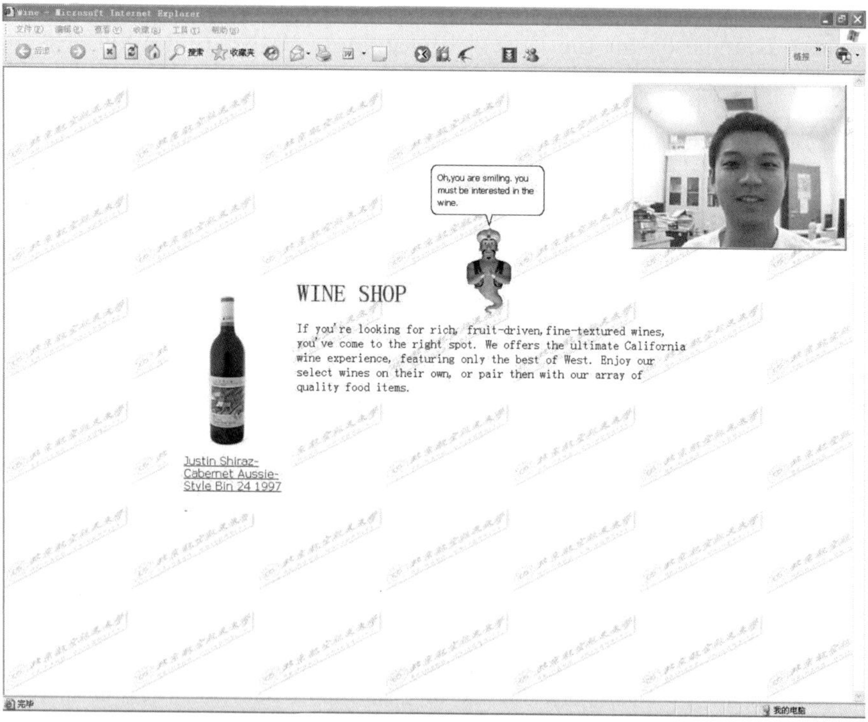

Fig. 6. Wine.html of E-commerce Recommendation System

emotional speech and facial expression can reach 90 percent. Users can click the mouse to jump to the right web page if there is a wrong recognition result.

4 Conclusion

Recent years have seen many efforts to include lifelike agents as a crucial component of application fields. That results in the growing number of describing languages for controlling the behavior of the lifelike agents. However, most of the languages are passive, i.e., all the verbal and non-verbal behavior described by these languages are predefined, and they can not describe and generate emotion detection behavior. In this paper, we have proposed our multi-modal interaction markup language. With the components and compiler we designed, the intelligent affective interaction can be embedded into web pages easily. In the future, we intend to design more tags and controllers to enrich the MIML architecture, such as textual emotion detection controller, eye-tracking controller, gesture and body detection controller (nodding or shaking head).

Acknowledgments. This work is supported by the National Nature Science Foundation of China (No.60572044), High Technology Research and Development

Program of China (863 Program, No.2006AA01Z135), the National Research Foundation for the Doctoral Program of Higher Education of China (No.20070006057). We would like to appreciate the Professor Mitsuru Ishizuka at University of Tokyo and the Associate Professor Helmut Prendinger at the National Institute of Informatics, who have provided many free resources on their home pages.

References

1. Nass, C., Tanber, E.: Computers are social actors. In: Proceeding of CHI 1994, Boston, pp. 72–78 (1994)
2. Picard, R.W.: Affective Computing. MIT Press, Cambridge (1997)
3. DeCarolis, B., Carofiglio: Apml: a mark-up language for believable behavior generation. In: Proceeding of AAMAS 2002 Workshop on ECA-Let's Specify and Evaluate Them, Italy (2002)
4. Marriott, A., Stallo, J.: Vhml - uncertainties and problems, a discussion. In: Proceeding of AAMAS 2002 Workshop on ECA-Let's Specify and Evaluate Them, Italy (2002)
5. Prendinger, H., Ishizuka, M.: Describing and generating multimodal contents featuring affective lifelike agents with mpml. New Generating Computing 24(2), 97–128 (2006)
6. Prendinger, H., Ishizuka, M.: Mpml: A markup language for controlling the behavior of life-like characters. Journal of Visual Languages and Computing 15(2), 183–203 (2004)
7. Huang, Z., Eliebs, A.: Step: a scripting language for embodied agent. In: Proceeding of PRICAI 2002 Workshop on Lifelike Animated Agent - Tools, Affective Functions and Applications,Tokyo (2002)
8. Badler, N.: Parameterized action representation for virtual human agents, Embodied Conversational Agents, pp. 256–284. MIT Press, Cambridge (2000)
9. Mao, X., Xue, Y.L.: Beihang university facial expression database and multiple facial expression recognition. In: Proceeding of Fifth International Conference on Machine Learning and Cybernetics, pp. 369–372 (2006)
10. Mao, X., Zhang, B.: Speech emotion recognition based on a hybrid of hmm/ann. In: Proceeding of the 7th WSEAS International Conference on Applied Informatics and Communications, pp. 3282–3287 (2007)

Scalable Content-Based Ranking in P2P Information Retrieval

Maroje Puh[1], Toan Luu[2], Ivana Podnar Zarko[1], and Martin Rajman[2]

[1] University of Zagreb, Faculty of Electrical Engineering and Computing, Zagreb, Croatia
[2] Ecole Polytechnique Fédérale de Lausanne (EPFL), Lausanne, Switzerland
{Maroje.Puh,Ivana.Podnar}@fer.hr,
{VinhToan.Luu,Martin.Rajman}@epfl.ch

Abstract. Numerous retrieval models have been defined within the field of information retrieval (IR) to produce a ranked and ordered list of documents relevant to a given query. Existing models are in general well-explored and thoroughly evaluated using traditionally centralized IR engines. However, the problem of producing global relevance scores to enable document ranking in peer-to-peer (P2P) IR systems has largely been neglected. Traditional ranking models in general require global document collection metrics such as document frequency, average document length, or the number of collection documents, which are not readily available in P2P IR systems. In this paper, we present a scalable solution for content-based ranking using global relevance scores in P2P IR systems that has been implemented as a part of ALVIS PEERS, a full-text IR engine developed for structured P2P networks. The provided experimental results show efficient and scalable performance of here proposed ranking implementation.

Keywords: P2P, Information retrieval, Content-based ranking.

1 Introduction

As the amount of web content is continuously growing and changing, it becomes more important to design and deploy widely-distributed and decentralized search engines that can efficiently operate in such dynamic environments. State-of-the art search engines are currently centralized and optimized for highly-responsive query answering using huge document indexes distributed over large proprietary clusters. Although such systems enable highly-efficient information access to millions of users, the amount of indexed documents currently represents a small fraction of the constantly growing web data. Centralized engines have difficulties to scale with the growing web size and constantly changing content [1]. Therefore, research efforts are currently directed to designing distributed and decentralized open-source retrieval systems [2]. Peer-to-peer (P2P) technology has become an appealing architecture for widely-distributed IR systems due to its properties, such as decentralization, self organization, and resource sharing [3]. Search engines are designed to efficiently find documents relevant to a user query, while the quality of retrieved documents depends on the ability of the information provided by retrieved documents to satisfy user information needs. Various IR models and ranking techniques have been developed over the years that aim at achieving a better quality of

I. Lovrek, R.J. Howlett, and L.C. Jain (Eds.): KES 2008, Part I, LNAI 5177, pp. 633–640, 2008.

retrieved ordered list of documents. The state-of-the art centralized search engines have successfully implemented existing models, however the process of adopting existing ranking techniques in distributed environments is not straightforward because of the unavailability of global collection statistics that are needed for ranking computation. Furthermore, P2P solutions can potentially induce high and unscalable traffic during the ranking process.

The paper presents the ranking technique implemented as part of ALVIS PEERS, a fully-functional IR search engine which uses a structured P2P overlay for building a distributed inverted index for large document collections [4]. Alvis uses a novel retrieval model based on indexing with *Highly Discriminative Keys* (HDKs)— terms and sets of terms occurring in a limited number of documents. HDKs may be seen as highly-selective multiterm queries associated with precomputed answer sets which enable efficient retrieval because of the short size of the associated posting.

Section 2 of this paper presents the related work. Section 3 describes the architecture of ALVIS search engine and the HDK indexing approach. The design and implementation of the contend-based ranking component integrated in ALVIS is presented in section 4, with its performance analysis given in section 5. Section 6 concludes the paper and presents future work.

2 Related Work

Distributed content-based ranking techniques depend on the indexing strategy used in the system. Two basic indexing strategies in P2P IR networks are federated local indexes used in unstructured P2P networks and the global index used in structured P2P networks [2].

In federated local indexes approach, disjunctive subsets of a global document collection are hosted on the peers and each peer is an independent search engine with its own local index. In such networks flooding is used to locate the data, resulting in high bandwidth costs and no guarantee that all relevant nodes will eventually be reached. To decrease the bandwidth consumption during the query phase, advanced approaches use two level querying: peers are independent search engines with local indexes, while the network or special nodes maintain the global peer index that is smaller and easier to maintain than global single-term index. The example search engine project using this combination indexing strategy is the Minerva project [5].

In global index approach, the overlay structured P2P network maintains a global index and each peer in the network is responsible for a maintaining a disjunctive part of the global index. Structured P2P networks enable efficient resource lookup process by employing different strategies, one of which is distributed hash table (DHT). Odissea, a P2P architecture for Web search [6] is an example of this indexing strategy.

3 The Alvis P2P Search Engine

This section explains the HDK-based indexing approach and presents the architecture of the P2P retrieval engine ALVIS PEERS [4]. The major obstacle for implementing

P2P full-text retrieval is unscalable network bandwidth consumption, caused by transmissions of long posting lists among peers when processing queries in a P2P system. To overcome this obstacle, the HDK-based indexing approach [7] has been introduced that results in shorter posting lists, but still achieves the retrieval quality comparable to the one in a centralized environment. Instead of indexing with single terms, this approach truncates large posting lists to a constant size, while compensating the resulting loss of information by indexing, in addition, carefully selected combinations of terms. Consequently, the index contains a larger number of index entries while all are associated with short posting lists. Sets of terms (we call them *keys*) forming an index entry have to occur simultaneously in a single document within a window of predefined size. If such keys occur in less than *DFmax* documents (*DFmax* is the parameter of our model), they are considered *discriminative* w.r.t document collection, i.e. such keys are HDKs. In case keys occur in more than *DFmax* documents, the index stores only top-*DFmax* ranked documents, while such key is a candidate to be extended with another term to create a new HDK.

We assume that each peer participating in the P2P IR engine contributes a set of local documents which constitute a part of the global document collection. From the local point of view, each peer indexes its local documents, i.e., peers compute keys and associated posting lists for its local collection and insert them into the global index. From the global point of view, peers build a DHT which is used to maintain a global inverted index. Each peer maintains a part of the global index assigned to it by the DHT, i.e. it stores a number of keys and associated posting lists. The peer also enables document retrieval by interacting with the DHT to retrieve the list of documents relevant to the submitted query and participates in the ranking procedures which are described in details in Section 4. The architecture of the Alvis P2P search engine is decomposed into layers presented in Figure 1.

The *P2P layer* builds a DHT for storing the global HDK index associating keys to document frequencies and posting lists. Each posting also includes statistics relevant to that document: term frequency of each key term in the given document, and document length, which are used for ranking.

The *HDK layer* is responsible for two tasks: during the key-based indexing task peers build the set of keys and associated posting lists from their local document collection, and during the querying task, a peer has to find relevant keys in the global index, retrieve the posting lists that are associated with such keys and merge them. The indexing task is triggered when the peer joins the network: first, a peer builds a standard single term index from its local collection and inserts it into the DHT. Next, it waits for messages from the DHT notifying them to expand certain single-term keys that appear in more than *Dfmax* global documents. Upon receiving such a request, the peer expands the key, and inserts it together with its posting list into the DHT. Note, however, that during the insertion of a key-posting pair into the DHT, a maximum of *Dfmax* postings will be inserted into the network. During the querying phase, a peer that received the query maps the query terms into keys that are stored in global index. The peer explores the lattice of query term combinations starting with the largest possible term set, which is limited either by the query size or the maximal key size. If this term set doesn't exist in the global index, term combinations of decreasing sizes are explored, and this process continues until all terms forming a query are covered

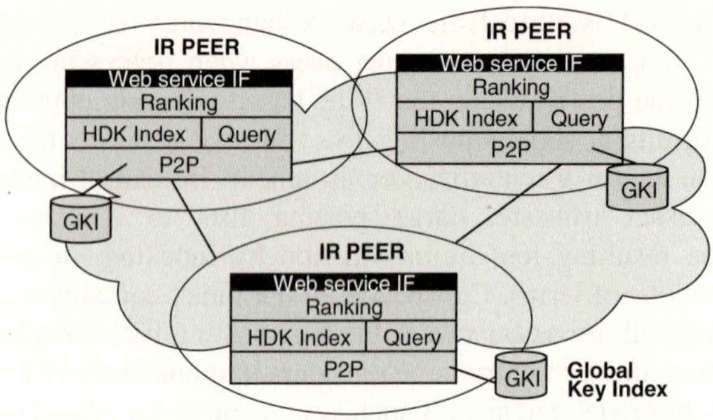

Fig. 1. Overview of the P2P search engine architecture

with retrieved keys. The resulting posting list is the union of postings associated with retrieved keys, and it is used as the input data to the Ranking layer.

The *Ranking layer* is responsible for producing a ranked and ordered set of documents during both the indexing and querying phases, and is described in detail in the following section.

4 Implementation of the Content Based Ranking Component

As mentioned, the Ranking layer implemented in Alvis prototype is in charge of computing document rankings during both the indexing and query processes. The score of a document w.r.t. a term is done using the well-known BM25 relevance function, which utilizes the following statistics:

Global values

- Term dependant: document frequency of the term in the global collection;
- Term independent: average document length, number of documents in the collection.

Local values

- Term dependant: term frequency in a document;
- Term independent: document length.

The score of a document w.r.t. a key (which is a set of terms) is calculated as the sum of individual scores of the document w.r.t. each term in the key. As mentioned before, local values are stored in the global index with each stored posting. Term independent global values are retrieved periodically by the ranking layer of each peer, and stored locally. Global document frequency of a key in the global collection is maintained by a peer responsible for that key, as determined by the DHT.

4.1 Ranking Layer Role in Indexing Process

When a peer joins the network, it starts to index its local document collection, and inserts its keys and associated postings containing local values into the P2P overlay. If the size of the posting list exceeds the *Dfmax* parameter, only *Dfmax* most relevant postings for the given key will be inserted into the network. Therefore, the ranking layer needs to rank all postings w.r.t. the key, produce an ordered set of postings, and insert top-*Dfmax* postings into the network. Local values needed for ranking are available as they are obtained during the indexing of the local collection. However, instead of using global document collection statistics which need to be retrieved from the network, the ranking layer uses local collection values. Ranking with local collection statistics (in particular document frequencies of key terms) produces the same postings order like when ranking with global statistics, but in addition saves bandwidth and improves performance. After producing the ordered set of postings for a key, top-*Dfmax* postings and inserted into the network.

As mentioned before, from the global view of the network, each peer is responsible for storing and maintaining a part of the global index. A peer stores a maximum of *Dfmax* postings associated with each key. As a peer continuously receives posting lists for a key, the cumulative number of postings may become greater than *Dfmax*. In this case, the ranking layer will need to rank both stored postings and newly received postings w.r.t. the key, and store only best ranked *Dfmax* postings. For ranking scores calculation, the ranking layer uses document-related statistics available in the global index and global statistics which are available at each peer (number of documents, average document length) as they are periodically requested from the network. Note that the global document frequency of a key is available locally at this peer only if the key is single-term. However, if the key consists of multiple terms, global document frequency of each key term needs to be retrieved from the peers which are responsible for maintaining them in the global index. After retrieving these document frequencies, the ranking layer ranks all documents, stores the top-Dfmax ranked documents, and discards the rest.

4.2 Ranking Layer Role in Querying Processing

The ranking layer produces the final ranked and ordered result set according to the relevance of a document w.r.t. a query Q. The following example illustrates the ranking function implemented in our prototype. First, it relies on a retrieval procedure to locate relevant documents. Second, the ranking of the retrieved documents is performed. Figure 2 illustrates the process of ranking a retrieved document set w.r.t. the query. Assume the query originator, $Peer_q$, produced a query Q consisting of terms t_1, t_2 and t_3. The HDK layer retrieved posting lists for keys $k_1 = \{t_1, t_2\}$ and $k_2 = t_3$ from the global index. Retrieved global index entries contain document-related statistics (term frequencies and document lengths), which are required for ranking. The ranking function also requires the global document frequency of each query term. For single term keys, the document frequencies are retrieved during the HDK search, while for multiple term keys they need to be separately retrieved. In our example, the global document frequency of key k_2 is retrieved during HDK retrieval, while global document frequency for terms t_1 and t_2 of key k_1 need to be additionally retrieved from peers responsible for storing these terms in the global index. Other global collection statistics such as the number of documents and average document length that are required for ranking are periodically retrieved from the network, so they are readily available at the querying peer.

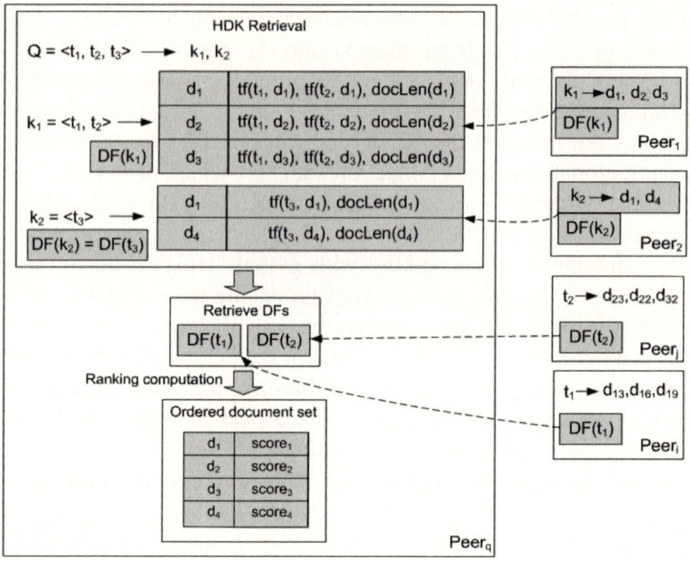

Fig. 2. Ranking a document set when answering a query

Having retrieved the required document frequencies from the global index, the querying peer is able to calculate the score of each retrieved document w.r.t. query, and produce a ranked and ordered set of retrieved documents. In order to present the retrieved and ranked documents to the user, document digests comprising a document title, snippet and URL, need to be retrieved from peers storing these documents. However, digests for only ten top-ranked documents will be retrieved, as the user is rarely interested in more than ten best-ranked documents. Besides, retrieving digests for all retrieved documents might result in contacting a very large number of peers, and would prove unscalable in terms of bandwidth consumption. If a user is interested in other documents besides the ten best, digests for these documents will be retrieved on demand, in steps of ten digests for ten documents currently viewed by the user.

5 Performance Analysis

Performance analysis of the ranking layer implemented in our prototype was performed in our lab environment. Each peer was running on a separate machine with the following characteristics: Intel Celeron, 2.66 GHz, 1024 MB of RAM. The document collection used in the setup was from the Reuters corpus, and each test included submitting 1000 queries from the Wikipedia query log were to the engine.

Figure 3 depicts the average distributed ranking time per query in a network of 10 peers and a growing document collection where each peer stored from 1000, up to 12000 documents. At first, the ranking time slightly increases since peers become loaded with the size of the index, but afterwards becomes constant with global collection growth.

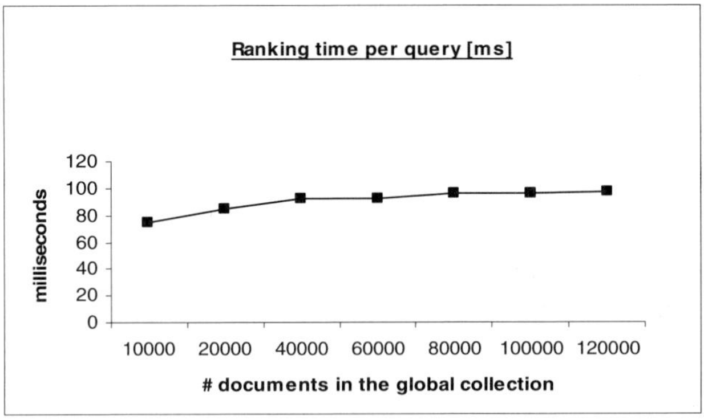

Fig. 3. Ranking time per query with a growing document collection

We experimentally measured the number of peers that need to be contacted by the querying peer in order to rank the resulting documents. Our network consisted of 4, 6, 8, and 10 peers, where each peer stored 5000 documents. The ranking layer needs to retrieve global document frequency for each term in the query (had they not been retrieved during the HDK search), and retrieve document digests for best ten documents.

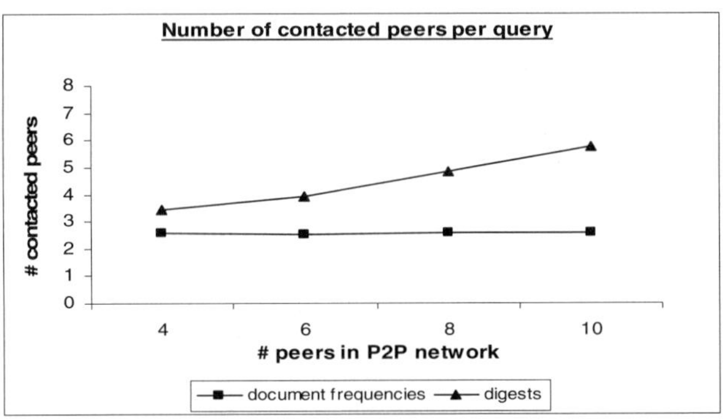

Fig. 4. Number of peers contacted by the ranking peer during a query

The results shown in figure 4 indicate that during a query the querying peer needs to contact in average 2,5 peers to retrieve global document frequencies, and that this number remains constant with network growth. This is reasonable, as this value depends mostly on the number of terms in a query, which is also an upper bound for this value. The results also indicate that the number of requests for document digests sent by the querying peer grows linearly when increasing the number of peers in the network. However, we only retrieve 10 document digests for best-ranked documents, so in the worst case scenario the number of contacted peers for digest retrieval is 10, no matter the network size. Experimental evaluation has shown that our ranking process is scalable and independent of the network size, as both ranking time and number of exchanged ranking messages between the peers when processing a query are bounded.

6 Conclusion

To satisfy user information needs in a P2P IR system, efficient ranking functionality is a necessity. Ranking implementation depends on the indexing strategy supported by the P2P search engine and the available collection statistic data. The Alvis P2P search engine maintains global collection statistics, and local document statistics that are needed for the computation of document ranking scores. With the available global statistics it is possible to use state-of-the-art ranking models that use both term dependant, and term independent statistic values statistic values for rank computation. The ranking mechanism implemented in Alvis search engine scales well in a growing P2P network in terms of ranking time per query and number of messages exchanged between the peers during the ranking process.

Future work will include the design of a link-based ranking module to additionally refine the content-based ranking scores. We also consider designing a community based ranking model to identify peers with similar interests and rank documents depending on the preferences of the community in which they are included.

References

[1] Baeza-Yates, R., Castillo, C., Junqueira, F., Plachouras, V., Silvestri, F.: Challenges in distributed information retrieval (invited paper). In: ICDE (2007)

[2] Yee, W.G., Beigbeder, M., Buntine, W.: SIGIR06 workshop report: Open Source Information Retrieval systems (OSIR06). SIGIR. Forum. 40(2), 61–65 (2006)

[3] Aberer, K., Alima, L.O., Ghodsi, A., Girdzijauskas, S., Haridi, S., Hauswirth, M.: The Essence of P2P: A Reference Architecture for Overlay Networks. In: Fifth IEEE International Conference on Peer-to-Peer Computing, pp. 11–20 (2005)

[4] Luu, T., Klemm, F., Podnar, I., Rajman, M., Aberer, K.: ALVIS Peers: A Scalable Full-text Peer-to-Peer Retrieval Engine. In: Workshop on Peer-to-Peer Information Retrieval (P2PIR 2006), ACM 15th Conference on Information and Knowledge Management Workshops, November 2006, pp. 41–48 (2006)

[5] Bender, M., Michel, S., Weikum, G., Zimmer, C.: The MINERVA Project: Database Selection in the Context of P2P Search. In: BTW 2005, Karlsruhe, Germany (2005)

[6] Suel, T., Mathur, C., Wu, J.-W., Zhang, J., Delis, A., Kharrazi, M.I., Long, X., Shanmugasundaram, K.: ODISSEA: A Peer-to-Peer Architecture for scalable Web Search and Information Retrieval. In: International Workshop on the Web and Databases (WebDB 2003), San Diego, California, USA (2003)

[7] Podnar, I., Rajman, M., Luu, T., Klemm, F., Aberer, K.: Beyond term indexing: A P2P framework for web information retrieval. Informatica 2(30), 153–161 (2006)

A New Algorithm for Community Identification in Linked Data

Nacim Fateh Chikhi, Bernard Rothenburger, and Nathalie Aussenac-Gilles

Institut de Recherche en Informatique de Toulouse
118, route de Narbonne
31062 Toulouse Cedex 4
{chikhi,rothenburger,aussenac}@irit.fr

Abstract. In this paper, we propose four specifications which can be used for the evaluation of community identification algorithms. Furthermore, a novel algorithm VHITS meeting the four established specifications is presented. Basically, VHITS is based on a two-step approach. In the first step, the Nonnegative Matrix Factorization is used to estimate the community memberships. In the second step, a voting scheme is employed to identify the hubs and authorities of each community. VHITS is then compared to the HITS and PHITS algorithms. Experimental results show that VHITS is more adapted than HITS and PHITS to the task of community identification in citation networks.

Keywords: authorities, hubs, authoritative documents, VHITS, HITS, PHITS, community identification, community mining, web communities.

1 Introduction

Since late nineties, identification of web communities has received much attention from researchers. HITS is a seminal algorithm in the community identification (CI) algorithms family. Since its invention, HITS has been followed by a multitude of CI algorithms. Some of them are just extensions of HITS, but some others use completely different approaches (like the graph based approaches) [1].

Unfortunately, the existence of a large variety of CI algorithms has caused a new problem, the problem of their evaluation and comparison. In fact, CI algorithms are usually evaluated by examining manually the extracted communities. Therefore, we propose four specifications which can be used for the evaluation of CI algorithms. Furthermore, a novel algorithm meeting the four established specifications is presented.

Although our algorithm may apply on different kinds of networks, we focus on the citation networks (or citation graphs). In a citation network, nodes correspond to web pages (resp. research papers), and edges represent hyperlinks (resp. bibliographic citations).

The rest of the paper is organized as follows. In section 2, we present the four proposed specifications for CI algorithms evaluation. These specifications are used in section 3 to analyze two popular CI algorithms. Section 4 describes our new CI algorithm. A case study in section 5 illustrates the behavior of three algorithms with

I. Lovrek, R.J. Howlett, and L.C. Jain (Eds.): KES 2008, Part I, LNAI 5177, pp. 641–649, 2008.

respect to some specifications. Experimental results are given in section 6 before concluding in section 7.

2 Requirements for Community Identification Algorithms

In this section, we propose the following requirements that a CI algorithm should meet. These specifications have been established after an analysis of existing CI algorithms.

- *Requirement 1*: Results of the CI algorithm should be as close as possible to the communities that may be identified by a human expert. This can be confirmed using conventional clustering assessment methods such as accuracy or F-measure.
- *Requirement 2*: CI algorithm should be able to identify the members of each community. The CI algorithm must be able to answer questions such as: "To which community does a document *d* belong?"
- *Requirement 3*: the hubs and authorities identified by the CI algorithm should have a straightforward interpretability.
- *Requirement 4*: CI algorithm should be able to handle overlapping communities.

3 A Critical Analysis of HITS and PHITS

In this study, we consider only a specific category of CI algorithms which deal with the identification of hubs and authorities. HITS [2] and PHITS [3] are two of the most influential algorithms in this field [1].

3.1 Hypertext Induced Topic Search

The HITS algorithm starts from a citation graph which is represented by an adjacency matrix **A**. A Singular Value Decomposition (SVD) is then performed on **A**, yielding three matrices such that: $\mathbf{A} = \mathbf{USV}$. In the HITS' terminology, matrix **U** is known as the hub matrix. It corresponds to the eigenvectors of the bibliographic coupling matrix $\mathbf{AA^T}$. Respectively, matrix **V** is called the authority matrix. It represents the eigenvectors of the co-citation matrix $\mathbf{A^TA}$.

The major drawback of the HITS algorithm is related to the interpretability issue of the discovered communities. More precisely, it is well-known that the dominant community found by HITS can be easily interpreted because, by definition, its left and right singular vectors contain only positive values. However, the interpretability problem arises when trying to interpret the non principal singular vectors since they contain both positive and negative values [3] [4]. To bypass this problem, Kleinberg suggests an empirical rule to identify the communities in such situations. His heuristic consists in manually examining the positive and negative parts of each hub or authority vector. In fact, this rule is based on the observation that the relevant communities are in some cases present in the positive part, and in other cases they are found in the negative part. Clearly, the Kleinberg's rule imposes a serious limitation since we cannot automate the community identification task.

By analyzing the HITS algorithm according to the requirements introduced in section 2, we can conclude the following:

- Interpretation of HITS's results is a tricky task (Req. 3).
- HITS is unable to extract overlapping communities (Req. 4) since computed components are orthogonal.
- Memberships of documents to communities cannot be obtained in a direct manner from matrices **U** and **V** (Req. 2). One has to use, subsequently, a clustering algorithm like K-means to get these memberships.

3.2 Probabilistic HITS

PHITS is a community identification algorithm based on the PLSA model [5]. PLSA is a latent variable model which was initially proposed for text analysis. More precisely, the PLSA's principle is that the relationship between documents and words can be explained by a small number of factors called topics. This model has been transposed by Cohn and Chang to the case of citation analysis by replacing words with citations.

Although PLSA has been successfully applied to text analysis [5] and was shown to be superior to the well-known Latent Semantic Analysis through many experiments, no comparative evaluation has been carried out to validate the performances of PHITS over other CI algorithms. A notable exception is [6] where authors compare classification accuracy of PHITS to PLSI (i.e. link versus content analysis) in many configurations. Authors report a significant superiority of PLSI over PHITS. PHITS' poor performances are actually due to its unsuitability for citation analysis.

Citation (bibliographic and web) data are particular and different from other data such as texts. Especially, citation data are characterized by their large *sparsity*. To illustrate this specificity, let's consider the Cora dataset used in Section 6. The dataset is composed of 3000 documents and 2500 unique words. While the total number of word occurrences is very large (170 000), the total number of links between documents is rather small (5 500). As it is well established in the discrete data analysis community, very sparse contingency tables poses many problems to techniques assuming a multinomial distribution of data and fitting models using the maximum likelihood estimation principle [7].

In section 5, we show that PHITS has poor clustering performance to find correct communities. Thus, we can claim that PHITS does not satisfy Req. 1.

4 Voting-Based HITS

The general approach used in VHITS is based on two steps:

Step A: Using an appropriate clustering algorithm, determine the members of each community. Here, "appropriate" means that the clustering algorithm should satisfy some requirements. These requirements include the ability to identify overlapping communities (Req. 4) and the efficiency in clustering citation data (Req. 1). In VHITS, we propose to use Nonnegative Matrix Factorization [8] which is a recent clustering algorithm meeting the aforementioned specifications.

Step B: Once the communities have been identified, VHITS determines hubs and authorities characterizing each community. A *voting scheme* is employed so that each community elects its hubs and authorities based on the degrees of membership.

4.1 Nonnegative Matrix Factorization

Nonnegative Matrix Factorization (NMF) is a recent statistical technique for matrix decomposition. Recently, NMF has been shown to be a powerful clustering technique. It has been explored in many data mining applications [9]. In text mining, for instance, it has been reported that NMF is able to elegantly extract the topics present in a collection of documents.

NMF is based on the idea that any positive matrix $\mathbf{M} \in \mathbb{R}_{+}^{p \times q}$ can be approximated by two positive matrices $\mathbf{F} \in \mathbb{R}_{+}^{p \times k}$ and $\mathbf{G} \in \mathbb{R}_{+}^{q \times k}$ such that: $\mathbf{M} \approx \mathbf{F}\mathbf{G}^{\mathrm{T}}$.

Basically, NMF is an optimization problem with positivity constraints. Many versions of NMF exist, but they all differ in their objective function [8]. Here, we propose to use the objective minimizing the Euclidean distance between the original matrix $\mathbf{M}$ and the approximation $\mathbf{F}\mathbf{G}^{\mathrm{T}}$. In [8], the authors propose two simple update rules to solve the Euclidean version of NMF. These rules are used in the first part of VHITS.

In our community identification task, NMF offers two important features. On the one hand, NMF clusters both rows and columns of the input data matrix. This characteristic is essential when identifying communities since rows (citing documents) and columns (cited documents) have different semantics. On the other hand, NMF is adapted to the analysis of overlapping clusters. This means that NMF allows a data point to belong to more than one cluster. In a nutshell, NMF is a soft clustering algorithm.

The complete VHITS algorithm is given in Table 1. The first part (steps 1-5) of the algorithm identifies the members of the communities according to inlinks and outlinks. Thus, each community can be regarded from two different points of view: one is relative to the citing documents and the other is relative to the cited documents.

In step 6 of VHITS, a normalization of communities' indicators is performed. This normalization makes the membership levels look like fuzzy values.

4.2 Voting Scheme

The voting scheme employed by VHITS aims at ordering the hubs (resp. authorities) of each community according to their importance. It is based on the idea that, for each community, a vote is held to choose the hubs and authorities of that community. This vote is organized according to the following rules:

- Each document has a *voting right VR* equal to one. However, this *VR* can be divided into portions in cases where the document belongs to more than one community. In such a situation, each community receives a *VR* part in proportion to the membership level of the document in that community. For example, if a document *D* belongs to communities *C1* and *C2* with membership degrees of 0.8 and 0.2 respectively, then, each member of *C1* which is linked by *D* will receive a *VR* of 0.8, and each member of *C2* which is linked by *D* will receive a *VR* of 0.2.

Table 1. VHITS algorithm

Algorithm: Voting-based HITS (VHITS).

Input: An adjacency matrix $\mathbf{D} \in \mathbb{R}^{N \times N}$ and the number of communities K.

Output: Authority matrix $\mathbf{A} \in \mathbb{R}^{N \times K}$, hub matrix $\mathbf{H} \in \mathbb{R}^{N \times K}$, membership matrices $\mathbf{I} \in \mathbb{R}^{N \times K}$ and $\mathbf{O} \in \mathbb{R}^{N \times K}$ according to inlinks and outlinks, respectively.

Steps:

1. Initialization: initialize $\mathbf{A}$ and $\mathbf{H}$ with random positive values, $t \leftarrow 0$;
2. Update membership value for every document in each community according to **inlinks**
   For i=1 to N and j=1 to K
   $$\mathbf{I}_{ij}^{(t+1)} = \mathbf{I}_{ij}^{(t)} \frac{(\mathbf{D}^{\mathrm{T}}\mathbf{O})_{ij}}{(\mathbf{I}\mathbf{O}^{\mathrm{T}}\mathbf{O})_{ij}};$$
3. Update membership value for every document in each community according to **outlinks**
   For i=1 to N and j=1 to K
   $$\mathbf{O}_{ij}^{(t+1)} = \mathbf{O}_{ij}^{(t)} \frac{(\mathbf{D}\mathbf{I})_{ij}}{(\mathbf{O}\mathbf{I}^{\mathrm{T}}\mathbf{I})_{ij}};$$
4. $t \leftarrow t+1$;
5. *If* a convergence criterion is not met *then* go to step 2;
6. Normalize rows of **I** and **O** to have unit L1 norm;
7. Identify authority candidates of each community
   $$\mathbf{AC}_{ij} = \begin{cases} 1 \text{ if } \mathbf{I}_{ij} > \mathbf{0} \\ 0 \text{ else} \end{cases}$$
8. Identify hub candidates of each community
   $$\mathbf{HC}_{ij} = \begin{cases} 1 \text{ if } \mathbf{O}_{ij} > \mathbf{0} \\ 0 \text{ else} \end{cases}$$
9. Compute authority score of every document in each community
   For i=1 to N and j=1 to K
   $$\mathbf{A}_{ij} = \begin{cases} \sum_{\mathrm{m=1}} \mathbf{D}_{\mathrm{mi}}\mathbf{O}_{\mathrm{mj}} & \textit{if } \mathbf{AC}_{ij} = 1 \\ 0 & \textit{else} \end{cases}$$
10. Compute hub score of every document in each community
   For i=1 to N and j=1 to K
   $$\mathbf{H}_{ij} = \begin{cases} \sum_{\mathrm{m=1}} \mathbf{D}_{\mathrm{im}}\mathbf{I}_{\mathrm{mj}} & \textit{if } \mathbf{HC}_{ij} = 1 \\ 0 & \textit{else} \end{cases}$$

- An *authority candidate* of a community *C* is any member of *C* according to inlinks. Here, being a member of a community according to inlinks means having a non null membership in the community's inlinks view.
- A *hub candidate* of a community *C* is any member of *C* according to outlinks. Here, being a member of a community according to outlinks means having a non null membership in the community's outlinks view.
- In a community *C*, *authority score* of an authority candidate *AC* is computed by counting the sum of the *VR*s received by *AC* from the members of *C* which link to *AC*.
- In a community *C*, *hub score* of a hub candidate *HC* is computed by counting the sum of the *VRs* received by *HC* from the members of *C* which are linked by *HC*.

5 Case Study

Let us consider the citation graph depicted in Figure 1. This graph is used to illustrate which of the four requirements established in section 2, are met by the three algorithms: HITS, PHITS and VHITS. A natural analysis of this citation graph would reveal three communities: one independent (community 1: nodes 1-2-3-4-5) and two overlapping (community 2: nodes 6-7-8-9-10 and community 3: nodes 8-11-12-13-14).

First, we apply HITS on the adjacency matrix corresponding to the graph of Figure 1. The obtained authority matrix (AHITS) is reported in Table 2.

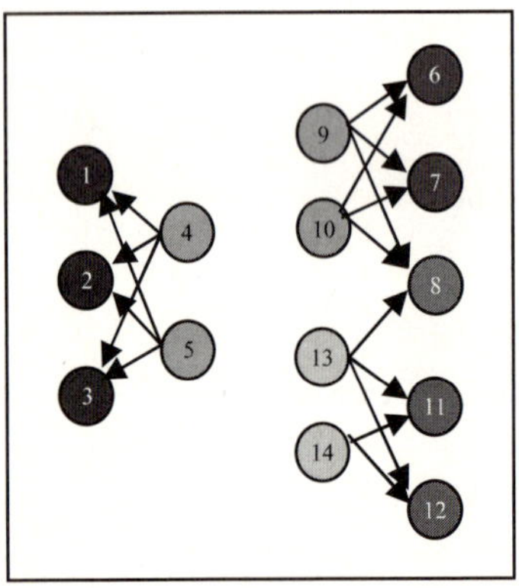

Fig. 1. An illustrative web graph

Table 2. Authority matrix returned by HITS

$\mathbf{A}_{HITS}$ =

0	0.577	0
0	0.577	0
0	0.577	0
0	0	0
0	0	0
0.481	0	-0.316
0.481	0	-0.316
0.650	0	0
0	0	0
0	0	0
0.240	0	0.633
0.240	0	0.633
0	0	0
0	0	0

From Table 2, we observe that the first component (i.e. the first column of $\mathbf{A}_{HITS}$) computed by HITS corresponds to a mix of community 2 and community 3. It is composed of nodes 6-7-8-11-12 as authorities. Apropos of this result, many studies have been carried out about the first component returned by HITS. It has been shown that HITS suffers from the Tightly Knit Community effect, which means that HITS returns in his first component the most dense structure in the graph [10]. In practice, however, this dense structure does not necessarily represent a correct community (which is the case in our example). The second component, containing only positive values, can be easily interpreted. The component corresponds exactly to the desired community 1. Now, considering the third component, it is not clear which part of the component one has to consider. The component is composed of both positive and negative values. Surprisingly, in this example, the two parts of the component make sense. The positive part (nodes 11-12) corresponds to a subset of the correct community 3 and the negative part (nodes 6-7) denotes a subset of community 2.

In this example, the HITS' results confirm its non adequacy to extract overlapping communities. The example has also elucidated its interpretability problem.

Due to space limitations, we report in Table 3 membership matrix according to inlinks and authority matrix returned by running VHITS on the graph of Figure 1. Moreover, results of PHITS, for this case study, are not reported because of their high similarity with those of VHITS.

Unlike HITS, VHITS successfully identifies the expected communities. Furthermore, the returned components by VHITS have a straightforward interpretability. For instance, the first membership component of VHITS (first column of $\mathbf{I}_{VHITS}$) corresponds exactly to community 3. It is also interesting to observe the ability of VHITS to localize overlapping communities. In matrix $\mathbf{I}_{VHITS}$, we note that node 8 is reported to belong to both community 2 and community 3.

Table 3. Membership matrix according to inlinks (I_{VHITS}) and authority matrix (A_{VHITS}) returned by VHITS

$$\mathbf{I}_{VHITS} = \begin{bmatrix} 0 & 0 & 1.000 \\ 0 & 0 & 1.000 \\ 0 & 0 & 1.000 \\ 0 & 0 & 0 \\ 0 & 0 & 0 \\ 0 & 1.000 & 0 \\ 0 & 1.000 & 0 \\ 0.322 & 0.678 & 0 \\ 0 & 0 & 0 \\ 0 & 0 & 0 \\ 1.000 & 0 & 0 \\ 1.000 & 0 & 0 \\ 0 & 0 & 0 \\ 0 & 0 & 0 \end{bmatrix} \qquad \mathbf{A}_{VHITS} = \begin{bmatrix} 0 & 0 & 2.000 \\ 0 & 0 & 2.000 \\ 0 & 0 & 2.000 \\ 0 & 0 & 0 \\ 0 & 0 & 0 \\ 0 & 2.000 & 0 \\ 0 & 2.000 & 0 \\ 0.842 & 2.158 & 0 \\ 0 & 0 & 0 \\ 0 & 0 & 0 \\ 1.842 & 0 & 0 \\ 1.842 & 0 & 0 \\ 0 & 0 & 0 \\ 0 & 0 & 0 \end{bmatrix}$$

Now, let's consider the authorities computed by VHITS for each community. For community 2 (corresponding to the second column of $\mathbf{A}_{VHITS}$), VHITS finds that node 8 is the most authoritative in that community. This result illustrates the difference between the authority and the membership concepts. Actually, according to column 2 of matrix $\mathbf{I}_{VHITS}$, nodes 6 and 7 are more likely than node 8 to be members of community 2. However, matrix $\mathbf{A}_{VHITS}$ indicates that node 8 is more authoritative than nodes 6 and 7 in that community.

6 Experimental Results

We report experimental results carried out to assess the clustering quality of HITS, PHITS and VHITS. We have used two pre-classified datasets. The first one is a subset of the well-known WebKb dataset [11]. This subset is a collection of 3100 web pages, where each page belongs to one of four predefined classes. The second one is the Cora dataset [12] which is composed of 3000 scientific papers. Each paper was manually classified into one of seven categories.

The clustering output of each CI algorithm is evaluated using the traditional *accuracy* and the *normalized mutual information*. Obtained results are presented in Figures 2 and 3. We observe from Figures 2 and 3 that the PHITS' results are poor comparatively to those of HITS and VHITS. This observation is even more emphasized by the normalized mutual information measure, which is a non biased clustering assessment measure. Results show also that HITS and VHITS have almost the same performances with a slight advance for VHITS on the WebKb dataset.

Let's notice also that, for the WebKb dataset, inlinks seem to have more importance than outlinks. However, the opposite phenomenon is observed with the Cora dataset. This observation is in accordance with previously reported studies about the importance of the inlinks for web page classification, and the usefulness of outlinks for bibliographic data classification.

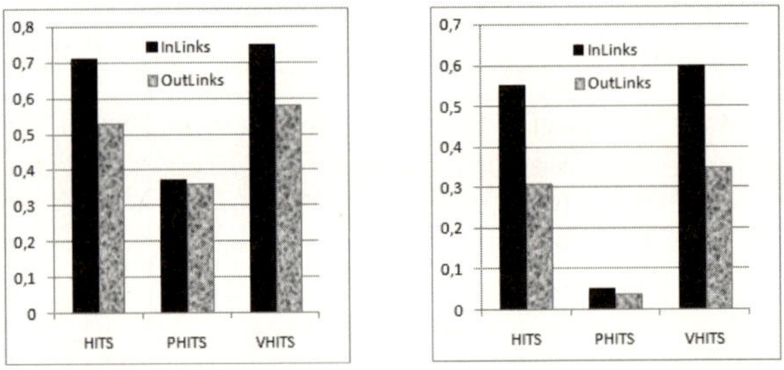

Fig. 2. Accuracy (left) and NMI (right) on WebKb

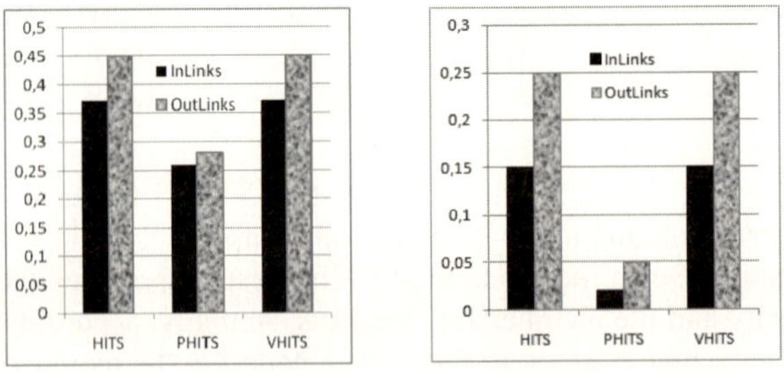

Fig. 3. Accuracy (left) and NMI (right) on Cora

7 Conclusion

In this paper, two contributions have been presented. On the one hand, four specifications have been proposed for the evaluation of community identification algorithms. These four requirements include: the clustering quality, the ability to assign memberships to the documents, the interpretability of identified hubs and authorities, and the ability to deal with overlapping communities. We believe that using such specifications can play an important role in the construction of new community identification algorithms.

On the other hand, an original algorithm namely, VHITS, has been described. VHITS is based on two techniques: Nonnegative Matrix Factorization and a voting scheme. Unlike HITS and PHITS, VHITS satisfied the four specifications.

Our further investigations include the use of additional sources of information in VHITS such as content information [13], author information or anchor text information.

References

1. Zhang, Y., Xu Yu, J., Hou, J.: Web communities: Analysis and construction. Springer, Heidelberg (2006)
2. Kleinberg, J.M.: Authoritative sources in a hyperlinked environment. Journal of the ACM 46(5), 604–632 (1999)
3. Cohn, D., Chang, H.: Learning to probabilistically identify authoritative documents. In: 17th International Conference on Machine Learning, pp.167–174 (2000)
4. Chikhi, N.F., Rothenburger, B., Aussenac-Gilles, N.: A comparison of dimensionality reduction techniques for web structure mining. In: IEEE/WIC/ACM International Conference on Web Intelligence, pp. 116–119 (2007)
5. Hofmann, T.: Probabilistic latent semantic analysis. In: 15th UAI Conference (1999)
6. Fisher, M., Everson, R.: When Are Links Useful? Experiments in Text Classification. In: Sebastiani, F. (ed.) ECIR 2003. LNCS, vol. 2633, pp. 41–56. Springer, Heidelberg (2003)
7. Agresti, A.: An Introduction to Categorical Data Analysis, 2nd edn. Wiley, Chichester (2007)
8. Lee, D., Seung, H.: Algorithms for non-negative matrix factorization. In: Neural Information Processing Systems, pp. 556–562 (2000)
9. Chu, M.: Data mining and applied linear algebra. In: International Conference on Informatics Education and Research for Knowledge-Circulating Society, pp. 20–25 (2008)
10. Lempel, R., Moran, S.: The stochastic approach for link-structure analysis (SALSA) and the TKC effect. Computer Networks 33(1-6), 387–401 (2000)
11. WebKB, `http://www.cs.cmu.edu/~webkb/`
12. McCallum, A., Nigam, K., Rennie, J., Seymore, K.: Automating the construction of internet portals with machine learning. Information Retrieval Journal 3, 127–163 (2000)
13. Zhu, D., Yu, K., Chi, Y., Gong, Y.: Combining content and link for classification using matrix factorization. In: 30th Annual Intl. ACM SIGIR Conference, pp. 487–494 (2007)

GrasSmart: An Intelligent Robotic System for Continuous Area Coverage

Miri Weiss-Cohen, Igal Sirotin, and Erez Rave

ORT Braude College, Karmiel, Israel
miri@braude.ac.il

Abstract. This work implements and simulates the problem known as, coverage of a continuous planar area by a mobile robot. The robot is given a bitmap of a known geometric area as an input, and derives an optimal path of coverage by implementing and improving the On-line Full Scan Spanning Tree Covering (STC) algorithm[1]. The path is calculated for the continuous area coverage by defining a DFS (Depth First Search) spanning tree. We improved the STC algorithm by optimizing U-Turns of the path, and optimizing the shifting of the path directions, moreover, we suggest using different sensor information, which reduces errors. The results of our work are presented by a 3D simulation program which mimics the grass robots' path and statistical calculations for testing optimality.

Keywords: Area coverage, spanning tree, robot motion.

1 Introduction

The coverage problem has been defined [2] as the maximization of the total area covered by a robot's motion. The static coverage problem is addressed as designed to deploy a robot in a static configuration, such that every point in the environment is known to the robot, and is analyzed to be covered [5,6]. Many tasks in the real world involve area coverage, such as, mapping and validation of topological maps [10]. Some tasks are possible only by non-human devices such as automated minesweeping which reduces human losses. As automated devices evolved, simpler and home-use tasks such as vacuum cleaning [3,4], snow removal, painting, grass-mowing[9], milling and pool-cleaning, are possible to be done automatically. In these cases, a robot is given a bounded work-area which, in many cases contain obstacles. The area coverage problem may be looked at as a geometric version of the Covering Salesman Problem [7]. Choset [2] surveys results in coverage path planning and organizes the coverage algorithms into four categories: heuristic, approximate, partial-approximate and exact cellular decompositions[1,8]. A recent work by Batalin et al [11,], presents an algorithm named Least Recently Visited (LRV), which solves the problems of coverage, by deploying a sensor network using a robot which carries a network of nodes as a payload, and emplaces the nodes into the environment based on certain local criteria. In this work we use the cell-decomposition approach [2] and improve some important features and measurement found in [1].

I. Lovrek, R.J. Howlett, and L.C. Jain (Eds.): KES 2008, Part I, LNAI 5177, pp. 650–657, 2008.

2 The Improved Full STC Coverage Algorithm

In this paper we use algorithms based on area coverage that uses a boundary of the work area known as off-line coverage [1]. We define the tool to be a square of size *D*. The work area is then approximately decomposed into cells, with each cell being a square of size *4D*. As with other approximate cell-decomposition approaches [2], cells that are partially covered by obstacles or outside of the bounds of the work area are discarded from consideration.

Our system, which we call GrasSmart, improves and simulates a spanning-tree coverage algorithm [1] to extract a path that visits all sub-cells. Previous work on generating such a path (called STC for Spanning-Tree Coverage) has shown it to be complete and non-backtracking. This work combines the off-line bitmap work area with on line Scan STC algorithm.

We analyzed the run-time, (which is linear to the size of the area), and examined the robustness and efficiency of the *Scan STC* algorithm [1]. It was improved by optimizing U-turns along the path and optimizing shifting of the path directions. Our method uses different sensor-acquired information that reduces the sensor errors (detailed later), and furthermore, eliminates accumulation of errors through robot motion.

The complete on-line *Scan* STC algorithm can be found in [1] which incrementally constructs a spanning tree for the free and partially occupied cells. The full algorithm inspects only sub-cells which are relevant to the robot. In Fig. 1 this stage is depicted. Let *p* be the point where the cells *x*, *y*, *y* + 45 and *y* 90 meet. Then the full algorithm inspects the four sub-cells surrounding *p*. These sub-cells are denoted *xp*, *yp*, *(y+45)p*, *(y+90)p*.

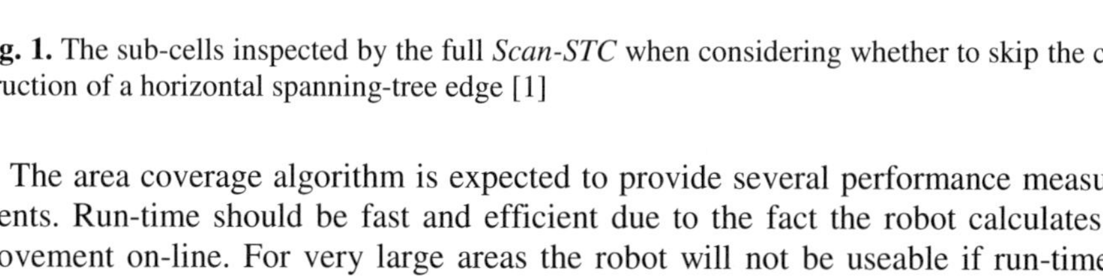

Fig. 1. The sub-cells inspected by the full *Scan-STC* when considering whether to skip the construction of a horizontal spanning-tree edge [1]

The area coverage algorithm is expected to provide several performance measurements. Run-time should be fast and efficient due to the fact the robot calculates its movement on-line. For very large areas the robot will not be useable if run-time is slower than linear. The over-cover measurement is the relative size of area that was covered more than once by the robot. This measurement should be as low as possible. As the over-cover measurement grows, the overall time of the coverage becomes longer, which increases the cost of using the robot (money, energy, time and resources). Turns measurement is the number of turns made by the robot along its movement. The number of turns should also be as low as possible for the same reasons as mentioned above.

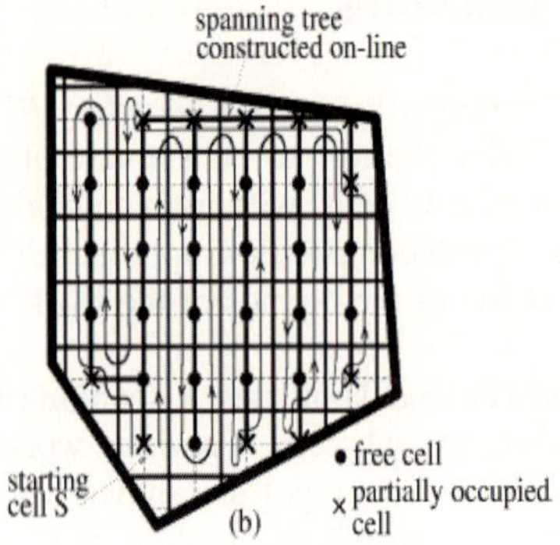

Fig. 2. Grid approximation of a given work-area and the spanning tree [1]

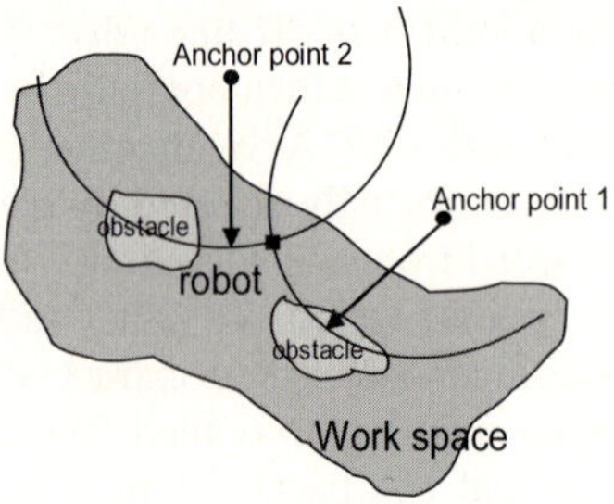

Fig. 3. Determining the location of the robot

In [1] the robot must know its exact location at the beginning of the run and calculates its new location as it moves from its movement history. This calculation is very problematic in practice due to error accumulation. The robot cannot accurately calculate its location at every moment of movement, because of the accumulated errors of the sensors and measuring devises. Without human assistance the robot not only cannot be calibrated to the original path, it cannot calculate or define the accumulated errors in its movement. A possible solution to this problem is for the robot to use a GPS device; however, this solution is not accurate enough under a cost limitation.

The GrasSmart system demonstrates the use of another positioning service. By allocating two poles, as shown in Fig. 3 (the simulation works on 2D space), which sends RF signals in a pre-defined frequency, the robot's sensors can measure its distance from each one of the poles. Assuming the poles are in a constant static, known location, the robot can triangulate and calculate its location based on the distance measurements. In most cases, only three poles are needed to determine an accurate location of a point. In cases where the anchor poles are analyzed, for robustness location, two poles are sufficient to determine the location of the robot. When the two poles are relatively close to one another, relative to the size of the work area, the circles constructed from R1 and R2 will have one intersection point relevant for the robot's location. This method for determining the location of the robot is not errorless, but the errors are not accumulated and each measurement is not dependent on a prior one, hence resulting in a local error in the worst case.

Our algorithm reduces the over-cover measurement locally, and consequently, the overall run-time. The improvements are local corrections for common scenarios that the Scan STC algorithm does not resolve. We depict the two major scenarios and demonstrate how the GrasSmart overcomes the problem:

2.1 U-Turn Optimization

In the progress of building the DFS (Depth Fist Search) graph, some optimization issues were discovered to be very meaningful and created better results for almost every coverage area problem which was analyzed.

When the robot moves through the cells for the first time, as shown in Fig. 4, it moves along line 1, in the upper line of cells. The blue line is a bypass of the non-coverable cells in the line above. When the robot returns from the DFS vertex to the previous position, it moves along line 2.

By moving along line 3, the robot can save two steps. It is important to note that this step-saving scenario occurs more than once in an area. Our improvement_to the STC [1] can be seen in the process of returning from a cell to its predecessor. The robot has to search for the 'next uncovered cell' and find the shortest path to it. This path can be found using BFS (Breadth First Search) on the cells. By implementing this optimization, the robot would move along line 3.

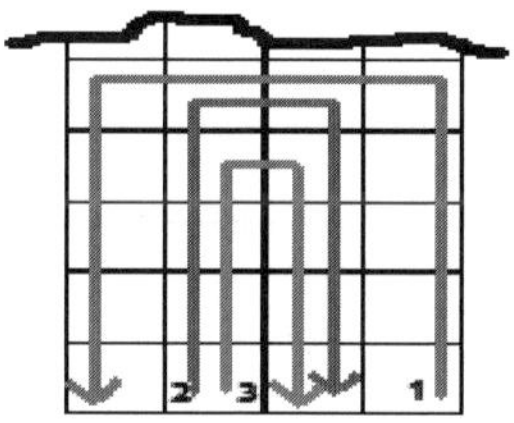

Fig. 4. U- Turn optimization

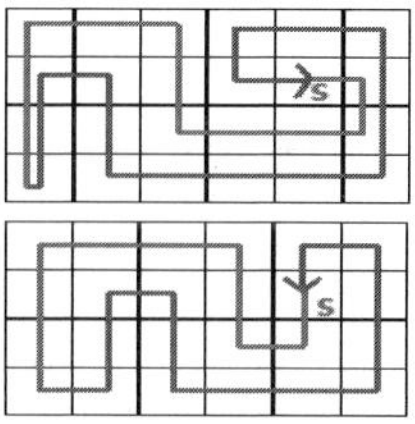

Fig. 5. Shift optimization

2.2 Shift Optimization

In the scenario described in Fig. 5, the area is identical as in Fig. 4. In the current implementation of the system, cell division is done automatically by the area's position in the bitmap file. As seen in the top half of the figure, the cells could be divided in a way that causes non-complete double cells on both sides of the work area (low edge-completeness). If this happens, the robot will cover five cells twice. This extra coverage, however, can be prevented by shifting the double-cell division one cell east or west. As shown in the bottom half of the figure, the double-cells are shifted to the east or to the west, with the result that no cells are over-covered.

In a bigger area, we need to know the number of similar scenarios, with or without shifting the double-cells division east-west or north-south. They are counted easily by measuring the edge-completeness of the four shifting options and selected by the maximal value.

The GrasSmart system is designed and implemented using a component model consisting of the algorithm model, which implements the improved STC [1] algorithm, and results a robust simulating system. The Image Processing Module processes bitmap files into area coverage map. The Simulation Module is used for presenting the algorithm results during run-time and for evaluating all performance measurements and statistics. The system supports different algorithm implementations with no changes and different user interfaces (or physical devices) with minimal changes. Three components were created: a "common" component, an "algorithm" component and a "main" (GUI) component. The "common" component is referenced by all other components. The algorithm component is in charge of all aspects of the area coverage tasks: definition of the area, analysis of its shape, execution and results.

The "main" component provides the user with the ability to load different lawn shapes and scenarios, view the progress of the mower at each step of its movement and displays the area covering progress on an on-line graphic 3D simulator.

3 Results and Examples

The result parameters, which have been determined and by which we analyzed our system, are:

* Over covers – Relative number of cells that were covered more than once
* Turns – Number of turns made by the robot
* Edge Completeness – Relative number of cells that are fully coverable.

Table 1. Some examples and their relevant parameter values

Input area	Results	Conclusions
	Over covers: 11.79% Turns: 119 Edge completeness: 74.53	On a non-regular shape, the algorithm provides very good results; shows that the number of steps does not depend on the shape of the area.
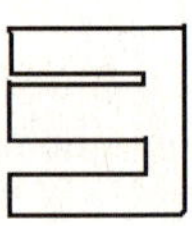	Over covers: 14.37% Turns: 165 Edge completeness: 63.19%	On a shape with straight lines, the number of turns depends on the lines' direction; the run shows that the number of turns is relatively high.
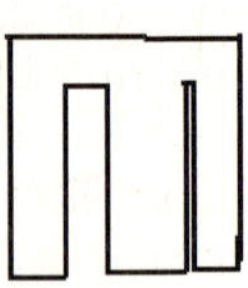	Over covers: 16.50% Turns: 75 Edge completeness: 67.80%	On the same shape as in the previous test, the run shows that with straight lines going north-south, turns are reduced to less than half.
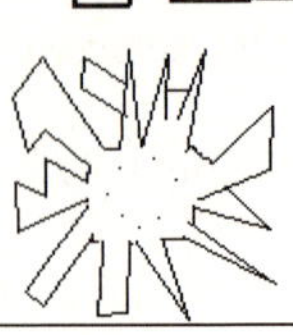	Over covers: 41.53% Turns: 271 Edge completeness: 39.71%	On a non-realistic area, the run shows that the relative number of re-covered cells and the number of turns are extremely high.

Over covers. The system demonstrates that the estimation that the number of cells being covered more than once will not exceed 17% of the total number of cells (with realistic inputs). Some results confirm this estimation and some cases show that the relative number of cells that are covered more than once can be extremely high when dealing with non-realistic area shapes. The most important issue that can be learned from the over covers result is that the number of over covers does not depend on the shape of the area at all but strongly depends on the relative size of the 'clean' area (double-cells that are fully coverable).

Turns. The result counts the number of turns made by the robot while covering the area. Viewing the results, it is possible to see that the number of turns depends on several factors. The most significant factor is the direction of the area. The robot moves in long straight lines. If work area runs north-south, the robot can move along the long straight lines from the beginning to the end without turning. If the direction of the area is east-west, the lines will be broken by the robot and more turns will be made.

Edge Completeness. For many possible implementations and uses of the algorithm it is very important to estimate the number of over-covered cells (e.g. mine sweeping in enemy territory, deep-sea search etc.). In order to estimate the number of over-covered cells before running the algorithm and the time and energy required to complete a task, the number of over-covered cells must be determined. Results were recorded on several system runs and showed that there is a linear dependency between the number of double-cells that are fully coverable (Edge Completeness) and the number of over-covered cells.. Table 1 summarizes some examples.

Example 1. Fig. 6 depicts the stages of the algorithm running with a non-regular shape. The results of the simulation are described in Table 1, The shape in this figure has the same work area as the first example as depicted in Table 1, but contains two obstacles. We can see the bitmap, the spanning tree and the motion and location of the robot.

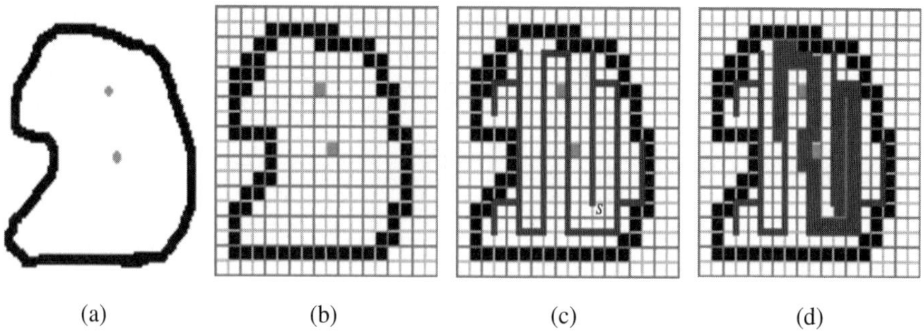

Fig. 6. (a) Input (bitmap file). (b) Image recognition (division to cells and sub-cells). (c) DFS graph with priorities (FULL- STC priorities allocation). The DFS graph is built while the robot moves through the area. (d) Movement on one side of the graph while bypassing obstacles on the other side of the graph.

Fig. 7 is the result of the simulation system. Cells that have been visited once are in pale green, and the ones that have been visited twice are in ocher green. To measure the number of double-cells that are fully coverable, the number of obstacles and the edges' delimiter length need to be estimated. Edge completeness defines the relative number of cells that are fully coverable. The full results of the statistical measurements are shown in Fig. 8.

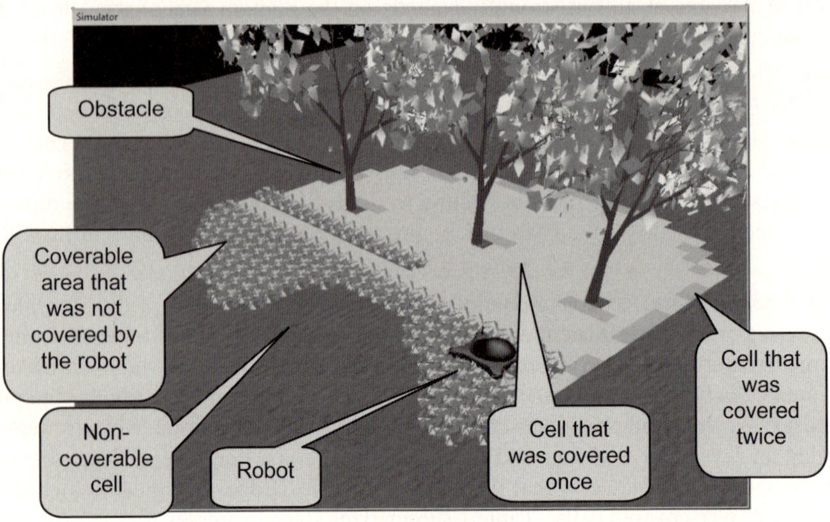

Fig. 7. Example for the non-regular shape with details of the 3D graphics

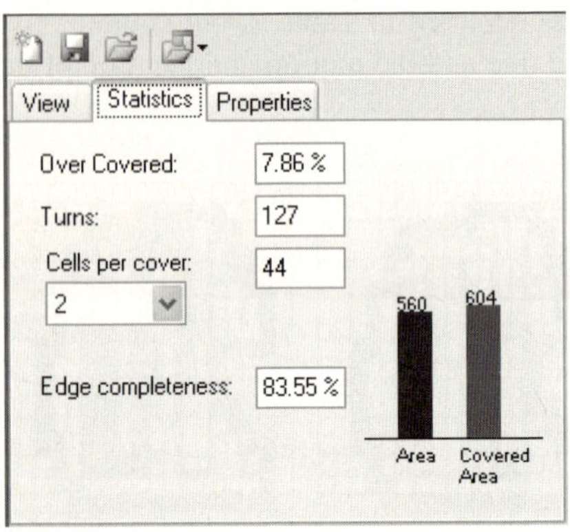

Fig. 8. Statistical measurements

Fig. 9. Final coverage mapping the area

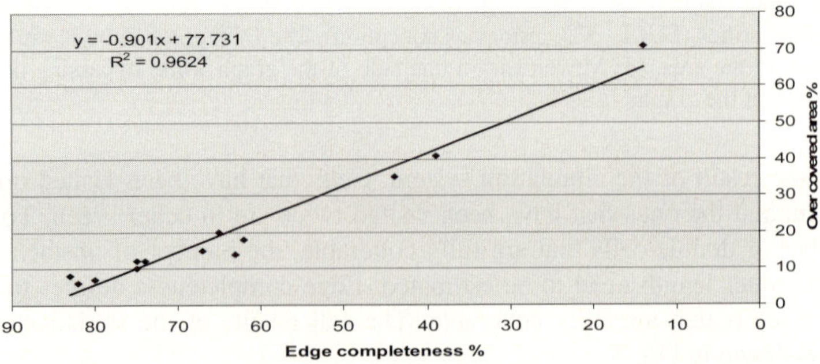

Fig. 10. Correlation between edge completeness and over covered area

Fig. 10 shows the linear connection between two values: edge-completeness and over-covered area. The scatter of the points on the graph indicates that in most of the cases, the edge-completeness value ranges from 60% to 85%. The slope of the linear dependency between edge completeness and over-covers is almost 1, which indicates that the dependency is strong.

4 Conclusions

In this work we improved the STC algorithm in which a mobile robot is given a bit-map of a known geometric area as an input and derives an optimal path of coverage. The results of our work are presented by a 3D simulation program that mimics the grass cutting robot's path and provides statistical calculations for testing optimality. Run-time results show that there is a linear correlation between algorithm optimality in terms of area over-coverage and edge-completeness in terms of the relative number of cells in the coverage area perimeter.

References

1. Gabriely, Y., Rimon, E.: Competitive On-line Coverage of Grid Environments by a Mobile Robot. Computational Geometry 24, 197–224 (2003)
2. Choset, H.: Coverage for Robotics. A Survey of Recent Results. Ann. Math. and AI. 31, 113–126 (2001)
3. Ulrich, I.R., Mondada, F., Nicoud, J.D.: Autonomous Vacuum Cleaner. Robotics and Autonomous Systems 19(3–4), 233–245 (1997)
4. Wong, S., Coghill, G., MacDonald, B.: Landmark-based World Model for Autonomous Vacuuming Robots. In: Proceedings of the International ICSC Congress on Intelligent Systems and Applications (ISA), Wollongong, Australia (2002)
5. Butler, Z.J., Rizzi, A.A., Hollis, R.L.: Contact Sensor-based Coverage of Rectilinear Environments. In: Proceedings of the IEEE International Symposium on Intelligent Control/Intelligent Systems and Semiotics, pp. 266–271 (1999)
6. Acar, E., Choset, H.: Critical Point Sensing in Unknown Environments. In: IEEE International Conference on Robotics and Automation, San Francisco, CA (2000)
7. Arkin, E.M., Refael, H.: Approximation Algorithms for the Geometric Covering Salesman Problem. Discrete Appl. Math. 55, 197–218 (1994)
8. Choset, H., Acar, E., Rizzi, A., Luntz, J.: Exact Cellular Decompositions in Terms of Critical Points of Morse Functions. In: IEEE International Conference on Robotics and Automation, San Francisco, CA (2000)
9. Arkin, E.M., Fekete, S.P., Mitchell, J.S.: The Lawnmower Problem. In: Proceedings of the 5th Canadian Conference on Computational Geometry, Waterloo, Canada, pp. 461–466 (1993)
10. Dudek, G., Jenkin, M., Milios, E., Wilkes, D.: Map Validation and Robot Self-location in a Graph-like World. Robotics Autonom. Syst. 22, 159–178 (1997)
11. Batalin, M., Sukhatme, G.S.: The Design and Analysis of an Efficient Local Algorithm for Coverage and Exploration Based on Sensor Network Deployment. IEEE Transactions on Robotics 23(4), 661–675 (2007)

Stator Resistance Tuning Based on a Neural Network in an Indirect Rotor Field Oriented Control System of an Induction Motor

Dinko Vukadinovic, Mateo Basic, and Ljubomir Kulisic

Faculty of Electrical Engineering Mechanical Engineering and Naval Architecture,
R. Boskovica bb, 21000 Split, Croatia
dvukad@fesb.hr, mateo.basic@gmail.com, ljubomir.kulisic@fesb.hr

Abstract. This paper presents an ANN-based (artificial neural network-based) method of stator resistance tuning in an IRFO (indirect rotor field oriented) control system of an induction motor. This method is based on the conventional two-layer ANN in which the rotor time constant is not a constant parameter and is identified using a model reference adaptive system (MRAS) - based procedure. During the training, rotor speed estimation of the induction motor is enabled. The difference between the actual and the estimated rotor speed is used as a signal for manual stator resistance tuning. Computer simulations and experimental results show the effectiveness of the described approach in a low rotor speed region.

Keywords: induction motor, indirect field-oriented control, neural network, adaptive control.

1 Introduction

An induction motor is the most commonly used electric motor in modern electric drives (e. g. an indirect rotor field oriented (IRFO) control system). In addition, induction motor control systems are known to be extremely non-linear control systems because of induction motor parameter variability under different conditions. Heating of motor windings depends on stator and rotor currents leading to variability of stator and rotor resistances. The accuracy and control quality of the IRFO control system is greatly influenced by the value of rotor resistance R_r used for control. In the past, several methods have been developed for rotor resistance identification. A brief review of the methods for rotor resistance estimation is in [11] and includes the following classification:

1. Spectral analysis techniques ([1])
2. Observer-based techniques ([4, 12, 5])
3. Model reference adaptive system-based techniques ([8])
4. Other methods

The other methods are based on the neural networks or fuzzy logic schemes. In recent years, the use of artificial neural networks (ANNs) in ac drives has been proposed ([5], [2]). Supervised learning methods, where the neural network is trained to learn the input/output pattern presented to it, are typically used ([6]). Two-layer

I. Lovrek, R.J. Howlett, and L.C. Jain (Eds.): KES 2008, Part I, LNAI 5177, pp. 658–665, 2008.

neural networks, where the training step is not required, are, therefore, preferable ([9], [10]). That type of ANN is utilized in this paper.

This paper deals with the IRFO system including inverse rotor time constant identification and stator resistance adjustment. In this paper, the rotor flux space vector is estimated using four different types of induction motor models (so called voltage and current models), and each is described in the stationary reference frame (α,β) or in an synchronously rotating reference frame (d,q). First, the rotor flux magnitude is estimated in the stationary reference frame by the voltage model (reference model) and by the current model (adaptive model). The error signal of the rotor flux magnitude of the two estimators is applied to drive a PI mechanism which provides correction of the inverse rotor time constant. Compared with the method described in [8], the stator resistance is not a constant parameter, but is identified by the two-layer neural network as described hereafter. Second, the rotor flux space vector is estimated in the d,q reference frame by the voltage model (reference model) and by the two-layer ANN model (adaptive model). The errors between rotor flux components are applied to drive a PI estimator which provides rotor speed estimation. The weights dependent on the inverse rotor time constant are tuned based on the inverse rotor time constant identification as described above. The weights dependent on the rotor speed are tuned based upon estimated rotor speed. Any mismatch between actual and estimated rotor speed will result from inaccurate stator resistance. Stator resistance is manually adopted in order to achieve zero error between the actual and estimated rotor speed.

2 IRFO Control System

Fig. 1 shows the IRFO control system of induction motor including both inverse rotor time constant identification and stator resistance identification.

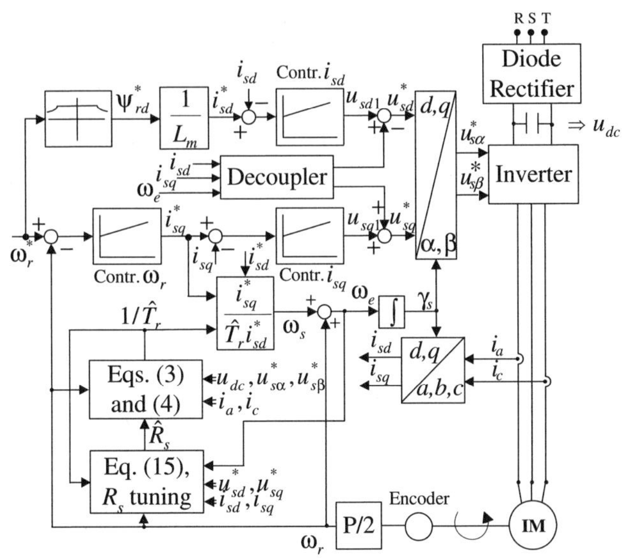

Fig. 1. ANN-based IRFO control system

An induction motor can be described by the following equations in a synchronously rotating (d,q) reference frame [7]:

$$0 = \frac{1}{T_r}\boldsymbol{\psi}_r - \frac{L_m}{T_r}\mathbf{i}_s + s\boldsymbol{\psi}_r + j(\omega_e - \omega_r)\boldsymbol{\psi}_r \tag{1}$$

$$\mathbf{u}_s = (R_s + (s + j\omega_e)\sigma L_s)\mathbf{i}_s + (s + j\omega_e)\frac{L_m}{L_r}\boldsymbol{\psi}_r . \tag{2}$$

where T_r is the rotor time constant, and s is the Laplace operator (=*d/dt*).

In equations (1) and (2) the space vectors are denoted in the bold face. Equation (1) is well known as the current model and (2) as the voltage model of the induction machine.

3 Identification of Inverse Rotor Time Constant

This paper utilizes the identification of inverse rotor time constant described in [8]. Equations (1) and (2) described in the α,β reference frame (ω_e=0) become

$$0 = \frac{1}{\hat{T}_r}\boldsymbol{\psi}_r - \frac{L_m}{\hat{T}_r}\mathbf{i}_s + s\boldsymbol{\psi}_r - j\omega_r\boldsymbol{\psi}_r \tag{3}$$

$$\mathbf{u}_s = (\hat{R}_s + s\sigma L_s)\mathbf{i}_s + s\frac{L_m}{L_r}\boldsymbol{\psi}_r . \tag{4}$$

A hat above a symbol in (3) and (4) denotes estimated parameters. Equation (3) (adaptive model) gives an estimation of the rotor flux space vector based upon easily measured stator currents and rotor speed. This estimation mainly depends on the accuracy of the inverse rotor time constant identification. Equation (4) is independent of the inverse rotor time constant and, accordingly, can be used as the reference model of the rotor flux space vector. An adaptive mechanism (PI) provides correction of the inverse rotor time constant (Fig. 2.).

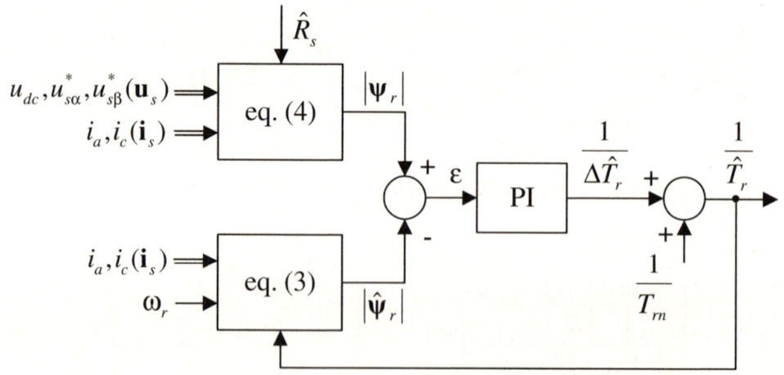

Fig. 2. Inverse rotor time constant identification

In our case, the inverse rotor time constant identification is valid beyond a rotor speed that is approximately 15 % of the rated speed.

4 Stator Resistance Tuning Based on ANN

The MRAS theory, as described in the section 1, has been utilized in order to estimate the rotor speed of induction motor. The rotor flux space vector is estimated in the d,q reference frame by the voltage model (reference model) and by the ANN-based model (adaptive model) of the induction motor. The difference between flux space vectors estimated using the two ways is then used in an adaptation mechanism that outputs the estimated value of the rotor speed and adjusts the adaptive model until good performances are obtained.

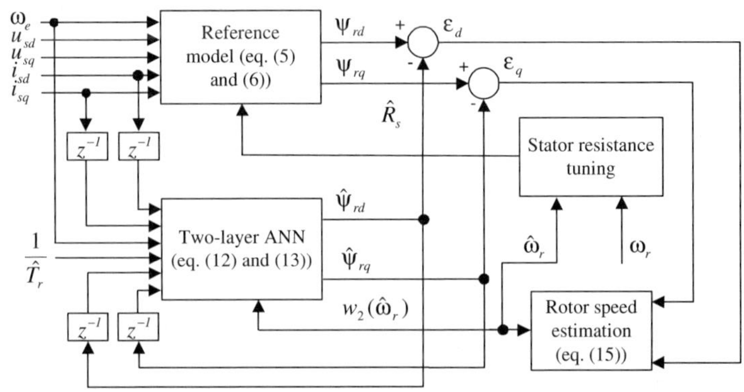

Fig. 3. Stator resistance tuning based on MRAS theory and ANN

The inputs to the reference model are the d- and q- axis stator voltages and currents of the induction motor and the angular stator frequency ω_e. The outputs of the reference model are the components of the rotor flux space vector in the d,q reference frame, which can be obtained from equation (2) as follows:

$$\frac{d\psi_{rd}}{dt} = \frac{L_r}{L_m}\left\{u_{sd} - \hat{R}_s i_{sd} - \sigma L_s \frac{di_{sd}}{dt} + \omega_e \sigma L_s i_{sq}\right\} + \omega_e \psi_{rq} \tag{5}$$

$$\frac{d\psi_{rq}}{dt} = \frac{L_r}{L_m}\left\{u_{sq} - \hat{R}_s i_{sq} - \sigma L_s \frac{di_{sq}}{dt} - \omega_e \sigma L_s i_{sd}\right\} - \omega_e \psi_{rd} \, . \tag{6}$$

These equations do not contain the rotor speed. However, equation (1) contains the rotor flux space vector and the rotor speed as well. This is the equation of the adaptive model. Rewriting (1) yields

$$\frac{d\hat{\psi}_{rd}}{dt} = \frac{1}{\hat{T}_r}\left(L_m i_{sd} - \hat{\psi}_{rd}\right) + \left(\omega_e - \omega_r\right)\hat{\psi}_{rq} \tag{7}$$

$$\frac{d\hat{\psi}_{rq}}{dt} = \frac{1}{\hat{T}_r}\left(L_m i_{sd} - \hat{\psi}_{rq}\right) - \left(\omega_e - \omega_r\right)\hat{\psi}_{rd} \,. \tag{8}$$

Equations (7) and (8) contain the rotor speed which, in general, is changing, and the intent is to estimate this speed by using an ANN. Consequently, equations (7) and (8) can be implemented by a two-layer ANN, which contains variable weights proportional to the rotor speed.

When there is no mismatch between the actual and estimated parameters of the induction motor, then the errors ε_d and ε_q (Fig. 3.) are zero in the steady state. In this case, the estimated rotor speed must be the same as the speed estimated by the ANN. The difference between the actual and the estimated rotor speed can be caused due to the following two reasons:

a) incorrect rotor resistance identification (incorrect inverse rotor time constant), and

b) incorrect stator resistance identification.

When the stator resistance is incorrectly identified, then the inverse rotor time constant is incorrectly identified as well. As a result, there is a mismatch between the actual rotor speed and the estimated rotor speed in the steady state. To obtain the weight adjustment in the ANN, the sampled data forms of equations (7) and (8) are derived. The actual rotor speed is now replaced by the estimated rotor speed. The rotor flux components can be described in the recursive form as follows:

$$\hat{\psi}_{rd}(k) = \hat{\psi}_{rd}(k-1)\left(1 - \frac{T}{\hat{T}_r}\right) + \left(\omega_e - \hat{\omega}_r\right)T\hat{\psi}_{rq}(k-1) + \frac{T}{\hat{T}_r}L_m i_{sd}(k-1) \tag{9}$$

$$\hat{\psi}_{rq}(k) = \hat{\psi}_{rq}(k-1)\left(1 - \frac{T}{\hat{T}_r}\right) - \left(\omega_e - \hat{\omega}_r\right)T\hat{\psi}_{rd}(k-1) + \frac{T}{\hat{T}_r}L_m i_{sq}(k-1), \tag{10}$$

where T is the sampling rate.

In equations (9) and (10) the following weights are introduced:

$$w_1 = 1 - \frac{T}{T_r}, w_2 = -\left(\omega_e - \hat{\omega}_r\right)T, w_3 = \frac{T}{T_r}L_m \,. \tag{11}$$

It can be seen that w_2 is a variable weight and is proportional to the speed. From the viewpoint of the training procedure of the ANN, the weights w_1 and w_3 do not depend on the ANN training, than on the inverse rotor time constant procedure described in the section 3.

Equations (9) and (10) can be expressed in the following forms:

$$\hat{\psi}_{rd}(k) = w_1\hat{\psi}_{rd}(k-1) - w_2\hat{\psi}_{rq}(k-1) + w_3 i_{sd}(k-1) \tag{12}$$

$$\hat{\psi}_{rq}(k) = w_1\hat{\psi}_{rq}(k-1) + w_2\hat{\psi}_{rd}(k-1) + w_3 i_{sq}(k-1). \tag{13}$$

Equations (12) and (13) present the two-layer ANN. There are four input nodes and two output nodes. The weight adjustment can be obtained from ([13])

$$w_2(k) = w_2(k-1) + \eta\left\{-\left[\psi_{rd}(k) - \hat{\psi}_{rd}(k)\right]\hat{\psi}_{rq}(k-1) + \left[\psi_{rq}(k) - \hat{\psi}_{rq}(k)\right]\hat{\psi}_{rd}(k-1)\right\} . \tag{14}$$

The estimated rotor speed can be obtained as follows ([13]):

$$\hat{\omega}_r(k) = \omega_e(k) + \frac{w_2(k-1)}{T} + \frac{\eta}{T}\left\{-\left[\psi_{rd}(k) - \hat{\psi}_{rd}(k)\right]\hat{\psi}_{rq}(k-1) + \left[\psi_{rq}(k) - \hat{\psi}_{rq}(k)\right]\hat{\psi}_{rd}(k-1)\right\}, \quad (15)$$

where η is the so-called learning rate.

Equation (15) presents the simple algorithm of the rotor speed estimation by the ANN. In comparison with the similar ANN described in reference [13], there are the two differences: the inverse rotor time constant is not a constant parameter but is identified, and the rotor flux is estimated in the d,q frame.

5 Simulation and Experimental Results

A simulation program of the complete control system in MATLAB-Simulink environment has been developed. In order to compare the theory with a test case, a control algorithm was executed on the dSpace DS1104 board. Parameters of the induction motor are given in Appendix.

5.1 Simulation Results

Fig. 4 demonstrates the dynamic performance of the IRFO system with half of the rated load torque.

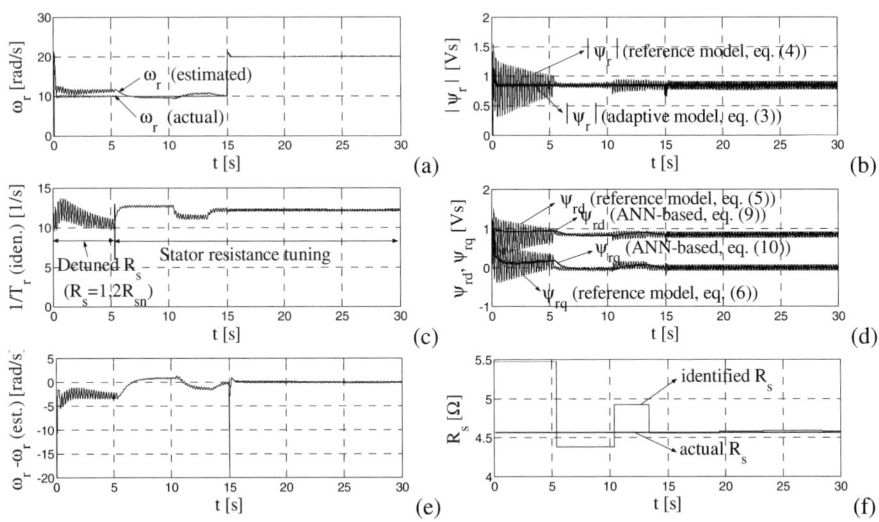

Fig. 4. Dynamic performance of the proposed control system (simulation)

The induction motor starts with the stator resistance initially overestimated by 20 % of the rated stator resistance. Stator resistance tuning starts at time 5.5 s, as shown in Fig. 4f. At time 15 s, the rotor speed reference value rapidly changes from 10 rad/s to 20 rad/s. The actual rotor speed and the estimated rotor speed are shown in Fig. 4a. As the identified rotor resistance reaches the actual resistance that the actual and estimated rotor speeds correspond closely. The estimated rotor flux magnitude is shown in Fig. 4b. As a consequence of this estimation procedure, the inverse rotor time constant identification is enabled (Fig. 4c). Fig. 4d shows the components of the rotor flux space vector in the d,q frame. The difference between the corresponding components enables rotor speed estimation as shown in Fig. 3. The sampling rate of 0.5 ms and the learning rate of 10^{-5} were chosen.

5.2 Experimental Results

Fig. 5 shows experimental results obtained at a rotor speed reference of 40 rad/s. At time t=1.2 s a step load of 5.5 Nm is applied. Fig. 5a shows the actual rotor speed. Fig. 5b shows the estimated rotor speed obtained by the ANN. The identified inverse rotor time constant is shown in Fig. 5c (it has an upper limit 15 s^{-1}).

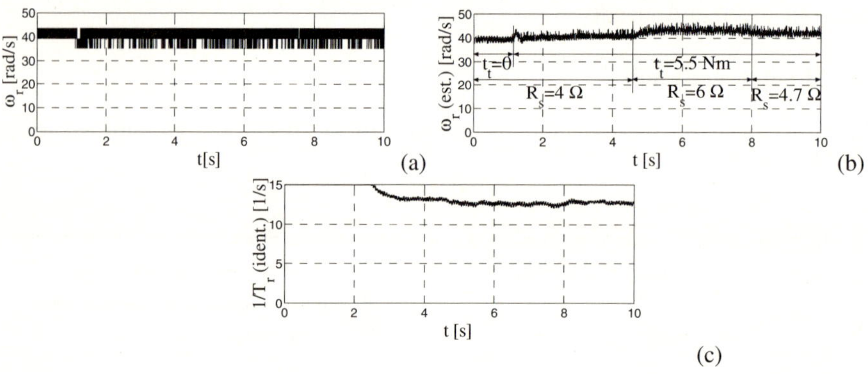

Fig. 5. Dynamic performance of the proposed control system (experiment)

By observing the performance shown in Fig. 5 the following could be concluded:

a) identification of the inverse rotor time constant does not work well at zero load torque,
b) when the identified stator resistance is higher than the actual resistance, then the estimated rotor speed is higher than actual speed, and vice versa.

6 Conclusion

A study of the IRFO control system of an induction motor including deviations in the stator resistance has been carried out. MRAS-based identification of the inverse rotor time constant has been included in the observed system. The identified inverse rotor time constant is an input parameter for a two-layer ANN. The ANN presents the adaptive model of the induction motor. Simultaneously, the components of the rotor flux space vector have been estimated by the reference model (voltage model) in the same reference frame. As a result of these simultaneous estimation procedures, the rotor speed estimation is enabled. The difference between the actual and estimated rotor speed converge to zero if the identified rotor resistance is near to the actual one. Therefore, the difference between the actual and estimated rotor speed has been utilized for manual stator resistance tuning. The overall control system exhibits excellent performances of operation over a low speed range (up to 15 % of the rated rotor speed). We expect to replace manual stator resistance tuning with an automated procedure.

References

1. Baghli, L., Al-Rouh, I., Rezzoug, A.: Signal analysis and identification for induction motor sensorless control. Control Engineering Practice 14(11), 1313–1324 (2006)
2. Ben-Brahim, L., Tadakuma, S., Akdag, A.: Speed control of induction motor without rotational transducers, IEEE Trans. on Ind. Appl. 35(4), 844–850 (1999)
3. Burton, B., Harley, R.G., Diana, G., Rodgerson, J.L.: Implementation of a neural network to adaptively identify and control VSI-fed induction motor stator currents. IEEE Trans. on Ind. Appl. 34(3), 580–588 (1998)
4. Du, T., Vas, P., Stronach, F.: Design and application of extended observers for joint state and parameter estimation in high-performance ac drives. IEE Proceedings—Electric Power Applications 142(2), 71–78 (1995)
5. Mondal, S.K., Pinto, J.O.P., Bose, B.K.: A neural network based space-vector PWM controller for a three voltage-fed inverter induction motor drive. IEEE Tran. on Ind. Appl. 38(3), 660–669 (2002)
6. Nguyen, H.T., Prasad, N.R., Walker, C.L., Walker, E.A.: A First Course in Fuzzy and Neural Control. A CRC Press Company (2003)
7. Novotny, D.W., Lipo, T.A.: Vector Control and Dynamics of AC Drives. Oxford University Press, New York (1996)
8. Radwan, E., Mariun, N., Aris, I., Bash, S.M., Yatim, A.H.: IRFOC induction motor with rotor time constant estimation modelling and simulation. COMPEL 24(4), 1093–1119 (2005)
9. Shimane, K., Tanaka, S., Tadakuma, S.: Vector controlled IM using neural networks. IEEJ Trans. on Elect. and Electronic Engineering 113-D(10), 1154–1161 (1993)
10. Takahashi, Y.: Adaptive control via neural networks. J. SICE 29(8), 729–733 (1990)
11. Toliyat, H.A., Levi, E., Raina, M.: A Review of RFO Induction Motor Parameter Estimation Techniques. IEEE Trans. on Energy Conversion 18(2), 271–283 (2003)
12. Verghese, G.C., Sanders, S.R.: Observers for Flux Estimation in Induction Machines. IEEE Trans. on Ind. Elect. 35(1) (1988)
13. Vas, P.: Electrical Machines and Drives, Application of Fuzzy, Neural, Fuzzy-Neural and Genetic-Algorithm-Based Techniques. Oxford Univ. Press, Inc., New York (1999)

Appendix (Induction Motor Parameters)

P_n=1.5 kW, U_n=380 V, P=4, Y, I_n=3.81 A, n_n=1391 r/min, L_m=0.3269 H, L_{sl}=0.01823 H, L_{rl}=0.02185 H, $R_s = 4.5633\ \Omega$, $R_r = 3.866\ \Omega$, t_n=10.5 Nm, J=0.0071 kgm^2.

A Simple Goal Seeking Navigation Method for a Mobile Robot Using Human Sense, Fuzzy Logic and Reinforcement Learning

Hamid Boubertakh[1,2,3], Mohamed Tadjine[2], and Pierre-Yves Glorennec[3]

[1] LAMEL, University of Jijel, BP. 98, Ouled Aissa, 18000, Jijel, Algeria
[2] LCP, Département de Génie Electrique, Ecole Nationale Polytechnique, 10 av. Hassen Badi, BP. 182, El Harrach, Alger, Algeria
[3] INSA de Rennes, 20 av. des Buttes de Coëmes, 35700 Rennes France
boubert_hamid@yahoo.com, tadjine@yahoo.fr, glorenne@irisa.fr

Abstract. This paper proposes a new fuzzy logic-based navigation method for a mobile robot moving in an unknown environment. This method endows the robot the capabilities of obstacles avoidance and goal seeking without being stuck in local minima. A simple Fuzzy controller is constructed based on the human sense and a fuzzy reinforcement learning algorithm is used to fine tune the fuzzy rule base parameters. The advantages of the proposed method are its simplicity, its easy implementation for industrial applications, and the robot joins its objective despite the environment complexity. Some simulation results of the proposed method and a comparison with previous works are provided.

Keywords: Fuzzy logic, Reinforcement learning, Mobile robot navigation, Obstacle avoidance.

1 Introduction

Path planning is an important topic in robotics, mainly for its practical applications, and can be classified on two types; the global path planning and the local path planning. The first one is done in off-line manner, and since the path is designed, the user can decide about the control method to use to follow the path by the robot. Many methods are developed in the literature; A* algorithm [1], potential field method [2, 3], cell decomposition method [3] and recently, the ant colony methods [4, 5]. The second one also named obstacle avoidance or navigation is done in on-line reactive manner; it imposes that the robot must be equipped by sensors to have a vision of its neighbourhood, and after, takes the adequate action to achieve its a priori defined task. We can find in the literature a large number of publications dedicated to this purpose where several methods are used; the force field [6], the reinforcement learning [7], and the one witch has known a great success in the last years is the fuzzy logic [1, 8-14]. The success of fuzzy logic in mobile robots navigation is due to its capability to represent the human reasoning and therefore the robot doesn't need an exact vision of its environment. Recently many research woks in vehicles navigation using fuzzy logic are published. In [8, 9], authors used a fuzzy rule base composed of

I. Lovrek, R.J. Howlett, and L.C. Jain (Eds.): KES 2008, Part I, LNAI 5177, pp. 666–673, 2008.

243 rules. The use of such a big rule base is unjustified and questions the applicability of the method in real time. In [10-13], the authors proposed simples fuzzy rule bases. But these methods have a common drawback; the robot can be stuck in a local minimum in complex environments. The problem of local minima was treated in [1]; the authors proposed for the goal seeking the fusion of two elementary behaviours; the convex obstacles avoidance behaviour is achieved by 25+25 fuzzy rules, and the wall following behaviour is achieved by 15 fuzzy rules. Many of the above works use the reinforcement learning for parameters adjustment of the fuzzy navigator. In [9] a reinforcement learning algorithm assisted by a supervised learning algorithm is used, at the beginning, the supervised learning is performed by a gradient descent-based algorithm witch is used to tune the premise parameters, and after, the reinforcement learning is used for fine tuning of both premise and the conclusion parameters of the fuzzy rule base. This learning process is very complicated and may lead to a non interpretable rule base. In [10, 12, 14], authors used the fuzzy Q-learning algorithm. The main advantage of this algorithm is its ease for real time implementation.

In the present work, we propose a new simple fuzzy logic-based navigation method for a mobile robot. We use a single fuzzy rule base composed of 8 rules inspired from the human reasoning where both obstacles avoidance and goal seeking behaviours are merged. To achieve the fine tune of the navigator conclusion parameters of the rule base, the reinforcement Fuzzy Q-Learning algorithm is used. The proposed navigation method is compared to previous works.

This paper is organized as follows. Section II describes the considered robot architecture, in section III, the proposed fuzzy navigator is presented with the reinforcement fuzzy Q-learning algorithm. Section IV presents the simulation results and finally, section V, concludes the paper.

2 Robot Model

We use a cylindrical omnidirectional mobile robot model with a radius of 20 cm [9]. The robot is equipped with 24 ultrasonic sensors evenly distributed in a ring as depicted in Fig.1(a). Each sensor, s_i for i=1,…,24, covers an angular view of 15° and gives the distance to the obstacle l_i in its field of view. To reduce the number of inputs for the navigator, sensors in the front of the robot are arranged into tree sensor groups; the left group *SL* consists of the 3 neighbouring sensors s_i (i=1,…,3), the face group *SF* consists of the 6 neighbouring sensors s_i (i=4,…,9), and the right group *SR* consists of the 3 neighbouring sensors s_i (i=10,…,12). The distances measured by the tree groups *SL*, *SF* and *SR* denoted respectively by *dl*, *df* and *dr* are expressed as follows:

$$\begin{cases} dr = R + \min\left(l_i \,/\, i = 1,2,3\right); \\ df = R + \min\left(l_i \,/\, i = 1,..,9\right); \\ dl = R + \min\left(l_i \,/\, i = 10,11,12\right) \end{cases} \tag{1}$$

We use two coordinate systems; the world coordinate system *XOY* and the mobile robot coordinate system *xoy* where *o* is in the center of the robot and the *x* axis goes in the middle between the two sensors s_6 and s_7 (see Fig. 1(b)). The robot actions are the

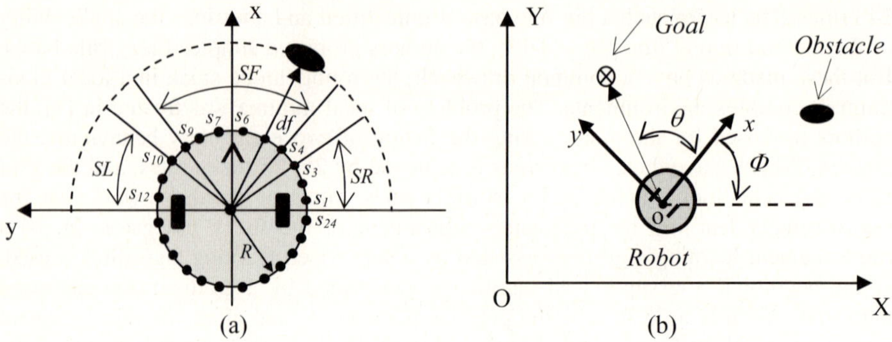

Fig. 1. (a) Mobile robot and the sensor arrangement, (b) The system coordinates

change of the heading angle $\Delta\Phi$ and the linear velocity v of the robot. For a goal seeking behaviour, the robot knows the position of its goal and θ defined as the angle between the orientation axis and the line connecting the center of the robot to the goal.

3 Fuzzy Navigator

The fuzzy navigator inputs are distances in the three directions; right *dr*, face *df* and left *dl*, and the outputs are the speed v of the robot centre and the change of the steering angle $\Delta\Phi$.

3.1 Fuzzification

Two fuzzy labels are used to describe each of the three distances; Near (N) and Far (F) as shown in Fig. 2. The fuzzy labels have the following membership functions:

$$\mu_N(d) = \min\left(\max\left(0, \frac{d-d_s}{d_m-d_s}\right), 1\right) \tag{2}$$

$$\mu_F(d) = \min\left(\max\left(0, \frac{d-d_m}{d_s-d_m}\right), 1\right) \tag{3}$$

Where d_m is the minimum permitted distance to an obstacle, and d_s is the safety distance beyond which the robot can move at high speed.

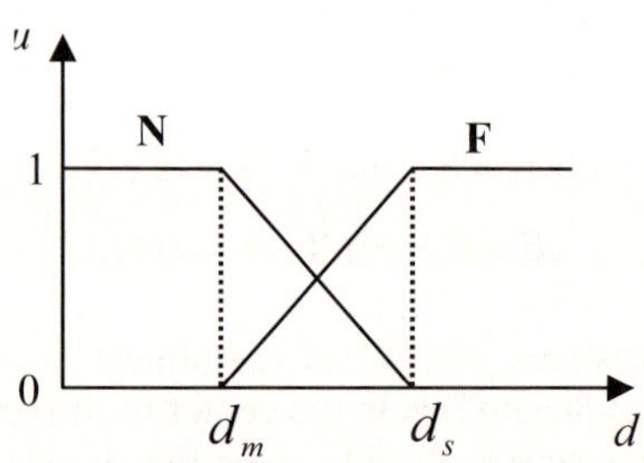

Fig. 2. Inputs membership functions

3.2 Fuzzy Rules Base

The rule base for both obstacles avoidance and goal seeking is constructed based on the human reasoning. It can be interpreted as:

- If the robot is far from obstacles in its three directions, then the robot steers to the goal and goes with its maximum speed;
- If the goal is not in the front of the robot and there exist obstacles, then the robot follows the nearest obstacle on its right or left, according to the smallest distance to the obstacle.
- If both the goal and obstacles are in the front of the robot, then the robot tries to steer to the goal and follows the nearest obstacle on its right or left, according to the smallest distance to the obstacle.

Let us define two parameters *a* and *b* as follows:

$$a = \begin{cases} 0 \ if \ \ g \le d \ et \ g \le d_M \\ 1 \ else \end{cases} , \text{ and } \quad b = \begin{cases} 1 \ if \ \ 90° < \theta < 270° \\ 0 \ else \end{cases} .$$

where d_M is the maximum distance that can be detected by the sensors.

The two parameters *a* and *b* allow the navigator to select its behaviour mode; if *p*=1 (*p*=0) then the right (left) wall following is selected, and if *q*=1 the goal seeking behaviour is activated else the wall following is activated.

Now, we group the three distances in a triplet (dl,df,dr). Then, the fuzzy rule base is expressed as:

Rule 1: If (NNN) Then v_1 is ZR and $\Delta\Phi_1$ is a*PB+(1-p)*NB
Rule 2: If (NNF) Then v_2 is ZR and $\Delta\Phi_2$ is NB
Rule3: If (NFN) Then v_3 is C*V_{max} and $\Delta\Phi_3$ is ZR
Rule 4: If (NFF) Then v_4 is V_{max} and $\Delta\Phi_4$ is a* NS+(1-p)*PS
Rule 5: If (FNN) Then v_5 is ZR and $\Delta\Phi_5$ is PB
Rule 6: If (FNF) Then v_6 is ZR and $\Delta\Phi_6$ is a*PB+(1-p)*NB
Rule 7: If (FFN) Then v_7 is V_{max} and $\Delta\Phi_7$ is a*NS+(1-p)*PS
Rule 8: If (FFF) Then v_8 is V_{max} and $\Delta\Phi_8$ is q*θ+(1-q)*(p*NM+(1-p)*PM)

where V_{max} is the maximum velocity of the robot, $\Delta\Phi_i$ and v_i are the fuzzy rule conclusions, and *C* is the speed decrease coefficient. In the simulation part we take V_{max}=1 m/s and C=0.1.

The rules conclusions are singletons expressed linguistically by:

PB (*Positive Big*), PM (*Positive Medium*), PS (*Positive Small*), ZR (*Zero*), NS (*Negative Small*), NM (*Negative Medium*), and NB (*Negative Big*).

Here, because of the geometrical symmetry of the robot, we take: NB=-PB, NM=-PM, NS=-PS. Hence, the number of parameters is reduced and the rule base becomes:

Rule 1: If (NNN) Then v_1 is ZR and $\Delta\Phi_1$ is (2*p-1)*PB
Rule 2: If (NNF) Then v_2 is ZR and $\Delta\Phi_2$ is NB
Rule 3: If (NFN) Then v_3 is C*V_{max} and $\Delta\Phi_3$ is ZR
Rule 4: If (NFF) Then v_4 is V_{max} and $\Delta\Phi_4$ is (2*p-1)*NS
Rule 5: If (FNN) Then v_5 is ZR and $\Delta\Phi_5$ is PB

Rule 6: If (FNF) Then v_6 is ZR and $\Delta\Phi_6$ is (2*p-1)*NB
Rule 7: If (FFN) Then v_7 is V_{max} and $\Delta\Phi_7$ is (2*p-1)*NS
Rule 8: If (FFF) Then v_8 is Vmax and $\Delta\Phi_8$ is q*θ+(1-q)*(2*p-1)*NM.

3.3 Inference

In order to determine control actions; the steering angle $\Delta\Phi$ and the robot linear speed v, we use the Sugeno fuzzy inference method [15].

$$\Delta\Phi = \frac{\sum \alpha_i . \Delta\Phi_i}{\sum \alpha_i} \tag{4}$$

$$v = \frac{\sum \alpha_i . v_i}{\sum \alpha_i} \tag{5}$$

Where α_i is the truth value of the ith rule calculated by the product method.

3.4 Fuzzy Reinforcement Learning

In reinforcement learning, or "learning with a critic", the received signal is a behaviour punishment (positive, negative or neutral). This signal indicates what you have to do without saying how to do it. The agent uses this signal to determine a policy permitting to reach a long-term objective. It exist several reinforcement learning algorithms. Some are based on policy iteration such as *Actor Critic Learning* and others on value iterations as *Q-learning* or *Sarsa*. In the present work, we are interested by the use of the fuzzy version of the Q-learning algorithms named Fuzzy Q-Learning algorithm presented on table 1. The basic theory of the algorithm can be found in [13]. Here, several competing conclusions are associated with each fuzzy rule. To each conclusion is associated a q-value that is incrementally updated. The learning process consists then in determining the best set of rules, the one that will optimize the future reinforcements. The initial rule base is composed therefore of N rules such as:

Table 1. Fuzzy Q-learning algorithm

1. t=0, observe the state x_t.
2. For each rule, i, compute $\alpha_i(x_t)$.
3. For each rule, i, choose $c[i]$ with an exploration /exploitation policy.
4. Compute the action $A(x_t)$ and its correspondence quality $Q(x_t,A(x_t))$.
5. Apply the action $A(x_t)$. Observe the new state, x_{t+1}.
6. Receive the reinforcement, r_t.
7. Compute $\alpha_i(x_{t+1})$.
8. Compute a new evaluation of state value.
9. Update parameters $q[i,j]$ using this evaluation.
10. t←t+1, go to 3

If x is S_i **Then** $y=c[i,1]$ ***with*** $q[i,1]=0$
or $y=c[i,2]$ ***with*** $q[i,2]=0$
...
or $y=c[i,J]$ ***with*** $q[i,J]=0$

To simplify the use of knowledge, we adopt the method in [16]. We fix a interval for each linguistic conclusion c(j,j)∈[a(i), b(i)] and the number J of the potential candidate conclusions is evenly distributed on this interval.

4 Simulation Results

The problem consists of equip the robot with the capability of obstacles avoidance and goal seeking without being stuck in local minima and without collusion with obstacles. For this purpose we use the mobile robot model presented in section II, we assume that the effective range of the ultrasonic sensors is 10 cm - 250 cm. We use the proposed fuzzy navigator presented in section III, and to learn the robot, we use the fuzzy Q-Learning algorithm presented in the subsection III-D.

4.1 Initialisation

The number of potential candidate conclusions is J=5. And the possibility interval of each fuzzy rule conclusion is given by Tab. 2.

Table 2. Fuzzy rule conclusions for ΔΦ

	NS	NM	NB	ZR	PS	PB
MIN	-40°	-40°	-80°	0°	5°	80°
MAX	-5°	-5	-40°	0°	40°	40°

As a reinforcement signal, we want to punish any conclusion of an activated rule witch can cause collusion with an obstacle:

$$r=\begin{cases}-1 & if\ \min(dl,df,dr)<d_m \\ 0 & otherwise\end{cases} \qquad (6)$$

4.2 Learning Phase

In the learning phase (Fig. 3(a)), only the wall following behaviour is selected (b=0), the robot is kept in unknown environment and after a sufficient learning time, 600 steps in the simulation, the reinforcement learning algorithm will be stopped.

4.3 Goal Seeking

After the learning stage, the rule conclusions with the best qualities are chosen and the algorithm is stopped. Fig. 3(b) shows the simulation results of a goal seeking task when the robot starts from different initial positions.

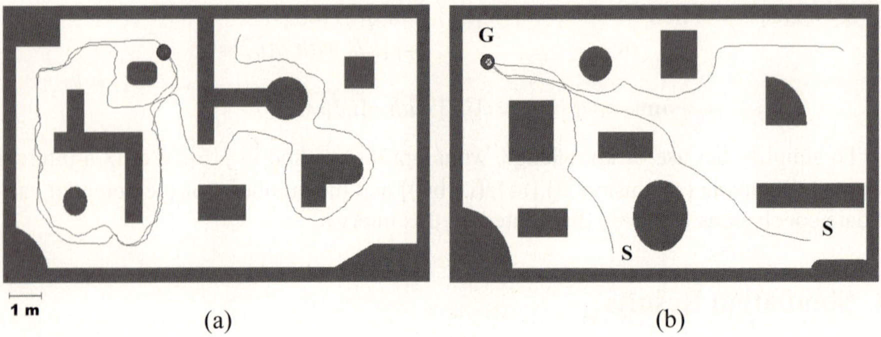

Fig. 3. (a) Learning phase, (b) Goal seeking

4.4 Comparison with the Related Methods

Fig. 4(a) shows the simulation results when one of the rule bases of [10, 11, 12] is used; the robot is suck in a local minimum in relatively complex environment. Fig.4(b) shows simulation results with the proposed navigation method; the robot can easily escape from the local minima.

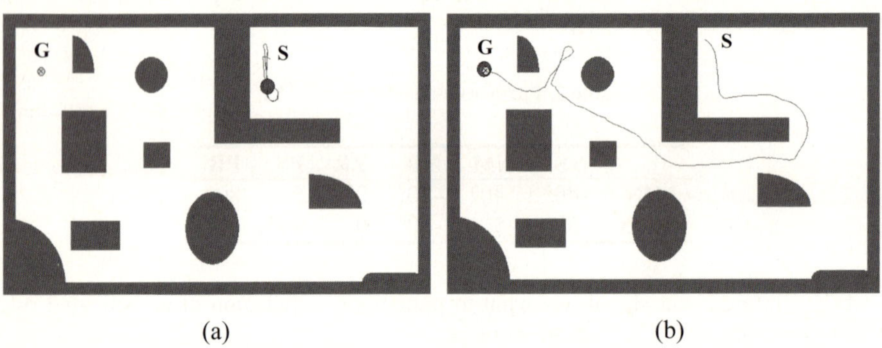

Fig. 4. Navigation in complex environment: (a) Local minima with some related works; (b) No local minima with the proposed method

5 Conclusion

In this paper, we have proposed a solution to the local minima problem in gaol seeking navigation for mobile robots in unknown environments. The proposed navigator is based on fuzzy logic. The rule base of the navigator is inspired from human reasoning and the parameters fine tuning is get by the Fuzzy Q-Learning algorithm. This method endows the robot with capabilities of obstacle avoidance and goal seeking without being stuck in local minima. The merits of the proposed method are its simplicity and its easy implementation for industrial applications. The efficiency of the proposed method is demonstrated trough simulation results.

References

[1] Maaref, H., Barret, C.: Sensor Based Navigation of an Autonomous Mobile Robot in an Indoor Environment. Control Engineering Practice 8, 757–768 (2000)

[2] Latombe, J.-C.: Robot Motion Planning. Kluwer Academic Publishers, Dordrecht (1991)

[3] Koren, Y., Borenstein, J.: Potential Field Methods and Their Inherent Limitations for Mobile Robot Navigation. In: The 1991 IEEE International Conference on Robotics and Automation, Sacramento, California (1991)

[4] Tan, G.-Z., He, H., Sloman, A.: Ant Colony System Algorithm for Real-Time Globally Optimal Path Planning of Mobile Robots. Acta Automatica 33(3) (2007)

[5] Liu, G., Li, T., Peng, Y., Hou, X.: The ant algorithm for solving robot path planning problem. In: The third IEEE International Conference on Information Technology and Applications, Sydney, vol. 2, pp. 25–27 (2005)

[6] Brooks, R.A.: A robust Layered Control System for a mobile robot. IEEE Trans. On Robotics and Automation RA-2(1), 14–23 (1986)

[7] Suwimonteerabuth, D., Chongstvatana, P.: Online robot learning by reward and punishment for a mobile robot. In: The 2006 IEEE/RSJ International Conference on Intelligent Robots and Systems, Lausanne, Swizerland (October 2002)

[8] Beom, H.B., Cho, H.S.: A Sensor Based Navigation for a Mobile Robot using Fuzzy Logic and Reinforcement Learning. IEEE Trans. On Systems, Man, and Cybernetics 25(3) (March 1995)

[9] Ye, C., Nelson, H.C.Y.: A Fuzzy Controller with Supervised Learning Assisted Reinforcement Learning Algorithm for Obstacle Ovoidance. IEEE Trans. On Systems, Man, and Cybernetics 33(1) (February 2003)

[10] Zavlangas, P.G., Tzafestas, S.G.: Motion control for Mobile Robot Obstacle Avoidance and Navigation: A Fuzzy Logic-Based Approach. Systems Analysis Modelling Simulation 43(12), 1625–1637 (2003)

[11] Dahmani, Y., Benyettou, A.: Fuzzy Reinforcement Rectilinear Trajectory Learning. Journal of Applied Science 4(3), 388–392 (2004)

[12] Kermiche, S., Saidi, M.L., Abbassi, H.A.: Gradient Descent Adjusting Takagi-Sugeno Controller for a Navigation of Robot Manipulator. Journal of Engineering and Applied Science 1(1), 24–29 (2006)

[13] Glorennec, P.Y., Jouffe, L.: A Reinforcement Learning Method For Autonomous Robot. In: The Fourth European Congress on Intelligent Techniques and Soft Computing, Aachen, Germany (September 1996)

[14] Gu, D., Hu, H.: Accuracy Based Fuzzy Q-Learning for Robot Behaviours. In: Proceedings of the IEEE International Conference on Fuzzy Systems (IEEE-FUZZY 2004), Budapest, 25-29 July (2004)

[15] Yager, R.R., Filev, D.P.: Essential of fuzzy modelling and control. John Wily & Sons inc (1994)

[16] Boubertakh, H., Glorennec, P.Y.: Optimization of a Fuzzy PI Controller Using Reinforcement Learning. In: The 2006 IEEE Internationnal Conference on Information and Communication Technologies: From Theory to Applications, Damascus, vol. 1, pp. 1657–1662 (2006)

Fuzzy Control of a Real Time Inverted Pendulum System

Selcuk Kizir, Zafer Bingul, and Cuneyt Oysu

Kocaeli University, Department of Mechatronics Engineering,
41380, Kocaeli, Turkey
{selcuk.kizir,zaferb,coysu}@kocaeli.edu.tr

Abstract. In this study, a real-time control of the cart-pole inverted pendulum system was developed using fuzzy logic controller. Swing-up and stabilization of the inverted pendulum were implemented directly in fuzzy logic controller. The fuzzy logic controller designed in the Matlab-Simulink environment was embedded in a dSPACE DS1103 DSP controller board. Swing-up algorithm brings the pendulum near to its inverted position in 10 seconds from downward position. In order to test the robustness of the fuzzy logic controller internal (changing model parameters) and external disturbances (applying external forces) were applied on the inverted pendulum. The inverted pendulum system was shown to be robust to the external and internal disturbances. The maximum errors of the pendulum angle to the impulse input were between 1.89° and 4.6449° in the robustness tests.

Keywords: Inverted pendulum, swing up, stabilization, fuzzy logic.

1 Introduction

Inverted pendulum is very common and interesting nonlinear system in the control applications. Nonlinear, unstable inverted pendulum system is used to test performance of the different control algorithms. Inverted pendulum problem is a very useful system to utilize in the control education due to simplicity of establishing the system. It is therefore very widely used laboratory tool in the control laboratories.

Fuzzy set theory is employed to model the amount of ambiguity or uncertainty subject to parameter inaccuracy and unmodeled dynamics. It offers a very attractive way to extract information from data for designing a controller. Comparison of classical set theory with fuzzy set theory helps one to understand the basic differences between them. The main difference is that with fuzzy set theory a set membership is represented as a possibility distribution, while the classical set theory defines sets with a clear-cut decision that a feature either belongs to a set or not [1].

In the literature, different inverted pendulum systems and their control methods have been studied. The inverted pendulum systems can be grouped structurally as cart-pole single pendulum systems [2-4], cart-pole double pendulum systems [5], rotary single and double pendulum systems [6-7]. The control methods applied to inverted pendulum systems can be summarized to linear methods such as PID, state

I. Lovrek, R.J. Howlett, and L.C. Jain (Eds.): KES 2008, Part I, LNAI 5177, pp. 674–681, 2008.

feedback [2,4,6,8,9] and nonlinear methods such as energy based, fuzzy logic, feedback linearization and etc. [2,3,5,6,10-14]. Muskinja and Tovornik [3] proposed energy based and fuzzy logic controllers for swing-up of inverted pendulum. Adaptive state controller is also suggested for the stabilization of inverted pendulum. Ji, Lei and Kin [12] presented simulation results of inverted pendulum using fuzzy logic controller for swing-up and stabilization routines.

2 The Structure of the Inverted Pendulum

In the cart-pole single inverted pendulum system, pendulum is attached to the cart which can move in the limited horizontal track. The rod is attached to an optical encoder, so it can rotate very small friction. Inverted pendulum system has two equilibrium points: downward and upright positions. Stable downward equilibrium point corresponds to $\theta = 0,\ \dot{\theta} = 0$. Unstable upright equilibrium point corresponds to $\theta = \pi,\ \dot{\theta} = 0$.

The inverted pendulum developed here consists of several parts such as servo motor providing movement of the cart and applying the desired force to the pendulum, the sensors measuring the state variables (cart position x and pendulum angle θ) and the controller supplying control signals. These parts are shown in Fig. 1. The control algorithm is embedded in dSPACE DS1103 DSP controller board. This board takes the sensor outputs and supply control signal. Photography of the inverted pendulum system developed here is shown in Fig. 1.

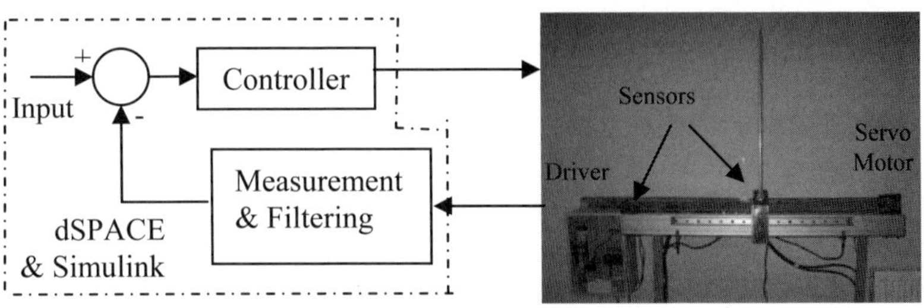

Fig. 1. Block diagram and setup of the inverted pendulum system

3 Controller Design

Generally, hybrid control approach is used in the control of inverted pendulum systems. The hybrid approach is implemented in two steps: swing-up and stabilization routines. The intention of swing-up routine is to bring up the pendulum from downward position to upright position. In stabilization routine, the pendulum is balanced in a limited track. In hybrid approach, the controller decides which routines switch on based on pendulum angle. However, the fuzzy logic control can handle both routines in rules base. Swing-up and stabilization routines are explained below.

3.1 Swing-Up Routine

In this study, a cart motion strategy was determined to swing up the pendulum based on work effect of a given acceleration [3]. The basic strategy is to move the cart in such a motion that energy is gradually pumped to the pendulum. How to add the best energy to the pendulum should be explored and the strategy should be determined based on the analysis. The pendulum model based on Newtonian approach is shown in Fig. 2-a. Throughout this analysis, bold characters are used to denote vector quantities, while non-bold characters represent scalar quantities.

The forces act on the pendulum is,

$$\begin{aligned} \mathbf{F_r} &= F_r(\sin\theta\ \hat{\mathbf{i}} + \cos\theta\ \hat{\mathbf{j}}) \\ \mathbf{F}_g &= -mg\ \hat{\mathbf{j}} \end{aligned} \quad . \tag{1}$$

Using Newton's law F_r is found,

$$F_r = m(\sin\theta\ \ddot{x} + l\dot{\theta}^2 + g\cos\theta)\,. \tag{2}$$

Using F_r the work is found,

$$\delta W = \mathbf{F_r} \bullet \partial x = F_r \sin\theta\ \partial x = m(\sin^2\theta\ \ddot{x} + \sin\theta\ l\dot{\theta}^2 + g\sin\theta\cos\theta)\partial x\,. \tag{3}$$

The term $\sin^2\theta\ \ddot{x}$ in Equation 3 is important, because it describes the work effect of a given acceleration at any angle. This term is shown in Fig. 2-b. For θ close to ±90°, a given acceleration and displacement does maximum work. For θ close to 0°, hardly any work is done. The work is positive when the acceleration and displacement are in the same direction. In this case, the energy is added to the system.

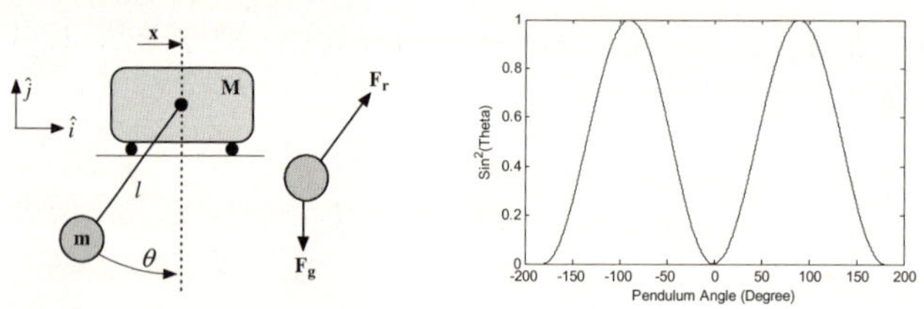

(a) (b)

Fig. 2. (a) Coordinate system and free body diagram of the system. **(b)** Work effect of the system.

Considering the work effect of the system, the strategy for a cart trajectory that is driven by pendulum angle is developed. To maximize the positive work done on the pendulum, the cart must be accelerated when θ near ±90° to achieve high work transfer and decelerated when θ near 0° since there is low work transfer. Acceleration and deceleration regions are shown in Fig. 3. The cart must be accelerated when the pendulum is in region I, cart must be decelerated in region II and it must be waited while pendulum reaches its maximum point in region III. This process is repeated until pendulum reaches its inverted position. Because the pendulum begins at rest, hanging

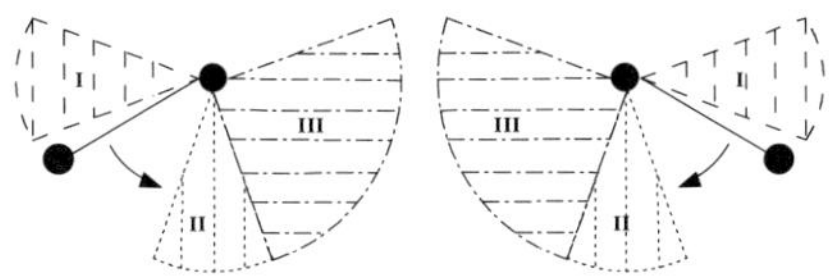

Fig. 3. Swing-up strategy

downward position in region II, this strategy does not produce an output, so swing-up controller should include a series of quick cart movements that begin the pendulum swinging. Finally, to bring the pendulum close to its inverted position and in low angular velocity, cart movement amplitude should be gradually reduced when the pendulum swings higher.

The rules mentioned above are very suitable to design a fuzzy controller. In the proposed fuzzy swing-up controller, two input variables are used: pendulum angle θ and pendulum angular velocity $\dot{\theta}$. The system has a limited track length but it is enough to apply this control strategy. Seven fuzzy subsets [NLS (Negative large), NBS (Negative big), SALN (Start swing negative), Z (Zero), SALP (Start swing positive), PBS (Positive big) and PLS (Positive large)] are used for the error of pendulum angle. Three fuzzy subsets are chosen for the error of pendulum angular velocity. The input membership functions are shown in Fig. 4, the rule base is given in Table 1 and the output membership functions are shown in Fig. 6. The output membership functions has nine fuzzy sets [NVVB (Negative very very big), NVB (Negative very big), NB (Negative big), N (Negative), Z (Zero), P (Positive), PB (Positive big), PVB (Positive very big), PVVB (Positive very very big)].

Response of the swing-up controller used here is shown in Fig. 5. Pendulum is swung-up from downward position to inverted position in about 10 seconds. The number of fluctuations and swing-up time can be reduced or increased by changing amplitude of the output membership functions.

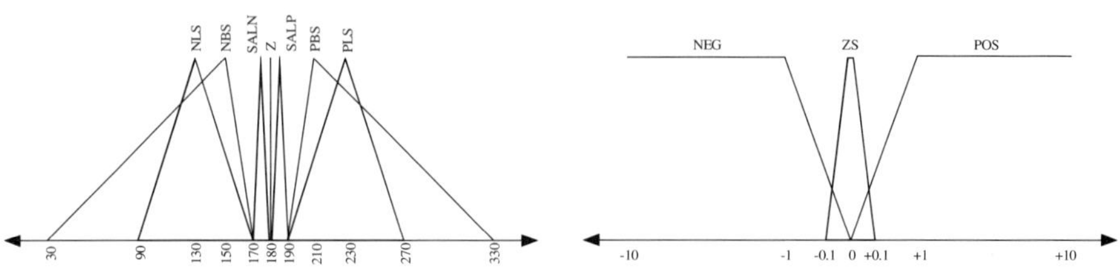

Fig. 4. Membership functions for θ and $\dot{\theta}$

Table 1. Fuzzy rule base for swing-up

	$e\theta$						
$e\dot{\theta}$	***NLS***	***NBS***	***SALN***	***Z***	***SALP***	***PBS***	***PLS***
NEG					P	Z	PB
ZS				P			
POS	NB	Z	N				

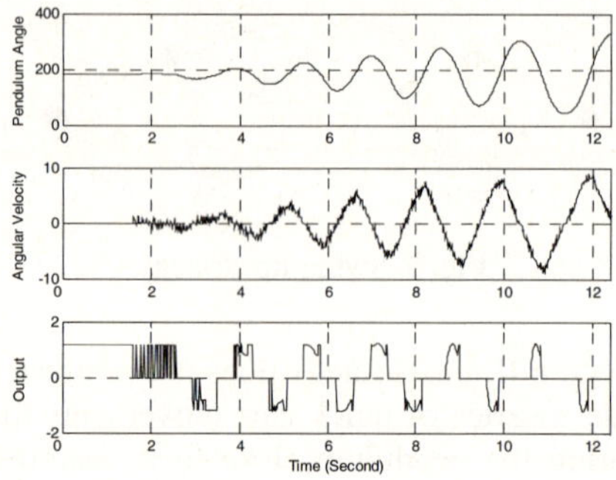

Fig. 5. Response of the system during swing-up

3.2 Stabilization

In the proposed fuzzy stabilization controller, four input variables are used: pendulum angle θ, pendulum angular velocity $\dot{\theta}$, cart position x and cart velocity $\dot{x}$. In order to balance the pendulum in the range of ±30°, seven fuzzy subsets [NVB, NB, N, ZO, P, PB, PVB] are chosen for the error of pendulum angle and five fuzzy subsets [NB, N, ZO, P, PB] are chosen for the error of angular velocity. Also, five fuzzy subsets [NBIG, NEG, Z, POS, PBIG] are chosen for the error of cart position and three fuzzy subsets [NEG, Z, POS] are chosen for the error of cart velocity to return the cart to its home position. Swing-up and stabilization rules use the same output fuzzy subsets. These membership functions are shown in Fig. 6. The rule bases for cart position control and pendulum angle control are given in Table 2 and Table 3, respectively. Totally, 49 rules are used to control the whole system.

Table 2. Fuzzy rule base for cart position

	ex				
$e\dot{x}$	***NBIG***	***NEG***	***Z***	***POS***	***PBIG***
NEG	PVVB	PVB	PB		
ZERO			Z		
POS			NB	NVB	NVVB

Table 3. Fuzzy rule base for stabilization

	$e\theta$						
$e\dot{\theta}$	***NVB***	***NB***	***N***	***ZO***	***P***	***PB***	***PVB***
NB	NVVB	NVVB	NVB	NB	N	Z	P
N	NVVB	NVB	NB	N	Z	P	PB
ZO	NVB	NB	N	Z	P	PB	PVB
P	NB	N	Z	P	PB	PVB	PVVB
PB	N	Z	P	PB	PVB	PVVB	PVVB

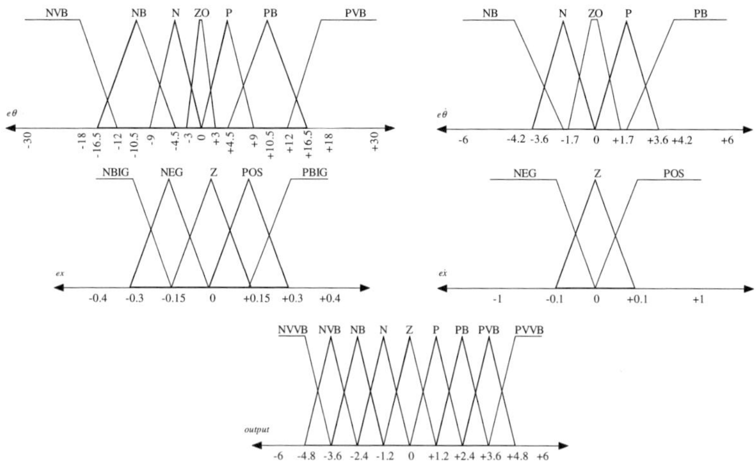

Fig. 6. Membership functions for $\theta, \dot{\theta}, x, \dot{x}$ and output

Response of the controller developed here is shown in Fig. 7. In the figure, the fuzzy logic controller based on the rules switches from swing-up to stabilization. After the switching stabilization routine, pendulum angle and cart position are brought to their desired values.

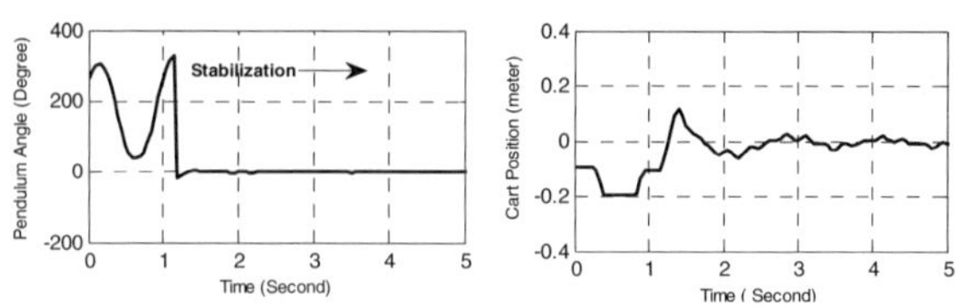

Fig. 7. Controller response during switching from swing-up to stabilization

4 Experimental Results

In order to test the robustness of the designed fuzzy controller, several experiments were implemented using the rods with different lengths and masses. In these experiments, two types of disturbances were applied to the system: external and internal disturbances. Internal disturbances were implemented by changing the system parameters: rod mass and rod length. Minimum and maximum values of the state variables are given in Table 4 as the system parameters change. The fuzzy controller was designed for the inverted pendulum with the rod shown in Fig. 1 (0.65m and 0.2 kg). This rod is called as a normal. Its physical properties were changed by:

- choosing heavier mass of the rod
- same mass and different length of the rod
- increasing length of the normal rod
- decreasing length of the normal rod
- attaching a moving mass to the end of the normal rod.

According to Table 4, longer and lighter pendulum has a minimum error, and shorter and heavier pendulum has a maximum error in state variables of the system. Natural frequency of inverted pendulum is based on length of the rod about equlibrium point so longer pendulum has a lower natural frequency. Longer pendulum is controlled more easily because of its low natural frequency. It causes a high rotational inertia giving a high resistance to a rotational change.

Table 4. Errors in the state variables

Rod Properties		θ peak	x peak	$\int\lvert\theta\rvert$	$\int\lvert x\rvert$	$\int\lvert u\rvert$
Pendulum length	Pendulum mass					
0.65m	0.2 kg	-2.6952	0.0655	0.0540	0.0444	0.0492
0.97 m	0.3 kg	-1.89	0.0384	0.0426	0.0065	0.0410
0.325 m	0.1 kg	-4.6449	0.1482	0.1568	0.0816	0.1388
0.42 m	0.2 kg	-3.8655	0.1881	0.0988	0.1131	0.0731
0.65 m	0.3 kg	-2.79	0.1027	0.0782	0.0680	0.0643
0.65 m	0.4 kg (moving mass)	-2.3400	0.0440	0.0575	0.0121	0.0435

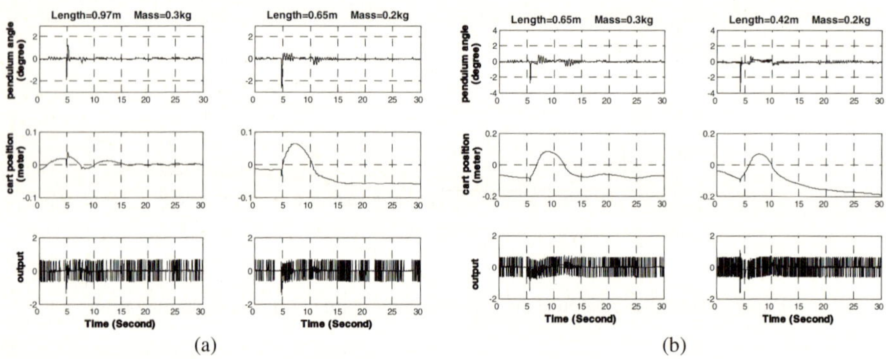

Fig. 8. (a) System responses for longest and normal rod experiments. **(b)** System responses for heavier and shorter rod experiments.

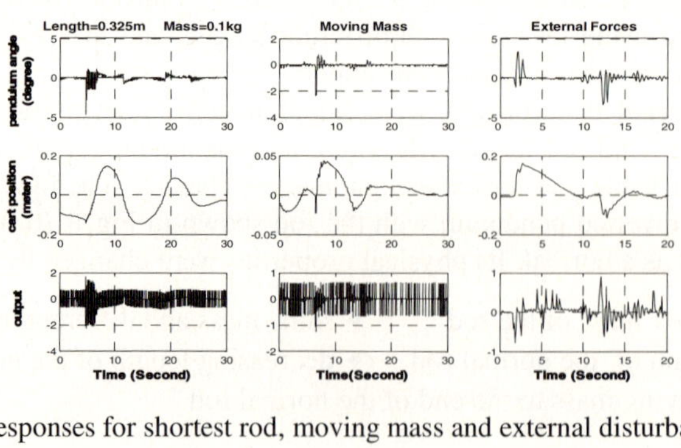

Fig. 9. System responses for shortest rod, moving mass and external disturbances experiments

5 Conclusion

The real-time fuzzy logic control of the cart-pole inverted pendulum system was developed for swing-up and stabilization. The controller was implemented in dSPACE-1103 development board. Several experiments were conducted to verify the robustness of the fuzzy controller under external and internal disturbances. Based on these experiments, the inverted pendulum system was seen to be robust to the disturbances. During disturbance experiments, the peak errors of the pendulum angle to the impulse input were obtained between 1.89° and 4.6449°.

Acknowledgements. One of the authors (S.K.) acknowledges the financial support of TÜBİTAK-BİDEB. This work is also partially supported by the Turkish State Planning Organization (DPT) under the Grant No. DPT-2003K120790.

References

1. Bingul, Z., Cook, G.E., Strauss, A.M.: Application of Fuzzy Logic to Spatial Thermal Control in Fusion Welding. IEEE Transactions on Industry Applications 36(6) (2000)
2. Bugeja, M.: Non-Linear Swing-Up and Stabilizing Control of an Inverted Pendulum System. In: EUROCON Ljubljana, Slovenia (2003)
3. Muskinja, N., Tovornik, B.: Swinging Up and Stabilization of a Real Inverted Pendulum. IEEE Transactions on Industrial Electronics 53(2) (2006)
4. Stimac, A.K.: Standup and Stabilization of the Inverted Pendulum., Massachusetts Institute of Technology (1999)
5. Zhong, W., Röck, H.: Energy and passivity based control of the double inverted pendulum on a cart. In: IEEE Conference on Control Applications (2001)
6. Krishen, J., Becerra, V.M.: Efficient Fuzzy Control of a Rotary Inverted Pendulum Based on LQR Mapping. In: IEEE International Symposium on Intelligent Control, Germany, pp. 2701–2706 (2006)
7. Craig, K., Awtar, S.: Inverted Pendulum Systems: Rotary And Arm-Driven A Mechatronic System Design Case Study. Mechatronics 12, 357–370 (2001)
8. Mirza, A.: Inverted Pendulum. Journal of AMSE 55(3,4), France (2000)
9. Nundrakwang, S., Benjanarasuth, T., Ngamwiwit, J., Komine, N.: Hybrid PD - Servo State Feedback Control Algorithm for Swing up Inverted Pendulum System. In: ICCAS 2005, KINTEX, Gyeonggi-Do, Korea, June 2-5 (2005)
10. Magana, M.E., Holzapfel, F.: Fuzzy-logic control of an inverted pendulum with vision feedback. IEEE Transactions on Education 41(2), 165–170 (1998)
11. Yasunobu, S., Mori, M.: Swing up fuzzy controller for inverted pendulum based on a human control strategy. In: Proceedings of the Sixth IEEE International Conference on Fuzzy Systems, vol. 3, pp. 1621–1625 (1997)
12. Ji, C.W., Lei, F., Kin, K.: Fuzzy logic controller for an inverted pendulum system. In: IEEE International Conference on Intelligent Processing Systems, ICIPS 1997, pp. 185–189 (1997)
13. Becerikli, Y., Celik, B.K.: Fuzzy control of inverted pendulum and concept of stability using Java application. Mathematical and Computer Modeling 46(1-2), 24–37 (2007)
14. Passino, K.M., Yurkovich, S.: Fuzzy Control. Addison Wesley Longman, Menlo Park (1998) (later published by Prentice-Hall)

Construction and Evaluation of a Robot Dance System

Kuniya Shinozaki[1], Akitsugu Iwatani[2], and Ryohei Nakatsu[3]

[1] Kwansei Gakuin University, School of Science and Technology
2-1 Gakuen, Sanda Japan, 669-1337
scbc0052@kwansei.ac.jp
[2] Universal Studios Japan
Osaka, Japan
[3] National University of Singapore, Interacive & Digital Media Institute
Blk E3A #02-04, 7 Engineering Drive 1, Singapore 117574
idmdir@nus.edu.sg

Abstract. Dance is one form of entertainment where physical movement is the key factor. The main reason why robots are experiencing a kind of "boom" is that they have a physical body. We propose a robot dance system that combines these two elements. First, various factors concerning entertainment and dance are studied. Then we propose the dance system by robot using motion unit and the synthetic rule referring the speech synthesis. Also we describe the details of the system by focusing on its software functions. Finally we show the evaluation results of robot dance performances.

1 Introduction

The research and development of various kinds of robots is actively being carried out, especially in Japan [1][2][3][4][5]. Several reasons explain the current robot boom. One main reason is that robots have physical bodies, and so human-robot interaction extends beyond human-computer interaction.

Although in the future these robots are expected to support various aspects of our daily life, so far their capabilities are very limited. At present, installing such a task in robots remains very difficult. To break through such a situation, entertainment might be a good application area for robots.

Developing a dancing robot would be remarkable from various points of view. First, it might become a new form of entertainment, activates both the body and brain. Watching humans dance is already one established type of entertainment. Second, we might develop a new type of communication with computers, because dance can be considered one of the most sophisticated nonverbal communication methods.

Based on the above considerations we started to research dancing robots. In this paper we clarify the relationship among entertainment, humans, and robots and propose a robot dance system by robot using motion unit and the synthetic rule referring the speech synthesis. Also we will describe an evaluation experiment carried out to test this basic concept's feasibility.

I. Lovrek, R.J. Howlett, and L.C. Jain (Eds.): KES 2008, Part I, LNAI 5177, pp. 682–691, 2008.

2 Dance Entertainment and Robots

2.1 Entertainment

The role of entertainment in our daily life is very important. It offers relaxation and thus contributes to our mental health. Many aspects concerning entertainment must be considered and discussed [6]. One of the most important may be the existence of two sides: entertainer and audience. Although these two sides change positions depending on the case, the existence of performers and spectators is an absolute prerequisite for entertainment. Many entertainments have both entertainer and spectator characteristics. In the case of dance, people sometimes go to theaters to watch good dance performances, and they sometimes go to dance clubs or discos to dance themselves.

Furthermore, when viewed from a different aspect entertainment can be classified into two types. One is a real-time type that includes performers or entertainers performing live in front of an audience. Good examples include plays and/or concerts. Another is the non-real-time type; reading books and watching movies are good examples.

Following this classification, dance basically belongs to the real-time type of entertainment. For robot dancing, however, as described later, its position is somewhat special.

2.2 Dance Robot

One main reason why we choose dance as an entertainment for robots is that dance is quite sophisticated [7]. Based on the considerations described above, what is the role of robots in dance entertainment? Dance robots allow us to become both entertainers and spectators. When watching a robot dance, we are spectators. On the other side, many people will probably want to install dance motions on their robots and show these actions to others. In this case they are entertainers. For the classification between real-time and non-real-time entertainment, dance robots also have significant characteristics. If we want to show people the robot dance, we have to install the dance actions beforehand, meaning that the robot dance is non-real-time entertainment. At the same time, by developing interactive capabilities, the robot would show impromptu dancing behaviors. For example, it could change the dance depending on audience requests. Or it could sense the audience mood and could adopt its dancing behaviors to reflect the sensor results. A dance robot could provide flexible entertainment that ranges between real-time and non-real-time entertainment.

3 Dance Robot System

3.1 Basic Concept

Based on the above considerations we want to develop a system that can generate various dance motions. Since different dance genres exist, it is necessary to restrict dance genres to a specific one. Then the system would generate various dance motions by selecting several basic dance motions and by concatenating them. This basic idea resembles text-to-speech synthesis (TTS) [8], where by restricting the language to be synthesized and by selecting a basic speech unit, any kind of text described by the language can be generated. The following is the basic concept adopted in TTS:

(1) Speech consists of a concatenation of basic speech units.
(2) Selection of the speech unit is crucial.
(3) Connection of speech units is also crucial.

As basic speech units, various basic units such as phonemes, phoneme pairs, CV (consonant-vowel concatenation), CVC, VCV and so on have been studied [8]. Based on research of the last several decades, phonemes including variations that depend on previous and following phonemes are widely used as speech units. Taking these situations into consideration, the basic concept of dance generation is as follows:

(1) We restrict the generated dance to a specific genre.
(2) All dance motions consist of a concatenation of several basic dance motions.
(3) Deciding what to select dance units as basic dance motions is very important.
(4) Connecting dance units is crucial.
(5) Also it is crucial how to express a dance unit as robot motion.

In the following sections, we answer the above questions.

3.2 Dance Genre

For basic dance motions, there are several researches on classic ballet [9]. The classification of ballet motions is based on several leg positions and movements called steps. Although each leg position and step has its own name, basically no rules describe the details of whole body motions. We chose hip-hop as the dance genre because all of its dance steps and motions are classified into several categories, so it is easier to handle the whole body motions of hip-hop than ballet.

3.3 Dance Unit

Next we must decide the basic unit for dance motions. As described above, since each hip-hop step/body motion has its own name, it can be selected as a dance unit. However, it is difficult for an amateur to extract them from continuous dance motions. Therefore we collaborated with a professional dancer to simplify the extraction of basic motions from continuous dance motions. In addition, when constructing robot motions based on human motions, we must deform complicated human motions into rather simple robot motions. In this deformation process, a professional dancer's advice is also of great help.

3.4 Concatenation of Dance Units

The next question is how to connect each motion unit. One method interpolates the last posture of the previous motion and the first posture of the next motion. The difficulty in the case of a dancing robot is how to connect these two motions and prevent the robot from falling down. We introduced a method in which a neutral posture represented by a standing still pose is used as a transition posture between two dance units. In this case developing an algorithm is unnecessary to generate a transitional motion that connects two different motions.

3.5 Realization of Robot Dance Motion

The next issue is transforming human dance motions into the motions of robots. One common method adopts a motion capture system that is used to generate the motion of CG characters. For a robot, however, due to the limitations of the degree of freedom at each joint, directly transforming the motion captured by the system into robot motion does not work well. Research that transforms captured motions into robot motions is described in [10] that treats a Japanese traditional dance whose motions include legs moving slowly and smoothly front/back and left/right instead of dynamically. In this case it is relatively easy to maintain balance. However, hip-hop motions include dynamic body motions, and therefore it is difficult to maintain balance. Taking these situations into considerations, we chose a method where each motion unit extracted from continuous motion is transformed manually.

3.6 System Architecture

Based on the above considerations, we constructed the first prototype of a robot dance system, as shown in Fig. 1, that consists of dance unit sequence generation, a dance unit database, and dance unit concatenation.

(1) Dance unit database

A large amount of dance units are stored here; each one corresponds to a basic short dance motion and is expressed as robot motion data.

(2) Dance unit sequence generation

An input data that expresses a dance motion is analyzed and converted into a sequence of dance units by this part. At the present stage a sequence of dance units is directly used as input data and fed into the system.

(3) Dance unit concatenation

As is described in 3.4, a neutral posture is introduced as an intermediate posture between two dance units, and therefore, they can be easily connected.

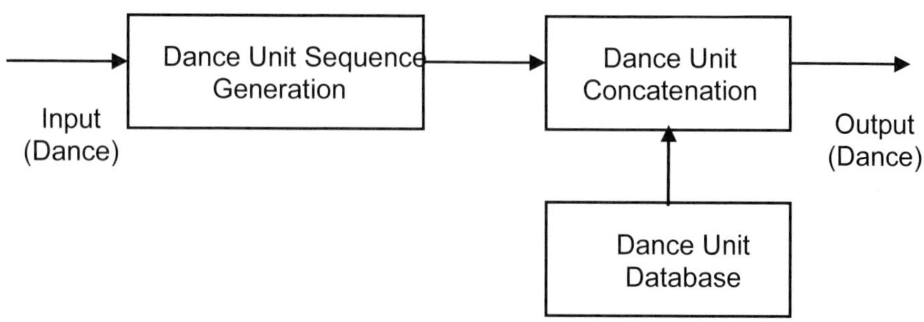

Fig. 1. Structure of dance robot system

4 System Development and Evaluation

4.1 Humanoid Robot

From the several humanoid robots already available on the market, we selected a humanoid robot developed by Nirvana Technology [11] and installed dance motions on it. Figure 2 shows its appearance, and Table 1 shows its basic specifications. Various robot motions can be designed and produced on PC using a "motion editor" realized by motion making and editing software.

Fig. 2. Humanoid robot

Table 1. Specifications of humanoid robot

Size/Weight	34 cm / 1.7 kg
Degree of flexibility	22 (12 legs, 8 arms, 1 waist, 1 head)
CPU	SH2/7047F
Motor	KO PDS-2144, FUTABA S3003, FUTABA S3102, FUTABA S3103
Battery	DC6V

4.2 Development of Dance Unit Database

As described above, we collaborated with a dancer to develop a dance unit database and conducted the following database generation:

(1) First, a typical hip-hop motion of several minutes long was recorded.
(2) Then we observed and discussed the dance sequence and selected about 60 motions as dance units that included almost all the representative hip-hop motions.
(3) We asked the dancer to separately perform each motion corresponding to each dance unit and recorded it. At the same time we asked him to start each dance motion from a "natural standing posture" and to finish in the same posture.
(4) By watching each dance motion being performed, we tried to create a robot dance motion that corresponds to human dance motion using motion editor.

4.3 Evaluation of Robot Dancing

Using the system described above we carried out simple evaluation experiments.

(1) Comparison of the two types of robot dance units
We evaluated the two types of dance units; one was generated by the professional dancer (type 1) and the other by non-experts (type 2). First we classified all the dance motions into three categories according to the complications of the motions; primary, intermediate, and advanced. And we selected one representative motion for each category. These dance motions are "Lock"(primary), "Rolling Arm" (intermediate), and "Club"(advanced). Then we generated two types of robot dance motions for each of these motions.

Ten subjects were asked to compare these two types of robot dance motions by giving a score ranging from 1 to 5 to each dance motion (1 is the worst and 5 is the best). Figure 3 shows the comparison between the two types of dance motions; robot dance motions developed by the dancer himself (type 1) and those developed by non-experts (type 2) for three kinds of motions; (a) Lock, (b) Rolling arm, and (c) Crab. Also the live dance motions performed by the dancer is shown as references. Figure 4 shows the evaluation results for each of the three kinds of motions. The evaluation result and the consideration for each motion are described below.

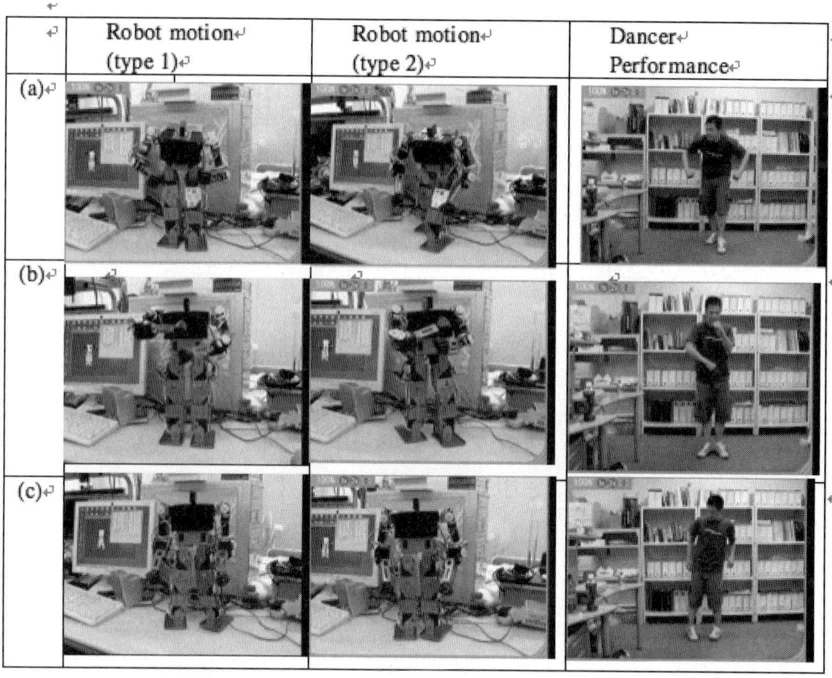

Fig. 3. Comparison of three dance motions

(a)Lock

This is a repeating motion of moving and stopping like being locked. In this move the sharpness of stopping motion is an important factor as a dance. For "sharpness," type 1 motion (motion designed by a professional dancer) obtained the higher score than type 2 (motions designed by non-experts) as expected. On the other hand, for such evaluation items as "exciting," "wonder," and "smooth," the type 2 motion got higher scores than the type 1 motion. It seems that the stop-and-go motion designed by the dancer was judged awkward by the subjects.

(b)Rolling arm

This is a motion of moving body while turning arms smoothly. For the sharpness, the type 1 motion obtained higher score than the type 2. But for other evaluation items, the type 2 motions generally got slightly higher scores. Especially for "smooth" type 2 received much higher scores against type 1. Originally this motion contains a step raising legs, and the type 1 motion precisely simulates this process and in the case of sharpness it worked well and obtained the high score. On the other hand, the type 2 motion achieves this move by sliding legs without raising legs. As a result, it was judged that the type 2 motion looked smoother than the type 1, and this gave a influence to the result of smoothness evaluation and others.

(c)Crab
This motion is a move peculiar to the Hip-hop dance. It includes a move of sliding legs sideways without raising them and fixing their backside on floor and thus moving the body sideways. The motion designed by the professional dancer (type 1) receives higher scores than the motion designed by non-expert (type 2) for almost all evaluation items. Especially, important evaluation items for this move such as "exciting," "wonder," and "smooth," the type 1 obtains fairly higher evaluation scores than the type 2.

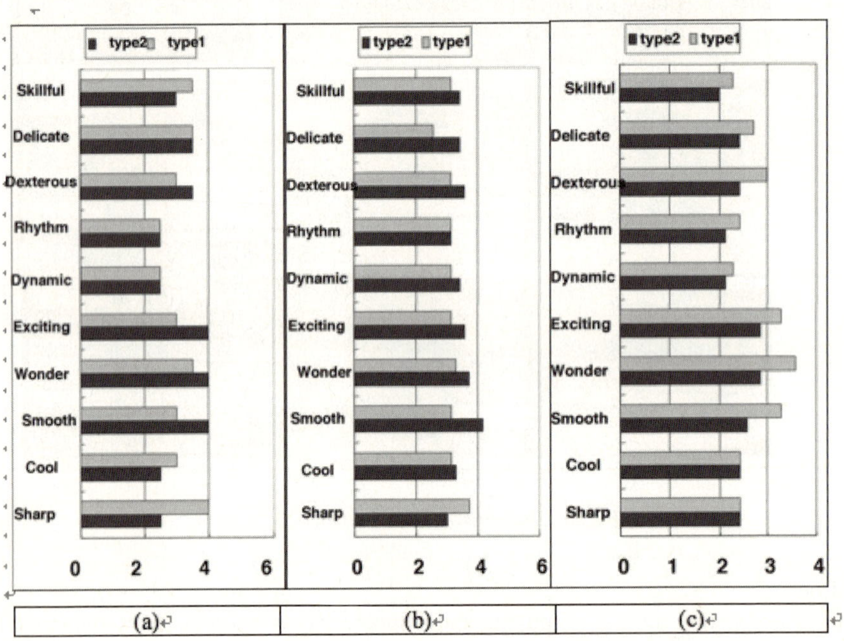

Fig. 4. Evaluation results for three kinds of motions

These result shows that as the robot dance motions become more complex, they can get higher scores. The reason for this would be that the professional dancer understands so well the characteristics of each dance motion and his knowledge and nowhow is reflected on the robot dance motion. Even though it does not appear so well in the case of simple motions, this characteristic reveals itself in the case of complicated motions. On the other hand, the motion designed by non-expert (type 2) obtained higher evaluation scores than the type 1 for simple motions. The explanation for this would be that the subjects got good impressions for the over-actions and the unstableness that the type 2 motions generally contain and express themselves. Contrarily, the type 1 motions designed by a professional dancer are sophisticated without containing such over-action nor unstableness. This characteristic sometimes leads to rather low evaluation scores as the subjects are non-expert of dances and thus could not understand the details of the dance motions where the knowledge and now-how of the professional are stored.

(4) Evaluation of the continuous dance motion

Then we carried out the experiment to evaluate the feasibility of the dance generation system. We compared two types of continuous dance motions. One is a continuous dance motion which is automatically generated by this system and has the length of about one minute (type 3). Another is the same dance motion where instead of automatic generation the professional dancer designed the whole continuous dance motion from scratch (type 4).

For evaluation twelve items generally used for the sensibility evaluation such as "stable," "soft", "smooth," and so on were selected. Each evaluation item has a seven level score ranging from -3 to 3. For example, for the evaluation item "stable" the 0 means neutral, 3 means very stable, and -3 is very unstable. Figure 5 shows the evaluation result. The type 4 obtained fairly good results for most of the evaluation

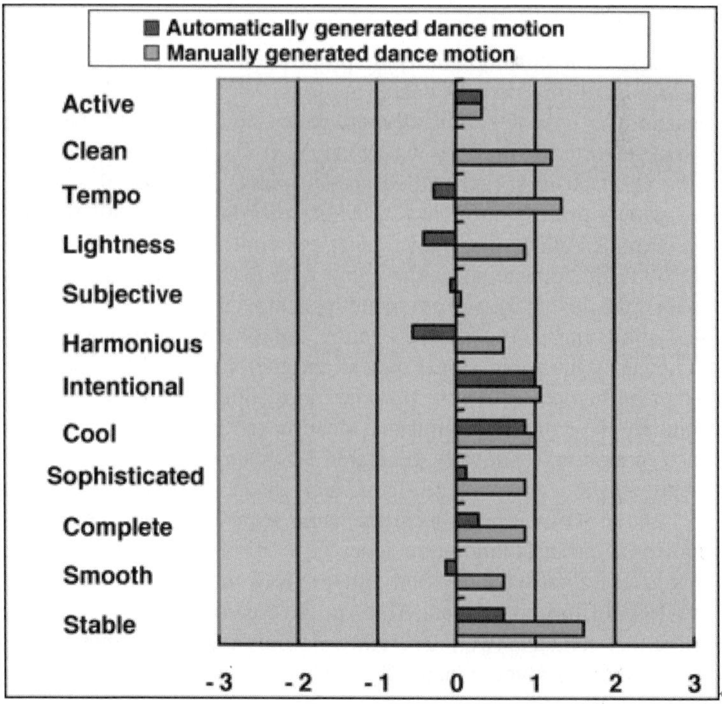

Fig. 5. Comparison between automatically generated motions and manually generated motions

items. This means that the evaluation items were fairly well selected. Generally the dance motion generated by this dance generation system (type 3) obtained lower evaluation scores than the type 4 motion. Especially, for such evaluation items as "harmony," "lightness," and "tempo, " the type 3 motion obtained minus evaluation scores. This is because the subject felt unnaturalness due to the neutral posture effect used to connect the two dance units. This means that the system still needs further improvement to generate continuous dance motion, especially for the connection of two dance units. At the same time, however, the type 3 motion got plus scores for "stability", "cool", and "intentional." Especially for "cool" and "intentional" the evaluation results are almost as high as the results of the type 4 motion. This shows that the continuous dance motion generated by this system would be effective as far as it is used as a performance even at the present stage.

The difference between type 3 and type 4 motions are that in the case of type 3 motion it goes back to a neutral position at the point of the dance unit connection. It is necessary to improve this point by introducing better neutral posture or introducing multiple neutral postures.

5 Conclusion

In this paper we proposed a dance robot system as a new application area for humanoid robots. We clarified several distinctive entertainment characteristics and investigated the role of robots in entertainment.

Based on these basic considerations we proposed a dance robot system in which a humanoid robot performs various dance motions. We hypothesized that any dance motion consists of a concatenation of short dance motions called dance units. This basic idea was imported from TTS, where any text can be converted into speech by concatenating short basic speech called speech units. Based on this basic idea, we collaborated with a professional dancer. After recording and analyzing his hip-hop dancing, we extracted about sixty dance units and converted them into the motions of a humanoid robot. By concatenating these dance units we found that a huge amount of dance variations for the hip-hop genre could be achieved.

Then we carried out two types of evaluation experiments. First we compared dance motions designed by the professional dancer and the ones by non-experts of dancing. We found that as the dance motions become more complicated and sophisticated, the dance motions by the dancer got higher evaluation results. Then we compared a continuous dance motion automatically generated by this system and one fully manually designed. Although the automatically generated dance got lower evaluation results, for some evaluation items it got almost the same scores. This means that this system is promising from a point of automatic dance generation. Further studies must address the following issues. First we have to investigate how many dance units are enough to generate any type of hip-hop dance. Also we have to investigate the feasibility of a neutral posture that connects two dance units. As only one type of neutral posture was used so far, still there is some unnaturalness for the automatically generated continuous dance motion. We expect that by introducing several other neutral postures, continuous dance motions achieved by the robot would become more natural.

References

1. Golubovic, D., Li, B., Hu, H.: A Hybrid Software Platform for Sony AIBO Robots. In: RoboCup 2003: Robot Soccer World Cup VII, pp. 478–486 (2003)
2. Ishimura, T., Kato, T., Oda, T., Ohashi, T.: An Open Robot Simulator Environment. In: RoboCup 2003: Robot Soccer World Cup VII, pp. 621–627 (2003)
3. Kerepesi, A., Kubinyi, E., Jonsson, G.K., Magnusson, M.S., Kiklosi, A.: Behavioural Comparison of Human-Animal (Dog) and Human-Robot (AIBO) Interactions. Behavioural Processes 73(1), 92–99 (2006)
4. Wama, T., Higuchi, M., Sakamoto, H., Nakatsu, R.: Realization of Tai-chi Motion Using a Humanoid Robot. Entertainment Computing. LNCS, pp. 14–19. Springer, Heidelberg (2004)
5. http://www.expo2005.or.jp/en/index.html
6. Callois, R.: Les Jeux et les Hommes. Callimard, Paris (1958)
7. Laban, R.: The Mastery of Movement. Macdonald and Evans, 4th, revised and enlarged edn. (1980)
8. Kleijn, W.B., Paliwal, K.K. (eds.): Speech Coding and Synthesis. Elsevier, Amsterdam (1995)
9. Lee, C.: Ballet in Western Culture: A History of Its Origins and Evolution. Routledge, London (2002)
10. Nakaoka, S., Nakazawa, A., Yokoi, K., Hirukawa, H., Ikeuchi, K.: Generating Whole Body Motions for a Biped Humanoid Robot from Captured Human Dances. In: IEEE 2003 International Conference on Robotics and Automation (2003)
11. http://www.nirvana.ne.jp/

Development and Evaluation of a Centaur Robot

Satoshi Tsuda[1], Kuniya Shinozaki[1], and Ryohei Nakatsu[2]

[1] Kwansei Gakuin University, School of Science and Technology
2-1 Gakuen, Sanda, 669-1337 Japan
{amy65823,scbc0052}@ksc.kwansei.ac.jp
[2] National University of Singapore, Interactive & Digital Media Institute
Blk E3A #02-04, 7 Engineering Drive 1, Singapore 117574
idmdir@nus.edusg

Abstract. Recently various types of robots are being studied and developed, which can be classified into two groups: humanoid type and animal type. Since each group has its own merits and demerits, a new type of robot is expected to emerge with greater strengths and fewer weaknesses. In this paper we propose a new type of robot called the "Centaur Robot" by merging the concepts of these two types of robots. This robot has a human-like upper body and a four-legged animal-like lower body. Due to this basic architecture, the robot has several merits, including human-like behaviors and stable walking even on non-smooth ground. We describe its hardware and software architectures. Also we describe the experiments to evaluate its walking capability.

1 Introduction

In recent years, various robots are being studied and developed in research institutes and companies that can be classified into two groups: a humanoid robot with two legs [1][2], an animal type robot with four or more legs [3][4][5]. Also a humanoid robot can be classified into those with two legs and those with wheels [6]. Each of these types has its own merits. The design of a humanoid robot with two legs is based on humans and can mimic such human motions as walking. Since this robot's behavior resembles human behavior, it might easily be introduced into society. In the future, such robots are expected to support us in various aspects of our daily life. At the same time, however, its walking capability still lacks stability, and it sometimes falls down, restricting its area of activity. Also tt has difficulty maintaining its balance on ground that is not flat. On the other hand, the merit of an animal type robot is its four legs, which allow it to walk stably even on uneven ground. Since it can also basically stand on three legs, it can adopt to various ground pattern changes. So far, however, the robot has mainly been developed as a pet to which useful applications have rarely been applied. A humanoid robot with wheels for locomotion, which we call a wheel type robot, can move very smoothly and stably on the ground. It rarely falls down. It can even move on slightly uneven ground. On the other hand, it has no ability to move on stairs, which greatly restricts its area of activity since houses usually contain stairs and other types of height differences.

I. Lovrek, R.J. Howlett, and L.C. Jain (Eds.): KES 2008, Part I, LNAI 5177, pp. 692–700, 2008.

One approach to overcome these problems is to develop new types of robots by merging the strengths of existing robots. In this paper we propose a new type of robot with a human-like upper body and an animal-like lower body that we call a "Centaur Robot." In the following sections, we describe its basic concept and then its detailed software/hardware architectures. Also we describe the experiments to evaluate how this robot can achieve a waking capability on non-smooth ground.

2 Related Works

Recently, especially in Japan, various kinds of robots have been studied and developed, particularly humanoid robots that are expected to support our daily life. For example, HONDA has developed a humanoid robot called ASIMO that has sophisticated walking capability [1]. For animal types of robots, on the other hand, most have been studied and developed as pets instead of supportive robots, including AIBO developed by Sony [3].

Although much research/development continues on humanoid and animal types of robots, little research has integrated these two types for several reasons. One reason is that since there are so many research themes for new functions and improvements for each of these types of robots, researchers have little incentive to concentrate on new types of robots that go beyond humanoid or animal types. Another is that even myths or folktales only contain a few examples of such creatures as centaurs, mermaids, and sphinxes in which humans and animals are integrated. Thus it is rather hard to imagine the functions and application areas that such a new type of robot might have.

Therefore, we developed a centaur robot because we believed by integrating two types of robots we could create a new type of robot with advantages over conventional robots.

3 Humanoid Robots

In our work, we are developing a robot that can stably achieve various motions by merging two types of robots: a humanoid and an animal.

There are two approaches for such integration: from the humanoid robot side and from the animal robot side. The former approach tries to realize a four-legged body as well as four-legged walk while maintaining a human-like upper body and achieving human-like motions. On the other hand, the latter approach achieves various human-like motions by adding a human upper body to a four-legged robot. In our study, we chose the former approach and modified the hardware and software of a humanoid robot to realize a centaur robot.

We adopted a humanoid robot developed by Nirvana Technology as a platform robot [7]. This robot has 22 servo motors that can express various human-like motions. Figure 1 shows its appearance, and Table 1 shows its specifications. Figure 2 illustrates the construction of its hardware. The control board, on which a microprocessor SH2 is attached, is connected to the servo motors, a gyro sensor, acceleration sensors, PC, and a battery. The program on the controller can achieve autonomous robot behaviors. At the same time, we can send commands to the robot by PC.

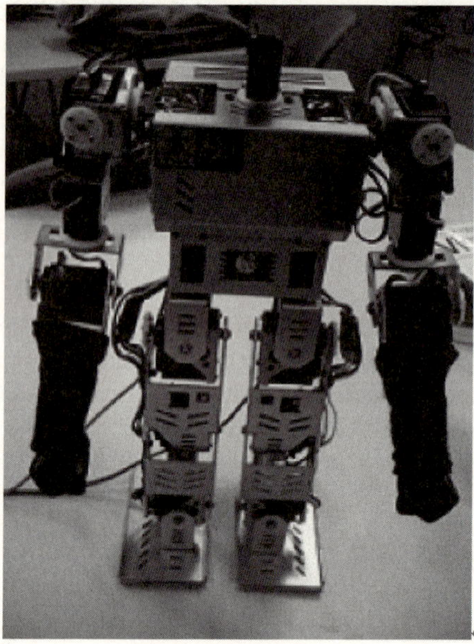

Fig. 1. Humanoid robot

Table 1. Specifications of humanoid robot

Size/Weight	34 cm/1.7 kg
Degree of flexibility	22 (12 legs, 8 arms, 1 waist, 1 head)
CPU	SH2/7047F
Motor	KO PDS-2144, FUTABA S3003, FUTABA S3102, FUTABA S3103
Battery	DC6V

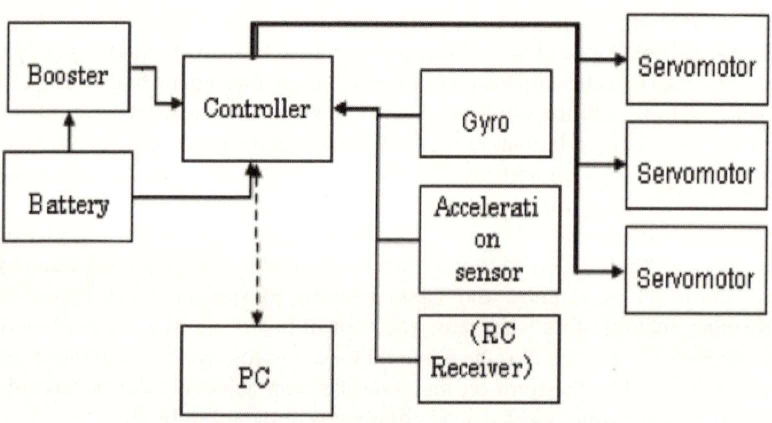

Fig. 2. Hardware construction of humanoid robot

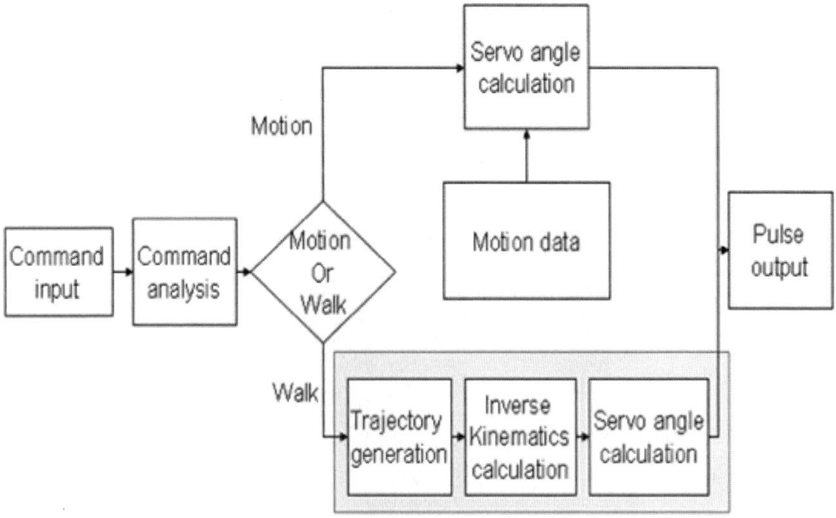

Fig. 3. Software construction of humanoid robot

Figure 3 illustrates the software construction. The calculation of the commands necessary to move each motor is carried out each fifteen milliseconds and sent to each servo motor. The instructions to the robot from the PC are first analyzed and based on results go through one of two processes: one command for walking and other commands for other motions. For other commands, the motion data corresponding to the command is read from memory and the control data for each motor is calculated, and then the control data is sent to each servo motor. On the other hand, if the input command is a command for walking, then the real time calculation of the control data for each servo motor is carried out and sent to each servo motor. Calculation consists of three processes: trajectory generation calculation, inverse kinematics calculation, and servo motor angle calculation. In trajectory generation calculation, the position of each ankle studied by observing human walking motion is calculated every fifteen seconds. Then by inverse kinematics calculation the rotation angle of each foot joint is calculated for the same timing. Based on these calculations, finally the angle of each servo motor is calculated. Thus the rotation angle to be achieved for each motor is sent every fifteen milliseconds.

4 Centaur Robot

4.1 Overview

We developed a centaur robot based on the humanoid robot described in the previous section. We prepared two humanoid robots and used one as a front body. For another robot, we only used its lower body as a back of the centaur robot. Then we connected these two parts by a flat plastic board that functions as the shoulder part. Figure 4 shows the centaur robot's appearance.

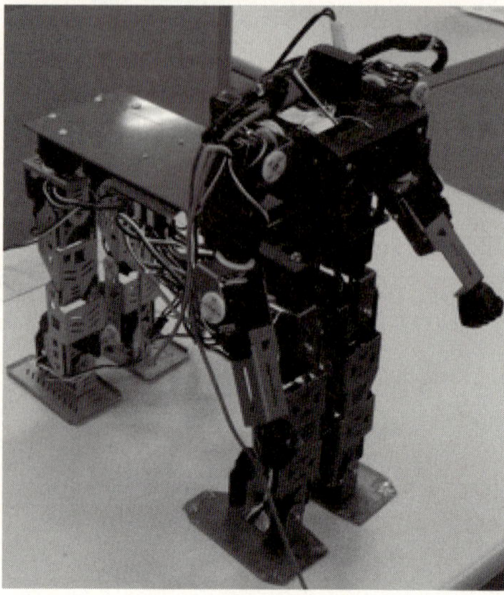

Fig. 4. Centaur robot

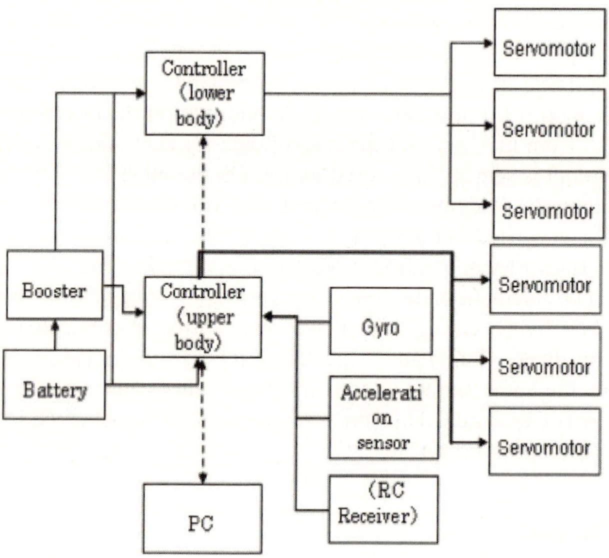

Fig. 5. Hardware construction of centaur robot

4.2 Hardware Construction

Now we explain the robot's hardware construction, as illustrated in Figure 6. Apparently for the front the hardware of the original humanoid robot was used, and for the back only the lower body was used. But a comparison of Figs. 3 and 5 shows that this robot's control structure is somewhat different from the original. Two controllers were used for complete control of the robot. One controls the servo motors required for upper body motions. The other controls the servo motors corresponding to the lower body. Since all the sensors are provided for the upper body, the controller corresponding to the upper body manages all sensor feedback. We adopted these two boards for several reasons. One, by using two boards, one of which controls the motions of the upper body and the lower body, it is possible to separately control the behaviors of the upper body as well as the lower body. For the power supply and battery, both controllers are connected to one battery. Also commands from PC are sent to both controllers.

4.3 Software Construction

Next, we explain the robot's software construction, as illustrated in Figure 6. The software of the original humanoid controls both the upper and lower bodies together. For the centaur robot, we checked all the original robot's software and separated the software codes into two groups: one that controls the upper body and another that controls the lower body. Thus we reconstructed the whole software. For the upper body, it is unnecessary to carry out calculations for walking. When commands other than a walking command are sent from the PC, it retrieves motion data stored in the memory and sends the necessary rotation angle data to each servo motor. On the other

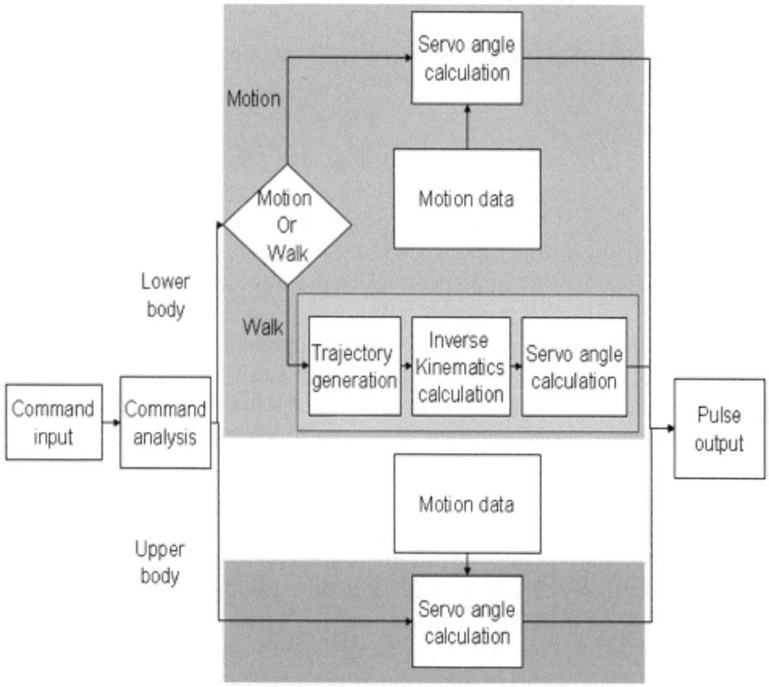

Fig. 6. Software construction of centaur robot

hand software corresponding to the lower body must treat two types of commands as in the case of the original humanoid robot: a command for walking and other commands for additional motions. Also we adopted a method of inserting an arbitrary phase shift between the servo motor control of the front and back legs so that the robot can adopt the most adequate walking motions depending on the walking speed. By adopting such basic software structure, robot control has the following merits:

(1) The upper and lower body motions can be controlled separately. So far all the motion data developed for achieving various types of humanoid robot motions must be developed to describe the whole body movement. Since the motions of the upper and lower bodies have been separated, we can separately develop two types of motions, and by combining these two types of data, we can generate various kinds of whole body movements for the robot. This idea can easily be applied to the original humanoid robot.
(2) The front and back body movements can be separately controlled. Although it seems natural to let the front lower body and back lower bodies perform identical motions, sometimes it is better to control the two bodies by different body motions. Especially in the case of walking and running motions there would be some differences between these two bodies. For example, for trot type walking there should be a 180° phase shift between the front and the back legs. In the case of gallop running, the front legs and the back legs should move synchronously.

4.4 Evaluation of the Robot

We carried out several experiments to evaluate the motion capability of our centaur robot.

(1) Walking capability
We inserted a phase shift of 0 degree, 90 degree, and 180 degree between the walking motion cycle of the front and back legs. These waling styles correspond to those of animals such as "pace," "gallop," and "trot." We confirmed that the robot could move smoothly with almost the same speed by adopting each of these walking styles. Therefore for walking stability and speed all of the three walking styles perform the same capability.

As a next step, as we expect that one of the applications of this robot would to carry light load, we evaluated the walking stability from a point of carrying a load. For this we measured the tilt angle of the shoulder when it walks by fixing a gyro sensor on its shoulder and obtaining tilt data from it. We observed the time sequence data of the tilt angle four times for each walking style and averaged the data. Figures 7, 8, and 9 show the obtained data for a phase shift of 0 degree (pace), 90 degree (gallop), and 180 degree (trot).

Figures 7, 8, and 9 show that the tilt angle for pace walking style is larger than other two styles especially at a first step. Thus this waking style is not adequate for carrying a load such as a teacup or a newspaper. When comparing gallop and trot waling styles, although for the first step the tilt angles are almost the same, for the following steps trot walking style shows far better result. In this case of trot walking the front and back legs move in opposite modes. For example, when the front left leg moves forward, so does the back right leg. These results show that the tilt angle for

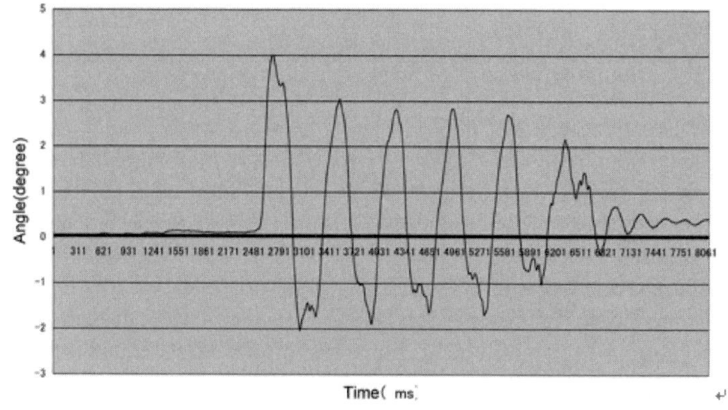

Fig. 7. Tilt angle for "pace" walking style

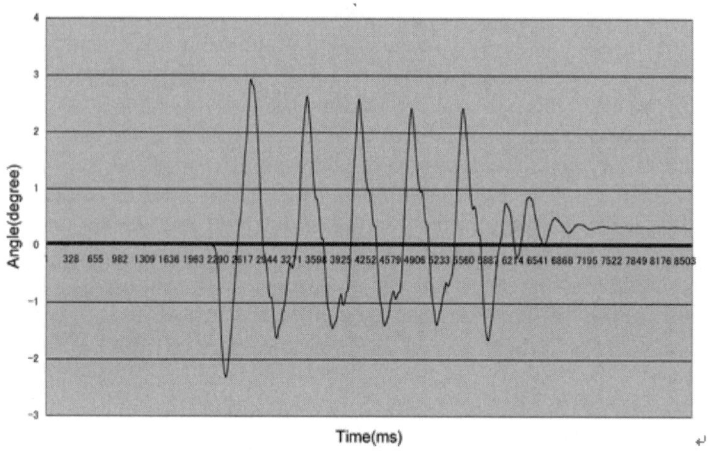

Fig. 8. Tilt angle for "gallop" walking style

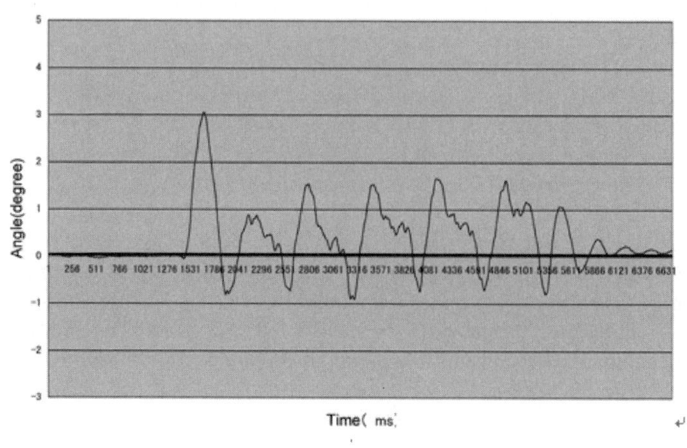

Fig. 9. Tilt angle for "trot" walking style

trot walking style is lower than the other two and thus more stable when carrying a load. At the same time even in the case of trot, the first step causes a little bit large tilt angle. And this is the issue we have to improve in the further study.

(2) Capability for other motions
We developed various types of human-like motions for the original humanoid robot [7]. An interesting question is which of these motions could work well on the centaur robot. We tried to transfer the humanoid robot motions to this robot and found that most of the motions worked fairly well on this robot. On the other hand, motions including such postures as bending and twisting did not work well or needed modifications. One interesting future research theme is automatically transferring the humanoid robot motions to the motions of four-legged robot such as this robot.

5 Conclusion

In this paper we proposed a new type of robot that is an integration of two types of robots: humanoid and four-legged. We adopted a humanoid robot with two legs and walking capability as a platform for this new robot. By integrating two of the humanoid robots we easily and successfully developed a centaur robot. We described its software and hardware and also its merits. We confirmed that by inserting a phase shift of 0 degree, 90 degree and 180 degree between the front and back leg motions the robot can stably achieve pace, gallop, and trot walking motions. Then we evaluated these three walking styles from a point of tilt angle of its shoulder and found that the trot walking style is more stable than other styles. Since this robot has merits of both humanoid and four-legged robots, we are also going to evaluate its new capabilities that neither of the two type robots could achieve by themselves.

References

1. http://www.honda/co.jp/ASIMO/
2. Friedmann, M., Kiener, J., Petters, S., Thomas, D., von Stryk, O., Sakakmoto, H.: Versatile, high-quality motions and behavior control of humanoid soccer robots. In: Workshop on Humanoid Soccer Robots of the 2006 IEEE International Conference on Humanoid Robots, pp. 9–16 (2006)
3. http://www.jp.aibo.com/
4. Golubovic, D., Li, B., Hu, H.: A Hybrid Software Platform for Sony AIBO Robots. In: RoboCup 2003: Robot Soccer World Cup VII, pp. 478–486 (2003)
5. Kerepesi, A., Kubinyi, E., Jonsson, G.K., Magnusson, M.S., Kiklosi, A.: Behavioural Comparison of Human-Animal (Dog) and Human-Robot (AIBO) Interactions. Behavioural Processes 73(1), 92–99 (2006)
6. Ishiguro, H., Ono, Y., Imai, M., Kanda, T.: Development of an interactive humanoid robot Robovie -An interdisciplinary approach. In: Jarvis, R.A., Zelinsky, A. (eds.) Robotics Research, pp. 179–191. Springer, Heidelberg (2003)
7. Wama, T., Higuchi, M., Sakamoto, H., Nakatsu, R.: Realization of Tai-chi Motion Using a Humanoid Robot. Entertainment Computing. LNCS, pp. 14–19. Springer, Heidelberg (2004)

Collaboration Options for Small and Medium Size Enterprises

Sylvia Encheva[1] and Sharil Tumin[2]

[1] Stord/Haugesund University College, Bjørnsonsg. 45, 5528 Haugesund, Norway
sbe@hsh.no
[2] University of Bergen, IT-Dept., P. O. Box 7800, 5020 Bergen, Norway
edpst@it.uib.no

Abstract. Collaboration options for small and medium size enterprises interested in entering new markets while providing fast, efficient, reliable and customized transport solutions are in the main focus of this work. For automated selection of the most desirable collaboration options we propose employment of a decision support system, where all collaboration options for a particular firm are evaluated by three experts with respect to two criteria.

Keywords: intelligent infrastructures and collaboration options.

1 Introduction

Alliances of small and medium size enterprises can solve many of the problems caused by the lack of direct transport. "Strategic alliances provide an effective means to improve both the economies of scale and scope offered by traditional modes of organization," [10].

This paper discuses collaboration options for small and medium size enterprises interested in entering new markets while providing fast, efficient, reliable and customized transport solutions. For automated selection of the most desirable collaboration options we propose employment of a decision support system, where all collaboration options for a particular firm are evaluated by three experts with respect to two criteria.

The rest of the paper is organized as follows. Related work and supporting theory may be found in Section 2. The evaluation model in Section 3. The system implementation is described in Section 4. The paper ends with a conclusion in Section 5.

2 Related Work

In this paper the term collaboration means a generic, cooperative interaction between firms to achieve some agreed upon objectives. Market considerations imply that early entry into large, growing markets is more likely to lead to success [14]. Theory and application of cost-benefit analysis are presented in [8] and [9].

I. Lovrek, R.J. Howlett, and L.C. Jain (Eds.): KES 2008, Part I, LNAI 5177, pp. 701–708, 2008.

Business decisions on projects with potential positive rate of return are discussed in [1] and the human side of a decision making process is considered in [7].

An object oriented model of a simulation based decision support for supply chain logistics is presented in [5]. A decision support system for administration of shipping enterprises is presented in [6]. Both cases involve statistical methods where our approach is based on classification methods from formal concept analysis and lattice theory.

Let L be a non-empty ordered set. If $sup\{x, y\}$ and $inf\{x, y\}$ exist for all $x, y \in L$, then L is called a *lattice*, [3]. In a lattice illustrating partial ordering of knowledge values, the logical conjunction is identified with the meet operation and the logical disjunction with the join operation.

By $\mathcal{N} - SubG$ we denote normal subgroups of the ordered subset of subgroups $< SubG; \subseteq >$ of a group G, and by $\mathbf{D_4}$ the groups of symmetries of a square.

Let L be a lattice with 0 and 1. An *orthocomplementation* on L is a unary operation $a \to a^{\perp}$ on L satisfying the following three conditions:

(i) $a \wedge a^{\perp} = 0$, $a \vee a^{\perp} = 1$, that is, a is a complement of $a^{\perp}$;

(ii) $a \leq b$ implies $b^{\perp} \leq a^{\perp}$;

(iii) $a^{\perp} \perp$ for every $a \in L$.

An orthocomplemented lattice is a lattice with 0 and 1 carring out an orthocomplementation [13].

A lattice is upper-semimodular if

$$a \wedge b \prec a \Rightarrow b \prec a \vee b, \forall a, b \in L,$$

where $a \prec b$ stands for b covers a, i.e. if $b \geq c > a$ implies $b = c$ for every $c \in L$, [2]. These lattices are referred to as semimodular in [13].

A nested lattice is the product of two concept lattices, graphically represented by a nested line diagram, sometimes referred to as inner and outer lattice.

A nested line diagram consists of an outer diagram that contains inner diagrams in each node. Inner diagrams are not necessarily congruent but only substructures of congruent diagrams. Congruent diagrams are shown as structures possessing some unrealized concepts.

3 Evaluation of Collaboration Options

We propose employment of a decision support system for selecting the most desirable collaboration options. All collaboration options for a particular firm are evaluated by three experts with respect to two criteria, importance and relevance. Importance refers to the degree of interest a collaboration option has regarding the firm's current goals and relevance refers to evaluation of the amount of resources a collaboration option requires considering results and organizational

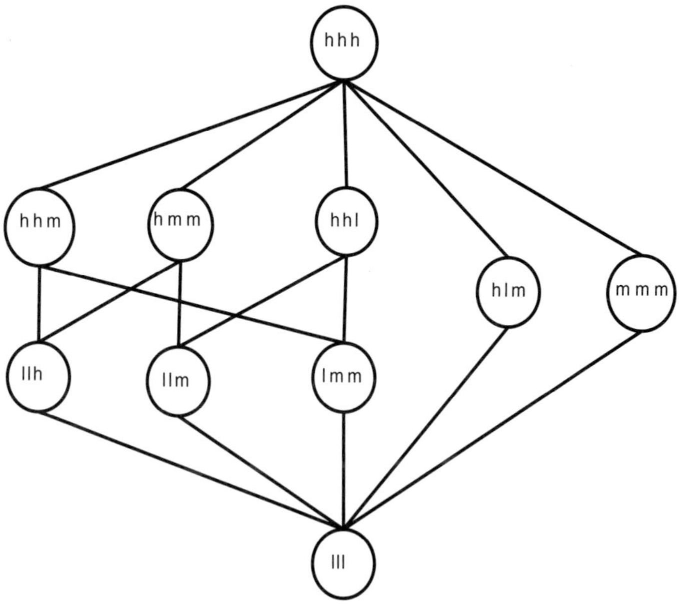

Fig. 1. Lattice illustrating response combination triplets related to a single criteria

constrains. To each criteria an expert can assign exactly one of the following recommendations -

- high priority abbreviated (h),
- medium priority abbreviated (m), and
- low priority abbreviated (l).

The experts' recommendations are not graded, i.e. they carry equal weight and their order does not effect the decision process. This results in ten response combination triplets per criteria -

- high, high, high, abbreviated (hhh)
- high, high, medium, abbreviated (hhm)
- high, high, low, abbreviated (hhl)
- high, medium, medium, abbreviated (hmm)
- high, medium, low, abbreviated (hml)
- high, low, low, abbreviated (hll)
- medium, medium, medium, abbreviated (mmm)
- medium, medium, low, abbreviated (mml)
- medium, low, low, abbreviated (mll)
- low, low, low, abbreviated (lll)

Every collaboration option is assigned an ordered pair of response combination triplets where the first one is related to the importance criteria and the second one to the relevance criteria. In order to simplify the visibility of all possible

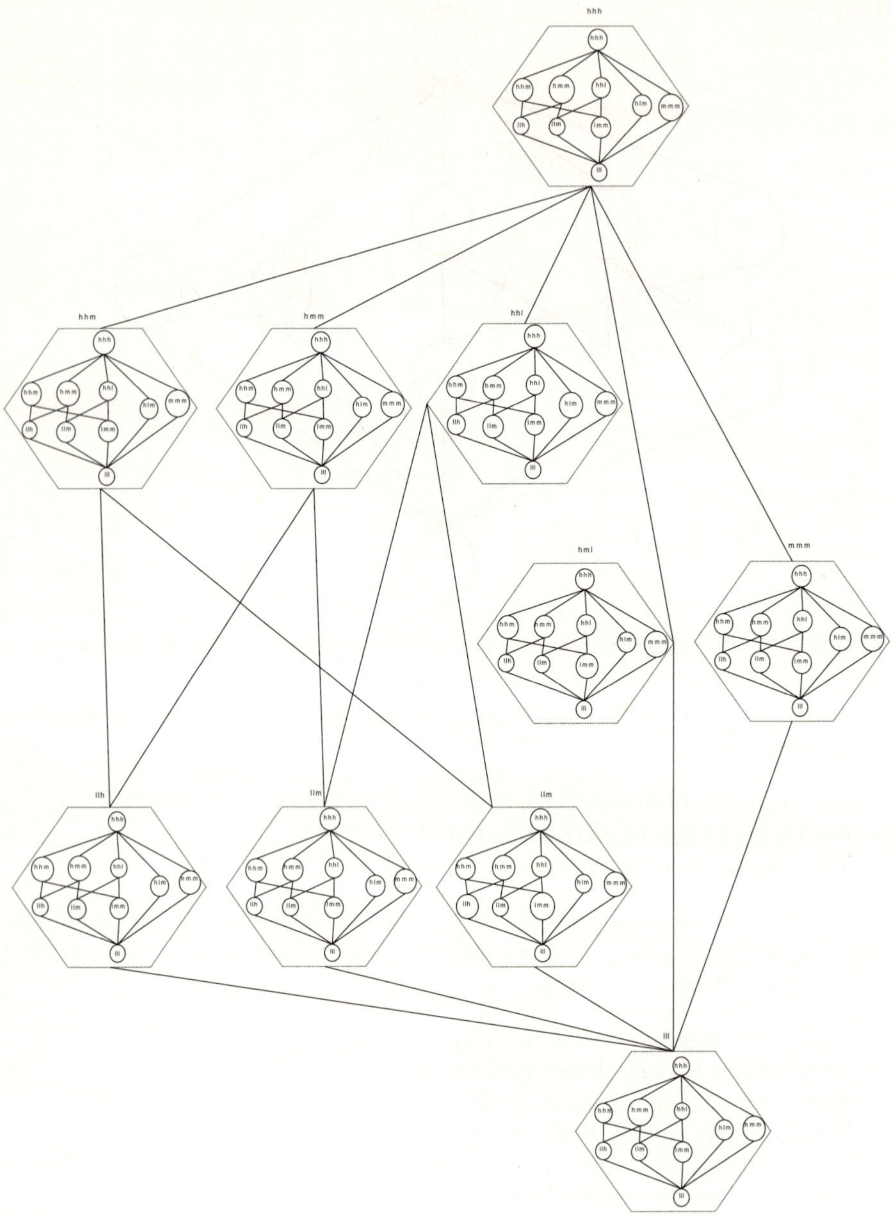

Fig. 2. Lattice illustrating response combination triplets related to two criteria

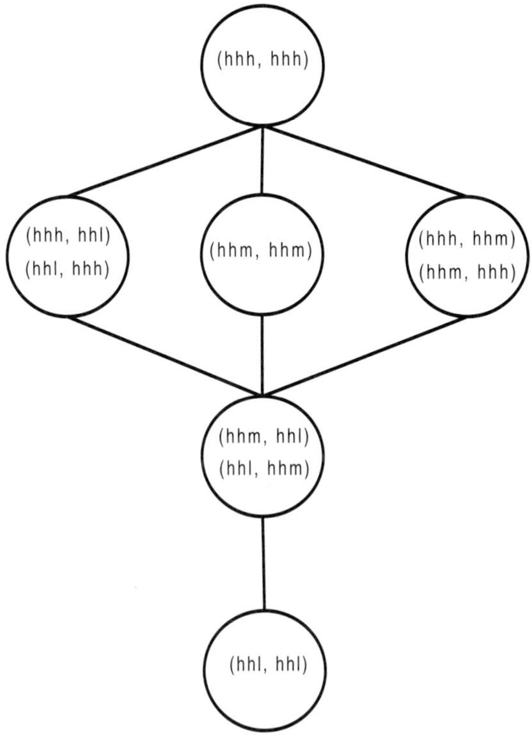

Fig. 3. Lattice with evaluations implying profitable projects

recommendations we suggest an arrangement of the recommendations in an orthocomplemented lattice as shown in Fig. 1. In this case the triplets (hhh) and (lll) correspond to the 1 and 0 elements in a lattice. Every node in the lattice in Fig. 1 is labelled with one of the ten response combination triplets. A collaboration option is placed in a node if the first criteria is assigned the same triplet as the node label. The lattice in Fig. 1 is actually the simplest orthocomplemented lattice that is not orthomodular [13].

The nested lattice in Fig. 2 shows all possible evaluation outcomes, where the outer lattice has nodes with labels corresponding to triplets assigned to the first criteria, and nodes in the inner lattice correspond to triplets assigned to the second criteria. The idea is to show where is the place of any single collaboration option with respect to the assigned evaluations.

We assume projects are considered as being profitable if they are assigned an ordered pair of triplets where both triplets belong to the set {hhh, hhm, hhl}. The resulting triplets are arranged in a lattice in Fig. 3. This is a lattice for the elements of $\mathcal{N} - Sub\mathbf{D_4}$ group of symmetries of a square, [3].

The lattice in Fig. 3 illustrates graphically projects' evaluations coming from a single request. Two collaboration options assigned symmetrical pair of triplets like f. ex. {hhh, hhm} and {hhm, hhh} are considered equally profitable and are therefore placed in one node. The lattice facilitates automated projects' evaluation process in case of a second request where new projects are added and or new experts are involved.

4 System Implementation

The prototype system (Fig. 4) is implemented using a typical Web application framework with the following system components

1. Apache Web server
2. Python programmable runtime environment
3. SQLite database

The Apache Web server provides system users with Web-based interface to the system. All users, administrators, experts, and clients interact with the system using Web browsers. Python is used to program the system middleware. All data are stored into multiple databases provided by SQLite database engine.

SQLite database is incoperated into the system by using a SQL interface compliant with the DB-API 2.0 specification provided by 'pysqlite' module. Web applications programming is done with the help of 'mod_python' extension module into Apache runtime environment All user requests and responses are processed by the Python programs running under Apache Web server main process.

Web application server supports the following sub-systems

- User authentication
- User authorization
- Data gathering
- Decision support
- Reports generation

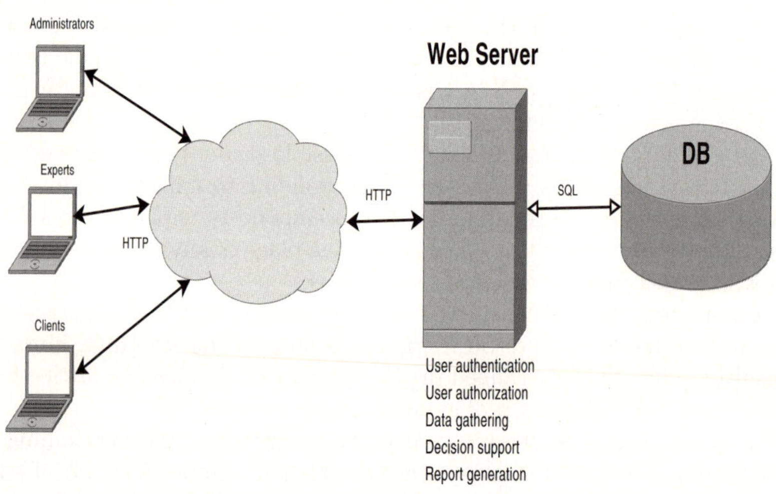

Fig. 4. The system

The clients are the normal users of the system who are interested in the expert evaluation of a specific collaboration option. Once such request is sent to the system, it will provide the client with expert analysis in the form of recommendation supported by analytical report and visualization aids in the form of lattice diagrams as shown in Fig. 2 and Fig. 3.

If a specific collaboration option is not yet being evaluated by three experts then the system will prepare Web-based questionnaires, gather relevant data and invite the experts to evaluate the collaboration option requested by the client. The client will be notified when the evaluation process is completed.

The lattice diagram shown in Fig. 3 is used as the basic rule as to whether to recommend a specific collaboration as being profitable or not. The experts' evaluations are not graded, thus combinations like for example, 'hhl', 'hlh', and 'lhh' are considered to be equivalent. Fig. 3 shows the lattice for profitable evaluations combinations where '(hhh, hhh)' being the most profitable and '(hhl, hhl)' being the least profitable.

Under the process of analysis the system converts recommendation levels into numerical characters

$$\begin{aligned} &\mathrm{h} \rightarrow 3 \\ &\mathrm{m} \rightarrow 2 \\ &\mathrm{l} \rightarrow 1 \end{aligned}$$

Any of the equivalent combinations is represented with the descending sort order of three digits numbers, for example 'lmh', 'mlh', 'mhl', 'hlm', and 'hml' is represented as '321'. Thus a recommendation level shown in Fig. 3 can be mapped into a number (recommendation index), for example '(hhl, hhl)' is mapped into $331 \times 331 = 109561$. The highest combination of which contains only one 'h' is 'hmm', and this is mapped into $322 \times 322 = 103684$. With this algorithm, profitable collaboration options are those with recommendation index greater then 109560.

Together with lattice diagrams and recommendation indexes the system can assists clients to decide on an optimal collaboration option calculated from the recommendations of the expert evaluators.

5 Conclusion

The proposed model facilitates an evaluation process of investment alternatives based on qualitative data and chooses those investments that are feasible within constrains of available capital.

References

1. Bierman, H., Smidt, S.: The Capital Budgeting Decision, Routledge, New York (2007)
2. Bordalo, G.H., Rodrigues, E.: Complements in modular and semimodular lattices. Portugalie Mathematica 55(3), 373–380 (1998)

3. Davey, B.A., Priestley, H.A.: Introduction to lattices and order. Cambridge University Press, Cambridge (2005)
4. Dyer, J.H., Kale, P., Singh, H.: How to Make Strategic Alliances Work. Sloan Management Review 42(4), 37–43 (2001)
5. Ganapathy, S., Narayanan, S., Srinivasan, K.: Logistics: simulation based decision support for supply chain logistics. In: Proceedings of the 35th conference on Winter simulation: driving innovation, New Orleans, Louisiana, pp. 1013–1020 (2003)
6. Jing, L., Jiafu, W.: Administration decision support system for shipping enterprises. In: Proceedings of 1999 IEEE International Conference on Systems, Man, and Cybernetics, vol. 3, pp. 1048–1053 (1999)
7. Klein, G.A.: Recognition-Primed Decision Making. In: Klein, G.A. (ed.) Sources of power: How people make decisions, pp. 15–30. MIT Press, Cambridge (1998)
8. Layard, R., Glaister, S.: Cost-Benefit Analysis, 2nd edn. Cambridge University Press, Cambridge (1994)
9. Nas, T.F.: Cost-benefit Analysis, Theory and Application. Sage Publications, Thousand Oaks (1996)
10. Rai, A., Borah, S., Ramaprasad, A.: Critical Success Factors for Strategic Alliances in the Information Technology Industry: An Empirical Study. Decision Sciences 27(1), 141–155 (1996)
11. Rosenbloom, R.S., Christensen, C.: Technological discontinuities, organizational capabilities and strategic commitments. Industrial and Corporate Change 3, 655–685 (1994)
12. Sim, K.M.: Bilattices and Reasoning in Artificial Intelligence: Concepts and Foundations. Artificial Intelligence Review 15(3), 219–240 (2001)
13. Stern, M.: Semimodular Lattices Theory and Applications. Cambridge University Press, Encyclopedia of Mathematics and its Applications (1999)
14. Zirger, B.J., Maidique, M.: A model of new product development: an empirical test. Management Science 36, 867–883 (1990)

Service Robot System Using Personal Attribute and Acquisition Preference Attribute with Networked Robots

Kota Nakamura, Yoshiharu Yoshida, Toru Yamaguchi,
and Eri-Shimokawara-Sato

Tokyo Metropolitan University,
6-6 Asahigaoka, Hino, Tokyo 191-0065 Japan
nakamura-kota@sd.tmu.ac.jp, yoshiharu@fml.ec.tmit.ac.jp,
yamachan@tmu.ac.jp, eri@sd.tmu.ac.jp

Abstract. In this research, the authors aimed to implement human-centered system, which understands user and provides a service which suits user. This paper consists of two systems. First, the authors implement service robot system, which provides information by employing user's personal attribute. "Personal attribute" consists of "Preference attribute" and "Body attribute" The system has networked robots share personal attribute information and provides useful information for user. Second, the authors focus on a method for getting user's preference attribute. The authors provide a method for getting user's preference attribute from user's buying history and watching TV history by SOM algorithm. Then, the authors implement service robot systems in a store and room by using the method. In a store and room, the system gets user's preference attribute and provides a service. Finally authors summarized our service robot systems.

Keywords: Preference attribute, body attribute, networked robot, SOM algorithm.

1 Introduction

Recently, the products related to IT have been developed, and have been high performance. But, the products related to IT have so many functions that people cannot use all functions. So, the systems have come to provide functions and service that suit each person. These systems are the "human-centered systems" that focus on the user mainly. Various systems like "human-centered systems" have come to be implemented by networking robots with spreading the word "ubiquitous network".

In this research, it is the authors' aim to implement human-centered systems that understand user and provide a service that suits user.

2 Networked Robot

Recently, the word "robot" is used in a broad sense than before. As illustrated in figure 1, ATR that researches the networked robot divides the robots into three types. The different types robots share information each other on the network, cooperate with each other, and provide a person with a service [1][2].

I. Lovrek, R.J. Howlett, and L.C. Jain (Eds.): KES 2008, Part I, LNAI 5177, pp. 709–716, 2008.

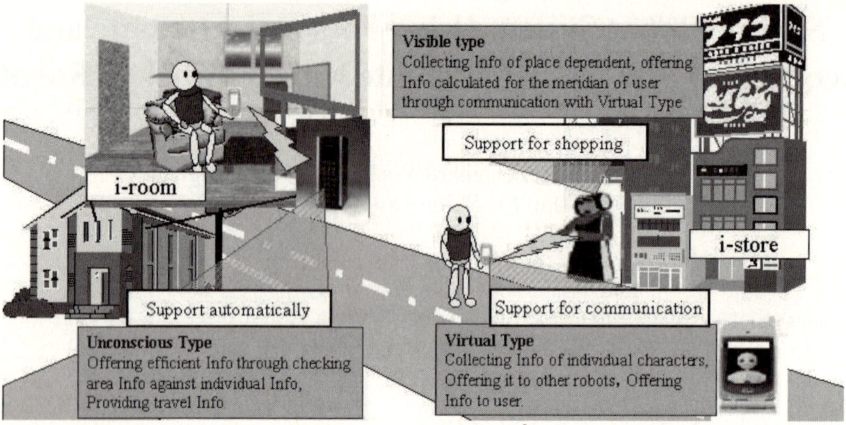

Fig. 1. Networked robots

3 Service Robot System Using Personal Attribute with Networked Robots

In this section, the authors defined "body attribute information" as information based on person's bodily features such as sex, "preference attribute information" as information that is been abele to get from a person's daily life such as favorite color. The authors defined "body attribute information" and "preference attribute information" as "personal attribute information". The authors implemented the system that had the robots share personal attribute information, and the system provided a service that suited the user. The authors implemented this system as a clothing store system.

3.1 Composition of System in Section 3

The system which the authors implement is the service robot system by indicating the show window. As illustrated figure 2, the authors have user's virtual type robot interact with networked robots. In this system the user's virtual type robot saves user's personal attribute information, shares user's personal attribute information with the networked robots, gets products' information that suits user's personal attribute.

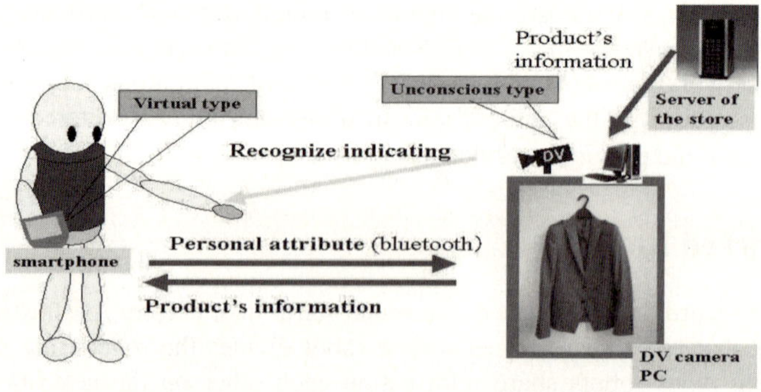

Fig. 2. Composition of system in section 3

3.2 Match and Inference of Experiment in Section 3

In this system, the authors use clothes data of sex and the size as body attribute information, and the authors use clothes data of the color and the category as preference attribute information [3].The authors think that it is meaningless that the system recommends a product which does not suit user's body attribute even if the product suits preference attribute. Therefore, first, the system infers from user's body attribute, and makes database of only products that are matched user's body attribute. Second, the system infers from user's preference attribute information in the database of only products that are matched body attribute. Finally, the system decides the recommended products which don't suit only body attribute but also preference attribute. As illustrated figure 3, the inference model's structure, which the authors use in section 3, is three layers. Between two layers are connected with each other. They compose bidirectional associative memory (BAM) [4].

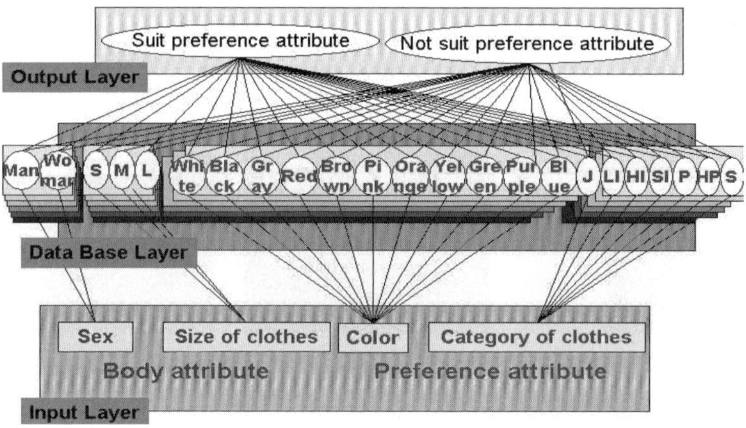

Fig. 3. Inference model in section 3's system

The bottom of the layer is an input layer and personal attribute information is input there. The system extracts two products' data from database of only products that are matched body attribute, and infers from its. The products' data of the sex, the size, the color, and the category of clothes is stored in middle layer, and then the system outputs a product which suits user's preference attribute information and a product which does not suit user's preference attribute information. The system extracts another product from database of only products that are matched body attribute, infers from the product which was extracted and the product which had suited user's preference attribute in the previous inference. By repeating this work, the system decides the most suitable product for user's preference attribute information in database of only products that are matched body attribute. Finally, the system provides information of it as a recommended product that suits user's personal attribute information.

3.3 Result of Experiment in Section 3

In section 3, the authors will show you result of eight subjects' experiment.

3.3.1 Experiment at a Show Window

As illustrated figure 4, the system tracks coordinates of subject's face and hand. The system recognizes whether the right's product is indicated or the left's product is indicated or the center's product is indicated by the difference between coordinates of face and hand. A subject sends personal attribute information from smart phone to PC with camera with indicating the product. When PC with camera gets personal attribute information, PC with camera begins to infer. As the result, PC with camera sent e-mail, which had recommended products' information that suited subject's personal attribute information with URL and the information of the product which the subject indicated, to the subject's smart phone. When a subject clicked URL in which subject is interested, the image of the clothes is displayed.

3.3.2 Evaluation of Experiment in Section 3

In section 3, the authors conducted a questionnaire. Number of subjects was eight. The questionnaire consisted of two questions. The questions are "How did you like a recommended products" and "utility of this system". The authors let subjects evaluate the system on a scale of one to five. The figure 5 shows the questionnaire's result of average. The utility of this system was shown by this figure.

Fig. 4. Image of Indicating, Tracking and, Show window

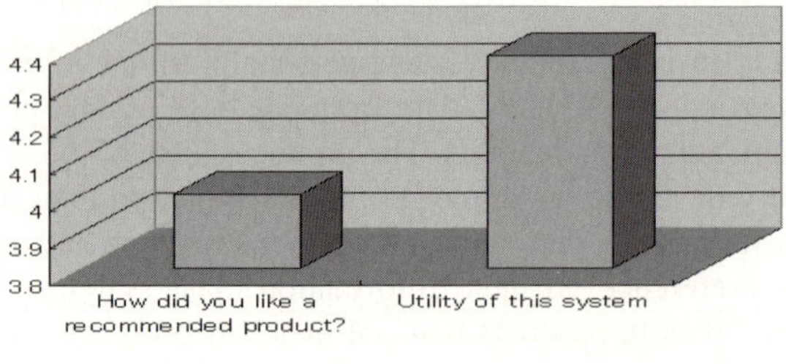

Fig. 5. The questionnaire's result of average in section 3

4 Acquisition Preference Attribute with Networked Robots

In this section, the authors show a method for getting preference attribute from user's buying history and watching TV history by SOM algorithm. Then, the authors propose service robot systems in a store and room by using the method. In a store and room, the system gets user's preference attribute and provides a service.

4.1 A Method for Getting Preference Attribute

In section 4, the authors get preference attribute from user's buying history and watching TV history. First, the system makes "general preference attribute profile" from buying history and watching TV history whose kinds are different from each other. The system gets preference attribute by inputting buying history and watching TV history into general preference attribute profile. Buying history is constructed based on Japan Standard Commodity Classification. Watching TV history is constructed based on ONTV JAPAN (http://www.ontvjapan.com/program/). ONTV JAPAN is the web site which has iEPG information. ONTV JAPAN divides TV shows into eleven types. In section 4, the authors got five subjects' buying history and watching TV history. The authors input frequency of five subjects' buying history and watching TV history as five dimensional vectors to input into Self-organizing map (SOM). Left one of figure 6 shows Self-organizing map before learning. Right one of figure 6 is Self-organizing map after learning. Right figure shows relation between categories.

As illustrated figure 7, distance between categories is strength of relation between categories. The smaller distance between categories is, the stronger relation between categories is. The system makes general preference attribute profile based on this strength of relation. In section 4, the author used only five dimensional vectors as attribute which is input into SOM because the authors wasn't able to get more subjects' buying history and watching TV history. But, SOM is able to be input multidimensional vectors. So, SOM is able to be input more large user's data.

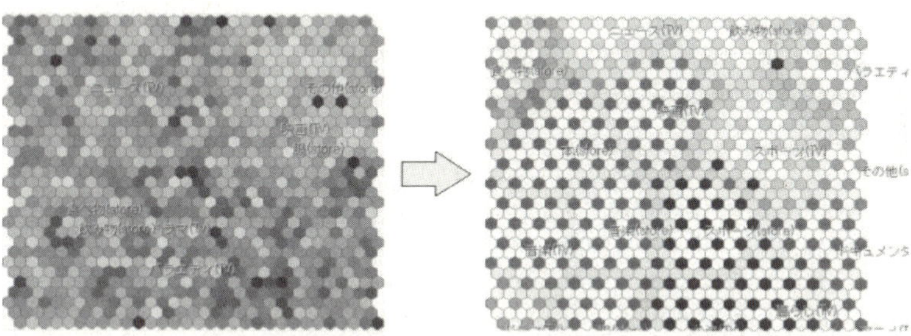

Fig. 6. Learning by SOM

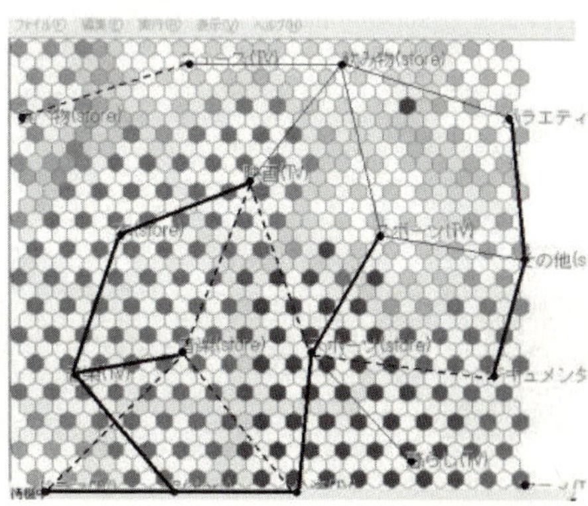

Fig. 7. Relation between categories

When the system inputs user's buying history and watching TV history into general preference attribute profile, a category which is related in a category, which user often buy or often watch, fires. Then, the system gets user's preference attribute information. By using this way, the system can get preference attribute information from buying history and watching TV history whose kinds are different from each other.

4.2 Result of Experiments in Section 4

In section 4, the system gets preference attribute information, and provides a service. In a store, the system gets preference attribute from product's data which user has, shows detailed information of products which user has. In a room, the system gets watching TV history. When user sends preference attribute, the system recommends TV show which suits user's preference attribute.

4.2.1 Experiment in a Store

Figure 8 shows flow of experiment in a store. Unconscious type robot tracks user's hand and face, recognizes user's motion. When user has a product, the system sends information of a product which user has to use's virtual type robot. The authors think that user has a product because user interested in a product. So the system gets new preference attribute information from data of a product which user has, updates preference attribute information.

In experiment in a store, the authors used data of product's color to recognize a product which user had. But the authors assume that products are going to be put RFID in the near future, the system will be able to recognize a product which user has with higher accuracy.

4.2.2 Experiment in a Room

Figure 9 shows flow of experiment in a room. In a room, the system gets watching TV history. When user sends preference attribute, the system recommends TV show which suits user's preference attribute. In this experiment, the authors built server

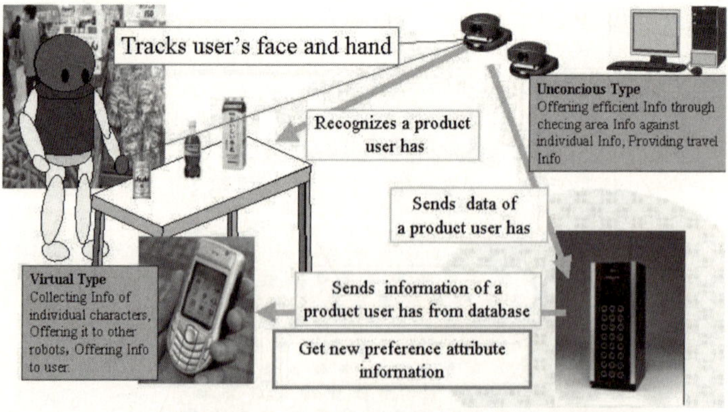

Fig. 8. Flow of experiment in a store

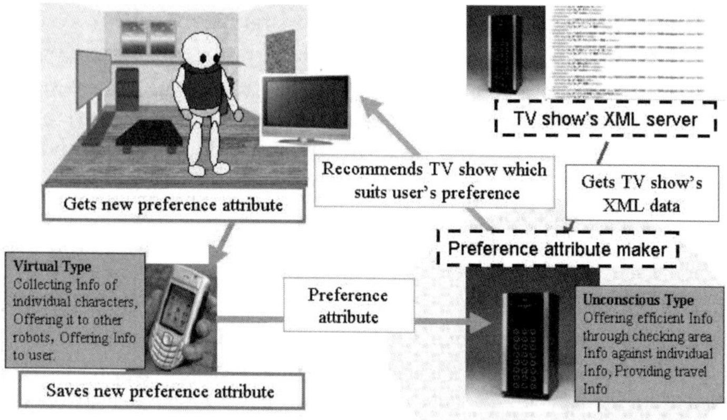

Fig. 9. Flow of experiment in a room

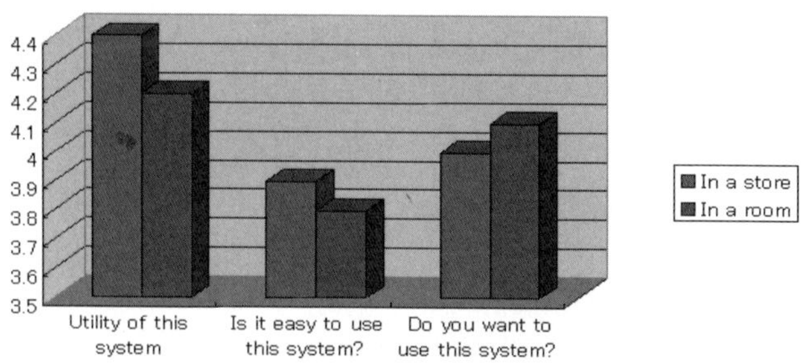

Fig. 10. The questionnaire's result of average in section 4

which makes XML from IEPG information which has TV show information. The system gets TV show's information from the TV show's XML server. When user sends preference attribute, the system displays recommended TV show on the TV's display. If user is interested in recommended TV show, user can watch recommended TV show by clicking tab.

4.2.3 Evaluation of Experiment in Section 4

In section 4, the authors conducted a questionnaire. Number of subjects was five. The questionnaire consisted of three questions. The questions are "utility of this system", "Is it easy to use this system?", and "Do you want to use this system?". The authors let subjects evaluate the system on a scale of one to five. The figure 10 shows the questionnaire's results of average. The utility of this system was shown by this figure.

5 Conclusion

In this research, the authors aimed to implement human-centered system, which understands user and provides a service which suits user. In section 3, the authors implemented the system, which provided a service by having networked robots share personal attribute information. The authors show the utility of this system. In section 4, the authors proposed a method of how to get and update preference which is one of the elements in personal attribute. Then the authors implemented the system in which the method is adopted. Next the authors showed the utility of the system. As a future task, the authors take it into account that the authors adopt the method of how to get and update preference in section 4 to the system in section 3.

References

[1] Yamaguchi, T., Sato, E., Takama, Y.: Intelligent Space and Human Centered Robotics. IEEE Transactions on Industrial Electronics 50(5), 881–889 (2003)

[2] Kawakatsu, J., Yamaguchi, T.: Networked Robots on Ontological Networks Using Humatronics1-st. Slovak – Japanese Seminar on Intelligent SystemsHerl'any, Slovak Republic (2005)

[3] Berlin, B., Kay, P.: Basic Color Terms Their Universality and Evolution. Univ of California Press, Barkley (1969)

[4] Hagiwara, M., Yamaguchi, T.: Collaboration: Neural Network and Fuzzy Signal Processing, pp. 79–94 (1998)

Self-Protecting Session Initiation Protocol Stack

Zoran Rusinovic[1] and Nikola Bogunovic[2]

[1] Ericsson Nikola Tesla, Krapinska 45,
10002 Zagreb, Croatia
zoran.rusinovic@ericsson.com
[2] Faculty of Computing and Electrical Engineering, Unska 3,
10000 Zagreb, Croatia
nikola.bogunovic@fer.hr

Abstract. We present the Self-Protecting Session Initiation Protocol Stack capable of recognizing malicious SIP messages and protecting itself in high-load conditions. The stack model is based on the two-step processing and uses hash lookup tables and cellular automata rules to identify SIP message either as a regular or malicious one. Results show that a presented SIP stack exhibits very promising results with respect to messages classification and enables regular operation of SIP stack under high-load consisting of highly malicious traffic.

Keywords: Self-protecting, SIP, Denial of Service, Cellular automata.

1 Introduction

Session Initiation Protocol (SIP) is a controlling protocol for initiating, managing and terminating sessions across packet networks, which is widely used for IP-based multimedia services. SIP is an application-layer control and can be used for sessions with one or more participants. These sessions include Internet telephone calls, multimedia distribution, and multimedia conferences. [1].

As an integral part of the IP Multimedia Subsystem (IMS) SIP is used as a controlling protocol between different call session control function (CSCF) nodes in the IMS network. All network elements processing SIP messages (i.e. SIP entities) require a SIP stack which is the software that interprets and generates SIP messages, typically consisting of SIP message parser and a protocol handling logic.

1.1 SIP Protocol Imposed Characteristics

We assume the reader to be familiar with SIP, and present here only SIP constraints to pinpoint important elements during SIP messages processing. For a full treatment of the SIP protocol we refer the reader to [1]. SIP is based on an HTTP-like request/response transaction model. Each transaction consists of a request from a server and at least one response from a client. This imposes significant load on message processing elements which must deal with SIP messages having following characteristics: Text-based messages, no fixed size of messages, no fixed order of message elements, allowed SIP specification extensions, case insensitive keywords and context-dependent grammar. All

I. Lovrek, R.J. Howlett, and L.C. Jain (Eds.): KES 2008, Part I, LNAI 5177, pp. 717–724, 2008.

of the aforementioned characteristics impact both memory and CPU on all intermediate proxies and on the receiver machine. Reading and parsing bytes of information to find the SIP message boundaries, SIP message header lines and all of the SIP message elements needed to convert a message from its external format (e.g. SIP format) to an internal data format, regardless of implementation, is relatively constant and takes up to 25% of overall processing time [2].

2 SIP Messages Differentiation

Even in case of careful network engineering, it might happen that SIP network element temporarily experience high-load condition. Such high-load situation will occur each time when the number of SIP messages in the system (which represents the load of the server) gets higher than a number of SIP messages that network can process within a given time interval. In such case network element cannot process all SIP messages, but nevertheless must retain control over how to handle all of them (e.g. which messages to process and which messages to reject). In case when network element cannot control anymore how to handle excess messages, it might arbitrarily drop any of the received messages in which case a high-load situation becomes an overload situation.

2.1 SIP Messages Flood Attack

Flooding a SIP network element with malicious SIP messages is the most difficult Denial of Service (DoS) attack to defend against. Due to a high degree of freedom allowed by SIP grammar, it is very difficult for the victim to distinguish regular and malicious traffic. At the same time malicious SIP messages can be constructed in such way that they consume as much as resources as possible during message processing. An attacker may create very long SIP messages, with many headers and parameters that at the same time have a big-sized message body that complicates message parsing and depletes processing power. SIP protocol specifies many requests and response codes used to create or end a session, redirect a call or update a session parameters. Any of those requests or response codes can be employed in flood attack which can cause DoS to the provided service. Our approach is very effective in terms of computing power since it is put in force before the actual decoding of the message.

3 Description of the Proposed Model

Our model of the SIP stack is based on a two-step processing applied in high-load condition. The idea behind the model is to avoid full-scale parsing, decoding and complete processing of the messages in situations when only by peeking at the very limited set of features, stack can conclude that a SIP message cannot be a regular one. Figure 1. illustrates the principle of the model which consists of table lookup/cellular automata calculation and the actual processing of a valid messages.

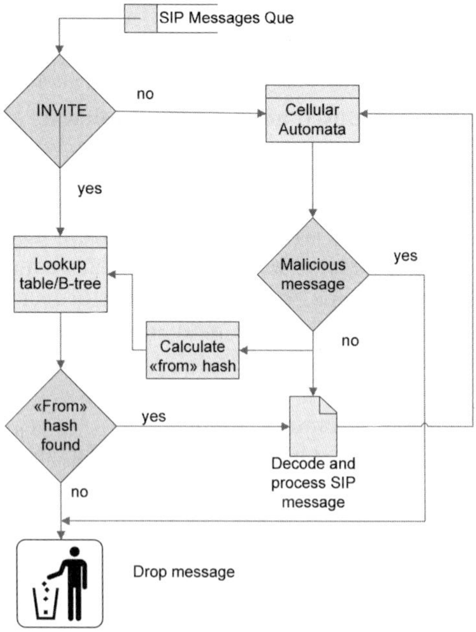

Fig. 1. Message processing steps in a Self-Protecting SIP Stack

3.1 Message Differentiation Based on the to Header URI

In the first step it is determined if the received message is an INVITE request, non-INVITE request or response. If it is an INVITE request we don't proceed to the full-scale parsing and complete processing, but rather look at the set of simple CAs to determine weather the particular INVITE can be considered a regular or a malicious one (the details of the automata rules are explained later). If the message is considered a regular one, a hash of the To header URI is calculated. It is crucial that a hash function used is quick (not CPU expensive) while maintaining a low collision rate. For this reason we use 32 bit FNV-1 hash value (which is used not just for hashing To header URIs, but whenever hash calculation of any element is performed). The calculated 32 bit value is xor-folded down to 16 bits (i.e. 16 excess high order bits are shifted down and xored with lower 16 bits) to reduce it to the size of the in-memory lookup table. The 16 bit value is then used as key for the search of from URI record in the lookup table. However, the original 32 bit hash value is not thrown away, but is stored as a value in the B-tree which has a lookup table cell as a root node (e.g. this is where we start to probe for the 32 bit hash).The property relied upon is that by storing hash values of the received INVITE messages we have extremely cheap way to prove invalidity of received non-INVITE requests or responses. For each received non-INVITE request or response we probe the hash table. If the hash is not found the message is dropped, otherwise we proceed to the decoding of the message. Using this approach, by calculating the hash value of a specific SIP message value and probing the hash table we can instantly determine if the message cannot be a regular one, which saves the budget by skipping a full-scale parsing, decoding and than searching for a dialog and transaction value. It is worth noticing that to calculate hash of the

From header URI we don't need to parse the entire message, but only to skip everything until the desired element is found.

3.2 Message Differentiation Based on Cellular Automata

Cellular automata are models of physical systems where space and time are discrete and interactions are only local. The overall behavior of the SIP node depends upon characteristics of SIP transactions which are initiated from the it's peer SIP entities. This corresponds to the local interaction between cellular automata's cells in a way that global behavior arises from the collective behavior of locally interacting components. We therefore argue that using CA is particularly efficient approach for capturing essential features of malicious SIP traffic.

3.2.1 Definition of Cellular Automata

Cellular automata, firstly introduced by Ulam an Von Neumann [3], are a special class of finite automata that can be described by the 3-tuple of Eq. (1) and they contain large numbers of simple identical components with only local interconnections.

$$A = (S, N, \delta) \tag{1}$$

In the above equation S is a nonempty set, called the state set, $N \subseteq \mathbb{Z}^2$ is the neighborhood, and $\delta : S^N \to S$ is the local transition rule.

A lattice of N identical finite-state machines (i.e. cells), each with an identical pattern of local connections to other cells for input and output, is called a cellular space. Each cell is denoted by an index i and its state at time t is denoted s_i^t (where $s_i^t \in S$). Cell i together with the cells to which cell i is connected is called the neighborhood η_i^t of the cell i.

Local transition rule $\delta : S^N \to S$ gives the update state s_i^{t+1} for each cell i as a function of η_i^t. Typically CA works in a discrete manner. That is to say time goes step by step and a global clock provides an update signal for all cells [5].

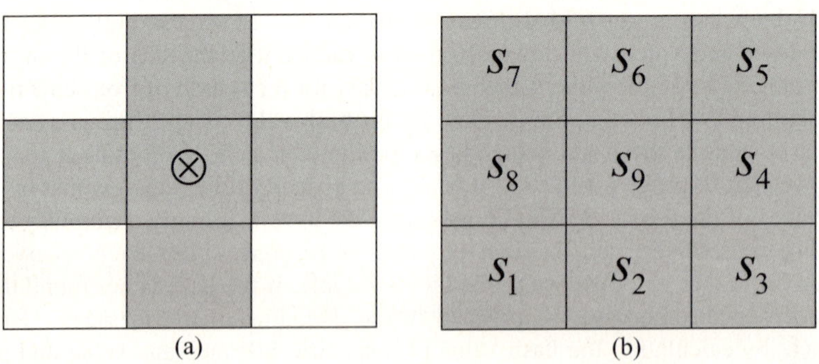

Fig. 2. (a) The von Neumann neighborhood. (b) The Moore neighborhood. In both cases the cell to be updated is marked with ⊗.

A two-dimensional CA is illustrated in Figure 2. The neighborhood consisting of the cell itself and the four bordering cells (four orthogonal neighbors) is called "von Neumann neighborhood" (Figure 2a). The two dimensional neighborhood consisting of the cell itself and the eight bordering cells is called the "Moore neighborhood" (Figure 2b).

3.2.2 Proposed CA Model

Our model is based on the combination of three two-dimensional cellular automata. Each two-dimensional automaton consists of exactly 3*3 set of cells. The goal is to have one set of cells describing observed behavior of a particular feature important to distinct between malicious and regular SIP message.

First set of cells describes behavior of SIP messages sent to the particular recipient, as observed based on the To header in the previous SIP message.

Second set of cells describes behavior of SIP messages sent from the particular IP address, as observed based either on the IP host address of the bottom-most Via header or based on the "received" parameter in the bottom-most Via header (if the host in the bottom-most Via header is given as a FQDN).

Third set of cells describes behavior of SIP messages received through the particular set of SIP proxies. Reason for heaving this additional value is to have a safety-check in the case that the remote party is sending SIP messages with a spoofed IP address through the outbound proxy that doesn't perform ingress filtering (SIP message sent from such UAC, although send with a spoofed IP address, will nevertheless most likely traverse the same set of SIP proxy servers).

Central cell represents a state of the particular feature at any given time t, whereas surrounding cells represent a state of the feature at times {t-1, t-2,…,t-8}. The form of CA is given by $A = (S, N, \delta)$ with $S = \{0,1\}$, where one indicates that a SIP message is marked as regular one with respect to the particular feature; whereas zero indicates that a SIP message is marked as a malicious one with respect to the feature. Initially all cells are in the state one. To define the neighborhood of the cell we will mark each cell of the automata as in Figure 2(b).

N is defined as follows (N(s_x) indicates the neighborhood of the cell x):

N(s_1) = s_9

N(s_n) = s_{n-1} for n = 2…8

N(s_9) = $s_1 + s_2 + s_3 + s_4 + s_5 + s_6 + s_7 + s_8$ (e.g. for the central cell, we observe the entire Moore's neighborhood). Because of the neighborhood being defined in such a way that each cell has its unique neighborhood, border condition problem is avoided (i.e. there is no need to, for example, stick opposite borders of automata together).

The local transaction function δ is based on comparison criteria of the central cell state with those of the cells from its neighborhood. Thus $\delta : S^N \rightarrow S$ is defined by (2). A theoretical explanation for using such transaction function is that in high-traffic conditions we try to identify INVITE messages which have same set of feature (i.e. were addressed to the same recipient and were sent by the same sender) as ones that have previously already failed to establish a session.

$$\delta(\eta_i^t) = \left\lfloor \frac{1}{2} + \frac{\sum_i^n s_i^t - \frac{1}{2}}{n} \right\rfloor \quad (2)$$

Equation (2) represents Majority function, which means that a new state of the cell equals zero if more cells from neighborhood are zero, otherwise a new state equals to one. In case of even n (such is the case for all cells, except the s_9) formula break ties in favor of zeros. After the new state for the cell s_9 has been calculated across all three automata, decision whether to mark the incoming SIP message as a regular or a malicious one is done by majority of all three results. If the INVITE message has been recognized as a regular one, it is sent to UAS, otherwise the message is dropped. The new decision is than replicated into s_9 cell of each automaton. It might happen that an INVITE message, despite being detected as a regular one, was in fact a malicious message in which case we must reflect this in CAs configuration. Through this feedback CA learns which calls were malicious, and can update its states making further discrimination each time more accurate. If neither ACK (that should be received on 200 OK sent by UAS) nor BYE was received from UAC, the status of the message is updated to malicious. In addition, if the session duration for the particular INVITE was shorter than 3 seconds (which is a typical time session may last without the call being charged), we update the message status to a malicious as well. In both cases this must be reflected in CAs configurations. For the first CA (receiver describing) and the second CA (sender describing) , this is achieved by updating the cell s_9 state to 0. Since the third CA might have already updated its configuration (due to the same route being used for multiple INVITE messages) this is reflected by updating an arbitrary cell in the $\{s_1,..,s_7\}$ set from the state 0 to the state 1 (to penalize this route for providing a malicious message). After returning from high-traffic condition back to regular traffic, states of all cells in all automata is set to one.

4 Test and Results

To test the cost of SIP message preprocessing and impact on high-load traffic handling we tested both the regular and the Self-Protecting SIP Stack implementations in our local lab environment. The test configuration setup is shown in the Figure 3. Two call generators are sending SIP messages according to the standard proxy 200 scenario. Messages from each call generator are equally distributed across two stateless proxies (e.g. 50% of messages from each generator is sent to one proxy and 50% to another). In addition to the regular proxy 200 scenario malicious messages are send from both call generators. Call generator one is used to send malicious INVITEs with spoofed IP addresses 40% of the time, whereas Call generator two is used to send random SIP messages 50% of the time. Regular holding time is 180 seconds, but for malicious INVITEs, BYE is sent by the call generator 2 seconds after 200 OK is received from the call receiver (i.e. wait time is 2 seconds).

Test was run on an isolated Ethernet network using Dual-Core AMD Opteron processor 2216 on 2.40 GHz for running Self-Protecting Proxy.

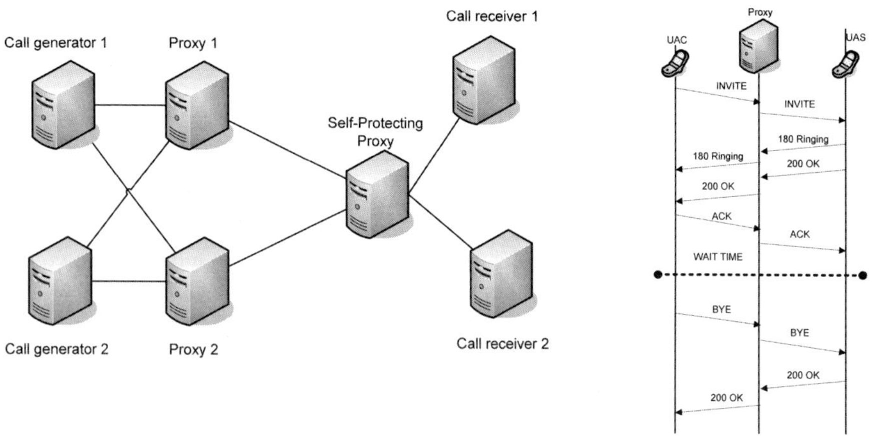

Fig. 3. Test Setup Scenario

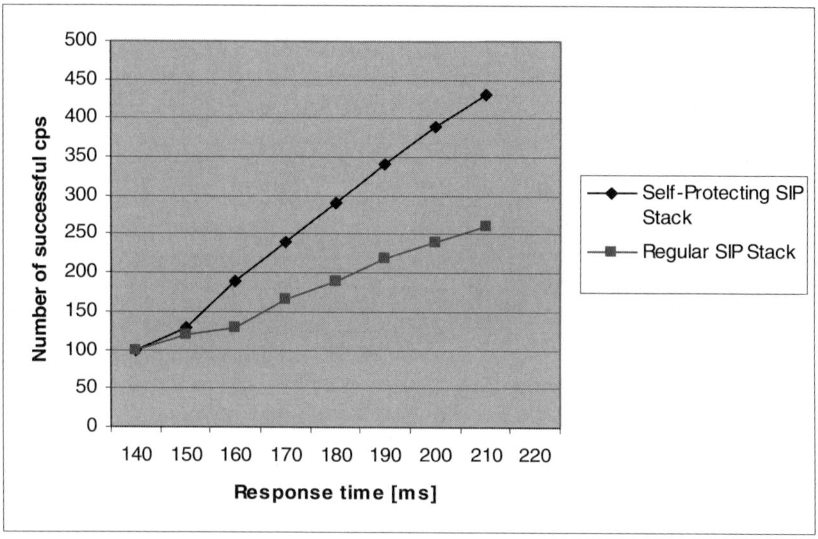

Fig. 4. Response time vs. calls per second

Figure 4. depicts effect of message differentiation to the number of calls being processed by the proxy. For proxy under test, initial load of 100 cps (equally distributed across both call generators) was ran, then we incremented call load by 50 cps (this includes only regular calls, without malicious SIP messages). In addition to this, 0,4*cps of malicious INVITEs was generated from a call generator 1 and 0.5*cps of malicious non-INVITEs was generated from a call generator 2.

Beside much higher values of realized cps seen in Figure 4, we have succeeded to drop virtually every malicious SIP message at the Self-Protecting Proxy node, without forwarding it to the UAS. Unlike for Self-Protecting SIP stack, each of these malicious messages was happily forwarded in case of regular SIP stack. In practice this would mean the difference between continuous SIP device activity (e.g. phone ringing) caused by malicious SIP messages traffic and a normal device operation in case of Self-Protecting SIP stack.

5 Conclusion

In this work we have presented a Self-Protecting SIP stack model. By experimental comparison we observe that the described method is very efficient in reducing the effect of malicious SIP traffic. It is shown that the proposed model is efficient in protecting itself from arbitrary malicious SIP messages, making it possible to reach almost double cps rate within the given time interval, compared to the regular SIP stack. For this reasons we can consider this approach to be a step forward in providing a robust attack-tolerant SIP service.

References

1. Rosenberg, J., Schulzrine, H., Camarillo, G.: SIP: Session Initiation Protocol., RFC 3261 (June 2002)
2. Cortes, M., Ensor, J., Esteban, J.: On SIP Performance. Bell Labs Tech. J., 155–172 (2004)
3. Neumann, J.V.: The Theory of Self-Reproducing Automata. In: Burks, A.W. (ed.). Univ. of Illinois Press, Urbana (1966)
4. Vichniac, G.Y.: Simulating physics with CA. Physica 10, 96–116 (1984)
5. Mitchell, M.: Computation in Cellular Automata: A Selected Review, Santa Fe Institute (1998)
6. Kuthan, J.: Accelerating SIP. In: SIP 2002, Paris,France (2002), `http://www.iptel.org/`
7. Batteram, H., Meeuwissen, E., van Bemmel, J.: SIP Message Prioretization. Bell Labs Tech. J. 11(1), 21–36 (2006)

Early Forest Fire Detection with Sensor Networks: Sliding Window Skylines Approach

Krešimir Pripužić, Hrvoje Belani, and Marin Vuković

University of Zagreb, Faculty of Electrical Engineering and Computing,
Department of Telecommunications, Unska 3, HR-10000 Zagreb, Croatia
{kresimir.pripuzic,hrvoje.belani,marin.vukovic}@fer.hr

Abstract. Wireless sensor networks are widely used in environmental applications, like forest fire detection. Although forest fires occur relatively rarely, their number is increasing in Europe in the last years, so their manifestation must be early detected in order to prevent higher damages. To minimize needless communication between the sensor nodes for this usage, new data suppression technique using sliding window skylines is described in this paper. We experimentally evaluate our algorithm for continuous sliding windows skylines computation, and show its usability in practice.

Keywords: Data suppression, forest fire detection, sensor network, sliding window skylines, experimental evaluation.

1 Introduction

Sensor network technology is considered as one of the key technologies for 21st Century [1]. It is a network of spatially distributed autonomous devices with various sensors that cooperatively monitor certain conditions (e.g. physical or environmental) at different locations. Communication between sensors is wireless in most cases. A wireless sensor network (WSN) is the term for Low-Rate Wireless Personal Area Networks employing no fixed infrastructure and having communication links less than 10 meters in length and sensors centered on a subject or individual or in the targeted area [2].

WSN find their spread usage in a variety of applications: military, environmental, health, etc. The amount of data sent over the network can be very low, and the message latency can be on the order of minutes. Yet, costs must be low, and power consumption must be low enough for the entire network to last an entire required timeframe (e.g. season). These low-data-rate applications involve sensing of one form or another and require short-range links without a preexisting infrastructure on site, except network of sensors. Sensors can monitor conditions at different locations, such as temperature, relative humidity, wetness, lightning conditions, smoke, atmospheric pressure, wind speed and direction, etc [3].

According to the application type and design requirements, there are two basic types of WSNs: the one for rare event detection (e.g. forest fire detection or intrusion detection), and the other for periodic data gathering (e.g. temperature monitoring or biosensors). Although forest fire occurs relatively rarely, its detection must be early

I. Lovrek, R.J. Howlett, and L.C. Jain (Eds.): KES 2008, Part I, LNAI 5177, pp. 725–732, 2008.

manifested in order to prevent higher damages. For network nodes to live long enough to fulfill their purpose, supply batteries consumption must be minimized. Hence, needless communication between the nodes must be lowered using different data aggregation and suppression techniques. New data suppression technique using sliding window skylines is described in this paper.

The next section discusses applicability of WSNs for forest fire detection, their advantages comparing to other solutions (e.g. satellite imagery) and issues that must be dealt with in order to improve reliability of the solutions and make readings of fire detection more accurate. The third section describes implementation of algorithm that uses sliding window skylines and simulation based on the case study of early forest fire detection. The fourth section presents the algorithm implementation and its simulation in WSNs usage for forest fire detection. The fifth section evaluates related work and the final section gives conclusion and future work.

2 Wireless Sensor Networks for Forest Fire Detection

As stated before, various applications deploy large number of small and inexpensive sensor nodes with one or more sensors attached and a fewer number of intermediate nodes, which aggregate and/or suppress sensing data forwarding them to the sink. Nowadays, WSNs are widely applied in environmental applications, such as habitat monitoring, agriculture research, earthquake monitoring, traffic control and fire detection, because of the nature of such applications that involve people, assets and environment in events of disasters, accidents and other needs of a today's society.

Wild fires, including forest and plant fires, are uncontrolled fires occurring in wild and rural areas which can cause significant damage to natural and human resources [4]. Common causes of forest fires are lightning strikes, human carelessness, and exposure of fuel to extreme heat and aridity.

In the last years heat waves in Europe caused magnified number of forest fires with devastating outcomes. Croatia also belongs to countries with increased risk of summer forest fires, especially on Dalmatian coast and islands in Adriatic Sea. For example, in period from January 1st to December 31st 2007 there were 8945 fires registered in Croatia, and 5455 of them (61%) were plant and forest fires that burnt the area of 67992 hectares [5]. Comparing these data with the ones for 2006, it can be seen that number of fires has increased for 25.7%, number of plant fires for even 52.6%, and the burnt area has increased 3.6 times. Burnt area index, given in hectares per fire, has increased even 2.4 times, which shows the enlarged threat of every fire occurred for landscape, human and animal life in affected areas. Therefore, early detection and suppression of forest fires is crucial for restriction of their propagation.

Most of existing forest fire detection systems rely on the satellite imagery. These approaches are limited due to the weather conditions (e.g. clouds) that can seriously decrease the detection accuracy. These issues become irrelevant when implementing forest fire detection using WSNs.

One of the most important systems aspects of a WSN for forest fire detection is its lifetime [6], because the fire detection network must operate for a very long period of time in order to detect such a comparatively rare event. Occurring of forest fires must not be confused with sensor battery exhaustion, as well as antennas being reoriented in the wrong direction by falling branches, curious animals, wind, etc. Another critical issue for such application is field coverage, since the WSN must identify the event quickly and accurately. In order to do so each node must be given a unique identification number (ID) which is associated with a node's precise position in the WSN or globally, with the use of GPS. With knowledge of the node ID and specific coverage area it is possible to precisely locate the source of the fire.

Fires can differ in size, shape, growth, frequency, and intensity. Nevertheless, all these parameters become irrelevant if it is possible to notice the fire at its very beginning. However, for early fire detection, coverage area of single node should be as low as possible; if the coverage area is too wide the sensor node will report fire once the fire reaches the sensor, which could be too late. Small size coverage areas would require more sensors which could be too expensive, so it is necessary to reach the optimum size of coverage area, which is out of the scope of this paper.

The sensors clustering using a distributed protocol, their physical distribution along the targeted area and data routing problems are outside the scope of this paper.

3 Sliding Window Skylines

Suppose that each sensor in our network measures a certain number of parameters to which we will refer as attributes. These attributes form multi-dimensional attribute space. Therefore, we assume that every sensor reading is a point in the attribute space.

For any two values of an attribute (e.g. temperature), it is easy to say which of the values is better indicator of forest fire (one with the higher temperature). In the case of sensor readings, which are multi-dimensional, it is not so easy to say which reading is better indicator of forest fire. For example, let us take a look at Fig. 1, where some readings in two-dimensional attribute space are shown. We have two attributes: temperature and wind speed. We want to know which readings are better indicators of forest fire than the others. This is typical example of ranking objects by more than one criterion. In our example, a reading is better indicator than another (i.e. we will say that it dominates) if it has larger temperature and higher wind speed. The top readings are the set of readings which are not dominated by any other reading. They are shown as gray dots in the figure. We see that they form line which bounds all other readings. This line is usually called the skyline, while the top readings are called skylines.

Each sensor in our network periodically measures attributes. For early detection of forest fire we do not have to collect all these readings at sink. Actually, at some point in time, it is sufficient to know skylines, because all other readings are bounded with them. To save power, we propagate only the skylines to the sink. At each inner sensor node on path to the sink, we aggregate both its own and received skylines, and send these aggregated skylines to next node on the path. Thereby, at the sink, we will receive skylines that will be aggregates of the whole WSN, such that all readings of all sensors will be bounded with them. This is an effective way of sensor readings aggregation, which reduces power consumption of WSN, because local processing and storage at node consumes less energy than transmitting data over the radio [19].

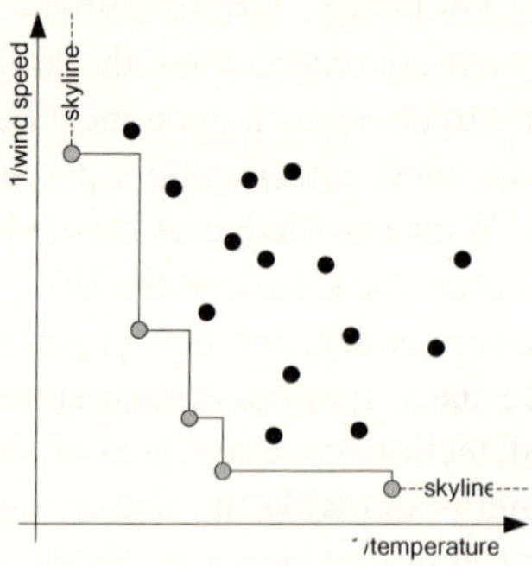

Fig. 1. Skyline of Sensor Readings

At the sink, we want to continuously monitor the environment, and therefore we have to know only most recent skylines, because the older ones do not reflect a present state of the monitored environment. For all readings (and skylines) we define lifetime in the form of time window. Every reading older than a size of the window will be deleted from the network. If the size is very little, every reading will be skyline in the window, and this will result in too many transmitted readings. On contrary, if the size is too large, aggregated skylines will not adequately reflect the present state. In practice, this size highly depends on dynamism of the phenomenon of interest (i.e. forest fire in our case).

4 Algorithm

At each sensor node we maintain an index of unexpired skylines and readings dominated by them. We propagate every change in the skylines to next sensor node on the path to the sink. In the rest of this section we present our algorithm for continuous maintaining of sliding window skylines. This algorithm is run at every sensor node, once for every of its own and received readings. The synchronization between nodes is not needed, because every node on the path has its own clock and for each propagated skyline it adds time spent on it. In this way, any node on the path may delete expired readings from its index.

```
method add new reading
  check oldest skyline
  foreach skyline
    if (skyline dominates reading)
      add reading to skyline
      return
    else if (reading dominates skyline)
      remove skyline
  add reading to skylines and propagate it.
```

The upper method adds new reading to the index. First we check if oldest skyline is expired. After that we go through present skylines and check if the reading is dominated by any of them. If this is the case, we add it to the set of readings dominated by the first such skyline we encountered. Otherwise, we remove from the index every skyline dominated by it, add it to the skylines and propagate it to next node on the path. We have to keep all younger readings that are dominated by any skyline (or any other reading) in the index, because they are potential skylines in future. Readings dominated by any younger reading (or skyline) cannot become skyline in future, and we may delete them from index. It is important to notice that we do not have to add a reading to all sets of all skylines that dominate it, because it may become skyline only if all such skylines are expired.

The following method adds a reading to the set of readings dominated by some other reading (i.e. parent). It is almost identical to the upper method, except we do not check if the parent is expired and we do not propagate reading after adding it to parent's children, because it is not a skyline. We do not have to check if the parent is expired, because it is not a skyline, and therefore it is younger than unexpired skyline which dominates it.

```
method add reading to parent
  foreach child of parent
    if (child dominates reading)
      add reading to child
      return
    else if (reading dominates child)
      remove child
  add reading to children of parent.
```

As we said before, each time we add new reading we have to check if the oldest skyline is expired. If it is expired we have to remove it and add all of its children from the index. The following two methods do this job. To improve performances of our algorithm we keep skylines (and children of other readings) in red-black trees sorted by their time of expiry.

```
method check oldest skyline
  if (oldest skyline is too old)
    remove oldest skyline from skylines
    foreach child of oldest skyline
      add child.

method add reading
  foreach skyline
    if (skyline dominates reading)
     if (skyline is younger than reading)
       foreach child of reading
         add child
       return
     else
       add reading to skyline
       return
  add reading to skylines and propagate it.
```

5 Experimental Evaluation

In this section we examine our algorithm and index structure on the performances of sliding window skyline computation. Following well-established methodology set by previous research on skyline algorithms [12, 13, 14] we choose to use the following three data distributions: independent, correlated and anti-correlated. In this section we want to see simulation runtimes for different cardinality (i.e. the number of readings) and different sizes of time windows. Shorter runtime means less processing at every sensor node in the network, and therefore smaller consuming of battery power which results in longer lifetime of the network.

The following setup was used in experiments: We created synthetic datasets and vary the cardinality from 10 thousands to 1 million. We fixed dimensionality to 2, and used 2 different sizes of the time window. The first size was 5000, and the second was 10 percent of the cardinality. Experiments were run in Java on a PC with 2.8 GHz Pentium Processor and 1GB of main memory.

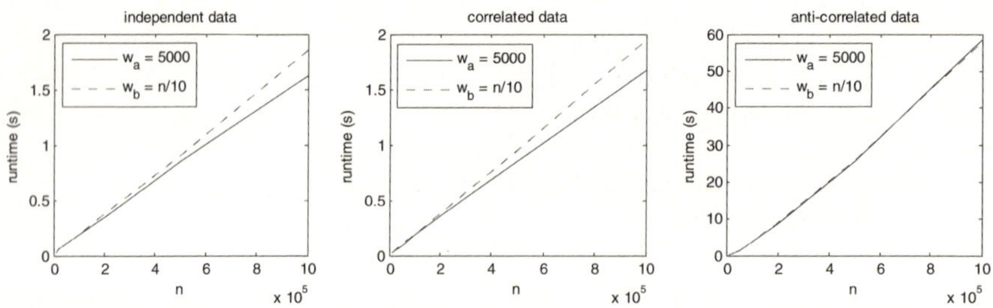

Fig. 2. Results of Experiments

In Fig. 2 we can see the results of the experiments, where n is the cardinality. We conclude that there is no significant difference in runtime for the independent and correlated datasets. We see that for larger cardinalities there is slight increase in the runtime for window w_b, which has the size of 10 percent of the cardinality. This is expected, because skylines live longer in the case of larger window sizes, and therefore the number of readings in the index is larger. For the anti-correlated dataset, we see that the runtime is more than one order of magnitude longer. This is expected, because many readings are skylines in the anti-correlated dataset.

Morse at al. [14] did an extensive evaluation of different algorithms for continuous skyline computation. For the independent and correlated datasets the runtime of our algorithm is one order of magnitude smaller (0.1 s comparing to their 1 s for cardinality 5000), while for the anti-correlated dataset our algorithm is equally fast to their algorithm. They are using R^*-trees and Quadtrees for indexing of skylines (and other readings). We suppose that our approach is faster for continuous skyline computation over sliding windows because we store skylines (and other readings) directly to the index as its nodes, while Morse at al. are storing them as end-nodes (i.e. leaves) of used structures. On the other hand, the memory consumption should be lower in our approach, because of the same reason.

6 Related Work

We have recognized two different topics of related work for our subject: usage of wireless WSNs in fire detection systems, and skylines computation in WSNs.

Some papers exploit the essence of artificial neural networks to perform simple calculations at many organized single nodes in order to conduct complex data processing [7]. This approach also, they claim, reduces communication overhead and energy consumption. Some recent, also domestic, works combine video and network sensor monitoring [8] [9], but their approach doesn't analyze the importance of data aggregation and suppression from sensors. They rely on algorithms for image analysis and looking of visual signs of forest fires, particularly forest fire smoke during the day and forest fire flames during the night. Some solutions employ the usage of mobile agents [8] [10] for efficient network exploration and fire detection, but the communication paths seem overused while agents tracking fire, which exhausts the sensor supplies (e.g. batteries). Some enhanced supply solutions exploit solar energy.

Skyline computation in WSNs has received much attention recently [15, 16, 17]. Kwon at al. [15] propose an algorithm for in-network processing of skyline queries. Chen at al. [17] use advanced approach for continuous in-network skyline computation that employs hierarchical thresholds at the nodes. Xin at al. [16] propose an energy-efficient algorithm for continuous skyline computation. There are also many other paper related to continuous skyline computation, which are not related to WSNs [12, 14, 18]. We compared our algorithm with [14] in Section 5.

7 Conclusion

This paper proposes enhanced algorithm for continuous skyline computation in order to suppress collected data from wireless sensors that are monitoring environment for potential forest fires. In experimental evaluation, we showed that our approach is faster than other algorithms for continuous skyline computation over sliding windows.

Because of the increased risk of forest fires in Croatia, especially during summer seasons, early detection and suppression of forest fires is crucial, in order to restrict their propagation. According to the Law of fire protection in the Republic of Croatia, National Protection and Rescue Directorate and its Fire Fighting Sector are obligated to propose Activity Program of special measures in fire protection on a yearly basis. One of the main goals of the program is development and implementation of new fire protection systems. This paper therefore represents useful and applicable contribution in that direction.

Acknowledgements

This work was carried out within the research project "Content Delivery and Mobility of Users and Services in New Generation Networks", supported by the Ministry of Science, Education and Sports of the Republic of Croatia.

References

1. Chong, C., Kumar, S.: Sensor Networks: Evolution, Opportunities and Challenges. Proceedings of the IEEE 91(8), 1247–1256 (2003)
2. Callaway, E.H.: Wireless Sensor Networks: Architectures and Protocols. Auerbach Publications, CRC Press LLC, Boca Raton (2004)
3. Akyildiz, I.A., Su, W., Sankarasubramaniam, Y., Cayirci, E.: A Survey of Sensor Networks. IEEE Communications Magazine 40(8), 102–114 (2002)
4. Hedeeda, M.: Forest Fire Modeling and Early Detection using Wireless Sensor Networks. Technical report CMPT 2007, Faculty of Applied Sciences, Simon Fraser University, Canada (2007)
5. Firefighter Gazette, No 3, Croatian Firefighter Association, Zagreb (2008) (in Croatian)
6. Tanenbaum, A.S., Gamage, C., Crispo, B.: Taking Sensor Networks from the Lab to the Jungle. IEEE Computer 39(8), 98–100 (2006)
7. Yu, L., Wang, N., Meng, X.: Real-Time Forest Fire Detection with Wireless Sensor Networks. In: Proceedings of International Conference on Wireless Communications, Networking and Mobile Computing, vol. 2, pp. 1214–1217. IEEE Press, New York (2005)
8. Stipanicev, D., Bodrozic, L., Stula, M.: Environmental Intelligence Based on Advanced Sensor Networks. In: Proceedings of 14th International Workshop on 2007 IWSSIP&EC-SIPMCS. Maribor, Slovenia (2007)
9. iForestFire – intelligent Forest Fire monitoring system, Faculty of Electrical Engineering, Machine Engineering and Naval Architecture, University of Split, `http://iforestfire.fesb.hr`
10. Fok, C.-L., Roman, G.-C., Lu, C.: Efficient Network Exploration and Fire Detection using Mobile Agents in a Wireless Sensor Network. ONR-MURI Review, Baltimore, MD (2004)
11. Bekara, C., Laurent-Maknavicius, M., Bekara, K.: SAPC: A Secure Aggregation Protocol for Cluster-Based Wireless. In: Zhang, H., Olariu, S., Cao, J., Johnson, D.B. (eds.) MSN 2007. LNCS, vol. 4864, pp. 784–798. Springer, Heidelberg (2007)
12. Lin, X., Yuan, Y., Wang, W., Lu, H.: Stabbing the Sky: Efficient Skyline Computation over Sliding Windows. In: Proceedings of the 21st International Conference on Data Engineering (ICDE 2005), pp. 502–513. IEEE Computer Society, Los Alamitos (2005)
13. Papadias, D., Tao, Y., Fu, G., Seeger, B.: An optimal and progressive algorithm for skyline queries. In: Proceedings of the ACM SIGMOD International Conference on Management of Data, San Diego, California (2003)
14. Morse, M., Patel, J.M., Grosky, W.I.: Efficient Continuous Skyline Computation. In: Proceedings of the 22nd International Conference on Data Engineering, IEEE Computer Society, Washington (2006)
15. Kwon, Y., Choi, J.-H., Chung, J.-D., Lee, S.K.: In-Network Processing for Skyline Queries in Sensor Networks. IEICE Transactions on Communications E90-B(12), 3452–3459 (2007)
16. Xin, J., Wang, G., Chen, L., Zhang, X., Wang, Z.: Continuously Maintaining Sliding Window Skylines in a Sensor Network. In: Kotagiri, R., Radha Krishna, P., Mohania, M., Nantajeewarawat, E. (eds.) DASFAA 2007. LNCS, vol. 4443, pp. 509–521. Springer, Heidelberg (2007)
17. Chen, H., Zhou, S., Guan, J.: Towards Energy-Efficient Skyline Monitoring in Wireless Sensor Networks. In: Langendoen, K.G., Voigt, T. (eds.) EWSN 2007. LNCS, vol. 4373, pp. 101–116. Springer, Heidelberg (2007)
18. Tao, Y., Papadias, D.: Maintaining Sliding Window Skylines on Data Streams. In: IEEE Transactions in Knowledge Data Engineering, pp. 377–391. IEEE Computer Society, Los Alamitos (2006)
19. Zeinalipour-Yazti, D., Kalogeraki, V., Gunopulos, D., Mitra, A., Banerjee, A., Najjar, W.: Towards In-Situ Data Storage in Sensor Databases. In: Bozanis, P., Houstis, E.N. (eds.) PCI 2005. LNCS, vol. 3746, pp. 36–46. Springer, Heidelberg (2005)

Analysing Flight Data Using Clustering Methods

Christopher Jesse, Honghai Liu, Edward Smart, and David Brown

Institute of Industrial Research, University of Portsmouth, Burnaby Building, Burnaby Road, Portsmouth, PO1 3QL, England
{christopher.jesse,honghai.liu,edward.smart,david.j.brown}@port.ac.uk

Abstract. This paper reviews existing forms of density-based, partitional and hierarchical clustering methods in the context of flight data analysis. Advantages and disadvantages are fully explored with a focus on proposing a clustering-based ensemble framework for monitoring flight data in order to search for anomalies during flight operation. Case studies in selected flight scenarios are provided to demonstrate the potential of clustering methods and their integration with reasoning techniques in detecting abnormal flights.

1 Introduction

Flight data monitoring is a powerful tool for analysing historical aircraft flight data for exceedances of normal operating procedures. The results are used by Airlines to further improve flight safety. Flight data monitoring is limited to detecting only pre-programmed events. This requires knowledge of a particular event (usually lessons learnt from historic flight safety incidents). Using data clustering of flight parameters, one can assume that the majority of flights are safe and can be classified as so. It can therefore be determined that those flights which do not fit the normal flight pattern could be unsafe and further investigation can determine the reasons for these abnormalities.

Snapshots are a single point in time measurement taken from the flight data parameters on every flight. A collection of multiple snapshot parameters measured at the same point in time will create a vector which describes the condition of the aircraft at that point in time. Snapshots of flight data parameters have been measured at altitudes from 12,000 feet to touchdown during the descent phases of flight. Expert domain knowledge has shown that the majority of incidents occur, or have precursors, at these points in the flight. Flights that differ from the norm follow a different approach pattern than usual and need to be highlighted for further investigation.

By using raw data and not being limited to pre-defined events, data clustering can find hidden trends and patterns within the data including the identification of precursors to flight safety incidents. By identifying these precursors early on, safety improvements can be put in place, stopping incidents before they occur. This paper is focused at exploring existing clustering methods for monitoring, separating and identifying different types of behaviours of flight data. Future

I. Lovrek, R.J. Howlett, and L.C. Jain (Eds.): KES 2008, Part I, LNAI 5177, pp. 733–740, 2008.

work is focused on constructing a clustering-based ensemble framework for detecting abnormal data behaviours in order to efficiently support everyday flight services.

2 Problem Formulation

To validate the application of clustering applications to flight data snapshots, it is first necessary to detect a known incident. Unstable approaches have been chosen as it is a *hot* topic in flight safety forums and there has been much work done to both define and identify unstable approaches. A stable approach is defined as an approach where the aircraft is established on a proper glide path and with a proper air speed, with a stable descent rate and engine power setting, and with a proper landing configuration (landing gear and flaps extended) below a set threshold altitude (usually 1,000 feet for instrumented flight rule (IFR) approaches and 500 feet for visual flight rules (VFR)). In order to detect an unstable approach, the appropriate parameters associated with the features of an unstable approach have been listed in Table 1. It is assumed that instrument flight rules are in use for the selected approaches (1,000 feet) and that the landing gear is extended.

Table 1. Parameters defining a stable approach

Approach Rule	Recorded Parameter	Unit
Established on the glide path	Glideslope deviation	dots
Proper air speed	Indicated Airspeed	knots
Stable descent rate	Inertial Vertical Velocity	feet/minute
Stable engine power setting	Engine 1 N1 Speed	%
Proper landing configuration	Flap Configuration	degrees

Measurement definition: The *glide path* or *glideslope* is the ideal approach path for an aircraft to land safely on the runway. *Glideslope deviation* is the vertical displacement (in dots) that the aircraft has deviated (above or below) from the projected glideslope for the runway. *Airspeed* is a measure of the aircraft velocity relative to the speed of the air outside. *Inertial Vertical Velocity* is the vertical speed of the aircraft. *Engine 1 N1* is the rotational speed of the intake fan at the front of the left-most engine on the aircraft (around 30% when idling and 100% maximum engine speed). *Flap Configuration* is the angle from horizontal that the flaps have been set to (between 0deg and 45deg).

3 Clustering Techniques

Data clustering is a method of representing a large dataset in a more manageable compact size. Clusters are defined by grouping objects with shared features together. Each object will belong to every cluster to a greater or lesser degree defendant on its similarity with other objects in the cluster. All objects within a cluster should be similar, whilst all clusters should in turn be dissimilar from one and another. The clusters are usually labelled to identify the data to the user. The representation of a dataset by fewer clusters inherently loses certain fine details during the compression, but achieves greater simplification. Detection of abnormal flights is a one-class classification problem. First one must define (classify) normal in order to determine what is abnormal. Abnormal observations shall be referred to as *outliers*.

It is evident that clustering is a powerful unsupervised learning method for data exploratory when training datasets are not available. For clustering, it is important that the compact dataset of clusters is an accurate representation of the original dataset. The measure of similarity is defined by the clustering technique in combination with the object representation in the feature space [1]. The most important measure of a good structure is the number of clusters, which is usually defined *a priori* by the user. This is often difficult in practical implementations and a process of trial-and-error is normally used to determine the best fit. Validity indices indicate the most appropriate number of clusters.

There are three main forms of clustering algorithms: Density-based, Partitional and Hierarchical Clustering. Firstly, hierarchical algorithms [2] establish clusters based on previously defined clusters. There are two forms of hierarchical clustering, Agglomerative and Divisive. The main differences between various hierarchical clustering methods are the way in which they update the similarity between existing clusters and the merged clusters. This can be done using a centroid or medoid like in k-means or k-medoid, but these suffer the same limitations of identifying clusters of spherical shape and similar sizes. Secondly, partitional algorithms assign all objects in the dataset membership to the clusters at once. The k-means algorithm [3] is by far the most popular clustering algorithm. The concept is to represent each of the k clusters by the mean (or weighted average) of its points, the so-called centroid. The distance measure is used as the objective function. k-means is a crisp or hard partitioning clustering method and is commonly used with a standard Euclidean distance norm. Thirdly, the most widely used fuzzy clustering algorithm is the fuzzy **c**-means (FCM) algorithm . FCM is an extension of k-means clustering. It applies a fuzzy assignment on the relative distances between one object and all cluster centroids. FCM is an objective function designed to find the optimal fuzzy partition to fit the clustering objective of the original dataset. The FCM algorithm uses an iterative technique until convergence is reached, other optimisation techniques have also been used to solve the objective function. Note that there have been many variants of the FCM algorithm to overcome some of the listed shortcomings. FuzzySOM was developed in order to improve FCM by arraying the cluster centroids into a regular grid [4]. Techniques exist to ignore and cut-off any spurious noise points, creating a noise cluster class [5] and applying a weighting to render noise / outliers less significant. Fuzzy Gustafson-Kessel [6] clustering method is a derivation of FCM with the ability to detect ellipsoidal clusters. It calculates distance using the squared Mahalanobis distance norm.

Finally, density-based clustering methods [7] use information about the proximity of objects to perform the clustering. The assumption is that all objects which are spatially proximate belong to the same cluster. Hierarchical algorithms create clusters by iteratively building up from individual objects (agglomerative) or breaking down the entire dataset (divisive) using a measure of similarity to determine the definition of clusters. Partitional methods apply a degree of membership to every item in the dataset straight away and iteratively change object memberships in order to solve an objective function. Additionally, Fuzzy clustering by Local Approximation of MEmberships (FLAME) [8] is a relatively new data clustering algorithm. It defines a cluster in dense regions of the dataset based solely on the neighbourhood relationships among objects. The neighbourhood relationships among neighbouring objects in the feature space are used to constrain the memberships of neighbouring objects in the fuzzy membership space. The FLAME algorithm first creates a k-Nearest Neighbours graph to identify objects with the highest local density (called Cluster Supporting Objects) and those objects with a local density lower than a threshold (Outliers). Fuzzy memberships are then assigned to objects with varying degrees of memberships to the cluster supporting objects. Outliers are assigned full membership to the outlier group. From the fuzzy results, clusters can be defined by those which exceed a pre-defined cluster threshold membership.

In clustering, outliers (observations which greatly deviate from other observations) are often considered to be noise observations which can exert undue influence on the results of the clustering process and for that reason they are either removed or ignored to make more reliable clustering. However in data mining the detection of anomalous patterns in data is more interesting than detecting inlier clusters [9].

The exact definition of an outlier depends on the context. Definitions fall roughly into five categories [10]:

1. Distribution-based, where outliers are observations which deviate from a standard distribution.
2. Depth-based which relies on the computation of different layers of k-d convex hulls.
3. Clustering-based methods define outliers as observations that do not fit to the overall clustering pattern.
4. Distance-based outliers define outliers as an observation that is $d_{\min}$ distance away from p percentage of observations in the dataset.
5. Density-based methods detect outliers as objects that are in a less-dense region of the feature space than the rest of the dataset. Objects can be outlying *relative to their local neighbourhoods*, particularly with respect to the densities of the neighbourhoods.

4 Case Study

To test the application of clustering upon the dataset, it is first necessary to split the dataset into its lowest level of granularity. Each dataset relates to a single

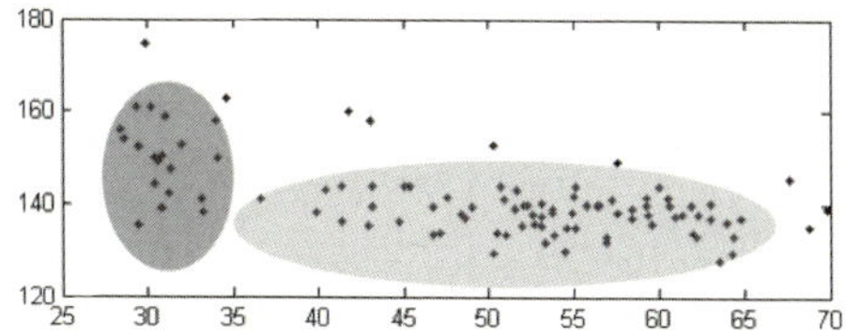

Fig. 1. Manual representation of the desired clusters $k = 2$. X axis : N1 (%). Y axis : Airspeed.

operator with data from 6 aircraft of the same aircraft type with approaches into a single airport runway. In accordance with confidentiality agreements, the data has been de-identified to remove any implications to any pilot, airline, airport or aircraft manufacturer. The dataset analysed consists of 100 datapoints.

The X axis represents the engine throttle N1 in %. The Y axis represents the airspeed in knots of the aircraft. Manual identification of clusters has been shown in Figure 1 The cluster highlighted in green to the lower right of the graph represents "good" flights where the airspeed is good and the engine power is not at idle (around 30% N1). The cluster in red represents a "poor" flight condition where the engine power is at or around idle. This is often caused when the aircraft has too much airspeed (also notable on the graph) so the pilots reduce throttle to slow the aircraft.

All the following clustering methods were performed after normalising the dataset with the exception of the FLAME algorithm. Cluster centroids, medoids are identified with a small red circle. The crisp partitional clustering methods k-means (Figure 2) and k-medoid (3) use colour coding to identify object membership to each of the k clusters. Where $k = 2$, both methods return almost identical results. However, when $k = 4$ k-medoid identifies 3 medoids in the centre of the larger "good" cluster and creates a clear segregation between the "good" and "bad" clusters.

Fuzzy clustering algorithms which apply a fuzzy membership to objects are represented by the use of contour lines. The FCM algorithm determines centroids in very similar locations to that of the k-means algorithm. The Gustafson-Kessel algorithm in Figure 5 displays an ellipsoid effect on upon the fuzzy contours which better encapsulates the clusters as per Figure 1 with a value of $k = 2$. When $k = 3$, the method identifies two clusters in the "good" cluster.

The FLAME algorithm is shown in two steps. First after the identification of the CSOs and Outlier and other objects in Figure 6. The second step of identifying the fuzzy membership values has been shown as a colourmap in Figure 7. Each CSO is assigned a colour and each object is assigned a new colour depending on its fuzzy membership to the CSOs around it. Using k-NN where $k = 5$, 8 CSOs are identified and 12 objects are identified as outliers (threshold $s = 1/4$ maximum density). Increasing the k nearest neighbours to 10 results in 2 CSOs and 8 outliers. The combination of the k-NN and the outlier threshold value s plays an important role on the effectiveness of cluster and outlier identification.

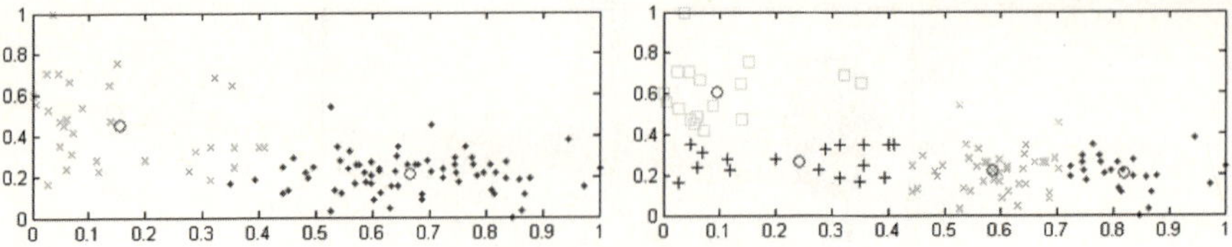

(a) (b)

Fig. 2. k-means algorithm. (a) $k = 2$. (b) $k = 4$.

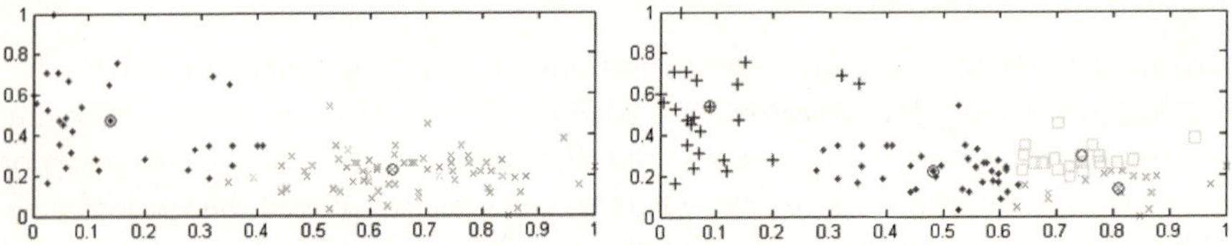

(a) (b)

Fig. 3. k-medoid algorithm. (a) $k = 2$. (b) $k = 4$.

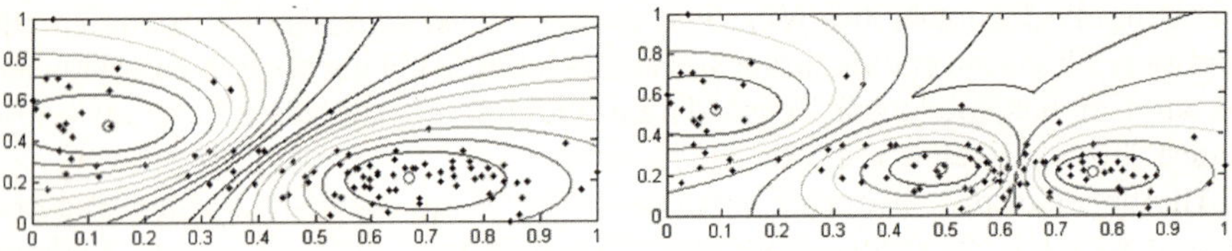

(a) (b)

Fig. 4. FCM algorithm. (a) $k = 2$. (b) $k = 3$.

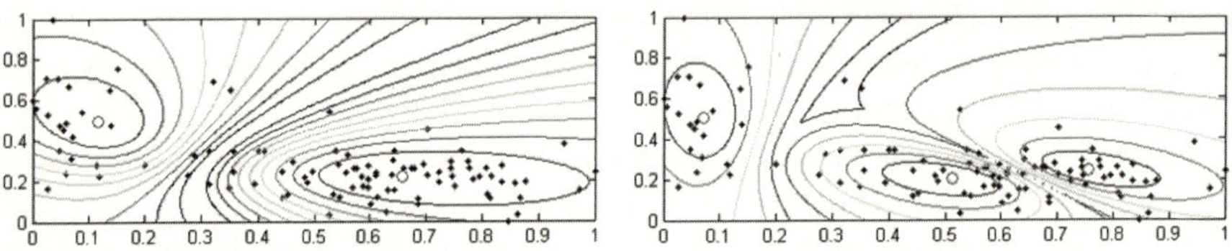

(a) (b)

Fig. 5. Gustfson-Kessel algorithm. (a) $k = 2$. (b) $k = 3$.

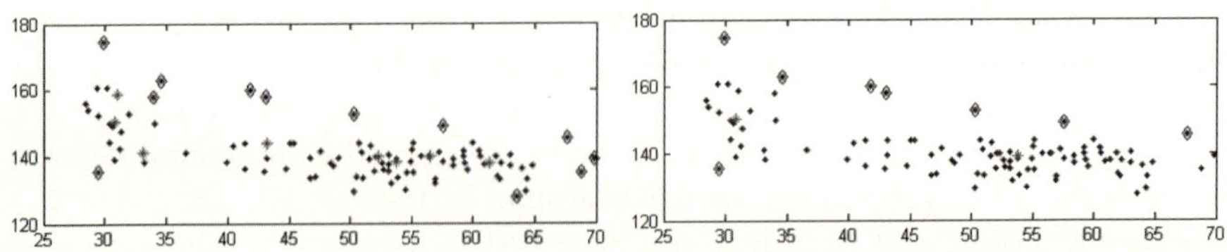

(a) (b)

Fig. 6. FLAME algorithm results highlighting the CSOs and Outliers. (a) k-NN = 5, $s = 1/4$ maximum density. (b) k-NN = 10, $s = 1/4$ maximum density.

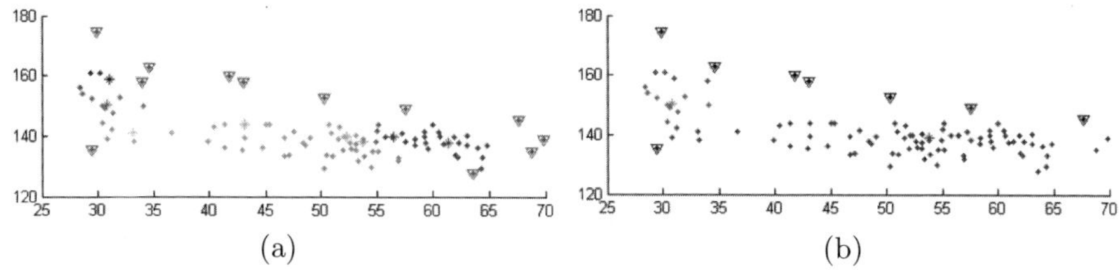

Fig. 7. FLAME algorithm results using a colourmap to represent the fuzzy membership to each of the CSOs or Outlier clusters. (a) k-NN = 5, $s = 1/4$ maximum density. (b) k-NN = 10, $s = 1/4$ maximum density.

5 Concluding Remarks

The results have shown that crisp clustering methods alone do not represent the real-life dataset accurately and reliably enough for data identification. Fuzzy clustering methods better represent an object's membership to multiple data clusters as is seen in many real-life applications. When combined with the density based method k-nearest neighbours (FLAME) the results have shown good promise. Currently there has been little application of hierarchical clustering techniques upon the dataset. Future work will include the implementation of Hierarchical clustering methods such as CURE, ROCK and CHAMELEON [11]. CHAMELEON measures the similarity of two clusters and only merges them if their inter-connectivity and proximity between each other is relatively high in comparison to the internal inter-connectivity and proximity of the objects within the clusters.

One of the important consideration when choosing a clustering algorithm for a real-world application is its ability to adapt to different datasets. This is particularly true in the domain of Flight Data Monitoring. Abilities such as scalability, detection of the best-fit value for k clusters and the method for identifying outliers. Validity indexes allow one to score the result of a clustering technique depending on a set of criteria. This criteria often includes the compactness and separation of the clusters identified. A similar measure would be helpful when scaling the FLAME algorithm which has shown sensitivity when scaling to different datasets with similar values of k nearest neighbours and the outlier threshold s. Due to the quantity of aircraft flight data received for analysis, it is necessary to create an optimisation technique to speed up the process of data clustering and identifying outliers. A cluster metrics will be formed in order to identify cluster membership regions in the feature space. New data objects can then be identified to a membership region which will drastically reduce the processing requirements compared to clustering.

References

1. Duin, R., Juszczak, P., Paclik, P., Pekalska, E., de Ridder, D., Tax, D.: Prtools4, a matlab toolbox for pattern recognition. Technical report, Delft University of Technology (2004)
2. Dubes, R.C., Jain, A.K.: Algorithms for Clustering Data, vol. 355. Prentice-Hall, Inc., Upper Saddle River (1988)

3. Hartigan, J.A., Wong, M.A.: Algorithm as 136: A k-means clustering algorithm. Applied Statistics 28(1), 100–108 (1979)
4. Pascual-Marqui, R.D., Pascual-Montano, A.D., Kochi, K., Carazo, J.M.: Smoothly distributed fuzzy c-means: a new self-organizing map. Pattern Recognition 34(12), 2395–2402 (2001)
5. Davé, R.N.: Characterization and detection of noise in clustering. Pattern Recognition Letters 12(11), 657–664 (1991)
6. Gustafson, D.E., Kessel, W.C.: Fuzzy clustering with a fuzzy covariance matrix. In: IEEE Conference on Decision and Control including the 17th Symposium on Adaptive Processes, 1978, vol. 17 (1978)
7. Ester, M., Kriegel, H.P., Sander, J., Xu, X.: A density-based algorithm for discovering clusters in large spatial databases with noise. In: 2nd International Conference on Knowledge Discovery and Data Mining (1996)
8. Fu, L., Medico, E.: Flame, a novel fuzzy clustering method for the analysis of dna microarray data. BMC Bioinformatics 8(3) (January 2007)
9. Hautamaki, V., Karkkainen, I., Franti, P.: Outlier detection using k-nearest neighbour graph. In: Proceedings of the 17th International Conference on Pattern Recognition, 2004. ICPR 2004, vol. 3 (2004)
10. Jin, W., Tung, A.K.H., Han, J.: Mining top-n local outliers in large databases. Knowledge Discovery and Data Mining, 293–298 (2001)
11. Karypis, G., Han, E.H.S., Kumar, V.: Chameleon: A hierarchical clustering algorithm using dynamic modeling. IEEE Computer 32(8), 68–75 (1999)

Visual-Based View-Invariant Human Motion Analysis: A Review

Xiaofei Ji[1,2], Honghai Liu[1], Yibo Li[3], and David Brown[1]

[1] The Institute of Industrial Research
University of Portsmouth, UK
[2] The Institute of Automation
Nanjing University of Aeronautics and Astronautics, China
[3] The Institute of Automation
Shenyang Institute of Aeronautical Engineering, China

Abstract. This paper provides a comprehensive survey of research on view-invariant human motion analysis. Recent research has shown that view-invariant related issues has been one of the bottlenecks for human motion understanding. The priority in this paper has been given to view-invariant pose representation and estimation, behaviour understanding. Research challenges and future directions are discussed in the end.

Keywords: Human motion analysis, View-invariant, Pose representation and estimate, Behaviour understanding.

1 Introduction

Visual-based human motion analysis is currently one of the most active research topics in computer vision. This strong interest is driven by a wide spectrum of applications in many areas such as visual surveillance, contend based video retrieval, precise analysis of athletic performance, etc. In the most of application actions are often observed from arbitrary viewpoint, so those application request the analysis methods exhibit some view invariance. As a novel direction of human motion analysis, the issue of viewpoint invariance in the representation and recognition of poses and actions has received more and more attentions [1, 2, 3]. This paper provides a comprehensive survey of research on view-invariant human motion analysis. The fundamental problems and applications of human motion analysis with a focus on view-invariance have been identified, main research groups and the contribution of this paper are outlined in the following.

1.1 Fundamental Problems and Application

1. Visual Surveillance: A human action recognition system can process image sequences captured by video cameras monitoring sensitive areas such as banks, department stores, parking lots and border to determine if one or more humans are engaging in suspicious or criminal activity.

I. Lovrek, R.J. Howlett, and L.C. Jain (Eds.): KES 2008, Part I, LNAI 5177, pp. 741–748, 2008.

2. Contend based Video Retrieval: A human behavior understanding system can scan through video taking as input, an action or event specified in high-level language as output. Such an application could be prove very useful for sportscasters to quickly retrieve important events in particular games.
3. Precise Analysis of Athletic Performance: Video analysis of athlete action is becoming an important tool for sports training, since it has no intervention to the athlete. For example, Li *et al.* present automatic analysis of complex individual actions in diving video [4].

1.2 Main Research Groups

Since there are growing requirements of its application in recent years, view-invariant human motion analysis has attract significant attention of research groups and several research groups have made some significant progresses to solve the view restriction in human motion analysis. Mainly research groups and their works are listed below.

1. Computer Science Department of Brown University: They have done a lot of research works in model-based tracking algorithm and 3D human pose estimate in video sequence. They provided a basic Matlab implementation for tracking a moving person in a sequence of multiple synchronized images and built a dataset(MOCAP) for testing the tracking algorithms [5].
2. Center for Automation Research (University of Maryland): It is a team early pay attention to the research of view-invariant human motion analysis. They focus on using 2D projective invariance theory and 3D mutual invariants to get the view-invariant human action representation and recognition [6].
3. INRIA PERCEPTION Team: It is a research group associated with INRIA. The scientific objective of the group is to map images and videos onto 3-D visual representations. They have obtained some improvement in view-invariant human pose representation and built a multi-view action database (IXMAS) [7,8].
4. Chinese Academy of Sciences, Institute of Automation: This group paid more attention on the research of human gait recognition [1], They built a multi-view gait database to study the relationship between the performance and the view angle.

1.3 Contribution of This Paper

The importance and popularity of human motion analysis has led to several previous surveys [9, 10, 11]. In contrast to the previous review, the current review focuses on the most recent developments in human motion analysis, *i.e.* view-invariant human motion analysis, especially on two issues: view-invariant pose representation and estimate, behavior understanding. It covers the latest researches ranging mainly from 2000 to 2007. This paper is organized as follows. Section 2 reviews the researches on view-invariant pose presentation and estimate, which is divided in three categories based on their use of a prior human model. The paper discuss the methods of human behaviour understanding in Section 3. Section 4 analyzes some challenges and possible directions for future research and concludes this paper.

2 View-Invariant Pose Representation and Estimate

Pose estimation refers to the process of estimating the configuration of the underlying kinematic or skeletal articulation structure of a person. There are infinite viewpoints from which a body in a given pose can be viewed, each leading to a different appearance of the body. A view-invariant method is specially necessary in this case. This paper only focuses on this direction and separates it into three categories based on their use of a prior human model.

2.1 3D Model-Based Pose Representation and Estimate

Many researcher are trying to depict the geometric structure of human body using 3D models. The simplest representation is the stick figure which consists of line segments linked by joints. More complex models include volumetric representations as generalized cones, elliptical, cylinders and spheres. The selection of the model employed usually depends on the application at hand.

Mostly 3D model-based pose estimate is an integral part of the tracking process. The introduction of stochastic sampling and search techniques has achieved whole-body pose estimation of complex movements from multiple views. The principal difficulty of their application is the dimensionality of the state space, it usually requires a relatively large number of samples to ensure a fair maximum likelihood estimation of the current state. Deutscher *et al.* presented the anneal particle filter which combines a deterministic annealing approach with stochastic sampling to reduce the number of samples required [12]. Kehl *et al.* proposed stochastic meta descent for whole-body pose estimation with 24 degrees-of-freedom from multiple views [13]. Balan *et al.* presented the first quantitative evaluation of Bayesian methods for the 3D tracking of humans in video [5].

Reconstruction of human pose from a single view image sequence is considerably more difficult than 3D pose estimate from multiple views. Sminchisescu and Triggs have achieved the most successful results to date in monocular markerless 3D human motion capture. Their algorithms are based upon propagating a mixture of Gaussians pdf, representing the probable 3D configurations of a body over time. These methods have proved effective on relatively short sequences [14]. Loy *et al.* presented a novel algorithm for the 3D reconstruction of human action in long monocular image sequences. A sequence is represented by a small set of automatically found representative keyframes. The skeletal joint positions are manually located in each keyframe and mapped to all other frames in the sequence. But this method need manual initialization [15]. Lee and Cohen combine a probabilistic proposal map representing the estimated likelihood of body parts in different 3D locations with an explicit 3D model to recover the 3D pose from single image frames. A data driven Markov chain Monte Carlo (MCMC) is used to search the high-dimensional pose space. Results demonstrate 3D pose estimation from single images of sports players in a variety of complex poses [16]. However, monocular reconstruction of complex 3D human movement remains an open problem. More attentions should be paid on automatic model initialization and re-initialization when tracking is lost.

2.2 3D Model-Free Pose Representation and Estimate

3D representation is a natural way to fuse multiple images information. A number of researchers have investigated direct 3D reconstruction of both model shape and motion from the visual-hull [7,17,18]. Visual-hull construction, also known as Shape-From-Silhouette(SFS), is a popular 3D reconstruction method which estimates the shape of an object from multiple silhouette images.

The representative work is the research of Mimic *et al.*, they presented an integrated system for automatic acquisition of the human body model and motion tracking from multiple synchronized video streams. The video frames are segmented and the 3D visual-hull reconstructions of the human body shape in each frame are computed from the foreground silhouettes. These reconstructions are then used as input to the model acquisition and tracking algorithms [17]. Cheung *et al.* proposed a SFS algorithm for articulated objects to recover the motion, shape and joints of an articulated object from silhouette and color images. The algorithm iteratively segments points on the silhouettes to each articulated part of the object and estimates the motion each individual part using the segmented silhouette. Once the motion/shape of each part is recovered, the joints are estimated by articulation constraints. Then applied articulated SFS algorithm to acquire the kinematic information of a person and used the model to track the person in new video sequences [18]. Those approaches exploit 3D reconstruction from multiple views to directly recover both shape and motion. They are suitable for multiple camera based systems allowing estimation of complex human movements, and can be used in real-time application because its computation price is relatively low.

2.3 Example-Based Pose Representation and Estimate

Example based methods store a database of example human figures with known 3D parameters and estimate 3D pose by searching for examples similar to the input image [19,20,21]. A potential advantage of example-based methods over model-based method is that the pose can be estimated independently at each frame, allowing pose estimation for rapid movements. Shakhnarovich *et al.* proposed an example-based approach for view-invariant pose estimation of upper-body 3D pose from a single image. Parameter-sensitive hashing is used to represent the mapping between observed segmented images from multiple views and the corresponding 3D pose [19]. Agarwal and Triggs presented a method that recovered 3D human body pose from monocular silhouettes by direct nonlinear regression of joint angles against histogram-of-shape-context silhouette shape descriptors [20]. Howe *et al.* used a direct silhouette look up using

Chamfer distance to select candidate poses together with a Markov chain for temporal propagation for 3D pose estimation of walking and dancing [21].

Recovering 3D poses from a single view is a more challenging problem, so upwards methods are usually multi-valued. Some alternative approaches don't recover the 3D parameters of the body joints but directly infer a high-level description of the type of key pose that the human is performing [2,22,8]. Comparing with known examples is certainly easier than inferring unknown parameters but it needs a large number of examples. The difficulty in getting enough examples makes the pose recovered not highly accurate.

3 Behaviour Understanding

Behaviour understanding aims to analyze and recognize the human motion patterns, and to produce high-level description of actions and interactions. They are large scale events that typically depend on the context of environment, objects, or interaction of humans. In this paper, classify behaviour understanding into two hierarchies: action recognition and behaviour description.

3.1 Action Representation and Recognition

In this section, we discuss action representation and recognition under the following groups of approaches similar to the survey [9].

Approaches Using Template Matching. Action recognition based on template matching, always converts an image sequence into a static shape pattern, and then compares it to prestored action prototypes during recognition. The advantage of template matching is low computational complexity and simple implementation. However, it is usually more sensitive to noise and the variance of movement duration.

RAO *et al.* presented a computational representation of human action to capture these dramatic changes using spatio-temporal curvature of 2D trajectory. This representation is compact, view-invariant [23]. Parameswaran and Chellappa considered the problem of view-invariant action recognition based on point-light displays by investigating 2D and 3D invariant theory. They employ a convenient 2D invariant representation by decomposing and combining the patches of a 3D scene [6]. Yilmaz and Shah proposed a novel representation for actions using spatio-temporal action volumes(STV). Set of these action descriptors define the action sketch which is invariant to the viewing angle of the camera [24]. Another view-invariant approach is the work of Weinland *et al.* [7], they use multiple cameras and shape from silhouette techniques to compute their visual hulls in every frame and accumulate visual hulls in a time period between primitive actions into motion history volumes(MHVs). MHVs fuse action cues, as seen from different viewpoints and over short time periods, into a single 3D representation. Using above view invariant features, they all can perform view invariant action recognition.

State-Space Approaches. Hidden Markov Models (HMMs) as a kind of sophisticated technique for analyzing time-varying data have been widely applied to express the temporal relationships inherent in human actions. Lv *et al.* decomposed the high dimensional 3D joint space into a set of feature spaces where each feature corresponds to the motion of a single joint or combination of related multiple joints. For each feature, the dynamics of each action class is learned with one HMM [25]. A novel view-invariant work [8] propose an exemplar-based hidden Markov Model(HMM) that accounts for dependencies between 3D exemplars, *i.e.* representative pose instances, and image cues.

Approaches using these HMMs usually apply intrinsic nonlinear models and do not have a closed-form solution. So it typically requires searching for a global optimum in the training process, which requires expensive computing iterations. In order to reducing the computing price, some new approaches infer human actions by taking advantage of the contextual constrains imposed by actions. Lv and Nevatia presented an example based action recognition system that explores the use of contextual constrains. Those constraints were inherently modeled using a novel action representation scheme called *Action Net* [22]. Though approaches based on state-space have been widely applied, selecting the proper number of states and dimension of the feature vector to avoid "underfitting" or "overfitting" still remains an issue.

3.2 Behaviour Semantic Description

The semantic description of human behavior has recently received considerable attention. Its purpose is to reasonably choose a group of motion words or shout expressions to report the behaviors of the moving objects in natural scenes. Ogale *etc.* presented a method for view-invariant action recognition using a probabilistic context-free grammar(PCFG). The PCFG construction process is completely automatic and uses multi-view data. The recognition process is also completely automatic, and parses a single viewpoint simultaneously [2]. Another work is [26], this paper proposes a new approach for recognition of task-oriented actions based on stochastic context-free grammar (SCFG). Giving one SCFG rule the multiple probabilities, one SCFG can recognize multiple actions. Park and Aggarwal presented a framework for recognizing human actions and interactions in video by using three levels of abstraction. The system provides a user-friendly natural-language description of several human interactions [27]. At present, human behaviour description is still restricted to simple and special action patterns and special scenes. Therefore, research on semantic description of human behaviors in complex unconstrained scenes still remains open.

4 Concluding Remarks

Although some work has been done in view-invariant human motion analysis, many issues are still open and deserve future research, especially in the following areas:

1. 3D model-based pose estimate: research in 3D pose estimate from monocular sequence remains an open problem, specially the problem of fully automatic initialization of model. Combining view-based and model-based tracking should be a direction to solve this problem.
2. Behaviour understanding: behaviour understanding is complex, as the same behaviour have several different meanings depending upon the scene and task context in which it is perform. According to unknown scenes, there are only a few relative research results are reported. So that patterns of behaviors are constructed by self-organizing and self-learning of image sequences should be a important direction to solve unknown scenes.

View-invariant human motion analysis is a new active research topics in computer vision. This strong interest is driven by a wide spectrum of promising applications in many areas. In this paper, we have presented an overview of recent developments in each key issue of view-invariant human motion analysis. At the end of this review, we have given some discussions on research difficulties and future directions in this direction.

References

1. Yu, S., Tan, D., Tan, T.: Modelling the effect of view angle variation on appearance-based gait recognition. In: Proc. of 7th Asian Conf. on Computer Vision, vol. 1, pp. 807–816 (2006)
2. Ogale, A., Karapurkar, A., Aloimonos, Y.: View-invariant modeling and recognition of human actions using grammars. In: Workshop on Dynamical Vision at ICCV, vol. 5 (2005)
3. Ong, E., Micilotta, A., Bowden, R., Hilton, A.: Viewpoint invariant exemplar-based 3D human tracking. Computer Vision and Image Understanding 104(2-3), 178–189 (2006)
4. Li, H., Lin, S., Zhang, Y., Tao, K.: Automatic Video-based Analysis of Athlete Action. In: 14th International Conference on Image Analysis and Processing, 2007. ICIAP 2007, pp. 205–210 (2007)
5. Balan, A., Sigal, L., Black, M.: A Quantitative Evaluation of Video-based 3D Person Tracking. In: International Workshop on 2nd Joint IEEE, pp. 349–356 (2005)
6. Parameswaran, V., Chellappa, R.: View Invariance for Human Action Recognition. International Journal of Computer Vision 66(1), 83–101 (2006)
7. Weinland, D., Ronfard, R., Boyer, E.: Free viewpoint action recognition using motion history volumes. Computer Vision and Image Understanding 104(2-3), 249–257 (2006)
8. Weinland, D., Grenoble, F., Boyer, E., Ronfard, R., Inc, A.: Action Recognition from Arbitrary Views using 3D Exemplars. In: Proceedings of the International Conference on Computer Vision (2007)
9. Wang, L., Hu, W., Tan, T.: Recent developments in human motion analysis. Pattern Recognition 36(3), 585–601 (2003)
10. Hu, W., Tan, T., Wang, L., Maybank, S.: A survey on visual surveillance of object motion and behaviors. IEEE Transactions on Systems, Man, and Cybernetics, Part C: Applications and Reviews 34(3), 334–352 (2004)

11. Moeslund, T., Hilton, A., Krüger, V.: A survey of advances in vision-based human motion capture and analysis. Computer Vision and Image Understanding 104(2-3), 90–126 (2006)
12. Deutscher, J., Blake, A., Reid, I.: Articulated body motion capture by annealed particle filtering. In: Proceedings. IEEE Conference on Computer Vision and Pattern Recognition, 2000. Proceedings, vol. 2 (2000)
13. Kehl, R., Bray, M., Van Gool, L.: Full Body Tracking from Multiple Views Using Stochastic Sampling. In: Proceedings of the 2005 IEEE Computer Society Conference on Computer Vision and Pattern Recognition, vol. 02, pp. 129–136 (2005)
14. Sminchisescu, C., Triggs, B.: Kinematic jump processes for monocular 3D human tracking. In: IEEE Computer Society Conference, vol. 1 (2003)
15. Loy, G., Eriksson, M., Sullivan, J., Carlsson, S.: Monocular 3D Reconstruction of Human Motion in Long Action Sequences. In: Computer Vision-ECCV 2004: 8th European Conference on Computer Vision, Prague, Czech Republic, May 11-14 (2004)
16. Lee, M., Cohen, I.: Proposal maps driven MCMC for estimating human body pose in static images. In: Proceedings of the 2004 IEEE Computer Society Conference, vol. 2 (2004)
17. Mikic, I., Trivedi, M., Hunter, E., Cosman, P.: Human Body Model Acquisition and Tracking Using Voxel Data. International Journal of Computer Vision 53(3), 199–223 (2003)
18. Cheung, K., Baker, S., Kanade, T.: Shape-from-silhouette of articulated objects and its use for human body kinematics estimation and motion capture. In: 2003 IEEE Computer Society Conference on Proceedings, vol. 1 (2003)
19. Shakhnarovich, G., Viola, P., Darrell, T.: Fast pose estimation with parameter-sensitive hashing. In: Proceedings of Ninth IEEE International Conference on Computer Vision, 2003, pp. 750–757 (2003)
20. Agarwal, A., Triggs, B.: Recovering 3D human pose from monocular images. IEEE Transactions on Pattern Analysis and Machine Intelligence 28(1), 44–58 (2006)
21. Howe, N.R.: Silhouette Lookup for Automatic Pose Tracking. In: Computer Vision and Pattern Recognition Workshop Conference on 2004, pp. 15–22 (2004)
22. Fengjun Lv, R.N.: Single View Human Action Recognition using Key Pose Matching and Viterbi Path Searching. In: IEEE International Conference on Computer Vision and Pattern Recognition (2007)
23. Rao, C., Yilmaz, A., Shah, M.: View-Invariant Representation and Recognition of Actions. International Journal of Computer Vision 50(2), 203–226 (2002)
24. Yilmaz, A., Shah, M.: Actions As Objects: A Novel Action Representation. In: Proc. IEEE Conf. Computer Vision and Pattern Recognition (CVPR 2005) (2005)
25. Lv, F., Nevatia, R.: Recognition and Segmentation of 3-D Human Action Using HMM and Multi-class AdaBoost. In: European Conf. on Computer Vision, vol. 4, pp. 359–372 (2006)
26. Yamamoto, M., Mitomi, H., Fujiwara, F., Sato, T.: Bayesian Classification of Task-Oriented Actions Based on Stochastic Context-Free Grammar. In: Proceedings of the 7th International Conference on Automatic Face and Gesture Recognition (FGR 2006), vol. 00, pp. 317–323 (2006)
27. Park, S., Aggarwal, J.: A hierarchical Bayesian network for event recognition of human actions and interactions. Multimedia systems 10(2), 164–179 (2004)

Generalized Extreme Value for Smooth Component Analysis in Prediction Improvement

Ryszard Szupiluk[1,2], Piotr Wojewnik[1,2], and Tomasz Ząbkowski[1,3]

[1] Polska Telefonia Cyfrowa Ltd, Al.Jerozolimskie 181, 02-222 Warsaw, Poland
[2] Warsaw School of Economics, Al.Niepodleglosci 162, 02-554 Warsaw, Poland
[3] Warsaw University of Life Sciences, ul.Nowoursynowska 159, 02-787 Warsaw, Poland
{rszupiluk,pwojewnik,tzabkowski}@era.pl

Abstract. In this paper we propose a new preprocessing method for smooth component analysis (SmCA). The smoothness measure used in SmCA depends on the signal extreme values directly. We propose the min/max transformation based on the extreme value distribution providing the more realistic and useful signal characteristic in terms of the smoothness. The full methodology is applied as an ensemble method for the energy load prediction improvement.

Keywords: predictive modeling, ensemble methods, statistical decomposition.

1 Introduction

The smooth component analysis is a method of the smooth components finding in a multivariate variable and can be treated as one of the blind signals separation (BSS) methods [2,3,6]. Those methods have variety of applications in such areas like telecommunication, medicine, finance [6]. The SmCA methods are addressed to the problems with temporal structure data [13]. In this paper we use the SmCA as ensemble method for the energy load prediction improvement. The crucial problem in this methodology is associated with the smoothness measure as a base for SmCA algorithms. This measure is very sensitive to the minimum and the maximum values of the signal. It can be a problem when our information about the signal characteristic is based on the different samples, and especially if such characteristic is used for prediction modelling. Therefore it is necessary to make such preprocessing that extreme values used in SmCA algorithm are adequate to general signal characteristic. To solve this problem we propose signal transformation according to the values obtained from extreme value distribution [4,7].

2 Smooth Component Analysis

Smooth component analysis (SmCA) is a method of the smooth components $\mathbf{y}_i$, $i=1,..,n$ finding in a multivariate variable $\mathbf{X}=[\mathbf{x_1},\mathbf{x_2},...,\mathbf{x_n}]^T$ [2,13]. The analysis of the signal smoothness or variability is strongly associated with the definitions and

I. Lovrek, R.J. Howlett, and L.C. Jain (Eds.): KES 2008, Part I, LNAI 5177, pp. 749–756, 2008.

assumptions about such characteristics [14]. For signals with temporal structure we propose a smoothness measure

$$K_0(\mathbf{y}) = \frac{\frac{1}{N}\sum_{k=2}^{N}|\mathbf{y}(k)-\mathbf{y}(k-1)|}{\max(\mathbf{y})-\min(\mathbf{y})+\delta(\max(\mathbf{y})-\min(\mathbf{y}))}, \tag{1}$$

where symbol $\delta(.)$ means Kronecker delta. Measure (1) has simple interpretation: it is maximal when the changes in each step are equal to the range (maximal change), and is minimal when the data are constant. The possible values of $K_0(\mathbf{y})$ vary from 0 to 1. The Kronecker delta term is introduced to avoid dividing by zero.

We assume the components are linear combination of signals $\mathbf{x}_i$ and they should be as smooth as possible. Our aim is to find such $\mathbf{W} = \lfloor \mathbf{w}_1, \mathbf{w}_2, ..., \mathbf{w}_j \rfloor$ that for

$$\mathbf{Y} = \mathbf{W}\mathbf{X}, \tag{2}$$

we obtain $\mathbf{Y} = [\mathbf{y}_1, \mathbf{y}_2, ..., \mathbf{y}_n]^T$ where $\mathbf{y}_1$ minimise $K_0(\mathbf{y}_1)$, so we can write

$$\mathbf{w}_1 = \arg\min_{||\mathbf{w}||=1}(K_0(\mathbf{w}^T\mathbf{x})) \quad , \tag{3}$$

Having estimated the first $j-1$ smooth components the next one is calculated as most smooth component of the residual obtained in Gram-Schmidt orthogonalization:

$$\mathbf{w}_j = \arg\min_{||\mathbf{w}||=1}(K_0(\mathbf{w}^T(\mathbf{x} - \sum_{i=1}^{j-1}\mathbf{y}_i\mathbf{y}_i^T\mathbf{x}))), \tag{4}$$

where $\mathbf{y}_i = \mathbf{w}_i^T\mathbf{x}, i = 1 \ldots j$. As the numerical algorithm for finding $\mathbf{w}_k$ we can take the quasi-Newton method with multiple starting points [11].

The main disadvantage of the measure (1) is its high sensitivity to the outliers due to minimum and maximum value in denominator. To avoid this problem and to make the measure more robust we can apply the estimation based on the generalized extreme value distribution (GEVD).

3 Extreme Value Distribution Preprocessing for SmCA

In this section we present the basic properties of generalized extreme value distribution GEVD and its application to the smoothness measurement. We show that the smoothness value estimated directly from a signal is less effective, than estimated from the signal regularized by the extreme value distribution. We show also that in some cases the regularization influences the smoothness measure much more than simple scaling.

There are signals, e.g. heavy tailed, with high probability that each particular observation will change the extremes a lot, and therefore the smoothness value K_0, too. It would be possible to stabilize the smoothness measure (1), if not the empirical but some representative extreme values were used. Therefore we propose to estimate the representative extremes using the extreme value distribution [4,7]. The probability density function of the generalized extreme value distribution with the location parameter μ, the scale parameter σ, the and shape parameter $\gamma \neq 0$ is described by

$$f(z)=\frac{1}{\sigma}\left(1+\gamma\frac{z-\mu}{\sigma}\right)^{-1-\frac{1}{\gamma}}\exp\left(-\left(1+\gamma\frac{z-\mu}{\sigma}\right)^{-1/\gamma}\right), \tag{5}$$

for $1+\gamma\frac{z-\mu}{\sigma}>0$, where $\gamma>0$ (Type II) or $\gamma<0$ (Type III). For $\gamma=0$ (Type I) GEVD is

$$f(z)=\frac{1}{\sigma}\exp\left(-\exp\left(\frac{z-\mu}{\sigma}\right)-\frac{z-\mu}{\sigma}\right). \tag{6}$$

From the method of moments we can estimate

$$\tilde{\mu}=\frac{s\sqrt{6}}{\pi}, \tag{7}$$

$$\sigma=\bar{z}-0.5772\tilde{\mu}, \tag{8}$$

where $\bar{z}$ and s are the sample mean and standard deviation, respectively.

The particular method for estimation smoothness of signal $\mathbf{Y}^{\alpha}_{i}$ is as follows:

1. From the signal generate the bootstrap samples and calculate their min's z,
2. Assume z's are realizations of $f(z)$ distribution and fit GEVD with μ_{-} as the location parameter,
3. For maximums calculate the location parameter μ_{+} of $f(-z)$, respectively.
4. Regularize the signal $\mathbf{Y}^{\alpha}_{i}$ to $\mathbf{Y}_{\mathrm{reg}}{}^{\alpha}_{i}$ by correction of outliers to range $[\mu_{-}, \mu_{+}]$,
5. Calculate the smoothness $K_1(\mathbf{Y}_{\mathrm{reg}}{}^{\alpha}_{i})$ for regularized signal

$$K_1(\mathbf{Y}_{reg_i}{}^{\alpha})=\frac{\sum_{j=1}^{n-1}\left|y_{j+1}^{\alpha(i)}-y_{j}^{\alpha(i)}\right|}{\mu_{+}-\mu_{-}}. \tag{9}$$

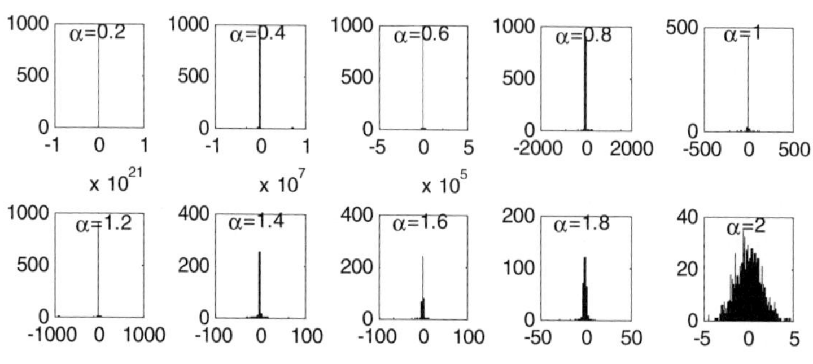

Fig. 1. Histograms of exemplar α-stable signals, α = 0.2,0.4,…,2

Now, we analyze the properties of K_0 and K_1 – the smoothness measure in primary form and after signal regularization with respect to the GEVD estimated outliers, $y_i \notin [\mu_-,\mu_+]$. We use the pseudo-random signals from alpha stable distributions with $\alpha = 0.2, 0.4, \ldots, 2$, [9,10]. In Fig. 1 we can observe the empirical distributions. We should note the decreasing range of signal as the parameter α grows.

The simulation study follows the scenario. For each level of α we generate 1000 pseudo-random signals $\mathbf{Y}^{\alpha}_i$ from α-stable distribution, $\alpha = 0.2, 0.4, \ldots, 2$, $i=1,\ldots,1000$, where each signal consist of 1000 elements. For each signal $\mathbf{Y}^{\alpha}_i$ we calculate directly the smoothness K_0 and estimate the extremes μ_- and μ_+ from GEVD's. We regularize the outliers to $[\mu_-,\mu_+]$, and thus the signal to $\mathbf{Y}_{reg}^{\alpha}$, and estimate K_1.

In Fig. 2 we present the histograms of empirical smoothness levels $K_0(\mathbf{Y}^{\alpha})$ and $K_1(\mathbf{Y}_{reg}^{\alpha})$ obtained for various $\alpha = 0.2, 0.4, \ldots, 2$.

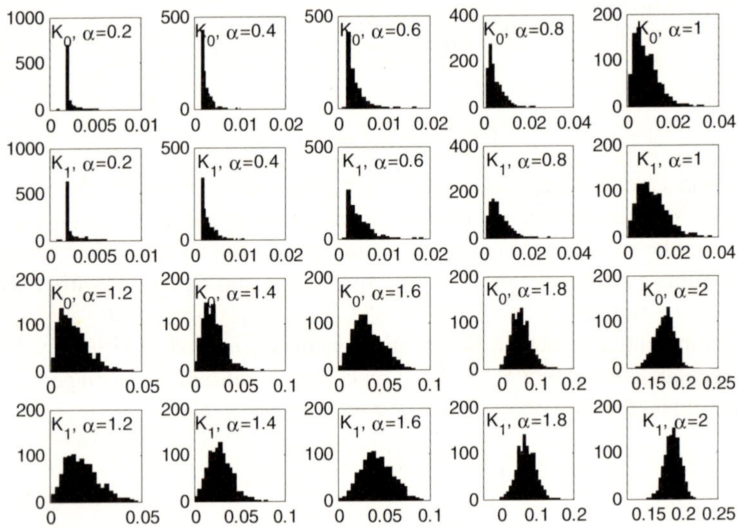

Fig. 2. Histograms of smoothness levels calculated for signals of different $\alpha = 0.2, 0.4, \ldots, 2$

We should note that for given α the empirical distributions of $K_0(\mathbf{Y}^{\alpha})$ and $K_1(\mathbf{Y}_{reg}^{\alpha})$ differ. In Tab. 1 we present the central moments of smoothness measures $K_0(\mathbf{Y}^{\alpha})$ and $K_1(\mathbf{Y}_{reg}^{\alpha})$ obtained for signals with various $\alpha = 0.2, 0.4, \ldots, 2$. As we can observe the K_1 converges better, because the skewness and kurtosis are smaller in comparison to K_0.

In Fig.3 we can observe the K_0/K_1 ratio versus range/$(\mu_+-\mu_-)$ ratio calculated for analysed signals. Small values of α characterizes the signals, where each observation can disturb the signal to high extend. This chaotic feature is visible also in the picture. The changes in range/$(\mu_+-\mu_-)$ ratio do not explain the whole change in K_0/K_1 ratio. Most of it depends on correction of outliers. For $\alpha=2$ (normal distribution), the outliers are not regularized frequently, and thus the range/$(\mu_+-\mu_-)$ ratio explains the most of change in K_0/K_1 ratio.

Table 1. Central moments of empirical smoothness levels $K_0(\mathbf{Y}^{\alpha})$ and $K_1(\mathbf{Y}_{\text{reg}}{}^{\alpha})$

α	Kurtosis		Skewness		Std. Deviation	
	K_0	K_1	K_0	K_1	K_0	K_1
0.2	11.00	7.80	2.69	2.19	0.0005	0.00068
0.4	8.63	5.87	2.05	1.55	0.0009	0.0013
0.6	9.28	6.08	1.99	1.44	0.002	0.002
0.8	5.61	4.88	1.46	1.16	0.003	0.004
1	4.58	3.67	1.15	0.84	0.005	0.006
1.2	3.64	3.03	0.93	0.64	0.008	0.009
1.4	3.67	3.13	0.76	0.43	0.011	0.012
1.6	2.69	2.59	0.50	0.26	0.016	0.017
1.8	3.20	3.02	0.42	0.13	0.024	0.025
2	2.77	2.76	-0.22	-0.09	0.013	0.011

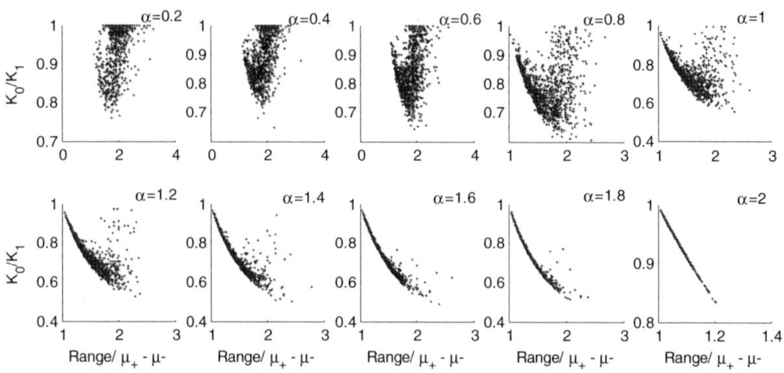

Fig. 3. K_0/K_1 ratio versus range/(μ_+-μ_-) ratio

In this paragraph we have shown, that estimation of the smoothness will be more effective with correction of the outliers based on extreme value distributions. We have shown that the impact differs for various signals. For Gaussian signal the smoothness is simply rescaled by range/(μ_+-μ_-) ratio, while for impulse signals (small α) the result is unpredictable.

4 SmCA in Ensemble Methods

In this paragraph we present SmCA utilized as an ensemble method in prediction improvement. The ensemble methods are addressed to prediction problems where many models can be applied. To improve the final prediction they integrate the information from different prediction methods. Usually such solutions propose the combination of a few models by mixing their results or parameters [1,5]. We develop an alternative concept based on the assumption that prediction results are mixture of the latent components common to all the model results [12].

We assume that after the learning process each prediction result includes two types of latent components: constructive, associated with the target, and destructive, associated with the inaccurate learning data, individual properties of models, missing data, not precise parameter estimation, distribution assumptions etc. The elimination of the destructive ones should improve the final results. To find the latent components we apply the SmCA method where the particular model prediction results are stored as $\mathbf{x}_i$, i=1,...,n, in $\mathbf{X}=[\mathbf{x}_1,\mathbf{x}_2,...,\mathbf{x}_n]^T$, $\mathbf{X}\in R^{n\times N}$, and set of latent components is $\mathbf{Y}=[\hat{\mathbf{y}}_1,\hat{\mathbf{y}}_2,...,\hat{\mathbf{y}}_k,\mathbf{y}_{k+1},\mathbf{y}_n]^T$, $\mathbf{Y}\in R^{n\times N}$, where N means the number of observations, $\hat{\mathbf{y}}_j$ denotes the constructive component and $\mathbf{y}_j$ is the destructive one [3]. Next we assume the relationship between the observed prediction results and the latent components as linear transformation

$$\mathbf{X}=\mathbf{AY}, \tag{10}$$

where the matrix $\mathbf{A}=\mathbf{W}^{-1}$ represents the mixing system. The (2) means the matrix $\mathbf{X}$ decomposition by the latent components matrix $\mathbf{Y}$ and the mixing matrix $\mathbf{W}$.

Our aim is to find the latent components and reject the destructive ones (replace them with zero). Next we mix the constructive components back to obtain the improved prediction results as

$$\hat{\mathbf{X}}=\mathbf{A}\hat{\mathbf{Y}}=\mathbf{A}[\hat{\mathbf{y}}_1,\hat{\mathbf{y}}_2,...,\hat{\mathbf{y}}_k,\mathbf{0}_{k+1},...,\mathbf{0}_n]^T. \tag{11}$$

In this paper we propose some transformation based on smooth component analysis to identify the latent signals. The whole algorithm of model aggregation by latent component identification can be written as:

1. For each prediction model $\mathbf{X}=[\mathbf{x}_1,\mathbf{x}_2,...,\mathbf{x}_m]^T$, $\mathbf{X}\in R^{m\times N}$, identify the extreme values $(\mu_{+(i)},\mu_{-(i)})$, i=1...m, basing on bootstrap and GEVD,
2. Regularize the prediction models $\mathbf{X}=[\mathbf{x}_1,\mathbf{x}_2,...,\mathbf{x}_m]^T$, $\mathbf{X}\in R^{m\times N}$, with ranges $[\mu_{+(i)},\mu_{-(i)}]$, i=1...m, to $\mathbf{X}_{reg}$,
3. Employing SmCA decompose the regularized models results $\mathbf{X}_{reg}\in R^{m\times N}$, into latent components $\mathbf{Y}=[\hat{\mathbf{y}}_1,\hat{\mathbf{y}}_2,...,\hat{\mathbf{y}}_k,\mathbf{y}_{k+1},...,\mathbf{y}_m]^T$ and for each component estimate smoothness value $K_1(\mathbf{y}_i)$, i=1...m,
4. Identify the destructive components as the most rough signals,
5. Eliminate destructive components $\hat{\mathbf{Y}}=[\hat{\mathbf{y}}_1,\hat{\mathbf{y}}_2,...,\hat{\mathbf{y}}_k,0,...,0]^T$,
6. Recompose the rest $\hat{\mathbf{X}}=\mathbf{A}\hat{\mathbf{Y}}$.

5 Practical Experiment

To verify whether the introduction of SmCA with the signals regularization improves the prediction results we analyze the real problem of the energy load prediction from the Polish market [8]. We want to predict the hourly energy consumption for 24 hours ahead based on the energy demand from the last 24 hours and the calendar variables:

month, day of the month, day of the week, and holiday indicator. The training data set included observations from 1988-1997 and the test was performed on the year 1998. As the primary prediction models we trained six MLP 28:q:1 neural networks with 28 inputs, q = 12, 18, 24, 27, 30, 33 – number of neurons in hidden layer, and single output. The error measures are: mean absolute percentage error, $MAPE = \frac{1}{N}\sum_{k=1\ldots N}\left|\frac{\mathbf{x}(k)-\mathbf{t}(k)}{\mathbf{t}(k)}\right|$, and mean square error, $MSE = \frac{1}{N}\sum_{k=1\ldots N}(\mathbf{x}(k)-\mathbf{t}(k))^2$, where k denotes the number of the observation, k=1…N, $\mathbf{t}(k)$ – the k-th real value, $\mathbf{x}(k)$ – the k-th prediciton, see Table 2-5.

Table 2. The results on MAPE for primary models

MLP 28:12:1	MLP 28:18:1	MLP 28:24:1	MLP 28:27:1	MLP 28:30:1	MLP 28:33:1
2.422	2.391	2.673	2.288	2.303	2.241

Table 3. The percentage improvements on MAPE after SmCA with preprocessing

MLP 28:12:1	MLP 28:18:1	MLP 28:24:1	MLP 28:27:1	MLP 28:30:1	MLP 28:33:1
1.8477	2.6278	4.478	2.376	3.844	5.821

In Table 2 we can observe that the MLP 28:33:1 performed the best among the primary models as to the MAPE criterion. After the introduction of SmCA with regularized signals the MAPE performance of the MLP 28:33:1 was improved by 5,821%, see Table 3.

Table 4. The results on MSE for primary models

MLP 28:12:1	MLP 28:18:1	MLP 28:24:1	MLP 28:27:1	MLP 28:30:1	MLP 28:33:1
1.127	1.115	1.038	1.147	1.094	1.121

Table 5. The percentage improvements on MSE after SmCA with preprocessing

MLP 28:12:1	MLP 28:18:1	MLP 28:24:1	MLP 28:27:1	MLP 28:30:1	MLP 28:33:1
-0.321	0.254	4.763	0.031	3.327	-0.196

In Table 4 we can observe that according to the MSE the network MLP 28:24:1 performed the best. After the SmCA with regularized signals the MSE performance was improved by 4,763%, see Table 5.

6 Conclusions

In this paper we derive a novel smooth component algorithm utilising the generalised extreme value distribution in the preprocessing stage. The maximal and the minimal values taken form this distribution allow to describe the smoothness characteristic of the signal in more reliable form than the range obtained directly from sample. In particular, the simulation study shows that the GEVD based SmCA decomposition enables improvement of the prediction accuracy. The practical experiment proved that the procedure can be successfully employed in the real-world problem. While the approach is time-consuming, it is recommended for the case where the computation is not crucial.

References

1. Breiman, L.: Bagging predictors. Machine Learning 24, 123–140 (1996)
2. Cichocki, A., Amari, S.: Adaptive Blind Signal and Image Processing. John Wiley, Chichester (2002)
3. Cichocki, A., Żurada, J.M.: Blind Signal Separation and Extraction: Recent Trends, Future Perspectives, and Applications. In: Rutkowski, L., Siekmann, J.H., Tadeusiewicz, R., Zadeh, L.A. (eds.) ICAISC 2004. LNCS (LNAI), vol. 3070, pp. 30–37. Springer, Heidelberg (2004)
4. Evans, M., Hastings, N., Peacock, B.: Statistical Distributions, 3rd edn. John Wiley and Sons, Chichester (2000)
5. Haykin, S.: Adaptive filter theory, 3rd edn. Prentice-Hall, Upper Saddle River (1996)
6. Hyvärinen, A., Karhunen, J., Oja, E.: Independent Component Analysis. John Wiley, Chichester (2001)
7. Johnson, N.L., Kotz, S., Kemp, A.W.: Univariate Discrete Distributions, 2nd edn. John Wiley and Sons, Chichester (1992)
8. Lendasse, A., Cottrell, M., Wertz, V., Verdleysen, M.: Prediction of Electric Load using Kohonen Maps – Application to the Polish Electricity Consumption. In: Proc. Am. Control Conf., Anchorage AK, pp. 3684–3689 (2002)
9. Nikias, C.L., Shao, M.: Signal Processing with Alpha-Stable Distributions and Applications. John Wiley &Son, Chichester (1995)
10. Samorodnitskij, G., Taqqu, M.: Stable non-Gaussian random processes: stochastic models with infinitive variance. Chapman and Hall, NY (1994)
11. Scales, L.E.: Introduction to Non-Linear Optimization. Springer, New York (1985)
12. Szupiluk, R., Wojewnik, P., Zabkowski, T.: Model Improvement by the Statistical Decomposition. In: Rutkowski, L., Siekmann, J.H., Tadeusiewicz, R., Zadeh, L.A. (eds.) ICAISC 2004. LNCS (LNAI), vol. 3070, pp. 1199–1204. Springer, Heidelberg (2004)
13. Szupiluk, R., Wojewnik, P., Zabkowski, T.: Smooth Component Analysis as Ensemble Method for Prediction Improvement. In: Davies, M.E., James, C.J., Abdallah, S.A., Plumbley, M.D. (eds.) ICA 2007. LNCS, vol. 4666, pp. 277–284. Springer, Heidelberg (2007)
14. Therrien, C.W.: Discrete Random Signals and Statistical Signal Processing. Prentice Hall, New Jersey (1992)

Context Cookies

Ramón Hervás[1], Gabriel Chavira[2], Salvador W. Nava[2], Vladimir Villarreal[3], and José Bravo[1]

[1] Castilla-La Mancha University, Paseo de la Universidad, 13071 Ciudad Real, Spain
{ramon.hlucas,jose.bravo}@uclm.es
[2] Autonomous University of Tamaulipas, Tampico-Madero, México
{gchavira,snava}@uat.edu.mx
[3] Technological University of Panama, Panama, Republic of Panama
vladimir.villarreal@utp.ac.pa

Abstract. In an ambient intelligence world, devices work in order to support people carrying out their everyday life activities in a natural way. Therefore, it is necessary to know the entities in the environment and to consider new interaction schemas with them. This paper proposes an autonomous and lively mechanism to refresh the context information while users interact with the environment using Near Field Communication technology. Context Cookies is the name of the mechanism that captures situation changes and provides awareness dynamically.

Keywords: Context Aware, Ambient Intelligence, NFC, Touching Interaction.

1 Introduction

Intelligent environments are responsive and sensitive to the presence of people that are integrated into a digital atmosphere which is adaptative to their needs, habits and emotions. In general, Ambient Intelligence (AmI), are the visions in which technology becomes invisible, embedded, present whenever we need it, enabled by simple interactions, attuned to all our senses and adaptive to users and contexts [1].

Only by understanding the world around us, the applications can be developed to achieve the Ambient Intelligence goals. Specifically, information that can be used to characterize the situation of significant users, places, or objects is considered C*ontext* and should be caught [2]. We need to know not only the object and people in the environment but also interact with them. Therefore, the traditional human-computer interaction is not appropriated for Ambient Intelligence.

In the upcoming years, we are going to think about Human-AmI interaction instead of Human-Computer interaction as a new perspective closer to human-human interaction [3]. Regarding human communication, there are three key issues: (a) Shared Knowledge between humans as an essential component to understand each other, but too extensive and not explicitly mentioned, (b) communication errors and recovery, including short term misunderstanding and ambiguities, and (c) situation and context [4]. We focus on

I. Lovrek, R.J. Howlett, and L.C. Jain (Eds.): KES 2008, Part I, LNAI 5177, pp. 757–764, 2008.

the last one; the physical environment, the situation, the role of the user, their relationship with others and the environment. Whenever a system is designed, we can change the current schemas of interaction using the above-mentioned information.

Some proposals regarding the interaction styles have to be considered. Spontaneous interaction is introduced in the Digital Aura project [5]. It is a model in which things interact with others within physical proximity. A similar approach is persistent interaction, "providing continuous interaction moves computing from a localized tool to a constant presence" [6]. Also, we can emphasize the embedded interaction. Within it, sensors and actuators are embedded into devices, tools, everyday objects and, also, interaction is embedded in the users' tasks [7]. Another related work is the Vazquez approaches; in [8] a model for knowledge sharing between devices in order to promote context-aware reactivity is proposed.

Most related works tackles the challenge of provide intelligence to digital object and interaction mechanisms between entities in the environment. We present a proposal for achieving a more natural interaction with the intelligent environment (with digital and non-digital objects, for example, a door, a table, etc.). Moreover this proposal provides an autonomous and dynamic mechanism in order to update the context information while users make their everyday task, in an implicit way. The mechanism is called "Context Cookies".

Section 2 of the paper analyzes innovative approaches to context modeling, looking for a simple way to identify objects in the environment and represent their behavior. Key issues of the Context Cookies, implementation details and applications, are shown in the section 3. Finally we provide the conclusions of our proposal.

2 Tagging the Context

Our mainly context source is the identification process, as an implicit and embedded input to the system, perceiving the entity identity, his profile and other kinds of dynamic data. Using NFC technology we can obtain awareness features in order to maintain a dynamic context model.

2.1 NFC: Technology and Architecture

It is obvious that we need a great variety of devices placed in the environment around us with wireless connection capabilities. Therefore, a new short-range wireless connectivity technology "Near Field Communications" (NFC), has appeared.

NFC systems consist of two elements: (a) The Initiator- as its name indicates it begins and controls the information exchange (called reader in RFID); and (b) The Target-the device that responds to the requirement of the initiator (called tag in RFID). In an NFC system, there are two modes of operation: Active and Passive. In the active mode, both devices generate their own field of radio frequency to transmit data (peer to peer). In the passive one, only one of these devices generates the radiofrequency field, while the other is used in order to load modulation for data transfers. It is important to mention that, although the NFC protocol can be installed in any electronic device, our interest will be centered on NFC-enabled cell phones. In [9] we analyze

RFID and NFC technologies, both with their corresponding models and we deem the advantages of the NFC approach.

2.2 Context-Awareness by Tagging and Touching

We need to identify relevant entities in the environment and distribute context information throughout the building. Every relevant entity wears a NFC tag storing context information (identification, location profile, etc.). When users touch a tag, three types of events can be thrown: (a) call up applications on the mobile phone (in this case, the idea of application server tent to disappear) (b) activation of services in the tagged object or in a nearby object (e. g a display, computer, etc.), and (c) redefinition of the tag information.

It is possible to classify the tag structure by two approaches. The first one is a categorization by the involved entity. Keeping in mind the definition of context, there are three important parts of the context: objects, places and users. Tag information is always about one of them. Figure 1 (left) shows an example of tagging a particular place: the door. On the other hand, thinking in the type of information, we can identify four categories (right)

- Tag Identification Number.
- Awareness Information: Static Contextual data about the user, object or place in which the tag is stuck. (e. g. identification, location, permissions, etc.).
- General Information: necessary information for the associated service. (e. g. cinema showtimes, product details, etc.). Using a metaphor, general information is like attached documents in emails.
- Cookext (Context Cookies): dynamic data that refresh and update the context model. It is explained in detail in section 3.

Each tag stores static information about the place, object or user and, additionally, contains dynamic information changing frequently. Whenever situation changes after a touching interaction, context cookies can capture and store the modification. This approach provides more awareness, mainly, about users and their needs and habits.

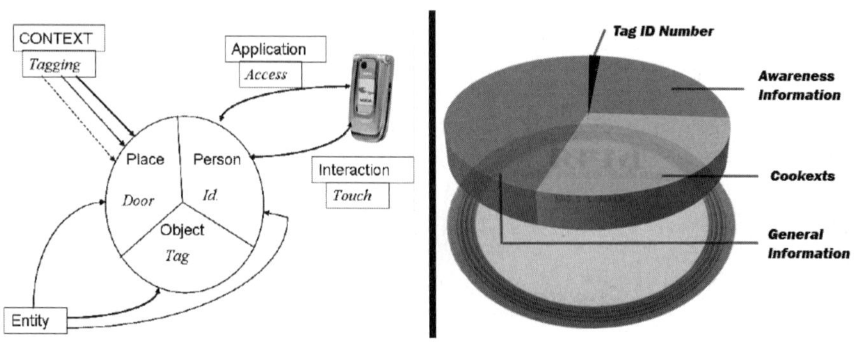

Fig. 1. Tag structure according to the involved entity (left) and according to the type of information (right)

3 Context Cookies

A Context-Cookie (Cookexts) is a parcel of data stored in a tag by a NFC device and then read back by the NFC device each time it touches the tag.

We propose using Cookexts by NFC systems in order to differentiate users and maintaining data related to the entities during the interaction with tagged environment, possibly across multiple visits. It is an approach inspired in HTTP cookies and applied to context-awareness environments.

3.1 Description

Technically, Cookexts are arbitrary pieces of data to describe the state about an object or a user. Therefore, there are two kinds of Cookexts: User Cookext and Object Cookext:

- User-Cookext ("I was here" metaphor): The Cookext is created by the NFC mobile and stored in the tag. Typically, these cookies provide awareness about the last uses of an object and allow personalized services reducing interactions. The personal information is kept into devices and returned when the NFC device touches the tag next time. Example: Continue with the last slide of the day-before presentation.
- Object-Cookext ("Take my presentation card" metaphor): The significant objects into the environment have a tag storing one or several Object Cookexts. When an NFC device touches a tag, it takes and saves a copy of the object Cookexts. Example: In a supermarket, save information about a product.

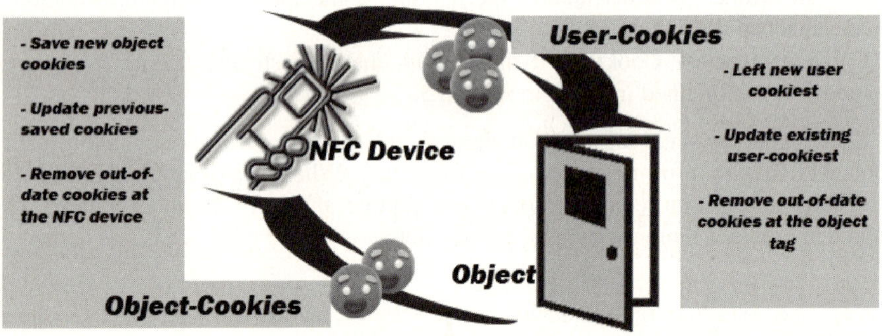

Fig. 2. General tasks for User-Cookies and Object-Cookies

When a NFC mobile device reads the context stored in a tag, first of all, it checks the deletion date and removes the out-of-date Cookext (Each Cookext must specify a deletion date). This date verification affects both User-Cookext in the tag as well as Object-Cookext in the NFC Mobile. After date verification, the Cookext content is read and refreshed (modifying information, increasing the deletion date or add new Cookext). Figure 2 shows the Cookexts exchange and the related actions.

3.2 Implementation

In our case of study, there are three main elements: (a) the user identification that determinates who stored this Cookext into the tag, (b) the deletion-date that indicates when the Cookext expires, and (c) the classes to update, i. e. the context attributes that should be modified in order to represent a new situation.

A context model is maintained and represented by ontologies, describing parts of the real world encompassing the users [10]. The context information is obtained from sensors embedded into the environment (tags and NFC devices) and from the static system information (stored in Data Bases). In addition, it is continuously inferred and selected according to the rules serialized into the Cookexts.

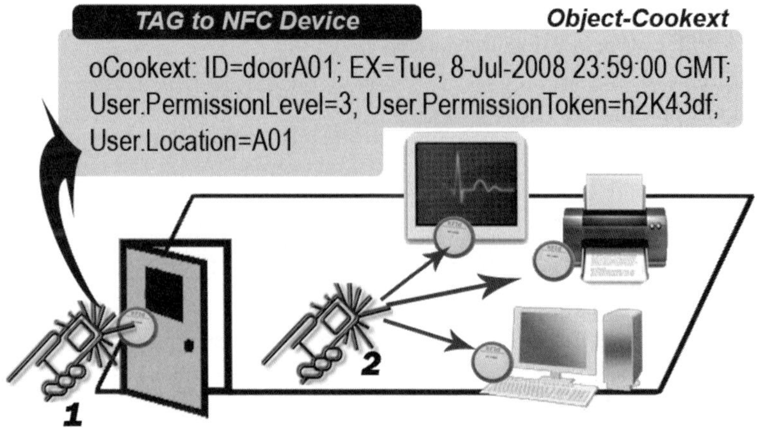

Fig. 3. Example of credential by means context cookies

When the context situation changes (typically when a user touches an NFC tag), the changes are evaluated. There may be changes in the context information that are not significant for our system (e.g. a new user in the room could launch a new service or not). The model decides which changes are important and either which are not (one change could be significant or not depending on the rest of the context information at this moment).

The Cookexts may be transformed into rules to determine which ontology attributes change. When a user touches a tag and reads back a Cookext, a new context graph is created to describe the new situation. In [11], we provide more details about the context model and our ontological approach.

The figure 3 shows an example: The user touches the laboratory-door tag and the system recognizes him after the user's login. The door is opened and an object Cookext is written back into the NFC device. The Cookext contain an authentication-token to recognize the user in the environment. This token allows user to use devices, such as printers, displays, personal computers, etc. Furthermore, the Cookext contains a rule to update the context information with the current user location

The Cookext at the figure is an Object-Cookext wrote back into the NFC device when user touches the door "doorA01". The expiration date is July, 8th and it includes a rule to update the user location. The PermissionLevel and PermissionToken are generated during the login process.

The user, her/his location, her/his permission level and her/his permission token are elements in our context model, i. e. the user is a concept or class in the ontology and location, permission level and permission token are user properties.

Due to the limited tag capacity, the content is not formatted. However, the NFC device parses the content to XML, check the consistency and validate the Cookext by means of the Document Type Definition. The code-behind shows the parsed XML code from the Cookext. Figure 4 shows the related RDF graph. The context information is updated taking the Cookext elements and making changes to the RDF graph.

Example of a XML code parsed from the Cookext in Figure 3.

```
<cookext>
    <ID> doorA01 </ID>
    <expireDate>Tue, 8-Jul-08 23:59:00 GMT</expireDate>
    <class>
        <classValue>User</classValue>
        <classPropoerty>
           <propertyName>PermisionLevel</propertyLevel>
           <propertyValue>3</propertyValue>
        </classProperty>
     </class> <class>
        <classValue>User</classValue>
        <classPropoerty>
            <propertyName>PermisionToken</propertyLevel>
            <propertyValue>h2K43df</propertyValue>
        </classProperty>
     </class> <class>
        <classValue>User</classValue>
        <classPropoerty>
            <propertyName>Location</propertyLevel>
            <propertyValue>A01</propertyValue>
        </classProperty>
     </class>
  </cookext>
```

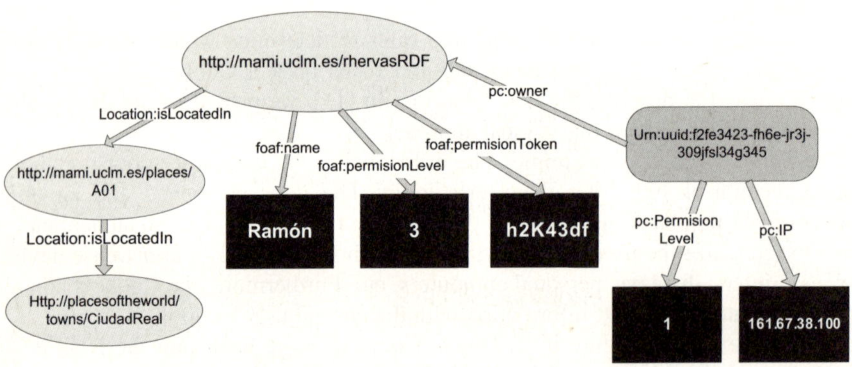

Fig. 4. RDF Graph representing the Cookext information in the figure 3

3.3 Awareness by Means of Context Cookies

There are many types of applications for Context Cookies:

- Authentication: Cookexts recognize previously-authenticated users.
- Customization: Cookext can be used in order to allow users to express devices' preferences, e.g. how and what contents are showed in public displays when the user activates them.
- Status: Cookexts facilitate to take up again a task. The current state of a task can be saved in order to retrieve this state in the future. For example, the session state of a personal computer, the current slide in an academic presentation, etc.
- Interaction: Usually, a tagged device (typically a display) has default behavior when a user touches it. As well as the main behavior, devices could have alternative interactive functions. Depending on the user, the Cookext facilitates to set the default behavior and the alternative ones.
- Tracking: Cookext can also be used for tracking the path of a person. It is useful to analyze the user experience and improve the available services in the environment. For example, in a tagged museum, the user can touch each work of art in order to obtain information. Each tag contains an object Cookext that is stored in the NFC device. This Cookext set allows the museum staff to know about the route of the visitor and his/her interests.
- Unexpected issues: Occasionally, it is necessary that the system has exceptional behaviors, for this reason a Cookext can define exceptions. For example, an unknown user visits an office or lab. An authorized user could save a Cookext in the unknown user NFC device in order to set temporal permissions, e.g printer permission
- Environment Resources: Cookext may content information about available resources in the environment. An illustrative application is the linking to active Bluetooth-points. NFC devices use closer Bluetooth devices to obtain connectivity. A cookiext stored in the door gives the MAC and UUID of available Bluetooth devices.

4 Conclusions

We have shown a technological adaptation supported by devices well known by the user (mobile phone) and offering some advantages over the traditional RFID for context awareness. The NFC approach is cheaper, allows us to provide storage capabilities to non-digital objects. Moreover, mobile devices reduce server dependency and ease the implementation of security and privacy strategies, because personal information is stored in the personal mobile and the distance of NFC transactions is about two centimeters.

NFC approach attains effortless and closer interaction, the infrastructure requirements are reduced and the schema of interaction is unified, i. e. users can interact with all objects (digital and non-digital) using touching interaction.

However, we think the most valuable contribution of this paper is the autonomous and dynamic mechanism to update the context information while users carry out their daily activities, in an implicit way. Moreover, Context Cookies allows us to offer new services according The Ambient Intelligent paradigm, adapted to the situation and, also, embedded in the environment and non-intrusive for users.

References

1. ISTAG, Scenarios for Ambient Intelligence in 2010 (February 2001), http://www.cordis.lu/ist/istag.htm
2. Dey, A.: Understanding and Using Context. Personal and Ubiquitous Computing 5(1), 4–7 (2001)
3. Reeves, B., Nass, C.: The Media Equation. C.U. Press, Cambridge (1996)
4. Schmidt, A.: Ubiquitous Computing - Computing in Context, in Com-puting Department. Lancaster University 1(294) (2002)
5. Ferscha, A., Hechinger, M., Mayrhofer, R., Dos Santos Rocha, M., Franz, M., Oberhauser, R.: Digital Aura. In: Advances in Pervasive Computing in Pervasive (2004)
6. Abowd, G., Mynatt, E.D.: Charting past, present and future research in ubiquitous computing. HCI in the new millennium 7(1), 29–58 (2000)
7. Schmidt, A., Kranz, M., Holleis, P.: Interacting with the Ubiquitous Computing - Towards Embedding Interaction. In: Smart Objects & Ambient Intel-ligence (sOc-EuSAI), Grenoble (2005)
8. Vazquez, I.: A Reactive Behavioral Model for Context-Aware Semantic Devices. Ph. D. Deusto University. Bilbao (2007)
9. Bravo, J., Hervás, R., Nava, S.W., Chavira, G., Sánchez, C.: Towards Natural Interaction by Enabling Technologies, A Near Field Communication Approach. In: The 1st Workshop on Human Aspect in Ambient Intelligence (European Conference on Ambient Intelligence), Frankfurt (2007)
10. Dogac, A., Laleci, G., Kabak, Y.: Context Framework for Ambient Intelligence. In: echallenges, Bologna (2003)
11. Hervás, R., Nava, S.W., Chavira, G., Sánchez, C., Bravo, J.: Towards implicit interaction in ambient intelligence through information mosaic roles. In: Engineering the User Interface: From Research to Practice - Invited papers from Interaccion 2006, Springer, Heidelberg (2006)

Smart Environment Vectorization
An Approach to Learning of User Lighting Preferences

Alejandro Fernández-Montes[1], Juan A. Ortega[1],
Luis González[2], Juan A. Álvarez[1], and Manuel D. Cruz[1]

[1] Department of Computer Science, University of Sevilla, Sevilla, Spain
{afernandez,ortega}@lsi.us.es, {jaalvarez,mancrudia}@us.es
[2] Department of Applied Economics, University of Sevilla, Sevilla, Spain
luisgon@us.es

Abstract. The automation of smart environment systems is one of the main goals of smart home researching. This paper focus on learning user lighting preference, considering a working field like a standard office. A review of the smart environment and devices setup is done, showing a real configuration for test purposes. Suitable learning machine techniques are exposed in order to learn these preferences, and suggest the actions the smart environment should execute to satisfy the user preferences. Learning machine techniques proposed are fed with a database, so a proposal for the vectorization of data is described and analyzed.

1 Introduction

Smart home technologies are often included as a part of ubiquitous computing. Mark Weiser [1] outlined some principles to describe Ubiquitous Computing (ubicomp) from which we emphasize that the purpose of a computer is to help you do something else.

Home technologies have tried to help home inhabitants since its creation. Nowadays, due to the popularization of computational devices, ubiquitous computing is called to be the revolution to develop smart systems with artificial intelligence techniques.

Domoweb [2] was a research project originally developed as a residential gateway implementation over the OSGi (Open Services Gateway Initiative) service platform. Nowadays Domoweb conform a great platform where researchers from different disciplines converges and where we can deploy, develop and test smart home related solutions, due to the component based model, and the service oriented architecture that Domoweb and OSGi supports.

This article focuses on modeling smart spaces to apply machine learning techniques. We have focused in the learning of user preferences for the lighting of a space. In order to interact with the space and retrieve these preferences, an office at the department of Computer Languages of the University of Seville has been provided of several devices to accomplish these tasks.

Artificial intelligent methods can be supported by this model like machine learning algorithms where is centered this article. Some techniques are presented

I. Lovrek, R.J. Howlett, and L.C. Jain (Eds.): KES 2008, Part I, LNAI 5177, pp. 765–772, 2008.

to accomplish this goal in section 3. Finally we propose some expansions which could be studied in order to cover other cases.

2 Experimental Environments

Two different environments are used for data collection. The first one is a simulated environment developed in Java that allows researchers to generate simulated and synthetic databases.

Second one is a standard office at the department of Computer Languages is used for data collection during the experiments. This office is intended to be used by a single person, who could be eventually visited by other work mates or students. It is illuminated by natural light from the window and four artificial fluorescent lights which can act as a complement of the natural light or like unique source of light.

Figure 1 shows the distribution and setup of the room.

As you can see, the orientation of the window is south, so it maximizes the quantity of light that receives during a day. This fact must be considered when analyzing results.

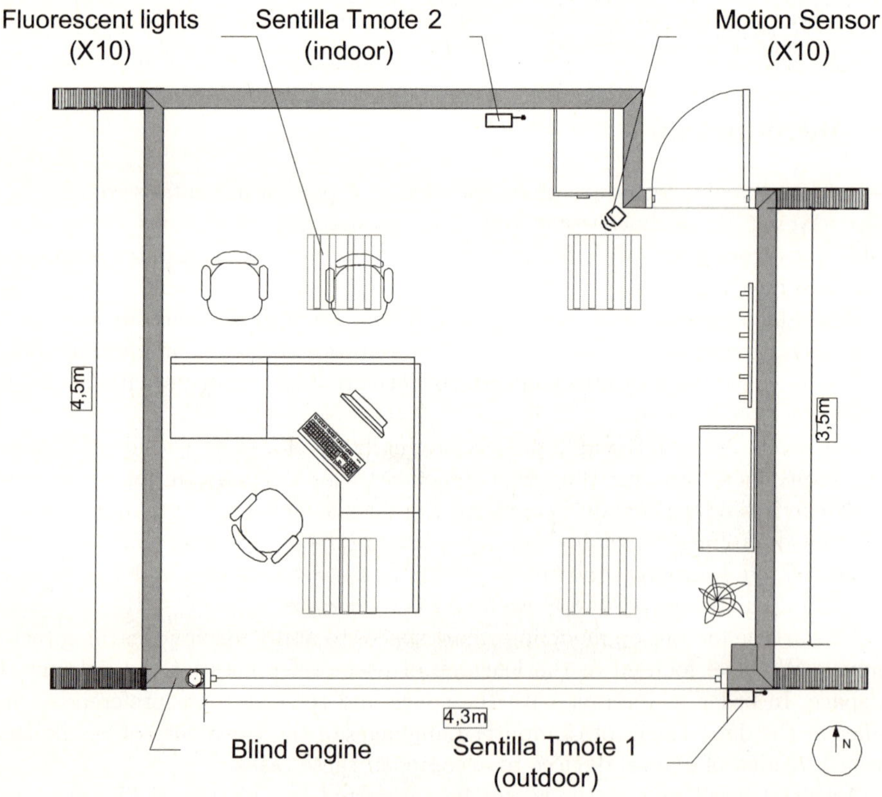

Fig. 1. Room setup

3 Related Devices

In this section we present the devices which will interact in the setup proposed. The concrete model and manufacturer of the devices are detailed although the learning system should be independent of these details.

3.1 Sentilla Tmote

Sentilla Tmotes are the devices which detects the quantity of light. In this setup we propose to install one mote indoor, and another outdoor. This way we can compare the preferences of the user indoor with the quantity of light outdoor. We'll be able to deduct some weather parameters in real time too. The dimensions are 8 cms. of width and 3.2 cm. of height, so it is quite small to suit ubiquitous applications and non intrusive systems. These devices also implement a humidity and temperature sensor, which could be requested for future improvement and expansions. The figure 2 shows the Sentilla *Tmote* module:

The connectivity with other motes and computers is done through the IEEE 802.15.4 (ZigBee) protocol, which minimizes battery consumption. Zigbee supports mesh networking so this is the topology we will adopt and this way the *motes* can create a network to share information and forward it to reach wider areas.

Nowadays Sentilla *Tmotes* are packaged in a beta development kit, including an IDE based on Eclipse 3.2 for developing. The hardware implements a Java Runtime Environment which can run different applications to retrieve, process and send information from sensors.

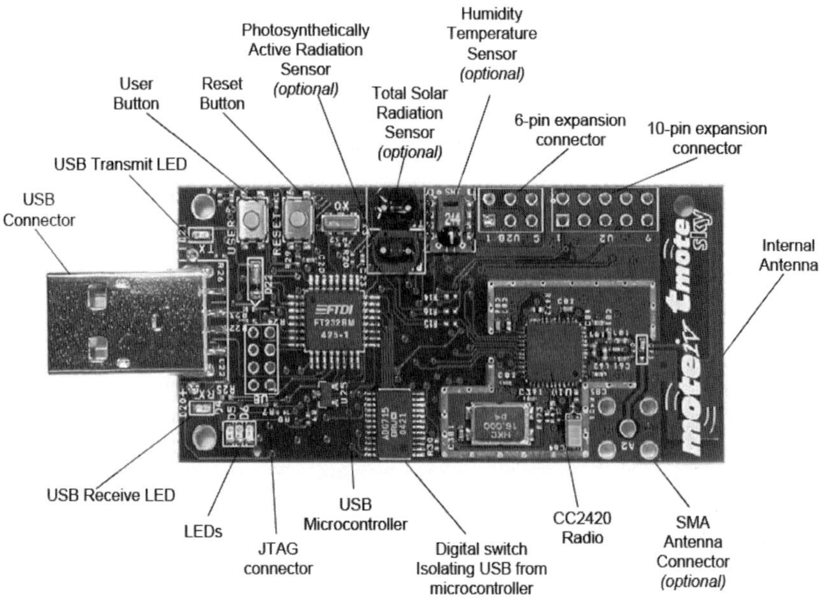

Fig. 2. Front of the *Tmote* module

3.2 Motion Sensor

A motion sensor is indispensable to determine when a user is at the room. This could act as a trigger of the learning algorithm to retrieve, process and send the information from sensors.

Domoweb project implemented an OSGi platform with software components to interact with X10 devices. The figure 3 shows the MS13A, a wireless device which interacts with a gateway that routes wireless messages over the electrical cable using X10 protocol. Notice that, although X10 protocol is an old-fashioned technology, it carries out its purpose perfectly.

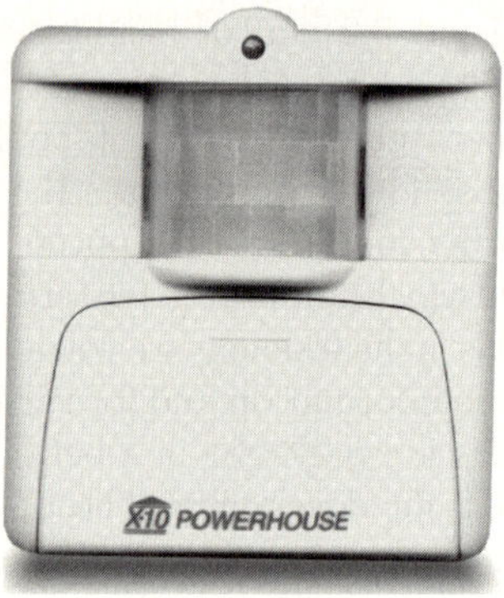

Fig. 3. Front of the MS13A. X10 Wireless Motion Sensor.

3.3 Fluorescent Light

These lights are activated with an X10 actuator, like the Appliance module AM486 in order to determine when a light has been switched on and off, and its current state.

3.4 Blind Engine

Lighting preferences are directly related with the quantity of light that goes through the window. Therefore the state of the blind or curtain will affect the lighting of the room. Some devices are under study but at the moment of writing this article none satisfied our requirements of wirelessly communication and standardized protocols.

4 Framework of Learning

As exposed before, our goal is the learning of the lighting preferences of a single user. This machine learning is done over the statistical data retrieved at the environment shown in sections 2 and 3. In general, we will retrieve a set of data X called *input*

$$X = \{x_1, x_2, \ldots, x_n\}$$

and we will have a set of data Y called *output.*

$$Y = \{y_1, y_2, \ldots, y_m\}$$

The process of learning tries to search a functional dependence between both sets of data. The framework is based on V.N. Vapnik [3] model of learning with examples. The model is composed by three elements as shown in the figure 4.

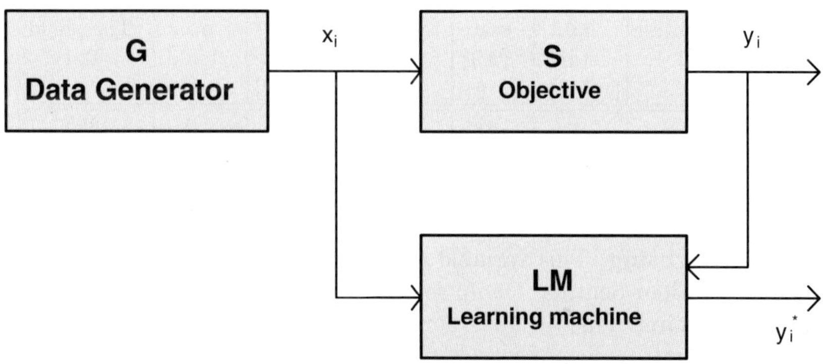

Fig. 4. Learning model

1. **Data generator.** Samples of X, retrieved by the infrastructure proposed.
2. **Objective** (aka supervisor). User preferences.
3. **LM.** Learning machine.

The learning can be carried out due to the dependence of the user preferences with his habits. Normally we have the same lighting preferences. These preferences must be learned at every environment, due to its dependence with the location, orientation of the window, devices setup and so on.

4.1 Techniques

Two families of algorithms, related with learning machine, can be considered although both are going to be supervised.

Support vector machine (SVM) and Neural networks (NN) are the selected techniques to learn user lighting preferences. The main advantage of these techniques is that always offer an output. On the other hand it is hard to interpret or understand their outputs which is an important feature that prediction algorithms should implement as expounded in [4].

The other family of algorithms considered is the machine learning techniques based on rules. These algorithms provide an output easier to understand and interpret, but their main disadvantage is that if no rule matches current state, these algorithms don't offer an output.

4.2 Input and Output

Table 1 shows the input variables X proposed and two sample input data:

Table 1. Input sample

Outdoor lighting $[0,1]$	Indoor lighting $[0,1]$	Indoor light state $\{0,1\}$	Blind state $[0,1]$	Motion $\{0,1\}$	Action over light $\{-1,0,1\}$	Action over blind $[-1,1]$	Threshold $[0,1]$
0.9	0	0	0	1	0	0	0.5
0.9	0.7	1	0.5	1	1	0.5	0.5

- **Outdoor lighting.** This variable represents the quantity of light received from the outdoor Sentilla *Tmote* sensor. Continuous variable from 0 to 1.
- **Indoor lighting.** This variable represents the quantity of light received from the indoor Sentilla *Tmote* sensor. Continuous variable from 0 to 1.
- **Indoor light state.** This variable represents the state of the lights of the room received from the X10 appliance module. Discrete variable, 0 for lights off, and 1 for lights on.
- **Blind state.** This variable represents the state of the blind or curtains. Continuous variable from 0 to 1. 0 for totally closed and 1 for totally open.
- **Motion.** This variable represents the detection of motion sent by the MS13A X10 device. Discrete variable, 0 for no motion, 1 for motion detected.
- **Action over light.** This variable represents the action done by the user over indoor light. Discrete variable, -1 lights switched off, 0 no action, +1 lights switched on.
- **Action over blind.** This variable represents the action done by the user over the blind/curtains. Continuous variable, -1 means the user closed it totally, 0 no action, +1 means the user opened it totally.
- **Threshold.** Represents the current user lighting preference. 0 represents minimum room lighting, 1 represents maximum room lighting.

Notice that Sentilla *Tmotes* offer the quantity of light received in luxes. We have to standarize this data to a $[0,1]$ interval.

Table 2 shows the output variables Y proposed and two sample output data:

- **Action over light.** Represents the action over indoor light predicted by the learning machine in order to satisfy user lighting preferences. Discrete variable, -1 lights switched off, 0 no action, +1 lights switched on.

Table 2. Output sample

Action over light {0,1}	Action over blind $[-1, 1]$	Action over threshold $[-1, 1]$
0	0	0
0	0	α

- **Action over blind.** Represents the action over the blind/curtains predicted by the learning machine in order to satisfy user lighting preferences. Continuous variable, -1 means the machine closed it totally, 0 no action, +1 means the machine opened it totally.
- **Action over threshold.** Represents the correction the machine must done in order to adapt current user threshold. Negative values reduce threshold, 0 represents no action, and positive values represent an increase correction over threshold. The value α of the correction must be determined with care, in order to avoid infinity jumps around user preference, and converge to the real user preference.

5 Future Work

Comparative results must be done between machine learning techniques proposed in section 4.1. Next step should focus in enlarge action field, to other rooms with different users, locations, orientations, and so on. This way we could compare results obtained with these techniques in different (but similar) environments, and create a wider *motes* mesh network.

Other field of action could be applying these techniques and algorithm to learn user preference over conditioning. Sentilla *Tmote* devices also include sensors to retrieve temperature and humidity, useful to learn conditioning preferences.

Acknowledgements. This research is supported by the MEC I+D project InCare. Ref: TSI2006-13390-C02-02 and the Andalusian Excellence I+D project CUBICO Ref: TIC2141.

References

1. Weiser, M.: The computer for the 21st century. SIGMOBILE Mob. Comput. Commun. Rev. 3(3), 3–11 (1999)
2. Alvarez, J.A., Cruz, M.D., Fernández, A., Ortega, J.A., Torres, J.: Experiencias en entornos de computación ubicua mediante arquitecturas orientadas a servicios. CEDI-JSWeb, 167–174 (September 2005)
3. Vapnik, V.: Statistical learning theory. Wiley, New York (1998)

4. Fernández-Montes, A., Álvarez, J.A., Ortega, J., Cruz, M., González, L., Velasco, F.: Modeling Smart Homes for Prediction Algorithms. In: Apolloni, B., Howlett, R.J., Jain, L. (eds.) KES 2007, Part II. LNCS (LNAI), vol. 4693, p. 26. Springer, Heidelberg (2007)
5. Cook, D., Youngblood, M., Das, S.: A multi-agent approach to controlling a smart environment. In: Augusto, J.C., Nugent, C.D. (eds.) Designing Smart Homes. LNCS (LNAI), vol. 4008, p. 165. Springer, Heidelberg (2006)
6. Li, J., Bu, Y., Chen, S., Tao, X., Lu, J.: FollowMe: On Research of Pluggable Infrastructure for Context-Awareness. In: Proceedings of the 20th International Conference on Advanced Information Networking and Applications (AINA 2006), vol. 01, pp. 199–204 (2006)
7. Das, S., Cook, D.: Designing Smart Environments: A Paradigm Based on Learning and Prediction. Mobile, Wireless, And Sensor Networks
8. Leake, D., Maguitman, A., Reichherzer, T.: Cases, Context, and Comfort: Opportunities for Case-Based Reasoning in Smart Homes. In: Augusto, J.C., Nugent, C.D. (eds.) Designing Smart Homes. LNCS (LNAI), vol. 4008, pp. 109–131. Springer, Heidelberg (2006)
9. Hagras, H., Callaghan, V., Colley, M., Clarke, G., Pounds-Cornish, A., Duman, H.: Creating an Ambient-Intelligence Environment Using Embedded Agents (2004)
10. Roy, N., Roy, A., Das, S.: Context-Aware Resource Management in Multi-Inhabitant Smart Homes: A Nash H-Learning based Approach. In: Proc. of 4th IEEE Int'l Conf. on Pervasive Computing and Communications (PerCom 2006) (2006)
11. Choi, J., Shin, D., Shin, D.: Research on Design and Implementation of the Artificial Intelligence Agent for Smart Home Based on Support Vector Machine. LNCS. Springer, Heidelberg
12. Yamazaki, T.: Beyond the Smart Home. In: Proceedings of the 2006 International Conference on Hybrid Information Technology, vol. 02, pp. 350–355 (2006)
13. Jiang, L., Liu, D., Yang, B.: Smart home research. In: Proceedings of 2004 International Conference on Machine Learning and Cybernetics, vol. 2 (2004)

Author Index

Abatangelo, Luca II-206
Abbas, Safia I-149
Abdelmoty, Alia I. II-689
Abe, Akinori II-750
Abe, Hidenao II-758
Abe, Jair Minoro II-466, II-474
Abe, Norihiro III-687
Abu Bakar, Rohani Binti III-136
Acito, Nicola III-424
Adachi, Yoshinori II-887, II-934, II-942
Adli, Alexander II-133
Akaishi, Mina II-823
Akama, Seiki II-474
Akman, Ibrahim II-879
Al-Fedaghi, Sabah S. I-294
Albertoni, Riccardo II-682
Aldana-Montes, José F. II-617, III-269
Alias, Mohd Fauzi III-541
Allwood, Carl Martin III-481
Álvarez, Juan A. I-765
Alves, Maria C.M. II-950
Amoiralis, Eleftherios I. II-9
An, Dong Chan II-346
Anagnostopoulos, John II-26
Ancona, Nicola II-206, III-556, III-564
Ancusa, Versavia II-297
Andonoff, Eric III-630
Annese, Vito III-564
Antunes, C. Henggeler I-90
Antunovic, Mladen I-49
Aoyama, Kouji II-814
Apelkrans, Mats III-788
Aritsugi, Masayoshi II-774
Arús, Carles II-173, II-189
Aufaure, Marie-Aude I-557
Aussenac-Gilles, Nathalie I-641
Avogadri, Roberto III-764
Avola, Danilo II-650
Awaji, Keita II-442
Axt, Margarete II-483
Azzini, Antonia II-371

Baba, Norio II-411
Babenyshev, Sergey II-711, III-210
Bacciu, Davide II-181
Bai, Yun I-278
Balachandran, Bala M. III-236, III-245
Balas, Marius M. III-658
Balas, Valentina E. III-658
Balic, Joze II-1
Balvig, Jens J. III-664
Banek, Marko I-456
Bankauskaite, Vaida II-50
Bao, Haiyan I-625
Bao, Yongguang II-926
Barbu, Tudor III-622
Barna, Cornel III-616
Barriga, Angel I-425
Basch, Danko I-384
Basic, Mateo I-658
Basile, Pierpaolo III-277
Baudrit, Cédric II-265
Beghini, Alessandro III-764
Belanche-Muñoz, Lluís A. II-197
Belani, Hrvoje I-725
Belean, Bogdan III-327
Benaben, Frédérick III-630
Biagetti, Giorgio I-580
Bienvenido, Fernando III-716
Bifulco, Ida III-724, III-732
Biganzoli, Elia II-181
Bingul, Zafer I-674
Blanes, Ian III-416
Bobillo, Fernando III-293
Bogdan, Razvan II-297
Bogdał, Adam I-262
Bogunovic, Nikola I-717
Bolinger, Joe III-799
Borda, Monica III-327
Borzemski, Leszek II-117
Boubertakh, Hamid I-666
Bouras, Christos II-593
Braga, Antõnio II-157
Bratina, Božidar I-25
Bravo, José I-757
Brioschi, Matteo III-764
Brkic, Marko II-73
Brocco, Michele II-806

Brown, David I-733, I-741
Bumbaru, Severin I-202
Bunke, Horst I-131
Butavicius, Marcus I-141

Cairó, Osvaldo II-81
Camastra, Francesco II-662
Camossi, Elena II-682
Cao, Cun Gen I-533
Cardoso, Gonçalo Nuno P. III-465
Carella, Massimo II-206
Castellano, Giovanna II-222
Castiello, Ciro II-230
Castilho, Wagner F. II-483
Castro, Félix II-173
Ceccarelli, Michele III-755
Çela, Arben II-157
Ceravolo, Paolo I-517
Chakravarthy, Sharma II-670
Chang, Jaewoo I-473
Chao, Kuei-Hsiang I-17
Chapurlat, Vincent III-630
Chavira, Gabriel I-757
Chen, Li-Che II-491
Chen, Yen-Wei II-141, II-149
Cheng, Jingde II-499, II-517, II-526
Chetty, Girija III-245
Chiang, Shu-Yin II-958, II-966
Chiesa, Stefano II-50
Chikhi, Nacim Fateh I-641
Chiodi, Giovanni I-362
Chiodi, Lorenzo I-362
Chniber, Othmane III-269
Cho, Jangsik III-50
Choi, Gyoung Soon II-727
Chowdhury, Nihad Karim I-473
Chung, Sheng-Luen II-491
Chyr, Wen-Li II-974
Cirlugea, Mihaela III-360, III-368
Cocu, Adina I-565
Colella, Roberto III-556, III-564
Corsini, Giovanni III-424
Costa, Marcelo II-157
Costa, Rui P. II-281
Costin, Mihaela III-622
Cottington, David II-727
Craciun, Marian I-565
Creanza, Teresa Maria II-206
Crippa, Paolo I-580
Cruz, Isabel F. II-354
Cruz, Manuel D. I-765
Csipkes, Doris III-335
Csipkes, Gabor III-302, III-335
Cui, Zhan I-517
Cuzzocrea, Alfredo II-670

D'Addabbo, Annarita II-206, III-556, III-564
d'Amato, Claudia II-656, III-252
d'Onofrio, Valentina III-556
Daemen, Anneleen II-165
Dalbelo Bašić, Bojana I-615
Damaševičius, Robertas I-98
Damiani, Ernesto I-517
de Freitas, Roberto Ivo C. III-465
De Hauwere, Yann-Michaël I-182
de la Calle, Guillermo II-50
de la Iglesia, Diana II-50
de Lima, Beatriz S.L.P. II-950
De Martino, Monica II-682
De Moor, Bart II-165
De Virgilio, Roberto II-650
Debenham, John I-220
Deguchi, Toshinori II-919
Del Nostro, Pierluigi II-650
Di Gesù, Vito III-747
Di Noia, Tommaso III-293
Di Palma, Annamaria III-556
Di Paolo, Salvatore III-556
Di Sciascio, Eugenio III-293
Diani, Marco III-424
Diao, Lu Hong I-533
Díaz-Parra, Ocotlán I-106
Dioşan, Laura III-376
Distaso, Angela II-206
Djedidi, Rim I-557
do Prado, Hércules A. II-483
Dornean, Irina III-343
Dourado, António II-281
Du, Qian III-392
Dumitriu, Luminita I-565
Duong, Trong Hai I-210
Dürst, Martin J. II-782

El-Bakry, Hazem M. I-33
El-Geresy, Baher A. II-689
El-Sisi, Ashraf II-607
Encheva, Sylvia I-304, I-701
Enda, Naoki II-887
Endo, Mamoru II-536

Eri-Shimokawara-Sato I-709
Esposito, Floriana II-656, III-252
Esposito, M. I-509
Etchells, Terence A. II-214
Evsukoff, Alexandre G. II-950

Falcoz, Paolo I-409
Fallucchi, Francesca III-260
Fanelli, Anna Maria II-230
Fanizzi, Nicola II-656, III-252
Farina, Daniel R. I-375
Faron-Zucker, Catherine I-341
Fazakas, Albert III-327, III-343
Fedullo, Carmine III-732
Feng, G. I-123
Feng, Jun II-543, II-551, II-567
Fernandes, Ana S. II-214
Fernández-Caballero, Antonio III-709
Fernandez-Canque, Hernando III-302, III-335
Fernández-Montes, Alejandro I-765
Ferneda, Edilson II-483
Ferrazzi, Fulvia III-764
Festila, Lelia III-335, III-351, III-360, III-368
Fiannaca, Antonino I-41
Ficko, Mirko II-1
Filho, Gentil J. Lucena II-483
Filko, Damir I-49
Finn, Anthony I-13
Fonseca, José M. II-214
Forster, Florian II-806
Frkovic, Fran I-166
Fu, Yu II-491
Fuchio, Tetsuo III-9
Fuji, Ren I-286
Fujinami, Tsutomu II-839
Fujita, Yoshikatsu II-994
Fukai, Hironobu II-458
Fukuda, Taro III-167
Fukui, Shinji II-887, II-903
Fukumi, Minoru II-403, II-450, II-458
Fukumura, Yoshimi III-499
Furumura, Takashi III-193
Furutani, Michiko II-750
Furutani, Yoshiyuki II-750
Fusco, Luigi III-260
Fuwa, Yasushi III-507

Gabbar, Hossam A. III-1
Gaglio, Salvatore I-41, III-740
Galiero, Giulio II-363
Gao, Kun II-289, II-306
Gao, Ying II-257
García-Remesal, Miguel II-42, II-50
Garofalakis, John II-238
Gentile, Anna Lisa III-277
Georgilakis, Pavlos S. II-9
Gesualdo, Loreto III-556
Gevaert, Olivier II-165
Gianfelici, Francesco I-580
Gianforme, Giorgio II-650
Giannakoudi, Theodoula II-238
Giannini, Franca II-682
Gjomemo, Rigel II-354
Glassey, Jarka II-257
Glavinić, Vlado I-245
Glöckner, Ingo I-606
Glorennec, Pierre-Yves I-666
Goda, Toshinao III-41
Gokceoglu, Candan I-74
Gomes, Alvaro I-90
González, Luis I-765
González-Navarro, Félix F. II-197
Goto, Kenji III-680
Goto, Takuya III-588
Goto, Yuichi II-499, II-526
Gottfried, Björn I-330
Graña, Manuel III-400
Grapsas, Vasilis II-26
Graves, Ian I-141
Greco, Gianluigi I-484
Greco, Mario III-424
Greco, Sergio II-695
Groza, Robert III-360, III-368
Guardati, Silvia II-81
Guidetti, Veronica III-260
Gusmini, Alex I-517

Hagita, Norihiro II-750
Håkansson, Anne I-312, III-779, III-788
Hamada, Mohamed I-33
Hamaguchi, Takashi III-9, III-17
Hamamoto, Yoshihiko I-433
Han, Xian-Hua II-149
Hanachi, Chihab III-630
Hanaue, Koichi II-790
Handa, Hisashi II-411
Hangos, Katalin M. I-501
Hara, Naoto I-417

Harada, Kouji III-432
Hartung, Ronald L. III-799
Hashimoto, Yoshihiro III-17
Hasida, Kôiti II-782
Hayashi, Yuki I-322
Hazeyama, Atsuo III-604
He, Yue II-324
Heldal, Ilona III-481
Hermosilla-Gimeno, Isabel II-66
Hervás, Ramón I-757
Hintea, Sorin III-302, III-335, III-368, III-384
Hiramatsu, Takuya III-507
Hirosaki, Makoto III-33
Hocenski, Zeljko I-49
Honda, Masaya III-25
Hori, Koichi II-823
Hori, Satoshi III-680
Horta, Euler II-157
Howlett, R.J. I-349, III-701
Hsiang, Tien-Ruey II-491
Huang, Xinyin II-141
Huang, Xu III-218, III-227
Hussein, Mahmoud II-607
Hussein, Nashaat M. I-425

Iaquinta, Leo III-277
Ichikawa, Teruhisa III-152
Ichimura, Hiroshi I-417, III-144
Ikeda, Mitsuru III-580
Ikeda, Tetsuo III-41
Ikezaki, Masakazu II-575
Imai, Shinya III-127
Incitti, Roberto II-157
Ingraffia, Nunzio III-285
Inuzuka, Nobuhiro II-911
Ionete, Cosmin II-273
Ishida, Hiroyuki II-911
Ishida, Yoshiteru III-432, III-440, III-448, III-456
Ishii, Naohiro II-895, II-919, II-926, II-942
Ishikawa, Norihiro III-159
Ishikawa, Yuta III-58
Ismail, Nabil II-607
Isokawa, Teijiro II-426
Isotani, Seiji III-523
Ito, Hideaki II-766
Ito, Kazunari II-782
Ito, Sadanori II-847
Ito, Shin-ichi II-450
Itoh, Hidenori II-903, III-50
Itokawa, Tsuyoshi II-774
Itou, Junko III-167, III-201
Iwahori, Yuji II-887, II-903, II-934
Iwata, Tomoharu III-588
Iwatani, Akitsugu I-682

Jain, Lakhmi C. III-642
Jarman, Ian H. II-214
Jascanu, Nicolae I-202
Jascanu, Veronica I-202
Jędrzejowicz, Piotr I-82
Jesse, Christopher I-733
Jevtic, Dragan II-418
Jezic, Gordan I-166, I-174
Ji, Xiaofei I-741
Jiang, J. I-123
Jimbo, Takashi II-895
Jin, Xin II-318
Jo, Geun Sik I-210
Johansson, Mikael III-481
Julià-Sapé, Margarida II-173, II-189
Jung, Kangsoo II-583
Jurasovic, Kresimir I-158
Jurić, Damir I-456
Juszczyszyn, Krzysztof I-194, II-97

Kagi, Tomonori III-183
Kanagawa, Shuya III-572
Kanda, Taki III-106
Kaplan, Emre II-642
Karacapilidis, Nikos I-525
Karasuyama, Masayuki III-76
Karatzas, George P. II-34
Karungaru, Stephen II-403
Kass, Alex I-375
Kato, Shohei III-50
Katsaggelos, Aggelos K. III-408
Katsumata, Yuji III-448
Kawabayashi, Hitoshi II-141
Kawaguchi, Masashi II-895
Kawai, Yuji II-766
Kawanaka, Haruki II-887
Kawaoka, Tsukasa I-286, I-549
Kazienko, Przemysław I-194, II-97
Kerzazi, Amine III-269
Kiewra, Maciej II-89
Kim, Jaehoon II-583
Kimura, Kakeru III-193

Kimura, Masahiro III-67
Kino, Yasunobu II-1018
Kipling, Katie II-257
Kirasić, Damir I-384
Kirei, Botond Sandor III-343
Kitajima, Teiji III-9
Kitasuka, Teruaki II-774
Kizir, Selcuk I-674
Kladas, Antonios G. II-9
Kobayashi, Kanami II-141
Koh, Takahiro II-499, II-526
Kojima, Masanori III-193
Kojima, Takashi II-903
Kojiri, Tomoko I-322, II-575, III-588
Kołaczek, Grzegorz I-237
Könönen, Ville II-379, II-391
Koshimizu, Hiroyasu II-766
Kouyama, Kenta I-417
Król, Dariusz I-262
Kulisic, Ljubomir I-658
Kumamoto, Tadahiko I-588
Kunifuji, Susumu II-847, II-855, II-863
Kunimune, Hisayoshi III-507
Kunstic, Marijan II-418
Kurahashi, Setsuya II-1026
Kuroda, Chiaki III-25
Kuroda, Ipei III-664
Kuroiwa, Toshiaki I-417
Kusek, Mario I-158, I-174

Lakner, Rozália I-501
Lambolais, Thomas III-630
Lamperti, Gianfranco I-362
Landero N., Vanesa I-106
Lányi, Cecília Sik III-473
La Rosa, Massimo III-740
Latiano, Anna III-564
Lazanas, Alexis I-525
Lee, Hsuan-Shih II-974, II-980
Lee, Huey-Ming III-92, III-98
Lee, Hyunjo I-473
Lee, Ren-Hao I-17
Lee, S.H. III-701
Legius, Eric II-165
Leida, Marcello I-517
Leitão, Bruno II-281
Leng, Jinsong III-642
Leunen, Karin II-165
Li, Jiuyong III-642
Li, Yibo I-741
Li, Zheng I-625
Lim, Chee Peng III-548
Lin, Chun-Mei II-332
Lin, Li-Hsueh III-92
Lin, Lily III-92, III-98
Lisboa, Paulo J.G. II-181, II-214
Liu, Honghai I-733, I-741
Liu, Lei I-533
Liu, Meiqun II-313
Liu, Wenting II-543
Lo Bosco, Giosuè III-747
Lopes, Helder F.S. II-466
López-Alonso, Victoria II-66
Lopez-Campos, Guillermo II-66
Lousame, Fabián P. II-248
Lovrek, Ignac I-174
Lu, C. I-123
Lu, Jiamin II-551
Luu, Toan I-633

Ma, YanLin II-318
Ma, Z.M. II-601, II-623
MacLeod, Veneta I-141
Maddalena, Lucia I-57
Maejima, Masashi III-515
Magaldi, Simona III-556
Maglietta, Rosalia II-206
Magnani, Lorenzo II-719
Mao, Xia I-625
Maojo, Víctor II-50
Maratea, Antonio III-755
Marrara, Stefania II-371
Martín-Sanchez, Fernando II-66
Martínez, Marcos II-58
Martins, A. Gomes I-90
Márton, Zoltán III-473
Marumoto, Haruna III-580
Mastrogiovanni, Fulvio I-541
Mat Isa, Nor Ashidi I-493, III-533, III-541
Mateos, Javier III-408
Matsuda, Motohiro II-782
Matsuda, Noriyuki III-673, III-687, III-695
Matsudaira, Kazuya III-152
Matsui, Nobuyuki II-426
Matsumoto, Hideyuki III-25
Matsuno, Keisuke II-919
Matsuoka, Rumiko II-750
Mencar, Corrado II-230

Meng, Xiangfu II-601, II-623
Michils, Geneviève II-165
Mineno, Hiroshi III-159
Mirchandani, Vinod I-220
Mirza, Hanane II-125, II-133
Misra, Sanjay II-879
Misue, Kazuo II-831, II-871
Mitani, Yoshihiro I-433
Mitsukura, Yasue II-403, II-450, II-458
Miura, Hirokazu III-673, III-687
Miura, Motoki II-847, II-855, II-863
Miyachi, Taizo III-664
Miyadera, Youzou III-604
Miyaji, Isao III-144
Miyazaki, Kohji I-440
Miyazaki, Koji III-491
Miyazaki, Shinya II-536
Mizoguchi, Riichiro III-523
Mizuno, Tadanori III-152, III-159, III-193
Mochizuki, Kazuki III-41
Moens, Marie-Francine I-615
Mohd Nordin, Zailani III-533
Mohd Yatid, Moonyati Binti III-167
Molina, Rafael III-408
Molinaro, Cristian II-695
Montague, Gary II-257
Monti, Marina II-682
Moradian, Esmiralda III-771
Moré, Gerard III-416
Morihiro, Koichiro II-426
Mount, Chloë I-141
Mukai, Naoto II-567
Mukherjee, Sayan II-206
Munemori, Jun III-167, III-201
Murino, Loredana III-724
Musa, Zalili Binti III-119
Musiał, Katarzyna I-194, II-97
Muškinja, Nenad I-25
Mutoh, Nobuaki III-41

Nadamoto, Akiyo I-588
Nagaoka, Hisao II-703
Nagasawa, Kenta II-411
Nakagami, Kayoko III-491
Nakagawa, Kenichi II-839
Nakamatsu, Kazumi II-466, II-474
Nakamura, Jun II-742
Nakamura, Koji II-536
Nakamura, Kota I-709
Nakamura, Shoich III-604
Nakamura, Tsuyoshi II-903
Nakanishi, Yasunori III-596
Nakano, Ryohei III-58, III-67, III-76
Nakano, Tomofumi II-911
Nakao, Toshiyuki II-1010
Nakao, Zensho II-125, II-133
Nakata, Kouta I-114
Nakatsu, Ryohei I-440, I-682, I-692
Nakayama, Hirotaka I-573, III-491
Napolitano, Francesco III-724, III-732
Nara, Yumiko II-734
Natowicz, René II-157
Nava, Salvador W. I-757
Navas-Delgado, Ismael II-617, III-269
Nebot, Àngela II-173
Negoita, Mircea Gh. III-310
Németh, Erzsébet I-501
Neupane, Ujjwal II-863
Ngah, Umi Kalthum III-533
Nguyen, Ngoc Thanh I-210, I-253
Nikolos, Ioannis K. II-34
Ninomiya, Daisuke I-440
Nishida, Yoshihiro III-687
Nishimoto, Kazushi II-863
Nishimura, Haruhiko II-426
Nowé, Ann I-182, II-379, II-391
Numa, Kosuke II-823

O'Grady, M.J. I-229
O'Hare, G.M.P. I-229
Oehlmann, Ruediger II-727
Ogiela, Lidia I-394
Ogura, Kazutaka III-687
Ogura, Masatoshi III-159
Ohsawa, Yukio II-742
Ohsugi, Isao J. I-417
Okamoto, Takeshi III-456
Okazaki, Yasuhisa I-598
Oliveira, Pedro II-281
Oltean, Gabriel III-351, III-384
Oltean, Mihai III-376
Ono, Akiko I-417
Orihara, Ryohei I-114
Orsini, Mirko II-354
Ortega, Juan A. I-765
Osano, Tomoyuki III-159
Otaka, Akira III-175
Ouziri, Mourad I-464
Oyama, Tadahiro II-403

Oysu, Cuneyt I-674
Ozaki, Masahiro II-934, II-942

Palmieri, Orazio III-564
Pandzic, Igor S. I-15, II-73
Paolozzi, Stefano II-650
Papadopoulou, Maria P. II-34
Papale, Massimo III-556
Papantonis, Dimitris II-26
Park, Seog II-346, II-583
Paulin, Dan III-481
Pazienza, Maria Teresa III-260
Pazos, Alejandro II-58
Pedersen, Thomas B. II-642
Peeters, Maarten II-379, II-391
Pejsa, Tomislav II-73
Pellow, Allan III-302
Pelzer, Björn I-606
Pereira, Javier II-58
Perrot, Nathalie II-265
Petre, Emil II-273
Petrosino, Alfredo I-57, II-662
Pham, Tuan D. I-401
Phillips-Wren, Gloria I-1
Piepoli, Ada II-206
Pinello, Luca III-747
Pinet, François II-106
Pirrò, Giuseppe II-631
Pirrone, Roberto III-285
Podnar Zarko, Ivana I-633
Podobnik, Vedran I-166
Pons, Xavier III-416
Popescu, Daniela E. III-650
Prada, Marcela F. III-650
Pripužić, Krešimir I-725
Prodan, Ante I-220
Puh, Maroje I-633
Purwitasari, Diana I-598

Qing, C. I-123
Quteishat, Anas III-548

Ragone, Azzurra III-293
Raiconi, Giancarlo III-724, III-732
Rajman, Martin I-633
Ramanathan, Jay III-799
Ranieri, Elena III-556
Rave, Erez I-650
Reusch, Bernd I-16
Riesen, Kaspar I-131
Rizzo, Riccardo I-41, III-740
Roccetti, Paolo II-363
Rodrigues, Guilherme II-281
Rodrigues, Thiago II-157
Romero, Enrique II-189
Rosić, Marko I-245
Rothenburger, Bernard I-641
Roupé, Mattias III-481
Rouzier, Roman II-157
Ruffino, Francesca III-764
Ruffolo, Massimo II-631
Ruiz-Vanoye, Jorge A. I-106
Rus, Cristian III-335
Rusinovic, Zoran I-717
Russo, Giuseppe III-285
Ruta, Dymitr I-448
Rybakov, Vladimir II-711, III-210

Saito, Kazumi III-41, III-67
Sakurai, Shigeaki I-114
Sampaio, Paulo N.M. III-465
Sánchez, Eduardo II-248
Sangiorgi, Pierluca III-285
Satake, Hiroki III-580
Sato, Shuji II-831
Savaş, Erkay II-642
Sawamura, Hajime I-149
Saygın, Yücel II-642
Scarpato, Noemi III-260
Schmidhuber, Jürgen I-11
Schmidt, Rainer I-270
Schneider, Michel II-106
Segal, Cristina I-565
Sekanina, Lukas III-310
Selişteanu, Dan II-273
Semeraro, Giovanni III-277
Şendrescu, Dorin II-273
Serra-Sagristà, Joan III-416
Seta, Kazuhisa III-580
Sgorbissa, Antonio I-541
Sharma, Dharmendra III-218, III-227, III-245
Shen, J.M. I-229
Shen, Pei-Di II-974
Shimada, Yukiyasu III-1, III-9, III-17
Shimazaki, Yosuke III-673
Shimohara, Katsunori II-434
Shinkai, Kimiaki III-572
Shinoda, Yuji I-573, III-491
Shinozaki, Kuniya I-682, I-692

Shojiguchi, Yoshitake III-695
Sibertin-Blanc, Christophe III-630
Sicard, Mariette II-265
Šilić, Artur I-615
Sinkovic, Vjekoslav I-174
Sipos, Emilia III-351, III-384
Sirotin, Igal I-650
Skakovski, Aleksander I-82
Skočir, Zoran I-456
Smart, Edward I-733
Smid, Karlo II-73
Smith, Huw III-302
Soga, Masato III-596, III-695
Sokolova, Marina V. III-709
Sprado, Jörn I-330
Stamatelos, Fotis II-26
Starita, Antonina II-181
Stellato, Armando III-260
Stergiadi, Maria II-34
Stoica, Adrian III-310
Straccia, Umberto III-293
Sugihara, Taro II-839
Sugimura, Yuuki I-433
Sugioka, Shingo III-9
Sulaiman, Siti Amrah III-541
Sumi, Kazuya III-515
Sunde, Jadranka I-141
Sunesson, Kaj III-481
Suzuki, Nobuo II-1002
Suzuki, Takeshi III-664
Suzuki, Taketoshi II-450
Szolga, Lorant Andras III-360
Szupiluk, Ryszard I-749

Tadjine, Mohamed I-666
Tagliaferri, Roberto III-724, III-732
Tahir, Omar III-630
Takagi, Hideyuki I-14
Takagi, Tsuyoshi III-175
Takahashi, Hiroshi II-986
Takahashi, Masakazu II-994, II-1010
Takahashi, Osamu III-175, III-183
Takahashi, Satoru II-994
Takashita, Taiki II-774
Takatsuka, Ryozo II-839, II-847
Takeda, Kazuhiro III-9, III-17
Takeuchi, Ichiro III-76
Taki, Hirokazu III-673, III-680, III-687, III-695
Takimoto, Hironori II-458
Takizawa, Takenobu III-572
Takizawa, Takeshi III-507
Talia, Domenico II-631
Tamura, Yasuhisa III-515
Tanaka, Jiro II-831
Tanaka, Katsuaki II-823
Tanaka, Katsumi I-588
Tanaka-Yamawaki, Mieko II-442
Terano, Takao II-986, II-1010
Terracina, Giorgio I-484
Thai, Hien II-125
Thanh, Nhan Le I-341
Thiel, Christian I-65
Titchener-Hooker, Nigel II-257
Tokoi, Kohe III-596
Tokumitsu, Masahiro III-440
Tolón-Becerra, Alfredo III-716
Tomita, Shigeyuki III-33
Tomita, Yohei II-450
Topa, Marina III-343
Topic, Gordan II-418
Toriumi, Kiyoko II-823
Torsello, Maria Alessandra II-222
Tovornik, Boris I-25
Tran, Trong Hieu I-253
Trzec, Krunoslav I-166
Tsili, Marina A. II-9
Tsogkas, Vassilis II-593
Tsourveloudis, Nikos C. II-18
Tsuchiya, Seiji I-286, I-549
Tsuda, Kazuhiko II-1002, II-1010, II-1018
Tsuda, Satoshi I-692
Tsuge, Satoru II-403
Tsukada, Koji III-673
Tsukahara, Tadashi II-1018
Tsumoto, Shusaku II-758
Tumin, Sharil I-304, I-701
Turchetti, Claudio I-580
Turli, Andrea II-363
Tyrrell, Andy I-12

Uchio, Fumitaka III-673
Uesu, Hiroaki III-84
Uesugi, Kentaro II-559
Ugai, Takanori II-814
Ukai, Yasuharu II-703
Umano, Motohide III-580
Ura, Masahiro II-536

Urso, Alfonso I-41, III-740
Ushiama, Taketoshi II-575, II-798

Valentini, Giorgio III-764
Van Segbroeck, Sven II-391
Vanspauwen, Vanessa II-165
Vast, Robyn I-141
Vázquez, José M. II-58
Vega, Miguel III-408
Veganzones, Miguel A. III-400
Vellido, Alfredo II-173, II-189
Verbeeck, Katja II-379, II-391
Vergote, Ignace II-165
Vescan, Andreea III-319
Vicari, Claudia III-285
Vieira, Ian N. II-950
Villarreal, Vladimir I-757
Vladutiu, Mircea II-297
Vopi, Agoritsa II-238
Vorobieva, Olga I-270
Vrancx, Peter I-182
Vukadinovic, Dinko I-658
Vuković, Marin I-725

Walters, S.D. III-701
Wang, Jing-Long II-958, II-966
Wang, Meng-Hui I-17
Wang, Shuming III-127
Wang, Su-Fen II-340
Wang, Zhijian II-543
Washio, Takashi II-703
Watabe, Hirokazu I-286, I-549
Watada, Junzo III-111, III-119, III-127, III-136
Watanabe, Kenzi I-598
Watanabe, Miho III-604
Watanabe, Toyohide I-322, II-551, II-559, II-567, II-575, II-790, II-798, III-588
Weiss-Cohen, Miri I-650
Westerdahl, Börje III-481
Willis, Mark II-257
Wojewnik, Piotr I-749
Wood, S. I-349
Woodham, Robert J. II-887, II-903
Wuillemin, Pierre-Henri II-265

Yaakob, Shamshul Bahar III-111
Yada, Katsutoshi II-703
Yamaba, Hisaaki III-33
Yamada, Kunihiro III-144, III-193
Yamada, Masashi II-536
Yamada, Takahiro II-926
Yamada, Takeshi III-588
Yamaguchi, Toru I-709
Yamamoto, Kosuke II-559
Yamamuro, Takeshi III-515
Yamashita, Hajime III-572
Yamashita, Seiu I-417
Yamashita, Yasuo II-986
Yamazaki, Atsuko K. III-680
Yan, Li II-601
Yan, Shu Ying I-533
Yasuda, Takami II-536
Yeh, Chen-Huei II-980
Yeh, Peter Z. I-375
Yokoyama, Kenzou III-507
Yokoyama, Setsuo III-604
Yoshida, Kenji I-573, III-491
Yoshida, Kouji III-144, III-193
Yoshida, Yoshiharu I-709
Yoshimura, Eriko I-286, I-549
Yoshino, Junichi I-417
Yoshiura, Noriaki II-508
Younan, Nicolas H. III-392
Younis, Mohammed I. I-493
Yu, Donggang I-401
Yukawa, Takashi III-499
Yurchyshyna, Anastasiya I-341

Zabala, Alaitz III-416
Ząbkowski, Tomasz I-749
Zaccaria, Renato I-541
Zamli, Kamal Zuhairi I-493, III-533, III-541
Zanella, Marina I-362
Zarli, Alain I-341
Zatwarnicki, Krzysztof II-117
Zelić, Marija I-245
Zhang, Sen I-533
Zhou, Xiaobo I-401
Zhou, Yuhong II-257
Zhu, Yuelong II-551
Žmak, Lovro I-615
Zorlu, Kivanc I-74

Printing: Mercedes-Druck, Berlin
Binding: Stein+Lehmann, Berlin

Lecture Notes in Artificial Intelligence (LNAI)

Vol. 5221: B. Nordström, A. Ranta (Eds.), Advances in Natural Language Processing. XII, 512 pages. 2008.

Vol. 5208: H. Prendinger, J. Lester, M. Ishizuka (Eds.), Intelligent Virtual Agents. XVII, 557 pages. 2008.

Vol. 5195: A. Armando, P. Baumgartner, G. Dowek (Eds.), Automated Reasoning. XII, 556 pages. 2008.

Vol. 5179: I. Lovrek, R.J. Howlett, L.C. Jain (Eds.), Knowledge-Based Intelligent Information and Engineering Systems, Part III. XXXVI, 817 pages. 2008.

Vol. 5178: I. Lovrek, R.J. Howlett, L.C. Jain (Eds.), Knowledge-Based Intelligent Information and Engineering Systems, Part II. XXXVIII, 1043 pages. 2008.

Vol. 5177: I. Lovrek, R.J. Howlett, L.C. Jain (Eds.), Knowledge-Based Intelligent Information and Engineering Systems, Part I. LVI, 781 pages. 2008.

Vol. 5144: S. Autexier, J. Campbell, J. Rubio, V. Sorge, M. Suzuki, F. Wiedijk (Eds.), Intelligent Computer Mathematics. XIV, 600 pages. 2008.

Vol. 5118: M. Dastani, A. El Fallah Seghrouchni, J. Leite, P. Torroni (Eds.), Languages, Methodologies and Development Tools for Multi-Agent Systems. X, 279 pages. 2008.

Vol. 5113: P. Eklund, O. Haemmerlé (Eds.), Conceptual Structures: Knowledge Visualization and Reasoning. X, 311 pages. 2008.

Vol. 5110: W. Hodges, R. de Queiroz (Eds.), Logic, Language, Information and Computation. VIII, 313 pages. 2008.

Vol. 5108: P. Perner, O. Salvetti (Eds.), Advances in Mass Data Analysis of Images and Signals in Medicine, Biotechnology, Chemistry and Food Industry. X, 173 pages. 2008.

Vol. 5097: L. Rutkowski, R. Tadeusiewicz, L.A. Zadeh, J.M. Zurada (Eds.), Artificial Intelligence and Soft Computing – ICAISC 2008. XVI, 1269 pages. 2008.

Vol. 5078: E. André, L. Dybkjær, W. Minker, H. Neumann, R. Pieraccini, M. Weber (Eds.), Perception in Multimodal Dialogue Systems. X, 311 pages. 2008.

Vol. 5077: P. Perner (Ed.), Advances in Data Mining. XI, 428 pages. 2008.

Vol. 5076: R. van der Meyden, L. van der Torre (Eds.), Deontic Logic in Computer Science. X, 279 pages. 2008.

Vol. 5064: L. Prevost, S. Marinai, F. Schwenker (Eds.), Artificial Neural Networks in Pattern Recognition. IX, 318 pages. 2008.

Vol. 5049: D. Weyns, S.A. Brueckner, Y. Demazeau (Eds.), Engineering Environment-Mediated Multi-Agent Systems. X, 297 pages. 2008.

Vol. 5043: N. Jamali, P. Scerri, T. Sugawara (Eds.), Massively Multi-Agent Technology. XII, 191 pages. 2008.

Vol. 5040: M. Asada, J.C.T. Hallam, J.-A. Meyer, J. Tani (Eds.), From Animals to Animats 10. XIII, 530 pages. 2008.

Vol. 5032: S. Bergler (Ed.), Advances in Artificial Intelligence. XI, 382 pages. 2008.

Vol. 5027: N.T. Nguyen, L. Borzemski, A. Grzech, M. Ali (Eds.), New Frontiers in Applied Artificial Intelligence. XVIII, 879 pages. 2008.

Vol. 5012: T. Washio, E. Suzuki, K.M. Ting, A. Inokuchi (Eds.), Advances in Knowledge Discovery and Data Mining. XXIV, 1102 pages. 2008.

Vol. 5009: G. Wang, T. Li, J.W. Grzymala-Busse, D. Miao, A. Skowron, Y. Yao (Eds.), Rough Sets and Knowledge Technology. XVIII, 765 pages. 2008.

Vol. 5003: L. Antunes, M. Paolucci, E. Norling (Eds.), Multi-Agent-Based Simulation VIII. IX, 141 pages. 2008.

Vol. 5001: U. Visser, F. Ribeiro, T. Ohashi, F. Dellaert (Eds.), RoboCup 2007: Robot Soccer World Cup XI. XIV, 566 pages. 2008.

Vol. 4999: L. Maicher, L.M. Garshol (Eds.), Scaling Topic Maps. XI, 253 pages. 2008.

Vol. 4994: A. An, S. Matwin, Z.W. Raś, D. Ślęzak (Eds.), Foundations of Intelligent Systems. XVII, 653 pages. 2008.

Vol. 4953: N.T. Nguyen, G.S. Jo, R.J. Howlett, L.C. Jain (Eds.), Agent and Multi-Agent Systems: Technologies and Applications. XX, 909 pages. 2008.

Vol. 4946: I. Rahwan, S. Parsons, C. Reed (Eds.), Argumentation in Multi-Agent Systems. X, 235 pages. 2008.

Vol. 4944: Z.W. Raś, S. Tsumoto, D.A. Zighed (Eds.), Mining Complex Data. X, 265 pages. 2008.

Vol. 4938: T. Tokunaga, A. Ortega (Eds.), Large-Scale Knowledge Resources. IX, 367 pages. 2008.

Vol. 4933: R. Medina, S. Obiedkov (Eds.), Formal Concept Analysis. XII, 325 pages. 2008.

Vol. 4930: I. Wachsmuth, G. Knoblich (Eds.), Modeling Communication with Robots and Virtual Humans. X, 337 pages. 2008.

Vol. 4929: M. Helmert, Understanding Planning Tasks. XIV, 270 pages. 2008.

Vol. 4924: D. Riaño (Ed.), Knowledge Management for Health Care Procedures. X, 161 pages. 2008.

Vol. 4923: S.B. Yahia, E.M. Nguifo, R. Belohlavek (Eds.), Concept Lattices and Their Applications. XII, 283 pages. 2008.

Vol. 4914: K. Satoh, A. Inokuchi, K. Nagao, T. Kawamura (Eds.), New Frontiers in Artificial Intelligence. X, 404 pages. 2008.

Vol. 4911: L. De Raedt, P. Frasconi, K. Kersting, S. Muggleton (Eds.), Probabilistic Inductive Logic Programming. VIII, 341 pages. 2008.

Vol. 4908: M. Dastani, A. El Fallah Seghrouchni, A. Ricci, M. Winikoff (Eds.), Programming Multi-Agent Systems. XII, 267 pages. 2008.

Vol. 4898: M. Kolp, B. Henderson-Sellers, H. Mouratidis, A. Garcia, A.K. Ghose, P. Bresciani (Eds.), Agent-Oriented Information Systems IV. X, 292 pages. 2008.

Vol. 4897: M. Baldoni, T.C. Son, M.B. van Riemsdijk, M. Winikoff (Eds.), Declarative Agent Languages and Technologies V. X, 245 pages. 2008.

Vol. 4894: H. Blockeel, J. Ramon, J. Shavlik, P. Tadepalli (Eds.), Inductive Logic Programming. XI, 307 pages. 2008.

Vol. 4885: M. Chetouani, A. Hussain, B. Gas, M. Milgram, J.-L. Zarader (Eds.), Advances in Nonlinear Speech Processing. XI, 284 pages. 2007.

Vol. 4874: J. Neves, M.F. Santos, J.M. Machado (Eds.), Progress in Artificial Intelligence. XVIII, 704 pages. 2007.

Vol. 4870: J.S. Sichman, J. Padget, S. Ossowski, P. Noriega (Eds.), Coordination, Organizations, Institutions, and Norms in Agent Systems III. XII, 331 pages. 2008.

Vol. 4869: F. Botana, T. Recio (Eds.), Automated Deduction in Geometry. X, 213 pages. 2007.

Vol. 4865: K. Tuyls, A. Nowe, Z. Guessoum, D. Kudenko (Eds.), Adaptive Agents and Multi-Agent Systems III. VIII, 255 pages. 2008.

Vol. 4850: M. Lungarella, F. Iida, J.C. Bongard, R. Pfeifer (Eds.), 50 Years of Artificial Intelligence. X, 399 pages. 2007.

Vol. 4845: N. Zhong, J. Liu, Y. Yao, J. Wu, S. Lu, K. Li (Eds.), Web Intelligence Meets Brain Informatics. XI, 516 pages. 2007.

Vol. 4840: L. Paletta, E. Rome (Eds.), Attention in Cognitive Systems. XI, 497 pages. 2007.

Vol. 4830: M.A. Orgun, J. Thornton (Eds.), AI 2007: Advances in Artificial Intelligence. XIX, 841 pages. 2007.

Vol. 4828: M. Randall, H.A. Abbass, J. Wiles (Eds.), Progress in Artificial Life. XII, 402 pages. 2007.

Vol. 4827: A. Gelbukh, Á.F. Kuri Morales (Eds.), MICAI 2007: Advances in Artificial Intelligence. XXIV, 1234 pages. 2007.

Vol. 4826: P. Perner, O. Salvetti (Eds.), Advances in Mass Data Analysis of Signals and Images in Medicine, Biotechnology and Chemistry. X, 183 pages. 2007.

Vol. 4819: T. Washio, Z.-H. Zhou, J.Z. Huang, X. Hu, J. Li, C. Xie, J. He, D. Zou, K.-C. Li, M.M. Freire (Eds.), Emerging Technologies in Knowledge Discovery and Data Mining. XIV, 675 pages. 2007.

Vol. 4811: O. Nasraoui, M. Spiliopoulou, J. Srivastava, B. Mobasher, B. Masand (Eds.), Advances in Web Mining and Web Usage Analysis. XII, 247 pages. 2007.

Vol. 4798: Z. Zhang, J.H. Siekmann (Eds.), Knowledge Science, Engineering and Management. XVI, 669 pages. 2007.

Vol. 4795: F. Schilder, G. Katz, J. Pustejovsky (Eds.), Annotating, Extracting and Reasoning about Time and Events. VII, 141 pages. 2007.

Vol. 4790: N. Dershowitz, A. Voronkov (Eds.), Logic for Programming, Artificial Intelligence, and Reasoning. XIII, 562 pages. 2007.

Vol. 4788: D. Borrajo, L. Castillo, J.M. Corchado (Eds.), Current Topics in Artificial Intelligence. XI, 280 pages. 2007.

Vol. 4775: A. Esposito, M. Faundez-Zanuy, E. Keller, M. Marinaro (Eds.), Verbal and Nonverbal Communication Behaviours. XII, 325 pages. 2007.

Vol. 4772: H. Prade, V.S. Subrahmanian (Eds.), Scalable Uncertainty Management. X, 277 pages. 2007.

Vol. 4766: N. Maudet, S. Parsons, I. Rahwan (Eds.), Argumentation in Multi-Agent Systems. XII, 211 pages. 2007.

Vol. 4760: E. Rome, J. Hertzberg, G. Dorffner (Eds.), Towards Affordance-Based Robot Control. IX, 211 pages. 2008.

Vol. 4755: V. Corruble, M. Takeda, E. Suzuki (Eds.), Discovery Science. XI, 298 pages. 2007.

Vol. 4754: M. Hutter, R.A. Servedio, E. Takimoto (Eds.), Algorithmic Learning Theory. XI, 403 pages. 2007.

Vol. 4737: B. Berendt, A. Hotho, D. Mladenic, G. Semeraro (Eds.), From Web to Social Web: Discovering and Deploying User and Content Profiles. XI, 161 pages. 2007.

Vol. 4733: R. Basili, M.T. Pazienza (Eds.), AI*IA 2007: Artificial Intelligence and Human-Oriented Computing. XVII, 858 pages. 2007.

Vol. 4724: K. Mellouli (Ed.), Symbolic and Quantitative Approaches to Reasoning with Uncertainty. XV, 914 pages. 2007.

Vol. 4722: C. Pelachaud, J.-C. Martin, E. André, G. Chollet, K. Karpouzis, D. Pelé (Eds.), Intelligent Virtual Agents. XV, 425 pages. 2007.

Vol. 4720: B. Konev, F. Wolter (Eds.), Frontiers of Combining Systems. X, 283 pages. 2007.

Vol. 4702: J.N. Kok, J. Koronacki, R. Lopez de Mantaras, S. Matwin, D. Mladenič, A. Skowron (Eds.), Knowledge Discovery in Databases: PKDD 2007. XXIV, 640 pages. 2007.

Vol. 4701: J.N. Kok, J. Koronacki, R. Lopez de Mantaras, S. Matwin, D. Mladenič, A. Skowron (Eds.), Machine Learning: ECML 2007. XXII, 809 pages. 2007.

Vol. 4696: H.-D. Burkhard, G. Lindemann, R. Verbrugge, L.Z. Varga (Eds.), Multi-Agent Systems and Applications V. XIII, 350 pages. 2007.

Vol. 4694: B. Apolloni, R.J. Howlett, L. Jain (Eds.), Knowledge-Based Intelligent Information and Engineering Systems, Part III. XXIX, 1126 pages. 2007.

Vol. 4693: B. Apolloni, R.J. Howlett, L. Jain (Eds.), Knowledge-Based Intelligent Information and Engineering Systems, Part II. XXXII, 1380 pages. 2007.